Laminar Flow and Convective Transport Processes

Laminar Flow and Convective Transport Processes

Scaling Principles and

Asymptotic Analysis

L. Gary Leal
Department of Chemical and Nuclear Engineering
University of California
Santa Barbara

Butterworth-Heinemann
Boston London Oxford Singapore Sydney Toronto Wellington

BUTTERWORTH-HEINEMANN SERIES IN CHEMICAL ENGINEERING

SERIES TITLES

REPRINT TITLES

Advanced Process Control *W. Harmon Ray*
Applied Statistical Mechanics *Thomas M. Reed and Keith E. Gubbins*
Elementary Chemical Reactor Analysis *Rutherford Aris*
Kinetics of Chemical Processes *Michel Boudart*
Reaction Kinetics for Chemical Engineers *Stanley M. Walas*

Library of Congress Cataloging-in-Publication Data

Leal, L. Gary.
    Laminar flow and convective transport processes : scaling principles and asymptotic analysis / L. Gary Leal.
        p.   cm. — (Butterworth-Heinemann series in chemical engineering)
    Includes bibliographical references and index.
    ISBN 0-7506-9117-4
    1. Laminar flow. 2. Transport theory. I. Title. II. Series.
QA929.L44   1992
532'.0525—dc20   91-29888
                                                            CIP

British Library Cataloguing in Publication Data

Leal, L. Gary.
    Laminar flow and convective transport processes : scaling principles and asymptotic analysis.
    I. Title
    620.106

    ISBN 0-7506-9117-4

Butterworth–Heinemann
313 Washington Street
Newton, MA 02158–1626

10 9 8 7 6 5 4 3 2

Printed in the United States of America

This book is dedicated to the many students who have taken my graduate classes in transport phenomena and fluid mechanics over the years, and thus contributed the insights, questions, and hard work that have led me to better understanding and a continued enthusiasm for teaching these subjects. I also thank Professor Andreas Acrivos who inspired me to seek an academic career, and taught me much of what I know. Finally, I thank my wife Mary and daughters, Heather, Kami, and Farrah, who did not complain over the time lost with them, and never gave up that the task would finally be completed.

Contents

CHAPTER 11 Thermal Boundary-Layer Theory at Large Reynolds Number   637

CHAPTER 12 Natural and Mixed Convection Flows   669

Preface

This book is based largely on lecture notes for a graduate-level course on fluid mechanics and thermal transport processes that I have taught at Caltech and UCSB. This course, like many in chemical engineering departments across the country, was originally built on the foundations of a course on the same topics taught for many years by Professor Andreas Acrivos at Stanford University. The emphasis is on asymptotic approximation techniques applied to classical problems involving laminar flows of incompressible, Newtonian fluids. However, there is also a significant amount of material on viscous-dominated creeping flows, which have historically been of special interest in chemical engineering applications but are now finding use across a broad spectrum of technologies involving fluid motions or transport process on short length scales.

Although the majority of the subject matter has been available in the research literature for 20 to 30 years, and has been taught in many graduate chemical engineering courses, there is no other textbook that I am aware of that covers a reasonable fraction of the same material. Thus, I hope that this book will meet an important need where this material is already being taught, and will also serve to make the material more easily accessible where it has not previously been incorporated into the curriculum. Although the book arises from chemical engineering coursework, and is written by a chemical engineer, I believe that many of the topics are of general interest to the fluid mechanics and heat transfer communities, and I will be disappointed if it does not serve a useful purpose within these groups as well.

Although the goals of a textbook should certainly speak for themselves once the book has actually been studied, it is perhaps useful to summarize briefly the intent of the book as I have conceived it in writing. First of all, I presume that the student using this book has already been exposed to fluid mechanics and convective transport processes from a fundamental point of view. For chemical engineers, a reasonable benchmark may be the well-known textbook of Bird, Stewart, and Lightfoot, though I do not intend to exclude more recent or comprehensive textbooks that cover similar material.

In spite of the intent to use this book for a second course in transport phenomena, it is written in a completely self-contained manner, and it should be possible for a graduate student with a reasonable background in classical methods of applied math, including an introduction to partial differential equations and vector analysis, to study it even as a first course in fluid mechanics and transport phenomena. In particular, the material from a typical undergraduate course is not required in the normal manner of a prerequisite. Instead, the need for undergraduate training is because there are a number of important subjects for engineering application that are not included in this presentation. Especially noteworthy in this regard is the lack of any discussion of macroscopic balances for momentum or mechanical energy, or of correlations and other material relevant to turbulent flows, multiphase flows, or complex fluids such as polymer solutions and melts.

Although the selection of problems and problem areas may seem somewhat narrow, my goal is not an exhaustive collection of problems. Rather, I seek a deep fundamental understanding of the basic principles of transport phenomena, and a theoretical framework that can be applied to a broad class of problems. In fact, the essence of much of what we shall do is just a sophisticated version of nondimensionalization and the use of characteristic scales to deduce approximations of the governing equations and boundary conditions. As we shall see, the successful application of characteristic scaling in this context requires a very thorough understanding of the basic physical phenomena, and of their relative importance in different parts of the domain. Thus there is an inherent emphasis on the relevance of physical understanding to the mathematical developments, and vice versa.

A most important message is that the key results for a particular problem, such as the form of correlations between independent and dependent dimensionless groups, can often be achieved as a consequence of simply formulating the problem in an asymptotic framework, without solving any differential equations. Nevertheless, it is likely that circumstances will arise where a more detailed knowledge of the velocity or temperature fields is desirable or even necessary. Unfortunately, in the research literature and in advanced textbooks, the steps between problem formulation and solution are often dominated by the phrase "it can be shown." While this may be quite acceptable in some instances, I have found that it can be extremely frustrating to the student, and I have attempted to avoid it as much as possible. The down side of this is that it leads to lengthy descriptions of the solutions for relatively few problems, and the criticism from at least one reviewer of "too many examples . . . and at too great a length." In a lecture, the criticism of too much detail is probably especially pertinent, but I believe that it is important for the student to have *access* to the analysis in more or less complete detail. Obviously, the student or the teacher using this text can choose to greatly curtail the details or study everything in complete detail or land somewhere between, depending on the needs of the students.

Finally, the reader may find it strange that I have gone into such a detailed account in Chapter 2 of the derivation and basis for the governing equations, given the fact that the majority of students will have already had some introduction to this material in a previous course. To some extent, this reflects the nature of my own introduction to fluid mechanics and transport phenomena at the graduate and undergraduate levels. Many undergraduate courses simply adopt the governing equations from the outset without attempting to either relate them to the material on classical mechanics that is usually encountered in earlier courses in physics, or derive them in a satisfactory, fundamental way. Although the various terms in the basic balance equations are often labeled according to the mechanism that they represent, this hardly provides a solid foundation for physical understanding to the uninitiated student. On the other hand, it is often assumed in graduate-level courses that the derivation of the basic equations is not necessary because it has been encountered before at the undergraduate level. The result is that the subject turns into mathematics only, for a class of nonlinear equations where classical techniques of analysis do not work, and the student never fully understands that the subject is really no more than Newton's laws of classical mechanics, applied to a body that deforms as a consequence of the fact that it moves, coupled with constitutive laws that specify the relationship between dynamical variables and force and between temperature gradients and heat flux.

I have generally offered a course based on this book in two quarters. To do this, I have sometimes found it necessary to be selective in deleting some examples or topics. Although the book and course are intended to provide a logical progression, and thus the discussion at each level builds on the material that came earlier, it is definitely possible to delete some material without serious damage to the general message that I have intended to convey. From my personal perspective, for a chemical engineer, the most essential topics are contained in Chapters 2, 3, and 4; the first section of Chapter 5; and Chapters 8, 9, 10, and 11. The choice of examples within any chapter is up to the discretion of the instructor.

There are, of course, many problems and problem areas within fluid mechanics and convective transport processes that are not covered in this text. Of particular importance to chemical engineers is the exclusion, after the introductory material of Chapter 2, of complex fluids (namely, polymeric solutions or melts, colloidal dispersions, and others that show non-Newtonian behavior); the exclusion of mass transfer, except for the sometimes useful analogy between convective transport of heat and convective transport of a single solute in a solvent; the lack of a comprehensive study of transport processes or flows that arise as instabilities from either a state of rest or some base motion; and, finally, the complete lack of any discussion of transport processes or flows in which the fluid motion is turbulent. These exclusions do not imply any suggestion that these problem areas are not of *vital* importance in engineering practice. In my own teaching (and in this book), I have simply run out of time to include these topics in any significant way in a two-term (quarter) course. Apart from mass transfer, the way that I have handled this myself is to teach one additional term of graduate-level transport phenomena, which is offered as a course on a rotating set of "special topics." Typically, the topics offered have included hydrodynamic stability theory, non-Newtonian fluid mechanics, suspension and biofluid mechanics, and an introduction to turbulent flow phenomena, all rotated on a three- to four-year cycle.

As far as mass transfer is concerned, the situation in my own experience has been somewhat less satisfactory. Once we generalize from a two-component (solute plus solvent) system and include the complications of homogeneous or heterogeneous chemical reaction, a systematic description of convective mass transfer processes does not exist at the same level and in the same spirit that is represented here. Descriptions of multicomponent mass transfer systems at the level of Bird, Stewart, and Lightfoot's well-known text are frequently encountered by chemical engineering students as undergraduates. Certainly, if mass transfer at the Bird, Stewart, and Lightfoot level has not been studied previously, I would highly recommend it either before or after the present course.

One additional topic that is *not* covered explicitly in the present book is vector or tensor calculus, though all of the presentation makes use of the efficiencies of vector/tensor notation whenever possible. I have found that vector and tensor analysis often represents a perceived liability of beginning graduate students. Indeed, I have often heard it suggested that every beginning course in fluid mechanics or convective heat transfer at the graduate level should provide formal training in this topic. This would, perhaps, be an excellent basis to suggest that every beginning graduate student should first have a course in applied mathematics, including a heavy dose of vector and tensor analysis. The main problem with this is that such a course has a considerable potential to seem "dry" and "without any obvious value" in the absence of a specific application. There is no doubt that courses on fluid mechanics and transport phenomena provide plenty of motivation and a useful framework to learn a great deal of applied mathematics. The real issue is one of emphasis. In my own experience, I have found that most of the vector analysis necessary can be picked up at an adequate level in the first few weeks of the course through a combination of an introductory lecture, supplementary sessions outside of the regular class with a graduate teaching assistant, and frequent use of one of the many texts and references that are devoted to this topic. The concept of tensors and dyadic products of vectors, on the other hand, is difficult for the student to grasp without specific discussion in lectures and/or the text. In my own teaching, I spend a significant amount of time explaining the concept of the stress tensor, with emphasis on its relationship to the surface force vector, but I do not attempt a broader, more general discussion of tensors and tensor algebra or calculus. This is primarily a pragmatic decision, based upon my concept of what is essential for this class, and the time that would be required to do more.

Acknowledgments

A number of people have been very helpful in the completion of this project, and I would like to acknowledge their contributions here. First, the preparation of the typecript was done by Kathy Lewis at Caltech and Bonnie Franzen at UCSB, and I thank both of them for their patience and expertise. Secondly, the camera-ready figures were prepared primarily by Yael Shapira, with expert assistance from Andrew Szeri, and I greatly appreciate their cooperation and willingness to help. Third, a large number of graduate students at Caltech and UCSB have been subjected to using all or major portions of this book, in manuscript form, as their "text" over the past several years. Many have made useful suggestions on presentation, and they have also found many of the small errors that inevitably creep into such a long manuscript. I would especially like to thank my own graduate and postdoctoral students, many of whom have taken a personal interest in this project and volunteered considerable time and effort in checking details of analysis. In this regard, I especially mention Howard Stone, Bill Milliken, In-Seok Kang, Seung-Man Yang, and Andrew Szeri. Fourth, several friends and former students have used portions of the manuscript in their classes at other universities and have sent useful corrections and comments. I especially thank Dave Leighton at Notre Dame, who sent a very interesting and innovative set of homework problems, some of which I have included in the book. Finally, I wish to express my most sincere appreciation to Andy Acrivos. As I said earlier, the course on this topic that he taught for many years at Stanford was the foundation for evolution of my own course. Beyond that, however, the education that he gave me as a Ph.D. student has served as the basis for the success that I have had in my own career. His example of quality, integrity, and innovativeness has been an inspiration to several generations of researchers and teachers of fluid mechanics within the chemical engineering community and has certainly had much to do with the current healthy state of fluid mechanics. Before Andy moved to his current post as Einstein Professor in the Levich Institute for Physico-chemical Hydrodynamics at CCNY, he had agreed to coauthor this book. I regret that his current duties did not leave him time to participate, for there is no doubt that the book would have been greatly improved with his help. Nevertheless, I thank him very much for the encouragement to begin. Of course, all errors of omission or commission are mine and mine alone.

Laminar Flow and Convective Transport Processes

CHAPTER 1

Introduction

This book is about fluid mechanics and the convective transport of heat (or any passive scalar quantity) for simple Newtonian, incompressible fluids, treated from the point of view of classical continuum mechanics. It is written primarily as a textbook for an introductory graduate course on these topics, and is based upon lecture notes that were developed over almost twenty years of teaching these topics to beginning graduate students in chemical engineering at Caltech and now at UCSB.

A. A Brief Historical Perspective

The earliest step toward the inclusion of specialized courses in fluid mechanics and heat or mass transfer processes within the chemical engineering curriculum probably occurred with the publication in 1923 of the pioneering text *Principles of Chemical Engineering* by Walker, Lewis, and McAdams.[1] This was the first major departure from curricula that regarded the techniques involved in the production of specific products as largely unique, to a formal recognition of the fact that certain physical or chemical processes, and corresponding fundamental principles, are common to many widely differing industrial technologies. A natural outgrowth of this radical new view was the gradual appearance of fluid mechanics and transport in both teaching and research as the underlying basis for many of the unit operations. Of course, many of the most important unit operations take place in equipment of complicated geometry, with strongly coupled combinations of heat and mass transfer, fluid mechanics, and chemical reaction, so that the exact equations could not be solved in a context of any direct relevance to the process of interest. Hence, insofar as the large-scale industrial processes of chemical technology were concerned, even at the unit operations level, the impact of fundamental studies of fluid mechanics or transport phenomena was certainly less important than a well-developed empirical approach (and this remains true today in many cases). Indeed, the great advances and discoveries of fluid mechanics during the first half of this century took place almost entirely without the participation (or even knowledge) of chemical engineers.

Gradually, however, chemical engineers began to accept the premise that the generally "blind" empiricism of the "lumped parameter" approach to transport processes at the unit operations scale should at least be supplemented by an attempt to understand the basic physical principles. This finally led, in 1960, to the appearance of the landmark textbook of Bird, Stewart, and Lightfoot,[2] which not only introduced the idea of detailed analysis of transport processes at the continuum level, but also emphasized the mathematical similarity of the governing field equations, along with the simplest constitutive approximations for fluid mechanics, heat, and mass transfer. The presentation of Bird, Stewart, and Lightfoot was overly simple in places, and the text itself primarily transmits simple results and solutions rather than methods of solution or analysis. Whatever the shortcomings of BSL, however, its impact has endured over the years. The combination of the more fundamental approach that it pioneered and the appearance of a few chemical engineers with very strong mathematics backgrounds who could understand the theoretical development of asymptotic methods by applied mathematicians produced the most recent major revolution in our way of thinking about and understanding transport processes. This is the application of the methods of asymptotic analysis to problems in transport phenomena, which began in the research literature of the late 1950s and early 1960s and is still a major contributor in the research arena.

B. Asymptotic Methods in Transport Processes

The major emphasis of this book is the use of asymptotic techniques in the analysis and understanding of transport processes from a fundamental point of view. At the most straightforward level, these methods provide a systematic framework to generate approximate solutions of the nonlinear differential equations of fluid mechanics, as well as the corresponding thermal energy (or species transport) equations. Perhaps more important than the detailed solutions enabled by these methods, however, is that they demand an extremely close interplay between the mathematics and the physics, and in this way contribute a very powerful understanding of the physical phenomena that characterize a particular problem or process. Of greatest importance in this regard is the concept of characteristic scales and nondimensionalization. It is the presence of large or small dimensionless parameters in appropriately nondimensionalized equations or boundary conditions that is indicative of the relative magnitudes of the various *physical* mechanisms in each case.

There is, in fact, an element of truth in the suggestion that asymptotic approximation methods are nothing more than a sophisticated version of dimensional analysis. Certainly it is true, as we shall see, that successful application of scaling/nondimensionalization can provide much of the information and insight about the nature of a given fluid mechanics or transport process without the need either to solve the governing differential equations or even be concerned with a detailed geometric description of the problem. The latter determines the magnitude of numerical coefficients in the correlations between dependent and independent dimensionless groups, but does not determine the *form* of the correlations. In this sense, asymptotic theory can reduce a whole class of problems, which differ only in the geometry of the boundaries and in the nature of the undisturbed flow, to the evaluation of a single coefficient. When the body or boundary geometry is simple, this can be done via detailed solutions of the governing equations and boundary conditions. Even when the geometry is too complex to obtain analytic solutions, however, the general asymptotic framework is unchanged, and the correlation between dimensionless groups is still reduced to determination of a single constant, which can now be done (in principle) via a single experimental measurement.

Thus, the power and impact of asymptotic techniques is not mainly because they provide approximate solutions of complex problems, but rather because they provide a framework to *understand* the essential features from a qualitative physical point of view. It is important, however, not to overstate what can be accomplished by asymptotic (and related analytic) techniques applied either to fluid mechanics or heat (and mass) transfer processes. At most, these methods can treat limiting regimes of the overall parameter domain for any particular problem. Furthermore, the approximate solutions obtained can be no more general than the framework allowed in the problem statement; that is, if we begin by seeking a steady axisymmetric solution, an asymptotic analysis will only produce an approximation for this class of solutions and, by itself, can neither guarantee that the solution is unique within this class nor that the limitation to steady and axisymmetric solutions is representative of the actual physical situation. For example, even if the geometry of the problem is completely axisymmetric, there is no guarantee that an axisymmetric solution exists for the velocity or temperature field, or if it does, that it corresponds to the motion or temperature field that would be realized in the laboratory. The latter may be either time dependent or fully three-dimensional or both. In this case, the most that we may hope is that these more complex motions may exist as a consequence of instabilities in the basic, steady, axisymmetric solution, and thus conditions for departure from this basic state predicted via the framework of classical stability theories. The important message is that analytic techniques, including asymptotic methods, are not sufficient by themselves to understand fluid mechanics or heat transfer processes. Such techniques would almost always need to be supplemented by some combination of stability analysis or, more generally, by experimental or computational studies of the full problem.

Of course, this situation is not unique. Generally, all of engineering science and/or mathematical physics is characterized by the essential interplay of mathematical analysis, and experiment, either from laboratory studies or, more recently, via computational studies of the full nonlinear problem. The existence of so-called "computational experiments," either as a supplement to the laboratory or as a replacement in some cases, is illustrative of another general feature of fluid mechanics and heat transfer processes. Provided that we limit ourselves to Newtonian fluids, and to flow domains involving only solid boundaries, there is no question of understanding the physical principles that govern the problem, at least from a continuum mechanical point of view. On a point-by-point basis, these are represented by the Navier-Stokes and thermal energy equations, with boundary conditions that are well established. The basic problem is to understand the *macroscopic phenomena* that are inherent in these physical principles. For example, if we consider a problem such as the translational motion of a cylindrical body through a fluid at low Mach number, the motion of the fluid in all circumstances is governed by the incompressible Navier-Stokes equations, yet the range of observable flow phenomena that occur as the velocity of the cylinder is increased relative to that of the fluid is quite amazing, beginning at low speeds with steady motion that follows the body contour and followed by a transition to asymmetric motion that includes a standing pair of recirculating vortices, which are at first steady and attached and then become unsteady and alternately detached as the velocity increases. Finally, at very high relative speeds the whole motion in the vicinity of the cylinder and downstream becomes three-dimensional and chaotic in the well-known regime of turbulent flow. *All* of these phenomena are inherent in the physical principles encompassed in the Navier-Stokes equations. It is "only" that the solutions of these equations (namely, the physical phenomena) become increasingly complex with increase in the relative velocity of the cylinder and the fluid. In principle, with the advent of increasingly powerful computers and numerical techniques, once the basic physical principles are established, the corresponding physical phenomena can be exposed equally well by solving the full problem numerically or by carrying out laboratory experiments.

It is unfortunate that a qualitative description, or even flow-visualization pictures, of complex phenomena do not translate immediately into "understanding." Obviously, if this were the case, it would be possible to provide students with a much more realistic picture of real phemonena than they can hope to achieve in the normal classroom (or textbook) environment. The difficulty with a qualitative description is that it can never go much beyond a case-by-case approach, and it would clearly be impossible to encompass all of the many flow or transport systems that will be encountered in technological applications. The present book does not provide anything like a catalog of physically interesting phenomena. Hopefully, the reader will have already encountered some of these in the context of undergraduate laboratory studies or personal experience. There is also at least one textbook[3] that attempts (with some success) to fill the gap between "analytic technique" and "physical phenomena" in fluid mechanics, and this can provide an important complement to the material presented here. In fluid mechanics, a very well-known film series, *Illustrated Experiments in Fluid Mechanics,*[4] is widely available also, and it is an *excellent* source of visual exposure to real phenomena, coupled with useful physical explanations as well. Finally, every student and teacher of fluid mechanics should examine the wonderful collection of photos in the book *An Album of Fluid Motion*[5] and in the series of articles "A Gallery of Fluid Motion,"[6] which was spawned from the book. The events depicted in these latter photographs provide a graphic reminder of the vast wealth of complex, important, and interesting phenomena that are encompassed within fluid mechanics. Clearly, the fluid mechanics and heat transfer presented in the classroom or textbook only scratches the surface of this fascinating subject.

C. Organization of the Book

The arrangement of the material in this book falls roughly into four parts: (1) an introduction to basic principles, including derivation of governing equations;

(2) linear problems, including unidirectional and creeping flows; (3) a transition toward fully nonlinear phenomena, including an introduction to asymptotic methods, which culminates in the addition of weak convection effects to creeping flow or pure conduction problems in heat transfer; and (4) boundary-layer theories for fluid flow and heat (or mass) transfer in both forced and natural convection problems. The intent of this organization is severalfold. First, by presenting the material roughly in order of mathematical complexity, there is an opportunity to learn or review classical methods of (linear) applied mathematics, as well as generally useful methods such as similarity transformations, before they become coupled with the complexities of asymptotic techniques. In this way, the emphasis in studying the latter chapters can focus on critical concepts of the asymptotic methodology, such as nondimensionalization or scaling or matching. Secondly, I have tried to order the material in a logical manner with respect to the complexity of the physical phenomena, with a special emphasis in the first part of the book on viscous-dominated phenomena that have been, historically, the special province of chemical engineers. However, the book does not attempt to provide tidbits of information about every possible problem of interest to chemical engineers. It does attempt to build a framework that can be applied to a much broader spectrum of problems than is actually discussed in the book.

Chapter 2 contains a comprehensive derivation of the governing equations from first principles, beginning with the concept of continuum mechanics as an average of a more fundamental, molecular description. Special emphasis is placed on the connection with classical mechanics, and on the microscopic origin of the stress and other surface contributions to the basic balance equations. The latter sections of the chapter are concerned primarily with constitutive equations. Most important is a clear statement of what is assumed in formulating these equations, and what is derived or deduced beyond these basic assumptions. Although the remainder of the book is about Newtonian fluids, I consider constitutive equations for both Newtonian and non-Newtonian fluids. In doing this I hope to underline the simplicity of the model for a Newtonian fluid and emphasize the fact that fluids can exist with much more complex behavior than those studied in this (and other) graduate texts in transport phenomena.

Chapter 3 is concerned with the class of motions that are known as unidirectional. For this class, the full nonlinear Navier-Stokes equations are reduced to a single, linear partial differential equation that governs the dependence of the single nonzero velocity component on spatial position and time. This, then, provides a very convenient and interesting framework to introduce the concept of similarity transformations and self-similar solutions. From a physical viewpoint, the evolution of flows via the diffusion of vorticity (alternatively, momentum) provides a basis to understand the characteristic time scale of the transients prior to establishment of a steady flow in a bounded domain.

Chapter 4 is the first of two chapters that address flows which are dominated by viscous effects, and can thus be described approximately via the so-called creeping flow equations. This class of problems differs from those discussed in Chapter 3 in that the nonlinear inertia terms are neglected because they are *small* under the prescribed flow conditions, rather than being precisely zero because of a geometric restriction as in unidirectional flows. The condition of approximation is that the Reynolds number is very (asymptotically) small. The material in Chapter 4 is split into three basic parts.

The first considers a number of general consequences of the linearity of the creeping flow equations. The second discusses two-dimensional and axisymmetric flows in which the creeping flow equations can be expressed in terms of the streamfunction and then solved via a general eigenfunction expansion. Specific flows considered are two-dimensional corner flows, uniform flow past a sphere (Stokes' flow), and a sphere in an axisymmetric extensional flow. The third topic introduces a more general solution methodology using superpositions of vector harmonic functions. This powerful technique allows solution of full three-dimensional problems.

Chapter 5 discusses additional topics in creeping flow theory. The most general of these is the motion of bubbles and drops, which begins with a discussion of boundary conditions at a fluid-fluid interface and then considers three specific problems using the solution methods from Chapter 4: the classic Hadamard-Rybcynski solution for translation of a spherical drop; the classic problem of bubble motion in a temperature gradient when surface-tension-driven Marangoni phenomena play a critical role; and the motion of a drop in a general linear-flow, including a first approximation to the flow-induced deformation of shape. The remainder of this chapter consists of one section on the use of fundamental solutions to construct solutions of general creeping flow problems, and a second section on special topics in creeping flow theory.

Chapter 6 begins the introduction to asymptotic approximations. After first returning to the unidirectional problem of pulsatile flow in a circular tube, where we now consider the asymptotic limits of high- and low-frequency pressure oscillations, there is a brief discussion of general considerations and principles for asymptotic approximations. Most important among these are the definition of convergence of an asymptotic series and the difference between regular and singular expansions. We then consider the problem of flow through a slightly curved tube, expressed as a regular asymptotic expansion in the magnitude of the curvature. This is followed by a discussion of the one-dimensional flow associated with the increase (or decrease) of the radius of a spherical gas bubble in a time-dependent pressure field. Analysis of various aspects of this problem involve regular perturbations, as well as an introduction to the method of domain perturbations and two-time scale approximations for resonant oscillations in a periodic pressure field.

Chapter 7 considers problems involving thin films, lubrication theory, and related topics. These generally represent the leading order approximation in a *singular* asymptotic expansion. Although the flow in the thin-film region is superficially similar to unidirectional flows, a critical difference is that the boundaries are no longer exactly parallel, and this leads to very important changes. There are three basic topics discussed: first, the motion between a pair of eccentric rotating cylinders, which reduces to the well-known "journal-bearing" lubrication geometry in the narrow gap limit; second, a general discussion of lubrication theory, including derivation of the famous Reynolds equation and its application to several specific lubrication problems; and third, an analysis of inertia effects in thin-film flows, including both the limit of high and low Reynolds number, where we encounter the concept of a boundary-layer.

The last five chapters of the book are devoted to the application of asymptotic approximation methods to the analysis of more general problems of fluid mechanics and convective heat transfer. The prototype problem involves the motion of a fluid in the vicinity of a body, which may either be at the same temperature or heated or cooled with respect to the ambient fluid temperature. The first four of these chapters, at least, should be studied as a unit. Together they provide a summary of forced convection heat transfer for both high and low Reynolds number flows.

Chapter 8 begins by considering the effects of small contributions of convection of heat and/or vorticity to fluid mechanics at small, but nonzero, Reynolds numbers and to heat transfer at small, but nonzero, Peclet number. The general features of these two classes of problem are very similar, but the heat transfer case is somewhat simpler in detail, and is therefore considered first, beginning with the *conduction* limit for Pe→0. The key idea, which is repeated from examples in earlier chapters, is that the temperature field is characterized by different length scales in different parts of the domain. Thus, asymptotic approximations, derived from nondimensionalized versions of the governing equations, must also be different in different parts of the domain, and this leads to the notion of *matched asymptotic approximations*. Following application to the weak convection heat transfer problem for a translating body, the same ideas are then applied to the calculation of heat transfer rates from a sphere in a simple shear flow, and to the inclusion of inertial corrections to the drag on a sphere at low, but nonzero, Reynolds number.

Chapter 9 then considers heat and mass transfer at low Reynolds numbers, but asymptotically large values of Peclet number. This leads naturally to the concept of a thermal boundary-layer for flows in the vicinity of heated (or cooled) bodies. Three types of problems are analyzed in the asymptotic framework of thermal boundary-layer theory: (1) solid bodies in streaming flows, (2) a solid sphere in a generalized shear flow, and (3) a bubble or drop that translates through a quiescent fluid. The last section in Chapter 9 is concerned with heat transfer across regions of closed streamline flow. This topic is not only of some practical importance but also involves novel (up to now) ideas for asymptotic analysis of this system, in which advection *along* streamlines is increasing in importance as the Peclet number increases, but the steady state temperature distribution is established by conduction *across* streamlines—with a time scale that becomes extremely long as the Peclet number increases (that is, thermal diffusivity decreases).

Chapter 10 then takes up the topic of laminar boundary-layer theory for fluid motion. This involves the potential flow approximation for the region that is not too near the body surface and the classical laminar boundary-layer equation for the remainder of the domain. Once the preliminaries are presented, we turn to a number of different types of solution for various physical problems. The first class of solutions are for two-dimensional bodies that allow the use of a similarity transformation. This includes the famous Blasius analysis for flow past a horizontal flat plate, and the equally well-known Falkner-Skan equation for flow past a semi-infinite wedge. Motivated by the behavior of the Falkner-Skan solution for "adverse" pressure gradients, we then consider boundary-layer separation. The analysis presented is for two-dimensional cylindrical bodies of arbitrary cross-sectional shape, via the so-called Blasius series. An important element of this section, however, is a discussion of the mechanism for separation and the critical role of viscous effects in this high Reynolds number process. In the next two sections, we consider an approximate integral technique known as the

Karman-Pohlhausen method to solve the boundary-layer equations, and generalization of the Blasius series to axisymmetric bodies. Finally, in the last section, we consider the boundary-layer on a bubble at high Reynolds numbers. The latter problem is not only of quantitative interest, but also the contrast in flow structure between the bubble with slip boundary conditions and solid bodies with no-slip conditions provides considerable insight into the nature of the boundary-layer approximation.

Chapter 11 returns to heat and mass transfer, this time for flow at large Reynolds numbers. The analysis follows the methodology established in the previous chapters to develop asymptotic solutions for arbitrary two-dimensional bodies at large and small Prandtl numbers.

Finally, Chapter 12 considers natural convection phenomena and mixed convection problems, which involve both forced and free convection simultaneously. One important distinction between natural and forced convection problems is that there is no longer an externally imposed velocity scale. Rather, the velocity scale depends upon the magnitude of the buoyancy force in the system. We consider two types of problems, both based upon the Boussinesq approximation of the Navier-Stokes equations. The first is buoyancy-driven motion in natural convection boundary-layers at the surface of a heated (or cooled) body at large Grashoff numbers. The second is the plumelike motion that occurs at large distances above (or below) a heated (or cooled) body, which appears in the analysis as either a *point* or *line* source of heat. The latter problem is representative of a larger class of problems that could be analyzed in a similar manner, namely, the wakes behind bodies in a flow, or a liquid jet far from a source of mass flow. The objective of our discussion of mixed convection problems is primarily to provide a rational basis for neglect of buoyancy effects in the forced convection approximation and/or neglect of forced convection effects in natural convection problems. However, an outgrowth of the asymptotic method is that a framework can be developed that allows a rigorous transition between these two limiting cases. Finally, we consider briefly the buoyancy-driven instability of a fluid layer that is heated from below. The fundamental distinction between this problem, which involves boundaries that are strictly horizontal (normal to **g**), and the natural convection problems of the preceding sections is that there is a steady-state configuration that is characterized by a nonzero temperature gradient but *no motion*. When buoyancy-driven motion does occur, it is a consequence of an instability in this basic state, which occurs only for temperature gradients above a certain critical value. In contrast, when the bounding surfaces or walls are not strictly horizontal, a buoyancy force of arbitrarily small magnitude will always produce motion, and this is characteristic of the problems analyzed earlier in the chapter.

References/Notes

1. Walker, W.H., Lewis, W.K., and McAdams, W.H., *Principles of Chemical Engineering*. McGraw-Hill: New York (1923).

2. Bird, S.R., Stewart, W.E., and Lightfoot, E.N., *Transport Phenomena*. Wiley: New York (1960).

3. Tritton, D.J., *Physical Fluid Dynamics*. Van Nostrand Reinhold: London (1977).

4. National Committee for Fluid Mechanics Films, *Ilustrated Experiments in Fluid Mechanics*. MIT Press: Cambridge (1972).

5. Van Dyke, M., *An Album of Fluid Motion*. Parabolic Press: Stanford, CA. (1982).

6. Reed, H.L., "A Gallery of Fluid Motion," *Physics of Fluids* 28:2631–2640 (1985); 29:2769–2780 (1986); 30:2597–2606 (1987); 31:2383–2394 (1988); 1:1439–1450 (1989); 2:1517–1528 (1990).

CHAPTER 2

Basic Principles

We shall be concerned in this book with a description of the motion of fluids under the action of some applied force and with convective heat transfer in moving fluids that are not isothermal. We assume that the reader is familiar with the basic principles and equations that describe these processes from a continuum mechanics point of view. Nevertheless, we begin our discussion with a review of these principles and the derivation of the governing differential equations. We aim to provide a reasonably concise and unified point of view. It has been our experience that the lack of an adequate understanding of the basic foundations of the subject frequently leads to a feeling on the part of the students that the whole subject is impossibly complex. However, the physical principles involved are actually quite simple and generally familiar to any student with a physics background in classical mechanics. Indeed, the main problems of fluid mechanics and of convective heat transfer are not in the complexity of the underlying physical principles, but rather in the attempt to understand and describe the fascinating and complicated phenomena that they allow. From a mathematical point of view, the main problem is not the derivation of the governing equations that is presented in this second chapter, but in their solution. The latter topic will occupy the remaining chapters of this book.

A. The Continuum Approximation

It will be recognized that one possible approach to the description of a fluid in motion is to examine what occurs at the microscopic level where the stochastic motions of individual molecules can be distinguished. Indeed, to a student of physical chemistry or, perhaps, chemical engineering, who has been consistently exhorted to think in molecular terms, this may at first seem the obvious approach to the subject. However, the resulting many-body problem of molecular dynamics is impossibly complex under normal circumstances where the fluid domain contains an enormous number of molecules. Attempts to simulate such systems with even the largest of present-day computers cannot typically handle more than a few hundred molecules of simple shape at one time.[1] Thus, efforts to provide a mathematical description of fluids in motion could not have succeeded without the introduction of sweeping model approximations. The most important among these is the so-called *continuum hypothesis*, according to which the fluid is modeled as completely lacking in microscale structure, with differences between materials being distinguished only through such averaged properties as the density, viscosity, or thermal conductivity.

The motivation for this approach, apart from an anticipated simplification of the problem, is that in most applications of applied science or engineering, we are concerned with fluid motions or heat transfer in the vicinity of bodies, such as airfoils, or in confined geometries, such as a tube or pipeline, where the physical dimensions are very much larger than any molecular or intermolecular length scale of the fluid. The desired description of fluid motion is, then, at this larger, *macrocscopic* level where, for example, an average of the forces of interaction between the fluid and the bounding surface may be needed, but not the instantaneous forces of interaction between this surface and individual molecules of the fluid.

Once the continuum hypothesis has been adopted, the usual macroscopic laws of classical continuum physics are invoked to provide a mathematical description of fluid motion and/or heat transfer in nonisothermal systems—namely, conservation of mass, conservation of linear and angular momentum (the basic principles of Newtonian mechanics), and conservation of energy (the first law of thermodynamics). Although the second law of thermodynamics does not contribute directly to the derivation of the governing equations, we shall see that it does provide constraints on the allowable forms for the so-called constitutive models that relate the velocity gradients in the fluid to the short-range forces that act across surfaces within the fluid.

The development of convenient and usable forms of the basic conservation principles and the role of the constitutive models in a continuum mechanics framework will occupy the remaining sections of this chapter. Before embarking on this discussion, with its implicit acceptance of the continuum hypothesis, we discuss the foundations and consequences of this hypothesis in somewhat more detail.

Foundations

In adopting the continuum hypothesis, we assume that it is possible to develop a description of fluid motion (or heat transfer) on a much coarser scale of resolution than on the molecular scale that is still *physically equivalent* to the molecular description in the sense that the former could be derived, in principle, from the latter by an appropriate averaging process. Thus, it must be possible to define any dependent macroscopic variable as an average of a corresponding molecular variable. A convenient average for this purpose is suggested by the utility of having macroscopic variables that are readily accessible to experimental observation. Now, from an experimentalist's point of view, any probe to measure velocity, say, whose dimensions were much larger than molecular, would automatically measure a *spatial* average of the molecular velocities. At the same time, if the probe were sufficiently small compared to the dimensions of the flow domain, we would say that the velocity was measured "at a point," in spite of the fact that the measured quantity was an average value from the molecular point of view. This simple example, suggests a convenient definition of the macroscopic variables in terms of molecular variables, namely as volume averages, for example,

$$\mathbf{u} \equiv \langle \mathbf{w} \rangle \equiv \frac{1}{V} \int_V \mathbf{w} \, dV, \qquad (2\text{-}1)$$

where V is the averaging volume.

It is important to remark that we shall never actually calculate macroscopic variables as averages of molecular variables. The purpose of introducing an explicit connection between the macroscopic and molecular (or microscopic) variables is that the conditions for $\langle \mathbf{w} \rangle$ to define a meaningful macroscopic point variable provide sufficient conditions for validity of the continuum hypothesis. In particular, if $\langle \mathbf{w} \rangle$ is to represent a statistically significant average, the typical linear dimension of the averaging volume, $V^{1/3}$, must be large compared to the scale δ that is typical of the microstructure of the fluid. Most frequently, δ represents a molecular length scale. However, multiphase fluids such as suspensions may also be considered, and in this case δ is the largest microstructural dimension—say, the interparticle length scale or the particle radius. If at the same time $\langle \mathbf{w} \rangle$ is to provide a meaningful point variable in the macroscopic description, it must have a unique value at each point in space at any particular instant, and this implies that the linear dimension, $V^{1/3}$, must be arbitrarily small compared with the macroscopic scale, L, that is determined by the geometry of the flow domain. Thus, with macroscopic variables defined as volume averages of corresponding microscopic variables, the existence of an equivalent continuum description of fluid motions or heat transfer processes (that is, validity of the continuum hypothesis) requires

$$\delta \ll V^{1/3} \ll L. \qquad (2\text{-}2)$$

In other words, it must be possible to choose an averaging volume that is arbitrarily small compared with the macroscale L while still remaining very much larger than the microscale δ. While the condition (2-2) will always be sufficient for validity of the continuum hypothesis, it is unnecessarily conservative, as a consequence of the choice (2-1) of volume averaging rather than the more fundamental ensemble average definition of macroscopic variables. Nevertheless, the preceding discussion is adequate for our present purposes.

Consequences

Two consequences of the continuum approximation will appear later in this chapter. The first is the necessity to hypothesize two independent mechanisms for heat or momentum transfer: one associated with the transport of heat or momentum via the continuum or macroscopic velocity field $\mathbf{u}$, and the other a "molecular" mechanism for heat or momentum transfer that will appear as a surface contribution to the macroscopic momentum and energy conservation equations. This split into two independent transfer mechanisms is a direct consequence of the coarse resolution that is inherent in the continuum description of the fluid system. If we revert to a microscopic or molecular point of view for a moment, it is clear that there is only a single class of mechanisms available for transport of any quantity—namely, those mechanisms associated with the motions and forces of interaction between the molecules (and particles in the case of suspensions). When we adopt the continuum or macroscopic point of view, however, we effectively split the molecular motion of the material into two parts: a molecular average velocity $\mathbf{u} \equiv <\mathbf{w}>$, and local fluctuations relative to this average. Since we define $\mathbf{u}$ as an instantaneous spatial average, it is evident that the local net volume flux of fluid across any surface in the fluid will be $\mathbf{u} \cdot \mathbf{n}$, where $\mathbf{n}$ is the unit *normal* to the surface. In particular, the local fluctuations in molecular velocity relative to the average value $<\mathbf{w}>$ yield no net flux of *mass* across any macroscopic surface in the fluid. However, these local random motions *will* generally lead to a net flux of heat or momentum across the same surface.

To illustrate this fact, we may adopt the simplest model fluid—the billiard-ball gas—and consider a surface in the fluid that is everywhere locally tangent to $\mathbf{u}$ so that $\mathbf{u} \cdot \mathbf{n} = 0$ along the surface. For the billiard-ball gas, the only microscopic or molecular vehicle for transport is the random translational motions of molecules across the surface. The significance of the condition $\mathbf{u} \cdot \mathbf{n} = 0$ is simply that for every gas molecule that passes across the surface in one direction, a second will, on average, pass across in the opposite direction. Thus, the net flux of mass across the surface will be zero if the surface is everywhere locally tangent to $\mathbf{u}$. On the other hand, the presence of mean temperature gradient (a gradient of the mean molecular kinetic energy) normal to the surface means that each interchange of gas molecules would transfer a net quantity of heat, even though the average *convective* flux of heat associated with $\mathbf{u}$ is zero (see Figure 2–1). It is this additional transport due to fluctuations in the molecular velocity about the continuum velocity $\mathbf{u}$ that must be incorporated in the continuum description as a local "molecular" surface heat flux contribution to the energy balance on a fluid element. We emphasize that the split of heat transfer into a convective contribution associated with $\mathbf{u}$, plus an additional molecular contribution, is due to the continuum description of the system. *Obviously, the sum of the convective and molecular heat flux contributions in the continuum description must be identical to the total flux due to molecular motions if the continuum description of the system is to have any value.* Similarly, the continuum description of momentum transfer across a surface must also include both a convective part associated with $\mathbf{u}$ and a molecular part due to random fluctuations of the actual molecular velocity about the local mean value $\mathbf{u}$. The billiard-ball gas again provides a convenient vehicle for descriptive purposes. In this case, if we consider a surface that is locally tangent to $\mathbf{u}$, it is evident that there can be no transport of mean momentum $\rho\mathbf{u}$ across the surface due to the macroscopic motion itself; the momentum flux due to this mechanism is $\rho\mathbf{u}(\mathbf{u} \cdot \mathbf{n}) \equiv 0$. On the other hand, if $(\nabla\mathbf{u}) \cdot \mathbf{n} \neq 0$, then the random interchange of gas molecules due to fluctuations in their velocity relative to $\mathbf{u}$ will lead to a net transport of momentum that must be included in the (macroscopic) continuum mechanics principle of linear momentum conservation as a local, molecular flux of momentum across any surface

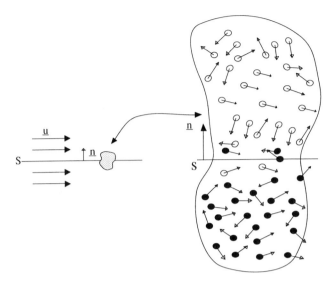

Figure 2–1 We consider a surface S drawn in a fluid that is modeled as a billiard-ball gas. Initially, when viewed at a macroscopic level, there is a discontinuity across the surface— the fluid above is white and the fluid below is black. The macroscopic (volume average) velocity is parallel to S so that $\mathbf{u} \cdot \mathbf{n} = 0$. Thus, there is no transfer of black fluid to the white zone, or vice versa, due to the macroscopic motion $\mathbf{u}$. At the molecular (billiard-ball) level, however, all of the molecules undergo a random motion (it is the average of this motion that we denote as $\mathbf{u}$). This random motion produces no net transport of billiard balls across S when viewed at the macroscopic scale since $\mathbf{u} \cdot \mathbf{n} = 0$. However, it does produce a net flux of *color*—on average there is a net flux of black balls across S into the white region and vice versa. In a macroscopic theory designed to describe the transport of white and black fluid, this net flux would appear as a surface contribution. The presence of this flux would gradually smear the initial step change in color until eventually the average color on both sides of S would be the same mixture of white and black.

element in the fluid. In this case, the molecular transport of mean momentum has the effect of decreasing any existing gradient of $\mathbf{u}$—crudely speaking, the slower moving fluid on one side of the surface appears to be accelerated, while the faster moving fluid on the other side is decelerated. Thus, from the continuum point of view, it appears that equal and opposite *forces* have acted across the surface, and the molecular flux of momentum is often described in the continuum description as a surface force per unit area and called the *stress vector*. Whatever it is called, however, it is evident that the continuum or macroscopic description of the system that results from the continuum hypothesis can only be successful if the flux of momentum due to fluctuations in molecular velocity about $\mathbf{u}$ is modeled in such a way that the sum of its contributions plus the transport of momentum via $\mathbf{u}$ is equal to the actual molecular average momentum flux. As we shall see, the attempt to provide models to describe the "molecular" flux of momentum or heat in the continuum formulation, without the ability to actually calculate these quantities from a rigorous molecular theory, is the greatest source of uncertainty in the use of the continuum hypothesis to achieve a tractable mathematical description of fluid motions and heat transfer processes.

A second similar consequence of the continuum hypothesis is an uncertainty in the boundary conditions to be used in conjunction with the resulting equations for motion and heat transfer. With the continuum hypothesis adopted, the conservation principles of classical physics, listed earlier, will be shown to provide a set of so-called field equations for molecular average variables such as the continuum point velocity **u**. To solve these equations, however, the values of these variables or their derivatives must be specified at the boundaries of the fluid domain. These boundaries may be solid surfaces, or the phase boundary between a liquid and a gas, or the phase boundary between two liquids. In any case, when viewed on the *molecular* scale, the "boundaries" are seen to be regions of rapid but continuous variation in fluid properties such as number density. Thus, in a molecular theory, boundary conditions would not be necessary. When viewed with the much coarser resolution of the macroscopic or continuum description, on the other hand, these local variations of density (and other molecular variables) can only be distinguished as discontinuities, and the continuum (or molecular average) variables such as **u** appear to vary smoothly on the scale L, right up to the boundary where some boundary condition is applied.

One consequence of the inability of the continuum description to resolve the region nearest the boundary is that the continuum variables extrapolated toward the boundary from the two sides may experience jumps or discontinuities. This is definitely the case at a fluid interface, as we shall see. Even at solid boundaries, the fluid velocity **u** may appear to "slip" when the fluid is a high molecular weight material or a particulate suspension.[2]

Here the situation is very similar to that encountered in connection with the need for continuum (constitutive) models for the molecular transport processes in that a derivation of appropriate boundary conditions from the more fundamental, molecular description has not been accomplished to date. In both cases, the knowledge that we have of constitutive models and boundary conditions that are appropriate for the continuum level description is largely empirical in nature. In effect, we make an educated guess for both constitutive equations and boundary conditions, and then normally judge the success of our choices by the resulting comparison between predicted and measured continuum velocity or temperature fields. Models derived from molecular theories are generally not available for comparison with the empirically proposed models. We shall discuss some of these matters in more detail later in this chapter, where specific choices will be proposed for both the constitutive equations and boundary conditions.

B. Conservation of Mass; The Continuity Equation

Once we adopt the continuum hypothesis and choose to describe fluid motions and heat transfer processes from a macroscopic point of view, the governing equations are derived by invoking the usual conservation principles of classical continuum physics. These are conservation of mass, conservation of linear and angular momentum, and conservation of energy. As usual in macroscopic mechanics, we apply these principles to a specific mass of the material that moves along through space. The only real conceptual difficulty is the fact that the chosen mass of material changes shape as a consequence of spatial gradients in the continuum point velocity field.

The simplest of the various conservation principles to apply is conservation of mass. It is instructive to consider its application relative to two different, but equivalent, descriptions of our fluid system. In both cases, we begin by identifying a specific macroscopic body of fluid that lies within an arbitrarily chosen volume element at some initial instant of time. Since we have adopted the continuum mechanics point of view, this volume element will be large enough that any flux of mass across its surface due to random molecular motions can be neglected completely. Indeed, in this continuum description of our system, we can resolve only the molecular average (or continuum point) velocities, and it is convenient to drop any reference to the averaging symbol < >. The continuum point velocity vector is denoted as **u**.

In the first description of mass conservation for our system, we consider the volume element (here called a *control volume*) to be fixed in position and shape as illustrated in Figure 2–2. Thus, at each point on its surface, there is a mass flux of fluid $\rho\mathbf{u}\cdot\mathbf{n}$. With **n** chosen as the outer unit normal to the surface, this mass flux will be negative at points where fluid enters the volume element and positive where it exits. There is no reason, at this point, to assume that the fluid density ρ is necessarily constant. Indeed, conservation of mass requires the density inside the volume element to change with time in such a way that any imbalance in the mass flux in and out of the volume element is compensated by an accumulation of mass inside. Expressing this statement in mathematical terms

$$\int_V \frac{\partial \rho}{\partial t} dV = -\int_S \rho\mathbf{u}\cdot\mathbf{n}dS, \tag{2–3}$$

where V denotes the arbitrarily chosen volume element of fixed position and shape, and S denotes its (closed) surface. Equation (2–3) is an integral constraint on the velocity and density fields within a given closed volume element of fluid. Since this volume element was chosen arbitrarily, however, an equivalent differential constraint at each point in the fluid can be derived easily. First, the well-known divergence theorem[3] is applied to the right-hand side of (2–3), which thus becomes

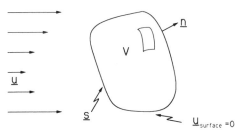

Figure 2–2 An arbitrarily chosen control volume of *fixed* position and shape, immersed in a fluid with velocity **u**. The velocity of the surface of the control volume is zero, and thus, there is a net flux of fluid through its surface. Equation 2-3 represents a mass balance on the volume V, with the left hand side giving the rate of mass accumulation and the right hand side the net flux of mass *into* V due to the motion **u**.

$$\int_V \left[\frac{\partial \rho}{\partial t} + \nabla \cdot (\rho \mathbf{u}) \right] dV = 0. \tag{2-4}$$

Then we note that this integral condition on ρ and $\mathbf{u}$ can be satisfied for an arbitrary volume element only if the integrand is identically zero, that is,

$$\boxed{\frac{\partial \rho}{\partial t} + \nabla \cdot (\rho \mathbf{u}) = 0.} \tag{2-5}$$

This is the famous *continuity equation,* which we now recognize as the pointwise constraint on ρ and $\mathbf{u}$ that is required by conservation of mass. In order to justify (2–5), we note that the only other way, in principle, to satisfy (2–4) would be if the integrand were positive within some portion of V and negative elsewhere in such a way that the nonzero contributions to the volume integral cancel. However, if this were the case, the freedom to choose an arbitrary volume element would lead to a contradiction. In particular, instead of the original choice of V, we could simply choose a new volume element that lies entirely within the region where the integrand is positive (or negative). Evidently, (2–4) is then violated, leading us to conclude that (2–5) must hold everywhere.

Although the derivation of the continuity equation using a fixed volume element is perfectly satisfactory, we consider a second approach to the same problem that will be conceptually preferable for application to the momentum and energy conservation principles. Here, instead of writing a mass balance for a volume element fixed in space, we allow an arbitrarily chosen volume element to translate and deform with time so that it always contains the same body of fluid, as in Figure 2–3. Thus, each point on the surface of this volume element moves with the local (continuum) velocity of the fluid. Such a volume element is sometimes called a *material control volume.* We shall denote a material control volume as $V_m(t)$. Note that $V_m(0)$ can be considered to be coincident with the fixed arbitrary volume element V that was used in the first derivation of the continuity equation. When applied to the material control volume, the mass conservation principle can be expressed in the form

$$\boxed{\frac{D}{Dt} \left[\int_{V_m(t)} \rho \, dV \right] = 0.} \tag{2-6}$$

Here, the symbol D/Dt stands for the so-called *convected* or *material* derivative with respect to time, which gives the rate of change of the mass of the specific material control volume of fluid that has been denoted as $V_m(t)$. The material control volume moves and deforms in such a way that the local flux of mass across its surface is identically zero for all time. Thus, the mass contained within the material control volume is constant, independent of time. The convected derivative of any quantity may be interpreted as giving its time rate of change in a reference frame that moves with the local velocity of the fluid. The ordinary partial time derivative that appears in (2–4), on the other hand, is the time rate of change at a fixed point in space. The integral relationship (2–6) is perfectly general and precisely equivalent to the relationship

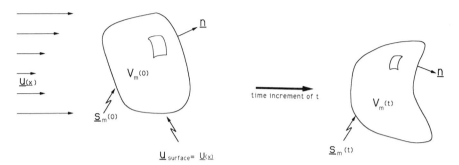

Figure 2–3 An arbitrarily chosen material volume element shown at some initial moment $t = 0$, and at a later time t, where it has translated and distorted in shape due to the fact that each point on its surface moves with the local fluid velocity **u**. Equation (2–6) represents a statement of mass conservation for this material volume element.

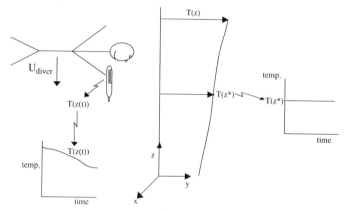

Figure 2–4 A sky diver falls with a velocity U_{diver} from a high altitude carrying a thermometer and a recording device that plots the instantaneous temperature, as shown in the lower left corner. During the period of descent, the temperature at any fixed point in the atmosphere is independent of time (i.e., the partial derivative $\partial T/\partial t \equiv 0$). However, the sky diver is in an inversion layer and the temperature decreases with decreasing altitude. Thus, the recording of temperature versus time obtained by the sky diver shows that the temperature decreases at a rate $DT/Dt^* = U_{diver}\partial T/\partial z$. This time derivative is known as the Lagrangian derivative, for an observer moving with velocity U_{diver}.

(2–4), but this cannot be seen easily without further discussion of the convected time derivative.

The first question that we may ask is the form of the relationship between D/Dt and the ordinary partial time derivative $\partial/\partial t$. The so-called "sky-diver" problem illustrated in Figure 2–4 provides a simple physical example that may serve to clarify the nature of this relationship without the need for notational complexity. A sky diver leaps from an airplane at high altitude and begins to record the temperature, T, of the atmosphere at regular intervals of time as he falls toward the earth. We denote his velocity $-U_{\text{diver}} i_z$, where i_z is a unit vector in the vertical direction, and the time derivative of the temperature he records as D^*T/Dt^*. Here, D^*/Dt^* represents the time rate of change (of T) measured in a reference frame that moves with the velocity of the diver. Evidently, there is a close relationship between this derivative and the convected derivative that was introduced in the preceding paragraph. Let us now suppose, for simplicity, that the temperature of the atmosphere varies with the distance above the earth's surface but is independent of time at any fixed point, say, $z = z^*$. In this case, the partial time derivative, $\partial T/\partial t$, is identically equal to zero. Nevertheless, in the frame of reference of the sky diver, D^*T/Dt^* is not zero. Instead,

$$\frac{D^*T}{Dt^*} = -U_{\text{diver}} \frac{\partial T}{\partial z}. \tag{2-7}$$

The temperature is seen by the sky diver to vary with time because he is falling relative to the earth's surface with a velocity U_{diver}, while T varies with respect to position in the direction of his motion.

The relatively simple concept represented by the sky-diver example is easily generalized to provide a relationship between the convected derivative of any scalar quantity B associated with a particular element of fluid that moves with velocity $\mathbf{u}$ and the partial derivatives of B with respect to time and spatial position in a fixed (inertial) reference frame. Specifically, B changes for a moving fluid element both because B may vary with respect to time at each fixed point at a rate $\partial B/\partial t$ and because the fluid element moves through space and B may be a function of spatial position in the direction of motion. The rate of change of B with respect to spatial position is just ∇B. The rate at which B changes with time for a fluid element with velocity $\mathbf{u}$ is then just the projection of ∇B onto the direction of motion multiplied by the speed, that is $\mathbf{u} \cdot \nabla B$. It follows that the convected time derivative of any scalar B can be expressed in terms of the partial derivatives of B with respect to time and spatial position as

$$\boxed{\frac{DB}{Dt} = \frac{\partial B}{\partial t} + \mathbf{u} \cdot \nabla B.} \tag{2-8}$$

However, this simple relationship cannot be applied directly to (2–6) because each point in the material control volume moves with a different local velocity $\mathbf{u}$. Further, the volume element itself changes as a function of time.

In order to evaluate the convected derivative of any integral quantity that is associated with a material control volume, we use a result from mathematics that is known as the Reynolds Transport Theorem.

Let us consider any scalar quantity $\alpha(\mathbf{x}, t)$ that is associated with a moving fluid. Then

$$\boxed{\frac{D}{Dt}\left[\int_{V_m(t)} \alpha(\mathbf{x}, t)\, dV\right] = \int_{V_m(t)}\left[\frac{\partial \alpha}{\partial t} + \nabla \cdot (\alpha \mathbf{u})\right] dV.} \tag{2-9}$$

The proof of this theorem is straightforward.[4] We first note that every point $\mathbf{x}(t)$ within a material control volume at any arbitrary instant of time can be specified completely by its position, say, $\mathbf{x}_0$, at some initial time and its velocity for $t \geq 0$. In particular,

$$\mathbf{x}(t) = \mathbf{x}_0 + \int_0^t \mathbf{u}(\tau, \mathbf{x}_0)d\tau,$$

where the notation $\mathbf{u}(\tau, \mathbf{x}_0)$ is used to denote the velocity of the particular element of fluid that was initially located at $\mathbf{x}_0$. Hence, once the (arbitrary) initial shape of the material control volume is chosen (so that all initial values of $\mathbf{x}_0$ are specified), a scalar quantity α associated with any point within the material control volume can be completely specified as a function of time only, that is, $\alpha[\mathbf{x}(t), t]$. Thus, the usual definition of an ordinary time derivative can be applied to the left-hand side of (2–9), and we write

$$\frac{D}{Dt}\left[\int_{V_m(t)}\alpha[\mathbf{x}(t),t]dV\right] \equiv \lim_{\delta t \to 0}\left\{\frac{1}{\delta t}\left[\int_{V_m(t+\delta t)}\alpha(t+\delta t)dV - \int_{V_m(t)}\alpha(t)dV\right]\right\}$$

(2–10)

To the quantity on the right-hand side of (2–10), we now add and subtract the term $\int_{V_m(t)}\alpha(t+\delta t)dV$ to give

$$\frac{D}{Dt}\int_{V_m(t)}\alpha dV = \lim_{\delta t \to 0}\left\{\frac{1}{\delta t}\left[\int_{V_m(t+\delta t)}\alpha(t+\delta t)dV - \int_{V_m(t)}\alpha(t+\delta t)dV\right]\right.$$

$$\left. + \frac{1}{\delta t}\left[\int_{V_m(t)}\alpha(t+\delta t)dV - \int_{V_m(t)}\alpha(t)dV\right]\right\}.$$

(2–11)

The second term is just

$$\lim_{\delta t \to 0}\int_{V_m(t)}\frac{\alpha(t+\delta t) - \alpha(t)}{dt}dV \equiv \int_{V_m(t)}\frac{\partial \alpha}{\partial t}dV.$$

(2–12)

The first term is simply rewritten as

$$\lim_{\delta t \to 0}\frac{1}{\delta t}\int_{V_m(t+\delta t) - V_m(t)}\alpha(t+\delta t)dV,$$

(2–13)

which shows that it is the integral of $\alpha(t)$ over the differential volume element $V_m(t+\delta t) - V_m(t)$. To evaluate (2–13), we notice that any differential element of surface dS_m of $V_m(t)$ will move a distance $\mathbf{u} \cdot \mathbf{n}\delta t$ in a time interval δt. Thus, for sufficiently small δt, the differential volume element dV in (2–13) can be approximated as $(\mathbf{u} \cdot \mathbf{n})\delta t dS$ and the volume integral over $V_m(t+\delta t) - V_m(t)$ converted to an integral over the surface of $V_m(t)$. Thus,

$$\lim_{\delta t \to 0}\left[\frac{1}{\delta t}\int_{V_m(t+\delta t) - V_m(t)}\alpha(t+\delta t)dV\right]$$

(2–14)

$$= \lim_{\delta t \to 0}\left[\frac{1}{\delta t}\int_{S_m(t)}\alpha(t+\delta t)\mathbf{u} \cdot \mathbf{n}\delta t dS\right] = \int_{S_m(t)}\alpha(t)\mathbf{u} \cdot \mathbf{n}dS,$$

where $S_m(t)$ is the surface of the material volume, and

$$\frac{D}{Dt} \int_{V_m(t)} \alpha dV = \int_{V_m(t)} \frac{\partial \alpha}{\partial t} dV + \int_{S_m(t)} \alpha(\mathbf{u} \cdot \mathbf{n}) dS. \tag{2-15}$$

The proof of the Reynolds transport theorem is completed by applying the divergence theorem to the surface integral in (2–15) to give the result (2–9).

We can now apply the Reynolds transport theorem to (2–6). In this case, the scalar property $\alpha(\mathbf{x},t)$ is just the fluid density, $\rho(\mathbf{x},t)$. Thus, the mass conservation principle (2–6) can be reexpressed in the form

$$\int_{V_m(t)} \left[\frac{\partial \rho}{\partial t} + \nabla \cdot (\rho \mathbf{u}) \right] dV = 0. \tag{2-16}$$

Since the initial choice of $V_m(t)$ is arbitrary, we obtain the same differential form for the continuity equation, (2–5), that was derived earlier using a fixed control volume. Of course, the fact that we obtain the same form for the continuity equation is not surprising. The two derivations are entirely equivalent. In the first, conservation of mass is imposed by requiring that the time rate of change of mass in a fixed control volume be exactly balanced by a net imbalance in the influx and efflux of mass through the surface. In particular, no mass is created or destroyed. In the second approach, we define the material volume element so that the flux through its surface is everywhere equal to zero. In this case, the condition that mass is conserved means that the total mass in the material volume element is constant. The differential form (2–5) of the statement of mass conservation, which we have called the continuity equation, is the main result of this section.

Before leaving the continuity equation and mass conservation, there are a few additional remarks that we can put to good use later. The first is that (2–5) can be expressed in a precisely equivalent alternative form:

$$\boxed{\frac{1}{\rho} \frac{D\rho}{Dt} + \nabla \cdot \mathbf{u} = 0.} \tag{2-17}$$

Or, since the specific volume of the fluid is $V \equiv 1/\rho$, we can also write (2–17) as

$$\boxed{\frac{1}{V} \frac{DV}{Dt} = \nabla \cdot \mathbf{u}.} \tag{2-18}$$

The left-hand side of (2–18) is sometimes referred to as the rate of expansion or the rate of dilation of the fluid and provides a clear physical interpretation of the quantity $\nabla \cdot \mathbf{u}$ (or div$\mathbf{u}$).

The forms of the continuity equation (2–17) or (2–18) also lead directly to a simpler statement of the mass conservation principle that applies when the fluid can be approximated as *incompressible*. The approximation of incompressibility in fluid mechanics is usually limited to the statement that the density is independent of the pressure. Thus, if an incompressible fluid is isothermal, its density must be constant, independent both of spatial position and time. In this case, the convected derivative $D\rho/Dt$ is identically equal to zero, and the continuity equation takes the simpler form

$$\boxed{\nabla \cdot \mathbf{u} = \text{div} \mathbf{u} = 0.} \tag{2-19}$$

Vector fields whose divergence vanishes are sometimes referred to as *solenoidal.* Since our subject matter will frequently deal with incompressible, isothermal fluids, we shall often make use of (2-19) in lieu of the more general form, (2-5). In the presence of heat transfer, however, the density will not be constant because of temperature variations in the fluid, and the simple form (2-19) cannot be used even for an incompressible fluid unless additional approximations are made. We shall return to this point later in our presentation.

C. Conservation of Linear and Angular Momentum

Now that we have shown how a pointwise differential equation can be derived from the macroscopic principle of mass conservation, it would appear to be relatively straightforward to apply the same methods to derive differential forms for the other basic conservation principles. We begin with the principle of linear momentum conservation.

As in the case of mass conservation, the principle of linear momentum conservation can be stated via either a fixed control volume or a material control volume. Fixed control volumes are often used in introductory textbooks because the details of deriving the corresponding differential form appear simpler. For example, there is no need for the Reynolds transport theorem. However, the connection between fluid mechanics and other branches of classical mechanics is more clearly emphasized by using a material control volume. In this case, the familiar form of the conservation principle is Newton's second law of classical mechanics, according to which

$$
\left\{ \begin{array}{c} \text{the time rate of change} \\ \text{of momentum} \\ \text{of a given body,} \\ \text{relative to an inertial} \\ \text{reference frame} \end{array} \right\} = \left\{ \begin{array}{c} \text{the sum of forces acting} \\ \text{on the body} \end{array} \right\} \qquad (2\text{-}20)
$$

We adopt this second point of view because it yields the simplest possible mathematical statement of the basic conservation principle, and because its subsequent use in conjunction with the energy conservation principle provides the most appealing justification for the necessary use of results from equilibrium thermodynamics.

Although the material control volume is constantly deforming as a consequence of the fluid's motion, it satisfies the necessary criteria for application of Newton's second law in that it consists, at the continuum level of approximation, of a fixed body of material. The fact that the material control volume deforms with time does not in itself lead to any complications of principle in deriving differential forms of the equations of motion for the fluid, though it does ultimately lead to nonlinear equations that are difficult to solve.

From the purely continuum mechanics viewpoint that we have now adopted, we recognize two kinds of forces acting on the material control volume. First are the body forces, associated with the presence of external fields, which are capable of penetrating to the interior of the fluid and acting equally on all elements (per unit mass). The most familiar body force is gravity, and we will be concerned exclusively with this single type of body force in this book. The second type of force is a surface force, which acts from the fluid outside the material control volume on the fluid inside and vice versa. In reality, there may exist short-range forces of molecular origin in the fluid. With the crude scale of resolution inherent in the use of the continuum approximation, these will appear as surface force contributions in the basic balance (2–20). In addition, the surface force terms will always include an *effective* surface force contribution to simulate the transport of momentum across the boundaries of the material control volume due to random differences between the continuum velocity **u** and the actual molecular velocities. As explained earlier, the necessity for these latter *effective* surface force contributions is entirely a consequence of the crude scale of resolution inherent in the continuum approximation, coupled with the discrete nature of any real fluid. Indeed, these latter contributions would be zero if the material were truly an indivisible continuum. We shall see that the main difficulties in obtaining pointwise differential equations of motion from (2–20) all derive from the necessity for including surface forces (whether real or effective) whose molecular origin is outside the realm of the continuum description.

With the necessity for body and surface forces thus identified, a mathematical statement of Newton's second law can be written with the material control volume chosen as the moving "body," that is,

$$\frac{D}{Dt}\int_{V_m(t)} \rho \mathbf{u} \, dV = \int_{V_m(t)} \rho \mathbf{g} \, dV + \int_{S_m(t)} \mathbf{t} \, dS. \tag{2–21}$$

The left-hand side is just the time rate of change of linear momentum of all the fluid within the specified material control volume. The first term on the right-hand side is the net body force due to gravity (other types of body force will not be considered in this book). The second term is the net surface force, with the local surface force per unit area being symbolically represented by the vector **t**. We call **t** the *stress vector*. It is the vector sum of all surface force contributions per unit area acting at a point on the surface of $V_m(t)$. We may recall that the flux of momentum across the surface of a material control volume due to the continuum level fluid motion **u** is identically zero, since $\mathbf{u} \cdot \mathbf{n} = 0$ by definition at each point on the surface. Thus, any change in linear momentum is due to the action of surface or body forces, as indicated in (2–21). This may be contrasted with the situation that arises for a fixed control volume, where momentum is transported through the surface by convection due to **u**, and an additional source of changes in momentum is thus any imbalance between the mean influx and efflux of momentum due to the fluid's overall motion.

We may now attempt to simplify (2–21) to a differential form, as we did for the mass conservation equation (2–6). The basic idea is to express all terms in (2–21) as integrals over $V_m(t)$, leading to the requirement that the sum of the integrands is zero since $V_m(t)$ is initially arbitrary. However, it is immediately apparent that this scheme will fail unless we can say more about the surface stress vector **t**. Otherwise, there is no way to express the surface integral of **t** in terms of an equivalent volume integral over $V_m(t)$.

We note first of all that **t** is not only a function of position and time, as is the case with **u**, but also of the orientation of the differential surface element through **x** on which it acts. The reader may well ask how this is known in the absence of a direct molecular derivation of a theoretical expression for **t** (the latter being outside the realm of continuum mechanics, even if it were possible in principle). The answer is that certain general properties of **t**, including its orientation dependence, can be either deduced or derived from (2–21) by considering the limit as the material control volume is decreased progressively toward zero while holding the geometry (shape) of V_m constant. Let us denote a characteristic linear dimension of V_m as l, with l^3 defined to be equal to V_m. An estimate for each of the integrals in (2–21) can be obtained in terms of l using the mean value theorem. A useful preliminary step is to apply the Reynolds transport theorem to the left-hand side. Although this might, at first sight, seem to present new difficulties since $\rho\mathbf{u}$ is a vector, whereas the Reynolds transport theorem was originally derived for a scalar, the result given by (2–9) carries over directly, as may be seen by applying it to each of the three scalar components of $\rho\mathbf{u}$, and then adding the results. Thus, (2–21) can be rewritten in the form

$$\int_{V_m(t)}\left[\frac{\partial(\rho\mathbf{u})}{\partial t} + \nabla\cdot(\rho\mathbf{uu}) - \rho\mathbf{g}\right]dV = \int_{S_m(t)}\mathbf{t}\,dS. \qquad (2\text{–}22)$$

Now, denoting the mean value over V_m or S_m by the symbol $<\ >$, we can express (2–22) in the symbolic form

$$<\ >l^3 = <\ >l^2.$$

It is evident that as $l\rightarrow 0$, the volume integral of the momentum and body force terms vanishes more quickly than the surface integral of the stress vector. Hence, in the limit as $l\rightarrow 0$, (2–22) reduces to the form

$$\boxed{\lim_{l\rightarrow 0}\frac{1}{S_m}\int_{S_m(t)}\mathbf{t}\,dS \rightarrow 0.} \qquad (2\text{–}23)$$

This result is sometimes called *the principle of stress equilibrium,* because it shows that the surface forces must be in local equilibrium for any arbitrarily small volume element centered at any point **x** in the fluid. This is true independent of the source or detailed form of the surface forces.

Now, it is clear that the stress vector at (or arbitrarily close to) a point **x** must depend not only on **x** but also on the orientation of the surface through **x** on which it acts, since otherwise the equilibrium condition (2–23) could not be satisfied. At first this dependence on orientation may seem to suggest that one would need a triply infinite set of numbers to specify **t** for all possible orientations of a surface through each point **x**. Not only is this clearly impossible, but (2–23) shows that it is not necessary. Let us consider a surface with completely arbitrary orientation, specified by a unit normal **n**, that passes near to point **x** but not precisely through it. Then using this surface as one side, we construct a tetrahedron, illustrated in Figure 2–5, centered around point **x**, whose remaining three sides are mutually perpendicular. In the limit as the volume of this tetrahedral volume element goes to zero, the surface stress equilibrium principle applies, and it is obvious that the surface stress vector on the arbitrarily-oriented sur-

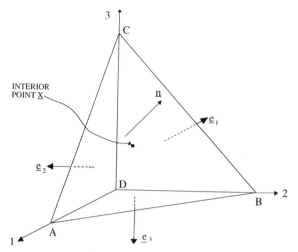

Figure 2–5 A tetrahedron, *ABCD*, is centered about the point **x**. The surface ABC is arbitrarily oriented with respect to the Cartesian axes 123 and its area is denoted ΔA_n. The unit outer normal to *ABC* is **n**. The areas of surfaces *BCD*, *ACD*, and *ABD* are denoted, respectively, as ΔA_1, ΔA_2, and ΔA_3, with outer unit normals e_1, e_2, and e_3 each parallel to the opposing coordinate axis but oriented in the negative direction.

face (which now passes arbitrarily close to **x**) must be expressible in terms of the surface stress vectors acting on the three mutually perpendicular faces (which also pass arbitrarily close to **x**). From this, we deduce that the specification of the surface force vector on three mutually perpendicular surfaces through a point **x** (nine independent components in all) is sufficient to completely determine the surface force vector acting on a surface of any arbitrary orientation at the same point **x**.

We can actually go one step further than this general observation and use the stress equilibrium principle applied to the tetrahedron to obtain a simple expression for the surface stress vector **t** on an arbitrarily oriented surface at **x** in terms of the components of **t** on the three mutually perpendicular surfaces at the point **x**. We denote the stress vector on a surface with normal **n** as **t(n)**. The area of the surface with unit normal **n** is denoted as ΔA_n. Then, applying the surface stress equilibrium principle to the tetrahedron, we have

$$<\mathbf{t(n)}> \Delta A_n - <\mathbf{t(e_1)}> \Delta A_1 - <\mathbf{t(e_2)}> \Delta A_2 - <\mathbf{t(e_3)}> \Delta A_3 = 0, \quad (2\text{–}24)$$

where the **e**'s are unit normal vectors for the three mutually perpendicular surfaces and $<>$ represents the mean value of the indicated stress vector over the surface in question. Now, ΔA_i is the projected area of ΔA_n onto the plane perpendicular to the $\mathbf{e}_i$ axis. Thus,

$$\Delta A_i = \Delta A_n (\mathbf{n} \cdot \mathbf{e}_i) \text{ for } i = 1, 2, \text{ or } 3,$$

and (2–24) can be expressed in the form

$$0 = [<\mathbf{t(n)}> - <\mathbf{t(e_1)}>(\mathbf{n} \cdot \mathbf{e}_1) - <\mathbf{t(e_2)}>(\mathbf{n} \cdot \mathbf{e}_2) - <\mathbf{t(e_3)}>(\mathbf{n} \cdot \mathbf{e}_3)] \Delta A_n.$$

It follows, in the limit as the tetrahedron collapses onto point **x**, that

$$t(n) = n \cdot [e_1 t(e_1) + e_2 t(e_2) + e_3 t(e_3)]. \tag{2-25}$$

The quantity in square brackets is a second-order tensor, formed as a sum of dyadic products of $t(e_i)$ and e_i for $i = 1, 2,$ and 3. This second-order tensor is known as the *stress tensor*

$$T = [e_1 t(e_1) + e_2 t(e_2) + e_3 t(e_3)]. \tag{2-26}$$

and is denoted here by the symbol **T**. It follows from (2–25) and (2–26) that the surface stress vector on any arbitrarily oriented surface through a point **x** is completely determined by specification of the nine independent components of the stress tensor, **T**, in the form

$$t(n) = n \cdot T \tag{2-27}$$

We see that the components of **T** can be interpreted as the components of the surface force vectors at **x** acting on the three perpendicular surfaces through **x** with unit normals e_1, e_2 and e_3. For example, the 21 component of **T** is the surface stress force component in the 1 direction acting on a surface with unit normal in the e_2 direction. Although the derivation of (2–27) has been carried out using the notation of Cartesian vector and tensor analysis, it is evident that the components of **T** can be defined in terms of stress vector components for any orthogonal coordinates through point **x**, and the result (2–27) is completely invariant to the choice of coordinate systems. The stress tensor **T** depends only on **x** and t, but not on **n**. Since a knowledge of **T** at a point enables us to determine the surface force acting on any surface through that point, **T** is said to represent *the state of stress* in the fluid at the given point.

With the relationship between the stress vector and the stress tensor in hand, differential equations of motion can be derived from the macroscopic form (2–22) of Newton's second law of mechanics. Substituting (2–27) into (2–22) and applying the divergence theorem to the surface integral in the form

$$\int_{S_m(t)} n \cdot T \, dS = \int_{V_m(t)} (\nabla \cdot T) \, dV,$$

we obtain

$$\int_{V_m(t)} \left[\frac{\partial(\rho u)}{\partial t} + \nabla \cdot (\rho u u) - \rho g - \nabla \cdot T \right] dV = 0. \tag{2-28}$$

Since the initial choice for $V_m(t)$ is arbitrary, it follows that the condition (2–28) can be satisfied only if the integrand is equal to zero at each point in the fluid, that is,

$$\frac{\partial(\rho u)}{\partial t} + \nabla \cdot (\rho u u) = \rho g + \nabla \cdot T. \tag{2-29}$$

Combining the first two terms in this equation with the continuity equation, the differential equations of motion can be written in the alternative form

$$\rho \left[\frac{\partial u}{\partial t} + u \cdot \nabla u \right] = \rho g + \nabla \cdot T. \tag{2-30}$$

This is known as *Cauchy's* equation of motion. It is clear from our derivation that it is simply the differential form of Newton's second law of mechanics, applied to a moving fluid.

It is, perhaps, well to pause for a moment to take stock of our developments to this point. We have successfully derived differential equations that must be satisfied by any velocity field that is consistent with conservation of mass and Newton's second law of mechanics (or conservation of linear momentum). However, a closer look at the results, (2–5) or (2–19) and (2–30), reveals the fact that we have far more unknowns than we have relationships between them. Let us consider the simplest situation in which the fluid is isothermal and approximated as incompressible. In this case, the density is a constant property of the material, which we may assume to be known, and the continuity equation (2–19) provides one relationship between the three unknown scalar components of the velocity $\mathbf{u}$. When Newton's second law is added, we do generate three additional equations involving the components of $\mathbf{u}$, but only at the cost of nine additional unknowns at each point: the nine independent components of $\mathbf{T}$. It is clear that more equations are needed.

One possible source of additional relationships between $\mathbf{u}$ and $\mathbf{T}$ that we have not yet considered is Newton's *third* law of motion, the law of *conservation of angular momentum*. We may state this principle, for a material control volume, in the form

$$\frac{D}{Dt}\int_{V_m(t)}(\mathbf{x}\wedge\rho\mathbf{u})dV = \text{sum of torques acting on the material control volume,}$$

(2–31)

where $\mathbf{x}$ is the position vector associated with points within the material volume $V_m(t)$. There are, in principle, four sources of torque acting on the material control volume. The first two are simply the moments of the surface and body forces that act on the fluid:

$$\int_{S_m(t)}\left[\mathbf{x}\wedge\mathbf{t}(\mathbf{n})\right]dS$$

(2–32a)

and

$$\int_{V_m(t)}(\mathbf{x}\wedge\rho\mathbf{g})dV.$$

(2–32b)

In addition, the possibility exists of body couples per unit mass $\mathbf{c}$, and surface couples per unit surface area $\mathbf{r}$, that are independent of the moments of surface and body forces. Thus, the full statement of Newton's third law takes the form

$$\frac{D}{Dt}\int_{V_m(t)}(\mathbf{x}\wedge\rho\mathbf{u})dV = \int_{S_m(t)}\left[\mathbf{x}\wedge(\mathbf{n}\cdot\mathbf{T})+\mathbf{r}\right]dS + \int_{V_m(t)}\left[\mathbf{x}\wedge\rho\mathbf{g}+\rho\mathbf{c}\right]dV.$$

(2–33)

Clearly, the presence of the surface torque terms in (2–33) contributes additional unknowns, and the imbalance between unknowns and equations is not improved in the most general case by consideration of Newton's third law. However, in practice, there is no evidence of significant surface torque contributions in real fluids, and we shall assume that $\mathbf{r} \equiv 0$. On the other hand, some fluids do exist where the influence of body couples is significant. One commercially available set of examples are the so-called *Ferrofluids*, which are actually a suspension of fine iron particles that have either permanent or induced magnetic dipoles.[5] When such a fluid flows in the presence of a magnetic field, there is a body *torque* applied to each particle, but in the continuum approximation this is described as a continuously distributed body couple per unit mass. Thus, a *Ferrofluid* is an example of a fluid in which $\mathbf{c} \neq 0$. In spite of the fact that fluids do exist in which body couples play a significant role, however, this is not true of the vast majority of fluids and none of the liquids or gases of common experience; water, air, oils, and so on. Let us suppose, therefore, that $\mathbf{c} = 0$. In this case, Newton's third law reduces to the simpler form

$$\frac{D}{Dt} \int_{V_m(t)} (\mathbf{x} \wedge \rho \mathbf{u}) dV = \int_{S_m(t)} \mathbf{x} \wedge (\mathbf{n} \cdot \mathbf{T}) dS + \int_{V_m(t)} \mathbf{x} \wedge \rho \mathbf{g} dV. \tag{2-34}$$

We can explore the consequences of this equation by converting it to an equivalent differential form. To do this, we first apply Reynolds transport theorem to the left-hand side. This gives

$$\frac{D}{Dt} \int_{V_m(t)} (\mathbf{x} \wedge \rho \mathbf{u}) dV = \int_{V_m(t)} \mathbf{x} \wedge \left(\frac{\partial(\rho \mathbf{u})}{\partial t} + \nabla \cdot (\rho \mathbf{u} \mathbf{u}) \right) dV. \tag{2-35}$$

Also, upon application of the divergence theorem, the surface integral in (2-34) becomes

$$\int_{S_m(t)} \mathbf{x} \wedge (\mathbf{n} \cdot \mathbf{T}) dS = \int_{V_m(t)} \left[\mathbf{x} \wedge (\nabla \cdot \mathbf{T}) - \boldsymbol{\epsilon} : \mathbf{T} \right] dV, \tag{2-36}$$

where $\boldsymbol{\epsilon}$ is the third-order alternating tensor and the symbol : indicates a double inner product of this tensor with $\mathbf{T}$.† Introducing (2-35) and (2-36) into (2-34), we thus obtain

$$\int_{V_m(t)} \left\{ \mathbf{x} \wedge \left[\frac{\partial(\rho \mathbf{u})}{\partial t} + \nabla \cdot (\rho \mathbf{u} \mathbf{u}) - \nabla \cdot \mathbf{T} - \rho \mathbf{g} \right] + \boldsymbol{\epsilon} : \mathbf{T} \right\} dV = 0. \tag{2-37}$$

In view of the differential form of the equation of motion, equation (2-29), and the fact that $V_m(t)$ is arbitrary, we see that Newton's third law of motion requires

$$\boxed{\boldsymbol{\epsilon} : \mathbf{T} = 0} \tag{2-38}$$

at all points in the fluid. Condition (2-38) requires

$$\boxed{\mathbf{T} = \mathbf{T}^T,} \tag{2-39}$$

that is, the stress tensor must be *symmetric*. Note that the condition of stress symmetry is not valid if there is a significant body couple per unit mass $\mathbf{c}$ in the field. In this case, it can be shown easily, following the same steps that were used in going from (2-33) to (2-38), that

---

†If $\mathbf{A}$ and $\mathbf{B}$ are second-order tensors, the double inner product is a scalar, defined in terms of Cartesian index notation as $A_{ij} B_{ji}$. The double inner product $\boldsymbol{\epsilon} : \mathbf{T}$ is thus a vector with components $\epsilon_{ijk} T_{kj}$.

$$\boxed{\boldsymbol{\epsilon}:\mathbf{T} - \rho\mathbf{c} = 0.}$$
(2–40)

This gives a relationship between **c** and the off-diagonal components of **T**, but the stress is clearly not symmetric. We shall hereafter restrict our attention to the case in which **c** = 0.

We see that application of Newton's third law of motion does reduce, somewhat, the imbalance between the number of unknowns and equations that derive from the basic principles of mass and momentum conservation. In particular, we have shown that the stress tensor must be symmetric. Complete specification of a symmetric tensor requires only six independent components rather than the full nine that would be required in general for a second-order tensor. Nevertheless, for an incompressible fluid we still have nine apparently independent unknowns and only four independent relationships between them. It is clear that the equations derived up to now—namely, the equation of continuity and Cauchy's equation of motion—do not provide enough information to uniquely describe a flow system. Additional relations need to be derived or otherwise obtained. These are the so-called constitutive equations. We shall return to the problem of specifying constitutive equations shortly. First, however, we wish to consider the last available conservation principle, namely, conservation of energy.

D. Conservation of Energy

We begin again by considering an arbitrary material control volume as it moves along with the fluid, and in this case consider the change in its total energy with respect to time. In the simplest molecular description based upon a hard-sphere gas model, this energy would be recognized as purely kinetic in nature, associated with the intensity of individual molecular motions. In the continuum approximation, however, we explicitly resolve only the molecular average velocity (denoted as **u** in the earlier developments of this chapter), and it is necessary to consider the total energy as consisting of two parts. First is the "kinetic" energy associated with the macroscopic or continuum velocity field **u**, and second is a so-called internal energy term that encompasses all additional contributions including those due to the differences between the continuum velocity **u** at a point and the actual velocities of the molecules that occupy the continuum averaging volume that is centered at that point. In this description, the *internal energy* is a measure of the intensity of random molecular motion relative to the mean, continuum velocity. Thus, from the continuum point of view, the total energy of an arbitrary material control volume is written as

$$\int_{V_{m(t)}} \left[\frac{\rho(\mathbf{u}\cdot\mathbf{u})}{2} + \rho e \right] dV,$$

where $\mathbf{u}\cdot\mathbf{u} = u^2$ is the local "speed" of the continuum motion and e is the internal energy (representing additional kinetic energy at the molecular level) per unit mass.

The rate at which the total energy changes with time is determined by the *principle of energy conservation* for the material volume element, according to which

$$\frac{D}{Dt}\int_{V_m(t)}\left[\frac{\rho\mathbf{u}^2}{2}+\rho e\right]dV = \begin{array}{c}\text{rate of work done}\\\text{on the material}\\\text{control volume by}\\\text{external forces}\end{array} + \begin{array}{c}\text{rate of energy flux}\\\text{across the boundaries}\\\text{of the material}\\\text{control volume.}\end{array}$$

$$(2\text{--}41)$$

We note that this conservation principle, for a closed system such as the material control volume, is precisely equivalent to the *first law of thermodynamics*, which can be obtained from it by integrating with respect to time, over some finite time interval.

The terms on the right-hand side of (2–41) can be expressed in mathematical form, based upon the following observations. First, work can be done on the material control volume only as a consequence of forces acting upon it. In our continuum description, these are body forces and surface forces associated with the stress vector $\mathbf{t(n)}$. We recall that the surface forces appear, in part, as a consequence of our inability to fully resolve momentum transfer at the molecular level in a continuum description. It is not surprising, therefore, that work done in the macroscopic description may lead to changes in either the macroscopic kinetic energy or the internal energy representing changes in the intensity of motions at the molecular level. The motivation for a term in (2–41) that is associated with energy flux across the boundaries of the material control volume is very similar to that associated with the appearance of a surface force (or stress) in the linear momentum principle. In particular, there would be no local flux of kinetic or internal energy across the surface of a material control volume if the fluid were actually a continuous, indivisible, and homogeneous medium, because the material control volume is defined as moving and deforming with the fluid in such a way that the local flux of mass across its surface is zero. However, random motions of molecules (which are not resolved explicitly in the continuum description) can contribute a net flux of internal energy across the surface, and this can only be included in the continuum energy balance (2–41) by the assumed existence of a *surface energy flux vector* $\mathbf{q}$. This surface energy flux is usually called the *heat flux vector*, in recognition of the fact that it is internal energy (or average intensity of molecular motion) that is being transferred across the surface by random molecular motion. Incorporating rate of working terms due to surface and body forces, as well as a surface flux of energy term, we can write (2–41) in the mathematical form

$$\frac{D}{Dt}\int_{V_m(t)}\left[\frac{\rho\mathbf{u}^2}{2}+\rho e\right]dV = \int_{S_m(t)}[\mathbf{t(n)}\cdot\mathbf{u}]\,dS + \int_{V_m(t)}(\rho\mathbf{g})\cdot\mathbf{u}\,dV - \int_{S_m(t)}\mathbf{q}\cdot\mathbf{n}\,dS.$$

$$(2\text{--}42)$$

Here, we have adopted the convention that a flux of heat into the material control volume is positive. The negative sign in the last term appears because $\mathbf{n}$ is the outer normal to the material control volume.

To obtain a pointwise differential equation from (2–42), we follow the usual procedure of applying the Reynolds transport theorem to the left-hand side and the divergence theorem to the right-hand side after first using (2–27) to express $\mathbf{t} \cdot \mathbf{u}$ as $\mathbf{n} \cdot (\mathbf{T} \cdot \mathbf{u})$. With all terms then expressed as volume integrals over the arbitrary material control volume $V_m(t)$, we obtain the differential form of the energy conservation principle

$$\rho \frac{D}{Dt} \left[\frac{\mathbf{u}^2}{2} + e \right] = \nabla \cdot (\mathbf{T} \cdot \mathbf{u}) + \rho \mathbf{g} \cdot \mathbf{u} - \nabla \cdot \mathbf{q}. \tag{2–43}$$

It appears, from (2–43), that contributions from any of the terms on the right-hand side will lead to a change in the sum of kinetic and internal energy, but may not contribute separately to one or the other of these energy terms. However, this is *not* true as we may see by further examination. First, we may note that the Cauchy's equation of motion provides an independent relationship for the rate of change of kinetic energy. In particular, if we take the inner product of (2–30) with $\mathbf{u}$, we obtain

$$\frac{\rho}{2} \frac{D\mathbf{u}^2}{Dt} = (\rho \mathbf{g}) \cdot \mathbf{u} + \mathbf{u} \cdot (\nabla \cdot \mathbf{T}). \tag{2–44}$$

This relationship is known as the *mechanical energy balance,* and is a direct consequence of linear momentum conservation. Substituting for $D\mathbf{u}^2/Dt$ in (2–43) using (2–44) and recalling that $\mathbf{T}$ is symmetric, we obtain the so-called *thermal energy balance*

$$\rho \frac{De}{Dt} = \mathbf{T}:\mathbf{E} - \nabla \cdot \mathbf{q}. \tag{2–45}$$

The second-order tensor $\mathbf{E}$ that appears in (2–45) is defined in terms of $\mathbf{u}$ as

$$\mathbf{E} \equiv \frac{1}{2} \left(\nabla \mathbf{u} + \nabla \mathbf{u}^T \right) \tag{2–46}$$

and is known as the *rate of strain tensor.* We shall later see the origins of this name. For now, we simply note that $\mathbf{E}$ is the symmetric part of the velocity gradient tensor, $\nabla \mathbf{u}$, that is,

$$\nabla \mathbf{u} \equiv \frac{1}{2} \left(\nabla \mathbf{u} + \nabla \mathbf{u}^T \right) + \frac{1}{2} \left(\nabla \mathbf{u} - \nabla \mathbf{u}^T \right) = \mathbf{E} + \mathbf{\Omega} \tag{2–47}$$

symmetric   antisymmetric

The antisymmetric contribution to $\nabla \mathbf{u}$, which we have denoted in (2–47) as $\mathbf{\Omega}$, is known as the *vorticity tensor.* Again, we shall have more to say about the vorticity tensor later in this chapter.

Returning to (2–45), the term $\mathbf{T}:\mathbf{E}$ represents a contribution to the internal energy of the fluid, due to the presence of mean motion (note $\mathbf{E} \equiv 0$ if $\nabla \mathbf{u} \equiv 0$); that is, it represents a conversion from kinetic energy of the velocity field, $\mathbf{u}$, to internal energy of the fluid—a process that is termed *dissipation* of kinetic energy to internal energy (or heat). The local rate of working due to body forces and surface forces may be seen from (2–44) to contribute directly to kinetic energy, but only to lead to changes in internal energy through dissipation. On the other hand, the surface energy (or heat) flux contribution to the total energy balance contributes directly to the change of internal energy, but only indirectly to the kinetic energy.

We may also note that the energy conservation principle (or, equivalently, the first law of thermodynamics) has not improved the balance between the number of unknown, independent variables and differential relationships between them. Indeed, we have obtained a single scalar equation but have introduced several new unknowns in the process, the three components of $\mathbf{q}$ and the internal energy e. A relationship between e and the thermodynamic state variables, say, density ρ and temperature θ, can be obtained provided that equilibrium thermodynamics is assumed to be applicable to a fluid element that moves with a velocity $\mathbf{u}$. In particular, a differential change in θ or ρ lead to a differential change in e for an equilibrium system

$$de = c_v d\theta - \left\{ \left(\frac{\partial p}{\partial \theta} \right)_\rho \theta - p \right\} \frac{d\rho}{\rho^2}.$$

Hence, for a fluid element moving with the fluid

$$\rho \frac{De}{Dt} = \rho c_v \frac{D\theta}{Dt} - \left\{ \theta \left(\frac{\partial p}{\partial \theta} \right)_\rho - p \right\} \frac{1}{\rho} \frac{D\rho}{Dt}, \tag{2-48}$$

and (2–45) can be expressed in terms of θ rather than e in the form

$$\boxed{\rho c_v \frac{D\theta}{Dt} = -\theta \left(\frac{\partial p}{\partial \theta} \right)_\rho \nabla \cdot \mathbf{u} + p(\nabla \cdot \mathbf{u}) + \mathbf{T} : \mathbf{E} - \nabla \cdot \mathbf{q}.} \tag{2-49}$$

This form of the thermal energy equation would appear to be particularly useful when the fluid can be approximated as incompressible since $\nabla \cdot \mathbf{u} = 0$ in that case, and (2–49) is considerably simplified. However, in practice, (2–49) is hardly ever used because it contains the constant volume heat capacity c_v, and this is extremely difficult to measure for liquids. Instead, an alternative form of the thermal energy equation is normally used that involves the heat capacity at constant pressure, c_p, since this is relatively easy to measure for both gases and liquids. To obtain this alternative form from (2–49), we introduce the general thermodynamic relationship

$$c_v = c_p + \frac{\theta}{\rho^2} \left(\frac{\partial p}{\partial \theta} \right)_\rho \left(\frac{\partial \rho}{\partial \theta} \right)_p.$$

This gives

$$\rho c_p \frac{D\theta}{Dt} = \frac{\theta}{\rho} \left(\frac{\partial p}{\partial \theta} \right)_\rho \left[\frac{D\rho}{Dt} - \left(\frac{\partial \rho}{\partial \theta} \right)_p \frac{D\theta}{Dt} \right] + p(\nabla \cdot \mathbf{u}) + \mathbf{T} : \mathbf{E} - \nabla \cdot \mathbf{q}.$$

Since a thermodynamic equation of state exists relating θ, p, and ρ, say, $\theta = \theta(p, \rho)$, the right-hand side can be simplified so that the equation takes the final form

$$\boxed{\rho c_p \frac{D\theta}{Dt} = -\frac{\theta}{\rho} \left(\frac{\partial \rho}{\partial \theta} \right)_p \frac{Dp}{Dt} + p(\nabla \cdot \mathbf{u}) + \mathbf{T} : \mathbf{E} - \nabla \cdot \mathbf{q}.} \tag{2-50}$$

Although the new term on the right-hand side is identically zero *only* for constant pressure conditions (that is, the material is a solid or it is stationary so that $Dp/Dt \equiv 0$), we shall see that it is frequently small compared with other terms in (2–50).

We have seen that the energy conservation principle, applied to a material control volume of fluid, is equivalent to the first law of thermodynamics. A natural question, then, is whether any additional useful information can be obtained from the *second law of thermodynamics*. In its usual differential form the second law states

$$dS \geq \frac{dQ}{\theta},$$

where dS is the entropy change for the thermodynamic system of interest, dQ is the change in its total heat content due to heat exchange with the surroundings, and θ is its temperature. When applied to a material control volume of fluid, this principle can be expressed in the form

$$\frac{D}{Dt} \int_{V_m(t)} (\rho s) \, dV + \int_{S_m(t)} \frac{\mathbf{n} \cdot \mathbf{q}}{\theta} \, dS \geq 0, \tag{2–51}$$

where s is the entropy per unit mass of the fluid. The only mechanism for heat transfer from the surrounding fluid is molecular transport represented by the heat flux vector $\mathbf{q}$. The sign in front of the second term is a consequence of the fact that $\mathbf{n}$ is the outer unit normal. A differential form of the inequality (2–51) is obtained easily by applying the Reynolds transport theorem to the first term and the divergence theorem to the second term to show

$$\int_{V_m(t)} \left[\rho \frac{Ds}{Dt} + \nabla \cdot \left(\frac{\mathbf{q}}{\theta} \right) \right] dV \geq 0.$$

This inequality can be satisfied for an arbitrary material control volume $V_m(t)$ only if

$$\rho \frac{Ds}{Dt} + \nabla \cdot \left(\frac{\mathbf{q}}{\theta} \right) \geq 0. \tag{2–52}$$

An inequality that is equivalent to (2–52) can be obtained using thermodynamics to express Ds/Dt in the form

$$\theta \rho \frac{Ds}{Dt} = \rho \frac{De}{Dt} - \frac{p}{\rho} \frac{D\rho}{Dt}$$

and then substituting for De/Dt from the energy conservation (2–45). The result for Ds/Dt is

$$\rho \frac{Ds}{Dt} = \frac{1}{\theta} \left[\mathbf{T} : \mathbf{E} + p (\nabla \cdot \mathbf{u}) - \nabla \cdot \mathbf{q} \right]. \tag{2–53}$$

Then, since

$$\nabla \cdot \left(\frac{\mathbf{q}}{\theta} \right) = \frac{1}{\theta} \nabla \cdot \mathbf{q} - \frac{1}{\theta^2} \mathbf{q} \cdot \nabla \theta,$$

the inequality (2–52) can be combined with (2–53) to obtain

$$\boxed{\frac{1}{\theta}\left(\mathbf{T}{:}\mathbf{E} + p\left(\nabla{\cdot}\mathbf{u}\right)\right) - \frac{\mathbf{q}{\cdot}\nabla\theta}{\theta^2} \geq 0.}$$
(2–54)

Although there is no immediately useful information that we can glean from (2–54), we shall see that it provides a constraint on allowable constitutive relationships for **T** and **q**. In this sense, it plays a similar role to the angular momentum principle which led to the constraint (2–39) that **T** be symmetric in the absence of body couples. In solving fluid mechanics problems, assuming that the fluid is isothermal, we will use the equation of continuity, (2–5) or (2–19), and the Cauchy equation of motion (2–30), to determine the velocity field, but angular momentum conservation and the second law of thermodynamics will appear only indirectly as constraints on allowable constitutive forms for **T**. Similarly, for nonisothermal conditions, we will use (2–5) or (2–19), (2–30), and either (2–49) or (2–50) to determine the velocity and temperature distributions, but neither angular momentum conservation nor the second law of thermodynamics will appear directly. However, we are getting ahead of our story.

So far, we have seen that the basic conservation principles of continuum mechanics lead to a set of five scalar differential equations—sometimes called the field equations of continuum mechanics—namely, (2–5) or (2–19), (2–30), and (2–49) or (2–50). On the other hand, we have identified many more unknown variables, **u**, **T**, θ, p, and **q**, plus various fluid or material properties such as ρ, c_v, (or c_p), $(\partial p/\partial\theta)_p$, or $(\partial\rho/\partial\theta)_p$, which generally require additional equations of state to be determined from p and θ if the latter are adopted as the thermodynamic state variables. Let us focus just on the independent variables **u**, **T**, θ, p, and **q**. Taking account of the symmetry of **T**, these comprise 14 unknown scalar variables for which we have so far obtained only the five independent "field" equations that were listed above. It is evident that we require additional equations relating the various unknown variables if we are to achieve a well-posed problem from a mathematical point of view. Where are these equations to come from? Why is it that the fundamental conservation principles of continuum physics do not, in themselves, lead to a complete, well-posed mathematical problem?

E. Constitutive Equations

We have seen that the basic field equations of continuum mechanics are not sufficient in number to provide a well-posed mathematical problem from which to determine solutions for the independent field variables **u**, **T**, θ, p, and **q**. It is apparent that additional relationships must be found, hopefully without introducing more independent variables. In the next several sections, we discuss the origin and form of the so-called constitutive equations that provide the necessary additional relationships.

We begin with some general observations. In the first place, the idea that additional equations are necessary has so far been based on the purely mathematical statement that the field equations by themselves do not lend to a well-posed problem. While this argument is powerful and certainly persuasive, it is also instructive to think about the problem from a more heuristic, physical point of view. In particular, if we first restrict ourselves to isothermal, incompressible conditions where the relevant field equations are continuity, in the form (2–19), and the Cauchy equations of motion (2–30), we see that the only material property that appears explicitly is the density ρ. That is, according to (2–19) and (2–30) in the form that they stand, it appears that the only material property that distinguishes the motion of one fluid from another is the density. This is clearly at odds with experimental observation—we can find (or create by blending) a variety of fluids that have the same density within experimental error, yet clearly demonstrate differences in flow properties. Consider, for example, the many grades of silicon oils that are sold commercially. These various grades differ in molecular weight, but their densities are all very nearly equal. Yet, if we were to simply pour a low- and a high-grade silicon oil from one container to another, we could not help but note a remarkable difference in the ease with which the fluids flow. The lowest grades would appear visually somewhat like water, whereas the highest grades would be more nearly akin to something like corn syrup. Quite apparently, there is something of the basic physics that is missing from the field equations alone. Similarly, if we consider a nonisothermal system in the absence of any mean motion, that is, $\mathbf{u} \equiv 0$, the thermal energy (2–50) reduces to the form

$$\boxed{\rho c_p \frac{\partial \theta}{\partial t} = -\nabla \cdot \mathbf{q}.} \tag{2–55}$$

Not only are there more independent variables (θ and the three components of $\mathbf{q}$) than equations (one!), but it would appear that the only material property relevant to energy transfer is ρc_p. Once again, simple observations would show that this is not enough to characterize the energy transfer processes in real materials.

Why is it that the basic conservation principles of continuum mechanics do not provide a complete problem statement, from either a mathematical or physical point of view? The answer is that the fluids or materials that we wish to consider are not actually indivisible and homogeneous as presumed in continuum mechanics, but rather they have a definite molecular structure. Although this structure is not directly evident at the scale of resolution relevant to continuum mechanics, we have seen in the derivation of the basic field equations that it cannot be ignored altogether even in a purely continuum mechanical formulation. Instead, the differences between the continuum velocity (which we have seen is really an average of the molecular velocities "at" a point) and the instantaneous, local molecular velocities are manifest as apparent *surface force or stress* and *surface energy or heat flux* contributions to the basic conservation principles of linear momentum and energy. Indeed, in the absence of the stress tensor $\mathbf{T}$ and the heat flux vector $\mathbf{q}$, as would be appropriate for a material with a completely continuous and homogeneous structure down to the finest possible scale of resolution, the basic field equations are completely adequate in number to determine all of the remaining independent field variables $\mathbf{u}$, θ, and p. It is the presence of $\mathbf{T}$ and $\mathbf{q}$, reflecting the existence of transport processes at the molecular scale, that causes the field equations to contain more apparently independent variables than there are equations. In view of this, we may anticipate that a full statement of the physics relevant to flowing fluids, whether isothermal or not, will require additional relationships between the surface stress and/or heat flux (representing molecular transport processes) and the macroscopic (or continuum) velocity and temperature fields. The relationships are known as the *constitutive equations* for the fluid.

But where do we get these additional equations? Since the underlying mechanisms responsible for the appearance of surface stress or surface heat flux in the continuum description are molecular, it is evident that continuum mechanics, by itself, can offer no basis to deduce what form these relationships should take. Thus, if we insist on a purely continuum mechanical approach, we must simply guess at the appropriate constitutive equations, and then judge the correctness of our guess by comparisons between theoretically predicted velocity, temperature, or pressure fields and experimental measurements of the same quantities. This is, in fact, the approach that was historically taken, and in some ways, it is still the most successful approach. Fortunately, just about the simplest possible guess of equations relating $\mathbf{T}$ and $\mathbf{u}$, or q and θ, turn out to provide an extremely good approximation for the large class of fluids (many liquids and all gases) that we know as Newtonian. We shall discuss the constitutive model for this class of fluids in more detail in Section **G** of this chapter. Regardless of the success of a particular constitutive equation, however, it is obvious that the status of constitutive equations in continuum mechanics is entirely different from the field equations that we derived in previous sections. The latter represent a deductive consequence of the basic laws of Newtonian mechanics and thermodynamics, whereas the former are never more than a guess, no matter how *educated*, in the absence of a fundamental molecular, statistical mechanical theory.

F. Fluid Statics-Constitutive Equations for a Stationary Fluid

Let us begin our quest for specific constitutive equations by considering the special case of a stationary fluid ($\mathbf{u} \equiv 0$). In this case, the acceleration of a fluid element is zero, and the linear momentum equation (2–30) reduces to a balance between body and surface forces,

$$\boxed{\nabla \cdot \mathbf{T} + \rho \mathbf{g} = 0,} \tag{2–56}$$

while the thermal energy equation reduces to the form (2–55). Although the equations are thus considerably simplified for a stationary fluid, the basic problem of requiring constitutive equations for $\mathbf{T}$ and $\mathbf{q}$ remains. However, the constitutive form for $\mathbf{T}$ in this case is very simple.

To deduce the form for the stress tensor $\mathbf{T}$ in a stationary fluid, we observe that the only surface forces in this case are pressure forces. These act normal to the surface with a magnitude that is independent of its orientation. That is, for a surface with orientation denoted by the unit normal vector $\mathbf{n}$, the surface force vector $\mathbf{t}(\mathbf{n})$ takes the form

$$\boxed{\mathbf{t}(\mathbf{n}) = -\mathbf{n}p.} \tag{2-57}$$

The minus sign in this equation is a matter of convention: $\mathbf{t}(\mathbf{n})$ is considered positive when it acts inward on a surface, while **n** is the outwardly directed normal, and p is taken as always positive. The fact that the magnitude of the pressure (or surface force) is independent of **n** is "self-evident" from its molecular origin but also can be proven on purely continuum mechanical grounds, since otherwise the principle of stress equilibrium, (2–23), cannot be satisfied for an arbitrary material volume element in the fluid. The form for the stress tensor **T** in a stationary fluid follows immediately from the general relationship (2–27) between the stress vector and stress tensor, (2–57) and namely,

$$\boxed{\mathbf{T} = -p\mathbf{I}.}$$
(2–58)

In other words, in this case **T** is strictly diagonal, that is,

$$\mathbf{T} = \begin{pmatrix} -p & 0 & 0 \\ 0 & -p & 0 \\ 0 & 0 & -p \end{pmatrix}.$$

Equation (2–58) is the constitutive equation for the stress in a stationary fluid.

Substituting (2–58) into the force balance (2–56), and noting that

$$\nabla \cdot \mathbf{T} = \nabla \cdot (-p\mathbf{I}) = -\nabla p,$$

we obtain the fundamental equation of fluid statics

$$\boxed{\rho\mathbf{g} - \nabla p = 0.}$$
(2–59)

It follows that the presence of a body force leads to a nonzero gradient of pressure parallel to the body force even in a stationary fluid. This is consistent with the well-known observation that the pressure increases with depth in a stationary fluid, with the difference in pressure between two horizontal levels being equal to the weight of a column of fluid with unit cross sectional area between the two levels.

The simplicity of the constitutive equation for stress in a stationary fluid is due to the fact that the only surface force is pressure. The constitutive equation for the surface heat flux vector **q** in a stationary fluid is not so easy to obtain. We adopt a continuum point of view in that we essentially attempt to guess the relevant form. However, we first consider the problem from a molecular viewpoint, with the goal of providing some motivation for this guess.

It may be helpful to recall the origin of **q**, which appears in the thermal energy balance to represent the flux of mean molecular kinetic energy (or heat) in a frame of reference moving with the mean continuum velocity **u** (here, equal to zero) due to random motions of individual molecules. Let us consider, for the moment, the simplest kinetic (or molecular) fluid model, namely, that of a hard-sphere (or billiard-ball) gas in which the only molecular interactions are due to collisions and the random molecular motions are pure translational in character. If we focus our attention on an arbitrary surface in this fluid, it is clear that the only possible mechanism for a flux of heat is the random interchange of hard-sphere molecules across the surface, and that this will lead to heat transfer only in the presence of a nonzero gradient of temperature (or average molecular kinetic energy) normal to the surface. The fact that a molecule passing across the surface in one direction is accompanied on average by a second molecule going in the opposite direction is guaranteed by the continuum approximation and conservation of mass. Furthermore, in this case of a hard-sphere gas, it is evident that the net flux of heat across a surface will be proportional to the product of the magnitude of the temperature gradient normal to the surface, the mean free path between successive molecular collisions, and the frequency of the molecular exchange process. Of course, the molecular transport process for real fluids will be much more complicated than for a billiard-ball gas. Nevertheless, we conclude from our considerations of that simple model that

$$\boxed{\mathbf{q} = \mathbf{q}\,(\nabla\theta, \text{ terms involving higher-order spatial derivatives of } \theta),} \qquad (2\text{-}60)$$

for all real fluids; that is, the rate of heat transport via molecular motions is dependent upon the magnitude of temperature gradients in the fluid (and quite possibly on higher-order spatial derivates of θ as well). The right-hand side of (2–60) represents, in this case, either a function of $\nabla\theta$ or, possibly, a functionale over the past "history" of $\nabla\theta$ for the fluid point of interest. Except for simple model materials like the hard-sphere gas, this is as much as we can deduce from our understanding of the molecular origins of $\mathbf{q}$. From this point, we must guess the constitutive form for $\mathbf{q}$ and ultimately judge the success of our guess by comparing between measured and predicted temperature fields in real fluids.

A reasonable initial guess is the simplest assumption that is consistent with the relationship (2–60); namely, that the heat flux vector depends linearly on $\nabla\theta$,

$$\boxed{\mathbf{q} = -\mathbf{K}\cdot\nabla\theta.} \qquad (2\text{-}61)$$

Here, $\mathbf{K}$ is a second-order tensor that is known as the *thermal conductivity tensor,* and the constitutive equation is known as the *generalized Fourier heat conduction model* for the surface heat flux vector $\mathbf{q}$. The minus sign in (2–61) is a matter of convention; the components of $\mathbf{K}$ are assumed to be positive while a *positive* heat flux is defined as going from regions of high temperature toward regions of low temperature (that is, in the direction of $-\nabla\theta$).

The reader may well be curious why the particular linear form (2–61) was chosen since at least one other vector function is linear in $\nabla\theta$, namely $\boldsymbol{\beta}\wedge(\nabla\theta)$, where $\boldsymbol{\beta}$ is a constant vector. To provide an answer, it is necessary to introduce two important principles that all constitutive relations are expected to obey. The first, which is frequently taken for granted, may be called *coordinate invariance.* This principle states simply that the *form* of a constitutive equation must be invariant under orthogonal coordinate transformations. The physical meaning is that the *form* of a constitutive equation must not change if, instead of specifying it with respect to some coordinate system, we choose to express it in terms of a second coordinate system derived from the first by a rigid rotation, or an inversion, of the coordinate axes, which is called an orthogonal transformation. Underlying this principle is the obvious fact that a change in orientation or sense of the coordinate system *cannot* influence the relevant physical processes, and thus should not influence the form of the constitutive equation. The second invariance requirement of a constitutive equation is that it must also remain unchanged under a transformation in the frame of reference of the observer, even if the frame of the observer (or the fluid) is accelerating with respect to an inertial frame. This is usually thought of as being a consequence of the intuitive notion that the mechanical or thermal properties of a material element cannot depend upon any motion of the person observing the material, and this is called the *principle of material objectivity.* Material objectivity is a stronger requirement than coordinate invariance for constitutive equations that involve dynamic variables, as we shall see shortly. However, for the constitutive equation for **q**, the requirement of invariance to rotation and inversion of the coordinate axes is sufficient to verify that (2–61) is the only acceptable constitutive form that is linear in $\nabla\theta$.

To demonstrate this fact, we proceed in two steps. First, we show that (2–61) is invariant under any orthogonal transformation, either proper or improper. Second, we show that the only other vector function that is linear in $\nabla\theta$ is *not* invariant to coordinate inversions.

In order to verify that the proposed constitituve form (2–61) is invariant to coordinate rotations or inversions, we introduce the transformation matrix **L**. This matrix relates the components of a general position vector **x** in some coordinate system to the components of the same vector in a second coordinate system that is obtained from the first by a rigid rotation or an inversion. If we denote the position vector in the second system as $\bar{\mathbf{x}}$, the transformation between **x** and $\bar{\mathbf{x}}$ is given by

$$\boxed{\mathbf{x}=\mathbf{L}\cdot\bar{\mathbf{x}} \text{ or } \bar{\mathbf{x}}=\mathbf{L}^{T}\cdot\mathbf{x}.} \tag{2-62}$$

The fact that the transformation from **x** to $\bar{\mathbf{x}}$ inverts in the manner shown is a reflection of the fact that **L**, corresponding to either a rigid rotation or an inversion of the coordinate axes, is orthogonal; that is, it exhibits the orthogonality property

$$\boxed{\mathbf{L}\cdot\mathbf{L}^{T}=\mathbf{L}^{T}\cdot\mathbf{L}=\mathbf{I}.} \tag{2-63}$$

The components of **L**, which we shall denote as l_{ij}, are just the direction cosines between the two coordinate axis systems—namely, the cosine of the angles between axis i of the untransformed system and axis j of the transformed system. Thus, as an example, for rigid rotation about the *3* axis, through an angle θ, as illustrated in Figure 2–6, the transformation matrix (which is sometimes called a rotation matrix in this case) is just

$$\mathbf{L} = \begin{pmatrix} \cos\theta & -\sin\theta & 0 \\ \sin\theta & \cos\theta & 0 \\ 0 & 0 & 1 \end{pmatrix},$$

while a complete inversion, illustrated in Figure 2–7 (that is, $\mathbf{i} \to -\bar{\mathbf{i}}, \mathbf{j} \to -\bar{\mathbf{j}}, \mathbf{k} \to -\bar{\mathbf{k}}$, where **i**, **j**, and **k** are Cartesian unit vectors) corresponds to

$$\mathbf{L} = \begin{pmatrix} -1 & 0 & 0 \\ 0 & -1 & 0 \\ 0 & 0 & -1 \end{pmatrix}.$$

Now, a general property for any vector **A** is that its components in the untransformed coordinate system may be expressed in terms of its components in the transformed system through the equations.

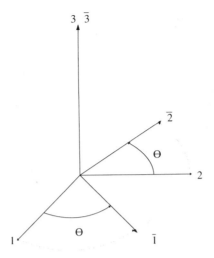

Figure 2–6 The coordinate axes $\bar{1}\,\bar{2}\,\bar{3}$ obtained by rotating the coordinate system *123* through an angle θ around the *3* axis. The corresponding rotation matrix **L** is given in the text.

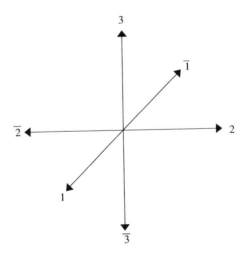

Figure 2–7 The coordinate axes $\bar{1}\,\bar{2}\,\bar{3}$ obtained from coordinate *123* by a complete inversion. The corresponding rotation matrix is given in the text.

$$\boxed{\mathbf{A} = \mathbf{L}\cdot\bar{\mathbf{A}} \text{ and } \bar{\mathbf{A}} = \mathbf{L}^{T}\cdot\mathbf{A}.} \tag{2-64}$$

Similarly, the transform rules for components of a second-order tensor, **B**, are just

$$\boxed{\mathbf{B} = \mathbf{L} \cdot \bar{\mathbf{B}} \cdot \mathbf{L}^T \text{ and } \bar{\mathbf{B}} = \mathbf{L}^T \cdot \mathbf{B} \cdot \mathbf{L}.} \tag{2-65}$$

With this brief background on orthogonal transformations of coordinate systems, we can verify that the proposed linear form for $\mathbf{q}$, given by (2–61), is invariant to such transformations. To do this, we begin with (2–61) and transform to a new coordinate frame using the "rules" (2–63) and (2–64). Thus, operating on (2–61) with $\mathbf{L}^T$ to transform from $\mathbf{q}$ to $\bar{\mathbf{q}}$, we find

$$\mathbf{L}^T \cdot \mathbf{q} = \mathbf{L}^T \cdot [-\mathbf{K} \cdot \nabla\theta].$$

The left-hand side is just $\bar{\mathbf{q}}$. Thus, if the constitutive model is to exhibit coordinate invariance, the right-hand side must equal $-\bar{\mathbf{K}} \cdot (\overline{\nabla\theta})$; that is, we should expect

$$\mathbf{L}^T \cdot [-\mathbf{K} \cdot \nabla\theta] \rightarrow -\bar{\mathbf{K}} \cdot (\overline{\nabla\theta}).$$

But, since $\mathbf{L} \cdot \mathbf{L}^T = \mathbf{I}$, the left-hand side can be reexpressed as

$$\mathbf{L}^T \cdot [-\mathbf{K} \cdot (\mathbf{L} \cdot \mathbf{L}^T) \cdot \nabla\theta],$$

or, rearranging slightly,

$$-[\mathbf{L}^T \cdot \mathbf{K} \cdot \mathbf{L}] \cdot (\mathbf{L}^T \cdot \nabla\theta).$$

However, according to the transformation rules, (2–64) and (2–65), this is just

$$-\bar{\mathbf{K}} \cdot \overline{\nabla\theta},$$

and we see that the proposed constitutive form (2–61) is invariant to arbitrary orthogonal transformations of the coordinate system.

However, it is important to recognize that other possible linear forms involving $\mathbf{q}$ and $\nabla\theta$ do not satisfy the requirement for coordinate invariance and thus are automatically ruled out as possible candidates for the constitutive equation for $\mathbf{q}$. For example, suppose we were to try the alternative linear form

$$\mathbf{q} = \boldsymbol{\beta} \wedge (\nabla\theta), \tag{2-66}$$

where $\boldsymbol{\beta}$ is a constant vector. Operating on both sides as before, with $\mathbf{L}^T$, we have

$$\mathbf{L}^T \cdot \mathbf{q} = \bar{\mathbf{q}} = \mathbf{L}^T \cdot [\boldsymbol{\beta} \wedge (\nabla\theta)].$$

But if the proposed form were coordinate invariant, the right-hand side should be equal to

$$\bar{\mathbf{q}} = \bar{\boldsymbol{\beta}} \wedge (\overline{\nabla\theta}).$$

Comparing the two preceding equations, we see that the condition of coordinate invariance requires

$$\mathbf{L}^T \cdot (\boldsymbol{\beta} \wedge \nabla\theta) \overset{?}{=} \bar{\boldsymbol{\beta}} \wedge (\overline{\nabla\theta}).$$

We put a question mark on the equal sign here because we have not yet established the equality. Indeed, if we proceed one step further by expressing $\bar{\boldsymbol{\beta}}$ and $\overline{\nabla\theta}$ in terms of $\boldsymbol{\beta}$ and $\nabla\theta$ using the transform rules (2–64), we find

$$\mathbf{L}^T \cdot (\boldsymbol{\beta} \wedge \nabla\theta) \overset{?}{=} (\mathbf{L}^T \cdot \boldsymbol{\beta}) \wedge (\mathbf{L}^T \cdot \nabla\theta). \tag{2-67}$$

Evidently, the two sides of (2–67) are not generally equal for an arbitrary orthogonal transformation of coordinates. For example, suppose we consider the coordinate inversion

$$\mathbf{L} = \begin{pmatrix} -1 & & \\ & -1 & \\ & & -1 \end{pmatrix}.$$

The left-hand side of (2–67) changes sign under this transformation. The right-hand side, however, remains invariant, that is,

$$(\mathbf{L}^T \cdot \boldsymbol{\beta}) \wedge (\mathbf{L}^T \cdot \nabla \theta) = (-\boldsymbol{\beta}) \wedge (-\nabla \theta) = \boldsymbol{\beta} \wedge \nabla \theta.$$

Clearly, then, the proposed linear form (2–66) is unacceptable as a constitutive approximation for **q** because it is not invariant to arbitrary orthogonal transformations of the coordinate system.

The reader who is experienced with vector and tensor analysis may immediately recognize that (2–66) is not an acceptable form for a constitutive equation without the need for the detailed manipulations that we have just described. The reason is that the entity on the right-hand side is a so-called *pseudovector*. The key property of a pseudovector is that it changes sign if one adopts a left-handed coordinate convention in evaluating the vector product instead of a right-handed system. However, the heat-flux vector **q** is a true vector that is invariant to such changes in the coordinate system: Clearly, the physical process represented by **q** is not dependent upon the coordinate system that we choose to represent **q**, and that is the underlying physical basis of the coordinate invariance principle.

The assumption of a *linear* relationship between **q** and $\nabla \theta$ is an *a priori* guess, which can only be verified, ultimately, by comparison between predicted and measured data for the heat flux and temperature gradient. However, among all possible linear forms, the requirement of coordinate invariance can be used to eliminate some possibilities [that is, (2–66)].

In addition to linearity between **q** and $\nabla \theta$, there are a number of additional assumptions inherent in the model (2–61) that need to be recognized here.

1. The surface heat transfer process is assumed to be *local*, in the sense that the flux associated with the fluid at some point depends only on the temperature gradient at the same point.
2. The surface heat transfer process is assumed to be *instantaneous*; the heat flux at a point depends only on the temperature gradient at that point at the same instant of time. In particular, there is no dependence on the thermal history of the fluid element that currently occupies the point in question.
3. The fluid is assumed to be *homogeneous*. The form of the relationship between the heat flux **q** and the temperature gradient $\nabla \theta$ is the same at all points. Furthermore, the only dependence of **q** on position **x** is due to the possible dependence of the so-called thermal conductivity tensor **K** on

the thermodynamic state variables (say, p and θ) or the dependence of $\nabla\theta$ on spatial position.

4. When there is no temperature gradient, the surface heat flux is identically zero.

It should be emphasized that all the preceding points are simply assumptions underlying the assumed constitutive form (2-61). We can make no claim *on the basis of continuum mechanics alone* that these assumptions or the basic linearity of (2-61) will necessarily be satisfied by any real fluid.

Fortunately, in view of the simplicity of (2-61), comparison between predicted and measured data for the heat flux and temperature gradient shows that the general linear form does work extremely well for many common gases and liquids. However, the majority of these materials exhibit one additional characteristic that leads to further simplification of the constitutive form (2-61)—they are *isotropic*. This means that the magnitude of the heat flux at any point is dependent only on the *magnitude* of the temperature gradient, not on its *orientation* relative to axes fixed in the material. A common material that is not isotropic in this sense is wood, because a temperature gradient of given magnitude in wood generally produces a larger heat flux if it is oriented along the grain than it does if it is oriented across the grain. In the absence of motion, almost all common fluids will be isotropic (an exception is a liquid crystalline material). If the fluid is made up of molecules and/or particles that are not spherical (or spherically symmetric), the orientations of these structural elements will generally be random as a consequence of random (Brownian) motions. Hence, when seen from the spatially averaged continuum viewpoint, such a fluid will be isotropic.

A mathematical statement of the *property of isotropy* is that the constitutive equation must be completely invariant to rotations of the coordinate reference system. In other words, in the present case, if we consider the components of $\mathbf{q}$ and $\nabla\theta$ defined relative to two coordinate systems that differ by rigid body rotation through some arbitrary angle (or angles) and then if

$$\mathbf{q} = \mathbf{F}(\nabla\theta),$$

it follows from the assumption of isotropy that

$$\bar{\mathbf{q}} = \bar{\bar{\mathbf{F}}}(\overline{\nabla\theta}),$$

where $\mathbf{F}$ and $\bar{\mathbf{F}}$ are *identical* functions.

We have already seen that the constitutive form (2-61) transforms under a general orthogonal transformation to the form

$$\bar{\mathbf{q}} = \bar{\mathbf{K}} \cdot \overline{\nabla\theta},$$

where

$$\bar{\mathbf{K}} = \mathbf{L}^T \cdot \mathbf{K} \cdot \mathbf{L}.$$

Hence, the condition of isotropy, in this case, requires that

$$\boxed{\bar{\mathbf{K}} \equiv \mathbf{K}.}$$

(2-68)

This condition on **K** will be satisfied for an arbitrary rotation tensor, **L**, if and only if **K** is a product of a scalar and the unit tensor **I**, that is,

$$\boxed{\mathbf{K} = k\mathbf{I},}$$
(2–69)

where k is a *scalar* property of the fluid that is known as the *thermal conductivity*.

Thus, for an *isotropic* fluid that exhibits a linear, instantaneous relationship between the heat flux and temperature gradient, the most general constitutive form for **q** is

$$\boxed{\mathbf{q} = -k\nabla\theta.}$$
(2–70)

This is known as *Fourier's law of heat conduction*. We may note that the inequality (2–54) imposes a restriction on the sign of the thermal conductivity k. In particular, in the absence of fluid motion, (2–54) reduces to the simple form

$$-\frac{\mathbf{q}\cdot\nabla\theta}{\theta^2} \geq 0.$$

It follows from this inequality and the constitutive form (2–70) that

$$k\left(\frac{\nabla\theta}{\theta}\right)^2 \geq 0.$$

Hence, assuming $k \neq 0$, we see that the thermal conductivity must be positive, $k > 0$.

Although this simplified version of Fourier's heat conduction law is well known to be an accurate constitutive model for many real gases, liquids, and solids, it is important to keep in mind that, in the absence of empirical data, it is no more than an educated guess, based upon a series of assumptions about material behavior that one cannot guarantee ahead of time to be satisfied by any real material. This status is typical of all constitutive equations in continuum mechanics, except for the relatively few that have been derived via a molecular theory.

G. Constitutive Equations for a Flowing Fluid— General Considerations

In the previous section, we discussed constitutive approximations for stress and surface heat flux in a stationary fluid, where $\mathbf{u} \equiv 0$. In view of the molecular origins of **q**, there is no reason to expect that the basic linear form for its constitutive behavior should be modified by the presence of mean motion, at least for materials that are not too complicated in structure. Of course this situation may be changed for materials such as polymeric liquids or suspensions, because in these cases the presence of motion may cause the structure to become anisotropic or changed in other ways that will affect the heat transfer process. We will return to this question in Section I.

The constitutive equation (2–58) for the stress, on the other hand, will be modified for all fluids in the presence of a mean motion in which the velocity gradient $\nabla\mathbf{u}$ is nonzero. To see that this must be true, we can again consider the simplest possible model system of a hard-sphere or billiard-ball gas, which we may assume to be undergoing a simple unidirectional shear flow, say, $\mathbf{u} = \gamma y \mathbf{i}$, where γ is known as the shear rate. Now, let us consider a surface in this fluid whose normal $\mathbf{n}$ is parallel to $\nabla \mathbf{u}$ (that is, $\mathbf{n} \equiv \mathbf{j}$). In this case, any interchange of molecules across the surface will result in a transfer of mean momentum; that is, the faster moving fluid on one side of the surface will appear to be decelerated as one of its molecules is exchanged for a slower moving molecule from the other side of the surface, while the slower moving fluid will appear to be accelerated by the same process. Hence, from the continuum point of view where this transfer of momentum is modeled in terms of equivalent surface forces (stresses), we see that the surface force vector in a moving fluid must generally have a component that acts *tangent* to the surface, and this is fundamentally different from the case of a stationary fluid in which the only surface forces are pressure forces that act *normal* to surfaces. We may also note, in the case of a hard-sphere gas, that the rate of momentum transfer via random molecular motions is proportional to $\nabla \mathbf{u}$. On the other hand, we know that $\mathbf{T}$ must reduce to the form (2–58) when $\nabla \mathbf{u} = 0$, both for a hard-sphere gas and other, more complicated (real) fluids.

We conclude, based upon the insight we have drawn from the hard-sphere gas model and our general understanding of the molecular origins of $\mathbf{T}$, that

$$\boxed{\mathbf{T} + p\mathbf{I} = \tau(\nabla \mathbf{u}, \text{ terms involving higher-order spatial derivatives})} \qquad (2\text{-}71)$$

for real fluids, where $\tau(\nabla \mathbf{u}, \ldots)$ could be either a function of present values of $\nabla \mathbf{u}$ or a functional that includes a dependence on both previous and present values. The second-order tensor τ is usually known as the *deviatoric* stress.

To obtain a specific form for $\tau(\nabla \mathbf{u}, \ldots)$, we again require a guess. However, some general properties of τ can be deduced that do not depend upon a specific constitutive form, and we discuss these general properties in the remainder of this section. First, it is obvious from (2–58) and (2–71) that $\tau(\nabla \mathbf{u}, \ldots) = 0$ for $\nabla \mathbf{u} = 0$ and $t \to \infty$. The requirement that $\tau \to 0$ asymptotically as $t \to \infty$ is necessary because all fluids are not "instantaneous" in the sense that τ depends only on the current values of $\nabla \mathbf{u}$. It is known, for example, that fluids exist where the deviatoric stress, $\tau(\nabla \mathbf{u}, \ldots)$, vanishes only if $\nabla \mathbf{u}$ has been zero for a finite period. Such fluids are said to possess a "memory" for past configurations and are typified by polymer solutions in which the molecular structure can return to an equilibrium state only via diffusion processes that require a finite period of time. A second (obvious) general property of τ is that is it symmetric in the absence of external body couples. This follows directly from (2–71) and the fact that $\mathbf{T}$ is symmetric in the same circumstances. A third general property is that τ must depend explicitly only on the symmetric part of $\nabla \mathbf{u}$, rather than $\nabla \mathbf{u}$ itself. We have already noted in (2–47) that the symmetric part of $\nabla \mathbf{u}$ is called the rate of strain tensor and is usually denoted as $\mathbf{E}$. It might seem, at first, that this third property would follow from the fact that τ is symmetric, but this is not the case.

There are two proper explanations, one intuitive and the other based upon the principle of material objectivity. The intuitive explanation is that nonzero contributions to the deviatoric stress cannot arise from rigid body motions—whether solid-body translation or rotation. Only if adjacent fluid elements are in relative (nonrigid-body) motion can random molecular motions lead to a net transport of momentum. We shall see shortly that the rate of strain tensor relates to the rate of change of length of a line element connecting two material points of the fluid (that is, to relative displacements of the material points), while the antisymmetric part of $\nabla \mathbf{u}$, known as the vorticity tensor, $\mathbf{\Omega}$, is related to its rate of (rigid-body) rotation. Thus, it follows that τ must depend explicitly on $\mathbf{E}$, but not on $\mathbf{\Omega}$:

$$\boxed{\tau = \tau(\mathbf{E}, \ldots).}\qquad(2\text{--}72)$$

The fact that the deviatoric stress can only depend on the symmetric part of $\nabla \mathbf{u}$ can also be deduced from the principle of material objectivity. As stated earlier, this principle states that the form of a constitutive law should be invariant to changes in the frame of reference and is based upon the belief that the behavior of a fluid must be independent of any motion of the observer. To show that material objectivity leads to a restriction in the independent variable for a general constitutive equation of the form (2–71), it is necessary to consider the consequences of a change of reference frame from $(\mathbf{x},t)$ to $(\mathbf{x}',t')$ where

$$\mathbf{x}' = \mathbf{U}(t') + \mathbf{L}^T(t') \cdot \mathbf{x}\qquad(2\text{--}73)$$

and

$$t' = t - a.\qquad(2\text{--}74)$$

Here, the vector $\mathbf{U}(t')$ represents a time-dependent translation of the "new" reference frame relative to the original frame, while $\mathbf{L}^T(t')$ is a rotation.

Now, it follows from (2–73) that

$$\mathbf{v}' = \mathbf{L}^T \cdot \mathbf{v} + \dot{\mathbf{U}}(t') + \dot{\mathbf{L}}^T \cdot \mathbf{x}\qquad(2\text{--}75)$$

and

$$d\mathbf{v}' = \mathbf{L}^T \cdot d\mathbf{v} + \dot{\mathbf{L}}^T \cdot d\mathbf{x},\qquad(2\text{--}76)$$

where $d\mathbf{v}$ and $d\mathbf{v}'$ represent the *relative* velocities between two points that are separated by $d\mathbf{x}$ in the original reference frame. By definition, the velocity gradient is related to $d\mathbf{v}$ and $d\mathbf{v}'$ according to

$$\nabla\mathbf{v} \cdot d\mathbf{x} = d\mathbf{v} \text{ and } \nabla\mathbf{v}' \cdot d\mathbf{x}' = d\mathbf{v}',\qquad(2\text{--}77)$$

where

$$d\mathbf{x}' = \mathbf{L}^T \cdot d\mathbf{x}.\qquad(2\text{--}78)$$

It follows from (2–76), (2–77) and (2–78) that

$$\nabla\mathbf{v}' \cdot \mathbf{L}^T \cdot d\mathbf{x} = \mathbf{L}^T \cdot (\nabla\mathbf{v} \cdot d\mathbf{x}) + \dot{\mathbf{L}}^T \cdot d\mathbf{x}.$$

Hence, since $\mathbf{L} \cdot \mathbf{L}^T = \mathbf{I}$ and $d\mathbf{x}$ is arbitrary, we find

$$\nabla\mathbf{v}' = \mathbf{L}^T \cdot \nabla\mathbf{v} \cdot \mathbf{L} + \dot{\mathbf{L}}^T \cdot \mathbf{L},\qquad(2\text{--}79)$$

and it is evident that the velocity gradient is *not* frame invariant because it does not transform like a regular tensor, that is, $\nabla\mathbf{v}' \neq \mathbf{L}^T \cdot \nabla\mathbf{v} \cdot \mathbf{L}$. On the other hand, if we examine the rate of strain tensor,

$$\mathbf{E} \equiv \frac{1}{2}(\nabla\mathbf{v} + \nabla\mathbf{v}^T),$$

we can easily show that

$$\mathbf{E}' = \mathbf{L}^T \cdot \mathbf{E} \cdot \mathbf{L} + \frac{1}{2}(\dot{\mathbf{L}} \cdot \mathbf{L}^T + \mathbf{L} \cdot \dot{\mathbf{L}}^T),$$

and since $\dot{\mathbf{I}} \equiv 0$, it follows that

$$\mathbf{E}' = \mathbf{L}^T \cdot \mathbf{E} \cdot \mathbf{L}. \tag{2–80}$$

Now, since the stress tensor is frame invariant according to the principle of material objectivity, it follows that the stress can depend only on the symmetric part of $\nabla\mathbf{v}$ (that is, on $\mathbf{E}$) but not on $\nabla\mathbf{v}$ itself. This is the same conclusion, (2–72), that we reached earlier via our intuitive arguments based upon the physical significance of the rate-of-strain and vorticity tensors.

As a final step before going on to consider special forms for the constitutive relationship, (2–72), we elaborate on our assertions regarding the physical significance of the rate of strain and vorticity tensors. To do this, we consider two nearby material points in the fluid P, initially at position $\mathbf{x}$ and Q, which is at $\mathbf{x} + \delta\mathbf{x}$. By material points, we mean points that move with the local velocity of the fluid. We denote the velocity of the material point P as $\mathbf{u}$, and that of Q as $\mathbf{u} + \delta\mathbf{u}$. Now, a Taylor series approximation can be used to relate $\mathbf{u}$ and $\mathbf{u} + \delta\mathbf{u}$, namely,

$$\mathbf{u} + \delta\mathbf{u} = \mathbf{u} + (\mathbf{E} + \mathbf{\Omega}) \cdot \delta\mathbf{x} + O(|\delta\mathbf{x}|^2). \tag{2–81}$$

Hence, the point Q moves *relative* to the point P with a velocity

$$\delta\mathbf{u} = \mathbf{E} \cdot \delta\mathbf{x} + \mathbf{\Omega} \cdot \delta\mathbf{x} + O(|\delta\mathbf{x}|^2). \tag{2–82}$$

Now, the length of the line element connecting P and Q is

$$|\delta\mathbf{x}| = (\delta\mathbf{x} \cdot \delta\mathbf{x})^{1/2},$$

and the rate of change in this length is thus proportional to

$$\delta\mathbf{x} \cdot \delta\mathbf{u} = \delta\mathbf{x} \cdot [\mathbf{E} \cdot \delta\mathbf{x} + \mathbf{\Omega} \cdot \delta\mathbf{x} + O(|\delta\mathbf{x}|^2)], \tag{2–83}$$

where $\delta\mathbf{u} = \dfrac{D}{Dt}\delta\mathbf{x}$. However,

$$\delta\mathbf{x} \cdot \mathbf{\Omega} \cdot \delta\mathbf{x} \equiv 0$$

since $\mathbf{\Omega}$ is antisymmetric, and it follows

$$\frac{1}{2}\frac{D}{Dt}(|\delta\mathbf{x}|^2) = \delta\mathbf{x} \cdot \mathbf{E} \cdot \delta\mathbf{x} + O(|\delta\mathbf{x}|^3). \tag{2–84}$$

Thus, the rate of change of the distance between two neighboring material points depends only on the symmetric part of $\nabla\mathbf{u}$, that is, on the rate of strain tensor $\mathbf{E}$. It can be shown in a similar manner that the contribution to the relative velocity vector $\delta\mathbf{u}$ due to the vorticity tensor, $\mathbf{\Omega}$, is the same as the displacement due to a (local) rigid-body rotation with angular velocity $\frac{1}{2}\omega$, where

$$\mathbf{\Omega} = \frac{1}{2}\,\boldsymbol{\epsilon}\cdot\boldsymbol{\omega}. \tag{2-85}$$

For future reference, we also note that

$$\boldsymbol{\omega} = \nabla\wedge\mathbf{u}. \tag{2-86}$$

The vector $\boldsymbol{\omega}$ is known as the vorticity vector. It can be calculated from $\mathbf{u}$ either via (2–85) or (2–86).

H. Constitutive Equations for a Flowing Fluid— The Newtonian Fluid

To proceed beyond the general relationship (2–72), it is necessary to make a guess of the constitutive behavior of the fluid. The simplest assumption consistent with (2–72) is that the deviatoric stress depends linearly on the rate of strain, that is,

$$\boxed{\mathbf{T} + p\mathbf{I} = \mathbf{A}{:}\mathbf{E}.} \tag{2-87}$$

Here, $\mathbf{A}$ is a fourth-order tensor that must be symmetric in its first two indices

$$A_{ijkl} \equiv A_{jikl} \tag{2-88}$$

since $\mathbf{T} + p\mathbf{I}$ is symmetric according to the constraint (2–39). The constitutive relation (2–87) is analogous to the generalized Fourier model (2–61), for the heat flux vector $\mathbf{q}$. Like the generalized Fourier model, it assumes that the fluid is *local, instantaneous, homogeneous*, and is invariant to rotations or inversions of the coordinate axes.

An additional physical assumption that is satisfied by many fluids is that the structure is *isotropic* even in the presence of motion. For an isotropic fluid, the constitutive equation is completely unchanged by rotations of the coordinate system. When applied to (2–72), for example, the condition of isotropy requires that

$$\mathbf{L}^T{\cdot}\boldsymbol{\tau}(\mathbf{E})\cdot\mathbf{L} = \boldsymbol{\tau}(\mathbf{L}^T{\cdot}\mathbf{E}{\cdot}\mathbf{L}). \tag{2-89}$$

Thus,

$$\mathbf{L}^T\cdot(\mathbf{A}{:}\mathbf{E})\cdot\mathbf{L} = \mathbf{A}{:}(\mathbf{L}^T\cdot\mathbf{E}\cdot\mathbf{L}) \tag{2-90}$$

for an isotropic fluid that satisfies the other assumptions inherent in (2–87). It can be shown, using the methods of tensor analysis, that the most general fourth-order tensor with the property (2–90) is

$$A_{ijpq} = \lambda\delta_{ij}\delta_{pq} + \mu(\delta_{ip}\delta_{jq} + \delta_{iq}\delta_{jp}) + \nu(\delta_{ip}\delta_{jq} - \delta_{iq}\delta_{jp}) \tag{2-91}$$

where λ, μ, and ν are arbitrary scalar constants and δ_{ij} is the ij component of the identity tensor, that is,

$$\delta_{ij} = \begin{cases} 1 & i=j \\ 0 & i\neq j \end{cases} \quad ij = 1,2,3. \tag{2-92}$$

Since the tensor $\mathbf{A}$ must also satisfy the symmetry condition (2–88), it follows that

$$\nu \equiv 0.$$

Substituting (2–91) into (2–87), we see that the most general constitutive equation that is consistent with the linear dependence of $\mathbf{T}$ on $\mathbf{E}$, and the assumption of isotropy, is

$$\boxed{\mathbf{T} = (-p + \lambda\,tr\mathbf{E})\mathbf{I} + 2\mu\mathbf{E}.} \tag{2-93}$$

Expressed in component form using Cartesian tensor notation, this equation is

$$T_{ij} = (-p + \lambda E_{pp})\delta_{ij} + 2\mu E_{ij}.$$

Fluids for which this constitutive equation is an adequate model are known as *Newtonian fluids*. We have shown that the Newtonian fluid model is the most general form that is linear and instantaneous in $\mathbf{E}$ and isotropic. If the fluid is also incompressible,

$$tr\mathbf{E} = \nabla \cdot \mathbf{u} = 0,$$

and the constitutive equation further simplifies to the form

$$\boxed{\mathbf{T} = -p\mathbf{I} + 2\mu\mathbf{E}.} \tag{2-94}$$

The coefficient μ that appears in this equation is known as the *shear viscosity* and is a property of the fluid.

We have, of course, said nothing about the physical reality of the initial assumptions of isotropy or of a linear, instantaneous dependence of $\mathbf{T}$ on $\mathbf{E}$. It is possible, insofar as continuum mechanics is concerned, that no fluid would be found for which these are adequate assumptions. Fortunately, in view of the simplicity of the resulting constitutive model, (2–93) or (2–94), experimental observation shows that the Newtonian constitutive assumptions are satisfied for gases in almost all circumstances and for the majority of low to moderate molecular weight liquids providing that $\|\mathbf{E}\|$ is not extremely large and does not change too rapidly with respect to time. Polymeric liquids, suspensions, and emulsions do not generally satisfy the Newtonian assumptions and require much more complicated constitutive equations for $\mathbf{T}$. We shall briefly discuss these latter fluids in the next section. An extremely important fact is that Newtonian fluids are also generally found to follow Fourier's law of heat conduction in the isotropic form (2–70).

We have emphasized repeatedly that continuum mechanics provides no guidance in the choice of a general constitutive hypothesis for either the heat flux vector $\mathbf{q}$ or the stress tensor $\mathbf{T}$. On the other hand, we have noted earlier that (2–38) and (2–54), derived, respectively, from the law of conservation of angular momentum and the second law of thermodynamics, must be satisfied by the resulting constitutive equations. It thus behooves us to see whether these two constraints are satisfied for the Newtonian and Fourier constitutive models that have been proposed in this and the preceding section. In the absence of external body couples (assumed to be true here), the constraint from angular momentum conservation requires only that the stress be symmetric, and this is obviously satisfied by the Newtonian model for any choice of λ and μ. The second constraint, from the second law of thermodynamics, requires

$$\mathbf{T}{:}\mathbf{E} + p(\nabla \cdot \mathbf{u}) - \frac{\mathbf{q} \cdot \nabla\theta}{\theta} \geq 0.$$

Substituting for $\mathbf{T}$ from (2–93) and for $\mathbf{q}$ from (2–70), we find

$$\left(\lambda + \frac{2}{3}\mu\right)(tr\mathbf{E})^2 + 2\mu\left(\mathbf{E} - \left(\frac{1}{3}tr\mathbf{E}\right)\mathbf{I} + \frac{k}{\theta}(\nabla\theta)^2 \geq 0. \tag{2-95}$$

If we consider the special case of an isothermal, incompressible fluid, the inequality (2–95) becomes

$$2\mu(\mathbf{E}:\mathbf{E}) \geq 0.$$

Obviously, if the Newtonian constitutive model for an incompressible fluid is to be consistent with the second law of thermodynamics, we require that the viscosity be nonnegative, that is,

$$\mu \geq 0. \tag{2–96}$$

This is, in fact, the most significant result that can be obtained for Newtonian fluids from the second-law inequality. If we do not restrict ourselves to incompressible or isothermal conditions, the inequality (2–95) can be satisfied for arbitrary motions and arbitrary temperature fields only if

$$\lambda + \frac{2}{3}\mu \geq 0, \qquad \mu \geq 0, \qquad \text{and} \qquad k \geq 0. \tag{2–97}$$

The quantity $(\lambda + \frac{2}{3}\mu)$ is commonly called the bulk viscosity coefficient. Besides the inequalities (2–97), no further information appears to be attainable for a Newtonian fluid from the constraint (2–54).

I. Constitutive Equations for a Flowing Field— Non-Newtonian Fluids

We have already noted that there are many fluids for which the Newtonian constitutive equation does not provide an adequate approximation of the relationship between stress and velocity gradients. One of the simplest experimental tests of Newtonian behavior is to measure, either directly or indirectly, the resistance to motion in a steady shear flow where only one component of the velocity gradient tensor is nonzero. It is not our purpose, here, to discuss the details of this type of measurement, which are really in the realm of experimental rheology or viscometry, subjects about which complete books have been written.[6] Instead, we note that a so-called "simple" unidirectional shear flow

$$u = \gamma y, \qquad v = w = 0, \tag{2–98}$$

in which (u, v, w) are the velocity components in the (x, y, z) directions, respectively, is a flow in which all components of the rate of strain tensor $\mathbf{E}$ are zero, except

$$E_{xy} = E_{yx} = \frac{1}{2}\gamma.$$

Thus, if a fluid were Newtonian, the nonzero components of the stress would be

$$T_{xx} = T_{yy} = T_{zz} = -p, \qquad T_{xy} = T_{yx} = \gamma\mu. \tag{2–99}$$

Evidently, if one could measure the shear stress, T_{xy} or T_{yx} and if the shear rate γ were either known or measured, we should find for a Newtonian fluid that the ratio T_{xy}/γ is a constant independent of the shear rate and equal to the shear viscosity μ. It should be cautioned, however, that a constant ratio T_{xy}/γ is not, in itself, a proof that the fluid is actually Newtonian. Some fluids are known, for example, that exhibit a constant shear viscosity but nonzero normal stress differences $T_{xx} - T_{yy}$ or $T_{yy} - T_{zz}$ that would be zero for a Newtonian fluid. If, on the other hand, the ratio T_{xy}/γ is not independent of the shear rate, this is definitive proof that the Newtonian model is not an adequate approximation for the particular fluid being tested.

Although the unidirectional, linear shear flow (2–98) is extremely simple in form, it cannot be realized easily in the laboratory since it ideally exists only between two infinite parallel plane boundaries that translate parallel to one another. A close approximation to a linear shear flow, however, can be achieved locally between two concentric cylinders that rotate relative to one another, known as a Couette flow device, or in a device favored by many rheologists that consists of a cone that is rotated relative to a flat plate. These two flow geometries are depicted in Figure 2–8. If the relative rates of rotation are known in either of these devices, the shear rate can be calculated. The corresponding shear stress can be determined by measuring the torque required to turn the cylinders in the Couette device, or to turn the cone in the cone-and-plate geometry.

Measurements in these devices show that many low molecular weight liquids closely satisfy the expected behavior for a Newtonian fluid, in that the torque required to rotate the cylinder or the cone varies linearly with the rate of rotation. On the other hand, other materials such as solutions or melts of high molecular weight polymer or suspensions frequently show marked deviations from linearity. For example, under most conditions polymeric liquids exhibit so-called *shear-thinning* behavior, in which the shear stress increases more slowly than linearly with shear rate so that the ratio of stress-to-strain rate (which would equal twice the viscosity in a Newtonian fluid) decreases with shear rate, as illustrated qualitatively in Figure 2–9. It is sometimes stated that such materials show a ''shear-thinning viscosity.'' In reality, the variation of T_{xy}/E_{xy} with shear rate means that this ratio is no longer equal to a material property

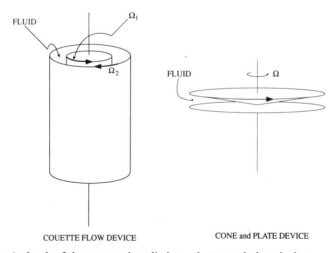

COUETTE FLOW DEVICE   CONE and PLATE DEVICE

Figure 2–8 A sketch of the concentric cylinder and cone-and-plate devices.

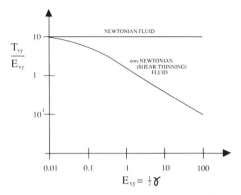

Figure 2-9 A qualitative sketch of the dependence of the ratio of shear stress to strain rate on the shear rate γ in a simple shear flow, $u_x = \gamma y$, (a) Newtonian fluid where the ratio T_{xy}/E_{xy} is twice the shear viscosity 2μ and (b) a non-Newtonian fluid that shows shear-thinning behavior. In both cases, the ratio is arbitrarily chosen to have an initial value of 10. The units for T_{xy}/E_{xy} and E_{xy} are arbitrary in this log-log plot.

since the latter would be expected to vary only with the thermodynamic state of the material (for example, its temperature and pressure), but to be independent of the nature or strength of the flow that the fluid may be undergoing. The concept of a shear-rate dependent viscosity arises only because we interpret the ratio T_{xy}/E_{xy} as defining a viscosity as though the fluid were Newtonian. At any rate, while many liquids exhibit apparent Newtonian behavior insofar as the shear stress T_{xy} and the shear rate (γ) are concerned, there are many others that do not. An immediate question then is how the constitutive hypothesis might be modified to accommodate these latter fluids. Again, our goal is not a comprehensive study of constitutive models for non-Newtonian fluids, but rather a brief introduction to give the reader a flavor for the subject. Many existing books do provide a comprehensive coverage, among which we may mention Bird et al. (1987), Tanner (1985), and Larson (1988).[7]

What then would seem to be a promising direction for generalizing the Newtonian assumptions of a linear and instantaneous relationship between the stress and rate of strain tensors? The fact that many non-Newtonian materials exhibit a nonlinear dependence of shear stress on strain rate in the case of a simple shearing flow would seem to suggest the possibility of a nonlinear constitutive model of the form

$$\boxed{\tau = \mathbf{T} + p\mathbf{I} = \mathbf{f}(\mathbf{E}),} \qquad (2\text{--}100)$$

where $\mathbf{f}$ is a nonlinear tensor function of $\mathbf{E}$. Although this would seem to be quite a general constitutive form, some of the assumptions of the Newtonian fluid model are preserved; namely, the stress is still assumed to depend only on $\mathbf{E}$ and to depend only on the instantaneous value of $\mathbf{E}$ at the same point in space. However, the functional relationship between $\mathbf{T}$ and $\mathbf{E}$ is not necessarily linear. Such a fluid has been called the *Stokesian fluid*. Evidently, the Newtonian fluid is a special case of the Stokesian fluid. As in the case of the general linear model, (2-87), which ultimately led to the Newtonian constitutive equation, the more general Stokesian fluid form represents a guess or hypothesis that may or may not encompass any real fluids beyond those already included in the Newtonian model.

A more specific form of (2–100) applies if the Stokesian fluid is assumed to be isotropic. Let us outline briefly the application of this condition to the general Stokesian form (2–100). We begin by noting that **E** is a symmetric second-order tensor; thus there exists a coordinate system in which **E** is diagonalized. We call the coordinate axes in this system the principle axes of strain-rate. Let us suppose that a coordinate system has been chosen that coincides with the principle axes of **E**. In this system, we denote the components of **E** as e_{ij} and write the general Stokesian constitutive form as

$$\boxed{\tau_{ij} = f_{ij}(e_{pg}).} \tag{2–101}$$

Now, let us consider a rotation of the coordinates in which one coordinate axis is held fixed and the other two are rotated about it through an angle of 180°. The corresponding rotation matrix with the *1* axis fixed is

$$\mathbf{L} = \begin{pmatrix} 1 & 0 & 0 \\ 0 & -1 & 0 \\ 0 & 0 & -1 \end{pmatrix}. \tag{2–102}$$

If the problem were carried out with the *2* or *3* axis fixed, the rotation matrix would take the forms

$$\mathbf{L} = \begin{pmatrix} -1 & & \\ & 1 & \\ & & -1 \end{pmatrix} \quad \text{or} \quad \mathbf{L} = \begin{pmatrix} -1 & & \\ & -1 & \\ & & 1 \end{pmatrix}, \tag{2–103}$$

respectively. In all these cases, the rate of strain tensor is unchanged by the rotation; that is, after rotation

$$\bar{e}_{ii} = e_{ii}$$
$$\bar{e}_{ij} = 0 \quad i \neq j.$$

The normal components τ_{11}, τ_{22}, and τ_{33} of the stress are similarly unchanged, and thus satisfy the isotropy requirement (2–89). The shear-stress components, on the other hand, change sign. For example, under the rotation (2–102) we find

$$\bar{f}_{13}(e_{pq}) = f_{13}(e_{pq}) l_{11} l_{33} = -f_{13}(e_{pq}).$$

Since the strain-rate tensor is unchanged, it is evident that the rotational invariance condition can be satisfied only if $f_{13} = 0$. Similarly, by considering the various rotations (2–102) and (2–103) it can be shown that $f_{31} = f_{13} = f_{12} = f_{21} = f_{23} = f_{32} = 0$. The important conclusion is that the stress tensor (and/or the function f_{ij}) must be diagonalized in the same coordinate system as **E** if the general Stokesian fluid form is to be invariant to a rotation of coordinate axes, as is required if the fluid is isotropic. In other words, the stress tensor and the rate of strain tensor must have the same principle axes if the condition of isotropy is to be satisfied.

This latter condition restricts considerably the form allowed for the tensor function $f_{ij}(e_{pq})$. In particular, a well-known result from tensor analysis is that the most general function of a symmetric second-order tensor $\mathbf{A}$ that has the property of having the same principle axes as $\mathbf{A}$ is

$$\alpha \mathbf{I} + \beta \mathbf{A} + \delta (\mathbf{A} \cdot \mathbf{A}),$$

where α, β, and δ can be functions of the three invariants of $\mathbf{A}$, namely,

$$\Psi = tr\mathbf{A},$$

$$\Phi = \frac{1}{2}(tr^2\mathbf{A} - tr\mathbf{A}^2)$$

and

$$\Delta = \det \mathbf{A}.$$

Hence, the most general constitutive form allowed for an *isotropic*, Stokesian fluid is

$$\boxed{\mathbf{T} = (-p + \alpha)\mathbf{I} + \beta \mathbf{E} + \delta(\mathbf{E} \cdot \mathbf{E}),} \tag{2-104}$$

where, α, β, and δ can depend on the thermodynamic state, as well as the invariants Ψ, Φ, and Δ (note that $\Psi = tr\mathbf{E} = 0$ for an incompressible fluid).

All of the development of the previous paragraph is aimed at determining the most general form, consistent with the original constitutive assumption, that satisfies the condition of isotropy. Nothing has been said vis-à-vis the physical reality of the initial assumption $\tau = \mathbf{f}(\mathbf{E})$, nor the subsequent assumption of isotropy. Apart from the fact that it is more general, there is no reason to expect the isotropic, Stokesian fluid to provide a better model for non-Newtonian fluids than the Newtonian model itself. In fact, no real material has been found where $\delta \equiv 0$ does not provide at least as good a fit to experimental data as $\delta \neq 0$. It should be noted, however, that the form

$$\boxed{\mathbf{T} = (-p + \alpha)\mathbf{I} + \beta \mathbf{E}} \tag{2-105}$$

is somewhat more general than the Newtonian fluid model because $\beta = \beta(\Delta, \Phi, \Psi)$ instead of being a constant. This generalization includes the so-called power-law fluid model and other models of the *generalized Newtonian* type. Although the function β can be chosen to provide a good empirical fit for the apparent viscosity in steady shear flows, the resulting model is then limited in its usefulness to essentially that single class of fluid motions.

Since the Stokesian fluid is no better, in general, than the Newtonian (or generalized Newtonian) model, it is evident that the original assumption of linearity in $\mathbf{E}$ was not the problem in and of itself. We have already indicated in one piece of experimental evidence that there can be dramatic departures from Newtonian behavior—namely, the apparent viscosity (that is, the ratio of shear stress to shear rate) in a simple shear flow may depend strongly on the shear rate, generally decreasing with increasing shear rate. This type of departure from Newtonian behavior is typical of macromolecular fluids (including technologically important examples such as molten plastics, polymer flooding liquids that are used in tertiary oil-recovery operations, and many biological fluids), suspensions, emulsions, dispersions, soap solutions, and many more. Since β

in the Stokesian model can depend upon the invariants of $\mathbf{E}$, it is not totally obvious that the effect of shear-dependent viscosity could not be captured by that model in either of forms (2–104) or (2–105). However, other deviations from Newtonian behavior definitely cannot be captured via the Stokesian class of constitutive models. Both to illustrate this fact and because it is possible that we may obtain some clue toward a more effective constitutive assumption, we briefly summarize other experimental observations for the class of materials listed above, which are collectively known as *viscoelastic*.

In the class of a steady, simple shear flow $u_1 = \gamma x_2$, we have already discussed the existence of a shear-rate dependent apparent viscosity. However, another even more characteristic observation for viscoelastic liquids in steady, simple shear flow is the fact that the normal stress components T_{11}, T_{22}, and T_{33} are not equal. It is not necessary here to give a detailed explanation of how these normal stress components might be measured except to note that it is always the differences $T_{11} - T_{22}$ and $T_{22} - T_{33}$ (or $T_{11} - T_{33}$) that can be determined directly. Sometimes the existence of nonzero normal stress differences can also be inferred without direct measurement from the existence of certain flow phenomena.[8] Examples of these flow phenomena, illustrated in Figure 2–10, include the famous Weissenberg effect, in which fluid is observed to rise up the shaft of a partially immersed rotating rod; the reversal of direction (compared to that observed in a Newtonian fluid) of the secondary-flow pattern when a circular disk rotates at the surface of liquid in a beaker; and the existence of an upward bulge at the center of the free surface of a liquid flowing down an open channel (or trough). We say that nonzero normal stresses are "more characteristic" of viscoelastic behavior than a shear-rate dependent apparent viscosity, because materials have been found in which the viscosity is constant over an appreciable range of shear rates, but the normal stress differences are strongly nonzero, while the reverse is rarely true. Nonzero normal stress differences cannot be accommodated within the Newtonian fluid model, because in that case $T_{11} = T_{22} = T_{33} = -p$ for steady, simple shear flow. Neither can realistic values of the normal stress differences be predicted via the most general version of the Stokesian fluid model. For $u_1 = \gamma x_2$, the Stokesian model gives

$$T_{22} - T_{33} = \left(\frac{\delta}{4}\right)\gamma^2$$

but

$$T_{11} - T_{22} = 0.$$

Experimentally, however, $T_{11} - T_{22}$ is actually found to be much larger than $T_{22} - T_{33}$, while $T_{22} - T_{33}$ is generally believed to be negative. We may also note that viscoelastic fluids almost always become anistropic in flow. This is a consequence of the fact that the microscale structure of such fluids is modified in the presence of a continuum velocity field. For example, in simple, steady shear flow, a polymer solution or a suspension, which is isotropic in the rest state, becomes optically anisotropic in the presence of flow, and this indicates the existence of a flow-induced anisotropic structure, which relaxes back to an isotropic, equilibrium state when the flow is removed.

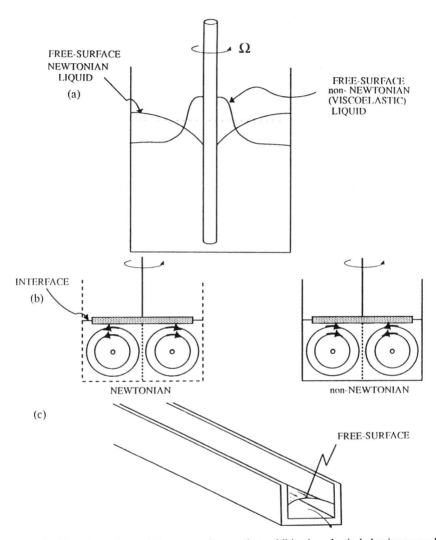

Figure 2–10   Illustrations of three experiments that exhibit viscoelastic behavior normally associated with the presence of normal stress differences in simple shear flow. (a) The rod-climbing (or Weissenberg) experiment, (b) the secondary-flow experiment, and (c) the tilted-channel experiment. In the Weissenberg experiment, a thin rod is rotated about its central axis in a beaker of fluid. For a Newtonian fluid, the fluid near the beaker walls rises slightly due to the outward thrust on the fluid associated with centrifugal effects. For a viscoelastic liquid, the fluid is observed to *climb* up the rod to a height that increases with Ω. In the secondary flow experiment, a thin disk is rotated in contact with the upper surface of the liquid. For a Newtonian fluid, centrifugal forces tend to drive fluid *outward* nearest the disk, and this creates a secondary circulation in the direction shown. For a viscoelastic fluid, the circulation is in the *opposite* sense. Finally, in the tilted-channel experiment, the fluid flows down the channel under the action of gravity. When the fluid is viscoelastic, the free surface is observed to bulge upward at the center of the channel.

Again, the reader may ask whether the differences between predictions of the Stokesian fluid model and the behavior of real viscoelastic fluids might not be due to the assumption of isotropy. The answer is that the assumption of isotropy appears *not* to be correct, but additional experimental observations described later indicate that this is not the only problem with the Stokesian fluid model.

Indeed, another set of experimental observations for viscoelastic fluids that cannot be modeled via any version of the Stokesian fluid is their behavior in unsteady shear flows, namely,

$$u_1 = \gamma(t)x_2.$$

One example of an unsteady shear flow is the so-called oscillatory shear flow in which $\gamma(t) = \gamma_0 \sin \omega t$.[9] For both Newtonian and Stokesian fluids, the stress response to changes in the strain rate is instantaneous. Thus, since $E_{12} = \gamma_0 \sin \omega t$, the stress must oscillate with the same phase if the fluid response is either Newtonian or Stokesian. What is observed experimentally for viscoelastic fluids, however, does not follow this pattern. Rather, as illustrated in Figure 2–11,

$$\left(\frac{T_{12}(t)}{\gamma_0} \right)_{\text{measured}} = A \sin \omega t + B \cos \omega t.$$

That is, there is both an in-phase and an out-of-phase component to the stress, suggesting the existence of a finite, rather than instantaneous response of the fluid to changing strain rates. Another manifestation of a finite response time for viscoelastic fluids occurs in the so-called stress-relaxation experiment in which a steady shear flow is suddenly turned off, that is,

$$u_1 = \gamma x_2 \quad \text{for} \quad t < t_0,$$
$$u_1 = 0 \quad \text{for} \quad t \ge t_0.$$

For a fluid with instantaneous response, as assumed in the Newtonian and Stokesian fluid models, the shear stress τ_{12} should follow the same pattern; that is, the stress

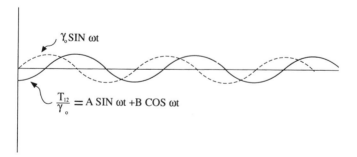

Figure 2–11 A schematic representation of the shear stress for a viscoelastic fluid that is subjected to a sinusoidal shear rate, shown *after* any start-up transients have decayed. The coefficients A and B are arbitrarily chosen for this illustration but would be determined by the rheological properties of a real fluid (for a Newtonian fluid, $A = 2\mu$ and $B = 0$, but for a viscoelastic fluid $B \ne 0$).

should drop instantaneously from its steady-state value for the shear rate γ to zero. Instead, what is observed for the class of materials that we have termed *viscoelastic* is that the shear stress drops to zero in a finite time interval. Finally, we may consider startup of simple shear flow

$$u_1 = 0 \qquad \text{for} \qquad t < t_0,$$

$$u_1 = \gamma x_2 \qquad \text{for} \qquad t \geq t_0.$$

In this case, instead of the stress jumping instantaneously from zero to its equilibrium value, it rises in a finite time scale, sometimes overshooting the final steady value and sometimes rising monotonically.

One final observation worth mentioning is the very strong sensitivity of the flow resistance of viscoelastic fluids to the type of motion that they are undergoing. For example, in a uniaxial extensional flow,

$$u_1 = E x_1, \qquad u_2 = -\frac{1}{2} E x_2, \qquad u_3 = -\frac{1}{2} E x_3,$$

the ratio of the normal stress difference $[2T_{11} - (T_{22} + T_{33})]$ to the strain rate E is observed in some viscoelastic fluids to exceed the shear viscosity (that is, the ratio of shear stress to shear rate) by many orders of magnitude and to be an increasing function of E. For a Newtonian fluid, on the other hand, the ratio

$$\frac{2T_{11} - (T_{22} + T_{33})}{E} = 3\mu$$

is independent of E. [10]

The most evident departure of real, viscoelastic fluid behavior from the basic assumptions of the Newtonian and Stokesian fluid models is that the stress response in transient flows is not instantaneous but is characterized by a finite time scale. A number of different constitutive assumptions consistent with the existence of finite relaxation times have been proposed over the last thirty to forty years, and efforts have also been made to derive constitutive equations from kinetic theories based upon "molecular" or *microstructural* models of polymeric fluids and suspensions. The resulting constitutive equations have met variable degrees of success, though by now the majority are capable of describing, at least qualitatively, the rheological or material behaviors that are described on the previous several pages. In spite of the apparent complexity of these phenomena, however, it must be noted that the flows considered are extremely simple, and the reliability of the many proposed constitutive models in other more complicated motions remains an unresolved research problem. For this reason, there does not yet exist an established branch of theoretical fluid mechanics for viscoelastic fluids, and we focus our attention in the remainder of this book on materials that can be approximated as Newtonian. We also do not attempt to list or catalog all of the apparently viable constitutive models since it remains to be seen which, if any, of the current proposals will survive the test of comparison between predicted and experimentally measured velocity, pressure, and stress fields in more complicated flows.

One useful general conclusion that can be drawn from the preceding discussion is that real fluids are characterized by a finite (relaxation) time scale that is representative of the rate at which the material can respond to achieve a new flow state when subjected to transient forces. We may call this time scale λ. If this time scale is very short compared with the rate of change of the strain rate experienced by an element of the fluid, the material will always tend to respond with a stress that is "in equilibrium" with the local, instantaneous strain rate. We may note that the time-dependent strain rate seen by a material fluid element can be a consequence of time-dependent variations in the flow, or of spatial gradients in the strain rate that are experienced by the fluid element as it moves through space. It is the material time derivative DE/Dt that is significant. From this point of view, fluids that can be approximated as Newtonian are ones where the material time scale λ is very much shorter than the time scale characteristic of the rate of change of $\mathbf{E}$. Let us denote this latter time scale as τ. In steady flows, an appropriate definition for τ is the ratio of the length scale L characteristic of gradients of $\mathbf{E}$ and the velocity U characteristic of the rate at which fluid elements move through space, that is,

$$\tau_{\text{steady flows}} = L/U. \tag{2-106}$$

A dimensionless parameter, known as the Deborah number, can be defined as the ratio

$$\boxed{\text{De} \equiv \frac{\lambda}{\tau}.} \tag{2-107}$$

Evidently, a real fluid can be approximated as Newtonian whenever

$$\text{De} \ll 1. \tag{2-108}$$

Obviously, the smaller the time scale λ of the fluid, the more likely it is that condition (2-108) will be satisfied for any particular flow. Or, to put it another way, fluids with a small material time scale λ would have to be subjected to flows with exceedingly rapid changes in $\mathbf{E}$ (small τ) before they would be expected to exhibit any deviation from the instantaneous response assumption that is inherent in the Newtonian fluid model. This is the reason why such common materials as water can almost always be approximated as Newtonian. On the other hand, high molecular weight polymers or materials such as suspensions with larger fluid elements tend to exhibit very much longer intrinsic time scales, λ, and for these materials the existence of so-called viscoelastic behavior will occur in all but the most slowly varying flows; that is, for these materials the relationship

$$\text{De} \geq 0(1)$$

is almost always true. The most important conclusions from all of this discussion is that it is not the material alone that determines whether a fluid may be considered as Newtonian but the combination of the material and the flow to which it is subjected. One fluid may be adequately modeled as Newtonian for some flow where $\text{De} \ll 1$, but not for a different flow where this condition is not satisfied. Indeed, the transition from Newtonian to viscoelastic fluid behavior can be realized, in principle, by simply increasing the flow rate (that is, the characteristic velocity U) so that the time scale characteristic of changes in **E** is reduced to a point where De $\sim 0(1)$. We shall assume throughout the remainder of this book that the fluids and flows considered are such that the Newtonian fluid constitutive model is a good approximation of material behavior. However, we must always keep in mind the fact that no real fluid will always satisfy these conditions independent of the flow.

J. The Equations of Motion for a Newtonian Fluid: The Navier-Stokes Equations

Let us now return to the case of a Newtonian fluid, which will be the subject of the remainder of this book. With the constitutive equation (2–93) [or (2–94) if the fluid is incompressible], the continuity equation (2–5) [or (2–19) if the fluid is incompressible], and the Cauchy equations of motion (2–30), we have achieved a balance between the number of independent variables and the number of equations for an isothermal fluid. If the fluid is not isothermal, we can add the thermal energy equation (2–50) and the thermal constitutive equation (2–70) and the system is still fully specified insofar as the balance between independent variables and governing equations is concerned.

In this section, we will combine the Cauchy equations and the Newtonian constitutive equation to obtain the famous *Navier-Stokes equations* of motion. First, however, we briefly reconsider the notion of pressure in a general, Newtonian fluid.

The physical significance of pressure, as it first appeared in the constitutive equation for stress in a stationary fluid, (2–58), is clear. This is the familiar pressure of thermodynamics. When a fluid is undergoing a motion, however, the simple notion of a normally directed surface force acting equally in all directions is lost. Indeed, it is evident upon examining the Newtonian constitutive equation (2–93) that the normal component of the surface force acting on a fluid element at a point will generally have different values depending upon the orientation of the surface. Nevertheless, it is often useful to have available a scalar quantity for a moving fluid that is analogous to static pressure in the sense that it is a measure of the local intensity of "squeezing" of a fluid element at the point of interest. Thus, it has become common practice to introduce a mechanical definition of pressure in a moving fluid as

$$\bar{p} \equiv -\frac{1}{3} tr\mathbf{T}. \qquad (2\text{--}109)$$

This quantity has the following desirable properties. First, it is invariant under rotation of the coordinate axes (unlike the individual components of **T**). Second, for a static fluid $-1/3 \cdot tr\mathbf{T} = p$, the thermodynamic pressure. And third, $\bar{p}$ has a physical significance analogous to pressure in a static fluid in the sense that it is precisely equal to the average value of the normal component of the stress on a surface element at position **x** over all possible orientations of the surface (alternatively, we may say that $1/3 tr\mathbf{T}$ is the average magnitude of the normal stress on the surface of an arbitrarily small sphere centered at point **x**). The definition (2–109) is a purely mechanical definition of pressure for a moving fluid, and nothing is implied directly of the connection between $\bar{p}$ and the ordinary static or thermodynamic pressure p. Although the connection between $\bar{p}$ and p can always be stated once the constitutive equation for **T** is given, one would not necessarily expect the relationship to be simple for all fluids since thermodynamics refers to equilibrium conditions, whereas the elements of a fluid in motion are clearly not in thermodynamic equilibrium. Applying the definition (2–109) to the general Newtonian constitutive model (2–93), we find

$$
\bar{p} = p - (\lambda + \frac{2}{3}\mu)\nabla\cdot\mathbf{u},
$$

$$
\bar{\tau} = \mathbf{T} + \left(\frac{1}{3}tr\mathbf{T}\right)\mathbf{I} = 2\mu\mathbf{E} - \frac{2}{3}\mu(\nabla\cdot\mathbf{u})\mathbf{I}. \tag{2–110}
$$

Only if the fluid can be modeled as incompressible does the connection between $\bar{p}$ and p simplify greatly for a Newtonian fluid. In that case,

$$
\bar{p} \equiv p; \qquad \bar{\tau} = \tau = 2\mu\mathbf{E}. \tag{2–111}
$$

So far, we have simply stated the Cauchy equations of motion and the Newtonian constitutive equations as a set of nine independent equations involving **u**, **T**, and p. It is evident in this case, however, that the constitutive equation (2–93) for the stress [or equivalently (2–110)], can be substituted directly into the Cauchy equations to provide a set of equations that involve only **u** and $\bar{p}$ (or p). These combined equations take the form

$$
\rho\left[\frac{\partial\mathbf{u}}{\partial t} + \mathbf{u}\cdot\nabla\mathbf{u}\right] = \rho\mathbf{g} - \nabla\bar{p} + \nabla\cdot(2\mu\mathbf{E}) - \frac{2}{3}\nabla\cdot(\mu\,\text{div}\,\mathbf{u}\mathbf{I}), \tag{2–112}
$$

where

$$
E \equiv \frac{1}{2}(\nabla\mathbf{u} + \nabla\mathbf{u}^T).
$$

If the fluid can be approximated as incompressible and if the fluid is isothermal so that the viscosity μ can be approximated as a constant, independent of spatial position (note that any dependence of μ on pressure is generally weak enough to be neglected altogether), these equations take the form

$$
\rho\left[\frac{\partial\mathbf{u}}{\partial t} + \mathbf{u}\cdot\nabla\mathbf{u}\right] = \rho\mathbf{g} - \nabla p + \mu\nabla^2\mathbf{u}. \tag{2–113}
$$

This is the famous *Navier-Stokes* equation of motion for an incompressible, isothermal Newtonian fluid.

Several comments are in order with regard to this latter equation. First, we shall always assume that ρ and μ are known—presumably by independent means—and attempt to solve (2–113) and the continuity equation (2–19) for **u** and p. Second the ratio μ/ρ, which is called *the kinematic viscosity* and denoted as ν, plays a fundamental role in determining the fluid's motion. In particular, the contribution to acceleration of a fluid element $(D\mathbf{u}/Dt)$ due to viscous stresses is determined by ν rather than μ.

Third, it is frequently convenient to introduce the concept of dynamic pressure into (2–113). This is a consequence of the utility of having the pressure gradient that appears explicitly in the Navier-Stokes equations act as a driving force for motion. In the form (2–113), however, a significant contribution to ∇p is the static pressure variation $\nabla p = \rho \mathbf{g}$, which has nothing to do with the fluid's motion. In other words, nonzero pressure gradients in (2–113) do not necessarily imply fluid motion. Because of this, it is convenient to introduce the so-called *dynamic pressure P*, such that $\nabla P = 0$ in a static fluid. This implies

$$\boxed{-\nabla P \equiv \rho \mathbf{g} - \nabla p.} \tag{2–114}$$

Notice that this equation requires that the sum $\nabla p - \rho \mathbf{g}$ be expressible in terms of the gradient of a scalar P. This is possible only if ρ is constant—that is, the fluid must be both incompressible and isothermal. In this case,

$$\boxed{\rho\left[\frac{\partial \mathbf{u}}{\partial t} + \mathbf{u}\cdot\nabla\mathbf{u}\right] = -\nabla P + \mu\nabla^2\mathbf{u}.} \tag{2–115}$$

Equation (2–115) would seem to imply that the gravity force $\rho\mathbf{g}$ has no direct effect on the velocity distribution in a moving fluid, provided the fluid is incompressible and isothermal so that the density is constant. This is generally true. An exception occurs when one of the boundaries of the fluid is a gas-liquid or liquid-liquid interface. In this case, the actual pressure p appears in the boundary conditions (as we shall see), and the transformation of the body force out of the equations of motion via (2–114) simply transfers it into the boundary conditions. We shall frequently use the equations of motion in the form (2–115), but we should always keep in mind that it is the dynamic pressure that then appears.

Finally, we stated previously that fluids which satisfy the Newtonian constitutive equation for the stress are often also well approximated by the Fourier constitutive equation (2–70) for the heat flux vector. Combining (2–70) with the thermal energy (2–50), we obtain

$$\boxed{\rho C_p \frac{D\theta}{Dt} = -\frac{\theta}{\rho}\left(\frac{\partial\rho}{\partial\theta}\right)_p \frac{Dp}{Dt} + p(\nabla\cdot\mathbf{u}) + (\mathbf{T}{:}\mathbf{E}) + \nabla\cdot(k\nabla\theta).} \tag{2–116}$$

For an incompressible Newtonian fluid this takes the simpler form

$$\boxed{\rho C_p \frac{D\theta}{Dt} = -\frac{\theta}{\rho}\left(\frac{\partial\rho}{\partial\theta}\right)_p \frac{Dp}{Dt} + 2\mu(\mathbf{E}{:}\mathbf{E}) + \nabla\cdot(k\nabla\theta),} \tag{2–117}$$

which we shall use frequently in subsequent developments.

K. Boundary Conditions

We shall be concerned in this book with the motion and transfer of heat in an incompressible, Newtonian fluid. For this case, the equations of motion, continuity, and thermal energy,

$$\rho \left[\frac{\partial \mathbf{u}}{\partial t} + \mathbf{u} \cdot \nabla \mathbf{u} \right] = \rho \mathbf{g} - \nabla p + \nabla \cdot (\mu \nabla \mathbf{u}), \qquad (2\text{--}118)$$

$$\nabla \cdot \mathbf{u} = 0, \qquad (2\text{--}119)$$

$$\rho C_p \left[\frac{\partial \theta}{\partial t} + \mathbf{u} \cdot \nabla \theta \right] = - \frac{\theta}{\rho} \left(\frac{\partial \rho}{\partial \theta} \right)_p \frac{Dp}{Dt} + 2\mu \, (\mathbf{E} : \mathbf{E}) + \nabla \cdot (k \nabla \theta), \qquad (2\text{--}120)$$

provide a complete set from which to determine the velocity $\mathbf{u}$, pressure p, and temperature θ. All that remains to have a completely specified problem is boundary conditions.

Boundary conditions occupy a position in continuum mechanics that is similar to the constitutive equations. At the molecular level, the fluid properties, such as density, vary rapidly but continuously near a solid wall or a fluid-fluid interface, and no special treatment is required. From the continuum point of view, however, these smooth but rapid variations in fluid properties cannot be resolved, and the boundaries appear as a surface of discontinuity upon which boundary values for the continuum variables must be specified. In this macroscopic description, the material properties that appear in (2–118), (2–119), and (2–120) are continuous right up to the plane of the boundary, where they undergo a discontinuity. The problem is that it is not clear, in the absence of an ability to describe the system at the molecular level, just what boundary conditions should be used. All that can be said for sure is that these macroscopic conditions must be equivalent to the underlying molecular physics in the sense that they yield the same macroscopic velocity and pressure fields in the interior of the domain as exist in the real fluid. Evidently, it is necessary to "guess" the appropriate form for the continuum boundary conditions and then judge the correctness of this guess by comparing the resulting velocity, pressure, and temperature fields with values measured experimentally. It is in this sense that the boundary conditions of continuum mechanics are seen to occupy a similar niche to the constitutive equations.

We focus our attention in this section only on boundary conditions for the velocity. Boundary conditions for the temperature field suitable for the solution of (2–120) will be discussed in Chapter 8, when we first consider fluid motion and heat transfer under nonisothermal conditions. Three types of velocity boundary conditions are encountered in theoretical analyses of isothermal motions of a Newtonian fluid. First, if the domain is unbounded, the form of the velocity field far from any boundaries is normally prescribed. A typical example is streaming flow past a sphere, where the velocity is prescribed as unidirectional and constant in magnitude sufficiently far from the sphere. Of course, the domain of a real fluid is never truly unbounded, but we often use this far-field condition as an approximation when the dimensions of the external boundaries are very much larger than the other length scales in the flow system. For example, we adopt this approximation in the case of streaming flow past a sphere when the diameter of the sphere is very much smaller than the dimensions of the external boundaries and the sphere is not close to any of these external boundaries. The second type of condition applies at either a solid wall or interface and is known as the *kinematic* condition. It states simply that the normal component of the velocity is continuous across any material boundary separating the fluid and another medium, that is,

$$\boxed{\mathbf{u} \cdot \mathbf{n} = \hat{\mathbf{u}} \cdot \mathbf{n}.}$$
(2-121)

Here, $\hat{\mathbf{u}}$ is the velocity of the "other medium" at the boundary. If the second medium is a solid wall, $\hat{\mathbf{u}} \cdot \mathbf{n} = 0$, in a frame of reference fixed to the wall, provided that the wall is impermeable. If the second medium is another fluid, then the kinematic condition (2-121) provides a relation between the normal components of velocity in the two fluids. Although the kinematic condition is strictly a guess, it must be true unless the boundary constitutes a source or sink for mass. The third type of boundary condition is the so-called *dynamic* condition at a solid wall or interface, which states that the tangential component of velocity is also continuous across any material boundary separating a viscous fluid and another medium. This condition is stated essentially as a consequence of empirical observation. Although it appears to be a good approximation under almost all circumstances for a Newtonian fluid, there are reasons to believe that it may break down locally in regions of ultrahigh stress—such as the vicinity of a moving three-phase contact line where the interface between two immiscible fluids contacts a solid boundary. The dynamic condition in mathematical form is

$$\boxed{\mathbf{u} - \mathbf{u} \cdot \mathbf{n}\mathbf{n} = \hat{\mathbf{u}} - \hat{\mathbf{u}} \cdot \mathbf{n}\mathbf{n}.}$$
(2-122)

Although the details of molecular phenomena near a solid wall or interface are unknown, it is evident, upon reflection, that the dynamic boundary condition on the velocity is a consequence of momentum transfer between the fluid and the second material. Thus, in the case of a Newtonian fluid it is due to the presence of a nonzero viscosity. If we consider an inviscid fluid (a fluid for which $\mu = 0$), there is no reason to expect the dynamic condition to be satisfied. In applying the dynamic boundary condition, there are two cases to consider. If the second medium is a solid, which has a known velocity, or is acted upon by a known force or torque so that its velocity can be calculated, then

$$\boxed{\hat{\mathbf{u}} = \mathbf{U}_{\text{solid}} \quad \text{(known)}.}$$
(2-123)

In this case, the dynamic condition is known as the *no-slip condition,* and the kinematic and dynamic conditions together are sufficient to completely specify the solution of (2-118) and (2-119). If, on the other hand, the second medium is another fluid, the situation is more complicated. In this case, $\hat{\mathbf{u}}$ and $\mathbf{u}$ are both unknown and the equations of motion and continuity must be solved in both fluids. The boundary conditions (2-121) and (2-122) then provide two relationships between the unknown velocity fields, but evidently supply only half enough boundary conditions to completely determine the two sets of solutions for $(\mathbf{u},p)$ and $(\hat{\mathbf{u}},\hat{p})$. Thus, when the boundary of a fluid domain is an interface, we need more boundary conditions, but we shall discuss this in Chapter 5.

Homework Problems

1. Prove (2–35). To do this, it will be convenient to show

$$\frac{D}{Dt}\int_{V_m(t)}\rho F\,dV = \int_{V_m(t)}\rho\frac{DF}{Dt}\,dV.$$

In addition, the identity

$$\frac{D\mathbf{x}}{Dt}=\mathbf{u}$$

is necessary, and this should also be demonstrated via the definition (2–8).

2. Show

a. $$\nabla\cdot(\rho\mathbf{u}\mathbf{u})\equiv\rho\mathbf{u}\cdot\nabla\mathbf{u}+\mathbf{u}(\nabla\cdot(\rho\mathbf{u}))$$

b. $$-\nabla\cdot(\mathbf{T}\wedge\mathbf{x})=\mathbf{x}\wedge(\nabla\cdot\mathbf{T})-\boldsymbol{\epsilon}\!:\!\mathbf{T}$$

c. $$\mathbf{u}\cdot\nabla\mathbf{u}=(\nabla\wedge\mathbf{u})\wedge\mathbf{u}+\nabla\left(\frac{\mathbf{u}\cdot\mathbf{u}}{2}\right)$$

d. $$\nabla\wedge(\nabla\wedge\mathbf{u})=\nabla(\nabla\cdot\mathbf{u})-\nabla^2\mathbf{u}$$

3. There has been a significant amount of theory developed for so-called *polar fluids*. These are fluids in which it is assumed that the surface couple, **r** and the body couple, **c**, are both nonzero. Show that the stress tensor will no longer, in general, be symmetric for such fluids, but will satisfy the relationship.

$$\boldsymbol{\epsilon}\!:\!\mathbf{T}-\rho\mathbf{c}+\nabla\cdot\mathbf{R}=0,$$

where

$$\mathbf{r}=\mathbf{n}\cdot\mathbf{R}.$$

To demonstrate the existence of the surface couple tensor **R**, you will need to first demonstrate the principle of surface couple equilibrium,

$$\lim_{l\to 0}\frac{1}{l^2}\int_{S_m(t)}\mathbf{r}\,dS\to 0,$$

in analogy with (2–23).

4. Magnetohydrodynamics is the study of the flow of an electrically conducting fluid in the presence of an electromagnetic field. The fundamental equations of magnetohydrodynamics are a combination of electromagnetic field equations and fluid flow equations, modified to include the effect of the interaction between the fluid motion and the electromagnetic field. The usual assumption in magnetohydrodynamics is that the induced current in the fluid will interact with the electromagnetic field in accordance with the classical laws governing electromagnetic interactions.

For a charged fluid particle moving in an electromagnetic field, the electromagnetic force per unit volume is

$$\mathbf{f} = \mathbf{J} \times \mathbf{B},$$

where $\mathbf{J}$ is the current density and $\mathbf{B}$ is the magnetic flux density.

Derive the appropriate form of the Navier-Stokes equation ror an electrically conducting Newtonian fluid with constant density and viscosity. This equation is called the magnetohydrodynamic momentum equation. Your derivation should begin from first principles. However, you may assume—but should state in appropriate places—the necessary properties of pressure, viscous stress, and convected time derivatives.

5. a. Assuming that we have a spherical coordinate system defined with respect to a Cartesian system according to $x = r \sin\theta \cos\phi$, $y = r \sin\theta \sin\phi$, and $z = r \cos\theta$,

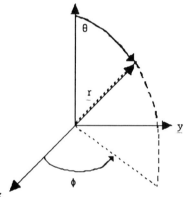

obtain an expression for the force on a sphere that is situated at the origin, in terms of the components of the stress tensor $\mathbf{T}$,

$$\mathbf{T} = T_{rr}\,\mathbf{e}_r\mathbf{e}_r + T_{r\theta}\,\mathbf{e}_r\mathbf{e}_\theta + T_{r\phi}\,\mathbf{e}_r\mathbf{e}_\phi + T_{\theta r}\,\mathbf{e}_\theta\mathbf{e}_r + T_{\theta\theta}\,\mathbf{e}_\theta\mathbf{e}_\theta$$
$$+ T_{\theta\phi}\,\mathbf{e}_\theta\mathbf{e}_\phi + T_{\phi r}\,\mathbf{e}_\phi\mathbf{e}_r + T_{\phi\theta}\,\mathbf{e}_\phi\mathbf{e}_\theta + T_{\phi\phi}\,\mathbf{e}_\phi\mathbf{e}_\phi.$$

   b. Suppose that the fluid is isothermal and stationary. Further, assume that the fluid and the sphere have the same density ρ. Show that the *net* force on the sphere is zero (that is, prove Archimedes's principle for this case).

6. a. If air can be treated as an ideal gas, and the temperature of the atmosphere varies linearly with height above the earth,

$$T(z) = T_0 - \alpha z \quad (T_0,\ \alpha \equiv \text{constants}),$$

derive an expression for the pressure as a function of z assuming that we can neglect fluid motion in the atmosphere, as well as the earth's rotation.

   b. A weather balloon, mass m_b, radius R_0, is filled with a mass m_g of some gas. To what altitude will the balloon rise? Assume that the radius of the balloon remains constant. Assume further that $\alpha R_0 \ll 1$ so that we can neglect density variations over the surface of the balloon.

   c. Suppose we have a fluid with the density variation given as

$$\rho(z) = \rho_0(1 + az^2 + bz^4),$$

where a, $b > 0$ and z increases in the direction of g. If a sphere, radius R, density ρ_s, is placed in this fluid, what will be the equilibrium position of the center of the sphere? If the sphere is replaced by a cube of side L, what will the equilibrium position be for the center of the cube?

7. a. It is proposed that the constitutive equation for stress in a moving fluid should take the form

$$\mathbf{T} = \boldsymbol{\beta} \wedge \nabla \mathbf{u},$$

where $\boldsymbol{\beta}$ is a constant "material" vector. Is this form allowable? If not, why not? Prove any assertions that you make.

b. There are many combinations of vectors, scalars, and tensors that are linear in $\mathbf{E}$ besides the form $\mathbf{A}:\mathbf{E}$ that was proposed as the most general form for $\mathbf{T}$ that is linear in $\mathbf{E}$. Consider the following possibilities, and determine whether they are viable alternative choices (note, in some cases, you may need only a few words of explanation).

i. $\mathbf{B} \cdot \mathbf{E}$	iv. $\alpha \nabla \cdot \mathbf{E}$
ii. $\mathbf{C} \wedge \mathbf{E}$	v. $\boldsymbol{\beta} \cdot \nabla \mathbf{E}$
iii. $\alpha \wedge \mathbf{E}$	

Assume that $\mathbf{B}$ is a third-order tensor, $\mathbf{C}$ is a second-order tensor, and α, $\boldsymbol{\beta}$ are vectors.

References

1. a. Hanson, J.P., and McDonald, I.R., *Theory of Simple Liquids*. Academic Press: London (1976).

   b. Boon, J.P., and Yip, S., *Molecular Hydrodynamics*. McGraw-Hill: New York (1980).

2. a. Schowalter, W.R., "The Behavior of Complex Fluids at Solid Boundaries," *J. Non-Newtonian Fluid Mechanics* 29:25–36 (1988).

   b. Goldstein, S. (ed.), *Modern Developments in Fluid Dynamics*, vol. 2, pp. 676–680. Oxford Univ. Press: London (1938).

3. Jeffreys, H., and Jeffreys, B.S., *Methods of Mathematical Physics*, chap. 5. Cambridge Univ. Press: Cambridge, UK (1966).

4. The proof given in this text closely follows that given by S. Whitaker in his book *Introduction to Fluid Mechanics*, chapt. 3. Prentice-Hall: Englewood Cliffs, NJ (1968). Another extremely useful discussion of convected derivatives and the Reynolds transport theorem is given in Aris, R., *Vectors, Tensors and the Basic Equations of Fluid Mechanics*. Prentice-Hall: Englewood Cliffs, NJ (1962).

5. a. Rosensweig, R.E., "Magnetic Fluids," *Ann. Rev. of Fluid Mech.* **19**:437–463 (1987).

   b. Rosensweig, R.E., *Ferrohydrodynamics*. Cambridge Univ. Press: Cambridge, UK (1985).

6. a. Walters, K., *Rheometry*. Chapman and Hall: London (1975).

   b. Whorlow, R.W., *Rheological Techniques*. John Wiley and Sons: New York (1980).

   c. Dealy, J.M., *Rheometers for Molten Plastics: A Practical Guide to Testing and Property Measurement*. Van Nostrand Rheinhold: New York (1982).

   d. Barnes, H.A., Hutton, J.R., and Walters, K., *An Introduction to Rheology*. Elsevier: Amsterdam (1989).

7. a. Bird, R.B., Curtiss, C.F., Armstrong, R.C., and Hassager, O., *Dynamics of Polymeric Liquids* (2 vols.), 2nd ed. John Wiley and Sons: New York (1987).

b. Tanner, R.I., *Engineering Rheology*. Oxford Univ. Press: Oxford, UK (1985).

c. Larson, R.G., *Constitutive Equations for Polymer Melts and Solutions*. Butterworth: Boston (1988).

8. A number of non-Newtonian and viscoelastic flow phenomena are described in chapter 3, "Flow Phenomena in Polymeric Liquids," in the first volume of Bird et al. (Reference 7).

9. The interested reader may refer to chapter 6 of Walters (Reference 6).

10. A summary of some of the recent efforts to measure extensional viscosity can be found in chapter 5 of Barnes et al. (Reference 6).

Unidirectional Flows

We are now finally in a position to begin to consider the solution of heat transfer and fluid mechanics problems using the equations of motion, continuity, and thermal energy, plus boundary conditions that were given in the preceding chapter. We will begin with the motion of an isothermal fluid. The thermal energy equation can be ignored in this case, and the fluid properties, such as viscosity and density, assumed to have constant values independent of spatial position. Nevertheless, the problem that remains is still extremely difficult. The trouble is that the Navier-Stokes equations are highly nonlinear, while all of the classical methods of solving partial differential equations analytically (for example, eigenfunction expansions via separation of variables, or Laplace and Fourier transform techniques, among others) are predicated upon the linearity of the governing equation(s). This is because they rely upon the construction of general solutions as sums of simpler, fundamental solutions of the differential equations. For a linear differential equation, the sum of any two (or more) solutions is also a solution. For nonlinear equations, however, this so-called superposition principle is not valid, and the analytical methods that have been developed for linear equations cannot be used. Of course, approximate solutions of partial differential equations can also be generated numerically via finite-difference or finite-element methods, and to a limited extent these methods can be carried over to nonlinear equations.[1] Even in this case, however, any nonlinearity of the basic equations leads to considerable difficulty, and it has not proven possible to extend solutions much beyond flows that are dominated by viscous effects; once the acceleration terms in the Navier-Stokes equations become too large compared with the viscous terms, all existing numerical methods fail.

Our main goal in this book is to develop and explain analytical methods for obtaining *approximate* solutions to Eqs. (2–118, 2–119, and 2–120). Prior to discussing approximate solution techniques, however, we consider in this chapter an exceptional class of flow problems where *exact* solutions of the Navier-Stokes equations are

71

possible. These are the so-called *unidirectional* flows, typified by simple shear flow and Poiseuille flow through a straight circular tube, in which the velocity is nonzero in only a single, fixed direction. Although these flows are extremely simple in form, they are often of considerable technical importance. Furthermore, they provide a convenient framework both for reviewing classical solution techniques and for illustrating a number of general principles about the motion of viscous fluids. The key to exact solution is the fact that the Navier-Stokes equations, for this particular class of flows, simplify to a very simple form in which the nonlinear terms are identically equal to zero.

A. Simplification of the Navier-Stokes Equations for Unidirectional Flows

We consider the fluid to be Newtonian, incompressible, and isothermal. Thus, in general, the governing equations for any flow are the Navier-Stokes and continuity equations (2–19) and (2–115), with the viscosity and density, μ and ρ, given.

The class of *unidirectional* motions that we consider in this chapter is defined simply as a flow in which the direction of motion is independent of position, that is,

$$\mathbf{u}(\mathbf{x},t) = u(\mathbf{x},t)\mathbf{k}, \tag{3-1}$$

where $u(\mathbf{x},t)$ is a scalar function of position and time and $\mathbf{k}$ is a constant (nonrotating) unit vector. We do not attempt to identify specific problems, at this stage, where the velocity field actually takes the form (3–1). Instead, we assume that such problems exist and show that the Navier-Stokes and continuity equations are greatly simplified whenever the form (3–1) can be applied. In the process, we derive a linear partial differential equation for the scalar function $u(\mathbf{x},t)$.

For convenience, we adopt a Cartesian coordinate system with the direction of motion coincident with the z axis. In this system,

$$\mathbf{u} = (0,0,u). \tag{3-2}$$

We first consider the continuity equation. Since the only nonzero velocity component is in the z direction, this equation becomes simply

$$\frac{\partial u}{\partial z} \equiv 0. \tag{3-3}$$

It follows that

$$u = u(x,y,t), \tag{3-4}$$

that is, the magnitude of u must be independent of position in the flow direction if the velocity field is unidirectional. Moreover, the conditions (3–2) to (3–4), in turn, imply that the nonlinear terms in the Navier-Stokes equations must be identically equal to zero, that is,

$$\mathbf{u}\cdot\nabla\mathbf{u} \equiv u\frac{\partial u}{\partial z} \equiv 0. \tag{3-5}$$

Finally, since the x and y components of $\mathbf{u}$ are zero, it follows from the Navier-Stokes equations that

$$\frac{\partial P}{\partial x} = \frac{\partial P}{\partial y} = 0. \tag{3-6}$$

Thus, the dynamic pressure is at most a function of z and t,

$$P = P(z,t) \text{ only.} \tag{3-7}$$

The only nonzero component of the Navier-Stokes equations is the z component that governs $u(x,y,t)$. Taking account of (3–5), this equation is

$$\rho\frac{\partial u}{\partial t} = -\frac{\partial P}{\partial z} + \mu\left(\frac{\partial^2 u}{\partial x^2} + \frac{\partial^2 u}{\partial y^2}\right). \tag{3-8}$$

But now it can be seen from (3–8) that $\partial P/\partial z$ must be a function of t only. Both the term on the left-hand side and the last two terms on the right are, at most, functions of x, y, and t. Thus, $\partial P/\partial z$ cannot depend on z, as suggested by (3–7), but only on t. Indeed, it is convenient to denote $\partial P/\partial z$ as

$$-\frac{\partial P}{\partial z} = G(t) \tag{3-9}$$

so that

$$\boxed{\rho\frac{\partial u}{\partial t} = G(t) + \mu\left(\frac{\partial^2 u}{\partial x^2} + \frac{\partial^2 u}{\partial y^2}\right).} \tag{3-10}$$

Equation (3–10), which we have derived from the Navier-Stokes equations, governs the unknown scalar velocity function for all unidirectional flows. However, instead of Cartesian coordinates (x,y,z), it is evident that (3–10) could have been derived using any cylindrical coordinate system (q_1,q_2,z) with the direction of motion coincident with the axial coordinate, z. In this case,

$$u = u(q_1,q_2,t) \tag{3-11}$$

and

$$\boxed{\rho\frac{\partial u}{\partial t} = G(t) + \mu(\nabla_2^2 u),} \tag{3-12}$$

where

$$\nabla_2^2 \equiv h_1 h_2\left[\frac{\partial}{\partial q_1}\left(\frac{h_1}{h_2}\frac{\partial}{\partial q_1}\right) + \frac{\partial}{\partial q_2}\left(\frac{h_2}{h_1}\frac{\partial}{\partial q_2}\right)\right]. \tag{3-13}$$

Here, h_1 and h_2 are the metrics (or scale factors) of the particular cylindrical coordinate system, defined such that

$$(ds)^2 \equiv \frac{1}{h_1^2}(dq_1)^2 + \frac{1}{h_2^2}(dq_2)^2 + (dz)^2,$$

where ds is the length of an arbitrary differential line element and dq_1, dq_2, and dz represent differential changes in q_1, q_2, and z.* The introduction of scale factors is necessary because differential changes in q_1 or q_2 of fixed size correspond to line elements of different length depending upon where we are in the coordinate domain. For example, the length of the arc associated with a differential change in the polar angle, $d\theta$, of a circular cylindrical coordinate system depends upon our proximity to the origin of the coordinate system. The scale factors h_1 and h_2 ensure that a differential change in q_1 or q_2 corresponds to a differential line element of fixed length independent of where we are relative to the origin of the coordinate system. The cylindrical coordinate system that will be of greatest interest here is the circular cylindrical system (r,θ,z). The metrics for this system are $h_1 = 1$ and $h_2 = 1/r$, and the two-dimensional Laplacian (3–13) in this case is

$$\nabla_2^2 \equiv \frac{1}{r}\frac{\partial}{\partial r}\left(r\frac{\partial}{\partial r}\right) + \frac{1}{r^2}\frac{\partial^2}{\partial\theta^2}. \tag{3-14}$$

To solve unidirectional flow problems, we must solve (3–10) or (3–12) subject to appropriate boundary conditions. Insofar as this equation is concerned, the pressure gradient $G(t)$ is considered to be a known function of time. It is either specified "a priori" or obtained from boundary values by solving (3–9). We will consider three types of unidirectional flows: steady flows driven either by boundary motion with $G=0$ or by a nonzero pressure gradient; start-up flows in which a steady boundary motion or a steady pressure gradient is suddenly imposed upon a stationary fluid; and transient flows driven by either time-dependent boundary motion or a time-dependent pressure gradient.

B. Steady Unidirectional Flows— Nondimensionalization and Characteristic Scales

We begin with the problem of steady flow between two infinite, parallel plane boundaries. We assume that the pressure gradient G is a nonzero constant and that the upper boundary moves in the same direction as the pressure gradient with a constant velocity U, while the lower boundary is stationary. A sketch of the flow configuration is given in Figure 3–1. It is convenient to describe the resulting motion of the fluid in terms of Cartesian coordinates x, in the direction of motion (parallel to the pressure gradient and the wall velocity), and y, normal to walls. We denote the constant gap width between the two walls as d. Then, starting from (3–10), we see that the equation governing the velocity distribution in the fluid is

$$\boxed{\frac{d^2u}{dy^2} = -\frac{G}{\mu},} \tag{3-15}$$

---

*The metrics (or scale factors) for a large number of orthogonal curvilinear coordinate systems can be found in the appendix by Happel and Brenner (1973).[2]

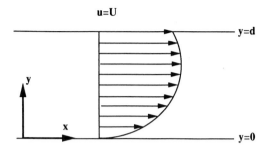

Figure 3–1 A schematic of the steady, unidirectional flow between two infinite parallel plane boundaries, with pressure gradient *G* acting from left to right in the *x* direction, and the upper plate translating with velocity $u = U$. The arrows represent the magnitude and direction of the fluid velocity at steady state.

which is to be solved subject to the no-slip boundary conditions (2–123).

$$\begin{array}{ll} u = 0 & \text{at} \quad y = 0, \\ u = U & \text{at} \quad y = d. \end{array} \tag{3-16}$$

The kinematic condition (2–121) is satisfied identically because the velocity component normal to the walls is identically equal to zero. Although this problem is extremely simple to solve, we temporarily postpone giving the solution.

Instead, we first illustrate the procedure of nondimensionalizing the governing differential equation and boundary conditions. For now, we do not attempt to motivate this procedure, except to say that it will result in a reduction in the number of parameters that characterize the problem (and its solution) from four dimensional parameters (G, μ, d, and U) to a single dimensionless parameter. To nondimensionalize, we define a dimensionless velocity, say, $\bar{u}$,

$$\bar{u} \equiv \frac{u}{u_c}, \tag{3-17}$$

and a dimensionless spatial coordinate $\bar{y}$,

$$\bar{y} \equiv \frac{y}{l_c}, \tag{3-18}$$

where u_c and l_c are known as the *characteristic* velocity and length scale, respectively, and have the same dimensions as u and y.

The use of the word characteristic in conjunction with the velocity is intended to imply that the magnitude of the velocity, anywhere in the flow, is proportional to u_c. Thus, in the present problem, a reasonable candidate for u_c would seem to be the boundary value U. If this choice is correct, we should expect that doubling the magnitude of U would lead approximately to doubling the magnitude of u everywhere in the fluid. Of course, this choice for u_c is not the only one that could have been made. In particular, another group of dimensional parameters exists that is proportional to the pressure gradient and has units of velocity

$$\frac{Gd^2}{\mu}, \tag{3-19}$$

and there is no evident reason why this quantity should not have been chosen for u_c instead of U. For the time being, let us stick with our original choice

$$u_c = U \tag{3-20}$$

but without forgetting the apparent arbitrariness of this choice relative to the other possibility (3–19).

The physical significance of a characteristic length scale, l_c, is that it is the distance over which u changes by an amount $0(u_c)$. In the present problem, this would seem to be the distance d between the two parallel plane surfaces,

$$l_c = d. \tag{3-21}$$

If this choice is correct, doubling of d should double the distance over which u changes by an amount $0(u_c)$.

With the definitions (3–20) and (3–21) adopted for u_c and l_c, we can express the original problem (3–15) and (3–16) in terms of the dimensionless variables (3–17) and (3–18), as

$$\boxed{\frac{d^2\bar{u}}{d\bar{y}^2} = -\frac{Gd^2}{\mu U},} \tag{3-22}$$

with

$$\boxed{\begin{array}{llll} \bar{u} = 0 & \text{at} & \bar{y} = 0, \\ \bar{u} = 1 & \text{at} & \bar{y} = 1. \end{array}} \tag{3-23}$$

Examining (3–22) and (3–23), we see that the dimensionless problem (and its solution) depends on a single dimensionless parameter,

$$\frac{Gd^2}{\mu U}.$$

This parameter is just the ratio of the two possible velocity scales Gd^2/μ and U, one characterized by the magnitude of the pressure gradient and the other by the magnitude of the boundary velocity. The solution of the problem, in its dimensionless form, is

$$\boxed{\bar{u} = -\left(\frac{Gd^2}{\mu U}\right)\left(\frac{\bar{y}^2}{2} - \frac{\bar{y}}{2}\right) + \bar{y}.} \tag{3-24}$$

A sketch showing profiles of $\bar{u}$ versus $\bar{y}$ for various values of $Gd^2/\mu U$ is shown in Figure 3–2.

Again, we note that the form of the velocity distribution (usually called the velocity profile) depends on the magnitude of the dimensionless ratio of velocities. When $Gd^2/\mu U \ll 1$,

$$\boxed{\bar{u} \sim \bar{y}.} \tag{3-25}$$

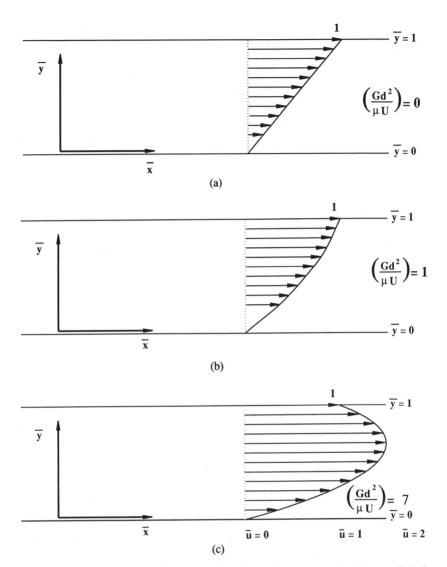

Figure 3–2 Typical velocity profiles for unidirectional flow between infinite parallel plane boundaries: (a) $Gd^2/\mu U = 0$ (simple shear flow), (b) $Gd^2/\mu U = 1$, and (c) $Gd^2/\mu U = 7$. The length of the arrows is proportional to the local dimensionless velocity, with $\bar{u} = 1$ at $\bar{y} = 1$ in all cases. The characteristic velocity scale in this case has been chosen as the velocity of the upper wall, $u_c = U$. The profiles are calculated from Equation (3–24).

In this case, the fluid motion is dominated by the motion of the boundary, and the velocity profile reduces to a linear (*simple*) shear flow. When $Gd^2/\mu U \gg 1$, on the other hand, it is apparent from (3–24) that the quadratic contribution dominates the velocity profile; indeed, the quadratic contribution to $\bar{u}$ becomes arbitrarily large as

$Gd^2/\mu U$ increases. This increase of $\bar{u}$ does not mean that the actual velocities in the system are blowing up, however. Indeed, $Gd^2/\mu U$ can be made arbitrarily large by simply taking U arbitrarily close to zero while holding Gd^2/μ constant. The increase in $\bar{u}$ simply means that the velocity contribution due to the pressure gradient G is increasing *relative* to the boundary velocity U that was used in the nondimensionalization. The problem is that the boundary velocity U is *not* an appropriate choice for the characteristic velocity when $Gd^2/\mu \gg U$. In this limit, the magnitude of the velocity in the flow domain is controlled primarily by Gd^2/μ rather than by U. Hence, it should be evident by now that

$$u_c = U, \quad \text{if} \quad Gd^2/\mu \ll U,$$

but
(3–26)

$$u_c = Gd^2/\mu, \quad \text{if} \quad Gd^2/\mu \gg U.$$

If we consider the governing equations and boundary conditions for $Gd^2/\mu \gg U$, where the second choice for u_c pertains, we find from (3–15) and (3–16) that

$$\frac{d^2\hat{u}}{d\bar{y}^2} = -1,$$ (3–27)

with $\hat{u} = 0$ at $\bar{y} = 0$,

and $\quad \hat{u} = \dfrac{U\mu}{Gd^2}$ at $\bar{y} = 1.$ (3–28)

In terms of dimensionless variables, the solution in this second case takes the form

$$\hat{u} = -\frac{1}{2}(\bar{y}^2 - \bar{y}) + \frac{U\mu}{Gd^2}\bar{y}.$$ (3–29)

Hence, as $U\mu/Gd^2 \to 0$, this solution reduces to

$$\hat{u} = -\frac{1}{2}(\bar{y}^2 - \bar{y}).$$ (3–30)

This parabolic velocity profile is characteristic of pressure-gradient-driven flow in the complete absence of boundary motion and is known as *two-dimensional Poiseuille flow*. A sketch of the velocity profile, (3–30), is shown in Figure 3–3. We see that it is the limiting form of the velocity profile when the motion driven by the pressure gradient dominates the boundary driven motion.

In the intermediate case, where neither of the limiting cases (3–26) pertains, neither choice for u_c is preferable over the other. In this case, the velocity profile is somewhere between the linear shear flow of $Gd^2/\mu \ll U$ and the parabolic profile of $Gd^2/\mu \gg U$. Either of the dimensionless solution forms (3–24) or (3–29) is satisfactory. Both show that the form of the velocity profile depends on the dimensionless ratio, $Gd^2/\mu U$. Indeed, when converted back to dimensional variables, the two dimensionless solution forms are identical [recall that $\bar{u} = u/U$ in (3–24), but $\hat{u} = u/(Gd^2/\mu)$ in (3–29)],

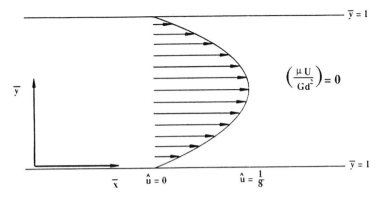

Figure 3–3 Velocity profile for unidirectional flow driven by a pressure gradient between two infinite parallel plane boundaries that are stationary [two-dimensional Poiseuille flow, Equation (3–30)]. The characteristic velocity in this case has been chosen as the centerline velocity divided by 8.

$$u = -\frac{Gd^2}{2\mu}\left(\frac{y^2}{d^2} - \frac{y}{d}\right) + U\frac{y}{d}. \tag{3-31}$$

Closely related to the pressure-driven unidirectional flow between two parallel plane surfaces is the pressure-driven motion in a straight tube of circular cross section. This is the famous problem first studied experimentally as a model for blood flow in the arteries by Poiseuille in 1840.[3] Although this problem could be solved using Cartesian coordinates with z being the axial direction, it is always much simpler to use a coordinate system in which the boundaries of the flow domain are coincident with a line or surface of the coordinate system. In this case, we use (3–12) with ∇_2^2 expressed in terms of the polar coordinates (r,θ) of a circular cylindrical coordinate system (r,θ,z), that is,

$$\frac{1}{r}\frac{\partial}{\partial r}\left(r\frac{\partial u_z}{\partial r}\right) + \frac{1}{r^2}\frac{\partial^2 u_z}{\partial \theta^2} = -\frac{G}{\mu}, \tag{3-32}$$

in which G is the constant pressure gradient. The boundary condition is

$$u_z = 0 \quad \text{at} \quad r = R, \tag{3-33}$$

where R is the radius of the tube. Since the boundary condition is independent of θ, we look for a solution $u_z = u_z(r)$ only, so that

$$\frac{1}{r}\frac{d}{dr}\left(r\frac{du_z}{dr}\right) = -\frac{G}{\mu}. \tag{3-34}$$

In this case, the only source of motion is the axial pressure gradient, and there is therefore only a single choice possible for the characteristic velocity

$$\bar{u}_z = \frac{u_z}{u_c}, \; u_c \equiv \frac{GR^2}{\mu},$$

and a single obvious choice for the characteristic length scale

$$\bar{r} = \frac{r}{l_c}, \quad l_c \equiv R.$$

In terms of the dimensionless variables $\bar{u}_z$ and $\bar{r}$, the Poiseuille flow problem (3–33) and (3–34) take the form

$$\frac{1}{\bar{r}} \frac{d}{d\bar{r}} \left(\bar{r} \frac{d\bar{u}_z}{d\bar{r}} \right) = -1, \tag{3–35}$$

$$\text{with } \bar{u}_z = 0 \quad \text{at} \quad \bar{r} = 1. \tag{3–36}$$

It is noteworthy that the Poiseuille flow problem in this dimensionless form is completely independent of the dimensional parameters G, R, and μ. This means that the form of the velocity profile that we seek is universal and does not depend on the values of these parameters. Integrating (3–35) twice with respect to $\bar{r}$, we obtain

$$\bar{u}_z = -\frac{\bar{r}^2}{4} + A \ln \bar{r} + B, \tag{3–37}$$

where A and B are two constants of integration. The boundary condition (3–36) gives us one condition from which to determine A and B. A second condition is that $\bar{u}_z$ is bounded for all $\bar{r}$, $0 \leq \bar{r} \leq 1$. This second condition can be satisfied only if $A \equiv 0$, and the solution satisfying the no-slip condition (3–36) is then

$$\bar{u}_z = \frac{1}{4}(1 - \bar{r}^2). \tag{3–38}$$

If this solution is expressed in terms of dimensional variables, it becomes

$$u_z = \frac{GR^2}{4\mu} \left(1 - \frac{r^2}{R^2} \right). \tag{3–39}$$

As we noted earlier, the form of the velocity profile is parabolic, independent of the dimensional parameters which serve only to determine the magnitude of u_z. If the velocity profile (3–39) is integrated over the cross section of the tube, we obtain an equation for the volumetric flux,

$$Q = \int_0^{2\pi} d\theta \int_0^R u_z(r) r \, dr \equiv \frac{\pi G R^4}{8\mu}. \tag{3–40}$$

This is the famous Hagen-Poiseuille law, which is the basis of capillary viscometry.

In the steady, unidirectional flow problems considered in this section, the acceleration of a fluid element is identically equal to zero. Both the time derivative, $\partial u / \partial t$, and the nonlinear "inertial" terms are zero so that $Du/Dt \equiv 0$. This means that the equation of motion reduces locally to a simple balance between forces associated with the pressure gradient and viscous forces due to the velocity gradient. Since this simple force balance holds at every point in the fluid, it must also hold for the fluid system

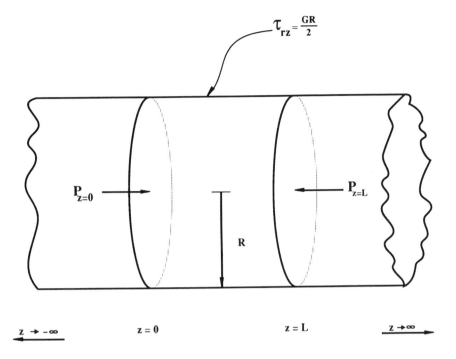

Figure 3–4 A pictorial representation of the force balance on the fluid within an arbitrarily chosen section of a circular tube in steady Poiseuille flow. Pressure forces act at the two cross-sectional surfaces as $z = 0$ and $z = L$, while a viscous stress acts at the cylindrical boundary and exactly balances the net pressure force.

as a whole. To illustrate this, we will use the Poiseuille flow solution. Let us consider the forces acting on a body of fluid in an arbitrary section of the tube, between $z = 0$, say, and a downstream point $z = L$, as illustrated in Figure 3–4. At the walls of the tube, the only nonzero shear stress components is τ_{rz}. The normal stress components at the walls are all just equal to the pressure and produce no *net* contribution to the overall forces that act on the body of fluid that we consider here. The viscous shear stress at the walls is evaluated using (3–39),

$$f_z = i_z \cdot (\mathbf{n} \cdot \mathbf{T}) = -\tau_{rz} = -\mu \frac{du_z}{dr}\bigg|_{r=R} = \frac{GR}{2}, \tag{3-41}$$

so the total viscous force at the walls between $z = 0$ and $z = L$ is

$$F_z = f_z(2\pi RL) = GR^2 \pi L. \tag{3-42}$$

But $G \equiv -\partial P/\partial z$ is a constant [see (3–9)], and so can be expressed in terms of P at $z = 0$ and $z = L$ as

$$G \equiv \frac{1}{L}(P_{z=0} - P_{z=L}). \tag{3-43}$$

Thus, substituting into (3–42), we obtain

$$F_z = (P_{z=0} - P_{z=L}) \pi R^2. \tag{3-44}$$

But, the quantity on the right-hand side is just the net pressure force in the z direction acting on the fluid between $z = 0$ and $z = L$: the force $P_{z=0} \pi R^2$ acting on the upstream surface at $z = 0$ minus the opposite-directed force $P_{z=L} \pi R^2$ acting on the downstream surface at $z = L$. Thus, we have demonstrated that the net pressure and viscous forces acting on any macroscopic body of fluid in steady, unidirectional Poiseuille flow must be in exact balance. The same conclusion is true for any steady, unidirectional flow.

A reader with some previous experience in fluid mechanics may have noted that some of the results that we have quoted in this section appear to be at odds with observation. The most obvious example of this is the universal form, independent of the flow rate, that we found for the Poiseuille flow velocity profile. Careful experiments on flow through a tube show quite different results. Up to some critical flow rate, a parabolic profile is found in agreement with the Poiseuille solution. For higher flow rates, however, the flow changes dramatically to a highly chaotic motion with a time-averaged velocity profile in the z direction that is very nearly a "plug-flow" form—that is, $(u_z)_{avg}$ is independent of r except very near the walls. In addition, if individual velocity components were measured, one would find not only rapid variations with respect to time but also nonzero values for all of the instantaneous components u_r, u_z, and u_θ. These experimental observations are a consequence of the transition from a simple, unidirectional laminar flow to a turbulent flow, which is always fully three-dimensional. How then are we to interpret the solution (3–39) which obviously satisfies the equations of motion and the boundary conditions, but has a universal form independent of the flow rate?

Our analysis shows that there is a steady, unidirectional flow solution of the Navier-Stokes equations for the Poiseuille flow problem for all flow rates. However, it says nothing about the stability of this solution. In order to examine this question, further analysis is necessary.[4] In principle, the simplest theoretical approach is to examine the stability of the solution to arbitrary "disturbances," that is, arbitrary departures from the laminar solution, (3–39). If these disturbances increase in magnitude with time, the basic laminar motion is said to be unstable, and it is clear that a new form for the velocity field will eventually be realized. If they decrease in magnitude, the flow will revert back to its "undisturbed" laminar form. It is generally possible to carry out such an analysis if the disturbance is assumed to be infinitesimal in magnitude, since then the equations of motion governing the disturbance can be linearized and an arbitrary disturbance analyzed by considering the growth or decay of its Fourier components (or normal modes). If the disturbance is initially of finite amplitude, however, its evolution in time will be governed by the full nonlinear Navier-Stokes equations (even if the undisturbed motion is unidirectional), and we can generally only study the growth or decay of specific disturbances numerically. Whatever the details of execution, this mathematical determination of instability is reflected in the laboratory by a transition at some critical flow condition from the basic laminar solution to a new, more complicated flow. In some instances this transition may result in turbulence, but in other cases it may lead only to a more complicated laminar flow. Generally, the motion that results from an instability of the basic laminar flow (or a bifurcation from it) is too complicated to be described via an analytical solution of the Navier-Stokes equations. (Even if we consider an infinitesimal initial disturbance via a linearized stability analysis, the growth of this disturbance will ultimately produce a disturbance of finite amplitude that can be described only via the full nonlinear Navier-Stokes equations.) An extreme example is the transition from a steady, unidirectional flow to a fully three-dimensional, time-dependent turbulent velocity field. All that we can say is that the simple, laminar solution will not be realized experimentally over the whole range of flow rates.

In the present book, we focus our attention on the solution of the Navier-Stokes equations for laminar flows, frequently without any attempt to analyze the stability (or realizability) of the resulting solutions. In using these solutions, it is therefore quite apparent that we must always reserve judgment as to the range of parameter values where they will exist in practice. In the case of pressure-driven unidirectional flows, experimental evidence shows that the laminar flow solutions are actually realized provided $u_c l_c / \nu$ is not too large. In the Poiseuille flow, transition to turbulence is generally observed experimentally for

$$\left(\frac{GR^2}{\mu} \right) R/\nu \sim 2000. \tag{3-45}$$

C. Start-up Flow in a Circular Tube— Solution by Separation of Variables

In the preceding section, we considered several time-independent unidirectional flows. It is evident that the resulting steady velocity profiles can be considered as the limiting solution for large times after the imposition of either a steady boundary motion or a steady pressure gradient. An interesting question that we can study in detail for unidirectional flows is how these steady-state velocity profiles evolve in time if the fluid is initially motionless at the instant of application of the boundary motion or pressure gradient.

In this section, we consider one example of this type of problem, namely, the start-up from rest of the flow in a circular tube when a nonzero pressure gradient is suddenly imposed at some instant (which we shall denote as $t = 0$) and then held at the same constant value thereafter ($t > 0$). In Section E, we consider the start-up of simple shear flow between two infinite, plane boundaries. Other relatively simple examples of start-up flows are left to the reader to solve (for example, see problem 6 at the end of this chapter).

In the present problem of the start-up of pressure-gradient-driven flow in a circular tube, we assume that the tube is very long and focus our attention only on the central region away from the ends so that the motion can be approximated as unidirectional, albeit time dependent. Thus, we assume, for $t \geq 0$, that

$$u_z = u_z(r,t), \quad u_r = u_\theta = 0, \tag{3-46}$$

and the governing differential equation (3–12), plus boundary and initial conditions are then

$$\rho \frac{\partial u_z}{\partial t} = G + \mu \left[\frac{1}{r} \frac{\partial}{\partial r} \left(r \frac{\partial u_z}{\partial r} \right) \right], \qquad (3-47)$$

with $u_z = 0$ at $r = R$, all t,

u_z bounded at $r = 0$, all t, $\qquad (3-48)$

and $u_z = 0$ at $t = 0$, all r.

It is clear that a much more complicated analysis would be required to consider the *entry* and *exit* regions near the ends of the tube, because the axial velocity component, u_z, in that region will be dependent upon the distance from the ends of the tube in violation of the basic unidirectional flow assumptions (see Section A).

We begin, again, by nondimensionalizing the governing equations. The characteristic length scale is the tube radius, R, and an appropriate choice for the characteristic velocity scale would seem to be GR^2/μ, since this quantity is proportional to the magnitude of the final steady-state velocities, cf. (3–39)

$$l_c = R, \quad u_c = GR^2/\mu. \qquad (3-49)$$

Although other combinations of the parameters (G,μ,ρ,R) can be constructed with dimensions of velocity, the choice (3–49) is the only one in which u_c is proportional to the pressure gradient G. Since the motion is driven solely by the pressure gradient G, it would indeed be surprising if the velocities in the system were not proportional to G.

One important implication of nondimensionalization that we shall use later is that spatial derivatives, such as $\partial \bar{u}_z/\partial \bar{r}$, must satisfy the condition

$$\frac{\partial \bar{u}_z}{\partial \bar{r}} = 0(1), \qquad (3-50)$$

if our choices for l_c and u_c are correct. The symbol $0(A)$ is introduced here (and will appear throughout this book) as a convenient shorthand notation to indicate that the magnitude of the quantity [for example, $\partial \bar{u}_z/\partial \bar{r}$ in (3–50)] has a numerical value that is proportional to A. If this "order of magnitude" symbol were applied to the dimensional velocity, for example, we would write

$$u_z = 0 \left(\frac{GR^2}{\mu} \right),$$

assuming that our choice of the characteristic velocity is correct. When we write $0(1)$, as in (3–50), we imply that the magnitude of $\partial \bar{u}_z/\partial \bar{r}$ is independent of the dimensional parameters of the problem.

In order to nondimensionalize (3–47) and the boundary conditions (3–48), it is also necessary to introduce a characteristic time scale, t_c, in order to define a dimensionless time, $\bar{t} = t/t_c$. The characteristic time scale t_c should be proportional to the time period over which the velocity profile evolves from its initial form, $u_z = 0$, to the final steady state (3–39). However, it is not immediately obvious how this time scale depends upon the dimensional parameters of the problem. Let us ▸ve t_c undefined for the moment and write (3–47) and (3–48) in dimensionless form in terms of t_c, that is,

$$\frac{R^2}{\nu}\frac{1}{t_c}\left(\frac{\partial \bar{u}_z}{\partial \bar{t}}\right) = 1 + \frac{1}{\bar{r}}\frac{\partial}{\partial \bar{r}}\left(\bar{r}\frac{\partial \bar{u}_z}{\partial \bar{r}}\right), \tag{3–51}$$

with $\bar{u}_z = 0$ at $\bar{r} = 1$, all $\bar{t}$,

$\bar{u}_z$ bounded at $\bar{r} = 0$, all $\bar{t}$, $\qquad$ (3–52)

and $\bar{u}_z = 0$ at $\bar{t} = 0$, all $\bar{r}$.

This problem can be converted to a slightly more familiar mathematical form by introducing $\bar{w}(\bar{r},\bar{t})$ as the difference between $\bar{u}_z(\bar{r},\bar{t})$ and the final steady-state velocity profile [which is given in dimensionless form in (3–38)], namely,

$$\bar{w}(\bar{r},\bar{t}) = \bar{u}_z(\bar{r},\bar{t}) - \frac{1}{4}(1 - \bar{r}^2). \tag{3–53}$$

Substituting (3–53) into (3–51) and (3–52), we obtain an equivalent problem for $\bar{w}$ in the form

$$\frac{R^2}{\nu}\left(\frac{1}{t_c}\right)\frac{\partial \bar{w}}{\partial \bar{t}} = \frac{1}{\bar{r}}\frac{\partial}{\partial \bar{r}}\left(\bar{r}\frac{\partial \bar{w}}{\partial \bar{r}}\right), \tag{3–54}$$

with $\bar{w} = 0$ at $\bar{r} = 1$, all $\bar{t}$,

$\bar{w}$ bounded at $\bar{r} = 0$, all $\bar{t}$, $\qquad$ (3–55)

and $\bar{w} = -\frac{1}{4}(1 - \bar{r}^2)$ at $\bar{t} = 0$, all $\bar{r}$.

The differential equation (3–54) is identical in form to the familiar *heat equation* for radial conduction of heat in a circular cylindrical geometry. Thus, we see that the evolution in time of the steady Poiseuille velocity profile is completely analogous to the conduction of heat starting with an initial parabolic temperature profile $-(1 - \bar{r}^2)/4$. In our problem, the final steady velocity profile is established by diffusion of momentum from the wall of the tube so that the initial profile for $\bar{w}$ eventually evolves to the asymptotic value $\bar{w} \to 0$ as $\bar{t} \to \infty$. The characteristic time scale for any diffusion process (whether it is molecular diffusion, heat conduction, or the present process) is (l_c^2/diffusivity), where l_c is the characteristic distance over which diffusion occurs. In the present process, $l_c = R$ and the kinematic viscosity ν plays the role of the diffusivity so that

$$t_c = \frac{R^2}{\nu}. \tag{3–56}$$

In retrospect, it should perhaps have been evident from (3–54) that this would be the appropriate characteristic time scale.* However, without the preceding discussion, the important observation of an analogy between the diffusion of momentum in start-up of a unidirectional flow and the conduction of heat would not have been evident. We shall discuss the nature of this process in more detail after we have solved (3–54) and (3–55) to obtain the time-dependent velocity profile.

Although the problem defined by (3–54) and (3–55) is time dependent, it is linear in $\bar{w}$ and confined to the bounded spatial domain, $0 \leq \bar{r} \leq 1$. Thus, it can be solved by the method of separation of variables. In this method we first find a set of eigensolutions that satisfy the differential equation (3–54) and the boundary condition at $\bar{r} = 1$, then determine the particular sum of those eigensolutions that also satisfies the initial condition at $\bar{t} = 0$. The problem (3–54) and (3–55) is one example of the general class of so-called Sturm-Louiville problems, for which an extensive theory is available that insures the existence and uniqueness of solutions constructed via eigenfunction expansions by the method of separation of variables. We assume that the reader is familiar with the basic technique and simply outline the solution of (3–54) and (3–55) without detailed proofs.[5] We begin with the basic hypothesis that a solution of (3–54) exists in the separable form

$$\bar{w}(\bar{r},t) = R(\bar{r})\Theta(\bar{t}). \tag{3–57}$$

Substituting (3–57) into (3–54), we obtain

$$\frac{1}{\Theta}\frac{d\Theta}{dt} = \frac{1}{R}\left[\frac{1}{\bar{r}}\frac{d}{d\bar{r}}\left(\bar{r}\frac{dR}{d\bar{r}}\right)\right]. \tag{3–58}$$

Now, since the left-hand side involves a function of $\bar{t}$ only, while the right-hand side involves a function of $\bar{r}$ only, it is evident that each can at most equal a constant. We call this constant $-\lambda^2$. Thence,

$$\frac{1}{\Theta}\frac{d\Theta}{dt} = -\lambda^2, \tag{3–59}$$

$$\frac{1}{R}\left[\frac{1}{\bar{r}}\frac{d}{d\bar{r}}\left(\bar{r}\frac{dR}{d\bar{r}}\right)\right] = -\lambda^2. \tag{3–60}$$

The solution of (3–59), for arbitrary λ^2, is

$$\Theta = e^{-\lambda^2\bar{t}}. \tag{3–61}$$

Equation (3–60) can be rewritten as

$$\bar{r}^2\frac{d^2R}{d\bar{r}^2} + \bar{r}\frac{dR}{d\bar{r}} + \lambda^2\bar{r}^2R = 0, \tag{3–62}$$

or, upon introducing a change of independent variables,

$$z = \lambda\bar{r}, \tag{3–63}$$

---

*Assuming that the characteristic length and velocity scales, (3–49), are correct, the right-hand side of (3–54) is $0(1)$—independent of the dimensional parameters of the problem. Thus, the parameter $R^2/\nu t_c$ must also be $0(1)$. For, if $R^2/\nu t_c \ll 1$, then the left-hand side would be negligible and the problem reduces back to the steady flow case. If, on the other hand, $R^2/\nu t_c \gg 1$, then $\partial\bar{w}/\partial\bar{t} \sim 0$, and $\bar{w}$ should be independent of time, thus adopting its initial form for all $\bar{t}$. But, this is inconsistent with the requirement that $\bar{w} \to 0$ as $\bar{t} \to 0$ in order that $\bar{u}_z$ achieve its known steady state form.

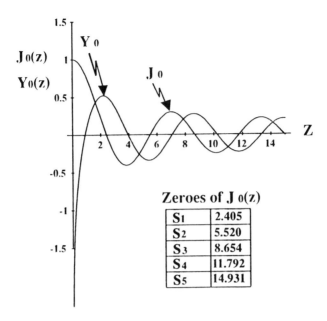

Figure 3-5 The Bessel functions of the first and second kind of order zero.

we have

$$z^2 \frac{d^2 R}{dz^2} + z \frac{dR}{dz} + z^2 R = 0. \tag{3-64}$$

This is Bessel's differential equation of order zero. It has two independent solutions,

$$J_0(z) \quad \text{and} \quad Y_0(z),$$

which are known as Bessel functions of the first and second kind of order zero.[6] A plot showing the behavior of these two functions is shown in Figure 3–5. Both oscillate back and forth across zero, but $Y_0(z) \to -\infty$ as $z \to 0$. Hence, the most general solution of the differential equation (3–54) of the form (3–57) that is bounded at $\bar{r} = 0$ is

$$\bar{w}_n = e^{-\lambda_n^2 \bar{t}} J_0(\lambda_n \bar{r}). \tag{3-65}$$

The subscript n is added in anticipation of the fact that there is an infinite, but discrete, set of values possible for λ such that the general solution, $e^{-\lambda^2 \bar{t}} J_0(\lambda \bar{r})$, satisfies the boundary condition $\bar{w} = 0$ at $\bar{r} = 1$. This set of values of $\lambda = \lambda_n$ are known as the eigenvalues of the problem, while the corresponding $\bar{w}_n$ are called the eigenfunctions. To determine the eigenvalues, λ_n, we apply the boundary condition at $\bar{r} = 1$ to (3–65), that is,

$$\bar{w}_n(1, \bar{t}) = e^{-\lambda_n^2 \bar{t}} J_0(\lambda_n) = 0, \quad \text{all } \bar{t}. \tag{3-66}$$

This condition requires

$$J_0(\lambda_n) = 0.$$

Hence, the eigenvalues, λ_n, are equal to the infinite set of zeroes for $J_0(z)$. Referring to Figure 3–5, we have denoted those zeroes as s_n, with the first crossing for the smallest value of z being s_1, namely,

$$\lambda_n = s_n, \quad n = 1, 2, 3, \ldots \infty. \tag{3-67}$$

Thus, since the differential equation (3–54) and boundary conditions are all linear, the most general form possible for $\bar{w}$ is

$$\bar{w} = \sum_{n=1}^{\infty} A_n \bar{w}_n(\bar{r}, \bar{t}) = \sum_{n=1}^{\infty} A_n e^{-s_n^2 \bar{t}} J_0(s_n \bar{r}), \tag{3-68}$$

where A_n are arbitrary, constant coefficients. This solution satisfies the differential equation (3–54) and the boundary condition $\bar{w} = 0$ at $\bar{r} = 1$ for any choice of the constant coefficients A_n.

The final step is to choose the A_n so that $\bar{w}(\bar{r}, \bar{t})$ satisfies the initial condition $\bar{w}(\bar{r}, 0) = -(1 - \bar{r}^2)/4$. The general Sturm-Louiville theory[7] guarantees that the eigenfunctions (3–68) form a complete set of orthogonal functions. Thus, it is possible to express the smooth initial condition $(1 - \bar{r}^2)$ via the Fourier-Bessel series (3–68) with $\bar{t} = 0$, that is,

$$\bar{w}(\bar{r}, 0) = \sum_{n=1}^{\infty} A_n J_0(s_n \bar{r}) = -\frac{1}{4}(1 - \bar{r}^2). \tag{3-69}$$

To determine the A_n, we multiply both sides of (3–69) by $\bar{r} J_0(s_m \bar{r})$ and integrate over $\bar{r}$ from 0 to 1, using the orthogonality properties of J_0,

$$\int_0^1 \bar{r} J_0(s_m \bar{r}) J_0(s_n \bar{r}) d\bar{r} = \begin{cases} 0 & n \neq m \\ \frac{1}{2} J_1^2(s_n) & n = m. \end{cases} \tag{3-70}$$

It follows from this process that

$$A_n = -\frac{1/4 \int_0^1 \bar{r}(1 - \bar{r}^2) J_0(s_n \bar{r}) d\bar{r}}{\int_0^1 \bar{r} J_0^2(s_n \bar{r}) d\bar{r}}. \tag{3-71}$$

Finally, using the general properties of the Bessel functions J_0 and J_1, it can be shown that

$$A_n = \frac{8}{s_n^3} [J_1(s_n)]^{-1}, \tag{3-72}$$

where J_1 is the Bessel function of the first kind of order 1. This completes the solution of (3–54) and (3–55) for the velocity function, $\bar{w}(\bar{r}, \bar{t})$.

Combining (3–72), (3–68) and (3–53), and reverting to dimensional variables, the solution of the full, original problem can be expressed in terms of the axial velocity profile

$$u_z = \frac{GR^2}{4\mu} \left[\left[1 - \left(\frac{r}{R}\right)^2 \right] - \sum_{n=1}^{\infty} \frac{32}{s_n^3} J_1(s_n)^{-1} e^{-s_n^2 \nu t / R^2} J_0\left(s_n \frac{r}{R}\right) \right].$$

$$\tag{3-73}$$

Obviously, as $t \to \infty$, this solution reverts to the steady state Poiseuille profile. In order to obtain other details of this velocity profile, it is necessary to evaluate the infinite series numerically for each value of t and r. A typical numerical example of the results is shown in Figure 3–6, where we have plotted $\bar{u}_z$ versus $\bar{r}$ for several values of $\bar{t}$. It can be seen that the initial profile for $\bar{t} = 0.05$ is flat, with $\bar{u}_z$ independent of $\bar{r}$ except for $\bar{r}$ very close to the tube walls. Right at the tube wall, $\bar{u}_z = 0$, and it can be seen that this manifestation of the no-slip condition gradually propagates across the tube via the diffusion process discussed earlier. The region in which the wall is felt increases in width at a rate proportional to $\sqrt{\nu t}$ as is typical of diffusion or conduction processes with "diffusivity" ν.

We shall see later that it is often advantageous to consider the temporal or spatial development of a velocity field in terms of diffusion and convection processes (though it is often more convenient to focus on vorticity rather than momentum transport). In this framework, the basic concept of diffusion over a scale $\sqrt{\nu t}$ in a time increment t plays a cricital role in helping to determine the extent of boundary influence on the flow. It should be recognized, however, that the class of unidirectional flows is a special case in that the direction of diffusion is always at right angles to the direction of motion. In most flows this will not be true, and the influence of the boundary geometry will be propagated via momentum transport by both diffusion and convection.

Finally, we wish to return briefly to the problem of choosing appropriate characteristic scales for nondimensionalization. In particular, it is important to recognize

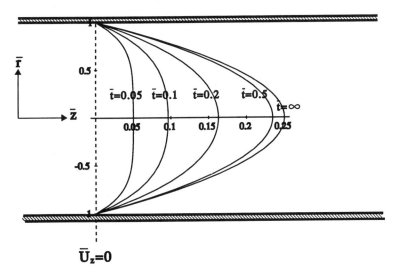

Figure 3–6 Velocity profiles at the different values of $\bar{t}$ depicting the start-up of pressure-gradient-driven flow in a circular tube (far from the ends of the tube). The profiles, from left to right, are for $\bar{t} = 0.05$, 0.1, 0.2, 0.5, and ∞. The velocities in all cases are scaled with the centerline velocity for $\bar{t} = \infty$ divided by 4.

the role played by nondimensionalization so that the consequences of an incorrect (or inappropriate) choice of characteristic scale is clear. The first and most important point is that the introduction of dimensionless variables does not change the problem at all, but simply renames variables. Suppose, for example, that we had used $l_c/u_c = \mu/GR$ as a characteristic time scale in (3–51) instead of R^2/ν. Then instead of

$$\frac{\partial \bar{u}_z}{\partial \bar{t}} = 1 + \frac{1}{\bar{r}} \frac{\partial}{\partial \bar{r}} \left(\bar{r} \frac{\partial \bar{u}_z}{\partial \bar{r}} \right) \tag{3-74}$$

with

$$\bar{t} = t \left(\frac{\nu}{R^2} \right) \tag{3-75}$$

as was actually used earlier in this section, we would have obtained

$$\left(\frac{GR^3 \rho}{\mu^2} \right) \frac{\partial \bar{u}_z}{\partial t^*} = 1 + \frac{1}{\bar{r}} \frac{\partial}{\partial \bar{r}} \left(\bar{r} \frac{\partial \bar{u}_z}{\partial \bar{r}} \right), \tag{3-76}$$

in which

$$t^* = t \left(\frac{GR}{\mu} \right). \tag{3-77}$$

The difference in scaling inherent in (3–74) and (3–76) is not significant in this case because the equations are linear. The exact solution must take the form (3–73) independent of how (or whether) the differential equation is nondimensionalized prior to its solution. Generally speaking, however, the governing equations in fluid mechanics will be nonlinear, and we can seek only approximate solutions obtained by some simplification of the equations. In this case, the relative importance of the various terms in the equations and boundary conditions is frequently determined on the basis of nondimensionalization. To illustrate the general ideas, let us reconsider (3–74) and (3–76), and recall that the basic concept of characteristic scales means that any term that involves only nondimensionalized variables should be $0(1)$ provided the scaling is correct. Hence, assuming that (3–77) is correct, the term on the left-hand side of (3–76) is $0(GR^3\rho/\mu^2)$, while the two on the right are $0(1)$. Likewise, if (3–75) is correct, all terms in (3–74) should be $0(1)$. As long as an exact solution of the full equation is achieved, the difference between (3–74) and (3–76) is not important. If we began with the time scaling of (3–77), however, we might be tempted to neglect the left-hand side of (3–76) whenever $GR^3\rho/\mu^2 \ll 1$. It is at this point of approximating the equation prior to solution that incorrect scaling could cause difficulties if one is not careful in the solution process. It is evident from the differential equation with correct scaling, (3–74), that the magnitude of the parameter $GR^3\rho/\mu^2$ has *nothing* to do with the relative importance of the various terms. But, suppose we had not previously derived (3–74), but had simply attempted to obtain an approximate solution of the problem for small $GR^3\rho/\mu^2$ by *neglecting* the left-hand side of (3–76). The resulting equation is identical to that for steady flow, (3–35), and the unique solution subject to the condition $\bar{u}_z = 0$ at $\bar{r} = 1$ is just the steady, parabolic Poiseuille profile. But this solution does not provide a uniformly valid first approximation in time to the actual solution because it does not satisfy the initial condition $\bar{u}_z = 0$ at $t^* = 0$. Hence, the incorrect scaling inherent in (3–76) has resulted in an incorrect approximation to the full equation for $GR^3\rho/\mu^2 \ll 1$ and an approximate solution that does not satisfy all of the boundary and initial conditions of the original problem. Provided that we recognize this last fact, we are forced to conclude that the scaling inherent in (3–76) is not correct—at least for short times, $t^* \ll 1$ or small values of $GR^3\rho/\mu^2$—and no harm is done provided that we go back and rethink the problem to finally achieve the correct scaling leading to (3–74).

Although we will typically be dealing with nonlinear rather than linear differential equations, the situation that we have outlined is fairly typical. If we attempt to simplify the governing equations by neglecting terms that appear to be small on the basis of nondimensionalization and if one (or more) of the characteristic scales inherent in the nondimensionalized equations is incorrect, we will always find that it is impossible to satisfy all of the boundary and/or initial conditions of the original problem. Indeed, this should be our signal to reevaluate the choice of characteristic scales. One final point with regard to (3–76) is to explain why the choice of an inappropriate characteristic time scale leads to the false conclusion that the time derivative in (3–47) can be neglected for $GR^3\rho/\mu^2 \ll 1$. The explanation is simple. As long as the nondimensionalization is correct, all terms involving only dimensionless variables are $0(1)$, and the relative magnitude of the various terms can be determined correctly by the magnitude of any dimensionless combination of parameters, such as $GR^3\rho/\mu^2$, that appear. On the other hand, if the characteristic scales are incorrect, then the magnitude of terms involving only dimensionless variables are not necessarily $0(1)$, that is, independent of the dimensional parameters of the problem. Thus, for example, since we know the appropriate scales for the present problem, we can see that as $GR^3\rho/\mu^2$ decreases, the dimensionless time derivative $\partial \bar{u}_z/\partial t^*$ must be increasing as $(GR^3\rho/\mu^2)^{-1}$, and the conclusion that

$$(GR^3\rho/\mu^2)\frac{\partial \bar{u}_z}{\partial t^*} \ll 1 \text{ for } \left(\frac{GR^3\rho}{\mu^2}\right) \ll 1$$

is incorrect.

D. The Rayleigh Problem—Solution by Similarity Transformation

In this section, we consider a second example of a transient, unidirectional flow. This is the famous problem, first studied in the 1800s by Lord Rayleigh, in which an initially stationary infinite flat plate is assumed to begin suddenly translating in its own plane with a constant velocity through an initially stationary unbounded fluid.

For convenience, let us adopt a Cartesian coordinate system in which the flat plate is assumed to occupy the **xz** plane, with the initially stationary fluid occupying the upper half space, $y > 0$. We denote the magnitude of the plate velocity as U so that

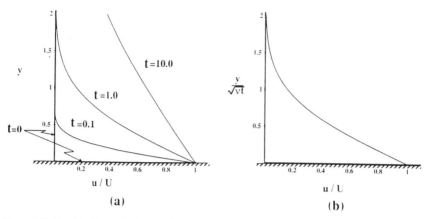

Figure 3–7 (a) The velocity profiles at various increasing times above an infinite flat plate that suddenly begins moving in its own plane at $t = 0$ with a constant velocity U (the Rayleigh problem). (b) The self-similar velocity profile for the same problem obtained by rescaling the distance from the plate y with the diffusion length scale, $\sqrt{\nu t}$ for all $t > 0$.

$$u_{plate} = \begin{matrix} 0 & t < 0 \\ Ui & t \geq 0. \end{matrix}$$

The problem is to determine the velocity distribution in the fluid as a function of time. In this problem, the fluid motion is due entirely to the motion of the boundary—the only pressure gradient is hydrostatic, and this does not effect the velocity parallel to the plate surface. At the initial instant, the velocity profile appears as a step with magnitude U at the plate surface and magnitude arbitrarily close to zero everywhere else, as sketched in Figure 3–7. As time increases, however, the effect of the plate motion propagates farther and farther out into the fluid as momentum is transferred normal to the plate by molecular diffusion and a series of velocity profiles is achieved similar to those sketched in Figure 3–7. In this section, we analyze the details of this motion and, in the process, introduce the concept of self-similar solutions that we shall use extensively in later chapters.

The governing differential equation for this problem is (3–10) with $G(t) \equiv 0$ and $u = u(y,t)$ only, thus

$$\rho \frac{\partial u}{\partial t} = \mu \frac{\partial^2 u}{\partial y^2}. \tag{3–78}$$

Initially,

$$u(y,0) = 0, \quad \text{all} \quad y > 0,$$

while

$$u(0,t) = U, \quad \text{all} \quad t \geq 0, \tag{3–79}$$

and

$$u(y, t) \to 0 \quad \text{as} \quad y \to \infty, \quad t \geq 0.$$

The second condition is just the no-slip condition at the plate surface, while the third condition arises as a consequence of the assumption that the fluid is unbounded. At any finite $t > 0$, a region will always exist sufficiently far from the plate that no significant momentum transfer will yet have occurred, and in this region the fluid velocity will remain arbitrarily small.

The problem posed by (3–78) with the boundary and initial conditions (3–79) is very simple to solve by either Fourier or Laplace transform methods.[8] Further, since it is linear, an exact solution is possible, and nondimensionalization need not play a significant role in the solution process. Nevertheless, we pursue the solution via the use of a so-called *similarity transformation,* whose existence is suggested by an attempt to nondimensionalize the equation and boundary conditions. Although it may seem redundant to introduce a new solution technique when standard transform methods could be used, the use of similarity transformations is not limited to linear problems (as are the Fourier and Laplace transform methods), and we shall find the method to be extremely useful in the solution of certain nonlinear differential equations later in this book.

Let us begin the solution process in the same way that we have adopted in the preceding sections of this chapter, by attempting to nondimensionalize the governing equation and boundary conditions. It is obvious from the problem formulation that an appropriate characteristic velocity for this purpose is just

$$u_c = U.$$

However, the geometry of the flow domain is just the upper half space bounded below by an infinite plane surface, and this does not offer any obvious characteristic length scale. Indeed, it is apparent from the governing equation (3–78) that the evolution of the velocity profile can be completely described in terms of a one-dimensional diffusion process transporting momentum from the fluid near the plate to the fluid in the rest of the domain, and this suggests that the velocity will vary from U to zero over a distance that increases continuously with time at a rate proportional to $\sqrt{\nu t}$. Thus, it appears that a physically relevant characteristic length scale simply does not exist in this case. On the other hand, it is obvious that it should be possible to express the solution of any physical problem in a dimensionless form that does not depend upon the system of dimensions that is being employed. In most cases, this is possible by simply expressing the problem in terms of dimensionless independent and dependent variables, which are obtained from the original dimensional variables by rescaling with respect to characteristic velocity, length, and time scales. In the present case, however, a characteristic length scale does not exist. The only other way in which a solution can be obtained that is independent of the system of dimensions is if it is a function of some *dimensionless combination* of the dimensional independent variables and parameters of the problem, y, t, ν, and U, rather than a function of y and t separately. We shall call this dimensionless combination the *similarity variable,* and denote it as η. The definition of η in terms of y and t is known as the *similarity transformation*

$$\eta = \eta(y,t,\nu,U). \tag{3–80}$$

A solution, such as $u(y,t)$ in the present problem, that depends on a single dimensionless variable η instead of y and t separately is said to be *self-similar,* and this designation is also the source of the names *similarity variable* and *similarity transformation.*[9] The basic idea of self-similarity is that the series of profiles $u(y,t)$ for various fixed times, t, will collapse into a single, universal form when u is plotted as a function of η rather than as a function of y.

We shall demonstrate later that a similarity transformation can always be generated by a systematic procedure that requires no guesswork or intuition, provided that it exists. We have suggested that the absence of any physically significant characteristic length scale in the present problem is a signal that such a transformation must exist, since otherwise it would not be possible to express the solution in a form that does not depend upon the arbitrary system of dimensions that has been adopted. It is clear, for example, that a similarity transformation (or a self-similar solution) would not exist if instead of an infinite, unbounded fluid above the moving plate we had a second, parallel stationary plate at a finite distance, say, d, away. To see this we need note only that the final steady profile in this later case will be a linear, simple shear flow, which cannot possibly be mapped into the initial ''step'' form of the velocity profile by any rescaling as would be required if a self-similar solution were to exist. Of course, the presence of a second plate means that a definite characteristic length scale is now present, namely, the separation distance between the plates $l_c = d$, and this in turn provides a definite indication that a similarity transformation should not be expected. Let us return to the solution of the classic Rayleigh problem when there is a single plate. We shall employ two approaches to the generation of a similarity transformation for this case: First, we shall consider a physical, intuitive approach that is based upon our understanding of the diffusion process by which the velocity profile evolves in time and space; second, we will present the same solution derived by a systematic procedure that requires no intuition or understanding whatever, other than the sense to look for a similarity transformation in the first place.

The physical, intuitive approach is very simple. We first recall that as t increases, the effect of the wall motion is propagated farther and farther out into the fluid as a consequence of the diffusion of momentum normal to the wall. Further, we have already seen that diffusion will occur across a region of characteristic dimension l_c in a characteristic time increment

$$t_c = \frac{l_c^2}{\nu}.$$

Thus, at any time t^* after the plate begins to move, the momentum generated at its surface will have diffused outward over a distance

$$l^* = \sqrt{\nu t^*}.$$

If we now consider two velocity profiles, one at time t_1^* and the other at time t_2^*, with $t_2^* > t_1^*$, we might expect them to reduce to a single ''universal'' (self-similar) profile if we were to scale the distance from the wall, y, with the length scale, $l^* = l^*(t^*)$, of the diffusion process, that is,

$$\boxed{\frac{u}{U} = F\left(\frac{y}{\sqrt{\nu t}}\right)} \tag{3-81}$$

for any y and t. If u/U does depend only upon $y/\sqrt{\nu t}$ rather than y and t independently, we shall say that u/U is self-similar, and the similarity transformation, (3–80), is

$$\eta = \frac{y}{\sqrt{\nu t}}. \tag{3-82}$$

It will be noted that the similarity variable suggested by our intuitive diffusion argument is dimensionless (though y and t are still dimensional). To determine whether a solution actually exists in the form (3–81), however, we must first substitute into (3–78) and (3–79) to see whether this form is consistent with the governing equation and boundary conditions. To do this, we first calculate

$$\frac{\partial u}{\partial t} = U\frac{dF}{d\eta}\frac{\partial \eta}{\partial t} = -U\left(\frac{1}{2t}\right)\eta\frac{dF}{d\eta} \tag{3-83}$$

and

$$\frac{\partial^2 u}{\partial y^2} = U\frac{\partial \eta}{\partial y}\frac{\partial}{\partial \eta}\left(\frac{\partial \eta}{\partial y}\frac{dF}{d\eta}\right) = U\frac{1}{\nu t}\frac{d^2 F}{d\eta^2}, \tag{3-84}$$

and then substitute into (3–78). The result is

$$\frac{d^2 F}{d\eta^2} + \frac{1}{2}\eta\frac{dF}{d\eta} = 0. \tag{3-85}$$

Thus, the transformation (3–81) yields an ordinary differential equation for F with coefficients that are either constant or a function of η only. This equation is, however, only second order in η, and this means that F can satisfy only two boundary conditions. On the other hand, the velocity u in the original problem satisfies three conditions, two boundary conditions and one initial condition. Evidently, if the similarity transformation (3–82) is to be successful, two of these original conditions must reduce to a single condition on F as a function of η. But, in the present case, this clearly happens; the boundary condition on u as $y \to \infty$ and the initial condition for $t = 0$ are both satisfied with the single condition on F,

$$F(\eta) \to 0 \quad \text{as} \quad \eta \to \infty. \tag{3-86}$$

The boundary condition on u at $y = 0$ requires

$$F(0) = 1. \tag{3-87}$$

Thus, the Rayleigh problem is reduced to the solution of the ordinary differential equation (3–85) subject to the two boundary conditions (3–86) and (3–87). When a similarity transformation works, this reduction from a partial differential equation to an ordinary differential equation is the typical outcome. Although this is a definite simplification in the present problem, the original partial differential equation was already linear, and the existence of a similarity transformation is not essential to its solution. When similarity transformations exist for more complicated, nonlinear partial differential equations, however, the reduction to an ordinary differential equation is often a critical simplification in the solution process.

Given the equation and boundary conditions (3–85), (3–86), and (3–87), the solution of the Rayleigh problem is very simple. Integrating (3–85) twice with respect to η, we obtain

$$F(\eta) = A \int_0^{\eta} e^{-t^2/4} dt + B, \qquad (3\text{–}88)$$

where t is a dummy variable of integration and A and B are arbitrary constants. Applying the boundary condition (3–87), we see

$$B = 1. \qquad (3\text{–}89)$$

In order to apply (3–86), we require the asymptotic value of the integral in (3–88) for $\eta \to \infty$. Instead of evaluating this directly, we first note that the integral is very nearly the error function

$$\text{erf}(z) \equiv \frac{2}{\sqrt{\pi}} \int_0^z e^{-y^2} dy. \qquad (3\text{–}90)$$

Using (3–90) and incorporating (3–89), we can express $F(\eta)$ in the form

$$F(\eta) = A\sqrt{\pi}\ \text{erf}\left(\frac{\eta}{2}\right) + 1. \qquad (3\text{–}91)$$

Now it is known* that

$$\text{erf}(z) \to 1 \quad \text{as} \quad z \to \infty.$$

Thus, the boundary condition (3–86) requires

$$A = -\frac{1}{\sqrt{\pi}}, \qquad (3\text{–}92)$$

and the solution for u can be expressed in the form

$$\boxed{\frac{u}{U} = 1 - \text{erf}\left(\frac{\eta}{2}\right),} \qquad (3\text{–}93)$$

where η is defined in (3–82). This velocity profile is plotted in Figure 3–7 as a function of η and as a function of y for several values of t. This emphasizes the nature of similarity solutions. We have argued that the lack of a physically meaningful characteristic length scale means that the solution of a problem should be expressible in terms of some dimensionless combination of dimensional, independent variables, such as $y/\sqrt{\nu t}$ in the present problem. However, this implies that the profiles for different values of the independent variables should collapse to a single, universal form when plotted as a function of the similarity variable rather than as a function of the individual independent variables. In the present case, $u = u(y,t)$ collapses to a single profile for all t when

*Cf. Abramowitz and Stegun, *Handbook of Functions*.

plotted as $u = u(y/\sqrt{\nu t})$. This internal *similarity* of the profiles for various t is responsible for the name *self-similar,* which is often used to describe solutions of this type.

Our discussion of the Rayleigh problem is essentially complete. Nevertheless, it is useful to briefly reconsider the solution methodology because we have used our intuitive, physical understanding of the problem to simply write down the solution form (3–81) and (3–82) that defines the similarity transformation, and it will not always be possible to do this. Indeed, despite of the general idea that the lack of meaningful characteristic length scales will generally imply the existence of a self-similar solution form, it may still not be clear in any particular case whether a self-similar solution actually exists. For example, it may be possible that a characteristic length scale exists, but we have not identified it. In the Rayleigh problem, we have argued that there is no characteristic length scale, but there is a combination of dimensional parameters with units of length, ν/U. Although there is no a priori reason to expect that ν/U is a length scale characteristic of changes of $0(U)$ in the velocity, it would be possible, in principle, that this is the case, and we would not then expect a similarity solution. Fortunately, a straightforward, systematic scheme can be developed from which one can determine both whether a similarity solution exists and, if so, the details of the similarity transformation. The starting point for this scheme is the assumption of a very general self-similar solution form,

$$\frac{u}{U} = F(\eta) \quad \text{with} \quad \eta = at^n y^m. \tag{3–94}$$

Of course, if a similarity solution exists, the variable η should be dimensionless, but beyond that general information we would not usually know the values of the constants a, n, and m, and these must be determined as part of the solution scheme. Necessary conditions for the existence of a self-similar solution are (1) the governing partial differential equation must reduce to an ordinary differential equation for F as a function of η alone, and (2) the original boundary and initial conditions must reduce to a number of equivalent conditions for F that are consistent with the order of the ordinary differential equation. Of course, a proof of *sufficient* conditions for existence of a self-similar solution would require a proof of existence of a solution to the ordinary differential equation and boundary conditions that are derived for F. In general, however, the problems of interest will be nonlinear, and we shall be content to derive a self-consistent set of equations and boundary conditions, and attempt to solve this latter problem numerically rather than seeking a rigorous existence proof. Let us see how the systematic solution scheme based on the general form (3–94) works for the Rayleigh problem.

The first step is always to ask whether the proposed solution form is compatible with the boundary conditions. This step is important because there is no point in spending the time and effort to try to determine values for n, m, and a that are compatible with the differential equation if the solution form could not possibly satisfy the boundary conditions in any case. It is evident that the reduction from a partial to an ordinary differential equation via the similarity transformation will decrease the number of boundary and initial conditions that can be satisfied by the function $F(\eta)$. Thus, two of the original conditions must be expressible in terms of a single condition on F.

Examining (3–79), it is obvious that the only possibility is that the conditions for $y \to \infty$ and $t = 0$ collapse into a single condition for F. This means that η must have the same value for $y \to \infty$ as it does for $t = 0$. Evidently, this should be possible with the general form (3–94) provided that n and m are opposite in sign. We do not have specific values for these coefficients yet [these will come from (3–78)], and so we do not know that the boundary conditions for $y \to \infty$ and $t = 0$ will actually reduce to one condition on F. All that we know is that the form (3–94) does not preclude this happening.

The second step is to ascertain whether the form (3–94) is compatible with the differential equation (3–78), and then determine specific values for a, m, and n and verify that m and n are of opposite sign. To do this, we differentiate (3–94) and substitute into (3–78). In this case

$$\frac{\partial u}{\partial t} = U \frac{dF}{d\eta} (na y^m t^{n-1}),$$

$$\frac{\partial^2 u}{\partial y^2} = U \left[\frac{d^2 F}{d\eta^2} (m^2 a^2 y^{2m-2} t^{2n}) + \frac{dF}{d\eta} am(m-1) y^{m-2} t^n \right],$$

and (3–78) can be expressed in the form

$$\frac{d^2 F}{d\eta^2} = \left(\frac{n}{m^2 va} y^{-m+2} t^{-n-1} \right) \frac{dF}{d\eta} - \left(\frac{m-1}{ma} y^{-m} t^{-n} \right) \frac{dF}{d\eta}. \tag{3–95}$$

If a solution of the similarity form actually exists, it is evident that the governing differential equation for u must yield an ordinary differential equation for F as a dimensionless function of η only. This means that the coefficients in (3–95) must be either dimensionless constants (independent of η, y, and t) or dimensionless functions of η alonê. But we can see that this will not be true for arbitrary values of a, n, and m. The most general case where the coefficients will be constant or functions of η alone is

$$m = -2n; \quad m > 0,$$

for which (3–95) takes the form

$$\frac{d^2 F}{d\eta^2} = \frac{1}{4nva} \left(\frac{a}{\eta} \right)^{\frac{n+1}{n}} \frac{dF}{d\eta} - \left(\frac{2n+1}{2n} \right) \frac{1}{\eta} \frac{dF}{d\eta}, \tag{3–96}$$

and

$$\eta = a \left(\frac{y^2}{t} \right)^{-n}.$$

In order to obtain n, we must determine which value yields a solution of (3–96) that satisfies the boundary conditions on F. The only value where this is true is $n = -1/2$. But the only method of solving the ordinary differential equation that results from a general similarity transformation will usually be numerical (the equation will usually be nonlinear). In this case, trying to establish the correct value of n will be like looking for a needle in a haystack. Fortunately, a much simpler ad hoc approach has been found that works in all cases in which it has been tried. This is to return to (3–95) and set the coefficient of one of the lower derivative terms identically equal to zero (after first arranging the equation so that the coefficient of the highest derivative term is unity). In the present case, this means either

$$n = 0,$$

or

$$m = 1.$$

However, if n were to be zero, there would be no similarity solution possible because then $\eta = \eta(y)$ only. Thus, we suppose $m = 1$, in which case

$$\frac{d^2F}{d\eta^2} = \frac{n}{va} yt^{-n-1} \frac{dF}{d\eta} \qquad (3\text{--}97)$$

and

$$\eta = ayt^n. \qquad (3\text{--}98)$$

Substituting for y in (3–97) using (3–98), we obtain

$$\frac{d^2F}{d\eta^2} = \frac{n}{va^2} \left(\frac{1}{t^{2n+1}} \right) \eta \frac{dF}{d\eta}. \qquad (3\text{--}99)$$

If there is to be a similarity solution, the coefficient in (3–99) can depend on η only, thus

$$2n + 1 = 0 \quad \text{or} \quad n = -\frac{1}{2}. \qquad (3\text{--}100)$$

Since $F(\eta)$ is dimensionless, we also see from (3–99) that

$$a^2v = \text{constant}.$$

The constant is arbitrary, since any choice will yield the same solution. We choose, for convenience,

$$a^2v = 1 \quad \text{or} \quad a = \frac{1}{\sqrt{v}}. \qquad (3\text{--}101)$$

With this choice, (3–97) becomes

$$\frac{d^2F}{d\eta^2} + \frac{1}{2}\eta \frac{dF}{d\eta} = 0, \qquad (3\text{--}102)$$

with

$$\eta = \frac{y}{\sqrt{vt}}; \quad \left(m = 1, \quad n = -\frac{1}{2}, \quad a = \frac{1}{\sqrt{v}} \right). \qquad (3\text{--}103)$$

Not surprisingly, this is the same similarity form of the problem that we obtained earlier, equation (3–85), by the physical diffusion argument. Here, apart from the general initial form (3–94), we did not require any real understanding of the physics of the problem in order to generate the same similarity transformation. The reader should note that m and n do indeed turn out to be opposite in sign, thus satisfying the condition on m and n that arose from our earlier consideration of boundary conditions. As already reported, the boundary conditions on F are (3–86) and (3–87).

E. Start-up of Simple Shear Flow

In the preceding section, we discussed the classical Rayleigh problem of start-up of a unidirectional flow in the unbounded half plane, $y > 0$, due to the sudden imposition of motion of an infinite plane wall at $y = 0$. Of course, no real fluid is truly unbounded in extent. Here, we consider the effect of a second stationary plane wall that is parallel to the first one and located at a distance $y = d$ away. It is evident from Section B that the solution at a sufficiently long time following imposition of the boundary motion will be simple shear flow, $u = U(1 - y/d)$. Thus, the solution for large times cannot be self-similar. Of course, this should not be a surprise in view of the discussion in the preceding section, since a definite characteristic length scale exists, namely, $l_c = d$. On the other hand, for sufficiently short times, $t \ll d^2/v$, when the effect of the boundary motion has only propagated a short distance relative to the gap width, d, the flow will not be affected significantly by the second wall at $y = d$, and in this case we should expect that the self-similar Rayleigh solution will give a good approximation to the actual velocity profile. In this section, we shall first derive an exact solution, via separation of variables, for the start-up flow between two parallel plane walls due to the sudden motion of one of them in its own plane with a constant velocity U. Following this, we shall explore the form of the solution for $t \ll d^2/v$ to show that it does indeed reduce to a self-similar form for sufficiently short times. Lastly, we shall see how this same conclusion can be reached (more simply) by considering the governing equations and boundary conditions rescaled for small times, $t = 0(\epsilon)$, where ϵ is a small parameter.

The problem is identical to that solved in the preceding section, (3–78) and (3–79), with the exception that the boundary condition as $y \to \infty$ is replaced by the boundary condition

$$u(d,t) = 0. \tag{3–104}$$

However, in this case, it is straightforward to nondimensionalize. The characteristic velocity scale is $u_c = U$, and an appropriate characteristic length scale is $l_c = d$. Since the velocity field is established via diffusion of momentum, the characteristic time scale is $t_c = d^2/v$. Using these characteristic quantities, the problem in dimensionless form is

$$\frac{\partial \bar{u}}{\partial \bar{t}} = \frac{\partial^2 \bar{u}}{\partial \bar{y}^2}, \tag{3–105}$$

with $\bar{u}(y,0) = 0$, all y,

$\bar{u}(0,\bar{t}) = 1$, all $\bar{t} \geq 0$, $\qquad$ (3–106)

and $\bar{u}(1,\bar{t}) = 0$, all $\bar{t} \geq 0$.

Since characteristic velocity, length, and time scales all exist, we do not expect a solution of self-similar form, as in the previous section, but instead seek a solution using separation of variables.

At steady state, we have already noted that the solution of (3–105) and (3–106) is just

$$\bar{u} = 1 - \bar{y} \quad (\bar{t} \to \infty),$$

and it is convenient, as is usual in start-up problems, to solve for the difference between this steady-state solution and the actual velocity field at any instant

$$u' = \bar{u} - (1 - \bar{y}). \tag{3-107}$$

Expressing (3–105) and (3–106) in terms of u', we have

$$\frac{\partial u'}{\partial \bar{t}} = \frac{\partial^2 u'}{\partial \bar{y}^2}, \tag{3-108}$$

with $u'(y,0) = -(1 - \bar{y})$,

$$u'(0,\bar{t}) = 0, \quad \text{all} \quad \bar{t} \geq 0, \tag{3-109}$$

and $u'(1,\bar{t}) = 0$, all $\bar{t} \geq 0$.

Now, we seek a solution for (3–108) and (3–109) in the form

$$u'(\bar{y},\bar{t}) = F(\bar{y})H(\bar{t}). \tag{3-110}$$

Substituting into (3–108), and solving the resulting ordinary differential equations for H and F, we find that the general solution which vanishes as $\bar{t} \to \infty$ is

$$u' = e^{-\lambda \bar{t}}(c_1 \sin \sqrt{\lambda}\bar{y} + c_2 \cos \sqrt{\lambda}\bar{y}). \tag{3-111}$$

Here, λ is the eigenvalue, which remains to be determined, and c_1 and c_2 are arbitrary constants. It follows from the boundary condition at $\bar{y} = 0$ that $c_2 = 0$, while the condition at $\bar{y} = 1$ yields the set of eigenvalues

$$\lambda = n^2 \pi^2 \quad \text{for} \quad n = 1,2, \ldots . \tag{3-112}$$

The corresponding eigensolution is then

$$u_n = e^{-n^2 \pi^2 \bar{t}} \sin n\pi \bar{y}, \tag{3-113}$$

and the most general solution of (3–108) that satisfies the boundary conditions at $\bar{y} = 0$ and $\bar{y} = 1$ and vanishes at $\bar{t} = \infty$ is

$$u' = \sum_{n=1}^{\infty} A_n e^{-n^2 \pi^2 \bar{t}} \sin n\pi \bar{y}. \tag{3-114}$$

All that remains is to determine the coefficients A_n so that the initial condition is satisfied:

$$-(1 - \bar{y}) = \sum_{n=1}^{\infty} A_n \sin n\pi \bar{y}. \tag{3-115}$$

Multiplying both sides of (3–115) by $\sin m\pi y$ and integrating over $\bar{y}$ from 0 to 1, we obtain

$$A_m = \frac{\int_0^1 (\bar{y}-1)\sin m\pi\bar{y}}{\int_0^1 \sin^2 m\pi\bar{y}d\bar{y}} \tag{3-116}$$

or

$$A_m = -\frac{2}{m\pi}. \tag{3-117}$$

Thus,

$$\boxed{\bar{u} = (1-\bar{y}) - \sum_{n=1}^{\infty} \frac{2}{n\pi} e^{-n^2\pi^2\bar{t}}\sin n\pi\bar{y}.} \tag{3-118}$$

This solution can also be expressed in terms of dimensional variables:

$$\boxed{\frac{u}{U} = \left(1 - \frac{y}{d}\right) - \sum_{n=1}^{\infty} \frac{2}{n\pi} e^{-n^2\pi^2\nu t/d^2}\sin n\pi\frac{y}{d}.} \tag{3-119}$$

The evolution of $\bar{u}$ with respect to $\bar{t}$ is plotted in Figure 3–8. As in previous examples of start-up flows, we see that the momentum from the lower wall propagates across the gap via diffusion in a dimensionless time interval $\bar{t} = 0(1)$; or in dimensional terms $t = 0(d^2/\nu)$.

It is evident, in examining (3–118) or (3–119), that the present exact solution is not generally self-similar in form. Nevertheless, we expect that the solution should reduce to the self-similar form for $t \ll 0(d^2/\nu)$ [or $\bar{t} \ll 0(1)$] since the upper wall at $y = d$ would not yet have any significant impact on the flow. However, it is by no means obvious in examining (3–118) or (3–119) that this will be the case.

Let us then consider the asymptotic form of (3–118) for very short times, $\bar{t} \ll 1$. A convenient way to do this is to introduce an arbitrary small parameter, ϵ, and to assume that $\bar{t} = 0(\epsilon)$ with $\epsilon \ll 1$ in the domain of interest. Thence, introducing the change of variable

$$\bar{t} = \epsilon\hat{t}, \tag{3-120}$$

we convert from a limit process involving the independent variable $\bar{t}$ directly to a formal process in which an arbitrarily small time interval in $\bar{t}$ is achieved by taking the asymptotic limit $\epsilon \to 0$ with the rescaled time $\hat{t} = 0(1)$. The main reason for introducing this formal limiting process is that one cannot consider the limit of small $\bar{t}$ in (3–118) without simultaneously considering the limit of small $\bar{y}$. This is a consequence of the fact that momentum diffusion only generates velocities that are significantly different from zero over a dimensionless distance $l/d \sim \sqrt{\bar{t}}$ starting from a stationary fluid at $\bar{t} = 0$. It follows that if we are to establish the asymptotic form of (3–118) for small times by taking the asymptotic limit $\epsilon \to 0$, we must simultaneously apply the limit $\bar{y} = 0(\epsilon^{1/2}) \to 0$ as $\epsilon \to 0$. This is most conveniently accomplished by introducing a rescaling of $\bar{y}$,

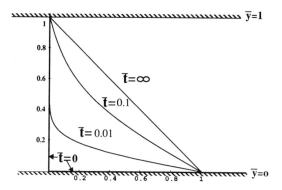

Figure 3–8 The velocity profiles at different times t for start-up of simple shear flow between two infinite, parallel plane walls.

$$\bar{y} = \epsilon^{1/2}\hat{y}, \tag{3-121}$$

and then taking the limit $\epsilon \to 0$ with both $\hat{t}$ and $\hat{y}$ fixed and $0(1)$.

Now let us consider the limiting form of (3–118) for small $\bar{t}$ by introducing (3–120) and (3–121) and taking the limit $\epsilon \to 0$. Upon substitution of (3–120) and (3–121) into (3–118), we obtain

$$\bar{u} = 1 - \epsilon^{1/2}\hat{y} - \sum_{n=1}^{\infty} \frac{2}{n\pi} e^{-n^2\pi^2\epsilon\hat{t}} \sin n\pi\epsilon^{1/2}\hat{y}. \tag{3-122}$$

Now, in the limit $\epsilon \to 0$, the term $\epsilon^{1/2}\hat{y}$ can be neglected compared to 1, and the infinite sum appears more and more as a continuous variation that can be approximated as an integral. To see this, define

$$\Delta S = \pi\sqrt{\epsilon},$$

rewrite the sum in (3–122) as

$$\sum_{n=1}^{\infty} \frac{2}{n\pi} \frac{1}{\Delta S} e^{-(n\Delta S)^2\hat{t}} \sin(n\hat{y}\Delta S)\Delta S, \tag{3-123}$$

and consider the limit as ϵ, and therefore ΔS, goes to zero. In this case,

$$\lim_{\Delta S \to 0} \sum_{n=1}^{\infty} \frac{2}{n\pi} \frac{1}{\Delta S} e^{-(n\Delta S)^2\hat{t}} \sin(n\hat{y}\Delta S)\Delta S = \frac{2}{\pi} \int_0^{\infty} \frac{1}{S} e^{-S^2\hat{t}} \sin(S\hat{y})dS. \tag{3-124}$$

But this integral is just

$$\frac{2}{\pi} \int_0^{\infty} \frac{1}{S} e^{-S^2\hat{t}} \sin(S\hat{y})dS = \mathrm{erf}\left(\frac{\hat{y}}{2\sqrt{\hat{t}}}\right). \tag{3-125}$$

Hence, we conclude that for small $\bar{t}$

$$\bar{u} \sim 1 - \text{erf}\left(\frac{\bar{y}}{2\sqrt{\bar{t}}}\right). \tag{3-126}$$

If $\bar{u}$, $\bar{y}$, and $\bar{t}$ are expressed in terms of dimensional variables, this expression becomes

$$\frac{u}{U} = 1 - \text{erf}\left[\frac{y/d}{2\sqrt{t\dfrac{v}{d^2}}}\right] = 1 - \text{erf}\left(\frac{y}{2\sqrt{vt}}\right). \tag{3-127}$$

This is precisely the self-similar solution of the Rayleigh problem (3–93), that was obtained in the previous section. Notice that the length scale d drops completely out of this limiting form of the solution (3–118). This is consistent with our earlier observation that the presence of the upper boundary should have no influence on the velocity field for sufficiently small times $\bar{t} \ll 1$ (or, equivalently, $t \ll d^2/v$).

Thus, we see, as expected, that the exact solution (3–118) reduces to a self-similar form for $t \ll 0(d^2/v)$. However, the analysis to demonstrate this fact was somewhat complicated, and we may ask whether a simpler approach might not be possible by direct examination of the governing equation and boundary/initial conditions, (3–105) and (3–106). To determine the asymptotic form for these equations for small $\bar{t}$, we again introduce the small parameter, ϵ, and rescale $\bar{y}$ and $\bar{t}$ in (3–105) and (3–106) according to (3–120) and (3–121). The result is

$$\frac{\partial \bar{u}}{\partial \hat{t}} = \frac{\partial^2 \bar{u}}{\partial \hat{y}^2}, \tag{3-128}$$

with

$$\bar{u}(\hat{y}, 0) = 0, \quad \text{all} \quad \hat{y} = 0(1),$$
$$\bar{u}(0, \hat{t}) = 1, \quad \text{all} \quad \hat{t} = 0(1), \tag{3-129}$$

and

$$\bar{u}(\epsilon^{-1/2}, \hat{t}) = 0 \,\text{all} \quad \hat{t} = 0(1).$$

Thus, in the limit $\epsilon \to 0$, the present problem reduces to the Rayleigh problem, with governing equation and boundary/initial conditions given by (3–78) and (3–79).

We have seen that the solution for start-up of simple shear flow between two parallel flat plates can be obtained as an eigenfunction expansion using the method of separation of variables. The solution is *not* self-similar in form, and this is consistent with the existence of a characteristic length scale d. On the other hand, when we consider the same problem with the upper stationary plate removed (that is, the Rayleigh problem), the solution *is* self-similar, and we have argued that this is to be expected in the absence of a characteristic length scale. Of course, no fluid domain is actually unbounded, and it is important to recognize the role of self-similar solutions in the context of real problems. In particular, we now see that the self-similar Rayleigh solution is the limiting, asymptotic form for small times compared to the time scale for diffusion of momentum across the gap separating the two plates, that is, $t \ll d^2/\nu$. It is important to note that it is much easier to obtain the self-similar solution directly—if one is really interested only in the small time domain—than it is to first derive the exact solution (3–118) or (3–119), and then obtain the self-similar form as a limit. The role of self-similar solutions in general is illustrated by the role of the self-similar solution here; since unbounded domains without characteristic length scales do not occur in reality, we must always expect that self-similar solutions will exist only as an approximation when the flow is sufficiently localized that any characteristic length scales are irrelevant to the problem.

F. Pulsatile Flow in a Circular Tube

In the preceding sections of this chapter, we have considered several examples of transient unidirectional flows. In each case, it was assumed that the flow started from rest with the abrupt imposition of either a finite pressure gradient or a finite boundary velocity, and we saw that the flow evolved toward steady state via diffusion of momentum with a time scale $t_c = l_c^2/\nu$. Here we consider a final example of a transient unidirectional flow problem in which time-dependent motion is produced in a circular tube by the sudden imposition of a periodic, time-dependent pressure gradient,

$$-\frac{\partial P}{\partial z} = G = G_0(1 + \epsilon \sin \omega t). \tag{3–130}$$

This problem of pulsatile flow in a circular tube has been studied extensively in the context of model studies of blood flow in the arteries,[10] though it is considerably simpler than the real problem where the vessel cross section is not circular, the walls are compliant, and the vessels quite short between branch or bifurcation points.[11] In the context of blood flow, and most other applications, we are usually not concerned with the start-up of pulsatile flow from rest, but only with the solution for sufficiently long times after imposition of the pulsatile pressure gradient that the velocity field is strictly periodic. We shall follow this lead in our present discussion and concentrate primarily on the asymptotic ($t \gg 1$) state where the velocity is periodic in time, though we shall begin by considering how the full transient problem can be solved. If a more complicated periodic pressure gradient is required, the solution for the velocity field can be achieved by Fourier decomposition, with the sinusoidal form, (3–130), being treated as a typical Fourier component.

In spite of its historical development in biomechanics, however, our main motivation for studying the problem of pulsatile flow in a circular tube is not in this application. Rather, our general goal is to expose the qualitative influence of the fluid's inertia (that is, of acceleration-deceleration effects) when there is an externally imposed time scale $t_c = 1/\omega$ that is characteristic of changes in the flow. In addition, we shall see in Chapter 6 that the problem provides a very convenient framework for illustrating the concepts of asymptotic analysis as a means of obtaining approximate solutions for large and small frequencies, $\omega \ll 1$ and $\omega \gg 1$.

We will begin in this section by deriving the exact solution. Later, in Chapter 6, the problem will be reexamined using asymptotic methods of analysis. The governing differential equation is just (3–12), with the pressure gradient function given by (3–130):

$$\rho \frac{\partial u_z}{\partial t} = G_0(1 + \epsilon \sin \omega t) + \mu \left[\frac{1}{r} \frac{\partial}{\partial r} \left(r \frac{\partial u_z}{\partial r} \right) \right]. \tag{3-131}$$

The boundary conditions are

$$u_z = 0 \quad \text{at} \quad r = R,$$

$$u_z \text{ bounded} \quad \text{at} \quad r = 0, \tag{1-132}$$

for all t. In addition, we assume that the fluid begins at rest for $t = 0$, that is,

$$u_z = 0 \quad \text{at} \quad t = 0, \quad \text{all} \quad r(0 \le r \le R).$$

The first step, as usual, is to introduce characteristic scales and nondimensionalize the problem. In this case, appropriate characteristic scales are

$$u_c = \frac{G_0 R^2}{\mu}, \quad l_c = R, \quad \text{and} \quad t_c = \frac{R^2}{\nu}. \tag{3-133}$$

The first two are obvious. The time scale R^2/ν is the time for diffusion of momentum across the tube and is certainly relevant to the initial start-up portion of the problem. We recognize the existence of a second time scale in this problem, however, and that is the period, $1/\omega$, of the imposed pressure gradient. We may anticipate that this second time scale will be especially relevant for larger times after the initial transients have died out. For the moment, we retain t_c as defined in (3–133). Then, introducing dimensionless variables

$$\bar{u}_z = \frac{u_z}{u_c}, \quad \bar{r} = \frac{r}{l_c}, \quad \text{and} \quad \hat{t} = \frac{t}{t_c}$$

into the governing equations and boundary conditions, we obtain

$$\frac{\partial \bar{u}_z}{\partial \hat{t}} = 1 + \epsilon \sin \left(\frac{\omega R^2}{\nu} \hat{t} \right) + \frac{1}{\bar{r}} \frac{\partial}{\partial \bar{r}} \left(\bar{r} \frac{\partial \bar{u}_z}{\partial \bar{r}} \right), \tag{3-134}$$

with $\bar{u}_z = 0$ at $\bar{r} = 1$,

$\bar{u}_z$ bounded at $\bar{r} = 0$, \hfill (3-135)

and $\bar{u}_z = 0$ for $\hat{t} = 0$.

It will be noted that a single dimensionless combination of parameters remains in the problem after nondimensionalization, namely,

$$R_\omega \equiv \frac{\omega R^2}{\nu}. \tag{3-136}$$

In the previous transient unidirectional flow problems that we have considered, the dimensionless governing equations and boundary/initial conditions were completely free of all dimensional parameters. In these problems, however, the only relevant time scale was the diffusion time, l_c^2/ν. Here, in contrast, the flow is characterized by a second imposed time scale due to the oscillatory pressure gradient, and the form of the resulting flow is predicted by (3–134) to depend upon the ratio of the diffusion time R^2/ν to the imposed time scale, $1/\omega$. The dimensionless ratio of time scales, R_ω, can also be considered to be a dimensionless frequency for the flow, and in that context is sometimes called the Strouhal number.

Before saying more about the physical significance of R_ω, it is convenient to proceed a few steps further with the solution of (3–134) and (3–135). Specifically, since the governing equation and boundary conditions are linear, it is clear that the solution can be expressed as the sum of two parts, one representing the response to the steady part of the pressure gradient and the other the response to the transient, oscillatory part. We thus denote $\bar{u}_z$ as

$$\bar{u}_z = u_z^{(0)} + \epsilon u_z^{(1)}, \tag{3–137}$$

where

$$\frac{\partial u_z^{(0)}}{\partial \hat{t}} = 1 + \frac{1}{\bar{r}}\frac{\partial}{\partial \bar{r}}\left(\bar{r}\frac{\partial u_z^{(0)}}{\partial \bar{r}}\right), \tag{3–138}$$

with

$$u_z^{(0)} = 0 \quad \text{at} \quad \bar{r} = 1,$$
$$u_z^{(0)} \text{ bounded} \quad \text{at} \quad \bar{r} = 0, \tag{3–139}$$
$$u_z^{(0)} = 0 \quad \text{at} \quad \hat{t} = 0,$$

and

$$\frac{\partial u_z^{(1)}}{\partial \hat{t}} = \sin\left(\frac{\omega R^2}{\nu}\right)\hat{t} + \frac{1}{\bar{r}}\frac{\partial}{\partial \bar{r}}\left(\bar{r}\frac{\partial u_z^{(1)}}{\partial \bar{r}}\right), \tag{3–140}$$

with

$$u_z^{(1)} = 0 \quad \text{at} \quad \bar{r} = 1,$$
$$u_z^{(1)} \text{ bounded} \quad \text{at} \quad \bar{r} = 0, \tag{3–141}$$
$$u_z^{(1)} = 0 \quad \text{at} \quad \hat{t} = 0.$$

Now the problem defined by (3–138) and (3–139) is just the transient start-up flow due to a suddenly imposed constant pressure gradient that was already solved in Section C of this chapter. The final steady-state solution is the steady Poiseuille flow profile

$$u_z^{(0)} = \frac{1}{4}(1 - \bar{r}^2).$$

Here we focus our attention on the transient pressure gradient problem defined by (3–140) and (3–141). We have seen in previous transient start-up problems that it is usually advantageous to solve for the difference between the actual instantaneous velocity distribution and the final steady-state profile, since this transforms the problem from a nonhomogeneous boundary value problem to a homogeneous initial value problem. In the present case, of course, the solution for large $\hat{t}$ will not be steady but will instead be a strictly periodic transient motion. Nevertheless, we propose adopting the same approach to solving (3–140) and (3–141) that we have used earlier: First, we will obtain the asymptotic solution for $\hat{t} \gg 1$ when all initial transients have died out and the solution has become independent of its initial state, and only then show that the complete initial value problem can be solved conveniently by looking for the difference between this periodic, asymptotic solution and the actual instantaneous solution $u_z^{(1)}$.

A key to solving (3–140) and (3–141), either exactly or for the large $\hat{t}$ asymptote, is to note that the velocity field will not generally be in-phase with the oscillating pressure gradient. We can see that this is true by a qualitative examination of the governing equation, but it is convenient for this purpose to temporarily rescale time according to

$$\bar{t} = \left(\frac{\omega R^2}{\nu}\right)\hat{t} \tag{3-142}$$

so that

$$\left(\frac{\omega R^2}{\nu}\right)\frac{\partial u_z^{(1)}}{\partial \bar{t}} = \sin\bar{t} + \frac{1}{\bar{r}}\frac{\partial}{\partial \bar{r}}\left(\bar{r}\frac{\partial u_z^{(1)}}{\partial \bar{r}}\right). \tag{3-143}$$

In this form, the pressure gradient oscillates on a dimensionless time scale of $0(1)$, independent of ω, and we see that the magnitude of R_ω determines the importance of acceleration effects in the fluid relative to the viscous effects (diffusion of momentum) that are represented by the second term on the right-hand side of (3–143). In this role, it is evident that R_ω can be considered as a Reynolds number, based upon a characteristic "velocity" ωR. Before proceeding with our discussion, it should be emphasized that the change of variable (3–142) does not alter the problem at all. Indeed, it is precisely equivalent to nondimensionalizing time with respect to $t_c = 1/\omega$ instead of $t_c = R^2/\nu$, as in (3–140). The choice of one characteristic time scale instead of the other is a matter of convenience only, although we may anticipate that the choice $t_c = 1/\omega$ inherent in (3–143) will be more convenient for discussing the long-time asymptotic behavior of $u_z^{(1)}$. In any case, the solution still depends upon the ratio of characteristic time scales, R_ω.

Now, examining (3–143), the qualitative nature of the phase relationship between $u_z^{(1)}$ and the pressure gradient can be seen by considering the two limiting cases, $R_\omega \gg 1$ and $R_\omega \ll 1$. Let us suppose first that $R_\omega \ll 1$. In this case, once the initial transient development of the flow has been accomplished, it is apparent that the acceleration term in (3–143) can be neglected compared to the viscous term, and this means that for large times, $u_z^{(1)}$ must be *quasi-steady* and completely in-phase with the pressure gradient; that is, if we examine (3–143) with the acceleration terms neglected, it is evident that the solution will be in the form

$$u_z^{(1)} \sim \sin\bar{t}\,F(\bar{r}) \quad \text{for} \quad \bar{t} \gg 1 \quad \text{and} \quad R_\omega \ll 1. \tag{3-144}$$

If $R_\omega \gg 1$, on the other hand, the viscous term in (3–143) is negligible compared to the acceleration term, and

$$u_z^{(1)} \sim -\cos\bar{t}\,\frac{1}{R_\omega}H(\bar{r}) \quad \text{for} \quad \bar{t} \gg 1 \quad \text{and} \quad R_\omega \gg 1. \tag{3-145}$$

The restriction to $\bar{t} \gg 1$ is clearly necessary in this case since the asymptotic solution form (3–145) does not satisfy the initial condition at $\bar{t} = 0$. In (3–145), the velocity lags $90°$ behind the pressure gradient. In other words, the first peak in G appears at $\bar{t} = \pi/2$, while the first peak in $u_z^{(1)}$ does not appear until $\bar{t} = \pi$. It will, perhaps, have been noted that the magnitude of the velocity component $u_z^{(1)}$ appears in (3–145) to be only R_ω^{-1} for $R_\omega \gg 1$. This result has been deduced from the governing equation (3–143) by the following reasoning. First, the time-dependent velocity field, $u_z^{(1)}$, exists only because of the time-dependent pressure gradient $\sin \bar{t}$. According to (3–143), this pressure gradient must be balanced either by a corresponding acceleration of the fluid or by viscous forces or, of course, some combination. Thus, for any value of R_ω, at least one of the terms corresponding to acceleration or viscous effects must have the same magnitude as the pressure gradient term, which is $0(1)$. When $R\omega \ll 1$, it is evident from (3–143) that the balance is between the viscous term and the pressure gradient and that both are $0(1)$ with $u_z^{(1)} = 0(1)$. When $R_\omega \gg 1$, on the other hand, the balance must be between acceleration and the pressure gradient, but it appears as though the former is $0(R_\omega)$, while the latter is only $0(1)$. This obviously cannot be true. Rather, as R_ω increases, the magnitude of the dimensionless velocity component $u_z^{(1)}$ must decrease as R_ω^{-1} so that the three terms in (3–143) are $0(1)$, $0(1)$, or $0(R_\omega^{-1})$, respectively, as we go from left to right.

The actual solution of the problem (3–140) and (3–141)—or the equivalent form (3–143) using $\bar{t}$ rather than $\hat{t}$—is straightforward once it is realized that the velocity field will not generally be in-phase with the pressure gradient. A convenient approach that allows a completely general phase relationship between $u_z^{(1)}$ and the pressure gradient, $\sin(\omega R^2/\nu)\hat{t}$, is to solve for the complementary complex velocity field, $\hat{u}_z$, which satisfies

$$\frac{\partial \hat{u}_z}{\partial \hat{t}} = e^{i\,(\omega R^2/\nu)\bar{t}} + \frac{1}{\bar{r}}\frac{\partial}{\partial \bar{r}}\left(\bar{r}\frac{\partial \hat{u}_z}{\partial \bar{r}}\right), \tag{3–146}$$

with $\hat{u}_z = 0$ at $\bar{r} = 1$, $\qquad$ (3–147)

$\hat{u}_z$ bounded at $\bar{r} = 0$,

and $\hat{u}_z = 0$ at $\hat{t} = 0$.

Comparing (3–146) and (3–140), it is evident that*

$$u_z^{(1)} = -\mathcal{R}e\,(i\hat{u}_z)$$

will satisfy (3–140) and (3–141) if $\hat{u}_z$ is a solution of (3–146) and (3–147).

We solve (3–146) and (3–147) in two steps. First, we obtain a solution for the periodic velocity field that will exist for $(\omega R^2/\nu)\hat{t} \gg 1$. The complete solution, including initial transients is then obtained by solving for the difference between $\hat{u}_z$ and this large-time periodic solution. The advantage of solving for the complementary complex velocity field is that the form of the solution for large times, satisfying (3–146) and the first two of the boundary conditions (3–147), can be guessed a priori to be

*Note $\mathcal{R}e$ ( ) denotes the real part of ( ).

$$\hat{u}_z = e^{i\,(\omega R^2/\nu)\hat{t}} G(\bar{r}).$$ (3–148)

Substitution into (3–146) and the first two conditions (3–147) shows that this is clearly a solution, provided $G(\bar{r})$ satisfies the ordinary differential equation and boundary conditions

$$\frac{d^2 G}{d\bar{r}^2} + \frac{1}{\bar{r}} \frac{dG}{d\bar{r}} - iGR_\omega = -1,$$ (3–149)

$$G(\bar{r}) = 0 \quad \text{at} \quad \bar{r} = 1,$$

and $G(\bar{r})$ bounded at $\bar{r} = 0.$ (3–150)

Let us denote the periodic solution (3–148) as $\hat{u}_z^\infty$ to remind us that it is the asymptotic form for $\hat{t}$ large enough that initial transients have died out. To obtain the complete solution of the initial value problem (3–146) and (3–147), we then seek a solution for the difference field

$$\hat{u}_z' = \hat{u}_z - \hat{u}_z^\infty.$$ (3–151)

Substituting (3–151) into (3–146), we obtain the governing equation and boundary conditions for $\hat{u}_z'$

$$\frac{\partial \hat{u}_z'}{\partial \hat{t}} = \frac{1}{\bar{r}} \frac{\partial}{\partial \bar{r}} \left(\bar{r} \frac{\partial \hat{u}_z'}{\partial \bar{r}} \right),$$ (3–152)

$$\text{with } \hat{u}_z' = 0 \quad \text{at} \quad \bar{r} = 1,$$

$$\hat{u}_z' \text{ bounded at } \bar{r} = 0,$$ (3–153)

and $\hat{u}_z' = -G(\bar{r})$ for $\hat{t} = 0.$

As in the previous transient problems that we have discussed, this problem for the difference between the actual velocity and its large-time form, $\hat{u}_z^\infty$, is reduced to a homogeneous, initial value problem that can be solved by standard methods such as separation of variables. Our interest, in the present problem, is primarily with the large-time behavior of the solution. Hence, we leave the details of solving (3–152) and (3–153) as an exercise (nontrivial) for the reader and concentrate our attention in the remainder of this section on the solution of (3–149) and (3–150).

We begin with the particular solution of (3–149) which we see, by inspection, is just

$$G_p = -\frac{i}{R_\omega}.$$ (3–154)

The homogeneous solution of (3–149) is a Bessel function. To see this, we note that the general Bessel's equation is

$$z^2 \frac{d^2 w}{dz^2} + z \frac{dw}{dz} + (z^2 - \nu^2) w = 0,$$ (3–155)

which has Bessel functions of the first and second kind of order ν as its solutions. The homogeneous equation for G is easily transformed into the standard form (3–155) by the change of independent variable

$$\bar{r} = \sqrt{\frac{i}{R_\omega}} R,$$

which yields

$$\frac{d^2 G_h}{dR^2} + \frac{1}{R}\frac{dG_h}{dR} + G_h = 0. \tag{3-156}$$

Thus, the two independent, homogeneous solutions for G are Bessel functions J_0 and Y_0 of order zero and

$$G = G_p + G_h = -\frac{i}{R_\omega} + AJ_0\left[\left(\frac{R_\omega}{i}\right)^{1/2}\bar{r}\right] + BY_0\left[\left(\frac{R_\omega}{i}\right)^{1/2}\bar{r}\right], \tag{3-157}$$

where A and B are arbitrary constants.

We have noted previously that $Y_0(z) \to -\infty$ as $z \to 0$. Thus $B = 0$ if G is to be bounded. To obtain A, we apply the condition $G(\bar{r} = 1) = 0$. The resulting solution for $\hat{u}_z^\infty$, obtained by substituting the solution for G in Equation (3–148), is

$$\hat{u}_z^\infty = e^{i\bar{t}}\left[-\frac{i}{R_\omega}\right]\left[1 - \frac{J_0\left[\left(\frac{R_\omega}{i}\right)^{1/2}\bar{r}\right]}{J_0\left[\left(\frac{R_\omega}{i}\right)^{1/2}\right]}\right]. \tag{3-158}$$

The corresponding velocity field $u_z^{(1)\infty}$ is

$$u_z^{(1)\infty} = -\mathcal{R}e\left[\frac{e^{i\bar{t}}}{R_\omega}\left\{1 - \frac{J_0\left[\left(\frac{R_\omega}{i}\right)^{1/2}\bar{r}\right]}{J_0\left[\left(\frac{R_\omega}{i}\right)^{1/2}\right]}\right\}\right]. \tag{3-159}$$

We use the superscript ∞ to remind us that this is the long-time periodic form of $u_z^{(1)}$ that pertains once the initial start-up transients have vanished. The result (3–159) is the general, exact solution for $u_z^{(1)\infty}$, valid for any value of R_ω. Although it has been relatively easy to obtain, it is very complicated to actually evaluate for any arbitrary R_ω. We therefore restrict ourselves here to the case of small $R_\omega \ll 1$.

To evaluate $u_z^{(1)\infty}$ for small R_ω from (3–159), we use the small z expansion of $J_0(z)$:[12]

$$J_0(z) = 1 - \frac{1}{4}z^2 + \frac{1}{64}z^4 + \cdots.$$

Let us denote $x \equiv \sqrt{R_\omega}\,\bar{r}$ and $X \equiv \sqrt{R_\omega}$. Then

$$J_0\left(\left(\frac{1}{i}\right)^{1/2}x\right) \sim 1 + \frac{1}{4}ix^2 - \frac{1}{64}x^4 + \cdots,$$

and

$$J_0\left(\left(\frac{1}{i}\right)^{1/2}X\right) \sim 1 + i\frac{1}{4}X^2 - \frac{1}{64}X^4 + \cdots .$$

Thus,

$$u_z^{(1)\infty} \sim -\mathscr{Re}\left\{\frac{e^{i\bar{t}}}{R_\omega}\left[1 - \frac{1 + i\frac{1}{4}x^2 - \frac{1}{64}x^4 + \cdots}{1 + i\frac{1}{4}X^2 - \frac{1}{64}X^4 + \cdots}\right]\right\}, \qquad (3\text{-}160)$$

with an error of $0(R_\omega^3)$ in both the numerator and denominator. The quantity in square brackets can be reexpressed in the form

$$u_z^{(1)\infty} \sim -\mathscr{Re}\left\{\frac{e^{i\bar{t}}}{R_\omega}\left[\frac{i\frac{1}{4}(X^2 - x^2) - \frac{1}{64}(X^4 - x^4) + \cdots}{\left(1 - \frac{1}{64}X^4\right) + i\frac{1}{4}X^2 + \cdots}\right]\right\}. \qquad (3\text{-}161)$$

Then multiplying both numerator and denominator by the complex conjugate of the denominator and retaining all terms of $0(x^4, X^4)$ or larger, we obtain

$$u_z^{(1)\infty} \sim -\mathscr{Re}\left[\frac{e^{i\bar{t}}}{R_\omega}\left\{i\frac{1}{4}(X^2 - x^2) + \frac{1}{16}X^2(X^2 - x^2) - \frac{1}{64}(X^4 - x^4)\right\}\right]. \qquad (3\text{-}162)$$

Finally, since

$$e^{i\bar{t}} = \cos\bar{t} + i\sin\bar{t},$$

we find that

$$u_z^{(1)\infty} \sim -\frac{1}{R_\omega}\left[-\sin\bar{t}\left\{\frac{1}{4}(X^2 - x^2)\right\} + \cos\bar{t}\left\{\frac{1}{16}X^2(X^2 - x^2) - \frac{1}{64}(X^4 - x^4)\right\}\right] + 0(R_\omega^2).$$

$$(3\text{-}163)$$

Or, substituting for x and X,

$$\boxed{u_z^{(1)\infty} = \frac{1}{4}(1 - \bar{r}^2)\sin\bar{t} - \frac{1}{16}(R_\omega)\left\{\frac{3}{4} - r^2 + \frac{r^4}{4}\right\}\cos\bar{t} + 0(R_\omega^2).} \qquad (3\text{-}164)$$

The error estimate $0(R_\omega^2)$ in (3-163) and (3-164) represents the terms of $0(R_\omega^3)$ that were neglected in the expansions of J_0 in (3-160) and in the simplification of (3-161) to obtain (3-162).

If we now revert to the full dimensional form for the velocity field $\bar{u}_z^\infty$, we have

$$\bar{u}_z^\infty = u_z^{(0)\infty} + \epsilon u_z^{(1)\infty},$$

and thus

$$u_z^\infty = \frac{G_0 R^2}{4\mu} \left[(1 + \epsilon \sin \omega t)(1 - \bar{r}^2) - \epsilon \frac{R_\omega}{4} \left(\frac{3}{4} - \bar{r}^2 + \frac{\bar{r}^4}{4} \right) \cos \omega t + 0(R_\omega^2) \right]. \qquad (3\text{-}165)$$

Examining the solution (3-165), we see that the first term at 0(1) is just the quasi-steady Poiseuille solution, with is identical to the steady solution (3-39) except that the instantaneous pressure gradient

$$G = G_0(1 + \epsilon \sin \omega t)$$

appears in place of the steady pressure gradient. This term dominates as $R_\omega \rightarrow 0$, and we see that all effects of inertia (or fluid acceleration) are negligible at this level of approximation. The second term is $0(R_\omega)$ and contains the first influence of fluid inertia. It can be seen that inertia causes the velocity field to lag behind the changes in pressure gradient. In addition, the form of the velocity profile is no longer parabolic. It is not surprising that the first influence of inertia is to generate a (small) phase lag between the velocity and the pressure gradient. Indeed, we had anticipated this effect in the approximate limiting forms of the velocity for $R_\omega \ll 1$ and $R_\omega \gg 1$ that were given by (3-144) and (3-145). One way of "explaining" the lag between the velocity and pressure for nonzero values of R_ω is simply to note that a finite mass of fluid ($\rho \neq 0$ since $R_\omega \neq 0$) cannot be accelerated instantaneously by the action of a finite force. However, a clearer picture of the significance of large and small values of R_ω emerges if we note that R_ω is the ratio of the time scale for momentum diffusion across the tube, R^2/ν, to the imposed time scale for changes in the flow, $1/\omega$. We have seen in previous unidirectional flow examples that the velocity field always evolves toward a new steady state by diffusion of momentum with a characteristic time scale R^2/ν. Thus, if G is changed at a rate, corresponding to a time scale ω^{-1}, which is slow compared with R^2/ν, the velocity field is able to remain arbitrarily close to its steady-state form for each instantaneous value of G. On the other hand, as G changes increasingly rapidly compared with the rate at which the diffusion process takes place (that is, as R_ω increases), the changes in the velocity field will fall further and further behind the changes in the pressure gradient.

We have shown that the first two terms in an asymptotic approximation for $u_z^{(1)\infty}$ can be generated for small R_ω with a modest degree of effort. With sufficient algebra, higher-order terms (that is, terms proportional to R_ω^n with $n > 1$) could be generated in exactly the same way. A much more difficult task is to determine the asymptotic form of (3-159) for $R_\omega \gg 1$. Indeed, we shall not pursue that task in this section. Instead, we shall see in Chapter 6 that the asymptotic form for both $R_\omega \ll 1$ and $R_\omega \gg 1$ is more easily achieved by direct asymptotic approximation of the governing equation, (3-149) for G, rather than first obtaining an exact solution and then attempting to evaluate it for large and small values for R_ω.

Homework Problems

1. We consider an initially motionless incompressible Newtonian fluid between two infinite solid boundaries, one at $y=0$ and the other at $y=d$. Beginning at $t=0$, the lower boundary oscillates back and forth in its own plane with a velocity $u_x = U\sin\omega t$ $(t>0)$.

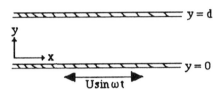

a. Identify characteristic velocity, length, and time scales, and nondimensionalize the governing differential equation and boundary conditions. You should note that there are two combinations of dimensional parameters that represent characteristic time scales, ω^{-1} and d^2/ν. What is the physical significance of each? The nature of the solution for the velocity field depends upon the magnitude of a single dimensionless parameter. What is it? What is its significance?

b. Assume that the boundary motion has been going on for a long period of time so that all initial transients have decayed away $(t \gg d^2/\nu)$ and the velocity field is strictly periodic. Solve for the velocity field in this case. Also, calculate the shear stress at the moving wall. Does fluid inertia increase or decrease the average shear stress?

c. Determine the limiting form of the solution for $d^2\omega/\nu \ll 1$ to determine the first effects of inertia. Explain the result in physical terms.

2. Consider an incompressible, Newtonian fluid that occupies the region above a single, infinite plane boundary. Beginning at $t=0$, this boundary oscillates back and forth in its own plane with a velocity $u_x = U\sin\omega t$ $(t>0)$.

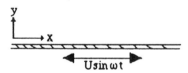

We wish to determine the velocity distribution in the fluid at large times, after any initial transients have decayed, so that the velocity field is strictly periodic. Normally we would proceed as in the related problem 1 by first nondimensionalizing and then solving the problem in dimensionless form. Here, however, there is not an immediate obvious characteristic length scale, so we follow a different procedure.

a. Solve for the velocity distribution using the full dimensional equations and boundary conditions. Does the solution exhibit a self-similar form? Should we have expected a self-similar solution? Based upon your solution, what is the characteristic length describing the distance the disturbance velocity penetrates into the fluid? Explain this penetration length scale in physical terms.

b. Based upon the characteristic penetration length scale identified in a, show that the governing equations and boundary conditions can be nondimensionalized to a form that contains no dimensionless parameters at all. This means that the *form* of the solution does not depend upon the frequency, ω, except as a scaling parameter for the independent variables. Does this make sense? Note that this feature in the present problem is fundamentally different from Problem 1, where the *form* of the solution was found to depend on the frequency of oscillation, ω. This difference may help you to understand the lack of dependence on ω in the present problem.

3. Consider the partial differential equation

$$\beta y \frac{\partial u}{\partial x} + \gamma y^2 \frac{\partial u}{\partial y} = \frac{\partial^2 u}{\partial y^2},$$

with boundary conditions

$$u = 1 \quad \text{at} \quad y = 0,$$

$$u \to 0 \quad \text{as} \quad y \to \infty,$$

$$\text{and } u = 0 \quad \text{at} \quad x = 0.$$

Show that a similarity solution exists of the form

$$u = F\left(\frac{y}{g(x)}\right),$$

and determine both the functions F and g.

4. Consider a semi-infinite body of fluid that is bounded below, at $y = 0$, by an infinite plane, solid boundary. For times $t < 0$, the fluid and the boundary are motionless. However, for $t \geq 0$, the boundary moves in its own plane under the action of a constant force per unit area, F. Assume that the boundary has zero mass so that its inertia can be neglected.

Determine the velocity distribution in the fluid by assuming a solution in the form

$$u = h(t)f(\eta),$$

where $\eta = y/g(t)$. Why does the simpler form $u = f(\eta)$ used in the Rayleigh problem not work in this closely related problem? If you had to program the motion of the boundary to produce the exact same solution, what would the velocity at the boundary need to be as a function of time? Does this present any difficulties?

5. Consider the motion of an incompressible, Newtonian fluid in the region between two parallel plane walls if the lower wall is subjected to a constant

force per unit area, F, in its own plane for all $t \geq 0$. The walls are separated by a distance d, there is no pressure gradient, and the fluid is initially at rest.

Determine the velocity distribution in the fluid. Show that your solution reduces for short times to form $u = h(t)f(\eta)$ obtained in Problem 4. What is the criteria for validity of this short-time approximation in this case?

6. Consider a circular cylinder (infinite length), radius R_1, immersed in a Newtonian fluid. The fluid may be treated as incompressible. The cylinder begins to rotate about its central axis at $t = 0$ with an angular velocity $\mathbf{\Omega}_1$ and continues with the same angular velocity for all $t \geq 0$. The fluid is initially motionless.

   a. Find the velocity distribution as a function of r and t (notice that $u_\theta \neq 0$, but $u_r = u_z = 0$—however, the flow field is *not* unidirectional). Is there a similarity solution? How would you calculate the pressure field? What happens in the limit as R_1 becomes arbitrarily small?

   b. Now, consider the same problem, but suppose that there is a second, outer cylinder, radius $R_2 > R_1$, that is held stationary for all $t \geq 0$. The outer cylinder is assumed to be *concentric* with the inner cylinder. The fluid is initially stationary.

   Determine the solution to the start-up problem in which the inner cylinder suddenly begins (at $t = 0$) to rotate with angular velocity $\mathbf{\Omega}_1$. How does this problem differ from a.? Do the two solutions agree for $t \ll (R_2 - R_1)^2/\nu$?

   c. At some time after the steady flow in part a has been established, the cylinder is instantaneously stopped. Calculate the torque required to keep the cylinder from rotating, as a function of time. Notice that the torque vanishes as $t \to \infty$.

References/Notes

1. Peyret, R., and Taylor, T.C., *Computational Methods for Fluid Flow.* Springer-Verlag (Springer Series in Computational Physics): Berlin (1983).

2. Happel, J., and Brenner, H., *Low Reynolds Number Hydrodynamics*, 2nd ed. Noordhoff International Publishing: Leyden (1973).

3. Poiseuille, J.L., "Recherches experimentales sur le mouvement des liquides dans les tubes de três petits diamétres," *Comptes Rendus* 11:961–967, 1041–1048 (1840); 12:112–115 (1841); *Mémoires des Seventes Étrangers* 9:433–543 (1846).

4. A large number of textbooks exist that are concerned with the stability of fluid motions. Among these, we recommend particularly

   a. Chandrasekhar, S., *Hydrodynamic and Hydromagnetic Stability.* Oxford Univ. Press: London (1961).

   b. Drazin, P.G., and Reid, W.H., *Hydrodynamic Stability.* Cambridge Univ. Press: Cambridge, UK (1981).

   c. Joseph, D.D., *Stability of Fluid Motions*, 2 vols. Springer-Verlag (Springer Tracts in Natural Philosophy 27 and 28): Berlin (1976).

   d. Swinney, H.L., and Gollub, J.P. (eds.), *Hydrodynamic Instabilities and the Transition to Turbulence.* Springer-Verlag (Topics in Applied Physics 45): Berlin (1981).

5. Although the following discussion is largely self-contained, we assume some prior experience in the solution of linear partial differential equations. For the reader who wishes to

review this material, there are, of course, many standard textbooks. Among these, we can recommend

a. Weinberger, H.F., *A First Course in Partial Differential Equations*. Blaisdell: New York (1965).

b. Boyce, W.E., and DiPrima, R.C., *Elementary Differential Equations and Boundary Value Problems*. John Wiley and Sons: New York (1977).

6. An excellent source of information about the special functions of mathematical physics is Abramowitz, M., and Stegun, I.A., *Handbook of Mathematical Functions*. Dover: New York (1965).

7. See Reference 5.

8. See chapters 10 and 11 of Reference 5a.

9. A general description of similarity transformations and related topics can be found in Ames, W.F., *Nonlinear Partial Differential Equations in Engineering,* chap. 4. Academic Press: New York (1965).

10. Womerslay, J.R., "Method for Calculation of Velocity, Rate of Flow, and Viscous Drag in Arteries When the Pressure Gradient is Known," *Journal of Physiology* 127:553–563 (1955).

11. Two review papers that can be used as a starting point to textbooks and research papers on blood flow and the circulation system are

a. Jones, R.T., "Blood Flow," *Ann. Rev. of Fluid Mech.* 1:223–244 (1969).

b. Skalak, R., Özkaya, N., and Skalak, T.C., "Biofluid Mechanics," *Annual Review of Fluid Mechanics* 21:167–204 (1989).

The interested reader may also wish to see

c. Pedley, T.J., *The Fluid Mechanics of Large Blood Vessels*. Cambridge Univ. Press: Cambridge, UK (1980).

12. See Reference 6.

CHAPTER **4**

Creeping Flows

The class of unidirectional flow problems considered in Chapter 3 was unusual in the sense that the nonlinear terms in the equations of motion were identically equal to zero so that exact solutions could be obtained using standard methods of analysis for linear partial differential equations. In virtually all circumstances besides unidirectional flows, the governing Navier-Stokes equations remain nonlinear and coupled, so that exact solutions cannot be found. The question then is whether methods exist to achieve approximate solutions for such problems. In fluid mechanics and in convective transport problems there are three possible approaches to obtaining approximate results from the nonlinear Navier-Stokes equations and boundary conditions.

First, we may discretize the differential equations and boundary conditions, using the formalism of either finite difference or finite element approximations, and thus convert them to a large but finite set of nonlinear algebraic equations that is suitable for attack via numerical (or computational) methods. It has become possible, especially with the advent of extremely large and fast computers, to obtain solutions of many flow and transport problems by these techniques. Indeed, the development of suitable numerical methods and their use in solving fluid mechanics problems has become sufficiently widespread and important that it is recognized as an independent subdiscipline known as *computational fluid dynamics* (CFD). In spite of this, we will not discuss numerical methods in this book. The interested reader will find many reviews of current techniques and results in such sources as *The Annual Review of Fluid Mechanics*.[1] For present purposes, we simply note that (1) existence proofs do not generally exist for solutions to the algebraic problem, or for convergence of the discretized solution to the continuous solution of the differential form of the Navier-Stokes equations; (2) numerical solutions of the finite difference or finite element equations are very difficult (or impossible) above a Reynolds number of 0(1000); and (3) it is often difficult to understand the physics of a problem from numerical solutions. (For example, the dependence of the solution structure on independent parameters of a problem is difficult to discern from a finite number of numerical solutions at discrete values of these parameters.) Thus, numerical solutions offer a very important and useful tool for many problems in fluid mechanics and convective heat or mass transfer, but they are not generally sufficient in themselves, and it is always helpful to supplement numerical analysis with analytic (or experimental) investigations of the same problem.

There are two distinct classes of analytic approximation that comprise the second and third approaches that were mentioned above. The first is based upon the use of so-called *macroscopic balances.* In this approach, we do not attempt to obtain detailed information about the velocity and pressure fields everywhere in the domain, but only to obtain results that are consistent with the Navier-Stokes equations in an overall (or macroscopic) sense. For example, we might seek results for the volumetric flow rates in and out of a flow system that are consistent with an *overall* mass or momentum conservation balance but not attempt to determine the detailed form of the velocity profiles. The macroscopic balance approach is described in detail in many undergraduate textbooks.[2] It is often extremely useful for derivation of quantitative relationships among the average inflows, outflows, and forces (or rates of working) within a flow system but is something of a "black-box" approach that provides no detailed information on the velocity, pressure, and stress distributions within the flow domain.

The third and final approach is to seek approximate, asymptotic solutions of the governing equations and boundary conditions for very large or very small values of the relevant dimensionless parameters. It is this approach that will be the primary focus of the present book. *Our general objectives are twofold: first, to develop general methods for obtaining approximate solutions of nonlinear differential equations via perturbation or asymptotic expansions, and, second, to show how the asymptotic framework for solution frequently can be used to determine the forms of correlations between independent and dependent dimensionless parameters without ever having to solve the resulting differential equations.*

This chapter represents a first step toward general asymptotic methods, in which we consider nondimensionalization of the Navier-Stokes and continuity equations for isothermal flows of an incompressible fluid. It was shown in Chapter 3 that nondimensionalization reveals the dimensionless combination(s) of independent parameters that control the form of the solution of a differential equation (or set of equations) that describes some physical process. In general, for isothermal flow of an incompressible, Newtonian fluid in a domain with solid, fixed boundaries, we shall see that there is a single dimensionless group, called the Reynolds number, that determines the form of solutions to the Navier-Stokes and continuity equations. When this parameter is very small, the (linear) viscous terms in the equation are dominant over the (nonlinear) inertial or acceleration terms, and a linear approximation of the equations is thus possible in the limit as the Reynolds number approaches zero. The class of fluid motions where this approximation can be used are known as *creeping flows,* and this chapter will focus on detailed analysis for many motions of this type. In subsequent chapters we shall consider flows and convective transport processes for heat or mass transfer where this linearized approximation is not valid. In so doing, we shall see that the creeping flow solutions from this chapter actually represent the first approximation of an asymptotic solution for the limit of arbitrarily small, but nonvanishing, Reynolds number.

A. Nondimensionalization and the Creeping Flow Equations

To begin, we restrict our attention to isothermal, laminar flow of an incompressible Newtonian fluid, in the absence of a gas-liquid or liquid-liquid interface, so that we can use the equations of motion and continuity in the forms (2–109) and (2–19), respectively, that is,

$$\rho\left[\frac{\partial \mathbf{u}'}{\partial t} + \mathbf{u}'\cdot\nabla'\mathbf{u}'\right] = -\nabla'p' + \mu\nabla'^2\mathbf{u}',$$

$$\nabla'\cdot\mathbf{u}' = 0.$$

These equations contain two independent dimensional parameters, the density ρ and the viscosity μ, which are both constant and assumed to be known. For convenience in what follows, we mark the dimensional variables that appear in these equations with a prime—$\mathbf{u}'$, p', and so on—and, in addition, denote the dimensional gradient operator as ∇'.

To completely characterize the flow, we require, in addition to these equations, the geometry of the flow domain and the boundary conditions that apply at the boundaries of the domain, including the form and magnitudes of the velocity distribution(s) on any inlet to the domain (or the form and magnitude of the velocity field far from any boundaries in an unbounded domain) and the velocities of the bounding surfaces if these are moving in the laboratory frame of reference. From these additional conditions, at least three independent dimensional parameters can normally be specified: a length scale, L_c; a velocity scale, U_c; and a time scale, T_c.

For purposes of nondimensionalization, the length scale L_c is intended to represent the physical distance over which the velocity changes by an amount proportional to U_c. For present purposes, it is enough to suggest that this length scale will typically be evident from the geometry of the flow domain. For example, for translation of a rigid sphere through a large body of liquid, the sphere radius would appear to be an appropriate choice. Similarly, the velocity scale U_c represents a typical magnitude of fluid velocities in the flow domain and generally will be set by either the boundary velocity or the magnitude of the velocity at large distances from the boundary in an unbounded domain; in the translating sphere problem, for example, an appropriate choice would be the sphere velocity. Finally, the time scale T_c must provide a measure of the rate of change of velocity as seen by a typical fluid element. Its definition will depend substantially upon the details of the particular flow. For example, if either the motion of a boundary or the flow at large distances is time dependent and periodic, then T_c is just proportional to the inverse of the frequency. If the geometry of the flow domain is changing with time, this may provide an appropriate characteristic time scale. Finally, if the flow and geometry are both steady from an Eulerian point of view, the time scale T_c may be just L_c/U_c, that is, the time for a "typical" fluid element to travel a distance L_c.

In some instances there may be more than a single length, velocity, or time scale inherent in the problem. In this case, we will assume for present purposes that it is not essential which choice is made for L_c, U_c, or T_c. For each added dimensional scale, we will simply get one more dimensionless parameter, in addition to the two that will appear based upon the five independent parameters ρ, μ, L_c, U_c, and T_c. These added parameters are most conveniently identified as ratios of length, velocity, or time scales.

In order to express (2–109) and (2–19) in dimensionless form, we introduce dimensionless dependent and independent variables as follows:

$$\mathbf{u} = \frac{\mathbf{u}'}{U_c},$$

$$t = \frac{t'}{T_c}, \tag{4–1}$$

$$\nabla = L_c \nabla',$$

and

$$p = \frac{p'}{\left(\dfrac{\mu U_c}{L_c}\right)}.$$

The reader may wish to verify that the dimensionless group $\mu U_c / L_c$ has dimensions of force per unit area and is thus appropriate as a choice for the characteristic pressure. Substituting (4–1) into (2–109) and (2–19), as given at the beginning of this chapter, we find

$$R\left[\frac{1}{S}\frac{\partial \mathbf{u}}{\partial t} + \mathbf{u}\cdot\nabla\mathbf{u}\right] = -\nabla p + \nabla^2\mathbf{u}, \tag{4–2}$$

$$\nabla\cdot\mathbf{u} = 0. \tag{4–3}$$

Two dimensionless parameters appear in (4–2), known, respectively, as the Reynolds number,

$$R \equiv \frac{\rho U_c L_c}{\mu}, \tag{4–4}$$

and the Strouhal number,

$$S \equiv \frac{T_c}{(L_c / U_c)}. \tag{4–5}$$

The fact that only two dimensionless combinations of dimensional parameters appear in (4–2) and (4–3) is one demonstration of the so-called *principle of dynamic similarity*. This principle is nothing more than the observation that the form of the velocity and pressure fields for any typical flow problem will depend upon the dimensionless parameters (4–4) and (4–5) rather than any of the dimensional parameters alone. It may not be obvious without examining the details of specific flow problems that additional dimensionless parameters will not appear in the boundary conditions. However, with the exception of problems that involve multiple scales as described above, no additional parameters will appear for flows that involve only solid boundaries.

For steady flows, examination of (4–2) shows that the Reynolds number determines the relative magnitudes of the acceleration terms on the left and the viscous and pressure gradient terms on the right. In the case of unsteady flows, the magnitude of the Eulerian acceleration is determined by the ratio of the Reynolds number to the Strouhal number, R/S (notice that for steady flows we have already suggested that $T_c = L_c/U_c$, and in this case $S \equiv 1$). In particular, in the limit as the Reynolds number becomes small (that is, $R \ll 1$ and $R/S \ll 1$), it appears that the acceleration (or inertia) terms in (4–2) will become negligible compared to viscous and pressure gradient terms. Although we will ultimately see that the situation can be more complicated than it first appears, the resulting linear equations,

$$\nabla^2 \mathbf{u} - \nabla p = 0, \tag{4–6a}$$

$$\nabla \cdot \mathbf{u} = 0, \tag{4–6b}$$

do represent a valid first approximation to the Navier-Stokes and continuity equations for the limit of very small Reynolds numbers. These equations are known as the *creeping flow equations*. Clearly, the creeping flow regime can be reached through a combination of very small velocity or length scales or very large kinematic viscosity μ/ρ. These conditions encompass a number of important flow problems, especially in chemical and biological applications.

The most important feature of the creeping flow equations is that they are linear and thus can be solved by a number of well-known methods for linear differential equations. In this limited sense, they are similar to the unidirectional flow equations of Chapter 3. However, unlike unidirectional flows, the velocity and pressure fields in the creeping flow limit can be fully three-dimensional. There is another extremely significant difference between unidirectional and creeping flows. For unidirectional flow, the nonlinear term $\mathbf{u} \cdot \nabla \mathbf{u}$ in the Navier-Stokes equation is identically equal to zero, simply because $\mathbf{u}$ and $\nabla \mathbf{u}$ are orthogonal. On the other hand, in the creeping flow limit, the nonlinear terms are not identically zero. They are simply neglected as an approximation for small values of R.

The creeping flow equations involve the dual limit $R \ll 1$ and $R/S \ll 1$ (or simply $R \ll 1$ when the flow is steady and $S = 1$), and the time derivative in the equations of motion is also neglected as an approximation. For this reason, creeping flows are sometimes called *quasi-steady*. Such a flow can depend on time only as a parameter rather than as a true independent variable. One way to see why this is true is to compare the time scales required for evolution of a velocity field from one steady state to another, relative to the time scale for a change in the boundary configuration due to, say, boundary movement with characteristic velocity U_c. If the boundary geometry is characterized by a length scale L_c, then the latter time scale is just L_c/U_c. On the other hand, at low Reynolds numbers the only mechanism for transport of momentum (or vorticity) is diffusion via the diffusivity ν. Hence, the transient process of going from one steady flow to another must be characterized by a diffusive time scale L_c^2/ν. The ratio of the two time scales is thus

$$\frac{L_c^2/\nu}{L_c/U_c} = \frac{U_c L_c}{\nu}.$$

But this is just the Reynolds number that, according to the creeping motion assumption, is arbitrarily small. It follows that the velocity and pressure fields adjust *instantaneously* relative to the rate at which the boundary configuration changes and therefore always appear to be at steady state with respect to the present configuration. Thus, time appears in a creeping flow solution only because the velocity field at any instant depends on the boundary geometry, and the latter may depend on time.

One final remark should be made about the nondimensionalization that led to (4–2) and (4–3), and thence for $R \ll 1$ to the creeping flow equation (4–6). This concerns the scale factor chosen to nondimensionalize the pressure. Assuming for simplicity that the geometry and boundary conditions lead to a single characteristic length, velocity, and time scale, the nondimensionalization of $\mathbf{u}'$, t', and ∇' in (4–1) is unambiguous. However, even for this simplest case there is a second possible choice for nondimensionalization of p', namely, ρU_c^2, which is another combination of dimensional parameters with dimensions of force per unit area. Let us then suppose that instead of (4–1) we had used a dimensionless pressure $\hat{p}$ defined as

$$\hat{p} = \frac{p'}{(\rho U_c^2)}. \tag{4–7}$$

In this case, instead of (4–2), we would have obtained

$$R\left[\frac{1}{S}\frac{\partial \mathbf{u}}{\partial t} + \mathbf{u}\cdot\nabla\mathbf{u}\right] = -R\nabla\hat{p} + \nabla^2\mathbf{u}, \tag{4–8}$$

and in the limit $R \to 0$, this would appear to yield

$$\nabla^2\mathbf{u} = 0 \tag{4–9}$$

in place of (4–6a). But the approximation (4–9) cannot be correct. To see this, we need only note that (4–9) and (4–3) now comprise a set of four scalar equations for only three unknowns, that is, the three scalar velocity components. We conclude that the scaling (4–7) is inappropriate, at least for the creeping flow limit $R \to 0$. It is important to recognize that the correct scaling will be obtained automatically even if we start with (4–7), assuming only that we require that the pressure be retained in the governing equations. Indeed, examination of (4–8) suggests that in order to retain pressure in the limit $R \to 0$, we should introduce a rescaled pressure in the form

$$p = R\hat{p}. \tag{4–10}$$

But this is identical to the original nondimensionalized form in (4–1). To see this, we can rewrite the right-hand side in the form as

$$p = \left(\frac{\rho U_c L_c}{\mu}\right)\frac{p'}{\rho U_c^2} = \frac{p'}{\left(\dfrac{\mu U_c}{L_c}\right)}.$$

In the remainder of this chapter we consider solutions of various flow problems based upon the approximate, linear creeping flow equations.

B. Some General Consequences of Linearity and the Creeping Flow Equations

A very important consequence of the approximation of the Navier-Stokes equations by the creeping flow equations is that the classical methods of linear analysis can be used to obtain exact solutions. Equally important, but less well known, is the fact that many important qualitative conclusions can be reached on the basis of *linearity* alone, without the necessity of obtaining detailed solutions. This, in fact, will be true of any physical problem that can be represented, or at least approximated, by a system of linear equations. In this section we illustrate some qualitative conclusions that are possible for creeping flows.

The Drag on Bodies that Are Mirror Images in the Direction of Motion

Let us consider two solid bodies that translate through a fluid in the creeping flow regime and are mirror images of one another about a plane orthogonal to the direction of motion. For simplicity in picturing the situation, we may consider the bodies as being held fixed relative to a uniform, undisturbed flow that has a velocity U. A typical example of this situation is illustrated in Figure 4–1. The two mirror-image configurations are denoted as (a) and (b). We wish to know whether the drag force D_A for configuration (a) is larger than, smaller than, or equal to the drag force D_B for configuration (b).

To determine either D_B (or D_A), we need to actually solve the creeping flow and continuity equations

$$\nabla^2 \mathbf{u} - \nabla p = 0, \qquad (4\text{–}11)$$

$$\nabla \cdot \mathbf{u} = 0,$$

subject to boundary conditions

$$\mathbf{u} = 0 \quad \text{on} \quad S(b),$$

$$\mathbf{u} = \mathbf{U} \quad \text{at} \quad \infty. \qquad (4\text{–}12)$$

However, once D_B has been determined, we can immediately obtain D_A.

To see that this is true, we simply note that the detailed flow problem for configuration (a) can be obtained directly from (4–11) and (4–12) by simply reversing the sign of the undisturbed flow at infinity. In particular, if the solution for configuration (b) is $(\mathbf{u}, p)$, the solution for (a) is just

$$\hat{\mathbf{u}} = -\mathbf{u},$$

$$\hat{p} = -p. \qquad (4\text{–}13)$$

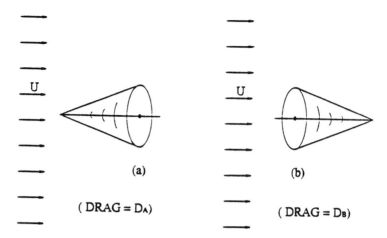

Figure 4–1 A schematic representation of two mirror-image configurations of cone-shaped bodies that are held fixed in a uniform, undisturbed flow. For the limit Re $\equiv 0$, $D_A \equiv D_B$.

This is a consequence of the linearity of governing equations and boundary conditions (4–11) and (4–12). It follows that the magnitude of the drag for configuration (a) must be identical (except for a sign change) to the drag for (b), that is,

$$D_A \equiv D_B. \tag{4–14}$$

Clearly, the result (4–14) is a consequence of linearity alone and requires no solutions of the governing equations and boundary conditions.

The Lift on a Sphere that is Rotating in a Simple Shear Flow

As a second example, let us consider a solid sphere rotating with angular velocity Ω about its center, with the center held fixed in space and the fluid "at infinity" undergoing a simple linear shear flow. The situation is sketched in Figure 4–2. We wish to determine whether there is any *lift* on the sphere in the creeping flow limit, that is, any force in the direction orthogonal to the fluid motion at infinity.

The governing nondimensionalized equations and boundary conditions for this problem are

$$\nabla^2 \mathbf{u} - \nabla p = 0; \quad \nabla \cdot \mathbf{u} = 0, \tag{4–15}$$

with

$$\mathbf{u} = \left(\frac{a}{U_c}\right)\Omega \wedge \mathbf{r} \quad \text{on the sphere } r = 1$$

and

$$\mathbf{u} = \left(1 + \frac{\gamma a}{U_c} \cdot y\right)\mathbf{i}_z \quad \text{at } \infty.$$

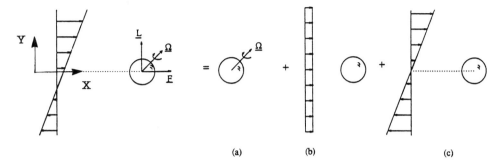

Figure 4–2 Illustration of the decomposition of the problem of a freely rotating sphere in a simple shear flow as the sum of three simpler problems: (a) a sphere rotating in a fluid that is stationary at infinity, (b) a sphere held stationary in a uniform flow, and (c) a nonrotating sphere in a simple shear flow that is zero at the center of the sphere. The angular velocity Ω in (a) is the same as the angular velocity of the sphere in the original problem. The translation velocity in (b) is equal to the undisturbed fluid velocity evaluated at the position of the center of the sphere. The shear rate in (c) is equal to the shear rate in the original problem.

Here, we have utilized the sphere radius as a characteristic length scale, and the undisturbed translational velocity evaluated at the sphere center U_c (see Figure 4–2) as a characteristic velocity. Because these equations and boundary conditions are linear, the problem can be solved as the sum of three simpler problems, namely,

1. a sphere rotating in a fluid that is stationary at infinity, that is,

$$\mathbf{u} = \frac{a}{U_c}(\mathbf{\Omega} \wedge \mathbf{r}) \quad \text{at} \quad r = 1 \quad \text{and} \quad \mathbf{u} = 0 \quad \text{at} \quad \infty;$$

2. a sphere held stationary in a uniform flow,

$$\mathbf{u} = 0 \quad \text{at} \quad r = 1 \quad \text{and} \quad \mathbf{u} = \mathbf{i}_z \quad \text{at} \quad \infty;$$

3. a sphere held stationary in a simple shear flow that is zero at the center of the sphere,

$$\mathbf{u} = 0 \quad \text{at} \quad r = 1 \quad \text{and} \quad \mathbf{u} = \left(\frac{\gamma a}{U_c}\right) y\mathbf{i}_z \quad \text{at} \quad \infty.$$

It is a simple matter to see that the lift is zero for each of these component problems. For a sphere rotating in a quiescent fluid, it is evident that all components of the force must be zero because there is no direction preferred over any other. The lift on a sphere in a uniform flow is clearly zero due to the symmetry of the problem. Finally, the lift on a sphere in simple shear flow can again be shown to be zero due to symmetry. If, for example, the lift were nonzero in the positive y direction (see Figure 4–2), then it would also have to be nonzero in the negative y direction because the symmetry of the flow problem is such that there is no distinction between $y > 0$ and $y < 0$. Thus, the lift must be zero. Finally, since the lift is zero for each of the component problems and the overall problem is linear, it follows that the lift is zero for the overall problem, (4–15), that is

$$\boxed{\text{Lift} \equiv 0.} \qquad (4\text{--}16)$$

The reader may find the result (4–16) surprising. For example, it is well known that a rotating and translating sphere in a stationary fluid will normally experience a sideways force (that is, *lift*) that will cause it to travel in a curved path—think, for example, of a curveball in baseball or an errant slice or hook in golf. The difference between these familiar examples and the problem analyzed above is that the Reynolds numbers are not small and the governing equations are the full, nonlinear Navier-Stokes equations rather than the linear creeping flow approximation. Thus, the decomposition to a set of simpler "component" problems cannot be used, and it is not possible to deduce anything about the forces on the body without actually solving the full fluid mechanics problem and calculating the force by integrating the stress vector **n·T** over the sphere surface.

Lateral Migration of a Sphere in Poiseuille Flow

One of the best known experimental results for particle motion in viscous flows is the observation by Segre and Silberberg (1962)[3] of *lateral migration* for a small, neutrally buoyant sphere ($\rho_{\text{sphere}} = \rho_{\text{fluid}}$) that is immersed in Poiseuille flow through a straight, circular tube or in the pressure-driven parabolic flow (sometimes called two-dimensional Poiseuille flow) between two parallel plane boundaries. The experiments of Segre and Silberberg, and many later investigators, show that a freely suspended sphere in these circumstances will slowly move perpendicular to the main direction of flow until it reaches an equilibrium position that is approximately 60 percent of the way from the centerline (or central plane) to the wall. Hence, a suspension of such spheres flowing in Poiseuille flow through a tube of radius R will tend to accumulate in an annular ring at $r = 0.6\,R$. Since the Reynolds number for many of the experimental observations was quite small, one might assume that a theoretical explanation could be achieved using detailed solutions of the creeping flow equations with suitable boundary conditions. However, in view of the complexity of the geometry (an eccentrically located sphere inside a circular tube), this theoretical problem is extremely complex and difficult to solve. Thus, prior to actually trying to solve this problem, it is prudent to determine whether lateral migration is possible at all in the creeping flow limit.

The fact is that a theory based entirely upon the creeping flow approximation *cannot* predict any lateral migration. To see that this is true, we can refer to Figure 4–3. Here, we have sketched the hypothetical situation of a sphere that is undergoing lateral migration in Poiseuille flow through a tube. The undisturbed flow in part (a) of Figure 4–3 is shown moving from left to right, and the sphere is assumed to be migrating radially inward toward the center of the tube. Now, however, if the creeping motion approximation is valid, the governing equations and boundary conditions are linear in the velocity and pressure, and we can change the signs of all velocities and the pressure and still have a solution of the same problem but with the direction of the undisturbed flow reversed as shown in Figure 4–3b. However, since all the velocities

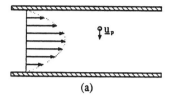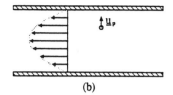

(a)   (b)

Figure 4–3 A schematic representation of the proof that a spherical particle cannot undergo lateral migration in either two-dimensional or axisymmetric *Poiseuille* flow if the disturbance flow is a creeping flow. In (a) we suppose that the undisturbed flow moves from left to right and the sphere migrates inward with velocity u_p. Then, in the creeping flow limit, if direction of the undisturbed flow is reversed, the signs of all velocities including that of the sphere would also have to be reversed, as shown in (b). Since the problems (a) and (b) are *identical* other than the direction of the flow through the channel or tube, we conclude that $u_p = 0$.

have the opposite sign, the *inward* migration velocity from configuration (a) must now become an *outward* migration velocity for configuration (b). But there is now a clear contradiction. The problems in (a) and (b) are clearly indistinguishable in all respects. Thus, if the sphere undergoes a lateral motion, it should be in the same direction in both cases. Since the preceding argument, based upon the linearity of the problem, shows that a nonzero migration velocity in case (a) must lead to a migration velocity of opposite sign in case (b), we can only conclude that the migration velocity must be zero in both cases. In other words, lateral migration *cannot* occur in the creeping flow limit.

The conclusion of the preceding paragraph does not, of course, mean that lateral migration cannot occur at all but only that the mechanism is not inherent in the creeping motion approximation. There have been, in fact, a number of theoretical and experimental studies that show that lateral migration of a rigid sphere in a Newtonian fluid is caused by inertial effects that remain nonzero, though small, for the small Reynolds numbers that were studied experimentally. The reader may find it instructive to think about how the flow reversal argument of the previous paragraph will fail if the nonlinear, inertial terms are retained in the Navier-Stokes equations.* The fact that the flow reversal argument cannot be applied does not prove that migration *will* necessarily occur at small, but finite, Reynolds numbers. We can only say that there is no simple argument, of the flow-reversal type, that is *inconsistent* with the existence of lateral migration. To prove theoretically that a sphere will experience a lateral force, and to evaluate that force, we must solve explicitly for the inertial contributions to the stress $n \cdot T$ on the sphere surface. This has been done by Ho and Leal (1974) and Vasseur and Cox (1976) for the case of a sphere in parallel flow between two parallel plane walls at small, but nonzero, Reynolds numbers, and for a sphere in a tube by Ishii and Hasimoto (1980).[4] Researchers have also shown that other departures from the solid sphere, Newtonian fluid assumptions can lead to lateral migration effects, even for

*If $(\mathbf{u}, p)$ is the solution of the Navier-Stokes equations and boundary conditions for case (a), is $(-\mathbf{u}, -p)$ a solution for case (b)?

flow at zero Reynolds number. For example, if the fluid is non-Newtonian, both experimental and theoretical studies have shown that lateral migration will occur, though in this case the sphere is found to travel all the way to the centerline or to the wall, depending on the nature of the fluid. For a non-Newtonian fluid, the governing equations are nonlinear because the constitutive equation is nonlinear (see Chapter 2), even if the nonlinear inertial terms are completely negligible. Another class of problems exhibiting lateral migration involves the motion of deformable particles (or drops), where migration can occur even for a Newtonian fluid at zero (or extremely small) Reynolds number. In this case, the shape of the particle is dependent upon the flow, and this leads (as we will see later) to nonlinear boundary conditions so that the flow-reversal argument does not apply. A summary of these various types of migration phenomena can be found in a paper by Leal (1980).[5]

The history of many failed theoretical attempts to explain the original observations of Segre and Silberberg using the creeping flow equations provides a noteworthy example of the importance of simple arguments based upon the linearity of the governing equations and boundary conditions in the creeping flow limit and a lesson about the need to think creatively at the outset instead of proceeding blindly with detailed analysis.

Resistance Matrices for the Force and Torque on a Body in Creeping Flow

Finally, linearity of the creeping flow equations and boundary conditions allows a great *a priori* simplification in calculations of the force or torque on a body of fixed shape that moves in a Newtonian fluid. To illustrate this assertion, we consider a solid particle of *arbitrary* shape moving with translational velocity $U(t)$ and angular velocity $\Omega(t)$ through an unbounded, quiescent viscous fluid in the creeping flow limit $R \ll 1$ and $R/S \ll 1$. The problem of calculating the force or torque on the particle requires a solution of the creeping flow equations, subject to boundary conditions; in *dimensional terms*, the problem is

$$\mu \nabla^2 \mathbf{u} - \nabla p = 0,$$

$$\nabla \cdot \mathbf{u} = 0, \tag{4-17a}$$

with

$$\mathbf{u} = \mathbf{U} + \mathbf{\Omega} \wedge \mathbf{x} \tag{4-17b}$$

on the particle surface S and

$$\mathbf{u} \rightarrow 0 \quad \text{as} \quad |x| \rightarrow \infty. \tag{4-17c}$$

Here, $\mathbf{x}$ is a position vector measured from the center of gravity of the particle. The force on the particle is

$$\mathbf{F} = \int_S (\mathbf{T} \cdot \mathbf{n}) \, dS, \tag{4-18a}$$

and the torque is

$$\mathbf{G} = \int_S (\mathbf{x} \wedge \mathbf{T} \cdot \mathbf{n}) dS, \tag{4-18b}$$

where $\mathbf{T}$ is the stress tensor and the integration is over the particle surface.

The critical difficulty with the problem, as stated above, is that the solution depends on the orientations of $\mathbf{U}$ and $\boldsymbol{\Omega}$ relative to axes fixed in the particle, as well as on the *relative* magnitudes of $\mathbf{U}$ and $\boldsymbol{\Omega}$. Thus, for every possible orientation of $\mathbf{U}$ and/or $\boldsymbol{\Omega}$, a new solution appears to be required in order to calculate $\mathbf{u}$, p, $\mathbf{F}$, or $\mathbf{G}$! Fortunately, however, this is not actually necessary in the creeping flow limit. Rather, to evaluate $\mathbf{u}$, p, $\mathbf{F}$, or $\mathbf{G}$ for any arbitrary choice of $\mathbf{U}$ and $\boldsymbol{\Omega}$, we will show that it is sufficient to obtain detailed solutions for three mutually orthogonal orientations of $\mathbf{U}$ with $\boldsymbol{\Omega} \equiv 0$ and for three mutually orthogonal orientations of $\boldsymbol{\Omega}$ with $\mathbf{U} \equiv 0$.

For convenience, we can consider the problem (4-17) as a superposition of two problems: The first is translation with arbitrary velocity $\mathbf{U}$ and $\boldsymbol{\Omega} = 0$, and the second is rotation with arbitrary angular velocity $\boldsymbol{\Omega}$ and $\mathbf{U} = 0$. We can denote the solutions of these two problems as $(\mathbf{u}_1, p_1)$ and $(\mathbf{u}_2, p_2)$, respectively, with the corresponding force and torque being $(\mathbf{F}_1, \mathbf{G}_1)$ and $(\mathbf{F}_2, \mathbf{G}_2)$. The *full* solution of (4-17) is then $\mathbf{u} = \mathbf{u}_1 + \mathbf{u}_2$ and $p = p_1 + p_2$, while the force and torque are $\mathbf{F} = \mathbf{F}_1 + \mathbf{F}_2$ and $\mathbf{G} = \mathbf{G}_1 + \mathbf{G}_2$.

Now, let us begin by considering the translation problem, again in dimensional terms,

with

$$\mu \nabla^2 \mathbf{u}_1 - \nabla p_1 = 0, \quad \nabla \cdot \mathbf{u}_1 = 0, \tag{4-19}$$

$$\mathbf{u}_1 = \mathbf{U} \quad \text{on } S,$$

$$\mathbf{u}_1 \rightarrow 0 \quad \text{at } \infty.$$

We see that the problem is linear in $\mathbf{U}$. Thus, the solution $(\mathbf{u}_1, p_1)$ can only depend linearly on $\mathbf{U}$, and in view of the relationships (4-18a) and (4-18b), this means that the force and torque must also be linear functions of $\mathbf{U}$. Since $\mathbf{F}_1$ and $\mathbf{U}$ are true vectors,* the most general linear relationship between them is

$$\mathbf{F}_1 = \hat{\mathbf{A}} \cdot \mathbf{U}, \tag{4-20a}$$

where $\hat{\mathbf{A}}$ is a true second-order tensor. Similarly, since $\mathbf{G}_1$ is a pseudo-vector, the most general linear relationship between $\mathbf{G}_1$ and $\mathbf{U}$ is

$$\mathbf{G}_1 = \hat{\mathbf{C}} \cdot \mathbf{U}, \tag{4-20b}$$

where $\hat{\mathbf{C}}$ is a second-order pseudo-tensor.

---

*A pseudo-vector differs from a true vector because it changes sign upon inversion of the coordinate axes. In particular, whereas a true vector transforms under an orthogonal transformation according to the rule (2-64), that is $\mathbf{A} = \mathbf{L} \cdot \bar{\mathbf{A}}$, a pseudo-vector transforms according to the rule $\mathbf{B} = \det(\mathbf{L})[\mathbf{L} \cdot \bar{\mathbf{B}}]$. Common examples of pseudo-vectors include the angular velocity, torque, the vector product of any two true (or pseudo-) vectors, the curl of a vector, and so on. Similarly, a pseudo-tensor differs from a true tensor because its components change sign upon coordinate inversion. Clearly, a pseudo-vector can only be equated to another pseudo-vector and vice versa for a true vector.

The components of the tensor $\hat{\mathbf{A}}$ and the pseudo-tensor $\hat{\mathbf{C}}$ have a simple interpretation. For example, the ij component of $\hat{\mathbf{A}}$ is just the i component of the force on the body for translation with unit velocity in the j direction. To see this, we may express (4–20a) in component form

$$(F_1)_i = A_{ij} U_j.$$

Then if $\mathbf{U} = \mathbf{i}_1$, for example,

$$(F_1)_1 = A_{11},$$

$$(F_1)_2 = A_{21},$$

$$(F_1)_3 = A_{31}.$$

Similarly, the ij component of $\hat{\mathbf{C}}$ is just the i component of the torque produced by translation of the body in the j direction. Thus if we solve the three problems of translation with unit velocity in the three coordinate directions and calculate the components of the force and torque on the body in each case, we completely determine all nine components of $\hat{\mathbf{A}}$ and $\hat{\mathbf{C}}$. But once $\hat{\mathbf{A}}$ and $\hat{\mathbf{C}}$ have been specified, the formulas in (4–20) can be used to evaluate the force and torque for translation of the body in any *arbitrary* direction.

An identical analysis also can be applied to the rotation problem that derives from (4–17), that is, rotation with arbitrary $\boldsymbol{\Omega}$ and $\mathbf{U} \equiv 0$. The result for the force and torque in this case is

$$\boxed{\begin{aligned} \mathbf{F}_2 &= \hat{\mathbf{B}} \cdot \boldsymbol{\Omega}, \\ \mathbf{G}_2 &= \hat{\mathbf{D}} \cdot \boldsymbol{\Omega}. \end{aligned}} \tag{4–21}$$

Again, specification of the components of $\hat{\mathbf{B}}$ and $\hat{\mathbf{D}}$ requires an evaluation of the three force and torque components for rotation about each of the three coordinate axes.

Thus, combining (4–20) and (4–21), as allowed by the linearity of the governing equations, we find that the force and torque on a particle that moves with arbitrary velocities $\mathbf{U}$ and $\boldsymbol{\Omega}$ through an otherwise quiescent fluid is

and

$$\boxed{\begin{aligned} \mathbf{F} &= \mu(\mathbf{A} \cdot \mathbf{U} + \mathbf{B} \cdot \boldsymbol{\Omega}) \\ \mathbf{G} &= \mu(\mathbf{C} \cdot \mathbf{U} + \mathbf{D} \cdot \boldsymbol{\Omega}). \end{aligned}} \tag{4–22}$$

Note that the factor μ will appear linearly in each of the second-order tensorial coefficients in (4–20) and (4–21) and thus has been factored out in (4–22)—that is, $\hat{\mathbf{A}} = \mu \mathbf{A}$, and so on. A most important property of the four remaining *resistance* tensors, $\mathbf{A}$, $\mathbf{B}$, $\mathbf{C}$, and $\mathbf{D}$, is that they depend only on the *geometry* of the particle and are independent of all other parameters of the problem including, of course, $\mathbf{U}$ and $\boldsymbol{\Omega}$.

To calculate the components of the resistance tensors, we have seen that it is necessary to solve three translation problems and three rotation problems. Of course, the orientation of the coordinate axes for specification of these components of $\mathbf{A}$, $\mathbf{B}$, $\mathbf{C}$, and $\mathbf{D}$ should be chosen to take advantage of any geometric symmetries that can simplify the fluid mechanics problems that must be solved. For example, if we wish to determine the force and/or torque on an ellipsoid for motions of arbitrary magnitude and direction (with respect to the body geometry), we should specify the components of the resistance tensors with respect to axes that are coincident with the principal axes of the ellipsoid, since this choice will simplify the fluid mechanics problems that are necessary to determine these components. If the arbitrary velocities $\mathbf{U}$ and $\mathbf{\Omega}$ are then specified with respect to these same coordinate axes, the equation (4–22) will yield force and torque components in the same coordinate system.

Although we will not carry our analysis further, there has been, in fact, significant progress in delineating the properties of the resistance tensors beyond the general formulas (4–22). The most important and general result is the symmetry conditions

$$\boxed{\mathbf{A} = \mathbf{A}^T, \ \mathbf{D} = \mathbf{D}^T, \text{ and } \mathbf{B} = \mathbf{C}^T.} \qquad (4\text{–}23)$$

It should be noted that there is a direct relationship between the two resistance matrices that represent the coupling between translation and rotation.

In addition to these general symmetry properties, considerable effort has been made to understand the relationships between symmetries in the geometry of the problem and the forms of the resistance tensors. It is beyond our present scope to discuss these relationships in a comprehensive manner; the interested reader can refer to Brenner (1972) or the textbook *Low Reynolds Number Hydrodynamics*, by Happel and Brenner (1973) for a detailed discussion of these questions.[6] Here we restrict ourselves to the results for several particularly simple cases. First, if we consider the motion of a body with spherical symmetry in an unbounded fluid, with the origin of coordinates at the geometric center of the body, it can be shown that

$$\boxed{\mathbf{A} = a\mathbf{I}, \ \mathbf{D} = d\mathbf{I}, \text{ and } \mathbf{C} = \mathbf{B} = 0.} \qquad (4\text{–}24)$$

On the other hand, for an ellipsoid of revolution with the origin of the coordinate system at the geometric center and coordinate axes parallel and perpendicular to the principal axes of the ellipse, it can be shown that

$$\mathbf{A} = \begin{pmatrix} a_\parallel & 0 & 0 \\ 0 & a_\perp & 0 \\ 0 & 0 & a_\perp \end{pmatrix}, \quad \mathbf{D} = \begin{pmatrix} d_\parallel & 0 & 0 \\ 0 & d_\perp & 0 \\ 0 & 0 & d_\perp \end{pmatrix}, \quad \text{and} \quad \mathbf{B} = \mathbf{C}^T = \begin{pmatrix} 0 & 0 & 0 \\ 0 & 0 & 0 \\ 0 & 0 & 0 \end{pmatrix}.$$

$$(4\text{–}25)$$

The most obvious example of a body that exhibits *coupling* between translation and rotation is a body with a screwlike structure that clearly *translates* if it is rotated about the screw axis (alternatively, if the particle is to be restrained from translating, a force equal to $-\mu \mathbf{B} \cdot \mathbf{\Omega}$ must be applied to *balance* the hydrodynamic force that is generated by its rotation, with

$$\mathbf{B} = \begin{pmatrix} b_1 & 0 & 0 \\ 0 & 0 & 0 \\ 0 & 0 & 0 \end{pmatrix}$$

if the 1 axis is coincident with the screw axis).

Although all the preceding development has been based upon a body moving through an *unbounded* fluid, the general expression (4–22) for the force and torque in terms of resistance tensors is also applicable to the motion of a spherical particle in a container or in the vicinity of a wall. Of course, the resistance tensors must be modified to account for the presence of the bounding walls. For example, to describe the force or torque on a sphere moving with arbitrary $\mathbf{U}$ and $\boldsymbol{\Omega}$ in the vicinity of an infinite plane wall, the obvious choice is to specify the components of the resistance tensors in a coordinate system whose axes are normal and parallel to the wall. This results in the simplest fluid mechanics problems for determining the components of the resistance tensors—that is, translation with unit velocity parallel and normal to the wall and rotation with unit angular velocity parallel and normal to the wall. We may anticipate intuitively that $\mathbf{A}$ and $\mathbf{D}$ will *not* be isotropic in this case, nor will $\mathbf{C}$ and $\mathbf{B}$ be zero. The fact that $\mathbf{A}$ and $\mathbf{D}$ are not isotropic is a consequence of the fact that a force or a torque of given magnitude will produce different translational or angular velocities depending upon the direction of motion relative to the wall. Thus, for example, the resistance to motion is different for motion parallel and perpendicular to the wall. Note, however, that we can use the arguments of Section A to show that the hydrodynamic force does not depend on the *sign* of $\mathbf{U}$; that is, the force generated by motion toward the wall is identical to the force on the same body at the same instantaneous position as it moves away from the wall. The fact that $\mathbf{C}$ and $\mathbf{B}$ are nonzero means that translational and rotational motions are *coupled*. For example, if a force parallel to the wall is applied to the sphere, the sphere not only will translate in that direction but also will rotate due to hydrodynamic interaction with the wall. Indeed, if we express the components of the resistance tensors in a Cartesian coordinate system with axes that are normal and tangential to the wall, it can be shown that

$$\mathbf{A} = \begin{pmatrix} a_\perp & 0 & 0 \\ 0 & a_\parallel & 0 \\ 0 & 0 & a_\parallel \end{pmatrix}, \quad \mathbf{D} = \begin{pmatrix} d_\perp & 0 & 0 \\ 0 & d_\parallel & 0 \\ 0 & 0 & d_\parallel \end{pmatrix}, \quad \text{and} \quad \mathbf{B} = \mathbf{C}^T = \begin{pmatrix} 0 & 0 & 0 \\ 0 & 0 & b \\ 0 & -b & 0 \end{pmatrix},$$

(4–26)

where we have taken the "1" direction to be normal to the plane wall. The components of $\mathbf{A}$, $\mathbf{D}$, $\mathbf{B}$, and $\mathbf{C}$ will depend on the ratio of the distance between the sphere and the wall and the sphere radius. Even nonspherical bodies in a container or near a wall will be characterized by a linear relationship between the force or torque and the velocities $\mathbf{U}$ and $\boldsymbol{\Omega}$. However, in the latter case, a formulation in terms of resistance tensors will be useful only if the body does not rotate and therefore maintains a fixed orientation relative to the boundaries. The reason is that the force (or torque) on the body generally depends on the orientation and position of the nonspherical body relative to the boundaries, as well as the orientation of $\mathbf{U}$ and $\boldsymbol{\Omega}$ relative to the body.

It is important to recognize the great simplification inherent in (4–22), which results from the linearity of the creeping flow problem for a body of fixed shape. In the absence of resistance tensors, the force or torque on an arbitrary body that translates and/or rotates with arbitrary velocities $\mathbf{U}$ and $\boldsymbol{\Omega}$ could be calculated only by completely re-solving the problem for each change in the relative magnitudes or orientation **U** and **Ω**, relative to axes that are fixed in the body. Indeed, for *nonzero* Reynolds number, where the governing equations are the full, nonlinear Navier-Stokes equations, this is precisely what must be done. Once the existence of the resistance tensors is recognized, however, we see that the force and torque for arbitrary **U** and **Ω** can be specified completely by solving a maximum of six fundamental problems, corresponding, respectively, to translation in three orthogonal directions with no rotation and rotation, about three orthogonal axes with no translation.

Solutions that Preserve the Symmetries of Boundary Data

In the preceding subsections we have seen several examples where the linearity of the creeping motion equations and boundary conditions have allowed important qualitative features of a flow to be deduced without resorting to detailed solutions of the flow problem. Here we show that even the process of solving the detailed flow problem may be greatly simplified as a consequence of linearity.

Of course, linearity allows one to utilize all of the classical tools for solving PDE's (Partial Differential Equations) that rely upon superposition, for example, series solutions based upon eigenfunction expansions and integral transforms such as Laplace and Fourier transforms. However, this is not the simplification that we refer to here. Rather, we exploit the fact that for linear systems, the spatial symmetry of the solution is often identical to the spatial symmetries of the boundary data, thus providing a major simplification in the details of obtaining the solution.

In order to demonstrate the basic idea, we consider two problems: First, the flow between a pair of rotating, parallel disks and second, the translation of a rigid sphere through an unbounded, stationary fluid.

The Rotating Disk

The rotating disk problem is illustrated in Figure 4–4. As shown, we consider two parallel disks of radius R that are separated by a gap H. For simplicity, we assume that the upper disk is stationary, while the lower disk rotates about the central axis with angular speed Ω. This relatively simple configuration is common[7] (or, at least repre-

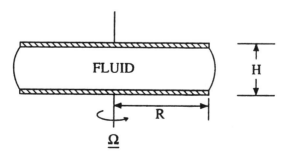

Figure 4–4 A sketch of a typical parallel disk configuration.

sentative of the actual geometry) in many applications, including lubrication, polymer processing, and the so-called parallel disk rheometer for measurement of viscosity. The claim made above is that the flow field in the creeping flow regime will often preserve the symmetry imposed on it through the boundary conditions. In the present case, it is convenient to discuss the problem in terms of a circular cylindrical coordinate system, as sketched in Figure 4-4, with the z axis oriented along the symmetry axis through the center of the disks. In this coordinate system, the velocity components on the stationary upper disk are

$$v_r = v_\theta = v_z = 0 \quad \text{for} \quad z = H, \quad r \le R, \tag{4-27a}$$

whereas the velocity on the lower rotating disk is

$$v_z = v_r = 0, \quad v_\theta = r\Omega \quad \text{for} \quad z = 0, \quad r \le R. \tag{4-27b}$$

When we say that the solution preserves the spatial symmetry of the boundary data, we mean that the solution everywhere for $r \le R$ should be in the form

$$\boxed{v_r = v_z = 0, \quad v_\theta = r\Omega F(z) \quad \text{for} \quad O \le z \le H, \quad r \le R,} \tag{4-28}$$

where $F(z)$ is a function that remains to be determined. We shall see shortly that it is very simple to determine F. First, however, we discuss several general features of the full nonlinear problem for v_r, v_θ, and v_z that arises when the creeping flow approximation cannot be used.

First, it is important to note that preservation of the spatial symmetry of the boundary data is strictly a consequence of the creeping flow approximation. For example, if we consider the rotating disk problem at finite Reynolds numbers, all three velocity components are coupled and nonzero! To see this, we can examine the steady-state Navier-Stokes equations in component form, assuming only that v_r, v_θ, and v_z are independent of θ, that is,

$$\frac{1}{r}\frac{\partial}{\partial r}(rv_r) + \frac{\partial v_z}{\partial z} = 0, \tag{4-29}$$

$$\rho\left(v_r\frac{\partial v_r}{\partial r} - \frac{v_\theta^2}{r} + v_z\frac{\partial v_r}{\partial z}\right) = -\frac{\partial p}{\partial r} + \mu\left[\frac{\partial}{\partial r}\left(\frac{1}{r}\frac{\partial}{\partial r}(rv_r)\right) + \frac{\partial^2 v_r}{\partial z^2}\right], \tag{4-30a}$$

$$\rho\left(v_r\frac{\partial v_\theta}{\partial r} + \frac{v_r v_\theta}{r} + v_z\frac{\partial v_\theta}{\partial z}\right) = \mu\left[\frac{\partial}{\partial r}\left(\frac{1}{r}\frac{\partial}{\partial r}(rv_\theta)\right) + \frac{\partial^2 v_\theta}{\partial z^2}\right], \tag{4-30b}$$

and

$$\rho\left(v_r\frac{\partial v_z}{\partial r} + v_z\frac{\partial v_z}{\partial z}\right) = -\frac{\partial p}{\partial z} + \mu\left[\frac{1}{r}\frac{\partial}{\partial r}\left(r\frac{\partial v_z}{\partial r}\right) + \frac{\partial^2 v_z}{\partial z^2}\right]. \tag{4-30c}$$

Clearly, if $v_\theta \ne 0$, we should expect $v_r \ne 0$ as a consequence of the coupling between v_r and v_θ in (4-30a), and if $v_r \ne 0$, it appears from (4-29) that v_z must also be nonzero. If, on the other hand, the nonlinear, inertia terms on the left-hand side of (4-30) are neglected, as in the creeping flow approximation, all coupling between v_r and v_θ vanishes, and, hence, the solution form (4-28) becomes a viable candidate.

It is also worthwhile to consider briefly the nondimensionalization of (4–29) and (4–30), since this leads to the necessary condition for validity of the creeping flow approximation. A unique feature that we have not seen previously is that the spatial derivatives in r and z are characterized by different length scales; that is, we assume

$$\frac{\partial}{\partial r} = \frac{1}{R}\frac{\partial}{\partial \bar{r}}, \qquad \frac{\partial}{\partial z} = \frac{1}{H}\frac{\partial}{\partial \bar{z}}. \tag{4-31}$$

A characteristic velocity scale for the v_θ component is $R\Omega$, that is,

$$v_\theta = R\Omega \bar{v}_\theta, \tag{4-32}$$

and because of the coupling between v_r and v_θ in (4–30a), we see that

$$v_r = R\Omega \bar{v}_r. \tag{4-33}$$

An appropriate estimate for the magnitude of v_z can then be obtained from the continuity equation. In particular, let us suppose

$$v_z = u_c \bar{v}_z$$

and express (4–29) in terms of dimensionless variables. The result is

$$\Omega\left(\frac{1}{\bar{r}}\frac{\partial}{\partial \bar{r}}(\bar{r}\bar{v}_r)\right) + \frac{u_c}{H}\frac{\partial \bar{v}_z}{\partial \bar{z}} = 0.$$

Hence,

$$v_z = \Omega H \bar{v}_z. \tag{4-34}$$

The only variable in (4–29) and (4–30) that does not have an obvious characteristic scale is p. In order to formally nondimensionalize (4–30a–c), let us denote the characteristic pressure scale as p_c, that is,

$$p = p_c \bar{p}.$$

Then, in dimensionless terms, the r component of (4–30) is

$$\frac{\rho\Omega H^2}{\mu}\left[\bar{v}_r\frac{\partial \bar{v}_r}{\partial \bar{r}} - \frac{\bar{v}_\theta^2}{\bar{r}} + \bar{v}_z\frac{\partial \bar{v}_r}{\partial \bar{z}}\right] = -\frac{p_c H^2}{\mu R^2 \Omega}\frac{\partial \bar{p}}{\partial \bar{r}}$$
$$+ \left[\left(\frac{H}{R}\right)^2\frac{\partial}{\partial \bar{r}}\left(\frac{1}{\bar{r}}\frac{\partial}{\partial \bar{r}}(\bar{r}\bar{v}_r)\right) + \frac{\partial^2 \bar{v}_r}{\partial \bar{z}^2}\right]. \tag{4-35}$$

It is evident that the relative magnitude of the inertia and viscous terms is proportional to the Reynolds number

$$\mathrm{Re} \equiv \frac{\rho\Omega H^2}{\mu},$$

which must therefore be vanishingly small for the creeping motion approximation to be valid. In this case, (4–35) reduces to a balance between viscous and pressure-gradient terms and we see that

$$p_c = \frac{\mu R^2 \Omega}{H^2}.$$ (4–36)

Assuming that the creeping flow approximation is valid, that is,

$$\text{Re} \equiv \frac{\rho \Omega H^2}{\mu} \ll 1,$$

the dimensionless forms of the governing equations thus become

$$\frac{1}{\bar{r}} \frac{\partial}{\partial \bar{r}} (\bar{r} \bar{v}_r) + \frac{\partial \bar{v}_z}{\partial \bar{z}} = 0,$$ (4–37)

$$0 = -\frac{\partial \bar{p}}{\partial \bar{r}} + \left[\frac{\partial}{\partial \bar{r}} \left(\frac{1}{\bar{r}} \frac{\partial}{\partial \bar{r}} (\bar{r} \bar{v}_r) \right) \left(\frac{H}{R} \right)^2 + \frac{\partial^2 \bar{v}_r}{\partial \bar{z}^2} \right],$$ (4–38a)

$$0 = \left(\frac{H}{R} \right)^2 \frac{\partial}{\partial \bar{r}} \left(\frac{1}{\bar{r}} \frac{\partial}{\partial \bar{r}} (\bar{r} \bar{v}_\theta) \right) + \frac{\partial^2 \bar{v}_\theta}{\partial \bar{z}^2},$$ (4–38b)

and

$$0 = -\frac{\partial \bar{p}}{\partial \bar{z}} + \left(\frac{H}{R} \right)^2 \frac{1}{\bar{r}} \frac{\partial}{\partial \bar{r}} \left(\bar{r} \frac{\partial \bar{v}_z}{\partial \bar{r}} \right) + \frac{\partial^2 \bar{v}_z}{\partial \bar{z}^2},$$ (4–38c)

with the dimensionless boundary conditions from (4–27a) and (4–27b),

$$\bar{v}_r = \bar{v}_z = \bar{v}_\theta = 0 \quad \text{at} \quad \bar{z} = 1, \ \bar{r} \le 1,$$

$$\bar{v}_r = \bar{v}_z = 0, \ \bar{v}_\theta = \bar{r} \quad \text{at} \quad \bar{z} = 0, \ \bar{r} \le 1.$$ (4–39)

Although completely general solutions of these linear creeping flow equations could be obtained, a substantial simplification is to assume that the symmetry of the boundary data is preserved, so that the solution takes the form (4–28), that is,

$$\bar{v}_r = \bar{v}_z = 0, \ \bar{v}_\theta = \bar{r} F(\bar{z}) \quad \text{at} \quad 0 \le \bar{z} \le 1, \ \bar{r} \le 1.$$ (4–40)

Then substituting (4–40) into (4–38a–c), we find that the whole problem reduces to the trivial form

$$\frac{\partial \bar{p}}{\partial \bar{r}} = \frac{\partial \bar{p}}{\partial \bar{z}} = 0$$

and

$$\frac{d^2 F}{d\bar{z}^2} = 0,$$ (4–41)

with boundary conditions

$$F(0) = 1, \quad F(1) = 0.$$ (4–42)

Thus, the pressure is just a constant, and the solution of (4–41) and (4–42) is

$$F = 1 - \bar{z}.$$

Substituting back into (4–40), we find that

$$\boxed{\bar{v}_r = \bar{v}_z = 0, \quad \bar{v}_\theta = \bar{r}(1 - \bar{z})}$$ (4–43)

or, in dimensional terms,

$$\boxed{v_\theta = r\Omega\left(1 - \frac{z}{H}\right).}$$ (4–44)

Stokes's Problem—Uniform Flow Past a Sphere

A second, somewhat more transparent, example of the simplification that may occur due to the preservation of the symmetry of the boundary conditions in the solution of creeping flow equations is the problem of a uniform flow (at infinity) past a solid sphere. This is known as Stokes's problem, in recognition of the British mathematician, Sir G. Stokes, who first worked out the solution in the late 1800s. A schematic of the flow configuration is shown in Figure 4–5. We may note that the problem, as shown, is identical to the steady translation of the sphere through a fluid that is stationary far from the sphere, described relative to a reference frame that translates along with the sphere.

We denote the magnitude of the velocity at infinity as U_∞ (notice that for translation of a sphere through a quiescent fluid, this is just the sphere velocity) and the radius of the sphere as a. Then the characteristic scales for nondimensionalization are

$$u_c = U_\infty, \quad l_c = a, \quad \text{and} \quad p_c = \mu\frac{U_\infty}{a}$$

[see equation (4–1)], and the governing dimensionless equations for the creeping flow limit are (4–6) with

$$\text{Re} \equiv \frac{U_\infty a}{\nu} \ll 1.$$ (4–45)

When expressed in terms of spherical coordinates, as sketched in Figure 4–5, these equations are

$$\frac{1}{r^2}\frac{\partial}{\partial r}(r^2 v_r) + \frac{1}{r\sin\theta}\frac{\partial}{\partial\theta}(\sin\theta\, v_\theta) = 0,$$ (4–46)

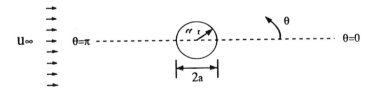

Figure 4–5 A schematic representation of uniform flow past a rigid sphere at Re ≪ 1.

$$\frac{1}{r^2}\left[\frac{\partial}{\partial r}\left(r^2 \frac{\partial v_r}{\partial r}\right) + \frac{1}{\sin\theta}\frac{\partial}{\partial\theta}\left(\sin\theta \frac{\partial v_r}{\partial\theta}\right) - 2v_r - 2\frac{\partial v_\theta}{\partial\theta} - 2v_\theta\cot\theta\right] - \frac{\partial p}{\partial r} = 0, \quad (4\text{--}47a)$$

and

$$\frac{1}{r^2}\left[\frac{\partial}{\partial r}\left(r^2 \frac{\partial v_\theta}{\partial r}\right) + \frac{1}{\sin\theta}\frac{\partial}{\partial\theta}\left(\sin\theta \frac{\partial v_\theta}{\partial\theta}\right) + 2\frac{\partial v_r}{\partial\theta} - \frac{v_\theta}{\sin^2\theta}\right] - \frac{1}{r}\frac{\partial p}{\partial\theta} = 0. \quad (4\text{--}47b)$$

In view of the symmetry of the problem, it is evident that the flow will be axi-symmetric. Thus, the azimuthal velocity component $v_\phi = 0$, and the other velocity components and the pressure are independent of ϕ. These conditions are reflected in (4–46) and (4–47).

The boundary conditions on the sphere surface are the kinematic and no-slip conditions

$$v_r = v_\theta = 0. \quad (4\text{--}48)$$

The motion at infinity is a uniform (constant) velocity in the z direction (see Figure 4–5),

$$\mathbf{u} \to \mathbf{i}_z \quad \text{as} \quad r \to \infty.$$

When expressed in terms of components in the spherical coordinate system using the identity

$$\mathbf{i}_z = \cos\theta\,\mathbf{i}_r - \sin\theta\,\mathbf{i}_\theta,$$

this *far-field* boundary condition is

$$v_r = \cos\theta, \quad \text{and} \quad v_\theta = -\sin\theta \quad \text{as} \quad r \to \infty. \quad (4\text{--}49)$$

Now, a general solution could be generated for (4–46) and (4–47) and the problem solved by applying the boundary conditions (4–48) and (4–49) to this solution.

In the present analysis, however, we take advantage of the linearity of the problem and suppose that the symmetry imposed upon the velocity field by its form at infinity is preserved throughout the domain. Thus, we look for a solution of the form

$$\boxed{\begin{aligned} v_r &= \cos\theta\, f(r), \\ v_\theta &= \sin\theta\, g(r), \end{aligned}} \quad (4\text{--}50)$$

If a solution of this form exists, the governing partial differential equations (4–46) and (4–47) should yield coupled ordinary differential equations for $f(r)$ and $g(r)$. There are two basic assumptions inherent in our formulation. First, there is no azimuthal or "swirl" component of velocity. This is due to the fact that there is no swirl in the motion at infinity, and the boundary geometry is also independent of ϕ (that is, it is axisymmetric). Second, the spherical shape preserves the fore-aft symmetry of the undisturbed flow, and thus allows the same θ symmetry everywhere in the domain.*

---

*The precise form (4–50) will work for a spherical body only. For example, if the body were an ellipsoid of revolution (that is, a spheroid) with its axis parallel to the undisturbed flow, the assumption $u_\phi = 0$ would still be applicable, but the expressions for u_r and u_θ could not be used because the surface of a spheroid is a function of both θ and r when expressed in terms of spherical coordinates. However, a decomposition similar to (4–50) would be possible if the solution and governing equations were expressed in terms

Before trying to determine the functions $f(r)$ and $g(r)$, it is important to consider the implications of (4–50) for the form of the pressure in the flow domain. To do this, we can substitute (4–50) into (4–47a) and (4–47b). The result of doing this for (4–47a) can be expressed in the form

$$\frac{\partial p}{\partial r} = H(r)\cos\theta,$$

where $H(r)$ is a function of r that involves $f(r)$ and $g(r)$. This demonstrates that the pressure must take the form

$$\boxed{p = P_o + P(r)\cos\theta} \qquad (4\text{–}51)$$

to be consistent with the proposed forms (4–50) for the velocity components. Assuming that a solution of the type (4–50) and (4–51) exists, the problem is reduced to determining the three functions $f(r)$, $g(r)$, and $P(r)$.

To obtain governing equations for these functions, we substitute (4–50) and (4–51) into the continuity and Navier-Stokes equations. From (4–46), we obtain a differential relationship between $f(r)$ and $g(r)$,

$$\boxed{f(r) + \frac{r}{2}\frac{df}{dr} + g(r) = 0,} \qquad (4\text{–}52)$$

while (4–47a) and (4–47b) yield

$$\boxed{-\frac{dP(r)}{dr} + \frac{1}{r^2}\frac{d}{dr}\left(r^2\frac{df}{dr}\right) - \frac{4f}{r^2} - \frac{4g}{r^2} = 0} \qquad (4\text{–}53)$$

and

$$\boxed{P(r) + \frac{1}{r}\frac{d}{dr}\left(r^2\frac{dg}{dr}\right) - \frac{2g}{r} - \frac{2f}{r} = 0.} \qquad (4\text{–}54)$$

Thus we obtain three coupled ordinary differential equations for the unknown functions $f(r)$, $g(r)$, and $P(r)$.

To solve these equations, it is convenient to combine them to achieve a single equation for the function $f(r)$. Thus, we can differentiate (4–54) with respect to r and combine the result with (4–53) to eliminate $P(r)$. Then, using (4–52), we can eliminate $g(r)$ in favor of $f(r)$ in the resulting equation. The result is a single fourth-order ordinary differential equation for f,

$$\frac{r^4}{8}\frac{d^4f}{dr^4} + r^3\frac{d^3f}{dr^3} + r^2\frac{d^2f}{dr^2} - r\frac{df}{dr} = 0. \qquad (4\text{–}55)$$

This is an *Euler* equation of the fourth order. It has four independent solutions of the form

$$f = r^k, \qquad (4\text{–}56)$$

---

of spheroidal, rather than spherical, coordinates. In fact, the same kind of substitution will reduce the creeping flow problem of uniform flow past an axisymmetric body to a pair of ordinary differential equations for any case where an analytic coordinate system exists in which the surface of the body is also a coordinate surface.

where the four values of k are solutions of the characteristic equation that is obtained by substituting (4–56) into (4–55),

$$-k^4 - 2k^3 + 5k^2 + 6k = 0. \tag{4–57}$$

This fourth-order algebraic equation has four distinct roots,

$$k = 0, \ -1, \ -3, \ \text{and} \ 2.$$

Thus, the general solution for $f(r)$ is

$$f(r) = c_1 + c_2 r^{-1} + c_3 r^{-3} + c_4 r^2, \tag{4–58}$$

where c_1, c_2, c_3, and c_4 are arbitrary constants that can be chosen so that $f(r)$ satisfies appropriate boundary conditions at $r = 1$ and as $r \to \infty$.

The far-field condition (4–49) requires that

$$f(r) \to 1 \quad \text{as} \quad r \to \infty. \tag{4–59}$$

Thus,

$$c_4 = 0 \quad \text{and} \quad c_1 = 1. \tag{4–60}$$

On the sphere surface $v_r = 0$, and thus

$$f(r) = 0 \quad \text{at} \quad r = 1. \tag{4–61}$$

In addition, $v_\theta = 0$ at $r = 1$, and this requires $g(r) = 0$ at $r = 1$. Since $f(r)$ and $g(r)$ are related through (4–52), we see that this condition on g is equivalent to

$$\frac{df}{dr} = 0 \quad \text{at} \quad r = 1. \tag{4–62}$$

The conditions (4–61) and (4–62) lead to a pair of algebraic equations for the remaining coefficients, c_2 and c_3. The solution of those equations is

$$c_2 = -\frac{3}{2}, \quad c_3 = \frac{1}{2}, \tag{4–63}$$

and, thus,

$$\boxed{f(r) = 1 - \frac{3}{2r} + \frac{1}{2r^3}.} \tag{4–64}$$

With $f(r)$ known, $g(r)$ can now be calculated directly from (4–52) as

$$\boxed{g(r) = -1 + \frac{3}{4r} + \frac{1}{4r^3},} \tag{4–65}$$

and it follows from (4–54) that

$$\boxed{P(r) = -\frac{3}{2r^2}.} \tag{4–66}$$

This completes the solution for the velocity and pressure fields. In terms of dimensional variables, we have shown that

$$
v_r = U_\infty \left[1 - \frac{3}{2}\left(\frac{a}{r}\right) + \frac{1}{2}\left(\frac{a}{r}\right)^3 \right] \cos\theta,
$$

$$
v_\theta = - U_\infty \left[1 - \frac{3}{4}\left(\frac{a}{r}\right) - \frac{1}{4}\left(\frac{a}{r}\right)^3 \right] \sin\theta, \tag{4-67}
$$

and

$$
p = P_0 - \frac{3\mu U_\infty}{2a}\left(\frac{a}{r}\right)^2 \cos\theta.
$$

One important prediction that we can achieve from the solution (4–67) is the hydrodynamic drag on the sphere due to its motion relative to the surrounding fluid. To do this, we apply the formula from Chapter 2 for the force on a surface in terms of the stress, and then integrate over the sphere surface. The result is

$$
F_{\text{drag}} = \int_A \mathbf{i}_z \cdot (\mathbf{i}_r \cdot \mathbf{T}) dA, \tag{4-68}
$$

or substituting for $\mathbf{i}_z$ and $\mathbf{T}$ in terms of their components in a spherical coordinate system, we have

$$
F_{\text{drag}} = \int_0^{2\pi} \int_0^\pi [\cos\theta \, T_{rr} - \sin\theta \, T_{r\theta}]|_{r=a} a^2 \sin\theta \, d\theta \, d\phi. \tag{4-69}
$$

Now, according to the constitutive equation (2–94) for an incompressible, Newtonian fluid, the T_{rr} component is

$$
T_{rr} = -p + 2\mu \left(\frac{\partial v_r}{\partial r}\right),
$$

while the $T_{r\theta}$ component is

$$
T_{r\theta} = \mu \left[r \frac{\partial}{\partial r}\left(\frac{v_\theta}{r}\right) + \frac{1}{r}\frac{\partial v_r}{\partial \theta} \right].
$$

However, at the sphere surface, we see from (4–67) that

$$
\left(\frac{\partial v_r}{\partial r}\right)_{r=a} = \left(\frac{\partial v_r}{\partial \theta}\right)_{r=a} = 0. \tag{4-70}
$$

Hence

$$
T_{rr}|_{r=a} = -P_0 + \frac{3\mu U_\infty}{2a} \cos\theta, \tag{4-71}
$$

while

$$
T_{r\theta}|_{r=a} = -\frac{3\mu U_\infty}{2a} \sin\theta. \tag{4-72}
$$

Substituting into (4–69), we thus have

$$F_{drag} = \int_0^{2\pi} \int_0^\pi \left[-P_0 \cos\theta + \frac{3\mu U_\infty}{2a}(\cos^2\theta + \sin^2\theta) \right] a^2 \sin\theta\, d\theta\, d\phi, \qquad (4\text{–}73)$$

and integrating, we obtain

$$\boxed{F_{drag} = 2\pi\mu a U_\infty + 4\pi\mu a U_\infty = 6\pi\mu a U_\infty.} \qquad (4\text{–}74)$$

This result is known as *Stokes's law*. Although $(\cos^2\theta + \sin^2\theta) = 1$, we display this term explicitly in (4–73) because it provides a basis for identifying the source of the two contributions to Stokes's law. Indeed, comparing (4–69)–(4–72) with (4–73), it is evident that the $\cos^2\theta$ contribution, which leads to the term $2\pi\mu a U_\infty$ in (4–74), is due to the dynamic pressure distribution on the sphere surface, while the $\sin^2\theta$ term corresponds to the viscous tangential stress contribution. It is common practice in fluid mechanics to identify the pressure contributions to the force on a body as the *form drag* and the viscous stress contributions as the *friction drag*. It is noteworthy that the form drag and friction drag contributions to (4–74) are of comparable magnitude for this low Reynolds number flow.

The result (4–74) can be used in conjunction with a force balance on the sphere to obtain an estimate for the sphere velocity for gravity-driven motion at steady state, where acceleration effects are not present. For this purpose, we need to know that the net buoyancy force on the sphere is

$$\boxed{F_{buoyancy} = \frac{4}{3}\pi a^3 (\rho_s - \rho)g.} \qquad (4\text{–}75)$$

This buoyancy force consists of two distinct contributions: $(4/3)\pi a^3 \rho_s g$ is the body force on a sphere of density ρ_s, and $4/3\,\pi a^3 \rho g$ is the net force contribution due to the hydrostatic pressure distribution at the sphere surface. At steady state, the two forces (4–74) and (4–75) must balance, that is,

$$6\pi a\mu U_\infty = \frac{4}{3}\pi a^3 (\rho_s - \rho)g.$$

Thus, solving for U_∞, we obtain

$$\boxed{U_\infty = \frac{2}{9}\frac{(\rho_s - \rho)ga^2}{\mu},} \qquad (4\text{–}76)$$

which is known as the *terminal* velocity for a sedimenting sphere in a quiescent fluid.

C. Representation of Two-Dimensional and Axisymmetric Flows in Terms of the Streamfunction

We have seen that the Navier-Stokes and continuity equations reduce, in the creeping motion limit, to a set of coupled, but linear, partial differential equations for the velocity and pressure, **u** and p. In the preceding section, we saw how the linearity of the creeping motion equations may be exploited to simplify the solution of problems when the form of the flow at the boundaries and the boundary geometry display similar symmetry. Although this ad hoc "preservation of symmetry" hypothesis is extremely useful when it works, it does not always work, and it is necessary to develop more general approaches for the solution of the creeping motion equations.

In the next three sections we consider the general class of two-dimensional and axisymmetric creeping flows. For this class of flows, it is possible to achieve a considerable simplification of the mathematical problem by combining the creeping flow and continuity equations to produce a single higher-order differential equation.

The starting point is a general representation theorem from vector calculus that states that any continuously differentiable vector field can be represented by three scalar functions ϕ, ψ, and χ in the form[8]

$$\mathbf{a} = \nabla\phi + \nabla\wedge(\psi\nabla\chi). \tag{4-77}$$

In effect, (4–77) represents a decomposition of the general vector field $\mathbf{a}$ into an irrotational part, associated with $\nabla\phi$, and a solenoidal (or divergence-free) part, represented by $\nabla\wedge(\psi\nabla\chi)$. It should be noted that general proofs exist that show not only that (4–77) can represent any arbitrary vector field $\mathbf{a}$ but also that an arbitrary, irrotational vector field can be represented in terms of the gradient of a single scalar function ϕ and that an arbitrary, solenoidal vector field can be represented in the form of the second term of (4–77). Since

$$\nabla\cdot[\nabla\wedge(\psi\nabla\chi)] = \nabla\cdot(\nabla\psi\wedge\nabla\chi) \equiv 0,$$

the representation (4–77) can be given also in terms of a solenoidal vector field $\mathbf{A}$, such that

$$\mathbf{a} = \nabla\phi + \nabla\wedge\mathbf{A} \quad \text{where} \quad \nabla\cdot\mathbf{A} = 0. \tag{4-78}$$

Now, since the velocity field $\mathbf{u}$ is a continuously differentiable vector field, it can be represented for any Reynolds number by either form (4–77) or (4–78), that is,

$$\mathbf{u} = \nabla\phi + \nabla\wedge(\psi\nabla\chi) \tag{4-79}$$

or

$$\mathbf{u} = \nabla\phi + \nabla\wedge\mathbf{A} \quad \text{where} \quad \nabla\cdot\mathbf{A} = 0, \tag{4-80}$$

and this is true for an arbitrary, three-dimensional motion. The function $\mathbf{A}$ that appears in (4–80) is known as the vector potential for the vorticity since

$$\nabla^2\mathbf{A} = -\nabla\wedge\mathbf{u} = -\boldsymbol{\omega}. \tag{4-81}$$

Now, the question is whether the representation (4–79) and/or (4–80) lead to any simplification of the mathematical problem for $\mathbf{u}$. To answer this question, let us for the moment stick to the creeping flow limit. For an incompressible fluid, the continuity equation requires that

$$\nabla^2\phi = 0, \tag{4-82}$$

while the equation of motion reduces to an equation for the vector potential function $\mathbf{A}$,

$$\nabla\wedge(\nabla^2\mathbf{A}) = \nabla p \tag{4-83}$$

or

$$\boxed{\nabla^4\mathbf{A} = 0.}$$
(4–84)

Clearly, $\mathbf{A}$ must be nonzero in general to satisfy (4–83). On the other hand, the contribution $\nabla\phi$ will only be nonzero if

$$(\mathbf{u} - \nabla\wedge\mathbf{A})\cdot\mathbf{n} \neq 0$$

at the boundaries. Generally, this will *not* be true, and $\mathbf{u}$ can then be represented simply as

$$\boxed{\mathbf{u} = \nabla\wedge\mathbf{A},}$$
(4–85)

with $\nabla\cdot\mathbf{A} = 0$ and $\mathbf{A}$ satisfying (4–83) or (4–84). In this case, introduction of the vector potential function $\mathbf{A}$ allows the problem to be simplified to the extent that the general form (4–85) satisfies continuity for any choice of $\mathbf{A}$, and the pressure can be eliminated to obtain a single fourth-order equation for $\mathbf{A}$. Nevertheless, for three-dimensional flows, the problem for $\mathbf{A}$ is not all that much simpler than the original problem because all three components of $\mathbf{A}$ are generally nonzero.

For two-dimensional and axisymmetric flows, however, the general representation results (4–79) and (4–80) do lead to a very significant simplification, both for creeping flows and for flows at finite Reynolds number where we must retain the full Navier-Stokes equations. The reason for this simplification is that the vector potential $\mathbf{A}$ can be represented in terms of a single scalar function,

$$\mathbf{A} = h_3\psi(q_1, q_2)\mathbf{i}_3,$$
(4–86)

where $\mathbf{i}_3$ is a unit vector that is orthogonal to the plane of motion for two-dimensional flows, and in the azimuthal direction for the axisymmetric case. We shall show shortly that this representation is sufficiently general for arbitrary two-dimensional and axisymmetric flows. First, however, we need to define all of the symbols that appear. The factor h_3 is the scale factor defined for a general, orthogonal curvilinear coordinate system

$$(q_1, q_2, q_3)$$
(4–87)

via the equation for the length of a differential line element

$$(ds)^2 \equiv \frac{1}{h_1^2}(dq_1)^2 + \frac{1}{h_2^2}(dq_2)^2 + \frac{1}{h_3^2}(dq_3)^2.$$
(4–88)

In general, for two-dimensional flows, q_3 can be identified with a Cartesian variable z, orthogonal to the plane of motion, and $h_3 \equiv 1$. However, for axisymmetric flows, q_3 represents the azimuthal angle ϕ about the axis of symmetry, and h_3 will generally be a function of q_1 and q_2. If we express $\mathbf{A}$ in terms of spherical coordinates, for example,

$$(q_1, q_2, q_3) \rightarrow (r, \theta, \phi)$$

then

$$h_1 = 1, \quad h_2 = \frac{1}{r}, \quad \text{and} \quad h_3 = \frac{1}{r\sin\theta}.$$

The scalar function ψ that appears in (4–86) is known as the *streamfunction*. The physical significance of ψ is best seen through the relationship (4–85). In particular, if we substitute (4–86) into (4–85), we obtain

$$\mathbf{u} = \left(h_2 h_3 \frac{\partial \psi}{\partial q_2}, \quad -h_1 h_3 \frac{\partial \psi}{\partial q_1}, \quad 0 \right). \qquad (4\text{–}89)$$

We see, from (4–89), that the form (4–86) is consistent with the existence of a two-dimensional or axisymmetric velocity field and, further, that the magnitudes of the two nonzero velocity components are directly related to the magnitudes of the spatial derivatives of ψ. In addition, if we calculate the derivative of ψ in the direction of motion, we see that

$$\frac{\mathbf{u} \cdot \nabla \psi}{|\mathbf{u}|} \equiv 0. \qquad (4\text{–}90)$$

Thus lines of constant ψ are everywhere tangent to the local velocity; that is, curves in space corresponding to constant values of ψ are coincident with the pathlines followed by an element of the fluid.

It is also worth noting that the volume flux of fluid across a curve joining any two arbitrary points, say, P and Q, in the flow domain is directly related to the difference in magnitude of the streamfunction at these two points. For simplicity, let us show that this is true for two-dimensional motions. The axisymmetric case follows in a very similar way. Thus, we consider a curve, as shown in Figure 4–6, that passes through the two points P and Q but is otherwise arbitrary. The volume flux of fluid across this curve, per unit length in the third direction, is

$$J = \int_l (\mathbf{u} \cdot \mathbf{n}) \, dl. \qquad (4\text{–}91)$$

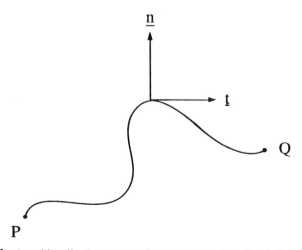

Figure 4–6 An arbitrarily chosen curve between two points, P and Q, with unit normal $\mathbf{n}$ and unit tangent $\mathbf{t}$.

Note that the integral is independent of the particular path between P and Q for an incompressible fluid where $\nabla \cdot \mathbf{u} = 0$. Now, the unit normal is

$$\mathbf{n} = \mathbf{t} \wedge \mathbf{i}_3,$$

where $\mathbf{i}_3$ is the positive unit normal in the third direction. Thus,

$$J = \int_l \mathbf{u} \cdot [\mathbf{t} \wedge \mathbf{i}_3] \, dl,$$

where

$$\mathbf{t} \, dl \equiv \left(\frac{1}{h_1} dq_1\right) \mathbf{i}_1 + \left(\frac{1}{h_2} dq_2\right) \mathbf{i}_2,$$

and thus,

$$(\mathbf{t} \wedge \mathbf{i}_3) \, dl \equiv \frac{1}{h_2} dq_2 \mathbf{i}_1 - \frac{1}{h_1} dq_1 \mathbf{i}_2.$$

Hence,

$$J = \int_P^Q \left[\frac{d\psi}{dq_2} dq_2 + \frac{d\psi}{dq_1} dq_1\right];$$

$$\boxed{J = \int_{\psi_P}^{\psi_Q} d\psi = \psi_Q - \psi_P,} \tag{4-92}$$

where ψ_Q and ψ_P denote the values of ψ at the points Q and P, respectively.

Introduction of the streamfunction for axisymmetric and two-dimensional problems simplifies their solution in two ways. First, in view of the definitions (4–85) and (4–86), it is evident that the continuity equation for an incompressible fluid

$$\nabla \cdot \mathbf{u} = 0$$

will be satisfied automatically for any function ψ that satisfies the other conditions of the problem. Second, the equations of motion are reduced from a coupled pair of equations relating $\mathbf{u}$ and p to a single higher-order equation for the scalar function ψ. In the creeping flow approximation this equation can be obtained directly from (4–84). Substituting from (4–86), we obtain

$$\boxed{E^4 \psi = 0,} \tag{4-93}$$

where

$$E^2 \equiv \frac{h_1 h_2}{h_3} \left[\frac{\partial}{\partial q_1}\left(\frac{h_1 h_3}{h_2} \frac{\partial}{\partial q_1}\right) + \frac{\partial}{\partial q_2}\left(\frac{h_2 h_3}{h_1} \frac{\partial}{\partial q_2}\right)\right].$$

In general, the E^2 operator is not the same as the more familiar $\nabla^2 (\equiv \nabla \cdot \nabla)$, which is defined as

$$\nabla^2 \equiv h_1 h_2 h_3 \left[\frac{\partial}{\partial q_1}\left(\frac{h_1}{h_2 h_3} \frac{\partial}{\partial q_1}\right) + \frac{\partial}{\partial q_2}\left(\frac{h_2}{h_1 h_3} \frac{\partial}{\partial q_2}\right)\right].$$

However, for two-dimensional flows, the third direction corresponds to a Cartesian coordinate direction and

$$h_3 = 1.$$

In this case, E^2 and ∇^2 are identical, and the governing equation for ψ in two dimensions is therefore normally expressed as

$$\nabla^4\psi = 0, \qquad (4\text{-}94)$$

which is the familiar biharmonic equation in two dimensions.

We shall discuss the solution of (4-93) and (4-94) shortly. First, however, it is worth noting the form of the full Navier-Stokes equations when expressed in terms of ψ. For this purpose, it is convenient first to take the curl of the equations to eliminate the pressure. In the two-dimensional and axisymmetric flows cases considered here, this gives

$$\text{Re}\left[\frac{\partial\omega}{\partial t} - \nabla\wedge(\mathbf{u}\wedge\omega)\right] + \nabla\wedge(\nabla\wedge\omega) = 0. \qquad (4\text{-}95)$$

Now,

$$\omega = \nabla\wedge(\nabla\wedge(h_3\psi\mathbf{i}_3)) = -h_3 E^2\psi\mathbf{i}_3, \qquad (4\text{-}96)$$

according to the definitions of ω and $\mathbf{u}$ (in terms of ψ). Thus substituting into (4-95),

$$\text{Re}\left[\frac{\partial}{\partial t}(E^2\psi) + \frac{h_1 h_2}{h_3}\left\{\frac{\partial}{\partial q_1}\left(h_3^2\frac{\partial\psi}{\partial q_2}E^2\psi\right) - \frac{\partial}{\partial q_2}\left(h_3^2\frac{\partial\psi}{\partial q_1}E^2\psi\right)\right\}\right] = E^4\psi.$$

$$(4\text{-}97)$$

Again, for the two-dimensional case, this reduces to

$$\text{Re}\left[\frac{\partial}{\partial t}(\nabla^2\psi) + h_1 h_2\left\{\frac{\partial}{\partial q_1}\left(\frac{\partial\psi}{\partial q_2}\nabla^2\psi\right) - \frac{\partial}{\partial q_2}\left(\frac{\partial\psi}{\partial q_1}\nabla^2\psi\right)\right\}\right] = \nabla^4\psi.$$

$$(4\text{-}98)$$

Clearly, in the limit $\text{Re}\to 0$, these equations reduced to the limiting forms (4-93) and (4-94).

D. Solutions via Eigenfunction Expansions for Two-Dimensional Creeping Flows

We saw, in the previous section, that problems of creeping motion in two dimensions can be reduced to the solution of the biharmonic equation (4-94), subject to appropriate boundary conditions. In order to actually obtain a solution, it is convenient to express (4-94) as a coupled pair of second-order partial differential equations,

$$\nabla^2\psi = -\omega, \qquad (4\text{-}99)$$

$$\nabla^2\omega = 0. \qquad (4\text{-}100)$$

The solution of these equations via standard eigenfunction expansions can be carried out for any curvilinear, orthogonal coordinate system for which the Laplacian operator ∇^2 is separable. Of course, the most appropriate coordinate system for a particular application will depend on the boundary geometry. In this section we briefly consider the most common cases for two-dimensional flows of Cartesian and circular cylindrical coordinates.

General Eigenfunction Expansions in Cartesian and Cylindrical Coordinates

For Cartesian coordinates, a solution of (4–100) exists in the separable form

$$\omega = X(x)Y(y). \tag{4-101}$$

Substituting into (4–100), we obtain

$$\frac{X''}{X} = -\frac{Y''}{Y} = \pm m^2, \tag{4-102}$$

where m is an arbitrary complex number. Hence,

$$X'' \pm m^2 X = 0, \quad Y'' \pm m^2 Y = 0, \tag{4-103}$$

and from this we deduce that

$$\boxed{\omega = e^{mx}e^{imy}} \tag{4-104}$$

for arbitrary complex m. Now, to obtain a general solution for ψ, we must solve (4–99) with the right-hand side evaluated using (4–104). Hence,

$$\nabla^2 \psi = -e^{mx}e^{imy}\gamma_m, \tag{4-105}$$

where γ_m is an arbitrary constant. The solution of (4–105) is the sum of a homogeneous solution of the form (4–104) plus a particular solution to reproduce the right-hand side. After some manipulation, we find

$$\boxed{\psi_m = \alpha_m e^{mx}e^{imy} + \beta_m x e^{mx}e^{imy} + \delta_m y e^{mx}e^{imy}.} \tag{4-106}$$

Hence, the most general solution for ψ expressed in Cartesian coordinates is

$$\boxed{\psi = \sum_m \psi_m} \tag{4-107}$$

with arbitrary complex values of m.

Starting with (4–99) and (4–100), we can also obtain a general solution for ψ in terms of a circular cylindrical coordinate system, and this solution is more immediately applicable to real problems. The governing equations in polar cylindrical coordinates are

$$\boxed{\frac{1}{r}\frac{\partial}{\partial r}\left(r\frac{\partial \psi}{\partial r}\right) + \frac{1}{r^2}\frac{\partial^2 \psi}{\partial \theta^2} = \omega} \tag{4-108}$$

and

$$\boxed{\frac{1}{r}\frac{\partial}{\partial r}\left(r\frac{\partial \omega}{\partial r}\right) + \frac{1}{r^2}\frac{\partial^2 \omega}{\partial \theta^2} = 0.} \tag{4-109}$$

We seek a solution of (4–108) and (4–109) in the separable form

$$\boxed{\omega = R(r)F(\theta) \quad \text{and} \quad \psi = S(r)H(\theta).} \tag{4-110}$$

Substitution for ω in (4–109) yields

$$\frac{r}{R}\frac{\partial}{\partial r}\left(r\frac{\partial R}{\partial r}\right) = -\frac{F''}{F} = \lambda_n^2,$$

where λ_n is an arbitrary constant, either real or complex. Hence, for $\lambda_n \neq 0$ there are two independent solutions for each of the functions R and F, namely,

$$F = \sin\lambda_n\theta, \quad \cos\lambda_n\theta, \quad \text{and } R = r^{\lambda_n}, r^{-\lambda_n}. \tag{4-111}$$

For $\lambda_n = 0$, on the other hand,

$$F = a_0, \theta \quad \text{and} \quad R = c_0, \ln r. \tag{4-112}$$

Hence, the most general solution for ω in the separable form of Equation (4-110) is

$$\omega = (a_0 + b_0\theta)(c_0 + d_0\ln r) + \sum_{n=1}^{\infty} (a_n\cos\lambda_n\theta + b_n\sin\lambda_n\theta)(c_n r^{\lambda_n} + d_n r^{-\lambda_n}).$$

$$\tag{4-113}$$

To determine ψ, we must solve (4-108) with the general form (4-113) substituted for ω. The general solution for ψ consists of a *homogeneous* solution of the same form as (4-113), plus a particular solution. To obtain terms of the particular solution corresponding to the summation in (4-113), we try

$$\psi_p = r^s\sin\lambda_n\theta, \quad r^s\cos\lambda_n\theta. \tag{4-114}$$

Substitution into the left-hand side of (4-108) yields

$$(s^2 - \lambda_n^2)r^{s-2}\sin\lambda_n\theta, \quad (s^2 - \lambda_n^2)r^{s-2}\cos\lambda_n\theta. \tag{4-115}$$

Thus, comparing (4-115) and (4-113), we see

$$s = \lambda_n + 2, \quad -\lambda_n + 2, \tag{4-116}$$

except for $\lambda_n = 1$. In this case, the second of the values for s is $s = 1$, and the particular solution form (4-114) reduces to a solution of the homogeneous equation. Thus, in this case, the corresponding particular solutions take the form

$$r^3\sin\theta, \quad r^3\cos\theta, \quad r\theta\sin\theta, \quad \text{and } r\theta\cos\theta. \tag{4-117}$$

Finally, for $\lambda_n = 0$, the particular solution is

$$\left[(\hat{a}_0\theta + \hat{b}_0)(\hat{c}_0 r^2 + \hat{d}_0 r^2(\ln r - 1))\right]. \tag{4-118}$$

Thus, the general solution for ψ is

$$\psi = (c_0 + d_0\ln r + \hat{c}_0 r^2 + d_0 r^2(\ln r - 1))(a_0\theta + b_0)$$

$$+ (c_1 r + d_1 r^{-1} + \hat{c}_1 r^3)(a_1\sin\theta + b_1\cos\theta) + \hat{d}_1 r(\hat{a}_1\theta\sin\theta + \hat{b}_1\theta\cos\theta)$$

$$+ \sum_{n=2}^{\infty}\left[c_{\lambda_n}r^{\lambda_n} + d_{\lambda_n}r^{-\lambda_n} + \hat{c}_{\lambda_n}r^{\lambda_n+2} + \hat{d}_{\lambda_n}r^{2-\lambda_n}\right](a_{\lambda_n}\sin\lambda_n\theta + b_{\lambda_n}\cos\lambda_n\theta).$$

$$\tag{4-119}$$

Application to Two-Dimensional Flow Near Corners

The general solution (4–119) can be applied to examine two-dimensional flows in the region between two plane boundaries that intersect at a sharp corner. This class of creeping motion problems was considered in a classic paper by Moffatt (1964),[9] and our discussion is similar to that given by Moffatt. A typical configuration is shown in Figure 4–7 for the case in which one boundary at $\theta = 0$ is moving with constant velocity U in its own plane and the other at $\theta = \alpha$ is stationary.

A noteworthy feature of this configuration is that it lacks a definite physical length scale. One rationalization of this fact is that the corner is generally a *localized* part of some more complicated global geometry. In any case, an obvious question is the range of validity of the creeping motion approximation for this situation in which a fixed characteristic length scale (and thus a fixed Reynolds number) does not exist. To answer this question, we need to obtain an estimate for the magnitudes of the inertial and viscous terms in the equations of motion. A starting point is the magnitude of the velocity, corresponding to the form (4–119) for the streamfunction, namely,

$$|\mathbf{u}| = 0(Ar^{\xi-1}), \tag{4-120}$$

where ξ is the real part of λ and A is a constant with dimension of velocity/(length)$^\xi$. Then, the magnitudes of the inertia and viscous terms in the equations of motion can be estimated as

$$|\mathbf{u}\cdot\nabla\mathbf{u}| = 0(A^2 r^{2\xi-3})$$

and

$$|\nu\nabla^2\mathbf{u}| = 0(\nu Ar^{\xi-3}).$$

Hence, the ratio of inertia to viscous terms is

$$\frac{|\mathbf{u}\cdot\nabla\mathbf{u}|}{|\nabla^2\mathbf{u}|} = 0\left(\frac{Ar^\xi}{\nu}\right), \tag{4-121}$$

and we see that the creeping motion approximation is valid provided

$$R \equiv \frac{Ar^\xi}{\nu} \ll 1, \tag{4-122}$$

where R is effectively a Reynolds number based upon the distance from the corner. Hence, for $\xi > 0$, inertia is negligible for sufficiently small values of r, whereas for $\xi < 0$, neglect of inertia requires r to be sufficiently large. We shall focus here on problems where $\xi > 0$. Because the resulting solution in this case is a *local approximation*, certain features of the flow will generally remain indeterminate. In reality, they are determined by the features of the flow at a large distance from the corner where the creeping flow approximation breaks down.

The simplest problem of the type considered here is the one sketched in Figure 4–7, which was originally solved by Taylor (1960).[10] The problem may be considered to be a local approximation for the action of a wiper blade on a solid surface that is completely covered by liquid. The boundary conditions in this case are

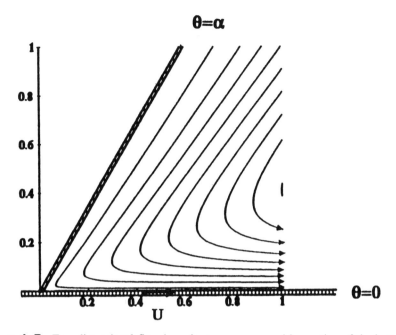

Figure 4–7 Two-dimensional flow in a sharp corner, caused by motion of the bottom surface (at $\theta = 0$) with the velocity **U**. The plot shows streamlines, $\bar{\psi} = \psi/U$, calculated from Equation (4–127) for $\alpha = \pi/3$. Contour values range from 0 at the walls in increments of 0.02.

$$u_r = U, \ u_\theta = 0 \quad \text{at} \quad \theta = 0,$$

$$u_r = u_\theta = 0 \quad \text{at} \quad \theta = \alpha, \tag{4-123}$$

where the r and θ components of velocity are related to the streamfunction via the definitions

$$u_r = \frac{1}{r}\frac{\partial \psi}{\partial \theta}, \quad u_\theta = -\frac{\partial \psi}{\partial r}. \tag{4-124}$$

In view of (4–124), it is clear that the requirement $u_r = U$ (constant) at $\theta = 0$ can only be satisfied by the solution form

$$\psi = rF(\theta).$$

Referring to the general solution (4–119), the terms that are linear in r are

$$\psi = r(A_1 \sin\theta + B_1 \cos\theta + C_1 \theta \sin\theta + D_1 \theta \cos\theta), \tag{4-125}$$

where A_1, B_1, C_1, and D_1 are constants. Applying the boundary conditions (4–123) to this solution, we find that

$$U = A_1 + D_1,$$

$$0 = (A_1 + D_1)\cos\alpha + (C_1 - B_1)\sin\alpha + C_1\alpha\cos\alpha - D_1\alpha\sin\alpha,$$

$$0 = B_1,$$

$$0 = A_1\sin\alpha + B_1\cos\alpha + C_1\alpha\sin\alpha + D_1\alpha\cos\alpha.$$

Thus, solving for A_1, B_1, C_1, and D_1, we have

$$B_1 = 0,$$

$$C_1 = \frac{U(\alpha - \sin\alpha\cos\alpha)}{\sin^2\alpha - \alpha^2},$$

$$D_1 = \frac{U\sin^2\alpha}{\sin^2\alpha - \alpha^2},$$

$$A_1 = \frac{U\alpha^2}{\sin^2\alpha - \alpha^2}, \tag{4-126}$$

and

$$\psi = -\frac{Ur}{(\sin^2\alpha - \alpha^2)}\left[\alpha^2\sin\theta + (\sin\alpha\cos\alpha - \alpha)\theta\sin\theta - (\sin^2\alpha)\theta\cos\theta\right]. \tag{4-127}$$

The streamlines corresponding to (4–127) are shown in Figure 4–7. Although the velocity components u_r and u_θ are perfectly well behaved, the shear stress $\tau_{r\theta}$ is *singular* in the limit $r \to 0$. Indeed, if we calculate $\tau_{r\theta}|_{\theta=0}$, we find that

$$\tau_{r\theta}|_{\theta=0} = -\frac{2U\mu}{(\sin^2\alpha - \alpha^2)}\frac{1}{r}(\sin\alpha\cos\alpha - \alpha). \tag{4-128}$$

Clearly, the solution breaks down in the limit $r \to 0$. In fact, according to (4–128), an *infinite force* is necessary to maintain the plane $\theta = 0$ in motion at a finite velocity U, and this prediction is clearly unrealistic. Presumably, one of the assumptions of the theory breaks down, although a definitive resolution of the difficulty does not exist at the present time. In our view, the most plausible explanation is that the no-slip boundary condition is inadequate in regions of extremely high shear stress, but we shall not attempt to carry the discussion of this point further at this time.

Closely related to the Taylor problem is the situation sketched in Figure 4–8, when a flat plate is drawn into a viscous fluid through a free surface (that is, an interface). In reality, or course, the interface will tend to deform as a result of the motion of the plate, but we assume here that the interface remains flat. Then the problem is identical to the previous Taylor problem except for the boundary conditions, which now become

$$u_r = U, \ u_\theta = 0 \quad \text{at} \quad \theta = -\alpha,$$

$$\tau_{r\theta} \equiv \frac{1}{r}\left(\frac{\partial u_r}{\partial\theta}\right) = 0, \ u_\theta = 0 \quad \text{at} \quad \theta = 0. \tag{4-129}$$

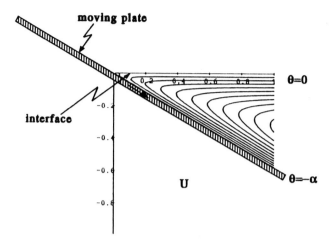

Figure 4–8 Two-dimensional flow in a sharp corner created when a flat plate is drawn into a fluid through a flat fluid interface. The plot shows streamlines, $\bar{\psi} = \psi/U$, calculated from Equation (4–130) for $\alpha = \pi/6$. Contour values range from 0 at the walls in increments of 0.0105.

The solution in this case is

$$\psi = Ur(\sin\alpha\cos\alpha - \alpha)^{-1}[\sin\alpha(\theta\cos\theta) - (\alpha\cos\alpha)\sin\theta].$$

$$(4\text{–}130)$$

In this case, it is of interest to calculate the tangential velocity on the free surface,

$$u_r|_{\theta=0} = -U\left\{1 - \frac{(\alpha - \sin\alpha)(+\cos\alpha + 1)}{-\sin\alpha\cos\alpha + \alpha}\right\}.$$

$$(4\text{–}131)$$

The term in brackets is positive and independent of r. Hence, the velocity on the free surface is constant but smaller than the velocity of the solid plate. The speed of a fluid particle that travels along the interface must therefore increase discontinuously as it reaches the plate and turns the corner—that is, it must undergo an infinite acceleration. This infinite acceleration is produced by an infinite stress and pressure, $0(r^{-1})$, on the plate in the limit $r \to 0$. Again, we conclude that the solution breaks down in the limit $r \to 0$.

A third interesting example of a flow in the vicinity of a corner is the motion of two hinged plates. As sketched in Figure 4–9, we assume that the plates rotate with angular velocities $-\omega$ and $+\omega$, respectively. Thus, the boundary conditions are

$$u_r = 0, \quad u_\theta = -\omega r \quad \text{at} \quad \theta = +\alpha,$$

and

$$u_r = 0, \quad u_\theta = \omega r \quad \text{at} \quad \theta = -\alpha.$$

$$(4\text{–}132)$$

Since $u_\theta = -\partial\psi/\partial r$, it is evident from the general solution (4–119) and the boundary conditions (4–132) that $\lambda = 2$, so that

$$\psi = r^2(A_2 + B_2\theta + C_2\sin2\theta + D_2\cos2\theta).$$

$$(4\text{–}133)$$

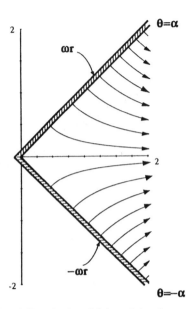

Figure 4–9 Two-dimensional flow in the vicinity of the sharp corner between two hinged, plane walls that are rotating toward one another with angular velocity $\omega(-\omega)$. The plot shows streamlines, $\bar{\psi} = \psi/\omega$, calculated from Equation (4–134) for $\alpha = \pi/44$. Contour values range from 0 at $\theta = 0$ in increments of 0.2105.

It is convenient to use the conditions (4–132) at $\theta = \alpha$ and the symmetry conditions $u_\theta = \partial u_r/\partial\theta = 0$ at $\theta = 0$ to determine the constants A_2, B_2, C_2, and D_2. After some manipulation, we find that the solution

$$\psi = \omega r^2 \frac{1}{2}(\sin 2\alpha - 2\alpha\cos 2\alpha)^{-1}(\sin 2\theta - 2\theta\cos 2\alpha). \tag{4–134}$$

In this case, both the velocity components and the stress are bounded in the limit $r \to 0$, but the pressure exhibits a $0(\log r)$ singularity.

Finally, it is of interest to consider the nature of the flow near a sharp corner that is induced by an arbitrary "stirring" flow at large distances from the corner. In general, there are two fundamental types of flow patterns that can be induced near the corner: an antisymmetric flow, as sketched in Figure 4–10a, and a symmetrical flow, as sketched in Figure 4–10b. The actual flow near a corner will generally be a mixture of antisymmetrical and symmetrical flow types, but it is permissible in the linear Stokes approximation to consider them separately (the more general flow can then be constructed as a superposition of the simpler fundamental flows). Here we consider only the antisymmetric case, which is the more interesting of the two. Thus, we consider the general antisymmetric form for ψ, namely,

$$\psi = \sum_{n=1}^{\infty} r^{\lambda_n} f_{\lambda_n}(\theta), \tag{4–135}$$

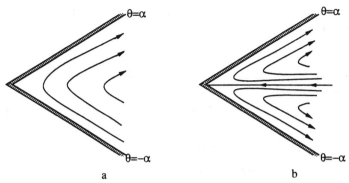

Figure 4–10 A sketch of the two-dimensional flow near a sharp corner that is induced by an arbitrary "stirring" flow at large distances from the corner: (a) antisymmetric, (b) symmetric.

where

$$f_{\lambda_n}(\theta) = A_n\cos\lambda_n\theta + C_n\cos(\lambda_n - 2)\theta. \qquad (4\text{-}136)$$

The boundary conditions at the walls require

$$f(\pm\alpha) = f'(\pm\alpha) = 0. \qquad (4\text{-}137)$$

We focus only on the dominant term in the expansion (4–135) for small r, that is, the term with the largest real part of λ_n,

$$\psi \sim r^{\lambda_1}f_{\lambda_1}(\theta). \qquad (4\text{-}138)$$

Thus, applying the boundary conditions (4–137) to (4–138), we find that

$$A_1\cos\lambda_1\alpha + C_1\cos(\lambda_1 - 2)\alpha = 0,$$

$$A_1\lambda_1\sin\lambda_1\alpha + C_1(\lambda_1 - 2)\sin(\lambda_1 - 2)\alpha = 0. \qquad (4\text{-}139)$$

Hence, in order to obtain a nontrivial solution, λ_1 must satisfy the condition

$$\det\begin{vmatrix} \cos\lambda_1\alpha & \cos(\lambda_1 - 2)\alpha \\ \lambda_1\sin\lambda_1\alpha & (\lambda_1 - 2)\sin(\lambda_1 - 2)\alpha \end{vmatrix} = 0. \qquad (4\text{-}140)$$

The resulting value of λ_1 is known as the eigenvalue for this problem, and the corresponding function $f_{\lambda_1}(\theta)$ is the eigenfunction. The coefficients A_1 and C_1 corresponding to (4–139) are

$$A_1 = K\cos(\lambda_1 - 2)\alpha,$$

$$C_1 = -K\cos\lambda_1\alpha,$$

so that

$$\psi = Kr^{\lambda_1}f_{\lambda_1}[\cos(\lambda_1 - 2)\alpha\,\cos\lambda_1\theta - \cos\lambda_1\alpha\cos(\lambda_1 - 2)\theta]. \qquad (4\text{-}141)$$

The coefficient K cannot be determined from the local analysis alone but only by matching the local solution to the stirring flow far from the corner. To obtain λ_1, we express (4–140) in the form

$$(\lambda_1 - 2)\sin(\lambda_1 - 2)\alpha\cos\lambda_1\alpha = \lambda_1\sin\lambda_1\alpha\cos(\lambda_1 - 2)\alpha$$

or, upon rearrangement,

$$-(\lambda_1 - 1)\sin 2\alpha = \sin[2(\lambda_1 - 1)\alpha]. \tag{4-142}$$

This equation has a *real* solution for λ_1 when $2\alpha > 146°$ but no real solutions for $0 \le 2\alpha < 146°$. In this range, λ_1 is complex.

Let us consider this latter case. If we denote $(\lambda - 1)$ as $p + iq$, then (4-142) can be expressed in the form

$$\sin\xi\cosh\eta = -k\xi,$$

$$\cos\xi\sinh\eta = -k\eta, \tag{4-143}$$

where $\xi = 2\alpha p$, $\eta = 2\alpha q$, and k is the positive parameter $k = \sin 2\alpha/2\alpha$. Any solution of these equations must satisfy the condition

$$(2n - 1)\pi < \xi_n < \left(2n - \frac{1}{2}\right)\pi,$$

where $\sin\xi_n$ and $\cos\xi_n$ are both negative. The corresponding eigenvalue is

$$\lambda_n = 1 + (2\alpha)^{-1}(\xi_n + i\eta_n). \tag{4-144}$$

The eigenvalue with the least positive real part, which dominates near the corner $(r < 1)$, obviously occurs for $n = 1$. Numerical values of ξ_1 and η_1 for $2\alpha < 146°$ were tabulated by Moffatt (1964). The specific values are not significant for present purposes. What is significant is the fact that λ_1 is complex. This feature of the solution implies the existence, for small r, of an infinite *sequence* of closed streamline eddies in the corner, the first two of which are sketched in Figure 4-11.

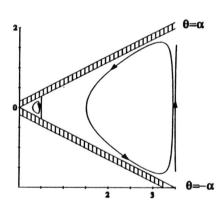

Figure 4-11 The infinite sequence of closed eddies in a sharp corner with an acute angle $2\alpha < 146°$. This sequence of closed streamline flows is commonly known as Moffatt eddies after the mathematician H.K. Moffatt (1964),[9] who first discovered their existence. Contours shown here are $\psi = 0$, -0.005, and 10, and we see the first two eddies in the sequence. The next eddy, closer to the corner, is too small to see with this resolution.

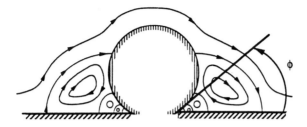

Figure 4–12 A qualitative sketch of the flow over a cylindrical ridge on a plane boundary (adapted from figure 1.3.9(a) in Reference 13).

To demonstrate that such a sequence of eddies does exist, we can show that there is an infinite sequence of dividing streamlines, $\psi = 0$, for successively smaller values of r. If we introduce (4–144) into (4–141), the dominant streamfunction can be written in the symbolic form

$$\psi_1 = r^{(1+p)}[\cos(q \cdot \ln r)g_1(\theta) - \sin(q \cdot \ln r)g_2(\theta)]. \qquad (4\text{--}145)$$

As demonstrated by Moffatt (1964), this streamfunction has infinitely many zeroes as r approaches zero, namely,

$$q \cdot \ln r = \tan^{-1}\left(\frac{g_1(\theta)}{g_2(\theta)}\right) - \frac{\pi}{2} - n\pi \quad (n = 0,1,2, \ldots). \qquad (4\text{--}146)$$

Hence, antisymmetric flow induced in a corner between two solid boundaries with an acute angle less than 146° may be expected to show an infinite sequence of increasingly small (and weak) eddies.

In this section we have considered the formation of a sequence of eddies near a corner between two plane boundaries with the suggestion that this type of motion could be driven by some external "stirring" motion at large distances from the corner. There have, in fact, been a number of investigations of problems in which the Moffatt eddies occur locally as part of an overall flow structure. Among these are the studies of Wakiya, O'Neill, and others[11] for simple shear flow over a circular cylindrical ridge on a solid, plane surface as depicted in Figure 4–12. In this case a sequence of eddies is found in the groove formed at the intersection between the cylinder and the plane wall when the angle of intersection, ϕ, is less than 146.3°. A closely related problem is the two-dimensional motion of two equal, parallel cylinders that are touching, either along the line connecting the centers of the cylinders or perpendicular to it, where a sequence of eddies appears in the neighborhood of the contact point.[12] A summary of these and related problems is available for the interested reader.[13]

Later, in Chapter 8, we shall briefly consider the additional two-dimensional problem of creeping flow past a circular cylinder (that is, a circle) in an unbounded fluid that undergoes a uniform motion at large distances from the cylinder. For the moment, however, we turn to general solution procedures for other classes of creeping flow problems.

E. Eigenfunction Expansion for Axisymmetric Creeping Flows in Spherical Coordinates

Solution by Separation of Variables

We saw in Section C that the creeping motion and continuity equations for axisymmetric, incompressible flow can be reduced to the single fourth-order partial differential equation for the streamfunction,

$$\boxed{E^4\psi = 0.} \tag{4-147}$$

An analytic solution for such a problem can thus be sought as a superposition of separable solutions of this equation in any orthogonal curvilinear coordinate system. The most convenient coordinate system for a particular problem is dictated by the geometry of the boundaries. As a general rule, at least one of the flow boundaries should coincide with a coordinate surface. Thus, if we consider an axisymmetric coordinate system (ξ, η, ϕ), then either $\xi = $ const or $\eta = $ const should correspond to one of the boundaries of the flow domain.

In this section we consider the general solution of (4–147) in the spherical coordinate system (r, θ, ϕ). This coordinate system is particularly useful for flows in the vicinity of a spherical boundary, but we shall begin by simply deducing the most general solution of (4–147) that is consistent with the constraint of axisymmetry, namely,

$$u_\theta = 0 \quad \text{at} \quad \theta = 0, \pi. \tag{4-148}$$

Rather than using the polar angle θ as an independent variable, it is more convenient to introduce

$$\eta \equiv \cos\theta$$

so that the coordinate variables are (r, η, ϕ) with $-1 \le \eta \le 1$, and the E^2 operator takes the simplified form

$$E^2 \equiv \frac{\partial^2}{\partial r^2} + \frac{(1 - \eta^2)}{r^2} \frac{\partial^2}{\partial \eta^2}. \tag{4-149}$$

The nonzero velocity components, expressed in terms of the streamfunction, are

$$\boxed{u_r = -\frac{1}{r^2} \frac{\partial \psi}{\partial \eta}, \quad u_\theta \sqrt{1 - \eta^2} = -\frac{1}{r} \frac{\partial \psi}{\partial r}.} \tag{4-150}$$

The symmetry condition (4–148) requires that

$$\frac{\partial \psi}{\partial r} = 0 \quad \text{at} \quad \eta = \pm 1 \tag{4-151a}$$

or, alternatively,

$$\psi = \text{const} \quad \text{at} \quad \eta = \pm 1. \tag{4-151b}$$

We seek a solution of (4-147) via the method of separation of variables. For this purpose, it is convenient to note that $E^4\psi = 0$ can be split into two second-order equations,

$$\boxed{E^2 w = 0} \tag{4-152}$$

and

$$\boxed{E^2 \psi = w.} \tag{4-153}$$

Obviously, substituting (4-153) into (4-152), we recover (4-147). In the present analysis, we first determine the most general solution for w by solving (4-152), and then we solve (4-153) for ψ with w given by this general solution.

To solve (4-152), we assume that

$$w = R(r)H(\eta). \tag{4-154}$$

Hence, substituting into (4-152), we obtain

$$\frac{r^2}{R}\frac{d^2 R}{dr^2} + \frac{(1-\eta^2)}{H}\frac{d^2 H}{d\eta^2} = 0. \tag{4-155}$$

The first term is a function of r only, while the second is a function of η. Hence, it follows that each must equal a constant. We denote this constant as $n(n+1)$, and thus (4-155) separates into two equations,

$$r^2\frac{d^2 R}{dr^2} - n(n+1)R = 0 \tag{4-156}$$

and

$$(1-\eta^2)\frac{d^2 H}{d\eta^2} + n(n+1)H = 0. \tag{4-157}$$

Equation (4-156) is a particular case of Euler's equation, for which a general solution is

$$R = r^s. \tag{4-158}$$

Substituting into (4-156), we obtain the "characteristic" equation for s,

$$s(s-1) - n(n+1) = 0. \tag{4-159}$$

The two roots of this quadratic equation are

$$s = n+1, \quad s = -n, \tag{4-160}$$

so that the two independent solutions of (4-156) are

$$\boxed{R = r^{n+1}, \; r^{-n}.} \tag{4-161}$$

To solve (4-157) for H, it is convenient to differentiate the whole equation with respect to η because this transforms it into an equation of well-known form, namely,

$$\frac{d}{d\eta}\left[(1-\eta^2)\frac{dY}{d\eta}\right] + n(n+1)Y = 0, \tag{4-162}$$

with

$$Y \equiv \frac{dH}{d\eta}. \tag{4-163}$$

Equation (4-162) is called Legendre's equation. In general, it has two independent solutions, known as Legendre's functions of the first and second kind. For the particular choice of the constant $n(n+1)$ in which n is an integer, the Legendre functions of the first kind are polynomials of degree n, which are known as Legendre polynomials and normally denoted as P_n. The first several Legendre polynomials are

$$P_0 \equiv 1,$$

$$P_1 \equiv \eta,$$

$$P_2 \equiv \frac{1}{2}(3\eta^2 - 1). \tag{4-164}$$

For each integer value of n in (4-162), there is a corresponding Legendre polynomial of degree n. The Legendre function of the second kind has a logarithmic singularity at $\eta = \pm 1$. The solution that we seek for Y (and thus for w) is regular in the domain $-1 \le \eta \le 1$, and so we eliminate this second solution of (4-162) from further consideration. Hence,

$$\boxed{Y(\eta) = P_n(\eta),} \tag{4-165}$$

for arbitrary integer values of n.

We have shown that the most general, regular solutions for Y is the Legendre polynomial $P_n(\eta)$, for arbitrary integer values of n. However, we require a solution for the function $H(\eta)$ rather than for $Y(\eta)$. To obtain $H(\eta)$ from $Y(\eta)$, we integrate according to (4-163). However, we want the solution for ψ (and thus also the solution for w) to satisfy the symmetry condition (4-151) at $\eta = \pm 1$. Since this requires that

$$H(\eta) = 0 \quad \text{at} \quad \eta = \pm 1, \tag{4-166}$$

a convenient choice for the integral transformation from $Y(\eta)$ to $H(\eta)$ is

$$\boxed{H(\eta) \equiv \int_{-1}^{\eta} Y(\eta)d\eta} \tag{4-167}$$

since $H(\eta)$, defined in this way, automatically satisfies the condition (4-166) at both $\eta = \pm 1$ for all n, other than $n = 0$. To see that the condition at $\eta = 1$ is satisfied, we note that the integral in (4-167) can be written in the form

$$\int_{-1}^{\eta} P_0(\eta) P_n(\eta) d\eta,$$

for any value of n, and that the Legendre polynomials are orthogonal functions, that is,

$$\int_{-1}^{1} P_n P_m d\eta = \begin{cases} 0 & \text{for } n \ne m \\ \dfrac{2}{2n+1} & \text{for } n = m. \end{cases} \tag{4-168}$$

We denote the polynomials that result from the definition (4–167) as $Q_n(\eta)$,

$$Q_n(\eta) \equiv \int_{-1}^{\eta} P_n(\eta)\, d\eta, \qquad (4\text{–}169)$$

of which the first several are

$$Q_0(\eta) \equiv \eta + 1,$$

$$Q_1(\eta) \equiv \frac{1}{2}(\eta^2 - 1),$$

$$Q_2(\eta) \equiv \frac{1}{2}(\eta^3 - \eta). \qquad (4\text{–}170)$$

These polynomial functions are closely related to the so-called Gegenbauer polynomials[14] and satisfy the orthogonality condition

$$\int_{-1}^{1} \frac{Q_n(\eta)Q_m(\eta)}{(1-\eta^2)}\, d\eta = \begin{cases} 0 & \text{for } n \neq m \\ \dfrac{2}{n(n+1)(2n+1)} & \text{for } n = m. \end{cases} \qquad (4\text{–}171)$$

Further, as noted earlier, the polynomials $Q_n(\eta)$ are all identically zero at $\eta = \pm 1$, except for $Q_0(\eta)$.

The polynomial functions $Q_n(\eta)$ for $n \geq 1$ are the *eigenfunctions* of (4–157) subject to the symmetry condition (4–151). The constants $\lambda_n = n(n+1)$ are the corresponding eigenvalues. It follows from the preceding discussion that the most general, nonsingular solution of $E^2 w = 0$ that satisfies the conditions (4–151) at $\eta = \pm 1$ is

$$w = \sum_{n=1}^{\infty} w_n, \qquad (4\text{–}172)$$

where

$$w_n = \left(\hat{A}_n r^{n+1} + \hat{C}_n r^{-n} \right) Q_n(\eta). \qquad (4\text{–}173)$$

We leave the constant coefficients undetermined (these require imposition of boundary conditions corresponding to some specific problem) and proceed to solve (4–153) for ψ.

Hence, we seek a solution of

$$E^2 \psi = \sum_{n=1}^{\infty} (\hat{A}_n r^{n+1} + \hat{C}_n r^{-n}) Q_n(\eta) \qquad (4\text{–}174)$$

for arbitrary $\hat{A}_n$ and $\hat{C}_n$. The solution consists of two parts, a homogeneous solution satisfying $E^2 \psi = 0$ and a particular solution that generates the right-hand side of (4–174). The homogeneous solution is clearly identical to w above, that is,

$$\psi_{\text{homog.}} = \sum_{n=1}^{\infty} \left[B_n r^{n+1} + D_n r^{-n} \right] Q_n(\eta). \qquad (4\text{–}175)$$

To obtain a particular solution, we try the form

$$\psi_{\text{part.}} = r^\lambda Q_n(\eta), \qquad (4\text{–}176)$$

which contains the same η dependence as the right-hand side of (4–174). With this choice for ψ, the left-hand side of (4–174) becomes

$$E^2\left[r^\lambda Q_n(\eta)\right] = \alpha r^{\lambda-2} Q_n(\eta), \tag{4–177}$$

where α is a constant that depends on λ and n.

Comparing (4–177) with the right side of (4–174), we see that we will obtain a particular solution of the form (4–176) for each of the terms in (4–174). The term involving r^{n+1} requires $\lambda = n + 3$, while the term involving r^{-n} requires $\lambda = 2 - n$. With these values, the particular solution for ψ is

$$\psi_{\text{part.}} = \sum_{n=1}^{\infty} \left[A_n r^{n+3} + C_n r^{2-n}\right] Q_n(\eta), \tag{4–178}$$

where A_n and C_n are *arbitrary* constants [related to the *arbitrary* constants $\hat{A}_n$ and $\hat{B}_n$ in (4–174)].

Combining (4–175) and (4–178), the general solution for arbitrary, axisymmetric flows, in terms of spherical coordinates, can be expressed in the form

$$\boxed{\psi = \sum_{n=1}^{\infty} \left[A_n r^{n+3} + B_n r^{n+1} + C_n r^{2-n} + D_n r^{-n}\right] Q_n(\eta).} \tag{4–179}$$

To apply (4–179) to a particular flow problem, we require boundary conditions to determine the four sets of arbitrary constants A_n, B_n, C_n, and D_n.

One extremely useful result can be proven that applies to any problem in which the streamfunction is expressed in the general form (4–179). This result can be formalized in the form of a *theorem*.

For any problem with ψ in the general form (4–179), the force exerted by the fluid on an arbitrary, axisymmetric body with its center of mass at $|\mathbf{x}| = 0$ is generally given by

$$F_z = 4\pi\mu U_c l_c C_1, \tag{4–180}$$

where U_c and l_c are the characteristic velocity and length scales, and z denotes the direction of the symmetry axis.

The result stated in this theorem is true independent of the values of the other coefficients A_n, B_n, C_n, and D_n. Further, although the numerical value of F_z might appear to depend upon the particular velocity and length scale chosen for U_c and l_c, this is not true because the combination $U_c l_c C_1$ remains constant for a body of fixed geometry in a particular flow, independent of U_c and l_c. One practical implication of the theorem is that to determine (or estimate) F_z, we require only an accurate numerical value for the coefficient C_1 in (4–179). The accuracy of the other coefficients is irrelevant for purposes of calculating (or estimating) the drag force.

The brute force way to calculate the force on a body is to integrate the surface stress vector over the body surface—in this case

$$\mathbf{F} = \mu U_c l_c \int_S (\mathbf{T} \cdot \mathbf{n}) \, dA, \tag{4–181}$$

where the integral is expressed in terms of dimensionless variables. We could, in principle, simply evaluate (4–181) by using the general solution (4–179) in order to prove (4–180). However, with this approach, we would need to restrict the class of allowable axisymmetric body shapes to ones where we could actually carry out the integration that is indicated in (4–181). To prove (4–180) for an axisymmetric body of *arbitrary* shape, we recall that the creeping motion equation, expressed in terms of the stress tensor, is just

$$\nabla \cdot \mathbf{T} = 0. \tag{4-182}$$

Thus, we can apply the divergence theorem to the volume of fluid that is contained between the body surface and any arbitrary closed surface that completely encloses the body to show

$$0 = \int_V (\nabla \cdot \mathbf{T}) dV = \int_S (\mathbf{T} \cdot \mathbf{n}) dS - \int_{S*} (\mathbf{T} \cdot \mathbf{n}*) dS*. \tag{4-183}$$

Thus, the surface integral on S that appears in (4–181) is precisely equal to the surface integral of $\mathbf{T} \cdot \mathbf{n}$ for any closed surface that encloses the body. Here, $\mathbf{n}*$ denotes the outer normal to the surface $S*$, and $\mathbf{n}$ denotes the outer normal to the body surface S. The basic idea of replacing the integral on S (which may be quite complicated for a body of complicated shape) with an integral over some other surface $S*$, is to choose $S*$ so that the integral is as simple as possible. A convenient choice for $S*$ is a spherical surface that is centered at the center of mass of the body.

Hence to prove (4–180), we begin with the expression for the z component of $\mathbf{F}$ (that is, the drag force) in terms of a surface integral over a large sphere of radius $R*$ that is circumscribed about the body and centered at the center of mass, namely,

$$\boxed{F_z = \mathbf{i}_z \cdot \mathbf{F} = \mu U_c l_c \int_0^{2\pi} \int_0^\pi \mathbf{i}_z \cdot (\mathbf{T} \cdot \mathbf{n}*)(R*)^2 \sin\theta \, d\theta \, d\phi.} \tag{4-184}$$

Now,

$$\mathbf{T} = -p\mathbf{I} + \mu(\nabla \mathbf{u} + \nabla \mathbf{u}^T). \tag{4-185}$$

Thus, for this class of axisymmetric problems where $T_{\phi r} = 0$,

$$\mathbf{T} \cdot \mathbf{n}* = \mathbf{T} \cdot \mathbf{i}_r = T_{rr}\mathbf{i}_r + T_{\theta r}\mathbf{i}_\theta, \tag{4-186}$$

where

$$T_{rr} = -p + 2\frac{\partial u_r}{\partial r} \tag{4-187}$$

and

$$T_{r\theta} = r\frac{\partial}{\partial r}\left(\frac{u_\theta}{r}\right) + \frac{1}{r}\frac{\partial u_r}{\partial \theta}. \tag{4-188}$$

Now, given the general form (4–179) for the streamfunction ψ, corresponding expressions for u_r and u_θ can be obtained by differentiating ψ using the definitions in (4–150), and thus general expressions can be obtained for T_{rr} and $T_{r\theta}$ that involve all of the coefficients A_n, B_n, C_n, and D_n. Substituting these expressions into (4–184) and carrying out the integration, the result (4–180) follows. We leave it to the reader to provide the details of this proof (see Problem 5).

Application to Uniform Streaming Flow Past an Arbitrary Axisymmetric Body

As an example of the application of (4–179), we consider creeping flow past an arbitrary axisymmetric body with a uniform streaming motion at infinity. For the case of a solid sphere, this is Stokes' problem. In the present case, we begin by allowing the geometry of the body to be arbitrary (and unspecified) except for the requirement that the symmetry axis be parallel to the direction of the uniform flow at infinity so that the velocity field will be axisymmetric. A sketch of the flow configuration is shown in Figure 4–14. We measure the polar angle θ from the axis of symmetry on the downstream side of the body. Thus $\eta = 1$ on this axis, and $\eta = -1$ on the axis of symmetry upstream of the body.

The asymptotic form of the streamfunction for $r \to \infty$ can be obtained from the uniform streaming condition at infinity, that is,

$$\mathbf{u} = \mathbf{i}_z \quad \text{as} \quad r \to \infty. \tag{4–189}$$

Since

$$\mathbf{i}_z = \mathbf{i}_r \cos\theta - \mathbf{i}_\theta \sin\theta = \mathbf{i}_r \eta - \mathbf{i}_\theta (1 - \eta^2)^{1/2}, \tag{4–190}$$

it follows from (4–189) that

$$\boxed{u_r = \eta, \ u_\theta = -(1 - \eta^2)^{1/2} \quad \text{for } r \to \infty.} \tag{4–191}$$

Thus, from the definitions (4–150), we see

$$\frac{\partial\psi}{\partial r} = r(1 - \eta^2), \quad \frac{\partial\psi}{\partial \eta} = -\eta r^2 \quad \text{as } r \to \infty. \tag{4–192}$$

or, integrating these expressions,

$$\psi \longrightarrow \frac{r^2}{2}(1 - \eta^2) + \text{const} \quad \text{for } r \to \infty. \tag{4–193}$$

The constant that appears in (4–193) is completely arbitrary. Notice from (4–150) that we can always add an *arbitrary* constant to ψ without changing the velocities at all. Since the constant is arbitrary, we set it equal to zero, so that

$$\boxed{\psi \longrightarrow \frac{r^2}{2}(1 - \eta^2) \quad \text{as } r \to \infty.} \tag{4–194}$$

Note that the choice const $= 0$ here also requires that the constant in (4–151b) is zero, so that

$$\boxed{\psi = 0 \quad \text{at} \quad \eta = \pm 1 \quad \text{(all } r).}$$

Since the general solution (4–179) is expressed in terms of the modified Gegenbauer polynomials $Q_n(\eta)$, it is convenient to also express (4–194) in terms of these func-

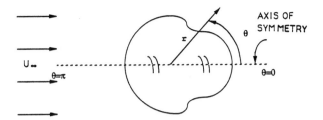

Figure 4–13 A schematic representation of the domain for a uniform flow past an arbitrary, axisymmetric body. For the case of a solid sphere, this is Stokes' problem.

tions. Referring to the definitions (4–170), we see that (4–194) can also be expressed in the form

$$\psi \longrightarrow -r^2 Q_1(\eta) \quad \text{as} \quad r \to \infty. \tag{4–195}$$

Thus, regardless of the details of the body geometry, the streamfunction must exhibit the asymptotic form (4–195) if there is a uniform streaming flow at large distances from the body. Comparing (4–195) and (4–179), it is therefore evident that many of the coefficients A_n and B_n in the general axisymmetric solution must be zero for this case. In particular, the asymptotic condition (4–194) requires that

$$A_n = 0, \quad \text{all } n,$$
$$B_n = 0, \quad n \geq 2, \tag{4–196}$$
$$B_1 = -1,$$

and thus the general solution (4–179) becomes

$$\psi = -r^2 Q_1 + \sum_{1}^{\infty} \left[C_n r^{2-n} + D_n r^{-n} \right] Q_n(\eta). \tag{4–197}$$

Again, we emphasize that this solution form is independent of the body geometry—this will only come into play when we try to apply boundary conditions at the body surface to determine C_n and D_n. For now we simply leave these constants unspecified.

By expressing the streamfunction in the form (4–197), we have essentially split the corresponding velocity field into two parts: the free stream (or undisturbed) flow and an as yet unspecified *disturbance flow* that is due to the presence of the body in the flow. Although the constants C_n and D_n remain to be determined, there are several general features of the disturbance flow that are worth mentioning here. First, it is clear from the theorem (4–180) that C_1 must be nonzero for any case in which the net force on the body is nonzero. For the problem considered here of uniform flow past a stationary body of axisymmetric, though otherwise arbitrary, geometry, the net force will always be nonzero, and thus $C_1 \neq 0$. Only the magnitude of C_1 will change from case to case. Second, since $C_1 \neq 0$, the largest possible contribution of the particle to the *far-field* (that is, $r \gg 1$) form of the disturbance flow is the term

$$\psi' = C_1 r Q_1(\eta).$$ (4–198)

In terms of velocity components, this contribution to the disturbance flow is

$$u_r' = -\frac{C_1}{r} P_1(\eta); \quad u_\theta' \sqrt{1 - \eta^2} = -\frac{C_1}{r} Q_1(\eta).$$ (4–199)

The most important feature of (4–199) is that the disturbance velocities fall off only *linearly* with distance from the body. Thus, for a body that exerts a net force on the fluid, the disturbance produced is extremely *long range* in a low Reynolds number flow. One important implication is that the motion of such a body will be extremely sensitive to the presence of another body, or other boundaries, even when these are a considerable distance away. For example, the velocity of a sphere moving through a quiescent fluid toward a plane wall under the action of buoyancy is reduced by approximately 12 percent when the sphere is 10 radii from the wall and by 35 percent when it is 4 radii away.

Now, we have expressed the general streamfunction (4–197), and the disturbance flow contribution in (4–198) and (4–199), in terms of spherical coordinates. However, we have *not* yet specified a body *shape*. Thus, the linear decrease of the disturbance flow with distance from the body must clearly represent a property of creeping flows that has nothing to do with specific coordinate systems. Indeed, this is the case, and the velocity field (4–199) plays a very special and fundamental role in creeping flow theory. It is commonly known as the *Stokeslet velocity field* and represents the motion induced in a fluid at Re ≡ 0 by a *point force* at the origin (expressed here in spherical coordinates). We shall see later that the Stokeslet solution plays an important role in many aspects of creeping flow theory.

Uniform Streaming Flow Past a Sphere— Stokes' Problem Revisited

The analysis of the preceeding two parts of this section was carried out using a spherical coordinate system, but the majority of the results are valid for an axisymmetric body of arbitrary shape. The necessity to specify a particular particle geometry only occurs when we apply boundary conditions on the particle surface (that is, when we evaluate the coefficients C_n and D_n in the spherical coordinate form of the solution). For this purpose, an exact solution requires that the body surface be a coordinate surface in the coordinate system that is used, and this effectively restricts the application of (4–197) to streaming flow past spherical bodies, which may be solid as considered below, or spherical bubbles or drops, as considered in Chapter 5.

Here, we briefly reconsider Stokes's problem of streaming motion past a stationary solid sphere, starting from equation (4–197). The boundary conditions on the body surface,

$$u_r = u_\theta = 0 \quad \text{for} \quad r = 1,$$ (4–200)

require that

$$\frac{\partial \psi}{\partial r} = \frac{\partial \psi}{\partial \eta} = 0 \quad \text{at} \quad r = 1. \tag{4-201}$$

It follows from the second of these conditions that $\psi = \text{const}$ on the sphere surface, and since $\psi = 0$ at $\eta = \pm 1$ for all r (because of symmetry), it follows that the appropriate version of the second condition in (4–201) is

$$\psi = 0 \quad \text{at} \quad r = 1. \tag{4-202}$$

Hence, we solve for C_n and D_n using the first of the conditions (4–201) and (4–202).

The *two* boundary conditions $\psi = (\partial \psi / \partial r) = 0$ at $r = 1$ lead to a pair of algebraic equations for C_n and D_n, namely,

$$0 = -Q_1 + \sum_{n=1}^{\infty} [C_n + D_n] Q_n(\eta) \tag{4-203}$$

and

$$0 = -2Q_1 + \sum_{n=1}^{\infty} [C_n(2 - n) - nD_n] Q_n(\eta) \tag{4-204}$$

Since the $Q_n(\eta)$ are a complete set of orthogonal functions, those equations can be solved for the constant coefficients C_n and D_n. The result is that

$$C_n = D_n = 0, \; n \geq 2,$$

$$C_1 = \frac{3}{2}, \quad D_1 = -\frac{1}{2}. \tag{4-205}$$

Hence, in dimensionless terms,

$$\boxed{\psi = -\left(r^2 - \frac{3r}{2} + \frac{1}{2r}\right) Q_1(\eta).} \tag{4-206}$$

This solution is identical to that obtained in Section B though the latter is expressed in terms of the velocity components u_r and u_θ instead of ψ.

Since $C_1 = 3/2$, the theorem (4–180) yields Stokes' law for the force

$$\boxed{F_z = 6\pi \mu a U}$$

as it should.

A plot showing lines of constant ψ in a plane that is parallel to the free stream and passing through the center of the sphere is shown in Figure 4–15. One striking feature is that the flow is completely symmetric about $\eta = 0$ (that is, $\theta = \pi/2$). Apart from the arrows that have been added to the figure, it would not be possible from the shape of the streamlines to tell whether the fluid is moving from right to left or left to right.

Another feature of the flow that we can calculate from ψ using the relationship (4–96) is the vorticity

$$\boldsymbol{\omega} = -h_3 E^2 \psi \mathbf{i}_\phi.$$

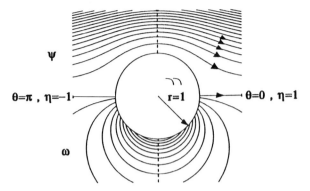

Figure 4–14 The streamlines and contours of constant vorticity for uniform streaming flow past a solid sphere (Stokes' problem). The streamfunction and vorticity values are calculated from equations (4–206) and (4–207). Contour values plotted for the streamfunction are in increments of 1/16, starting from zero at the sphere surface, while the vorticity is plotted at equal increments equal to 0.04125.

It may be recalled that the vorticity, defined as $\boldsymbol{\omega} \equiv \nabla \wedge \mathbf{u}$, provides a measure of the local rate of rotation of fluid elements. In the present problem, the vorticity becomes

$$\boldsymbol{\omega} = -\frac{1}{2r^2}(1-\eta^2)^{1/2}\mathbf{e}_\phi. \qquad (4\text{–}207)$$

A plot showing lines of constant $\omega\ (\equiv |\boldsymbol{\omega}|)$ is also shown in Figure 4–15. Again, the fore-aft symmetry of the flow is clearly evident. The fact that the constant vorticity lines all terminate on the body surface is indicative of the fact that the *source* of vorticity in the flow is the sphere surface. Indeed, the governing equation for vorticity in the creeping flow limit can be shown from (4–6) to be*

$$\nabla^2\omega = 0.$$

Further, there is no vorticity in the undisturbed flow. Hence, vorticity is created at the body surface and then transported into the fluid via diffusion (we shall see later that it is the no-slip condition that is responsible for production of vorticity). In the creeping flow limit, there is no convection of vorticity. We shall see later that it is the convection of vorticity from upstream toward the downstream direction that is responsible for asymmetry in the flow at finite Reynolds numbers, including the development of strong disturbances in the downstream wake. Since vorticity only diffuses in the creeping flow limit, the disturbance produced by the sphere is fore-aft symmetric. There is, in fact, a strong similarity between the transport of vorticity in two-dimensional and axisymmetric flows and the transport of heat from a heated body. Not surprisingly, constant vorticity lines are often similar in appearance to isotherms in the thermal problem. In the simplest thermal problem, however, the surface of the body would be assumed

---

*To obtain this result, we simply take the curl of both terms in (4–6a) and use the well-known result $\nabla \wedge (\nabla\phi) = 0$ for any scalar ϕ.

to be at constant temperature so that the temperature field in the case of pure conduction from a sphere would be spherically symmetric. The sphere surface is not a uniform source of vorticity, however; in regions where the velocity gradients are largest, the surface vorticity is highest. As a consequence, even for creeping flow with a spherical body, the vorticity distribution is not spherically symmetric, and this is evident in Figure 4–15.

A Rigid Sphere in Axisymmetric, Extensional Flow

Stokes' problem of a rigid sphere in a steady, uniform flow is relevant to the steady translation of the sphere through an unbounded quiescent fluid due to the action of a body force on the sphere. The most common example of such a problem is the sedimentation of a sphere due to the action of gravity, when the sphere density is not equal to the density of the suspending fluid. For many applications, however, we would like to understand the fluid motion in the vicinity of a body when the fluid at infinity is undergoing some nonuniform motion. For purposes of studying such a problem, it is convenient to consider a body whose density is equal to the density of the fluid so that the only motion in the fluid is due to the flow at infinity and its interaction with the body. Such bodies are called *neutrally buoyant,* for obvious reasons. In creeping flow, the solution for a nonneutrally buoyant body in some general flow at infinity can be constructed by superposition from the solution for a neutrally buoyant body in the same undisturbed flow and Stokes' solution for sedimentation/translation through a quiescent fluid. Of course, it will generally be difficult to consider the fluid motion in the vicinity of a body for some completely arbitrary flow at infinity. For creeping flow, where solutions can be constructed by superposition, a reasonable alternative is to consider a hierarchy of nonuniform undisturbed flows beginning with flows in which the magnitude of the velocity varies *linearly* with spatial position.

A more direct motivation for studying linear flows is that we are frequently interested in applications of creeping flow results for particles that are very small compared with the length scale, L, that is characteristic of changes in the undisturbed velocity gradient for a general flow. In this case, we may approximate the undisturbed velocity field in the vicinity of the particle by means of a Taylor series approximation, namely,

$$\mathbf{u}'_\infty = \mathbf{u}'_0 + (\nabla'\mathbf{u}')_0 \cdot \mathbf{x}' + \cdots,$$

where $(\nabla\mathbf{u})_0$ is the velocity gradient evaluated at the position of the center of mass of the particle, $\mathbf{x} = 0$. The next term involving the gradient of $\nabla\mathbf{u}$ is smaller by a factor proportional to the ratio of the length scale of the body l to the macroscopic scale L. We shall see later that the net hydrodynamic force on a sphere in a linear undisturbed flow is proportional to the difference between the velocity of the center of mass of the sphere and the velocity of the undisturbed flow evaluated at the position occupied by the center of mass. Thus, since the external force on the sphere is zero (recall that it is neutrally buoyant), it follows that

$$\mathbf{u}_p = \mathbf{u}'_0, \tag{4-208}$$

where $\mathbf{u}_p$ is the velocity of the center of mass of the body.

Thus, the undisturbed flow that is seen in a frame of reference that translates with the particle is just

$$\mathbf{u}_\infty = (\nabla \mathbf{u})_0 \cdot \mathbf{x} + \cdots. \tag{4-209}$$

Now, however, this can be expressed as the sum of a linear straining flow and a purely rotational flow, that is,

$$\mathbf{u}_\infty = \mathbf{E}_0 \cdot \mathbf{x} + \mathbf{\Omega}_0 \cdot \mathbf{x} + \cdots, \tag{4-210}$$

where $\mathbf{E}_0$ and $\mathbf{\Omega}_0$ are the rate of strain and vorticity tensors defined in terms of $(\nabla \mathbf{u})_0$, and we see that the motion of a small particle in a general linear flow can be described as a superposition of its motion in a pure straining flow and a purely rotational flow. In general, a particle in the purely rotational flow will simply rotate with the vorticity of the undisturbed flow, though we shall not prove this result here. The response of the fluid and particle to a pure straining flow is generally more complicated. In this section, we consider the special case of a rigid sphere in an axisymmetric pure straining flow,

$$\mathbf{u}_\infty = -E[x\mathbf{i} + y\mathbf{j} - 2z\mathbf{k}], \tag{4-211}$$

where

$$\mathbf{E} = E \begin{pmatrix} -1 & 0 & 0 \\ 0 & -1 & 0 \\ 0 & 0 & 2 \end{pmatrix}.$$

A plot showing streamlines for this flow with a spherical body at the origin (calculated below) is shown in Figure 4–15. For $E > 0$, there is flow outward away from the

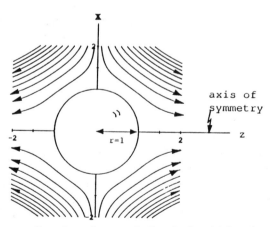

Figure 4–15 The streamlines for axisymmetric flow in the vicinity of a solid sphere with *uniaxial* extensional flow at infinity. When the direction of motion is reversed at infinity, the undisturbed flow is known as *biaxial* extensional flow. The streamfunction values are calculated from equation (4–220). Contour values are plotted in equal increments equal to 0.5.

sphere along the axis of symmetry and flow inward in the plane orthogonal to this axis. This flow is called *uniaxial extensional flow*. For $E < 0$, the direction of fluid motion is reversed and the undisturbed flow is known as *biaxial extensional flow*. In either case, with an axially symmetric body at the origin $x = 0$, with its symmetry axis parallel to the symmetry axis of the flow, it is evident that the complete velocity field will be axisymmetric. Hence, the general eigenfunction expansion (4–179) should be applicable in creeping flow, and if the body is spherical as we assume here, it should be possible to determine all of the coefficients A_n, B_n, C_n, and D_n from the form of the undisturbed flow and the boundary conditions on the sphere surface.

To do this, we must first determine the form of streamfunction at large distances from the sphere when the velocity field has the asymptotic form (4–211). Specifically, if we transform (4–211) to spherical coordinates using the general relationships

$$\mathbf{i} = \mathbf{i}_r \sin\theta \cos\phi + \mathbf{i}_\theta \cos\theta \cos\phi - \mathbf{i}_\phi \sin\phi,$$

$$\mathbf{j} = \mathbf{i}_r \sin\theta \sin\phi + \mathbf{i}_\theta \cos\theta \sin\phi + \mathbf{i}_\phi \cos\phi,$$

$$\mathbf{k} = \mathbf{i}_r \cos\theta - \mathbf{i}_\theta \sin\theta, \tag{4–212}$$

and

$$x = r\sin\theta\cos\phi,$$

$$y = r\sin\theta\sin\phi,$$

$$z = r\cos\theta, \tag{4–213}$$

we find that

$$\mathbf{u}_\infty = -Er(1 - 3\cos^2\theta)\mathbf{i}_r - Er(3\cos\theta\sin\theta)\mathbf{i}_\theta. \tag{4–214}$$

A relevant characteristic velocity in this case is thus

$$u_c = Ea, \tag{4–215}$$

where a is the sphere radius, and in dimensionless terms we therefore have

$$\mathbf{u}_\infty = -r(1 - 3\cos^2\theta)\mathbf{i}_r - r(3\sin\theta\cos\theta)\mathbf{i}_\theta, \tag{4–216}$$

where r is now the dimensionless radial variable that is equal to unity ($r = 1$) at the sphere surface.

From the form (4–216) and the definition (4–150) of the streamfunction in terms of the axisymmetric velocity components u_r and u_θ, we can easily obtain the asymptotic form for ψ. The result is

$$\boxed{\psi = r^3\eta(1 - \eta^2) = -2r^3 Q_2(\eta) \quad \text{for } r \to \infty.} \tag{4–217}$$

Thus, if we impose this far-field condition on the general axisymmetric solution (4–179), we must require that

$$A_n = 0, \quad n \geq 1,$$

$$B_n = 0, \quad n \geq 3,$$

$$B_2 = -2,$$

$$B_1 = 0. \tag{4–218}$$

The latter condition simply avoids the presence of a uniform streaming flow in the far field.

To complete the solution, we must apply boundary conditions at the surface of the solid sphere. Thus, again, we require that

$$u_r = u_\theta = 0 \quad \text{at} \quad r = 1,$$

and this translates to the two conditions on ψ

$$\psi = \frac{\partial \psi}{\partial r} = 0 \quad \text{at} \quad r = 1.$$

Hence, since the $Q_n(\eta)$ are a complete set of orthogonal eigenfunctions, it follows from (4–179) and (4–218) that

$$C_2 = 5, \quad D_2 = -3, \tag{4–219}$$

and

$$C_n = D_n = 0 \quad \text{for} \quad n \neq 2.$$

Hence, the solution for axisymmetric straining flow past a solid sphere is

$$\boxed{\psi = 2\left(-r^3 + \frac{5}{2} - \frac{3}{2}\frac{1}{r^2}\right)Q_2(\eta).} \tag{4–220}$$

A sketch of the streamlines corresponding to this solution is given in Figure 4–16.

There are a few brief, but useful, observations about the form (4–220) and about the solution in general. First, it may be noted that $C_1 = 0$. Hence, it can be shown that the net hydrodynamic force on the sphere is zero. Second, the sign of E plays no role in solution of this problem. Hence, for a solid sphere, the solution (4–220) is valid for both uniaxial and biaxial flows and the form of the velocity field is unchanged by a change from $E > 0$ to $E < 0$, though, of course, the direction of motion does reverse.

The velocity field given by (4–220) will be used later to estimate heat transfer rates for spherical particles in a straining flow. Here, we focus on a different application of (4–220), namely, its use in predicting the effective viscosity of a dilute suspension of solid spheres. In order to carry out the calculation, it is first necessary to briefly discuss the properties of a suspension in a more general framework.

Let us then consider a suspension of identical, neutrally buoyant solid spheres of radius a. We are interested in circumstances where the length scale of the suspension at the particle scale (that is, the particle radius) is very small compared with the characteristic dimension L of the flow domain so that the suspension can be modeled as a continuum with properties that differ from the suspending fluid due to the presence of the particles. Our goal is to obtain an a priori prediction of the macroscopic rheological properties when the suspension is extremely dilute, a problem first considered by Einstein (1905) as part of his Ph.D. dissertation.[15] A suspension is called *dilute* if the volume fraction of particles is small enough that each particle moves independently of all the others, with no hydrodynamic interactions possible.

To discuss the macroscopic (or "bulk") properties of a suspension, it is necessary to specify the connection between local variables at the particle scale and macroscopic variables at the scale L. One plausible choice, in view of the relationship between continuum and molecular variables in Chapter 2, is to assume that the macroscopic variables are just volume averages of the local variables. In particular, we shall assume in the discussion that follows that the macroscopic (or bulk) stress can be defined as a volume average of the local stress in the suspension, namely,

$$<\sigma_{ij}> \equiv \frac{1}{V}\int_V \sigma_{ij}dV, \qquad (4\text{-}221)$$

where σ_{ij} may pertain either to a point in the suspending fluid or to a point inside one of the particles.* The volume V is an averaging volume whose linear dimensions are large compared with the characteristic microscale of the suspension (so that V will contain a statistically significant number of spheres) but still arbitrarily small compared with the dimension L of the flow domain. Similarly, we define a bulk or macroscopic strain rate as

$$<e_{ij}> \equiv \frac{1}{V}\int_V e_{ij}dV, \qquad (4\text{-}222)$$

and it follows that

$$<\sigma_{ij}> - 2\mu <e_{ij}> \equiv \frac{1}{V}\int_V (\sigma_{ij} - 2\mu e_{ij})dV. \qquad (4\text{-}223)$$

Now, in the solid particles $e_{ij}=0$, and within the suspending fluid σ_{ij} differs from $2\mu e_{ij}$ by the pressure multiplied by the unit tensor δ_{ij}. Thus,

$$<\sigma_{ij}> - 2\mu <e_{ij}> = -p^*\delta_{ij} + \frac{1}{V}\sum_p \int_{V_p} \sigma_{ij}dV, \qquad (4\text{-}224)$$

where $p^*\delta_{ij}$ is an isotropic term that is related to the bulk pressure, and the summation is over all of the particles within V.

The expression (4-224) for the bulk stress is, of course, valid for arbitrary concentrations of particles, but the volume integrals over V_p are exceedingly difficult to evaluate in general because the value for a particular particle depends upon the complete configuration of particles in the suspension. For a *dilute* suspension of identical particles, on the other hand, the problem simplifies immensely, because the integral over V_p is *exactly the same* for all particles, and the expression (4-224) can be replaced by

$$<\sigma_{ij}> - 2\mu <e_{ij}> = -p^*\delta_{ij} + n\int_{V_p} \sigma_{ij}dV, \qquad (4\text{-}225)$$

where n is the number of particles per unit volume and the integration is now over the volume of a single sphere.

In order to evaluate the integral over V_p, it is convenient to convert it to an integral over the *surface* of the sphere by means of the divergence theorem. To do this, we note that

---

*Note that it is *not* necessary for the initial developments leading to (4-224) to assume that the suspension is dilute.

$$\frac{\partial \sigma_{ij}}{\partial x_j} \equiv 0 \tag{4-226}$$

within a freely suspended sphere, and thus

$$\sigma_{ij} = \frac{\partial}{\partial x_k}(x_j \sigma_{ik}). \tag{4-227}$$

It follows that

$$\boxed{\int_{V_p} \sigma_{ij} dV = \int_{S_p} x_j \sigma_{ik} n_k dS,} \tag{4-228}$$

where S_p is the sphere surface and $\mathbf{n}$ is the unit outer normal to S_p. Hence, to evaluate the bulk stress for a dilute suspension of identical spheres, we must evaluate the surface integral (4–228) for some specified flow.

In the present circumstances, we consider a *dilute* suspension that is undergoing an axisymmetric pure straining flow, $<e_{ij}>x_j$. For this case, the velocity field in the vicinity of each of the spherical particles is given by (4–220), namely,

$$u_r = -\frac{1}{r^2}\frac{\partial \psi}{\partial \eta} = \left(r - \frac{5}{2r^2} + \frac{3}{2r^4}\right)(3\cos^2\theta - 1),$$

$$u_\theta = -\frac{1}{r}\frac{1}{\sqrt{1-\eta^2}}\frac{\partial \psi}{\partial r} = -3\left(r - \frac{1}{r^4}\right)\sin\theta\cos\theta, \tag{4-229}$$

and the integral on S_p can be evaluated for the nonzero components of $<e_{ij}>$. The result, after some manipulation, is

$$\int_{S_p} x_j \sigma_{ik} n_k dA = \frac{4\pi}{3} a^3 5 <e_{ij}> \mu. \tag{4-230}$$

Since the volume fraction of spheres is just

$$C = \frac{4\pi a^3}{3} n,$$

it follows that

$$\boxed{<\sigma_{ij}> = -p^*\delta_{ij} + 2\mu\left(1 + \frac{5}{2}C\right)<e_{ij}>.} \tag{4-231}$$

It can, in fact, be proven that a dilute suspension of rigid spheres will *always be Newtonian* at the first $0(C)$ correction to the bulk stress, with an effective viscosity given by (4–231) of the form

$$\boxed{\mu^* = \mu\left(1 + \frac{5}{2}C\right).} \tag{4-232}$$

This is the famous result obtained by Einstein for the viscosity of a dilute suspension of spheres.* Although the integral (4–230) leading to this result was evaluated by using the velocity field (4–229) for the specific case of an axisymmetric extensional flow, the same result, in fact, is obtained for any arbitrary linear flow, including simple shear flow.

It is obvious from (4–231) that the effective viscosity of a dilute suspension of rigid spheres exceeds the viscosity of the suspending fluid by an amount that is proportional to the volume fraction of the particles. The increase in effective viscosity can be understood to be a consequence of the fact that the rigid particles cannot deform in response to the straining motion in the fluid and thus generate a disturbance flow that increases the rate of viscous dissipation relative to that which would occur in the absence of the particles. In this sense, the particles produce a macroscopic effect that is equivalent to an increase of the viscosity of the suspension when it is considered as a homogeneous continuum. It may be noted from (4–231) that the bulk stress for a given bulk strain rate, $<e_{ij}>$, is completely independent of the addition of vorticity to the flow. This is because the particle response to vorticity in the linear, creeping motion approximation is to simply rotate with the local angular velocity of the fluid with no additional disturbance of the fluid's motion and hence no additional contribution to the rate of viscous dissipation.

A general discussion of the rheology of suspensions and other materials that are *heterogeneous* at the microscale is beyond the scope of the present book. However, the interested reader may wish to refer to Schowalter (1978)[16] for a more comprehensive presentation of this material.

F. Solutions via Superposition of Vector Harmonic Functions— General Three-Dimensional Problems

Although the solution methods of the preceeding sections are very useful, they are restricted to two-dimensional and axisymmetric creeping flow problems. Clearly, it is extremely important to develop a more general solution procedure that can be applied to fully three-dimensional creeping flow problems. We pursue that goal in this section.

Generally speaking, there are two approaches that have been used successfully to solve general three-dimensional creeping flow problems. The first is a natural extension of the procedures of the two preceding sections, in the sense that it is based upon an eigenfunction expansion using a general solution of the creeping flow equations in terms of harmonic functions originally due Lamb (1932).[17] A detailed description of Lamb's solution and its application to the solution of creeping flow problems can be found in *Low Reynolds Number Hydrodynamics* by Happel and Brenner (1973),[18] among a number of other available sources. However, in spite of the fact that

---

*It is interesting to note that Einstein's original published result was incorrect, with the coefficient 3/2 instead of 5/2. Although Einstein quickly published a correction to the result, there is perhaps some "comfort" for the student in realizing that even brilliant researchers can make published mistakes.

Lamb's general solution does provide an effective procedure for solving creeping flow problems, we shall not pursue it here. Instead, we describe a second approach that is more general and powerful. In particular, we shall see that some basic properties of the solutions of linear systems can be used to simultaneously represent (solve) the solutions for a complete class of related problems, rather than solving one specific problem at a time, as is necessary using standard eigenfunction expansion techniques. For this purpose, we must begin by discussing some important (preliminary) concepts.

Preliminary Concepts

The first important concept is the significance of the apparently trivial equality

$$A = B \tag{4-233}$$

when used in a general vector analysis, where A and B may represent scalars, vectors, or tensorial quantities of any order. In particular, if (4–233) derives from some physical law or principle, it implies "equality" on a number of different levels:

1. A and B must have the same physical dimensions.
2. A and B must have the same tensorial rank (that is, if A is a scalar, then B must be a scalar; if A is a vector, then B must be a vector; and so on).
3. A and B must have the same parity (that is, if A is a pseudo-vector, then B is a pseudo-vector; if A is a true vector, then B is a true vector; and so on).
4. A and B have the same tensorial symmetry.

We shall see below that the second and third requirements play a critical role in the representation of solutions to the creeping flow equations.

A general discussion of the concept of pseudo-vectors (in general, pseudo-scalars, pseudo-tensors, and so on) can be found in most books on vector analysis. However, generally speaking, a pseudo-vector differs from a true vector because it changes sign upon inversion of the coordinate axes (alternatively, we may think of inversion as changing from a right- to a left-handed coordinate reference frame). Thus, if we consider an orthogonal transformation, specified by the transformation matrix $\mathbf{L}$, a pseudo-vector transforms according to the rule

$$\mathbf{B} = (\det \mathbf{L})\mathbf{L} \cdot \bar{\mathbf{B}},$$

whereas a true vector transforms according to

$$\mathbf{A} = \mathbf{L} \cdot \bar{\mathbf{A}},$$

as we have already seen in Chapter 2. Common examples of pseudo-vectors that will be relevant later include the angular velocity vector $\mathbf{R}$, the torque $\mathbf{T}$, the vorticity vector $\boldsymbol{\omega}$, or the vorticity tensor $\boldsymbol{\Omega}$ (or the curl of any true vector), the cross product of two vectors. A second-order pseudo-tensor can be formed from a general position vector $\mathbf{x}$; namely, $\boldsymbol{\epsilon} \cdot \mathbf{x}$, where $\boldsymbol{\epsilon}$ is the third-order, alternating tensor but no pseudo-vector can be formed via any standard vector operation that is linear in $\mathbf{x}$. We may also note that quantities that involve "products" of two pseudo-quantities will generally become true scalars, vectors, or tensors. For example, if $\mathbf{v} = \mathbf{V} \cdot \mathbf{\Omega}$, where $\mathbf{V}$ is a pseudo-tensor and $\mathbf{\Omega}$ is a pseudo-vector, then $\mathbf{v}$ is a true vector. Another true vector is the curl of a pseudo-vector, $\nabla \wedge \boldsymbol{\omega}$, say, or the vector product of a true vector and a pseudo-vector $\mathbf{v} \wedge \boldsymbol{\omega}$. A simple test to determine whether a scalar, vector, or tensor formed as a product of vectors and tensors is a true- or pseudo-quantity is to see whether its components change sign upon transformation from a right- to left-handed coordinate system. Thus, for example, $\mathbf{v} \wedge \boldsymbol{\omega}$ is a true vector because $\boldsymbol{\omega}$ and the vector product both change sign upon coordinate transformation so that the product $\mathbf{v} \wedge \boldsymbol{\omega}$ is invariant.

Another necessary preliminary involves a few words about the solution of the creeping flow equations (4–6). However, it is convenient for some problems involving bubbles or drops (to be considered later) to formulate the equations in terms of the disturbance pressure $p' = p - p_\infty$, where p_∞ is a constant, arbitrary reference pressure. One important fact is that the disturbance pressure (and thus also the pressure itself) is a harmonic function. To see this, we can simply take the divergence of the creeping flow equation (4–6a), expressed in terms of p', subject to the condition from continuity that $\mathbf{u}$ is a solenoidal vector function,

$$\nabla \cdot [\mu \nabla^2 \mathbf{u} - \nabla p'] = 0 \longrightarrow \boxed{\nabla^2 p' = 0.} \tag{4-234}$$

A second, even more important fact, is that a general solution of the creeping (dimensional) flow equations is

$$\boxed{\mathbf{u} = \frac{\mathbf{x}}{2\mu} p' + \mathbf{u}^{(H)},} \tag{4-235}$$

where $\mathbf{u}^{(H)}$ is also a harmonic function, that is,

$$\nabla^2 \mathbf{u}^{(H)} = 0.$$

This can be demonstrated easily by substituting (4–235) into the creeping flow equations. In order to satisfy the continuity equation, the harmonic function $\mathbf{u}^H$ is required to satisfy the condition

$$\boxed{\nabla \cdot \mathbf{u}^{(H)} = -\frac{1}{2\mu} \{3p' + \mathbf{x} \cdot \nabla p'\}.} \tag{4-236}$$

Although it is not essential, we shall see that it is convenient in the developments that follow to maintain the general solution (4–235) in dimensional form.

Finally, to develop a general representation procedure for solutions of the creeping flow equations, we require a brief introduction to harmonic functions, expressed in a coordinate-independent vector form using the general position vector $\mathbf{x}$. It is convenient to group the harmonic functions into two categories: *decaying* harmonics whose magnitude decreases with increase in $r \equiv |\mathbf{x}|$ and *growing* harmonics whose magnitudes increase with r. The decaying harmonics are conveniently represented by means of higher-order derivatives of $1/r$, which is the fundamental solution of Laplace's equation. The first decaying harmonic beyond $1/r$ is thus

$$\nabla\left(\frac{1}{r}\right) \longrightarrow \frac{x_i}{r^3},$$

while the second is

$$\nabla\left(\frac{\mathbf{x}}{r^3}\right) \longrightarrow \left(\frac{x_i x_j}{r^5} - \frac{\delta_{ij}}{3r^3}\right),$$

and so forth. The first several *decaying* harmonics are thus

$$\frac{1}{r}, \frac{x_i}{r^3}, \left(\frac{x_i x_j}{r^5} - \frac{\delta_{ij}}{3r^3}\right), \left(\frac{x_i x_j x_k}{r^7} - \frac{x_i \delta_{jk} + x_j \delta_{ki} + x_k \delta_{ij}}{5r^5}\right), \text{ etc.} \quad (4\text{-}237)$$

or

$$\phi_{-(n+1)} \equiv \frac{(-1)^n}{1 \cdot 3 \cdot 5 \cdots (2n-1)} \frac{\partial^n\left(\dfrac{1}{r}\right)}{\partial x_i \partial x_j \partial x_k \cdots}, \quad n = 0, 1, 2, \ldots.$$

The growing harmonics, on the other hand, are just

$$r^{2n+1}\phi_{-(n+1)}$$

or

$$1, \; x_i, \; \left(x_i x_j - \frac{r^2}{3}\delta_{ij}\right), \; \left(x_i x_j x_k - \frac{r^2}{5}(x_i \delta_{jk} + x_j \delta_{ki} + x_k \delta_{ij})\right), \text{ etc.} \quad (4\text{-}238)$$

It is important to note that these *vector harmonic functions* involve only the general position vector $\mathbf{x}$ and its magnitude $r \equiv |\mathbf{x}|$ and thus can be represented in any coordinate system that is convenient for a particular problem.

Now, let us see how the preliminary concepts in this section can be put together to achieve a general representation procedure for the solution of general classes of creeping flow problems. We restrict our discussion to the simplest case of particles, which are spherical.

Vector Representations of Solutions for Creeping Flows

The Rotating Sphere in a Quiescent Fluid

We begin with some simple problems to illustrate the basic ideas. The simplest is the flow induced by rotation of a sphere of radius a with angular velocity $\mathbf{\Omega}$ in an unbounded, quiescent fluid. Although this problem could be formulated in dimensionless terms using Ωa as a characteristic velocity (where $\Omega \equiv |\mathbf{\Omega}|$) and the sphere radius as a characteristic length scale, it is more convenient simply to solve it in dimensional form.

The starting point is the pressure p'. The pressure is a harmonic function and a true scalar. Further, in the present problem, it must be a decaying function that is linear in $\mathbf{\Omega}$. Finally, it must be "constructed" solely from $\mathbf{\Omega}$ and the general position vector $\mathbf{x}$ in the form of the vector harmonic functions.

Now, there is only a single function that can be constructed from $\mathbf{\Omega}$ and the decaying harmonics (4–237) that is linear in $\mathbf{\Omega}$, namely,

$$\mathbf{\Omega} \cdot \frac{\mathbf{x}}{r^3}. \tag{4–239}$$

The inner product of $\mathbf{\Omega}$ with the higher-order harmonics produces either a vector or a tensor, while all vector products of $\mathbf{\Omega}$ and the vector harmonic functions produce at least a vector. Thus, the only candidate form for p' is (4–239). However, p' cannot be of this form because p' is a true scalar, while $\mathbf{\Omega} \cdot \mathbf{x}$ is a pseudo-scalar. To see this, we simply note that $\mathbf{\Omega}$ is a pseudo-vector, so that the product $\mathbf{\Omega} \cdot \mathbf{x}$ changes sign upon inversion from a right- to a left-handed system. Thus, the only possibility consistent with all of the conditions on p' is

$$p' \equiv 0. \tag{4–240}$$

It follows, therefore, from (4–235) that $\mathbf{u} = \mathbf{u}^{(H)}$ must be a decaying harmonic function, linear in $\mathbf{\Omega}$ and a true vector. The only combination of $\mathbf{\Omega}$ and the vector harmonic functions that satisfies these conditions is

$$\mathbf{u} = c\mathbf{\Omega} \wedge \left(\frac{\mathbf{x}}{r^3} \right). \tag{4–241}$$

Here, c is an arbitrary constant.*

To determine c, and thus complete our solution of the rotating sphere problem, we apply the boundary condition

$$\mathbf{u} = \mathbf{\Omega} \wedge (a\mathbf{e}_x) \quad \text{at} \quad r = a. \tag{4–242}$$

Here, we denote the position vector $\mathbf{x}$ at the sphere surface as $a\mathbf{e}_x$ where $\mathbf{e}_x$ is a unit vector in the direction of $\mathbf{x}$ and $r = a$ is the sphere radius. Now, evaluating (4–241) at the sphere surface,

$$\mathbf{u} = c\mathbf{\Omega} \wedge \left(\frac{a\mathbf{e}_x}{a^3} \right), \tag{4–243}$$

and we see by comparing (4–242) and (4–243) that

$$c = a^3.$$

Hence, the final solution for the flow produced by a rotating sphere is

$$\boxed{\mathbf{u} = \mathbf{\Omega} \wedge \mathbf{x} \left(\frac{a^3}{r^3} \right).} \tag{4–244}$$

Uniform Flow Past a Sphere

A second simple example, solved previously by other means, is Stokes' original problem of uniform flow past a stationary sphere. To apply the methods of the preceding subsection to this problem, it is convenient to transform to the *disturbance flow* problem,

---

*Application of the same technique to bodies of nonspherical geometry would generally require an integral sum of such contributions, rather than a single point function.

$$\mathbf{u}' = \mathbf{u} - \mathbf{U}, \tag{4-245}$$

where $\mathbf{U}$ is the undisturbed, uniform fluid velocity and $\mathbf{u}'$ satisfies

$$\nabla^2\mathbf{u}' - \nabla p' = 0, \quad \nabla\cdot\mathbf{u}' = 0, \tag{4-246}$$

with

$$\mathbf{u}' = -\mathbf{U} \text{ on the sphere } r = a, \tag{4-247}$$

$$\mathbf{u}' \longrightarrow 0 \quad \text{at} \quad \infty.$$

Thus, $\mathbf{u}'$ and p' are decaying functions of r.

To construct a solution for $(\mathbf{u}', p')$, we again begin with the pressure p'. In this case, p' is a decaying harmonic function, linear in $\mathbf{U}$ and a true scalar, that must be constructed solely from $\mathbf{U}$ and $\mathbf{x}$. Examination of the decaying harmonics (4–237) shows that there is a single combination of $\mathbf{U}$ and the vector harmonics that satisfy the above conditions, namely,

$$p' = c\frac{\mathbf{U}\cdot\mathbf{x}}{r^3}. \tag{4-248}$$

Again, c is an arbitrary constant, and in this case the quantity on the right is a true scalar, and thus has the same parity as p'.

Now, with the form (4–248) for p', we see from (4–235) that

$$\mathbf{u}' = \frac{\mathbf{x}}{2\mu}\left(c\frac{\mathbf{U}\cdot\mathbf{x}}{r^3}\right) + \mathbf{u}^{(H)}. \tag{4-249}$$

The function $\mathbf{u}^{(H)}$ is harmonic, decaying, linear in $\mathbf{U}$ and a true vector. Thus, the most general form for $\mathbf{u}^{(H)}$ is

$$\mathbf{u}^{(H)} = \alpha\left(\frac{\mathbf{xx}}{r^5} - \frac{\mathbf{I}}{3r^3}\right)\cdot\mathbf{U} + \beta\frac{\mathbf{U}}{r}, \tag{4-250}$$

where α and β are arbitrary constants. The continuity condition (4–236) requires that

$$\beta = \frac{c}{2\mu}. \tag{4-251}$$

Thus, the general form for $\mathbf{u}'$ that satisfies continuity is

$$\mathbf{u}' = \frac{\mathbf{x}(\mathbf{U}\cdot\mathbf{x})}{r^3}\left(\frac{c}{2\mu} + \frac{\alpha}{r^2}\right) + \frac{\mathbf{U}}{r}\left(\frac{c}{2\mu} - \frac{\alpha}{3r^2}\right). \tag{4-252}$$

The boundary condition in this case requires that

$$\mathbf{u}' = -\mathbf{U} \quad \text{at} \quad r = a. \tag{4-253}$$

Hence

$$-\mathbf{U} = \frac{\mathbf{e_x}(\mathbf{U}\cdot\mathbf{e_x})}{a}\left(\frac{c}{2\mu} + \frac{\alpha}{a^2}\right) + \frac{\mathbf{U}}{a}\left(\frac{c}{2\mu} - \frac{\alpha}{3a^2}\right). \tag{4-254}$$

The two types of terms appearing in this relation are $\mathbf{U} \cdot \text{const}$ and $\mathbf{e}_x(\mathbf{U} \cdot \mathbf{e}_x) \cdot \text{const}$. Clearly, these terms exhibit a different dependence on position on the sphere surface. Thus, to satisfy (4–254), we must require that the coefficients c and α satisfy

$$1 + \frac{c}{2a\mu} - \frac{\alpha}{3a^3} = 0$$

and

$$\frac{c}{2a\mu} + \frac{\alpha}{a^3} = 0.$$

Hence,

$$\alpha = \frac{3a^3}{4}, \quad c = -\frac{3a\mu}{2}, \tag{4–255}$$

and thus substituting back into (4–248) and (4–252), we have

$$\mathbf{u} = \mathbf{U} - \left[\frac{3}{4} \frac{a}{r} + \frac{1}{4} \frac{a^3}{r^3} \right] \mathbf{U} - \left[\frac{3a}{4r^3} - \frac{3a^3}{4r^5} \right] \mathbf{x}(\mathbf{U} \cdot \mathbf{x}) \tag{4–256}$$

and

$$p = -\frac{3a\mu}{2} \frac{\mathbf{U} \cdot \mathbf{x}}{r^3}. \tag{4–257}$$

This is nothing more than Stokes' solution, obtained earlier in Sections **B** and **E** by other (more cumbersome) techniques.

A Sphere in a General Linear Flow

In the preceding two subsections, we demonstrated the construction of solutions of the creeping flow equations via the superposition of vector harmonic functions. However, in both cases, the problem could have been solved by the techniques described earlier, and though the present methodology requires less effort, its very important fundamental advantages have not really been exposed.

Here, however, we consider the problem of a nonrotating sphere in a general linear flow of an unbounded fluid, namely,

$$\mathbf{u}_\infty = \mathbf{\Gamma} \cdot \mathbf{x} = \mathbf{E} \cdot \mathbf{x} + \mathbf{\Omega} \cdot \mathbf{x}, \tag{4–258}$$

where $\mathbf{\Gamma}$ is the velocity gradient tensor and $\mathbf{E}$ and $\mathbf{\Omega}$ are the rate of strain and vorticity tensors, respectively. In this case, apart from the single special example of axisymmetric pure straining flow, where

$$\mathbf{\Omega} \equiv 0 \quad \text{and} \quad \mathbf{E} = \pm \begin{pmatrix} E & & \\ & E & \\ & & -2E \end{pmatrix},$$

the velocity and pressure fields will be fully three-dimensional and the eigenfunction expansion of the preceding section for axisymmetric problems cannot be used. In addition, instead of focusing upon a specific example of the general class of undisturbed flows (4–258), as would be required by any of the methods of solution discussed previously, we propose to formulate a solution directly for *arbitrary* **E** and **Ω**, that is, for arbitrary flows of the general class (4–258).

In order to proceed, we again formulate the problem in terms of the disturbance velocity and pressure fields $(\mathbf{u}', p')$, namely,

$$\mathbf{u}' = \mathbf{u} - \mathbf{\Gamma} \cdot \mathbf{x}, \tag{4–259}$$

which satisfies (4–246), but with boundary conditions

$$\mathbf{u}' = -\mathbf{\Gamma} \cdot (a\mathbf{e}_x) \quad \text{on} \quad r = a, \tag{4–260}$$

$$\mathbf{u}' \to 0 \quad \text{as} \quad r \to \infty.$$

To construct a solution, we begin, as in the previous examples, with pressure p'. In this case, p' is a decaying harmonic, which is linear in **E** and **Ω** and a true scalar. The only true scalar that can be formed by **E**, **Ω**, and the decaying harmonic functions (4–237) is

$$p' = c_1 \mathbf{E} : \left(\frac{\mathbf{xx}}{r^5} - \frac{\mathbf{I}}{3r^3} \right). \tag{4–261}$$

Since $\mathrm{tr}\,\mathbf{E} = 0$, this becomes

$$\boxed{p' = c_1 \frac{\mathbf{x} \cdot \mathbf{E} \cdot \mathbf{x}}{r^5}.} \tag{4–262}$$

No true scalar can be formed that is linear in **Ω** and a decaying harmonic, because **Ω** is a pseudo-tensor. The constant c_1 will be determined shortly via boundary conditions on the velocity $\mathbf{u}'$.

A general form for the homogeneous contribution $\mathbf{u}^H$ to the velocity can be obtained in a similar manner. However, it is convenient for this purpose to express the undisturbed flow in terms of the vorticity vector rather than **Ω**, that is,

$$\mathbf{u}_\infty = \mathbf{E} \cdot \mathbf{x} + \frac{1}{2}(\boldsymbol{\omega} \wedge \mathbf{x}),$$

where $\mathbf{\Omega} \equiv -(1/2)\boldsymbol{\epsilon} \cdot \boldsymbol{\omega}$. Now, $\mathbf{u}^H$ must consist of a sum of decaying harmonics, all of which are linear in either **E** or $\boldsymbol{\omega}$ and are true vectors since $\mathbf{u}^{(H)}$ is a true vector. The most general form for $\mathbf{u}^{(H)}$, satisfying these constraints, is

$$\boxed{\mathbf{u}^{(H)} = c_2 \mathbf{E} \cdot \frac{\mathbf{x}}{r^3} + c_3 \mathbf{E} : \left(\frac{\mathbf{xxx}}{r^7} - \frac{2\mathbf{xI}}{5r^5} \right) + c_4 \boldsymbol{\omega} \wedge \frac{\mathbf{x}}{r^3}.} \tag{4–263}$$

The constants c_2, c_3, and c_4 are again arbitrary, apart from the constraint (4–236), which insures that the continuity equation is satisfied. In this case, the condition (4–236) yields $c_2 = 0$.

Thus, combining (4–262) and (4–263) via the general solution form (4–235), we find that the disturbance velocity field is given by

$$\mathbf{u}' = \mathbf{x}(\mathbf{x} \cdot \mathbf{E} \cdot \mathbf{x})\left[\frac{c_1}{2 r^5 \mu} + \frac{c_3}{r^7}\right] + \mathbf{E} \cdot \mathbf{x}\left(-\frac{2 c_3}{5 r^5}\right) + c_4 \boldsymbol{\omega} \wedge \frac{\mathbf{x}}{r^3} \qquad (4\text{-}264)$$

for arbitrary $\mathbf{E}$ and $\boldsymbol{\omega}$. The constants c_1, c_3, and c_4 are obtained by applying the boundary condition (4-260) with $\boldsymbol{\Gamma} = \mathbf{E} + \boldsymbol{\Omega}$ and $\boldsymbol{\Omega} = -(1/2)\boldsymbol{\epsilon} \cdot \boldsymbol{\omega}$. The result is

$$-\mathbf{E} \cdot (a \mathbf{e}_x) - \frac{1}{2}\boldsymbol{\omega} \wedge (a \mathbf{e}_x) = \mathbf{e}_x(\mathbf{e}_x \cdot \mathbf{E} \cdot \mathbf{e}_x)\left[\frac{c_1}{2 a^2 \mu} + \frac{c_3}{a^4}\right] + \mathbf{E} \cdot \mathbf{e}_x\left(-\frac{2 c_3}{5 a^4}\right) + c_4\left(\frac{\boldsymbol{\omega} \wedge \mathbf{e}_x}{a^2}\right),$$

$$(4\text{-}265)$$

and thus

$$c_3 = \frac{5 a^5}{2}, \quad c_4 = -\frac{a^3}{2}, \text{ and } c_1 = -5 a^3 \mu. \qquad (4\text{-}266)$$

Combining (4-259), (4-264), and (4-266), we therefore obtain

$$\boxed{\mathbf{u} = \mathbf{E} \cdot \mathbf{x}\left(1 - \frac{a^5}{r^5}\right) + \frac{1}{2}(\boldsymbol{\omega} \wedge \mathbf{x})\left(1 - \frac{a^3}{r^3}\right) - \mathbf{x}(\mathbf{x} \cdot \mathbf{E} \cdot \mathbf{x})\left(\frac{5}{2}\frac{a^3}{r^5} - \frac{5 a^5}{2 r^7}\right).}$$

$$(4\text{-}267)$$

The corresponding pressure distribution is

$$\boxed{p = -5\mu(\mathbf{x} \cdot \mathbf{E} \cdot \mathbf{x})\left(\frac{a^3}{r^5}\right) + p_\infty.}$$

$$(4\text{-}268)$$

The expressions (4-267) and (4-268) represent a complete, exact solution of the creeping flow equations for a completely *arbitrary* linear flow. Among the linear flows of special interest are axisymmetric pure strain, which was solved via the eigenfunction expansion for axisymmetric flows in the previous section, and simple shear flow

$$\mathbf{u} = G x_2 \mathbf{i}_1, \text{ that is, } \mathbf{E} = \begin{pmatrix} 0 & G/2 & 0 \\ G/2 & 0 & 0 \\ 0 & 0 & 0 \end{pmatrix} \text{ and } \boldsymbol{\omega} = -G \mathbf{i}_3. \qquad (4\text{-}269)$$

It is of interest, in this latter case, to express the general solution (4-267) in terms of components in a spherical coordinate system. For this purpose, we note that the general position vector can be expressed in the form

$$\mathbf{x} = x_1 \mathbf{e}_1 + x_2 \mathbf{e}_2 + x_3 \mathbf{e}_3, \qquad (4\text{-}270)$$

with

$$\mathbf{e}_1 = \mathbf{i}_r \sin\theta \cos\phi + \mathbf{i}_\theta \cos\theta \cos\phi - \mathbf{i}_\phi \sin\phi,$$

$$\mathbf{e}_2 = \mathbf{i}_r \sin\theta \sin\phi + \mathbf{i}_\theta \cos\theta \sin\phi + \mathbf{i}_\phi \cos\phi,$$

$$\mathbf{e}_3 = \mathbf{i}_r \cos\theta - \mathbf{i}_\theta \sin\theta, \qquad (4\text{-}271)$$

and

$$x_1 = r\sin\theta\cos\phi,$$

$$x_2 = r\sin\theta\sin\phi,$$

$$x_3 = r\cos\theta. \tag{4-272}$$

Note that the spherical coordinates (r,θ,ϕ) are defined, as sketched in Figure 4–16, with the polar angle θ measured from the x_3 axis and the azimuthal angle ϕ measured around the x_3 axis with $\phi = 0$ on the x_1x_3 plane. For the specific case of simple shear flow, given by (4–269), the general solution (4–267) takes the form

$$\mathbf{u} = \frac{G}{2}(x_2\mathbf{e}_1 + x_1\mathbf{e}_2)\left(1 - \frac{a^5}{r^5}\right) - \left(\frac{G}{2}x_1\mathbf{e}_2 - \frac{G}{2}x_2\mathbf{e}_1\right)\left(1 - \frac{a^3}{r^3}\right)$$

$$- \frac{5}{2}Gx_1x_2(x_1\mathbf{e}_1 + x_2\mathbf{e}_2 + x_3\mathbf{e}_3)\left(\frac{a^3}{r^5} - \frac{a^5}{r^7}\right). \tag{4-273}$$

Thus, if we substitute (4–271) and (4–272) into (4–273) and collect all of the terms corresponding to each vector component, we find that

$$\mathbf{u} = \mathbf{i}_r\left(Gr\sin^2\theta\sin\phi\cos\phi\left(1 - \frac{5}{2}\left(\frac{a^3}{r^3}\right) + \frac{3}{2}\left(\frac{a^5}{r^5}\right)\right)\right)$$

$$+ \mathbf{i}_\theta\left(Gr\sin\theta\cos\theta\sin\phi\cos\phi\left(1 - \frac{a^5}{r^5}\right)\right)$$

$$+ \mathbf{i}_\phi\left(-\frac{G}{2}r\sin\theta + \frac{Gr}{2}\sin\theta\frac{a^3}{r^3} + \frac{Gr}{2}(\sin\theta\cos2\phi)\left(1 - \frac{a^5}{r^5}\right)\right). \tag{4-274}$$

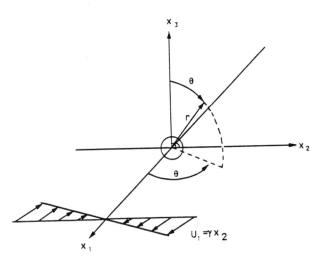

Figure 4–16 The undisturbed flow and spherical coordinate system for simple shear flow, $u = \gamma y$, in the vicinity of a sphere.

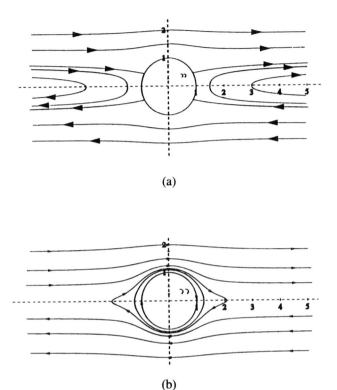

(a)

(b)

Figure 4–17 Fluid pathlines in the $x_1 x_2$ plane (see Figure 4-16) for simple shear flow past (a) nonrotating and (b) rotating spheres. The nonrotating case was obtained from equation (4–274), and the rotating case was obtained by adding (4–277) to (4–274).

This is the velocity field for a stationary, nonrotating sphere in a simple shear flow. A sketch of the fluid pathlines in the $x_1 x_2$ plane for this case is reproduced in Figure 4–17a. If the sphere is allowed to rotate with some angular velocity $\boldsymbol{\Omega}$, the corresponding velocity field can be obtained as a direct superposition of the present solution for a nonrotating sphere in simple shear, and the solution (4–244) for rotation with velocity $\boldsymbol{\Omega}$ in a quiescent fluid. If the sphere rotates as a consequence of the shear flow, it is evident that $\boldsymbol{\Omega}$ must be in the same direction as the vorticity vector $\boldsymbol{\omega}$, that is,

$$\boldsymbol{\Omega} = -\Omega \mathbf{i}_3. \tag{4–275}$$

In that case, the velocity field (2–244) becomes

$$\mathbf{u} = (-\Omega x_1 \mathbf{e}_2 - \Omega x_2 \mathbf{e}_1)\left(\frac{a^3}{r^3}\right), \tag{4–276}$$

and when this is expressed in terms of spherical coordinates, it simplifies dramatically to the form

$$\mathbf{u} = -\Omega\left(\frac{a}{r}\right)^3 r\sin\theta\,\mathbf{i}_\phi. \tag{4-277}$$

The composite velocity field for a sphere rotating with the angular velocity (4–275) in a simple shear flow, equation (4–269), is thus the sum of (4–274) and (4–277). The fluid pathlines in the x_1x_2 plane for this case are shown in Figure 4–17b. The corresponding torque on the sphere is

$$\mathbf{T} = -8\pi a^3 \mu\left(\frac{G}{2} - \Omega\right)\mathbf{i}_3. \tag{4-278}$$

To determine the angular velocity Ω, we must balance the hydrodynamic torque against the magnitude of any externally applied torque on the sphere. If the sphere is free to rotate—that is, $\mathbf{T} = 0$—it follows that

$$\Omega = \frac{G}{2}. \tag{4-279}$$

Thus, the sphere rotates with an angular velocity that is just 1/2 the vorticity of the undisturbed flow. The velocity field for a freely rotating sphere in a simple shear flow is identical to equation (4–274), except that

$$u_\phi = -\frac{G}{2}r\sin\theta + \frac{Gr}{2}\sin\theta\cos2\phi\left(1 - \frac{a^5}{r^5}\right). \tag{4-280}$$

Clearly, for either a nonrotating or freely rotating sphere, the velocity field in a simple shear flow is fully three-dimensional, so that the solution methods of earlier sections would necessarily fail.

The technique outlined in this section is extremely simple and powerful, allowing solutions for a complete class of undisturbed flows to be obtained at the same time. The reader should notice that all the eigenfunction-based methods of the preceding sections are included within the present methodology, which may thus seem to supersede them. To some degree, this is in fact true. However, the earlier methods are still frequently used, often because of the preference of the researcher or engineer for the older, more classical techniques. It should also be noted that instances will arise, in using the present techniques, where application of the boundary conditions to determine constants is extremely complicated if it is done (as above) for the most general form of the undisturbed flow. However, since the *general* form of the solution is the same for all members of a particular family of flows, it is often possible to use the detailed solution for one specific flow to determine the constants in the general solution form. In this case, the solution for the specific flow problem might be determined by the methods of previous sections (or by the generalization using Lamb's general solution that was mentioned at the beginning of this section).

Homework Problems

1. Consider a rigid sphere of radius a, which executes a rectilinear oscillatory motion with velocity

$$U(t) = U \sin \omega t$$

in an unbounded, quiescent body of fluid. We consider the asymptotic limit

$$\frac{Ua}{\nu} \ll 1, \quad \frac{a^2\omega}{\nu} \sim 0(1).$$

In this case, nonlinear inertia terms in the Navier-Stokes equation can be neglected to leading order of approximation, but the acceleration term, $\partial u/\partial t$, must be retained.

a. Derive dimensionless forms for the governing equations and boundary conditions in the limit described above for the case where t is sufficiently large that initial transients associated with start-up flow at $t=0$ can be neglected. For this purpose, it is convenient to adopt a spherical coordinate system whose origin is fixed at the position initially occupied by the center of the sphere. Show that the displacement of the sphere can be neglected at the first level of approximation so that boundary conditions can be applied at $r=a$ for all times t.

b. Solve for the velocity and pressure fields for the problem defined above. Hints: Do not assume that the motion of the fluid is in-phase with the motion of the sphere. In addition, this problem is an example of a case where we can anticipate the form of the solution based upon the symmetry of the boundary conditions. In particular, if the motion of the sphere is along the z axis as depicted here, and the spherical coordinate system is set up with the polar angle θ measured from the z axis, you may wish to try

$$u_r = e^{-i\bar{t}}\,\bar{f}(\bar{r})\ \cos\theta,$$

$$u_\theta = e^{-i\bar{t}}g\,(\bar{r})\ \sin\theta,$$

where, for algebraic convenience,

$$\bar{f}(\bar{r}) = -\frac{2}{r}\frac{df}{dr}.$$

You should find that the governing equation takes the form

$$\frac{d}{dr}\left[\nabla_r^2\,(\nabla_r^2 f) + i\left(\frac{\omega a^2}{\nu}\right)\nabla_r^2 f\right] = 0,$$

where

$$\nabla_r^2 \equiv \frac{1}{r^2}\frac{d}{dr}\left(r^2\frac{d}{dr}\right).$$

Hence, since **u** and all its derivatives vanish as $r \to \infty$, it follows that

$$\nabla_r^2\left(\nabla_r^2 f\right) + i\left(\frac{\omega a^2}{\nu}\right)\nabla_r^2 f = 0.$$

c. Solve the equation from part (b) for f and calculate the drag on the sphere. In dimensional terms, you should find that

$$F = 6\pi\mu a\left[1 + \left(\frac{a^2\omega}{2\nu}\right)^{1/2}\right]U\,e^{-i\omega\tau}$$

$$+ 3\sqrt{2}\pi a^2\sqrt{\frac{\mu\rho}{\omega}}\left(1 + \frac{2}{9}\left(\frac{a^2\omega}{2\nu}\right)^{1/2}\right)(-i\omega U e^{-i\omega\tau}).$$

Discuss the limit $\omega \to 0$. Show that there is a contribution to F for $\omega \gg 1$ that is independent of the viscosity. This term is known as the added mass contribution and is identical to the force on an accelerating sphere in an inviscid fluid (that is, a fluid with $\mu = 0$). It predicts that there is a contribution to the force that has the effect of adding an additional mass to the sphere that is equal to $1/2$ of the sphere volume times the fluid density.

d. Given the result in part (c) for the force on an oscillating sphere, we can obtain the corresponding result for the force on an accelerating sphere with velocity $u_s(t)$. The simplest way is to note that $u_s(t)$ can be represented in terms of a Fourier transform,

$$u_s(t) = \int_{-\infty}^{\infty} u_\omega e^{-i\omega t}d\omega,$$

where

$$u_\omega = \frac{1}{2\pi}\int_{-\infty}^{\infty} u_s(\tau)e^{i\omega\tau}d\tau,$$

so that $F \equiv F_\omega$ from part (c) can be interpreted as the force on a sphere with velocity $u_\omega e^{-i\omega t}$. Then $F(t)$ corresponding to $u_s(t)$ is just

$$F = \int_{-\infty}^{\infty} F_\omega d\omega.$$

Show that

$$F = 2\pi\rho a^3\left\{\frac{1}{3}\frac{du_s}{dt} + \frac{3\nu u_s}{a^2} + \frac{3}{a}\sqrt{\frac{\nu}{\pi}}\int_{-\infty}^{t}\frac{du_s}{d\tau}\frac{d\tau}{\sqrt{t-\tau}}\right\}.$$

Here, the first term is the added mass contribution, the second is Stokes' law, and the third is known as the Basset memory integral contribution. Evaluate this expression for

$$u_s = \begin{cases} 0; & t < 0 \\ u_\infty, & t \geq 0. \end{cases}$$

2. a. A jellyfish swims by slowly drawing fluid into its umbrella-like body (see the sketch below) and then ejecting it with high velocity, propelling itself forward. Would this still work if the jellyfish were of microscopic dimensions (consider the dimensionless velocity of the jellyfish here)? Briefly explain your answer.

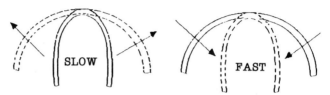

b. A straight, slender rodlike particle sediments through a quiescent fluid under the action of gravity ($\rho_{particle} \neq \rho_{fluid}$). The Reynolds number is very small so that the creeping motion approximation is relevant.

Show that the particle will fall vertically if it is oriented with $\alpha = 0$ or $\alpha = \pi/2$. What can you say for $\alpha \neq 0$, $\pi/2$? Justify your conclusions.

c. A neutrally buoyant drop (that is, $\rho_{drop} = \rho_{suspending\ fluid}$) in simple shear flow near a plane wall is observed to migrate in the direction normal to the wall. Is this possible in creeping flow? Discuss. (Note that the shape of the drop will not generally remain spherical.)

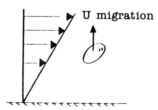

3. A rigid sphere is translating with velocity **U** and rotating with angular velocity **Ω** in an unbounded, incompressible Newtonian fluid. The position of the sphere center is denoted as $\mathbf{x}_p$ (that is, $\mathbf{x}_p$ is the position vector).

At large distances from the sphere, the fluid is undergoing a simple shear flow (this is the undisturbed velocity field). We may denote this flow in the form

$$\mathbf{u}_\infty = \mathbf{\Gamma} \cdot \mathbf{x},$$

where

$$\mathbf{\Gamma} = \gamma \begin{bmatrix} 0 & 1 & 0 \\ 0 & 0 & 0 \\ 0 & 0 & 0 \end{bmatrix}$$

and x is the general position vector associated with any arbitrary point in the fluid. (Note that $\mathbf{x}_p \neq 0$.) If the appropriate Reynolds number is very small, so that the creeping motion approximation is valid, it can be shown that the hydrodynamic force and torque acting on the body are expressible in the form

$$\mathbf{F} = a\mu(\mathbf{A} \cdot \mathbf{U} + \mathbf{B} \cdot \boldsymbol{\Omega} + \mathbf{D} : \boldsymbol{\Gamma}),$$

$$\mathbf{T} = a\mu(\mathbf{C} \cdot \mathbf{U} + \mathbf{E} \cdot \boldsymbol{\Omega} + \mathbf{G} : \boldsymbol{\Gamma}),$$

where **A**, **B**, **C**, **E**, **D**, and **G** are constant tensors. Here, a is the particle radius and μ the fluid viscosity.

a. Demonstrate that the above result for **F** and **T** is true.
b. Indicate the forms for **D** and **G** (that is, what terms are zero or nonzero and what can we say beyond that?).
c. Demonstrate the validity of the general formula for **F** and **G** for any linear flow of the type

$$\mathbf{u}_\infty = \boldsymbol{\Gamma} \cdot \mathbf{x}.$$

4. A spherical gas bubble rises through a stationary liquid that is sufficiently viscous that the creeping flow approximation is valid. Assume that the boundary conditions for the liquid motion at the bubble surface are

$$\mathbf{u} \cdot \mathbf{e}_r = 0, \, \mathbf{e}_\theta \cdot (\mathbf{e}_r \cdot \mathbf{T}) = 0,$$

where $\mathbf{e}_r$ and $\mathbf{e}_\theta$ are unit tangent vectors normal and parallel to the bubble surface. Further, assume that the gas inside the bubble has negligible viscosity and density relative to the liquid.

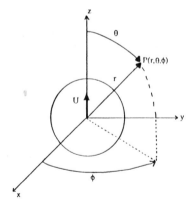

Determine the velocity and pressure fields in the liquid as well as the velocity of the bubble using:

a. the concept of "preservation of boundary symmetry,"
b. the full-eigenfunction expansion for ψ in spherical coordinates.

5. a. Prove that the hydrodynamic force exerted by the fluid on a solid sphere in *any* axisymmetric creeping flow is

$$F_z = 4\pi\mu U a C_1,$$

where

$$\psi = \sum_{n=1}^{\infty} \left[A_n r^{n+3} + B_n r^{n+1} + C_n r^{2-n} + D_n r^{-n} \right] Q_n(\eta).$$

b. Use the divergence theorem to demonstrate that the force on an *arbitrarily shaped* axisymmetric body in creeping flow is proportional to C_1.

6. Use the general eigenfunction expansion for axisymmetric creeping flows, in spherical coordinates, to determine the velocity and pressure fields for a solid sphere of radius a that is held fixed at the central axis of symmetry of an unbounded parabolic velocity field,

$$\mathbf{u}_\infty = u_z(r) \mathbf{i}_z,$$

where

$$u_z(r) = U \left(1 - \frac{r^2}{d^2} \right).$$

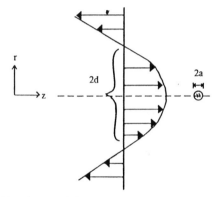

Use the result from Problem 5a. to calculate the drag force on the sphere. You should find that

$$F_z = 6 \pi \mu a U \left[1 - \frac{2a^2}{3d^2} \right].$$

This result should be checked against *Faxen's law*, which will be proven in Chapter 5. Faxen's law for a solid sphere holds that the force and torque due to an *arbitrary undisturbed flow*, $\mathbf{u}_\infty = \mathbf{u}(r)$, that satisfies the creeping motion equations can be calculated from the following formula:

$$\mathbf{F} = 6 \pi \mu a \left[(\mathbf{u}_\infty)_0 + \frac{a^2}{6} (\nabla^2 \mathbf{u}_\infty)_0 \right]$$

and

$$\mathbf{T} = 4 \pi \mu a^3 (\nabla \wedge \mathbf{u}_\infty)_0,$$

where the subscript zero means that the indicated quantities are to be evaluated at the point in space occupied by the center of the sphere.

7. Consider a solid sphere of radius a in a general quadratic undisturbed flow

$$\mathbf{u}_\infty = \mathbf{U} + \mathbf{A} : \mathbf{xx},$$

where $\mathbf{U}$ is a constant vector and $\mathbf{A}$ is a constant third-order tensor.

a. Determine the velocity and pressure fields in terms of $\mathbf{U}$ and $\mathbf{A}$, assuming that the sphere is stationary and its center lies at the origin of the coordinate reference system (that is, $\mathbf{x}$ is a general position vector defined with respect to the same origin of coordinates).

b. Show that problem 6 is a special case of this problem (what form do I need to assume for $\mathbf{A}$?). See if the solutions match if the general solution here is specialized to Problem 6.

References

1.a. Roache, P.J., *Computational Fluid Dynamics*. Hermosa: Albuquerque, NM (1976).

b. Baker, A.J., *Finite Element Computational Fluid Mechanics*. Hemisphere (McGraw-Hill): Washington, DC (1983).

c. Orszag, S.A., and Israeli, M., "Numerical Simulations of Viscous Incompressible Flows," Ann. Rev. of Fluid Mech. 6:281–318 (1974).

d. Shen, S.F., "Finite-Element Methods in Fluid Mechanics," *Ann. Rev. of Fluid Mech.* p:421–445 (1977).

2. a. Whitaker, S., *Introduction to Fluid Mechanics*. Prentice-Hall: Englewood Cliffs, NJ (1968).

b. Denn, M.M., *Process Fluid Mechanics*. Prentice-Hall: Englewood Cliffs, NJ (1980).

3. Segre, G., and Silberberg, A., "Behavior of Macroscopic Rigid Sphere in Poiseuille Flow, Part 1, Determination of Local Concentration by Statistical Analysis of Particle Passages Through Crossed Light Beams," J. Fluid Mech. 14:115–136 (1962).
"Part 2, Experimental Results and Interpretation," J. Fluid Mech. 14:136–157 (1962).

4. a. Ho, B.P., and Leal, L.G., "Inertial Migration of Rigid Spheres in Two-Dimensional Unidirectional Flows," J. Fluid Mech. 65:365–400 (1974).

b. Vasseur, P., and Cox, R.G., "The Lateral Migration of a Spherical Particle in Two-Dimensional Shear Flows," J. Fluid Mech. 78:385–413 (1976).

c. Ishii, K., and Hasimoto, H., "Lateral Migration of a Spherical Particle in Flows in a Circular Tube," J. Phys. Soc. of Japan 48:2144–2153 (1980).

5. Leal, L.G., "Particle Motion in a Viscous Fluid," *Ann. Rev. of Fluid Mech.* 12:435–476 (1980).

6. See Reference 2, Chapter 3. Also see Brenner, H., "Dynamics of Neutrally Buoyant Particles in Low Reynolds Number Flows," *Prog. In Heat and Mass Transfer* 6:509–574 (1972).

7. The rotating parallel disk problem has been studied by many investigators. One of the most interesting features is that there are at least 19 non-unique solutions at high enough Reynolds number. The reader who is interested in pursuing the literature on this topic may wish to consult the following references as a starting point:

a. Parter, S.V., "On the Swirling Flow Between Coaxial Disks: A Survey," in *Theory and Application of Singular Perturbations.* Springer (Lecture Notes in Mathematics 942): (1982).

b. Mellor, G.L., Chapple, P.J., and Stoes, V.K., "On the Flow Between a Rotating and a Stationary Disk," J. Fluid Mech. 31:95–112 (1968).

c. Lai, C.Y., Rajagopal, K.R., and Szeri, A.Z., "Asymmetric Flow Between Parallel Rotating Disks," *J. Fluid Mech.* 146:203–225 (1984).

8. An excellent sourcebook for vector and tensor calculus is Aris, R., *Vectors, Tensors and the Basic Equations of Fluid Mechanics.* Prentice-Hall: Englewood Cliffs, NJ (1962).

9. Moffatt, H.K., "Viscous and Resistive Eddies Near a Sharp Corner,'" *J. Fluid Mech.* 18:1–18 (1964).

10. Taylor, G.I., "Deposition of Viscous Fluid in a Plane Surface," *J. Fluid Mech.* 9:218–224 (1960).

11. a. Wakiya, S., "Application of Bipolar Co-ordinates to the Two-Dimensional Creeping Motion of a Liquid. I. Flow Over a Projection or a Depression on a Wall," *J. Phys. Soc. Japan* 39:1113–1120 (1975).

b. O'Neill, M.E., "On the Separation of a Slow Linear Shear Flow From a Cylindrical Ridge or Trough in a Plane," Z. Angew. Math. Phys. 28:438–448 (1977).

c. Davis, A.M.J., and O'Neill, M.E., "Separation in a Stokes Flow Past a Phase with a Cylindrical Ridge or Trough," *Quart. J. Mech. Appl. Math* 30:355–368 (1977).

12. Dorreapaal, J.M., and O'Neill, M.E., "The Existence of Free Eddies in a Streaming Stokes Flow," *Quart. J. Mech. Appl. Math* 32:95–107 (1979).

13. O'Neill, M.E., and Ranger, K.B., "Particle-Fluid Interaction," in Hetsron: (ed), *Handbook of Multiphase Flow.* Hemisphere: New York, pp. 1–96 (1982).

14. See Reference 6, Chapter 3.

15. Einstein, A., *Investigations on the Theory of the Brownian Movement.* Dover: New York (1956).

16. Schowalter, W.R., *Mechanics of non-Newtonian Fluids.* Pergamon Press: New York (1978).

17. Lamb, H., *Hydrodynamics,* 6th ed. Dover: New York (1945): reprint of Camb. Univ. Press edition (1932).

18. See Reference 2, Chapter 3.

CHAPTER 5

Further Results in the Creeping Flow Limit

The theory of flows dominated by viscous effects has long been the special province of chemical engineers. In large part, this is because the materials that we deal with are often very viscous, and also because we are frequently concerned with fluid dynamical phenomena on very small length scales—small particles, microdomains, and so forth. In Chapter 4 we outlined some of the classical solution methods for the creeping flow equations. In this chapter we continue this presentation, with a focus on additional solution techniques and on the application of creeping flow theory to a topic of special concern to chemical engineers: the motions of bubbles and drops in a viscous fluid. While all of the material presented here is very important, the reader who wishes to deemphasize low Reynolds number fluid mechanics in favor of later topics may wish to study those portions of the first section that pertain to general boundary conditions at a fluid interface and then proceed to Chapter 6.

A. The Motions of Bubbles and Drops

Many of the fluid mechanics problems of greatest interest and concern to scientists and engineers involve the motion of two (or more) contiguous fluids that are immiscible and separated by an interface. These problems exhibit many complicated phenomena and are especially difficult to solve because the position and shape of the interface is not known a priori and may change in response to the motions of the fluids. Thus, although the Navier-Stokes equations are relevant to the motions of the two fluids (each being Newtonian with its own viscosity and density), the boundary conditions must be applied at the interface, and the shape of this interface cannot be specified ahead of time. Such problems are known as *free-boundary problems.*

One class of free-boundary problems of particular interest to chemical engineers is the motion of bubbles or drops in a second immiscible fluid. Of course, bubble and drop motions may occur over a broad spectrum of Reynolds numbers, not only in the creeping flow limit which is the special province of this chapter.[1] Nevertheless, many problems involving small bubbles or drops in viscous fluids do fall into this class, and in this section we consider some typical examples. Before doing this, however, it is necessary to discuss the boundary conditions at a fluid interface. It should be emphasized that these conditions do not depend in any way on the creeping motion approximation. However, this section is the first discussion of this class of problems, and it is appropriate to introduce the boundary conditions here.

Boundary Conditions at a Fluid Interface

Let us then consider the boundary conditions at an interface separating two immiscible liquids, or a liquid and a gas. It is convenient to discuss these conditions in the context of a generalized drop dynamics problem, as described below, although we emphasize that the interface boundary conditions are not specific to the bubble/drop problem. Thus, we consider two Newtonian fluids which we shall denote as fluid 1 and 2, separated by an interface that is denoted symbolically as S. A typical bubble/drop configuration is sketched in Figure 5–1. Fluids 1 and 2 are characterized by viscosities μ_1 and μ_2 and densities ρ_1 and ρ_2, and the Navier-Stokes equations apply within the two fluids. We presume that boundary conditions are available at any solid boundary that is in contact with only fluid 1 or 2, as well as conditions at infinity if one of the fluid domains is unbounded as in the configuration of Figure 5–1. Our concern here is the conditions at the interface S.

It should be recognized at the outset that these conditions represent an attempt to impose continuum-mechanical conditions at an interface that are macroscopically equivalent to the molecular scale physics, but without the benefit of an a priori molecular theory. In particular, we attempt to replace the real interfacial region, in which physical properties vary extremely rapidly but continuously, by a macroscopic surface, with properties such as density or viscosity that are modeled as retaining their bulk values right up to the surface and then changing discontinuously. As a consequence of the approximation inherent in this simplified view, it is necessary to make some guesses about the macroscopic properties to ascribe to the surface S. The goal is to render it equivalent to the real interfacial region in the sense that the predicted macroscopic velocity, pressure, and stress fields in the two bulk fluids are the same as in the real system.

In effect, in order to formulate boundary conditions for an interface, we must hypothesize that it is characterized by certain *constitutive* properties. In the absence of a molecular theory, the validity of this hypothesis can ultimately be tested only by comparison between predictions obtained using the constitutive properties as boundary conditions and measurements of the macroscopic fields in real systems. It is fair to say that the appropriate interface formulation remains a topic of ongoing research, and the correct macroscopic conditions remain uncertain, particularly for fluid systems which involve surface-active solutes in addition to the two primary fluids. However, in this section and elsewhere throughout this book, we adopt the simplest possible description that is consistent with equilibrium thermodynamics—namely, we suppose that the interface can be characterized entirely by a *surface or interfacial tension* which may be a function of the local thermodynamic state, such as temperature or pressure, and may also depend on the concentration of any solutes, but is independent of whether the

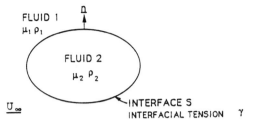

Figure 5–1 A typical configuration of a viscous drop of viscosity μ_2 and density ρ_2 immersed in a second immiscible fluid of viscosity μ_1 and density ρ_1. The interface, denoted as S, is deformed to a nonspherical shape, and it is characterized by the interfacial tension γ.

interface is undergoing any macroscopic motion or deformation. We shall shortly discuss the significance of surface or interfacial tension in the context of a macroscopic, continuum-mechanical description of a two-fluid system. However, to conclude this discussion, we should note that the simple model adopted here assumes implicitly that the interface has no macroscopically significant dynamical properties of its own, in spite of the fact that some research suggests that the interface may also have dynamical properties, such as a surface or interface viscosity, that are independent of the corresponding bulk-phase dynamical properties. Although the simple interface description adopted here may be incomplete, we do not believe that current knowledge provides a sufficiently strong motivation to adopt a more complex model. The reader who is interested in a more general prescription of interface properties may wish to refer to the classic paper by Scriven (1960).[2]

What conditions, then, apply at a fluid-fluid interface? All currently accepted interface models agree on the conditions that directly involve the velocity components in the two fluids.

Continuity of Velocity
This is essentially the *no-slip* boundary condition first introduced at the end of Chapter 2 and takes the form

$$\boxed{\mathbf{u} = \hat{\mathbf{u}} \quad \text{on} \quad S,} \tag{5-1}$$

where we denote the velocity in fluid 1 as $\mathbf{u}$ and the velocity in fluid 2 as $\hat{\mathbf{u}}$.

The Kinematic Condition
This is essentially the assumption that the interface is neither a source nor a sink of mass. According to this condition, the normal component of velocity at S in either of the two fluids must be equal to the normal velocity of the surface itself. Again, we introduced this condition at the end of Chapter 2. The only difference between its application in the problems we considered previously and in the present context is that the surface S not only may translate, as for rigid boundaries, but also may deform to new shapes as a function of time. In this case, the kinematic condition can be expressed in

the form of a relationship between $\mathbf{u} \cdot \mathbf{n}$ or $\hat{\mathbf{u}} \cdot \mathbf{n}$ (where $\mathbf{n}$ is the unit normal to S) and the time rate of change of the interface shape.

In order to express this relationship in mathematical form, let us introduce a function $F(\mathbf{x}, t)$ to represent the interface shape as the set of points $\mathbf{x}$ where $F(\mathbf{x}, t) \equiv 0$. Now the unit normal $\mathbf{n}$ to S can be defined in terms of F as

$$\mathbf{n} = \pm \frac{\nabla F}{|\nabla F|}, \tag{5-2}$$

where the sign is chosen by convention so that $\mathbf{n}$ is a positive unit vector when it points from S into the exterior fluid 1 (see Figure 5-1). Now, since F is a scalar function which is always equal to zero at any point on the fluid interface, its time derivative following any material point on the interface [which has velocity $\mathbf{u} = \hat{\mathbf{u}}$ on S according to (5-1)] is obviously equal to zero, that is

$$\frac{\partial F}{\partial t} + \mathbf{u} \cdot \nabla F = 0 \quad \text{for any point on } S. \tag{5-3}$$

Thus, rearranging slightly,

$$\boxed{\frac{1}{|\nabla F|} \frac{\partial F}{\partial t} + \mathbf{u} \cdot \mathbf{n} = 0 \quad \text{on } S.} \tag{5-4}$$

This is the most useful form of the kinematic condition. The reader should note that if the shape of the interface is independent of time and the interface is not translating relative to the origin of the coordinate system, $\mathbf{x} = 0$, then this condition becomes simply

$$\boxed{\mathbf{u} \cdot \mathbf{n} = 0 \quad \text{on } S.} \tag{5-5}$$

The Stress Conditions

For motions of a single fluid involving solid boundaries, we have seen that the no-slip and kinematic boundary conditions are sufficient to determine completely a solution of the equations of motion, provided the motion of the boundaries is specified. In problems involving two fluids separated by an interface, however, these conditions are not sufficient because they only provide relationships between the velocity components in the fluids and the interface shape, all of which are unknowns. The additional conditions necessary to completely determine the velocity fields and the interface shape come from a *force equilibrium* condition on the interface. In particular, since the interface is viewed as a surface of zero thickness, the volume associated with any arbitrary segment of the interface is zero, and the sum of all forces acting on this interface segment must be identically zero (to avoid infinite acceleration).

In order to formally express this force equilibrium condition, it is necessary to consider the nature of the forces which act on an interface. According to the simple interface description which involves only interfacial tension, these consist of two kinds: First, there are the bulk pressure and stresses which act on the faces of the interface element and produce a net effect that is proportional to the surface area; and second, there is a force due to surface or interfacial tension which acts in the plane of the interface at the edges of the surface element and is specified via the magnitude of the surface or interfacial tension as a force per unit length. If we assume that these are the only forces acting on the interface (this is akin to assuming that there are no dynamic forces in the interface itself, as explained above), it is not difficult to derive a differential, pointwise form for the force equilibrium condition.

Prior to actually doing this, however, it is useful to remind ourselves briefly of the origin of surface tension, as seen from thermodynamics, and explain why it appears in a continuum mechanics theory as a tensile force per unit length in the interface. First, from a thermodynamic point of view, we may recall that interfacial tension is introduced as a measure of a free energy per unit surface area. Thus, an increase in the area of the interface requires an increase in the free energy of the system (work). The necessity of doing work to create new interfacial area is a consequence of the fact that the molecules in the immediate vicinity of the interface experience a net force which tends to pull them back into the bulk liquid. This is particularly clear in the case of a gas-liquid interface, where the density of molecules in the gas is very much lower than in the bulk liquid, but it is generally true also at liquid-liquid interfaces. However, in a macroscopic mechanical theory, the only way to include the required rate of working for changes in interface area is to assume the existence of a force per unit length which acts at the edges of an interfacial element. The magnitude of this force is γ, the so-called *interfacial tension*.

To express the force equilibrium condition in a mathematical form, we can now consider a force balance on an arbitrary surface element of a fluid interface, which we denote as A. A sketch of this surface element is shown in Figure 5-2, as seen when

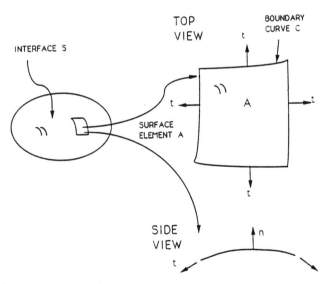

Figure 5-2 A sketch showing the top and side views of an arbitrary surface element, denoted as A, of a fluid interface. The unit normal vector is denoted as $\mathbf{n}$, and the unit tangent that is perpendicular to the boundary curve C is denoted as $\mathbf{t}$.

viewed along an axis that is normal to the interface at some arbitrary point within A. We do *not* imply that the interface is flat (though it could be)—indeed, we shall see that curvature of an interface almost always plays a critical role in the dynamics of two-fluid systems. We denote the unit normal to the interface at any point in A as $\mathbf{n}$ (to be definite, we may suppose that $\mathbf{n}$ is positive when pointing upward from the page in Figure 5-2) and let $\mathbf{t}$ be the unit vector that is normal to the boundary curve C and *tangent* to the interface at each point (see Figure 5-2). With these conventions, the force equilibrium condition applied to A requires that

$$\iint_A (\mathbf{T} - \hat{\mathbf{T}}) \cdot \mathbf{n} \, dA + \int_C \gamma \mathbf{t} \, dl = 0. \qquad (5\text{-}6)$$

Here, $\mathbf{T}$ is the total bulk stress in the flow above the surface element A evaluated in the limit as we approach the interface, while $\hat{\mathbf{T}}$ is the stress in the second fluid, evaluated as we approach the interface from below. The negative sign associated with $\hat{\mathbf{T}} \cdot \mathbf{n}$ is a consequence of the fact that the unit normal from the interface into the second fluid is $-\mathbf{n}$.

Although the expression (5-6) is a perfectly general statement of the force balance at an interface, it is not particularly useful in this form because it is an overall balance on a macroscopic element of the interface. In order to be used in conjunction with the differential Navier-Stokes equations which apply pointwise in the two bulk fluids, we require a condition equivalent to (5-6) that applies at each point on the interface. For this purpose, it is necessary to convert the line integral on C to a surface integral on A. To do this, we use an exact integral transformation that can be derived as a generalization of Stokes's theorem,

$$\int_C \gamma \mathbf{t} \, dl = \iint_A (\text{grad} \, \gamma) \, dA - \iint_A \gamma \mathbf{n} (\nabla \cdot \mathbf{n}) \, dA. \qquad (5\text{-}7)$$

The proof of this result is straightforward but will not be pursued here. What should be noted, however, is the qualitative physical implication of (5-7). To do this, it is useful to think of A as corresponding to an arbitrarily small surface element centered on some point P, with $\mathbf{n}$ thus being the unit normal to the interface at (or arbitrarily near) P. Then we see that the tensile force associated with the action of interfacial tension [represented by the integral over C in (5-6)] contributes a net force at P in the tangential direction that is proportional to the gradient of γ at P, plus a net force normal to the interface that is proportional to γ times the curvature of the interface at P (that is, $\nabla \cdot \mathbf{n}$). Physically, if the surface element A is flat, the action of surface tension can only produce a *net* contribution to the force balance if $\nabla \gamma \neq 0$. On the other hand, if the surface is curved, the simple sketch in Figure 5-2 shows that surface tension acting at the edges of A can produce a net force contribution that is in the direction of the normal at P, and this contribution will be nonzero even if $\nabla \gamma = 0$. These facts are reflected in the theorem (5-7) from vector calculus.

Now, combining (5-6) and (5-7), we obtain

$$\iint_A \left[(\mathbf{T} - \hat{\mathbf{T}}) \cdot \mathbf{n} + \text{grad} \, \gamma - \gamma \mathbf{n} (\nabla \cdot \mathbf{n}) \right] dA = 0. \qquad (5\text{-}8)$$

Thus, since A is an arbitrary surface element in the interface, it follows that

$$(\mathbf{T} - \hat{\mathbf{T}}) \cdot \mathbf{n} + \operatorname{grad} \gamma - \gamma \mathbf{n} (\nabla \cdot \mathbf{n}) = 0 \qquad (5\text{-}9)$$

at each point on the interface. This is the differential stress balance that we seek. For application to the solution of flow problems, it is convenient to discuss separately the components in the normal and tangential directions.

To obtain the normal component, which is generally referred to as the *normal stress balance*, we take the inner product of (5-9) with $\mathbf{n}$. Recalling that $\mathbf{T} = -p\mathbf{I} + \tau$ and $\hat{\mathbf{T}} = -\hat{p}\mathbf{I} + \hat{\tau}$, this gives

$$\hat{p}_{\text{tot}} - p_{\text{tot}} + \left[((\tau - \hat{\tau}) \cdot \mathbf{n}) \cdot \mathbf{n} \right] - \gamma (\nabla \cdot \mathbf{n}) = 0. \qquad (5\text{-}10)$$

Here, p_{tot} and $\hat{p}_{\text{tot}}$ represent the actual total pressure in the exterior and interior fluids, including both dynamic and hydrostatic contributions. In crossing an interface, we see that the normal component of the total stress undergoes a jump equal to $\gamma(\nabla \cdot \mathbf{n})$. In the limiting case of no motion in the fluids, this implies that

$$\hat{p}_{\text{tot}} - p_{\text{tot}} = \gamma (\nabla \cdot \mathbf{n}). \qquad (5\text{-}11)$$

Hence, the pressure *inside* a curved interface at equilibrium is larger than that *outside* by an amount which depends on the curvature ($\nabla \cdot \mathbf{n}$) and γ.

Now, let us apply (5-11) to a bubble or drop that is stationary due to the absence of an external force. A stationary bubble or drop will be spherical in shape. This can be explained either as a consequence of the fact that the sphere minimizes surface area (and thus surface-free energy for a given γ) or, equivalently, that the sphere is the shape that is consistent with a *constant* hydrostatic pressure difference between the two fluids. [According to (5-11), any other shape would require pressure gradients in addition to hydrostatic pressure gradients in one or both fluids, and thus the fluids could not remain in a static state—the ensuing motion would, in fact, drive the shape toward spherical.] In general, the curvature term $\nabla \cdot \mathbf{n}$ can be expressed as the sum of the inverse principal radii of curvature, that is,

$$\nabla \cdot \mathbf{n} = \frac{1}{R_1} + \frac{1}{R_2}. \qquad (5\text{-}12)$$

However, for a sphere $R_1 = R_2 = R$, the sphere's radius. Thus, for a spherical bubble or drop,

$$\hat{p}_{\text{tot}} - p_{\text{tot}} = \frac{2\gamma}{R}, \qquad (5\text{-}13)$$

and the internal pressure is seen to exceed the exterior pressure by $2\gamma/R$. Of course, this result is also well known from equilibrium thermodynamics.

The process described above, by which a nonspherical drop would be driven toward the spherical equilibrium shape by pressure gradients associated with variations in interface curvature, is but one example of a large number of situations in which fluid motions are actually caused by pressure gradients that are produced by variations in the curvature of a fluid interface. Collectively, these motions are known as *capillary flows*. Generally, a system of two immiscible fluids (with $\gamma \neq 0$) of the same density

Figure 5–3 Photographs of the relaxation of an initially deformed viscous drop back to a sphere under the action of surface tension. The characteristic time scale for this surface tension-driven flow is $\hat{\mu} a / \gamma = 4$ sec, with the viscosity of the drop $\hat{\mu} = 115$ Poise, the undeformed radius of the drop $a = 0.19$ cm and the interfacial tension $\gamma = 5.4$ dyne/cm. The ratio of interior to exterior viscosity is 11.2. The bottom four photos were taken at $t = 0$, 2.87, 7.38, and 17.4 seconds. The corresponding dimensionless times $\bar{t} = t / (\hat{\mu} a / \gamma)$ are 0, 0.72, 1.8, and 4.3. The top photo is at steady state.

can only exist in an *equilibrium* state if the configuration of the interface between them corresponds to a surface of constant curvature. The simplest such surfaces would be a sphere, an infinite cylinder of constant radius, and an infinite flat boundary for which $\nabla \cdot \mathbf{n} \equiv 0$. The spherical interface shape is not only an allowable equilibrium configuration, but it also is a configuration that is stable to small perturbations of shape. If, for example, a spherical drop is deformed slightly to an ellipse, the variations of interface curvature will produce capillary motions within the drop which drive it back toward a spherical shape. An illustrative example may be seen in Figure 5–3, where we show a drop that is initially deformed into the shape of a prolate spheroid. As a consequence the curvature at the ends is increased relative to that in the middle, and the internal pressure is correspondingly higher at the ends. But, this capillary-induced pressure gradient induces a motion of the fluid from the ends of the drop toward the middle, and this motion causes the drop to return to the equilibrium spherical shape. An infinitely long cylindrical thread of constant radius is also an equilibrium configuration as indicated previously. However, it is *unstable* to infinitesimal perturbations, which tend to cause it to break into a large number of small spherical drops. The source of this instability is easily seen qualitatively by considering an initial shape that has a small varicose wavelike departure from the cylinder of constant radius. In this case, depending upon the wavelength of the perturbations of shape compared to the cylinder radius, there is an obvious mechanism for growth of the initial wavelike disturbance of shape. If the total surface curvature is such that the points of minimum radius produce a local maximum in $\nabla \cdot \mathbf{n}$, then there will be a local maximum in the internal pressure at the same point, which tends to drive fluid away from the region of minimum radius toward the regions of maximum radius. But this motion further decreases the radius at the waists and increases the radius between the waists, and this produces a varicose shape of increasing amplitude with time which ultimately terminates when the waists approach zero radius and break, causing the original thread to disperse into a line of small drops. This process is known as capillary wave instability and is an important mechanism for breakup of liquid threads, elongated drops, and so forth. The interested reader may wish to refer to texts on hydrodynamic stability (see Chapter 3 References), in which a quantitative analysis of the capillary-wave instability can be found, that leads to detailed predictons of growth rate for initial disturbances as a function of wave length.

There will be, in general, two tangential components of (5–9), which are obtained by taking the inner product with the two orthogonal unit tangent vectors that are normal to $\mathbf{n}$. If we denote these unit vectors as $\mathbf{t}_i$ (with $i = 1$ or 2), the so-called *shear stress balances* can be written symbolically in the form

$$\boxed{0 = ((\boldsymbol{\tau} - \hat{\boldsymbol{\tau}}) \cdot \mathbf{n}) \cdot \mathbf{t}_{(i)} + (\mathrm{grad}\,\gamma) \cdot \mathbf{t}_{(i)}.} \tag{5–14}$$

Again, we see that the shear stress components are discontinuous across the interface—in this case, the jump is proportional to $\mathrm{grad}\,\gamma$. Conversely, in cases where γ is uniform on the interface, the shear stress components are continuous! An important implication of (5–14) is that systems in which $\nabla\gamma \neq 0$ must undergo motion. To see this, we note that if $\nabla\gamma \neq 0$, one or both of $\boldsymbol{\tau} \cdot \mathbf{n} \cdot \mathbf{t}_{(i)}$ and $\hat{\boldsymbol{\tau}} \cdot \mathbf{n} \cdot \mathbf{t}_{(i)}$ must be nonzero. But, $\boldsymbol{\tau}$ or $\hat{\boldsymbol{\tau}}$ can only be nonzero if the fluid is in motion. Thus, any mechanism which maintains $\nabla\gamma \neq 0$ will necessarily drive motion in the fluids. Such motions are often known as *Marangoni flows,* and we shall discuss some examples later in this section.

First, however, we consider the problem of a bubble or drop that translates through a quiescent fluid due to a difference in density between the two fluids.

Translation of a Drop Through a Quiescent Fluid at Low Re

Let us now consider the steady, buoyancy-driven motion of a bubble or drop through a quiescent fluid. We denote the viscosities and densities of the two fluids as μ, $\hat{\mu}$, ρ, and $\hat{\rho}$ with the capped variables corresponding to the fluid inside the drop. For present purposes, we assume also that the interfacial tension at the drop surface is uniform and that the Reynolds numbers for both the interior and exterior flows are sufficiently small that the creeping motion approximation can be applied for both fluids. A convenient choice for nondimensionalization is to use the radius of the undeformed drop as a characteristic length scale, $l_c = a$ and the (as yet unknown) translational velocity of the drop as a characteristic velocity scale, $u_c = U$.

The governing equations and boundary conditions in dimensionless form are thus

$$\nabla^2\mathbf{u} - \nabla p = 0,$$
$$\nabla\cdot\mathbf{u} = 0,$$

$\qquad\qquad$ (5–15)

and

$$\nabla^2\hat{\mathbf{u}} - \nabla\hat{p} = 0,$$
$$\nabla\cdot\hat{\mathbf{u}} = 0,$$

$\qquad\qquad$ (5–16)

with the far-field boundary condition being

$$\mathbf{u} \to \mathbf{i}_z \quad \text{as} \quad |\mathbf{x}| \to \infty$$

$\qquad\qquad$ (5–17)

for a coordinate reference frame fixed at the center of mass of the drop. Inside the drop, we require that the velocity and pressure fields are bounded at the origin [which is a singular point for the spherical coordinate system that we use to solve (5–16)]. Finally, at the drop surface, the boundary conditions from the preceeding subsections must be applied. However, a complication in using these boundary conditions is that the drop shape is actually unknown (and, thus, so too are the unit normal and tangent vectors **n** and **t** and the interface curvature $\nabla\cdot\mathbf{n}$). In order to simplify the problem, we will initially assume that the shape is spherical. Clearly, however, this is a severe assumption that must be checked in the course of the analysis.

If the drop is spherical, as assumed, it is convenient to solve the problem using a spherical coordinate system, with the axis of symmetry for the flow corresponding to $\theta = 0$ (and $\theta = \pi$), as illustrated for the corresponding solid sphere problem in Figure 4–5. In this case, the unit normal to the drop surface is just a unit vector in the r direction, namely,

$$\mathbf{n} = \mathbf{i}_r.$$

Further, the velocity and pressure fields are axisymmetric, and thus the only relevant unit tangent vector is

$$\mathbf{t} = \mathbf{i}_\theta,$$

and, since the shape is assumed to be fixed (spherical), the kinematic condition takes the form (5–5). It follows that the boundary conditions (5–1), (5–5), (5–10), and (5–14) at the interface of the drop take the simple form

$$u_\theta = \hat{u}_\theta \quad \text{at} \quad r = 1, \tag{5–18}$$

$$u_r = \hat{u}_r = 0 \quad \text{at} \quad r = 1, \tag{5–19}$$

$$e_{r\theta} = \lambda \hat{e}_{r\theta} \quad \text{at} \quad r = 1, \tag{5–20}$$

and

$$\hat{p}_{\text{tot}} - p_{\text{tot}} + 2(e_{rr} - \lambda \hat{e}_{rr}) = \frac{2}{\text{Ca}} \quad \text{at} \quad r = 1. \tag{5–21}$$

Here, e_{ij} is the ij component of the rate-of-strain tensor, λ is the viscosity ratio, $\hat{\mu}/\mu$, and

$$\text{Ca} \equiv \frac{\mu U_\infty}{\gamma} \tag{5–22}$$

is known as the *capillary number* and provides a measure of the relative magnitude of viscous and capillary (or interfacial tension) forces at the interface. The characteristic pressure used to nondimensionalize (5–21) is $\mu U_\infty / a$. The condition (5–18) derives from the continuity of velocity (5–1), while (5–19) comes from (5–1) plus the kinematic condition (5–5) for a steady interface (and drop) shape. The conditions (5–20) and (5–21) correspond to the tangential and normal components of the stress continuity condition (5–9).

The solution of the problem outlined above is straightforward and can be approached via either the eigenfunction expansion for the Stokes equations in spherical coordinates or the general tensor method. Here, we adopt the eigenfunction expansion procedure. Since the flow both inside and outside the bubble will be axisymmetric, we can employ the equations of motion and continuity, (5–15) and (5–16), in terms of the streamfunctions ψ and $\hat{\psi}$, that is,

$$E^4 \psi = 0 \quad \text{and} \quad E^4 \hat{\psi} = 0. \tag{5–23}$$

Thus, in both the inner and outer fluids, the general solution of (5–23) in spherical coordinates has the form derived earlier,

$$\psi = \sum_{n=1}^\infty \left[A_n r^{n+3} + B_n r^{n+1} + C_n r^{2-n} + D_n r^{-n} \right] Q_n(\eta),$$

and

$$\hat{\psi} = \sum_{n=1}^\infty \left[\hat{A}_n r^{n+3} + \hat{B}_n r^{n+1} + \hat{C}_n r^{2-n} + \hat{D}_n r^{-n} \right] Q_n(\eta). \tag{5–24}$$

The set of eight constants A_n, B_n, C_n, D_n, $\hat{A}_n$, $\hat{B}_n$, $\hat{C}_n$, and $\hat{D}_n$ must be determined from the boundary conditions.

As in the case of streaming flow past a solid sphere, the far-field condition (5–17) requires that

$$\psi \to -r^2 Q_1 \quad \text{as} \quad r \to \infty,$$

and thus $A_n = B_n = 0$ for all n, except $B_1 = -1$, and

$$\psi = -r^2 Q_1 + \sum_{n=1}^{\infty} \left[C_n r^{2-n} + D_n r^{-n} \right] Q_n(\eta). \tag{5-25}$$

Inside the drop, on the other hand, we require that the velocity components u_r and u_θ be bounded at $r = 0$. Since

$$u_r = -\frac{1}{r^2} \frac{\partial \psi}{\partial \eta} \quad \text{and} \quad u_\theta \sqrt{1 - \eta^2} = -\frac{1}{r} \frac{\partial \psi}{\partial r}$$

according to (4–150), it follows that

$$\hat{C}_n = \hat{D}_n = 0 \quad \text{all } n, \tag{5-26}$$

and thus

$$\hat{\psi} = \sum_{n=1}^{\infty} \left[\hat{A}_n r^{n+3} + \hat{B}_n r^{n+1} \right] Q_n(\eta). \tag{5-27}$$

It remains to determine the four sets of constants C_n, D_n, $\hat{A}_n$, and $\hat{B}_n$. For this, we still have the boundary conditions (5–18)–(5–21). However, it may be noted that there are actually five independent boundary conditions and only four sets of constants to be determined. Thus, it would appear that the problem is overdetermined. In reality, however, the shape of the drop is unknown, in addition to the velocity and pressure fields [namely, the four constants in (5–25) and (5–26)]. Since we have assumed that the shape is known (namely, spherical), one of the conditions (5–18)–(5–21) must be dropped. The obvious candidate is the normal stress condition (5–21), which involves the drop shape in an explicit way. The remaining four conditions (5–18), (5–19), and (5–20) are sufficient to completely determine the unknown coefficients in (5–25) and (5–26). Indeed, for any given (or prescribed) drop shape, the four conditions of velocity continuity, tangential stress continuity, and the kinematic condition are sufficient along with the far-field condition to completely determine the velocity and pressure fields in the two fluids. The normal stress condition can then be considered as a *consistency condition* to determine whether the proposed shape is actually correct. For the present problem, we first apply the conditions (5–18)–(5–20) at $r = 1$ (thus assuming the shape to be spherical) and then use (5–21) to check on the validity of this assumption.

We leave it to the reader to provide the details of algebra necessary to determine the constants in (5–25) and (5–27) from the boundary conditions (5–18)–(5–20). The result is

$$C_n = D_n = \hat{A}_n = \hat{B}_n = 0, \ n \neq 1, \tag{5-28}$$

and

$$C_1 = \frac{3\lambda + 2}{2(\lambda + 1)}, \quad D_1 = -\frac{\lambda}{2(\lambda + 1)}, \quad -\hat{A}_1 = \hat{B}_1 = \frac{1}{2(\lambda + 1)}, \tag{5-29}$$

where

$$\lambda \equiv \frac{\hat{\mu}}{\mu}.$$

Thus,

$$\psi = -Q_1 \left\{ r^2 - \frac{3\lambda + 2}{2(\lambda + 1)} r + \frac{\lambda}{2(\lambda + 1)} \frac{1}{r} \right\}, \tag{5-30}$$

$$\hat{\psi} = \frac{Q_1}{2} \frac{r^2 - r^4}{(\lambda + 1)}. \tag{5-31}$$

This result is known as the *Hadamard-Rybczyński* solution.

The drag on the drop can be predicted from (4–180) and the result (5–29) for C_1, namely,

$$\text{Drag} = 4\pi a \mu U C_1 = 4\pi a \mu U \left(\frac{3\lambda + 2}{2(\lambda + 1)} \right). \tag{5-32}$$

In the limiting case, $\lambda \to \infty$, this expression for the drag becomes

$$D = 6\pi \mu a U, \tag{5-33}$$

which is simply Stokes's law for the drag on a rigid, no-slip sphere that moves with relative velocity U through a quiescent fluid. On the other hand, in the limit $\lambda \to 0$, the expression (5–32) becomes

$$D = 4\pi a \mu U, \tag{5-34}$$

which is the drag on a spherical bubble at Re $\ll 1$. It is interesting that the drag on a solid sphere exceeds the drag on a spherical bubble by only a factor of 3/2. This result is illustrative of a general observation that the drag on a body of fixed shape at low Reynolds number is remarkably insensitive to the boundary conditions at the body surface.

Although the limit $\lambda \to 0$ in (5–30)–(5–32) may at first seem straightforward, it is not obvious that we can take this limit without violating the condition

$$\hat{\text{Re}} \equiv \frac{\hat{\rho} U a}{\hat{\mu}} \ll 1.$$

Surprisingly, however, the solution (5–30)–(5–32) yields the correct result for $\lambda \to 0$, without regard to whether or not $\hat{\text{Re}}$ remains small. The reason is that as $\lambda \to 0$, the motion of the fluid outside the drop becomes increasingly insensitive to the motion inside. Thus, provided that we focus on the motion of the external fluid and on results like (5–34) that depend on this motion, the details of the inside flow are unimportant. One clear way to see that this is true is to examine the boundary conditions at the drop surface for $\lambda \to 0$. In particular, the boundary condition (5–20) can be written in the form

$$r \frac{\partial}{\partial r}\left(\frac{u_\theta}{r}\right) = \lambda \left(r \frac{\partial}{\partial r}\left(\frac{\hat{u}_\theta}{r}\right)\right) \quad \text{at} \quad r = 1. \tag{5-35}$$

Thus, in the limit $\lambda \to 0$, this reduces to

$$r \frac{\partial}{\partial r}\left(\frac{u_\theta}{r}\right) = \tau_{r\theta} = 0 \quad \text{at} \quad r = 1. \tag{5-36}$$

But insofar as the flow in the *exterior* fluid is concerned, this condition

$$\boxed{\tau_{r\theta} = 0 \quad \text{at} \quad r = 1} \tag{5-37}$$

plus the kinematic condition

$$\boxed{u_r = 0 \quad \text{at} \quad r = 1} \tag{5-38}$$

is *completely sufficient to determine the exterior flow* (that is, the constants C_n and D_n) without any need to consider the fluid motion inside the drop! Indeed, theories of bubble motion in viscous liquids are almost always obtained by applying free-shear condition (5–37) and the kinematic condition (5–38) (or their generalizations for bodies of nonspherical geometry) to directly determine the velocity and pressure fields in the liquid phase, rather than first solving the full flow problem for a drop with arbitrary λ and then letting $\lambda \to 0$.

The remaining point is to determine whether the drop is actually spherical as assumed and, if not, to obtain an improved guess for the shape. If the shape is spherical, then the Hadamard-Rybczyński solution should be consistent with the normal stress balance (5–21). Since the right-hand side of (5–21) is constant, 2/Ca, for a sphere, then if this guessed shape were actually correct, the left-hand side of (5–21) should also reduce to a constant, independent of position on the bubble surface. To check on this possibility, we must calculate

$$\left(\hat{p}_{\text{tot}} - p_{\text{tot}} + 2 \frac{\partial u_r}{\partial r} - 2\lambda \frac{\partial \hat{u}_r}{\partial r}\right)_{r=1} \tag{5-39}$$

using the Hadamard-Rybczyński solution. Here $\hat{p}_{\text{tot}}$ and p_{tot} represent the sum of hydrostatic and dynamic (flow-induced) pressure variations in the fluid.

To calculate $\partial u_r / \partial r|_{r=1}$, we must first calculate u_r using the definition

$$u_r \equiv -\frac{1}{r^2} \frac{\partial \psi}{\partial \eta}$$

and (5–30), and then differentiate with respect to r. Similarly, we can evaluate $\partial \hat{u}/\partial r$ from (5–31). The pressure in the exterior fluid, on the other hand, is

$$p_{\text{tot}} = -\sum_{n=1}^{\infty} \frac{2(2n-1)}{n+1} \frac{P_n(\eta)}{r^{n+1}} C_n + p_{\text{hydrostatic}}. \tag{5-40}$$

To evaluate the first term on the right-hand side, we use the numerical value for C_1 from (5–29). The hydrostatic pressure contribution to (5–40) can be written in non-dimensionalized form as

$$P_{\text{hydrostatic}} = \left(\frac{\mu U}{a}\right)^{-1}[-\rho g(za) + c], \tag{5–41}$$

where c is an arbitrary constant. Since

$$z = -r\eta = -rP_1(\eta),$$

it follows that

$$P_{\text{hydrostatic}} = \frac{\rho g a^2}{\mu U}P_1(\eta) + C^* \quad \text{at} \quad r = 1. \tag{5–42}$$

Hence, at the sphere surface ($r = 1$),

$$P_{\text{tot}} = -\frac{P_1}{2}\left(\frac{3\lambda+2}{\lambda+1}\right) + \frac{\rho g a^2}{\mu U}P_1(\eta) + C_1^*. \tag{5–43}$$

Finally, the pressure inside the drop $\hat{p}_{\text{tot}}$ can be evaluated in a similar manner. When the results for $\partial u_r/\partial r$, $\partial \hat{u}_r/\partial r$, p_{tot}, and $\hat{p}_{\text{tot}}$ are combined in the form (5–39), we obtain

$$\frac{3}{2}\left(\frac{3\lambda+2}{\lambda+1}\right)P_1(\eta) + \frac{\rho g a^2}{\mu U}(\kappa-1)P_1(\eta) + \pi = \frac{2}{\text{Ca}}, \tag{5–44}$$

where $\kappa \equiv \hat{\rho}/\rho$ and $\pi \equiv C^* - C$ is a constant that can be equated to the pressure difference between the internal and external fluid due to the capillary pressure jump, 2/Ca.

The objective in deriving (5–44) was to determine whether the assumed spherical shape was correct. Since the capillary pressure term is a constant for a sphere (that is, independent of position on the sphere surface), it follows that the viscous pressure and stress difference on the left-hand side of (5–21) also must be a constant when evaluated with the solution for a sphere, if this shape is actually correct. If we examine (5–44), however, it appears at first that $T_{rr} - \hat{T}_{rr}$ on the left side is not a constant as required, but rather a function of η. The fact of the matter is that the two terms involving $P_1(\eta)$ actually cancel identically. To see that this is possible, we need to remind ourselves that the velocity U, which represents the translational velocity of the drop, has not yet been determined. For this purpose, we need to consider a force balance on the spherical drop. At steady state, this requires that the net buoyancy force be exactly balanced by the hydrodynamic drag, given by (5–32). Thus,

$$\frac{4}{3}\pi a^3 g\rho(1-\kappa) = 4\pi a\mu U\left(\frac{3\lambda+2}{2(\lambda+1)}\right) \tag{5–45}$$

and, rearranging,

$$U = \frac{2}{3}\left(\frac{\lambda+1}{3\lambda+2}\right)\frac{a^2 g\rho(1-\kappa)}{\mu}. \tag{5–46}$$

Comparing (5-46) with the left-hand side of (5-44), we see that the two terms involving $P_1(\eta)$ in the normal stress balance cancel exactly, and (5-44) reduces to the simple form

$$\pi = \frac{2}{\mathrm{Ca}}. \tag{5-47}$$

Thus, the normal stress balance is precisely satisfied for the spherical shape that we assumed initially, and the capillary (or interfacial tension) contribution is simply to produce a jump in pressure equal to $\pi = 2/\mathrm{Ca}$ across the drop surface. This pressure jump is, in fact, precisely the result (5-13) that was derived earlier, simply written in dimensionless terms based on a characteristic pressure, $p_c = \mu U/a$.

The fact that the normal stress balance is satisfied exactly proves that the sphere is an equilibrium shape for a drop that is translating at low Reynolds number through a quiescent fluid. This may, at first, seem to be a rather unremarkable result. After all, the equilibrium shape for a stationary drop would clearly be spherical as a consequence of the tendency to minimize interface area (surface free energy) for any immiscible system where $\gamma \neq 0$. However, in the result that we have obtained, there is an exact balance between viscous and pressure forces for a drop of spherical shape and no tendency for the drop to deform at any arbitrary capillary number, including $\mathrm{Ca} = \infty$ where $\gamma \equiv 0$. This is, in fact, a unique result! For virtually any other problem involving the motion of bubbles or drops in a viscous fluid, the shape would be nonspherical. Even for the buoyancy-driven motion of a bubble or drop in a quiescent fluid, the shape becomes nonspherical if the Reynolds number is nonzero. This particular problem, for small, but nonzero, Reynolds number, was solved originally by Taylor and Acrivos (1964).[3] In the creeping flow solution obtained above, the sphere is an equilibrium shape and the magnitude of the capillary number (interfacial tension) only determines the difference in pressure between the inside and outside of the drop.

One point that we have not emphasized is that all of the preceding analysis and discussion pertains only to the steady-state problem. From this type of analysis, we cannot deduce anything about the stability of the spherical (Hadamard-Rybczyński) shape. In particular, if a drop or bubble is initially nonspherical or is perturbed to a nonspherical shape, we cannot ascertain whether the drop will evolve toward a steady, spherical shape. The answer to this question requires additional analysis that is not given here. The result of this analysis[4] is that the spherical shape is stable to infinitesimal perturbations of shape for all finite capillary numbers but is unstable in the limit $\mathrm{Ca} = \infty$ ($\gamma = 0$). In the latter case, a drop that is initially elongated in the direction of motion is predicted to develop a tail. A drop that is initially flattened in the direction of motion, on the other hand, is predicted to develop an indentation at the rear. Further analysis is required to determine whether the magnitude of the shape perturbation is a factor in the stability of the spherical shape for arbitrary, finite Ca.[5] Again, the details are not presented here. The result is that finite deformation can lead to instability even for finite Ca. Once unstable, the drop behavior for finite Ca is qualitatively similar to that predicted for infinitesimal perturbations of shape at $\mathrm{Ca} = \infty$; that is, oblate drops form an indentation at the rear, and prolate drops form a tail.

Marangoni Effects in the Motion of Bubbles and Drops

The analysis of the preceding section is based on the assumption that the interfacial tension at the drop surface is uniform so that grad$\gamma \equiv 0$, and the condition (5–14) reduces to the requirement that the tangential components of stress in the two fluids are equal at the drop interface. However, the interfacial tension at any point on the interface depends on the thermodynamic state (p and T) as well as the concentrations of any solute molecules that are adsorbed at the interface. Hence, in a real flow system, we must often expect γ to vary from point to point on the interface, and it is important to consider how gradients of γ may influence the flow.

This question is particularly pertinent to the motion of gas bubbles (or drops) in water, or in any liquid which has a large surface tension (the surface tension of a pure air-water interface is approximately 70 dynes/cm). Experiments on the motion of gas bubbles in water at low Reynolds numbers show the perplexing result illustrated in Figure 5–4. For bubbles larger than about 1 millimeter in diameter, the translation velocity is approximately equal to the predicted value for a spherical bubble with zero shear stress at the interface, that is,

$$U = 1/3 \left(\frac{\Delta \rho g a^2}{\mu} \right).$$

However, for smaller bubbles the translation velocity decreases until it is approximately two-thirds of this value; that is, the small bubbles move through the liquid with a velocity that is consistent with a no-slip boundary condition at the interface (hence, yielding Stokes result $6\pi\mu aU$ for the drag) rather than a free surface condition (leading

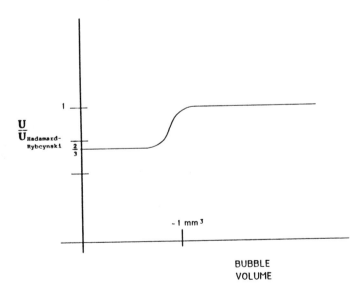

Figure 5–4 A qualitative sketch representing the terminal velocity of a gas bubble in water at low Reynolds number.

to the value $4\pi\mu aU$, according to (5–34)). A qualitative explanation for this observation was discovered many years ago by Frumkin and Levich (1947),[6] who carried out experiments on buoyancy-driven motion of gas bubbles in very highly purified liquids under a close approximation to isothermal conditions. These experiments led to two main conclusions. First, as the liquid became increasingly purified, the transition from the theoretically predicted bubble motion to motion of the bubble as a no-slip body occurred for smaller and smaller bubbles (or drops). Second, experiments in water showed extreme sensitivity to even minute concentrations of contaminant (even 1 ppm by weight of some contaminants led to an observable effect on bubble motion).

The explanation for these observations and for the existence of a range of bubble sizes where the bubble seems to act as a no-slip body is that real fluid systems almost invariably contain small quantities of contaminant, and this material can modify boundary conditions at the bubble (or drop) surface. This is especially true if the contaminant is a surfactant—that is, a solute that is preferentially adsorbed at the interface, since this means that infinitesimal amounts of surfactant in the bulk-phase fluids can yield very significant concentrations of surfactant at the interface. For air-water or oil-water systems, the chemical structure of a typical surfactant will be made of two distinct moieties: One part is polar (and thus hydrophilic), and the other is nonpolar (and thus hydrophobic). In general, a surfactant can modify the physical character of the interface in two quite distinct ways. When sufficient concentrations of several different surfactants are present, the surfactant mixture can form a semipermanent film at the interface, with a definite two-dimensional structure. For example, this is typical of surfactant mixtures that are used as *emulsifiers* designed to stabilize an emulsion by creating a physical barrier to droplet coalescence. On the other hand, for low surfactant concentrations or when a single surfactant species is present, the adsorbed surfactant can remain highly mobile on the interface and then influences dynamical behavior by producing spatial variations in the interfacial tension. The dynamic effects of interfacial tension gradients are known as Marangoni effects. It is these latter effects that we discuss now.

In almost all liquid-gas or liquid-liquid systems, the interfacial tension is known to decrease as the local concentration of surfactant increases at the interface. The importance of Marangoni effects depends upon the *sensitivity* of the interfacial tension to changes in surfactant concentration and also the magnitude of the interfacial tension in the absence of surfactant, since this controls the maximum possible variation in γ over the surface. Since the interfacial tension of the pure air-water interface is very large (~ 70 dynes/cm), such systems are particularly sensitive to the presence of surfactant at the interface.

If the concentration of surfactant is uniform at the interface, the effect is a uniform decrease in surface tension, which may generally change the bubble shape but otherwise has no influence on the bubble dynamics or on the hydrodynamic drag. The fact is, however, that the surfactant concentration at a bubble surface does not remain uniform as the bubble translates through the liquid. The transport processes for surfactant are extremely complicated, and we thus do not attempt to carry out an analysis here. Instead, we rely on the qualitative picture, sketched in Figure 5–5. The essential feature is that convection of surfactant molecules sweeps them from the front toward

θ=π

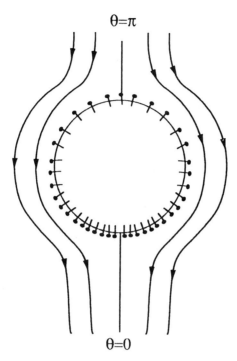

θ=0

Figure 5–5 A pictorial representation of the distribution of surfactant on the surface of a rising gas bubble. The small "sticks" at the bubble interface are intended to represent surfactant, which adsorbs preferentially at the gas-liquid interface. The fluid motion, from the top of the bubble toward the bottom, convects surfactant toward the rear of the bubble where it tends to accumulate. This tendency is counteracted to some extent by diffusion which tends toward a uniform surfactant distribution.

the rear of the bubble, and a gradient of surfactant concentration is established as a balance between this convective effect and surface diffusion which drives the system back toward a uniform surfactant concentration. But a nonuniform surfactant concentration implies that the surface tension must also vary over the bubble surface. Since surface tension goes down as the concentration of surfactant goes up, the net result is the development of a gradient of surface tension at the bubble surface, from a maximum value at the front of the bubble ($\theta = \pi$) where the surfactant concentration is lowest, to a minimum value at the rear ($\theta = 0$) where surfactant concentration is largest. Hence, referring to Figure 5-5, we expect that

$$\frac{d\gamma}{d\theta} > 0. \qquad (5\text{–}48)$$

However, such a gradient of γ modifies the boundary condition at the bubble (or drop) surface, from a zero shear stress condition ($\tau_{r\theta} = 0$) to the condition (5-14), which becomes

$$\tau_{r\theta} = \frac{1}{a} \frac{d\gamma}{d\theta} \tag{5-49}$$

for this axisymmetric problem. Clearly, the existence of a gradient of γ is equivalent to a tensile force in the interface which acts in the direction of increasing γ (in the direction to oppose the external flow). Thus, as $d\gamma/d\theta$ increases, the bubble surface becomes increasingly resistant to tangential motions and the bubble acts increasingly as a no-slip body. The limit to this process is complete immobilization of the interface. As this limit is approached, the convection of surfactant becomes weaker due to the decrease in interfacial velocity. Examination of (5-49) shows that, for a given gradient of γ, the effect on the shear stress increases as the bubble radius decreases. This is in qualitative accord with the experimental observation that the bubble reverts to no-slip behavior for small radii below some critical threshold value. This effect of nonuniform surfactant concentrations is one of the most easily observed manifestations of Marangoni effects in the motion of bubbles or drops.

A detailed analysis of the physical effects described above would be extremely difficult due to the complexity of the surfactant transport problem and the coupling between surfactant transport and fluid motion. However, some approximate model calculations have been carried out. Two examples are the early analyses of Davis and Acrivos (1966)[7] and the more recent work of Holbrook and LeVan (1983).[7] The interested reader also may refer to the textbook by Levich (1962)[8] for a summary of some of the earlier analyses.

In this section, we turn our attention to a second example of Marangoni effects on the motion of a bubble (or drop) which illustrates the important qualitative physical phenomena but is somewhat easier to analyze. In this example, a bubble is placed in a body of liquid in which a vertical temperature gradient is developed by some external means (for example, heating the fluid from below), so that the temperature decreases with increasing height. In this system, the temperature gradient in the fluid produces a surface-tension gradient at the bubble surface. Since the surface tension is found to decrease as the temperature increases, that is,

$$\frac{\partial \gamma}{\partial T} = -\beta < 0, \tag{5-50}$$

the surface tension will increase from the bottom toward the top of the bubble. Hence, a tensile force is developed in the bubble surface which tends to produce a fluid motion from the bottom toward the top of the bubble. If this Marangoni effect is strong enough, it may completely counteract the tendency of the bubble to rise in the liquid. The mechanism which makes this possible is most easily visualized by analogy to the downward motion produced by a swimmer in response to a sweeping upward motion of his arms. This is illustrated qualitatively in Figure 5-6. Experimental observation, in fact, demonstrates that the motion of the bubble can be arrested if the temperature gradient becomes sufficiently large. Surface-tension–induced flows generated by temperature gradients are usually called *thermocapillary motions*. Interest in this class of fluid motions has been particularly strong in recent years as a consequence of the potential for materials processing applications in the microgravity environment of space

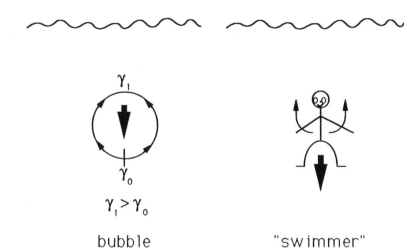

$$\gamma_1 > \gamma_0$$

bubble   "swimmer"

Figure 5–6 A sketch to illustrate the qualitative analogy between a gas bubble with a surface-tension gradient that causes fluid to be swept upward near the bubble surface and a swimmer who sweeps fluid upward by the action of his arms. In both cases, if the induced upward motion of fluid is sufficiently strong, the response is that the body will be thrust downward against the effects of buoyancy.

flight, where buoyancy-driven motions are nonexistent and thermocapillary phenomena can play a dominant role. One application of thermocapillary-driven motions is to provide a mechanism for removal of small gas bubbles from specialty glasses and other space-processed materials.[9]

In the remainder of this section we provide an approximate analysis of the thermocapillary problem of motion of a bubble in a gravitational field.[10] Thus, we suppose that we have a spherical bubble, which translates vertically through a clean stationary fluid, in which we impose a linear temperature distribution at large distances from the bubble, that is,

$$T_\infty(z) = T_\infty(0) - \alpha z, \tag{5–51}$$

where $T_\infty(0)$ is a reference value at $z = 0$, which we take for convenience to correspond to a horizontal plane that coincides with the bubble center. With the linear temperature field at large distances from the bubble surface, an exact analysis of the problem would require a solution of the thermal energy equation in order to determine the temperature distribution in the fluid, and hence the temperature distribution at the bubble surface. For example, if convection of heat could be neglected compared with conduction, and viscous dissipation also were negligible, we should solve $\nabla^2 T = 0$ with (5–51) as the asymptotic boundary condition at infinity, and $\partial T/\partial r = 0$ at the bubble surface (if we assume $k_{\text{bubble}}/k_{\text{fluid}} \ll 1$). However, for present purposes, we adopt an ad hoc approximate approach and simply assume that the temperature distribution (5–51) is maintained throughout the fluid, including the fluid immediately adjacent to the bubble surface. This approximation to the temperature field would be exact if heat transfer were dominated by conduction ($\nabla^2 T = 0$ in the fluid) and the thermal conductivities of the liquid and gas were equal ($k_{bubble}/k_{fluid} = 1$). However, we expect, in any case, that it is qualitatively correct in the sense that the temperature decreases monotonically on the bubble surface as we move from the bottom toward the top.

The problem, then, is to solve the creeping flow equations in the fluid, subject to the condition

$$\mathbf{u} \to -\mathbf{i}_z \quad \text{for} \quad |\mathbf{r}| \to \infty, \tag{5-52}$$

plus boundary conditions at the bubble surface. Here, we have adopted a frame of reference that is fixed at the center of the bubble and nondimensionalized with respect to the (as yet unknown) translational velocity, U. The conditions at the bubble surface are the kinematic condition (5–5),

$$u_r = 0 \quad \text{at} \quad r = 1, \tag{5-53}$$

and the tangential stress balance (5–14), which can be written in the dimensionless form

$$\tau_{r\theta} + \frac{1}{\mu U}\frac{\partial \gamma}{\partial \theta} = 0 \quad \text{at} \quad r = 1. \tag{5-54}$$

In order to evaluate $\partial \gamma / \partial \theta$ in (5–54), we use the temperature distribution (5–51), which is now assumed to hold throughout the fluid, and the relationship (5–50) between the surface tension and temperature. With the z axis being vertically upward, as shown in Figure 5–7 and the polar angle θ measured from the bottom ($\theta = 0$) to the top ($\theta = \pi$), the temperature distribution can be expressed in terms of spherical coordinates as

$$T_\infty(z) = T_\infty(0) + \alpha r \cos\theta. \tag{5-55}$$

Hence, at $r = 1$,

$$\frac{\partial \gamma}{\partial \theta} = \frac{\partial \gamma}{\partial T}\frac{\partial T}{\partial \theta} = -\beta(-\alpha \sin\theta)$$

$$= \alpha\beta(1 - \eta^2)^{1/2}, \tag{5-56}$$

and the tangential stress condition becomes

$$\tau_{r\theta} = -\frac{\alpha\beta(1 - \eta^2)^{1/2}}{\mu U} \quad \text{at} \quad r = 1. \tag{5-57}$$

Since the problem is axisymmetric about the z axis, we can solve the creeping flow equation in terms of the streamfunction ψ. Hence, after applying the asymptotic condition (5–52), the solution can be expressed in the form

$$\psi = +r^2 Q_1(\eta) + \sum_{n=1}^\infty [C_n r^{2-n} + D_n r^{-n}] Q_n(\eta). \tag{5-58}$$

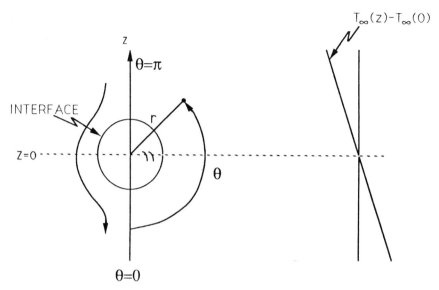

Figure 5–7   Schematic diagram of the coordinate systems and undisturbed temperature distribution for thermocapillary-driven motion of a gas bubble. Note, in this case $\mathbf{i}_z = -\eta\,\mathbf{i}_r + (1 - \eta^2)^{1/2}\mathbf{i}_\theta$.

The coefficients C_n and D_n must be determined by applying the boundary conditions (5–53) and (5–57) at the bubble surface $r = 1$. The first of these requires that $\psi = 0$ at $r = 1$. Hence,

$$0 = +Q_1(\eta) + \sum_{n=1}^{\infty}\left[C_n + D_n\right]Q_n(\eta). \tag{5–59}$$

The second condition requires that

$$r\frac{\partial}{\partial r}\left(\frac{u_\theta}{r}\right) = -\frac{\alpha\beta}{\mu U}(1 - \eta^2)^{1/2} \quad\text{at}\quad r = 1, \tag{5–60}$$

where

$$u_\theta = -\frac{1}{\sqrt{1 - \eta^2}}\frac{1}{r}\frac{\partial\psi}{\partial r}.$$

Hence,

$$+2Q_1 + 2C_1Q_1 - 4D_1Q_1 + \sum_{n=2}^{\infty}\left[(2 - n)(n + 1)C_n - n(n + 3)D_n\right]Q_n = 2\frac{\alpha\beta}{\mu U}Q_1. \tag{5–61}$$

Solving (5–59) and (5–61), we find that

$$C_1 = -1 + \frac{1}{3}\frac{\alpha\beta}{\mu U},$$

$$D_1 = -\frac{1}{3}\frac{\alpha\beta}{\mu U},$$

$$C_n = D_n = 0 \quad \text{for} \quad n \geq 2. \tag{5-62}$$

It follows from (4–189) that the hydrodynamic drag on the bubble is

$$F = 4\pi\mu a U \left[-1 + \frac{1}{3}\frac{\alpha\beta}{\mu U} \right]. \tag{5-63}$$

Thus, the hydrodynamic force is larger than it would be in the absence of the thermo-capillary (Marangoni) contribution to the shear stress at the bubble surface. As a consequence, the bubble moves slower. Indeed, at steady state, the bubble velocity can be calculated from the overall force balance

$$\frac{4}{3}\pi a^3 \rho g = 4\pi\mu a U \left[-1 + \frac{1}{3}\frac{\alpha\beta}{\mu U} \right]. \tag{5-64}$$

The result is

$$U = -\frac{1}{3}\frac{a^2 \rho g}{\mu}\left(1 - \frac{\alpha\beta}{a^2 \rho g} \right). \tag{5-65}$$

It would appear from (5–65) that the critical value of α to completely arrest the bubble motion is

$$\alpha_{\text{crit}} = \frac{a^2 \rho g}{\beta}. \tag{5-66}$$

The correct result, if we determine the actual temperature distribution at the bubble surface by solving $\nabla^2 T = 0$, as outlined earlier, was shown by Young, Goldstein, and Block (1959)[11] to be

$$\alpha_{\text{crit}} = \frac{2}{3}\frac{a^2 \rho g}{\beta}. \tag{5-67}$$

Clearly, the simplification of the problem that we have used does not change this result except in the quantitative sense.

Several brief comments should be made regarding the solution just outlined. First, the derivation assumes, implicitly, that $U \neq 0$; remembering the far-field condition (5–52) and the form of the streamfunction (5–58), the reader may well wonder about the efficacy of trying to use this solution to deduce conditions when $U = 0$. The simplest response is to notice that the analysis we have given is clearly appropriate for any $U \neq 0$, no matter how small, and we rely on this fact to hypothesize that the

INTERFACE

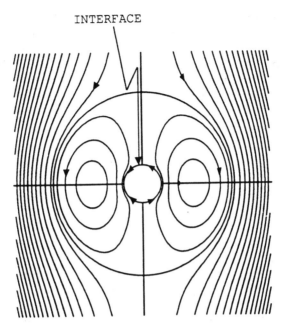

Figure 5–8 Streamlines for thermocapillary motion of a gas bubble for $\alpha = 0.8 \ (a^2 \rho g / \beta)$, where the bubble velocity is reduced to 20 percent of its value in the absence of thermocapillary effects. The streamfunction values are calculated from equation (5–58) with coefficients C_n and D_n from Equation (5–62). Contour values are plotted in increments of 0.7681.

limiting process $U \to 0$ is nonsingular. However, it is a useful exercise to go back and reformulate the problem starting with the presumption $U = 0$, and we suggest this as a supplement to the problems at the end of this chapter. Second, we remind the reader that it is only the bubble that is not moving when $U = 0$. The fluid is, in fact, undergoing a surface-tension–driven recirculating flow of the form sketched qualitatively in Figure 5–8. Indeed, it is the downward thrust on the bubble that results from this flow that is responsible for balancing the buoyancy force and maintaining the bubble in a stationary position.

Deformation of a Drop in a General Linear Flow

We noted in an earlier section that a drop will deform in almost any viscous flow, other than in steady translation through a stationary fluid. In all other cases, at zero Reynolds number, the magnitude of the deformation for a given flow primarily depends on the capillary number (and to a lesser extent upon the ratio of internal to external viscosity $\lambda \equiv \hat{\mu}/\mu$). The type of motion also influences the degree of deformation. The qualitative idea is illustrated by thinking of two limiting cases. In the first, we imagine a drop immersed in a fluid that is undergoing a purely rotational two-dimensional flow, and in the second, the fluid undergoes a two-dimensional pure straining flow with the same magnitude for the velocity gradient. It is intuitively obvious that the drop will not deform in the pure rotational flow (it will simply rotate) but will deform substantially in the extensional flow. If we have a two-dimensional flow that is somewhere between these extremes, the degree of deformation will be decreased by the addition of vorticity (holding $|\nabla \mathbf{u}| = $ const).

A qualitative prediction of the role of the capillary number in drop deformation can be obtained from the normal stress balance (5–10), written in a dimensionless form that is suitable to the low Reynolds number limit that we consider here. Thus,

$$\hat{p}_{\text{tot}} - p_{\text{tot}} + ((\boldsymbol{\tau} - \lambda \hat{\boldsymbol{\tau}}) \cdot \mathbf{n}) \cdot \mathbf{n} = \frac{1}{\text{Ca}} (\nabla \cdot \mathbf{n}), \qquad (5\text{–}68)$$

where $\text{Ca} = \mu u_c / \gamma$. Now, if the shape is spherical, the capillary term on the right-hand side is a constant, and thus the viscous pressure and stress contributions on the left must also produce a constant value that corresponds to the jump in pressure across the interface due to surface tension. In general, however, the pressure and stress differences will not reduce to this very simple form, but will instead vary as a function of position on the surface of the drop. Such a *nonuniform* distribution of pressure and stress will tend to deform the drop. In fact, in this case, the normal stress balance can only be satisfied if the drop deforms to a shape where the interface curvature $(\nabla \cdot \mathbf{n})$ varies in precisely the same way as the surface pressure and stress difference.

The *magnitude* of the change in shape that is required to satisfy the normal stress balance can be seen from (5–68) to depend on Ca. For $\text{Ca} \ll 1$, very small deviations from a spherical shape (where $\nabla \cdot \mathbf{n} = $ const) can produce sufficient variation in $(1/\text{Ca})(\nabla \cdot \mathbf{n})$ to balance the pressure or stress variations over the drop surface. Hence, for $\text{Ca} \ll 1$, a drop will remain approximately spherical in shape. This is consistent with the interpretation of Ca as the ratio of characteristic viscous (and pressure) forces to capillary forces. Thus, the limit $\text{Ca} \ll 1$ corresponds to dominant surface-tension effects, and this accounts for the tendency of the drop to remain almost spherical. On the other hand, for $\text{Ca} \gg 1$, pressure and stress variations in (5–68) can only be balanced by large deformations of shape to produce large variations in the curvature term $(1/\text{Ca})(\nabla \cdot \mathbf{n})$. In fact, experimental observation shows that steady shapes are not always possible for very large values of Ca. Essentially, in this case, no steady shape exists that allows a balance between the two sides of (5–68), and the drop must undergo a time-dependent change of shape, leading possibly to breakup into two or more pieces. Such breakup effects are extremely important in processing equipment such as mixers, or blenders, that are designed to disperse one continuous fluid in another, but an analysis of them is beyond the scope of this book.

Instead, in the present section, we consider the deformation of a drop in a general linear flow, for $\text{Ca} \ll 1$, where the shape remains approximately spherical. We assume that the density of the drop is equal to that of the suspending fluid and that the surface tension is constant on the drop surface—that is, there are no thermal gradients and no surfactants present. The fact that the drop is neutrally buoyant means that it does not translate relative to the surrounding fluid. Thus, at large distances from the drop, we can assume that the fluid undergoes a general linear flow of the form

$$\mathbf{u}_\infty = G(\mathbf{E} + \mathbf{\Omega}) \cdot \mathbf{x}', \quad |r| \to \infty,$$

where G represents the magnitude of $\nabla\mathbf{u}$, and $\mathbf{E}$ and $\mathbf{\Omega}$ are the dimensionless rate of strain and vorticity tensors, respectively. Hence, in dimensionless terms, with $u_c = Ga$,

$$\boxed{\mathbf{u} \to (\mathbf{E} + \mathbf{\Omega}) \cdot \mathbf{x}, \quad |r| \to \infty.} \qquad (5\text{--}69)$$

We have already seen that this general linear flow includes simple shear, uniaxial extension, and other flows of interest.

In the following analysis, we assume that the Reynolds numbers for the flows inside and outside the drop are both extremely small, that is,

$$\mathrm{Re} \equiv \frac{\rho Ga^2}{\mu}, \quad \hat{\mathrm{Re}} \equiv \frac{\hat{\rho}Ga^2}{\hat{\mu}} \ll 1,$$

so that the governing equations are again the creeping flow equations, namely, (5–15) and (5–16). The far-field boundary condition in this case is (5–69). In addition, the boundary conditions (5–1), (5–4), and (5–9), with $\mathrm{grad}\,\gamma \equiv 0$, must be applied at the surface of the drop, which we represent as the set of points $\mathbf{x}$ where the function $F(\mathbf{x},t) \equiv 0$. The stress balance (5–9), written in dimensionless form, is

$$\boxed{(\mathbf{T} - \lambda\hat{\mathbf{T}})\cdot\mathbf{n} = \frac{1}{\mathrm{Ca}}\,\mathbf{n}(\nabla\cdot\mathbf{n}) \quad \text{at} \quad F = 0.} \qquad (5\text{--}70)$$

Here, the characteristic pressures are μG and $\hat{\mu}G$, respectively. The capillary number is $\mathrm{Ca} \equiv \mu Ga/\gamma$, and λ is the viscosity ratio, $\lambda \equiv \hat{\mu}/\mu$.

In general, the problem defined above is nonlinear, in spite of the fact that the governing, creeping flow equations are linear. This is because the drop shape, and thus $\mathbf{n}$ and F, is unknown, so that the boundary conditions (5–1), (5–4), and (5–70) are nonlinear. Thus, for arbitrary Ca, where the deformation may be quite significant, the problem can only be solved numerically. Here, however, we consider the limiting case $\mathrm{Ca} \ll 1$. As we have discussed already, the deformation in this limit is very small, and we shall see that an approximate analytic solution can be obtained. The basic reason is that the drop shape is only slightly nonspherical, and the boundary conditions at the drop surface can be linearized about the boundary conditions for an exactly spherical drop. This approach is an example of a general technique known as the *method of domain perturbations*.

Thus, for convenience, we express the drop shape in the form

$$F(\mathbf{x},t) \equiv r - \left[1 + \mathrm{Ca}\, f(\mathbf{x},t)\right] \equiv 0, \qquad (5\text{--}71)$$

which is consistent with the fact that a first approximation to the drop shape is just a sphere,

$$r = 1.$$

The deviation from sphericity is contained in the function $f(\mathbf{x},t)$, and we have anticipated, from the arguments at the beginning of this subsection, that the magnitude of the deviation from sphericity is proportional to Ca. Thus, for small Ca, a first approximation to the unit normal vector **n** is just $\mathbf{i}_r$, the unit vector in the radial direction of a spherical coordinate system. A second approximation to **n** can be obtained by application of the definition of **n** in terms of F, namely,

$$\mathbf{n} \equiv \frac{\nabla F}{|\nabla F|} = \mathbf{i}_r - Ca\nabla f + 0(Ca^2). \tag{5-72}$$

The symbol $0(Ca^2)$ in (5-72) signifies that there are additional corrections to **n**, but that these are proportional to Ca^2 or smaller for $Ca \ll 1$. Substituting (5-71) and (5-72) into the boundary conditions (5-1), (5-4), and (5-70), we find that

$$\mathbf{u} = \hat{\mathbf{u}} \quad \text{at} \quad r \approx 1, \tag{5-73a}$$

$$u_r = Ca\frac{\partial f}{\partial t} + 0(Ca^2) \quad \text{at} \quad r \approx 1, \tag{5-73b}$$

and

$$(\mathbf{T} - \lambda\hat{\mathbf{T}})\cdot\mathbf{i}_r = \frac{\mathbf{i}_r}{Ca}(2 - Ca\nabla^2 f) + 0(Ca^2) \quad \text{at} \quad r \approx 1. \tag{5-73c}$$

Again, there are additional corrections that are proportional to Ca^2 or smaller for $Ca \ll 1$. Examination of (5-73) shows that the boundary conditions at this level of approximation are linear in the unknowns **u**, $\hat{\mathbf{u}}$, p, $\hat{p}$, and f. Thus, at least for a first approximation for $Ca \ll 1$, the overall problem is linear and can be solved using the linear methods outlined in Chapter 4.

In order to obtain a solution for the *complete class* of flows given by (5-69), we revert to the general method of Section **F** in Chapter 4 for the construction of solutions of the creeping motion equations via the superposition of vector harmonic functions. The development of a general form for the pressure and velocity fields in the fluid exterior to the drop follows exactly the arguments of Section **F** in Chapter 4 and the solution therefore takes the same general form obtained for a solid sphere in a general linear flow [see Equations (4-262) and (4-264)], that is,

$$p' = c_1 \frac{\mathbf{x}\cdot\mathbf{E}\cdot\mathbf{x}}{r^5} \tag{5-74}$$

and

$$\mathbf{u} = \mathbf{E}\cdot\mathbf{x} + \frac{1}{2}(\boldsymbol{\omega}\wedge\mathbf{x}) + \mathbf{x}(\mathbf{x}\cdot\mathbf{E}\cdot\mathbf{x})\left(\frac{c_1}{2r^5} + \frac{c_3}{r^7}\right) - \frac{2c_3}{5r^5}(\mathbf{E}\cdot\mathbf{x}) + c_4\left(\frac{\boldsymbol{\omega}\wedge\mathbf{x}}{r^3}\right). \tag{5-75}$$

Similarly, a general solution can be obtained for the fluid inside the drop (notice, however, that we must use growing harmonics to avoid any singularity at the origin of the coordinate system). The results are

$$\hat{p} = \mathbf{E} : (\mathbf{xx} - \frac{r^2}{3}\mathbf{I})d_2 \qquad (5\text{-}76)$$

and

$$\hat{u} = d_3(\mathbf{E}\cdot\mathbf{x}) + d_4(\boldsymbol{\omega}\wedge\mathbf{x}) - \frac{2}{21}d_2\mathbf{x}(\mathbf{x}\cdot\mathbf{E}\cdot\mathbf{x}) + \frac{5}{21}d_2r^2\mathbf{E}\cdot\mathbf{x}. \qquad (5\text{-}77)$$

Finally, since the shape function f is linearly related to the dynamic variables $\mathbf{u}$, $\hat{\mathbf{u}}$, and $(\mathbf{T} - \lambda\hat{\mathbf{T}})\cdot\mathbf{n}$, it follows that f must be expressible in invariant form as a linear function of $\mathbf{E}$ or $\boldsymbol{\Omega}$. Since f is a true scalar, it follows that

$$f(\mathbf{x},t) = b_1(\mathbf{x}\cdot\mathbf{E}\cdot\mathbf{x}). \qquad (5\text{-}78)$$

To complete the solution, we must apply the boundary conditions (5–73) at the interface to determine the coefficients c_i, d_i, and b_i in (5–74)–(5–78). The first condition, (5–73a), requires that the velocity components match at the drop surface, which at first approximation is just $r = 1$. Thus, the normal component of (5–73a) yields

$$\mathbf{n}\cdot\left[1 + \left(\frac{c_1}{2} + c_3\right) - \frac{2c_3}{5}\right]\mathbf{E}\cdot\mathbf{n} = \mathbf{n}\cdot\left[d_3\mathbf{E} + \frac{3}{21}d_2\mathbf{E}\right]\cdot\mathbf{n}$$

or

$$1 + \left(\frac{c_1}{2} + \frac{3c_3}{5}\right) = \left(d_3 + \frac{3}{21}d_2\right). \qquad (5\text{-}79)$$

Similarly, the tangential component of (5–73a) gives

$$1 - \frac{2}{5}c_3 = d_3 + \frac{5}{21}d_2 \qquad (5\text{-}80)$$

and

$$\frac{1}{2} + c_4 = d_4. \qquad (5\text{-}81)$$

To apply the stress balance (5–73c), we must first calculate the viscous stresses associated with $\mathbf{u}$, $\hat{\mathbf{u}}$, p, and $\hat{p}$. For the fluid exterior to the drop, the stress field at $r = 1$ is found to be

$$\mathbf{T}\cdot\mathbf{n}|_{r=1} = -p\mathbf{n} + 2\left\{\mathbf{E}\cdot\mathbf{n} + \frac{c_1}{2}(\mathbf{E}\cdot\mathbf{n} - 4\mathbf{n}(\mathbf{n}\cdot\mathbf{E}\cdot\mathbf{n}))\right.$$

$$\left. - \frac{c_3}{5}(-8\mathbf{E}\cdot\mathbf{n} + 20\mathbf{n}(\mathbf{n}\cdot\mathbf{E}\cdot\mathbf{n})) - 3c_4(\boldsymbol{\omega}\wedge\mathbf{x})\right\},$$

while the stress field inside the drop is

$$\hat{\mathbf{T}}\cdot\mathbf{n}|_{r=1} = -\hat{p}\mathbf{n} + 2\left\{d_3\mathbf{E}\cdot\mathbf{n} + \frac{d_2}{21}\left(8\mathbf{E}\cdot\mathbf{n} - \frac{19}{2}(\mathbf{n}\cdot\mathbf{E}\cdot\mathbf{n})\mathbf{n}\right)\right\}.$$

Hence, matching the *tangential* components of stress, we find that

$$c_4 = 0 \tag{5-82}$$

and

$$1 + \frac{c_1}{2} + \frac{8c_3}{5} = \lambda \left(d_3 + \frac{8}{21} d_2 \right). \tag{5-83}$$

To evaluate the *normal* component of the stress balance, we must first evaluate $\nabla \cdot \mathbf{n}$. With $f = b_1 (\mathbf{x} \cdot \mathbf{E} \cdot \mathbf{x})$, the normal $\mathbf{n}$ becomes

$$\mathbf{n} = \frac{\mathbf{x}}{r} + \left(2 \frac{\mathbf{x} \cdot \mathbf{E} \cdot \mathbf{x}}{r^2} \mathbf{x} - \mathbf{E} \cdot \mathbf{x} - \mathbf{x} \cdot \mathbf{E} \right) b_1 \mathrm{Ca},$$

and thus

$$\nabla \cdot \mathbf{n} = 2 + 4 (\mathbf{x} \cdot \mathbf{E} \cdot \mathbf{x}) b_1 \mathrm{Ca} + 0(\mathrm{Ca}^2).$$

It follows from Equation (5-73c) that

$$\hat{p}_0 - p_0 = \frac{2}{\mathrm{Ca}}$$

and

$$\left(1 - 3 \left(\frac{c_1}{2} \right) \right) - \frac{12c_3}{5} - \lambda \left(d_3 - \frac{1}{14} d_2 \right) = 2b_1. \tag{5-84}$$

We thus have six equations that can be solved for the unknowns c_1, c_3, c_4, d_2, d_3, d_4 in terms of b_1. The results are

$$c_4 = 0, \tag{5-85}$$

$$d_4 = \frac{1}{2}, \tag{5-86}$$

$$c_1 = -\frac{5(\lambda - 1)}{2\lambda + 3} (2) - \frac{8}{2\lambda + 3} b_1, \tag{5-87}$$

$$c_3 = 5 \left(\frac{\lambda - 1}{2\lambda + 3} \right) + \frac{20(3\lambda + 2)}{(2\lambda + 3)(19\lambda + 16)} b_1, \tag{5-88}$$

$$d_2 = \frac{84}{19\lambda + 16} b_1, \tag{5-89}$$

and

$$d_3 = \frac{5}{2\lambda + 3} - \frac{4(16\lambda + 19)}{(2\lambda + 3)(19\lambda + 16)} b_1. \tag{5-90}$$

The rate of change of b_1 describes the response of the interface to any rate of change of the flow. To obtain an equation for b_1, we apply the kinematic condition (5-73b). The result is

$$\boxed{\text{Ca}\,\frac{db_1}{dt} = \frac{5}{2\lambda+3} - \frac{40(\lambda+1)}{(2\lambda+3)(19\lambda+16)}b_1.}$$ (5-91)

Although Ca $\ll$ 1, the analysis above does not neglect Ca(db_1/dt), which may, in fact, be 0(1).

The solution of (5-91) for b_1 can be obtained easily. The two terms on the right-hand side of this equation represent the stretching of the drop due to the strain rate **E** and the restoring force due to interfacial tension. The dimensionless relaxation rate, for return to a spherical shape in the absence of a flow, is $(5/6)(1/\text{Ca})$ when the viscosity ratio is small, and $(20/19\lambda)(1/\text{Ca})$ for $\lambda \gg 1$. Hence, for a relatively inviscid drop, the relaxation rate is determined by the external fluid viscosity, while the relaxation rate for a viscous drop is determined by the internal fluid viscosity. The equilibrium solution of (5-91) is

$$b_1 = \frac{19\lambda+16}{8(\lambda+1)},$$ (5-92)

and the corresponding drop shape is

$$\boxed{r = 1 + \text{Ca}\left(\frac{19\lambda+16}{8(\lambda+1)}\right)(\mathbf{x}\cdot\mathbf{E}\cdot\mathbf{x}).}$$ (5-93)

Thus, for a simple shear flow $(u_1 = x_2)$,

$$\mathbf{E} = \begin{pmatrix} 0 & 1/2 & 0 \\ 1/2 & 0 & 0 \\ 0 & 0 & 0 \end{pmatrix} \quad \text{and} \quad r_s = 1 + \text{Ca}\left(\frac{19\lambda+16}{8(\lambda+1)}\right)(x_1 x_2),$$

whereas for a pure uniaxial elongation $\left(u_3 = x_3, u_1 = -\frac{1}{2}x_1, u_2 = -\frac{1}{2}x_2\right)$,

$$\mathbf{E} = \begin{pmatrix} -1/2 & 0 & 0 \\ 0 & -1/2 & 0 \\ 0 & 0 & 1 \end{pmatrix} \quad \text{and} \quad r_s = 1 + \text{Ca}\left(\frac{19\lambda+16}{8(\lambda+1)}\right)\left(x_3^2 - \frac{1}{2}(x_1^2+x_2^2)\right),$$

The corresponding drop shapes are sketched in Figure 5-9. If we express these equations for drop shape in terms of spherical coordinates as shown in Figure 5-10 $(x_1 = r\sin\theta\cos\phi, x_2 = r\sin\theta\sin\phi, x_3 = r\cos\theta)$, we find that

$$\boxed{r_s = 1 + \text{Ca}\left(\frac{19\lambda+16}{8(\lambda+1)}\right)\sin^2\theta\sin\phi\cos\phi}$$ (5-94)

for simple shear flow, and

$$\boxed{r_s = 1 + \text{Ca}\left(\frac{19\lambda+16}{8(\lambda+1)}\right)\left(\frac{3}{2}\cos^2\theta - \frac{1}{2}\right)}$$ (5-95)

for uniaxial extension. The major axis of deformation for simple shear flow lies in the

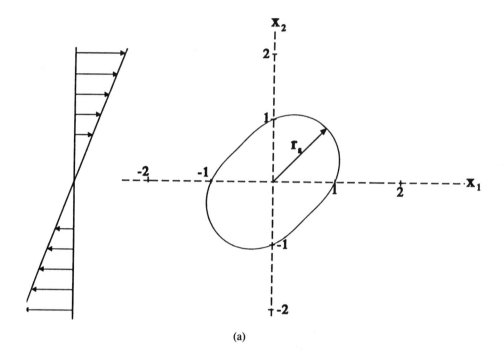

(a)

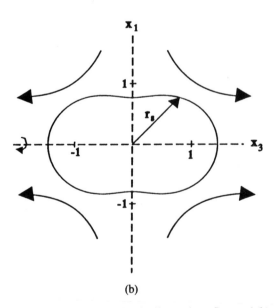

(b)

Figure 5–9 Shape of a viscous drop in (a) simple, linear shear flow and (b) uniaxial extension for $\lambda = 1$ and $Ca = 0.2$.

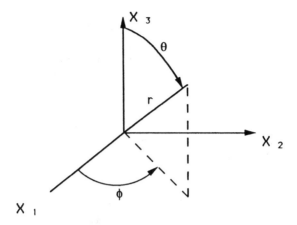

Figure 5–10 Relationship between Cartesian coordinates (x_1, x_2, x_3), and sperical coordinates (r, θ, ϕ).

$x_1 x_2$ plane at $45°$ from the x_1 axis, and the *ratio* of the longest to the shortest axis in this case is

$$\alpha = \frac{1 + \dfrac{1}{2} \mathrm{Ca} \left(\dfrac{19\lambda + 16}{8(\lambda + 1)} \right)}{1 - \dfrac{1}{2} \mathrm{Ca} \left(\dfrac{19\lambda + 16}{8(\lambda + 1)} \right)}. \tag{5-96}$$

In the extensional flow, the principal axis of deformation is x_1, and the shape is axisymmetric about this axis. The ratio of major to minor axis in this case is

$$\alpha = \frac{1 + \mathrm{Ca} \left(\dfrac{19\lambda + 16}{8(\lambda + 1)} \right)}{1 - \dfrac{1}{2} \mathrm{Ca} \left(\dfrac{19\lambda + 16}{8(\lambda + 1)} \right)}. \tag{5-97}$$

The drop in the extensional flow is slightly more elongated than in the simple shear flow. Although the deformation is small in all cases for the limit Ca $\ll 1$, the slight difference in shape indicated by (5–96) and (5–97) is illustrative of the general principle that extensional flows are more efficient at stretching deformable bodies than flows that contain vorticity, like the simple shear flow considered above.

B. Fundamental Solutions of the Creeping Flow Equations

All of the solution techniques in Chapter 4 were based upon the representation of solutions of the creeping flow equations in terms of harmonic functions and/or eigenfunction expansions. These techniques, and their generalization to other coordinate systems besides spherical, provide a powerful arsenal of methods to attack problems in which the geometry of the solid boundaries or interfaces either coincides with a coordinate surface(s) in some orthogonal coordinate system* or else lies close to such a surface, as in the last problem of the previous section. However, the linearity of the creeping flow equations allows another class of solution methods that is more readily applied to problems with *complicated boundary geometries*. These methods for solving Stokes equations are based upon *fundamental* solutions, corresponding to the flow produced by a point force in space. For problems involving *solid* boundaries, a general solution of the creeping flow equations can be obtained in integral form, corresponding to a *distribution of point forces* over all of the boundaries. For problems involving a fluid interface, a similar integral formulation is obtained easily by generalization to a distribution of singularities at the interface.

The solution of a specific problem is then transformed to determining the distribution of surface forces that is necessary to satisfy boundary conditions. If the boundary (or interface) shapes are complex, or at least not "close" to coordinate surfaces in some coordinate system, the solution of this distribution problem can only generally be accomplished numerically, and this leads to so-called *boundary-integral techniques* that have been used widely in the research literature[13] and are particularly well suited to free-boundary problems involving large steady or transient deformations of an interface.[14] We will discuss the basic principles of these boundary-integral methods at the end of this section, though we do not discuss the actual numerical implementation. If the boundary shapes are simpler, an *analytic* approximation for the distribution of point forces can sometimes be achieved by means of equivalent, *internal* distributions of force and force multipole singularities. One particularly important example is the solution for very elongated, slender bodies where the integral representation can be reduced, at least approximately, to a line distribution of point forces along the centerline of the body; the resulting theory is known as *slender-body theory* for creeping flows.

We organize this section of the chapter along the following lines: First, we discuss the fundamental solution of the creeping flow equations for a point force in an unbounded fluid and show that it leads to a general integral representation for the solution of the creeping flow equations; we then consider the special cases, including slender bodies, when the integral equation that results from applying boundary conditions can be solved analytically; finally, we return to the full integral equation and discuss its use in the context of the numerical, boundary-integral technique.

A Fundamental Solution for the Creeping Flow Equations

We begin by considering the solution of the creeping flow equations for a point force **f** in an unbounded fluid. This solution is a *fundamental solution* in the sense that it can be obtained formally from the governing equations

---

*Many such coordinate systems exist. Besides cylindrical, spherical, and Cartesian, the most common are probably elliptical, elliptical cylindrical, bispherical, and bicylindrical. A very useful compilation of the properties of vectors and vector operators, in a large number of orthogonal coordinate systems, may be found in an appendix to the book by Happel and Brenner (1973).[12]

$$\nabla \cdot \mathbf{u} = 0, \qquad (5\text{-}98a)$$

and

$$\nabla^2 \mathbf{u} - \nabla p = \delta(\mathbf{x})\mathbf{f}, \qquad (5\text{-}98b)$$

for a point force $\mathbf{f}$ applied to the fluid at $\mathbf{x} = 0$. Rather than solving (5–98) directly, via Fourier transforms or some similar technique, we resort to the method of superposition of harmonic functions as described in Chapter 4.

This method is particularly simple to apply in the case of a point force $\mathbf{f}$. The (disturbance) pressure must be a decaying harmonic function that is linear in $\mathbf{f}$ and a true scalar. Hence,

$$p = c_1 \frac{\mathbf{f} \cdot \mathbf{x}}{r^3}. \qquad (5\text{-}99)$$

The (disturbance) velocity is thus

$$\mathbf{u} = \frac{\mathbf{x}}{2}\left(\frac{c_1 \mathbf{f} \cdot \mathbf{x}}{r^3}\right) + \mathbf{u}^H, \qquad (5\text{-}100)$$

where $\mathbf{u}^H$ is harmonic, decaying, and linear in $\mathbf{f}$ and a true vector, that is,

$$\mathbf{u}^H = \alpha\left(\frac{\mathbf{xx}}{r^5} - \frac{\mathbf{I}}{3r^3}\right) \cdot \mathbf{f} + \beta\frac{\mathbf{f}}{r}, \qquad (5\text{-}101)$$

where α and β are coefficients that must still be determined.

If we apply the continuity Equation (5–98a), we find that

$$\frac{c_1}{2} = \beta$$

so that

$$\mathbf{u} = \frac{c_1}{2}\left[\frac{\mathbf{f}}{r} + \frac{\mathbf{x}(\mathbf{f} \cdot \mathbf{x})}{r^3}\right] + \alpha\left(\frac{\mathbf{xx}}{r^5} - \frac{\mathbf{I}}{3r^3}\right) \cdot \mathbf{f}. \qquad (5\text{-}102)$$

To this point, the solution is essentially identical to that obtained for uniform flow past a sphere in Section **F** of Chapter 4. Beyond this, however, the two differ. To satisfy boundary conditions at the surface of a sphere, a nonzero value of α is necessary. But, for a point force $\alpha = 0$. The simplest way to see this is to note that c_1 is dimensionless, while α must have dimensions of (length)^2. But a point force has no characteristic length scale, so $\alpha \equiv 0$. Thus,

$$\mathbf{u} = \frac{c_1}{2}\left(\frac{\mathbf{f}}{r} + \frac{\mathbf{x}(\mathbf{f} \cdot \mathbf{x})}{r^3}\right). \qquad (5\text{-}103)$$

To determine the scalar constant c_1, we make use of an integral relationship between the force $\mathbf{f}$ and the resultant force exerted on the fluid outside a control surface S.

First, consider the fluid region V between a body of arbitrary geometry ∂D and a control surface S of arbitrary shape that completely encloses it. Since the creeping motion approximation is being used,

$$\nabla \cdot \mathbf{T} = 0 \qquad (5\text{–}104)$$

everywhere in the fluid region V, and thus

$$\int_V \nabla \cdot \mathbf{T} \, dV = 0. \qquad (5\text{–}105)$$

Thus, by applying the divergence theorem to (5–105), we see that

$$\int_{\partial D} \mathbf{T} \cdot \mathbf{n} \, dA = \int_S \mathbf{T} \cdot \mathbf{n}_s \, dA, \qquad (5\text{–}106)$$

Here the integral on the left is taken over the surface of the body ∂D and $\mathbf{n}$ is the *outer normal from the body into the fluid*. The integral on the right is taken over the control surface S with outer normal $\mathbf{n}_S$. But, the integral on the left is just the hydrodynamic force acting on the body (that is, the negative of the force acting from the body onto the fluid). It thus follows that

$$\mathbf{F} = \int_S \mathbf{T} \cdot \mathbf{n}_s \, dA. \qquad (5\text{–}107)$$

In the present case, we have a point force at the origin, and the same expression holds but with $-\mathbf{f}$ replacing $\mathbf{F}$.

To obtain c_1, we apply (5–107) to the solution (5–99) plus (5–103), with S chosen for convenience as a *sphere* centered at the origin. The surface stress, evaluated using (5–99) and (5–103), is

$$\mathbf{T} \cdot \mathbf{n} = \left[-p\mathbf{I} + (\nabla\mathbf{u} + \nabla\mathbf{u}^T) \right] \cdot \mathbf{n}$$

$$= -3c_1 \frac{\mathbf{x}\mathbf{x} \cdot \mathbf{f}}{r^4}. \qquad (5\text{–}108)$$

Hence, integrating over the spherical surface S (arbitrary radius), we obtain from (5–107)

$$\mathbf{f} = 4\pi c_1 \mathbf{f}. \qquad (5\text{–}109)$$

Hence,

$$c_1 = \frac{1}{4\pi} \qquad (5\text{–}110)$$

so that

$$\boxed{\mathbf{u} = \frac{1}{8\pi} \left(\frac{\mathbf{f}}{r} + \frac{\mathbf{x}(\mathbf{f} \cdot \mathbf{x})}{r^3} \right)} \qquad (5\text{–}111)$$

and

$$\boxed{p = \frac{1}{4\pi} \left(\frac{\mathbf{f} \cdot \mathbf{x}}{r^3} \right).} \qquad (5\text{–}112)$$

The associated expression for the stress tensor $\mathbf{T}$ is

$$\boxed{\mathbf{T} = -\frac{3}{4\pi} \frac{\mathbf{x}\mathbf{x}(\mathbf{x} \cdot \mathbf{f})}{r^5}.} \qquad (5\text{–}113)$$

The solutions (5–111) and (5–112) are called the *Stokeslet* solution, apparently a name coined by Hancock (1953).[15] It is customary to choose

$$\boxed{\mathbf{f} = 8\pi\alpha\delta(\mathbf{x})}$$
(5–114)

so that the point force singularity is characterized in magnitude and direction by α.

We shall see that the Stokeslet solution plays a fundamental role in creeping flow theory. We have already seen in Section **E** of Chapter 4, that it describes the disturbance velocity far away from a body of any shape that exerts a nonzero force on an unbounded fluid. Indeed, when nondimensionalized and expressed in spherical coordinates, it is identical to the velocity field (4–199). In the next section we use the Stokeslet solution to derive a general integral representation for solutions of the creeping flow equations.

A General Integral Representation for Solutions of the Creeping Flow Equations

In order to obtain a general integral representation for solutions of the creeping flow equations, it is necessary first to derive a general integral theorem reminiscent of Green's theorem from vector calculus.

Let us then consider the space outside a closed surface ∂D, and let $\mathbf{u}$ and $\hat{\mathbf{u}}$ be any smooth vector fields (such as velocity fields) in this exterior domain that satisfy the conditions

$$\nabla\cdot\mathbf{u} = 0, \quad \nabla\cdot\hat{\mathbf{u}} = 0$$
(5–115)

(such vector fields are called *solenoidal*) and vanish at infinity as r^{-1} or faster. In addition to $\mathbf{u}$ and $\hat{\mathbf{u}}$, we also define the two tensor (stress) functions

$$\mathbf{T} \equiv -p\mathbf{I} + (\nabla\mathbf{u} + \nabla\mathbf{u}^T)$$
(5–116)

and

$$\hat{\mathbf{T}} \equiv -\hat{p}\mathbf{I} + (\nabla\hat{\mathbf{u}} + \nabla\hat{\mathbf{u}}^T),$$
(5–117)

where p and $\hat{p}$ are any smooth scalar fields that vanish as r^{-2} or faster. Then a general vector identity between these functions is

$$\boxed{\hat{\mathbf{u}}\cdot\nabla\cdot\mathbf{T} - \mathbf{u}\cdot\nabla\cdot\hat{\mathbf{T}} = \nabla\cdot\{\hat{\mathbf{u}}\cdot\mathbf{T} - \mathbf{u}\cdot\hat{\mathbf{T}}\} + \{\nabla\mathbf{u}:\hat{\mathbf{T}} - \nabla\hat{\mathbf{u}}:\mathbf{T}\}.}$$
(5–118)

But, if we utilize (5–116) and (5–117), then it can be shown that the last term in (5–118) is identically equal to zero. Hence, integrating over V,

$$\int_V [\hat{\mathbf{u}}\cdot(\nabla\cdot\mathbf{T}) - \mathbf{u}\cdot(\nabla\cdot\hat{\mathbf{T}})]dV = \int_V \nabla\cdot[\hat{\mathbf{u}}\cdot\mathbf{T} - \mathbf{u}\cdot\hat{\mathbf{T}}]dV,$$
(5–119)

and applying the divergence theorem, we obtain

$$\int_V [\hat{\mathbf{u}}\cdot(\nabla\cdot\mathbf{T}) - \mathbf{u}\cdot(\nabla\cdot\hat{\mathbf{T}})]dV =$$
$$\int_S [\hat{\mathbf{u}}\cdot\mathbf{T} - \mathbf{u}\cdot\hat{\mathbf{T}}]\cdot\mathbf{n}dA - \int_{\partial D} [\hat{\mathbf{u}}\cdot\mathbf{T} - \mathbf{u}\cdot\hat{\mathbf{T}}]\cdot\mathbf{n}dA,$$
(5–120)

where S is any surface enclosing ∂D. Finally, if we let the surface $S \to \infty$, we see that the integral over S is $0(r^{-1})$ and thus vanishes. It follows that

$$\int_V \{\hat{\mathbf{u}} \cdot (\nabla \cdot \mathbf{T}) - \mathbf{u} \cdot (\nabla \cdot \hat{\mathbf{T}})\} dV = -\int_{\partial D} \mathbf{n} \cdot [\hat{\mathbf{u}} \cdot \mathbf{T} - \mathbf{u} \cdot \hat{\mathbf{T}}] dA, \qquad (5\text{-}121)$$

where V is now the whole space exterior to ∂D.

The integral theorem (5-121) leads directly to a general integral representation for solutions of the creeping flow equations. To see this, let $\mathbf{u}, \mathbf{T}$ represent a solution of Stokes's equations—arbitrary except that they must be $0(r^{-1})$ and $0(r^{-2})$, respectively, for $r \to \infty$, as assumed in deriving (5-121). Further, let $\hat{\mathbf{u}}, \hat{\mathbf{T}}$ be the fundamental solution of Stokes's equations for a point force at a point $\boldsymbol{\xi}$, that is, referring to (5-111),

$$\hat{\mathbf{u}} = \frac{1}{8\pi} \left(\frac{(\mathbf{x} - \boldsymbol{\xi})(\mathbf{x} - \boldsymbol{\xi})}{R^3} + \frac{\mathbf{I}}{R} \right) \cdot \mathbf{e}, \qquad (5\text{-}122)$$

where the point force is assumed to be of unit magnitude $\mathbf{f} = \mathbf{e}$, and

$$R \equiv |\mathbf{x} - \boldsymbol{\xi}|$$

is the distance between a point $\mathbf{x}$ and the point of application of the force, $\boldsymbol{\xi}$. The corresponding stress $\hat{\mathbf{T}}$ is

$$\hat{\mathbf{T}} = -\frac{3}{4\pi} \frac{(\mathbf{x} - \boldsymbol{\xi})(\mathbf{x} - \boldsymbol{\xi})(\mathbf{x} - \boldsymbol{\xi})}{R^5} \cdot \mathbf{e}. \qquad (5\text{-}123)$$

Now, substituting (5-122) and (5-123) into (5-121) and noting that

$$\nabla \cdot \mathbf{T} \equiv 0 \qquad (5\text{-}124a)$$

and

$$\nabla \cdot \hat{\mathbf{T}} = \delta(\mathbf{x} - \boldsymbol{\xi})\mathbf{e}, \qquad (5\text{-}124b)$$

we obtain

$$\int_V (-\delta(\mathbf{x} - \boldsymbol{\xi})\mathbf{u} \cdot \mathbf{e}) dV_\xi = -\int_{\partial D} \mathbf{n} \cdot [(\hat{\mathbf{U}} \cdot \mathbf{e}) \cdot \mathbf{T} - \mathbf{u} \cdot (\hat{\boldsymbol{\Sigma}} \cdot \mathbf{e})] dA_\xi, \qquad (5\text{-}125)$$

where we have introduced the shorthand notation

$$\hat{\mathbf{u}} = \hat{\mathbf{U}} \cdot \mathbf{e}, \quad \hat{\mathbf{T}} = \hat{\boldsymbol{\Sigma}} \cdot \mathbf{e} \qquad (5\text{-}126)$$

in place of (5-122) and (5-123). The symbols dV_ξ and dA_ξ indicate that the variable of integration is $\boldsymbol{\xi}$, and the fixed point is thus $\mathbf{x}$. The tensor component of $\hat{\mathbf{U}}$ (i.e. $\hat{U}_{ik}$) is the i component of the velocity field generated by a point force of unit magnitude in the k direction. Similarly, the component Σ_{ijk} is the ij component of the stress tensor corresponding to the flow produced by a point force in the k direction.

To simplify (5-125), we can factor out the constant unit vector $\mathbf{e}$ from all terms and use the integral property of the delta function to evaluate the term on the left. The result, after substituting for $\hat{\mathbf{U}}$ and $\hat{\boldsymbol{\Sigma}}$ from (5-122), (5-123), and (5-126) is

$$\mathbf{u}(\mathbf{x}) + \frac{3}{4\pi} \int_{\partial D} \frac{(\mathbf{x} - \boldsymbol{\xi})(\mathbf{x} - \boldsymbol{\xi})(\mathbf{x} - \boldsymbol{\xi})}{R^5} \cdot \mathbf{u} \cdot \mathbf{n} \, dA_{\xi} =$$
$$-\frac{1}{8\pi} \int_{\partial D} \left[\frac{\mathbf{I}}{R} + \frac{(\mathbf{x} - \boldsymbol{\xi})(\mathbf{x} - \boldsymbol{\xi})}{R^3} \right] \cdot (\mathbf{T} \cdot \mathbf{n}) \, dA_{\xi}.$$

$$(5\text{--}127)$$

This is the famous integral representation for the solution of the creeping flow equations that is usually attributed to Ladyzhenskaya (1963).[16] Since the derivation requires $\mathbf{u} \sim r^{-1}$ for large r, we recognize that $\mathbf{u}$ must be interpreted as the disturbance velocity field if we wish to apply (5–127) to a problem that involves an undisturbed velocity field $\mathbf{u}^{\infty}(\mathbf{x})$ at large distances from the body or boundary that is denoted as ∂D. To apply (5–127) directly to the actual velocity field, we let $\mathbf{u} = \mathbf{u}'$ in (5–127) where $\mathbf{u}' = \mathbf{u} - \mathbf{u}^{\infty}$; $\mathbf{u}$ is now the true velocity, and $\mathbf{u}^{\infty}$ is the undisturbed velocity. This gives

$$\mathbf{u}(\mathbf{x}) + \frac{3}{4\pi} \int_{\partial D} \left(\frac{(\mathbf{x} - \boldsymbol{\xi})(\mathbf{x} - \boldsymbol{\xi})(\mathbf{x} - \boldsymbol{\xi})}{R^5} \cdot \mathbf{u} \right) \cdot \mathbf{n} \, dA_{\xi} =$$
$$\mathbf{u}^{\infty}(\mathbf{x}) - \frac{1}{8\pi} \int_{\partial D} \left[\frac{\mathbf{I}}{R} + \frac{(\mathbf{x} - \boldsymbol{\xi})(\mathbf{x} - \boldsymbol{\xi})}{R^3} \right] \cdot \mathbf{T} \cdot \mathbf{n} \, dA_{\xi}.$$

$$(5\text{--}128)$$

The term

$$\int_{\partial D} \mathbf{n} \cdot \left[\hat{\mathbf{U}} \cdot \mathbf{T}^{\infty} - \mathbf{u}^{\infty} \cdot \hat{\boldsymbol{\Sigma}} \right] dA_{\xi} \qquad (5\text{--}129)$$

is identically equal to zero, as the reader may wish to verify.

The formula (5–127) [or (5–128)] provides a formal solution of the creeping motion equations in a compact form. The integral on the left-hand side is denoted as the *double-layer potential* and has a density function that is just the velocity $\mathbf{u}$ on the boundaries ∂D of the flow domain. The integral on the right is termed the *single-layer potential,* and its density function is the surface stress vector $\mathbf{t} = \mathbf{T} \cdot \mathbf{n}$. Of course, (5–127) and (5–128) do not provide a solution for any specific problem until the density functions $\mathbf{u}$ and $\mathbf{T} \cdot \mathbf{n}$ are specified on ∂D. In fact, all that we really have done is to obtain an integral formula for $\mathbf{u}$ that is equivalent to the differential form of the creeping flow equation (4–6). To obtain a solution for any particular problem, we must determine the density functions so that the velocity field $\mathbf{u}$ satisfies the boundary conditions on ∂D. In general, this requires numerical solution of the integral equations that result from applying boundary conditions to (5–127) or (5–128). In fact, this is the essence of the so-called *boundary-integral method* for solution of creeping flow equations; this technique has been used widely in research and is especially suitable for free-surface and other Stokes flow problems with complicated boundary geometries. At the end of this section, we discuss some principles of the boundary-integral technique. First, however, we discuss some alternative techniques for the *analytic* solution of problems involving flow exterior to a solid body.

Solutions for Solid Bodies via Internal Distributions of Singularities

One important class of Stokes flow problems involves motion past a stationary solid surface. In this case, the no-slip boundary condition is

$$\mathbf{u} = 0 \quad \text{for all} \quad \mathbf{x} \text{ on } \partial D, \tag{5–130}$$

and the integral formula (5–128) can be applied for $\mathbf{x}$ on ∂D to obtain

$$\mathbf{u}^{\infty}(\mathbf{x}) = \frac{1}{8\pi} \int_{\partial D} \left[\frac{\mathbf{I}}{R} + \frac{(\mathbf{x} - \boldsymbol{\xi})(\mathbf{x} - \boldsymbol{\xi})}{R^3} \right] \cdot \mathbf{t}(\boldsymbol{\xi}) \, dA_{\boldsymbol{\xi}}. \tag{5–131}$$

The solution of this integral equation gives us directly the unknown surface stress vector $\mathbf{t}(\boldsymbol{\xi}) = \mathbf{T}(\boldsymbol{\xi}) \cdot \mathbf{n}$. Then, the general solution of the creeping flow equations is

$$\mathbf{u}(\mathbf{x}) = \mathbf{u}^{\infty}(\mathbf{x}) - \frac{1}{8\pi} \int_{\partial D} \left[\frac{\mathbf{I}}{R} + \frac{(\mathbf{x} - \boldsymbol{\xi})(\mathbf{x} - \boldsymbol{\xi})}{R^3} \right] \cdot \mathbf{t}(\boldsymbol{\xi}) \, dA_{\boldsymbol{\xi}}, \tag{5–132}$$

where $\mathbf{x}$ is now a fixed point in the flow field and $\mathbf{t}(\mathbf{x})$ is the distribution of surface stress on the boundary ∂D that is obtained by solving (5–131). We do not discuss the solution of the integral equation (5–131) here. The main objective in deriving (5–132) is to show that a general solution of the creeping flow equations for flow past stationary solid surfaces can be expressed completely as a superposition of surface forces (Stokeslets) at the boundaries ∂D. In fact, a solution of the creeping flow equations can always be written solely as a distribution of Stokeslets over the bounding surfaces, even if these are not solid and stationary, but the simple identity of the Stokeslet density function with the actual surface stress is only valid for this special case. To retain the simple physical interpretation of the density functions for other kinds of boundaries, we must use the more general form (5–128).

In this section, we pursue the basic idea that the solution for creeping flow past a body, in terms of a *surface distribution* of point forces (Stokeslets) on the surface of that body, can sometimes be replaced by an *internal distribution* of point forces and higher-order singularities. This is based upon two key points. First, if we begin with a solution of the creeping flow equations, then a derivative of any order of that solution is still a solution of the creeping flow equations. In particular, if we start with the Stokeslet solution (5–111), (5–112), and (5–114), which we denote here as $(\mathbf{u}_s, p_s)$,

$$\mathbf{u}_s(\mathbf{x}; \boldsymbol{\alpha}) = \frac{\boldsymbol{\alpha}}{r} + \frac{(\boldsymbol{\alpha} \cdot \mathbf{x}) \mathbf{x}}{r^3},$$

$$p_s(\mathbf{x}; \boldsymbol{\alpha}) = 2 \frac{\boldsymbol{\alpha} \cdot \mathbf{x}}{r^3}, \tag{5–133}$$

then a derivative of any order of $\mathbf{u}_s$ and p_s is also a solution of the creeping flow equations, corresponding to a point singularity that is a derivative of the same order of a point force $\mathbf{f}$. Second, a Stokeslet solution for a point force at $\mathbf{x} = \boldsymbol{\xi}$ can be expressed in terms of a formal *multipole* expansion in the form of a generalized Taylor series about **x**,

$$\mathbf{u}_s(\mathbf{x} - \boldsymbol{\xi}) = \mathbf{u}_s(\mathbf{x}) - (\boldsymbol{\xi} \cdot \nabla)\mathbf{u}_s(\mathbf{x}) + \frac{1}{2}(\boldsymbol{\xi} \cdot \nabla)^2 \mathbf{u}_s(\mathbf{x}) + \ldots \qquad (5\text{--}134)$$

with a similar series expression for p_s. But $(\boldsymbol{\xi} \cdot \nabla)\mathbf{u}_s$ and $(\boldsymbol{\xi} \cdot \nabla)^2 \mathbf{u}_s$ are the velocity fields generated by a *force dipole*, $(\boldsymbol{\xi} \cdot \nabla)\mathbf{f}_s$, and a *force quadrapole*, $(\boldsymbol{\xi} \cdot \nabla)^2 \mathbf{f}_s$, respectively. Thus, for flow past a solid body, we always can use the generalized Taylor series to replace the surface distribution of Stokeslets in the solution (5–132) by an equivalent internal distribution of Stokeslets and higher-order singularities (inside the body).

The obvious question is whether there is any advantage to be gained, especially in view of the fact that we must replace a surface distribution of Stokeslets only, by a whole hierarchy of higher-order singularities inside. One possibility is that we may be able to replace the surface distribution by an internal distribution of lower spatial order. Thus, for example, for axisymmetric bodies, it may be possible to express the solution in terms of a *line distribution* of singularities along the axis of symmetry of the body. In this case, the two-dimensional integral equation that arises from application of boundary conditions [equation (5–131)] would be replaced by a one-dimensional, though more complicated, integral equation. Another possibility is that one may be able to replace a Stokeslet distribution on a body which has a very complicated surface (for example, lots of "bumps") with an internal distribution of singularities on a nearby surface that has a much simpler geometry. In any case, however, the use of internal distributions of singularities will be an advantage only if the number of terms in the multipole expansion can be limited to a relatively small set. Intuitively, for an exact solution, this will require bodies of simple geometry in relatively simple flows. Alternatively, if the internal surface (or line) is close enough to the surface of the body, it should be possible to approximate the multipole expansion by a small number of terms (or even one term), since the higher-order terms will decrease rapidly in magnitude. Beyond these generalities, we cannot offer more definitive criteria for recognizing problems where internal distributions of singularities offer an advantage over the solution in terms of a surface distribution of Stokeslets [equation (5–132)]. This is, in fact, a subject of current research.

In the remainder of this section, we outline results for two types of problems where internal distributions of singularities have been used to advantage. The first, originally pursued by Chwang and Wu,[17] considers bodies of very simple shape— spheres, prolate ellipsoids of revolution (spheroids), and similar cases where *exact* solutions can be obtained for some flows either by a point or a line distribution involving only a few singularities. The second class of problems is for very slender bodies, where an *approximate* solution can be obtained via a distribution of Stokeslets along the particle centerline.[18]

To pursue the use of internal singularities for bodies of simple geometry, we begin by discussing the solutions of the creeping flow equations for force dipoles and higher-order multipole singularities. These may be obtained formally by differentiation of the basic Stokeslet solution. For example, the so-called Stokes dipole solution is

and

$$\mathbf{u}_{SD}(\mathbf{x};\boldsymbol{\alpha},\boldsymbol{\beta}) \equiv -(\boldsymbol{\beta}\cdot\nabla)\mathbf{u}_s(\mathbf{x};\boldsymbol{\alpha}) = \frac{(\boldsymbol{\beta}\wedge\boldsymbol{\alpha})\wedge\mathbf{x}}{r^3} - \left[\frac{(\boldsymbol{\alpha}\cdot\boldsymbol{\beta})\mathbf{x}}{r^3} - \frac{3(\boldsymbol{\alpha}\cdot\mathbf{x})(\boldsymbol{\beta}\cdot\mathbf{x})\mathbf{x}}{r^5}\right]$$

$$p_{SD}(\mathbf{x};\boldsymbol{\alpha},\boldsymbol{\beta}) \equiv -(\boldsymbol{\beta}\cdot\nabla)p_s(\mathbf{x};\boldsymbol{\alpha}) = 2\mu\left[-\frac{\boldsymbol{\alpha}\cdot\boldsymbol{\beta}}{r^3} + 3\frac{(\boldsymbol{\alpha}\cdot\mathbf{x})(\boldsymbol{\beta}\cdot\mathbf{x})}{r^5}\right].$$

(5–135a) (5–135b)

This corresponds to the velocity and pressure fields generated by a pair of forces, one at $\mathbf{x} = \boldsymbol{\beta}/2$ with strength $\boldsymbol{\alpha}$ and the other at $\mathbf{x} = -\boldsymbol{\beta}/2$ with strength $-\boldsymbol{\alpha}$, in the limit as $|\boldsymbol{\beta}| \to 0$ (see, for example, the sketch in Figure 5–11). Similarly, the Stokes quadrapole solution is

and

$$\mathbf{u}_{S4}(\mathbf{x};\boldsymbol{\alpha},\boldsymbol{\beta},\boldsymbol{\delta}) = (\boldsymbol{\delta}\cdot\nabla)(\boldsymbol{\beta}\cdot\nabla)\mathbf{u}_s(\mathbf{x};\boldsymbol{\alpha}) \qquad (5\text{–}136a)$$

$$p_{S4}(\mathbf{x};\boldsymbol{\alpha},\boldsymbol{\beta},\boldsymbol{\delta}) = (\boldsymbol{\delta}\cdot\nabla)(\boldsymbol{\beta}\cdot\nabla)p_s(\mathbf{x};\boldsymbol{\alpha}), \qquad (5\text{–}136b)$$

and we can obtain solutions for even higher-order singularities by continued differentiation as indicated in (5–135) and (5–136).

It is useful to express the Stokes dipole solution as the sum of two component parts, because each has a clear physical significance. These components are defined in a manner that is analogous to the symmetric and antisymmetric components of a second-order tensor, that is,

$$\mathbf{u}_{SD}(\mathbf{x};\boldsymbol{\alpha},\boldsymbol{\beta}) = \mathbf{u}_{SS}(\mathbf{x};\boldsymbol{\alpha},\boldsymbol{\beta}) + \mathbf{u}_R(\mathbf{x};\boldsymbol{\alpha},\boldsymbol{\beta}), \qquad (5\text{–}137)$$

where

$$\mathbf{u}_{SS}(\mathbf{x};\boldsymbol{\alpha},\boldsymbol{\beta}) \equiv \frac{1}{2}(\mathbf{u}_{SD}(\mathbf{x};\boldsymbol{\alpha},\boldsymbol{\beta}) + \mathbf{u}_{SD}(\mathbf{x};\boldsymbol{\beta},\boldsymbol{\alpha})) \qquad (5\text{–}138)$$

and

$$\mathbf{u}_R(\mathbf{x};\boldsymbol{\alpha},\boldsymbol{\beta}) \equiv \frac{1}{2}(\mathbf{u}_{SD}(\mathbf{x};\boldsymbol{\alpha},\boldsymbol{\beta}) - \mathbf{u}_{SD}(\mathbf{x};\boldsymbol{\beta},\boldsymbol{\alpha})). \qquad (5\text{–}139)$$

The symmetric part, given by (5–138), is known as the *stresslet* solution, and the antisymmetric part, (5–139), is known as the *rotlet* solution.

The detailed form of the rotlet can be obtained easily by substituting the Stokes dipole solution (5–135a) into the definition (5–139). The result is

$$\frac{1}{2}(\mathbf{u}_{SD}(\mathbf{x};\boldsymbol{\alpha},\boldsymbol{\beta}) - \mathbf{u}_{SD}(\mathbf{x};\boldsymbol{\beta},\boldsymbol{\alpha})) = \frac{(\boldsymbol{\beta}\wedge\boldsymbol{\alpha})\wedge\mathbf{x}}{r^3}. \qquad (5\text{–}140)$$

Hence, if we define $\boldsymbol{\gamma} \equiv \boldsymbol{\beta}\wedge\boldsymbol{\alpha}$, the rotlet velocity field is

$$\mathbf{u}_r(\mathbf{x};\boldsymbol{\gamma}) = \frac{\boldsymbol{\gamma}\wedge\mathbf{x}}{r^3}, \qquad (5\text{–}141a)$$

and the corresponding pressure is

$$p_r(\mathbf{x};\boldsymbol{\gamma}) = 0. \qquad (5\text{–}141b)$$

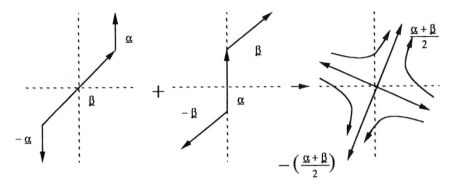

Figure 5–11 Stresslet velocity field constructed as a sum of the Stokes' dipoles $u_{SS}(\mathbf{x};\alpha,\beta)$ and $u_{SD}(\mathbf{x};\beta,\alpha)$.

The physical significance of the rotlet solution is that it is the flow due to a singular, point torque at the origin. To see this, we can calculate the moment exerted on the fluid outside an arbitrary control surface S centered at the origin by a rotlet of strength $\boldsymbol{\gamma}$. Thus,

$$\mathbf{M} = \int_S \mathbf{x} \wedge (\mathbf{T}_R \cdot \mathbf{n}) dA, \qquad (5\text{-}142)$$

and the result after some straightforward manipulation is

$$\mathbf{M}_R = 8\pi\mu\boldsymbol{\gamma}. \qquad (5\text{-}143)$$

The stresslet solution (5-138) takes the form

$$\mathbf{u}_{SS}(\mathbf{x};\alpha,\beta) = \left[-\frac{\alpha\cdot\beta}{r^3} + \frac{3(\alpha\cdot\mathbf{x})(\beta\cdot\mathbf{x})}{r^5} \right] \mathbf{x} \qquad (5\text{-}144\text{a})$$

and

$$p_{SS}(\mathbf{x};\alpha,\beta) = 2\mu \left[-\frac{\alpha\cdot\beta}{r^3} + \frac{3(\alpha\cdot\mathbf{x})(\beta\cdot\mathbf{x})}{r^5} \right]. \qquad (5\text{-}144\text{b})$$

The stresslet exerts zero net force or torque on the fluid, that is,

$$\mathbf{F}_{SS} = 0 \text{ and } \mathbf{M}_{SS} = 0 \qquad (5\text{-}145)$$

and can be thought of as a straining motion of the fluid that is symmetric about the α, β plane with principal axes of strain $\alpha + \beta$, $\alpha - \beta$, and $\alpha \wedge \beta$; see Figure 5-11.

The Stokes quadrupole solutions are more complicated; however, one component turns out to be particularly useful in the solution of Stokes flow problems, and that is the *potential dipole* solution, namely,

$$\mathbf{u}_D(\mathbf{x};\delta) \equiv -\frac{1}{2}\nabla^2 \mathbf{u}_s(\mathbf{x};\delta). \qquad (5\text{-}146)$$

This solution has the simple form

$$\mathbf{u}_D = -\frac{\delta}{r^3} + \frac{3(\delta \cdot \mathbf{x})\mathbf{x}}{r^5},$$

$$p_D = 0. \tag{5-147}$$

Again, the potential dipole exerts zero net force on the fluid. Physically, it corresponds to a mass dipole at the origin—that is, the flow generated by a mass source at $\mathbf{x} = \delta/2$ and a mass sink at $\mathbf{x} = -\delta/2$ in the limit $|\delta| \to 0$.

Unfortunately, the use of fundamental solutions to solve Stokes flow problems via internal distributions of singularities has not yet become completely systematized. Many (and perhaps all) problems involving flow in the vicinity of a solid sphere can be solved by a superposition of point force and/or higher-order singularities at the center of the sphere. Some problems involving prolate axisymmetric ellipsoids (spheroids) can be solved by a superposition of point force and higher-order singularities on the symmetry axis between the two foci of the ellipse. However, general rules that are guaranteed to guide the choice of singularities or the choice of internal surface geometry for all possible situations cannot be offered. If there is a net force on the body, the leading-order term must be a Stokeslet or Stokeslet distribution, and experience suggests that the Stokeslet is always accompanied by the potential dipole. If there is a net torque on the body, we require a rotlet or rotlet distribution, and this will be the leading term if the net force is zero. Finally, if the net force and torque are both zero, the leading term often (though not always) involves the stresslet. Several examples may illustrate these points.

Translation of a Sphere in a Quiescent Fluid

The translation of a sphere through a quiescent fluid produces a net force on the sphere. Thus, to construct a solution via internal singularities we require a Stokeslet of strength α located at the sphere center. However, by itself, the Stokeslet flow does not satisfy the no-slip and kinematic boundary conditions at the surface of the sphere. In particular, if we assume that velocities have been nondimensionalized with the magnitude of the sphere's velocity U, the boundary condition at the sphere surface, $r = 1$, is

$$\mathbf{u} = \mathbf{i}_u \quad \text{at} \quad r = 1, \tag{5-148}$$

where spatial variables are nondimensionalized using the sphere radius a. The Stokeslet velocity field at $r = 1$ takes the form

$$\mathbf{u}_s = \alpha + (\alpha \cdot \mathbf{i}_r)\mathbf{i}_r. \tag{5-149}$$

Hence, it is clear that the Stokeslet solution alone cannot satisfy (5-148). However, we have already suggested that the Stokeslet is most often accompanied by the potential dipole $\mathbf{u}_D$. At the sphere surface, the potential dipole field becomes

$$\mathbf{u}_D = -\delta + 3(\delta \cdot \mathbf{i}_r)\mathbf{i}_r. \tag{5-150}$$

Hence, comparing (5-149) and (5-150), it is obvious that a superposition of the Stokeslet and potential dipole solutions will satisfy the creeping flow equations and also the boundary condition (5-148). Specifically,

$$\boxed{\mathbf{u} = \mathbf{u}_S(\mathbf{x};\boldsymbol{\alpha}) + \mathbf{u}_D(\mathbf{x};\boldsymbol{\delta}),} \tag{5-151}$$

where

$$\boldsymbol{\alpha} - \boldsymbol{\delta} = \mathbf{i}_u \tag{5-152}$$

and

$$\boldsymbol{\alpha} + 3\boldsymbol{\delta} = 0. \tag{5-153}$$

Thus,

$$\boxed{\boldsymbol{\delta} = -\frac{\mathbf{i}_u}{4} \quad \boldsymbol{\alpha} = \frac{3}{4}\mathbf{i}_u.} \tag{5-154}$$

It may be noted that the force on the sphere in dimensionless terms, as given by (5-114), is $\mathbf{f} = 6\pi\mathbf{i}_u$. In dimensional terms this becomes $\mathbf{f}_D = (6\pi\mu Ua)\mathbf{i}_u$. Not surprisingly, we recover Stokes's law.

Sphere in Linear Flows: Axisymmetric Extensional Flow and Simple Shear

We have previously obtained solutions by other techniques for the problems of a rigid sphere immersed in axisymmetric straining or simple shear flow of an unbounded fluid (see Chapter 4). In this subsection, we show that those two problems also can be solved very simply by means of a superposition of fundamental singularities at the center of the sphere.

We begin with the problem of a solid sphere centered at the stagnation of an axisymmetric straining flow in which the undisturbed velocity at infinity takes the dimensionless form

$$\mathbf{u}^\infty = \mathbf{E} \cdot \mathbf{x}, \tag{5-155}$$

with

$$\mathbf{E} = \begin{bmatrix} -1 & 0 & 0 \\ 0 & -1 & 0 \\ 0 & 0 & +2 \end{bmatrix}.$$

In this technique, we seek the *disturbance* velocity field,

$$\boxed{\mathbf{u}' = \mathbf{u} - \mathbf{u}^\infty,} \tag{5-156}$$

so that

$$\mathbf{u}' \to 0 \quad \text{as} \quad |\mathbf{x}| \to \infty \tag{5-157}$$

and

$$\boxed{\mathbf{u}' = -\mathbf{E} \cdot \mathbf{x} \quad \text{for} \quad \mathbf{x} \epsilon \mathbf{x}_s \text{ (sphere surface).}} \tag{5-158}$$

We have already noted that a systematic procedure does not yet exist for choosing the singularities necessary to solve any particular problem. Thus, we rely on a combination of intuition and a knowledge of the properties of the basic singularities to make an educated guess. In the present case, it is clear from symmetry considerations alone that there is no net force or torque on the sphere. Thus, the solution cannot involve either a Stokeslet or a rotlet. Furthermore, we have already seen that the flow field associated with the stresslet is an axisymmetric extensional motion, and the form of the undisturbed flow at the sphere surface is also an axisymmetric extension. Thus, an educated guess in this case is that the dominant singularity at the sphere center should be a stresslet. Then, since the Stokeslet is always accompanied by a potential dipole and the stresslet is just a derivative of the Stokeslet, it seems reasonable to suppose that the stresslet should be accompanied by a derivative of the potential dipole—namely, the potential quadrapole.

Thus, we suggest trying to construct a solution that is a superposition of a stresslet and a potential quadrapole $\mathbf{u}_{D4}$, both located at the center of the sphere, that is,

$$\boxed{\mathbf{u}' = \mathbf{u}_{SS}(\mathbf{x}; \alpha, \beta) + \mathbf{u}_{D4}(\mathbf{x}; \delta, \gamma),} \tag{5-159}$$

where

$$\boxed{\mathbf{u}_{D4}(\mathbf{x}; \delta, \gamma) \equiv \gamma \cdot \nabla(\mathbf{u}_D(\mathbf{x}; \delta)) = \frac{6(\delta \cdot \gamma)\mathbf{x}}{r^5} - \frac{15(\delta \cdot \mathbf{x})(\gamma \cdot \mathbf{x})\mathbf{x}}{r^7} + \frac{3(\delta \cdot \mathbf{x})\gamma}{r^5}.} \tag{5-160}$$

Hence, at the surface of the sphere

$$\mathbf{u}' = \left[-\alpha \cdot \beta + 6(\delta \cdot \gamma)\right]\mathbf{i}_r + 3(\delta \cdot \mathbf{i}_r)\gamma + \left[3(\alpha \cdot \mathbf{i}_r)(\beta \cdot \mathbf{i}_r) - 15(\delta \cdot \mathbf{i}_r)(\alpha \cdot \mathbf{i}_r)\right]\mathbf{i}_r. \tag{5-161}$$

On the other hand, we have previously shown that the boundary condition (5-158) can be expressed in terms of spherical coordinates in the form (4-216), namely,

$$\mathbf{u}' = (1 - 3\cos^2\theta)\mathbf{i}_r + 3\sin\theta\cos\theta\,\mathbf{i}_\theta. \tag{5-162}$$

It follows, by equating (5-161) and (5-162), that

$$\alpha = \alpha\mathbf{i}_z, \ \beta = \beta\mathbf{i}_z, \ \gamma = \gamma\mathbf{i}_z, \quad \text{and} \quad \delta = \delta\mathbf{i}_z, \tag{5-163}$$

where

$$\beta\gamma = -1 \text{ and } \alpha\beta = -5. \tag{5-164}$$

Hence, we can write the solution (5-159) in the form

$$\mathbf{u}' = -5\mathbf{u}_{SS}(\mathbf{x}; \mathbf{i}_z, \mathbf{i}_z) - \mathbf{u}_{D4}(\mathbf{x}; \mathbf{i}_z, \mathbf{i}_z) \tag{5-165}$$

and

$$\mathbf{u} = \mathbf{u}^\infty + \mathbf{u}'. \tag{5-166}$$

The closely related problem of a rigid sphere in a linear shear flow is very easy to solve now by analogy to the solution for an axisymmetric straining flow. We consider the problem in the form

$$\mathbf{u}^\infty = \mathbf{\Gamma} \cdot \mathbf{x}, \tag{5-167}$$

with

$$\mathbf{\Gamma} = \begin{bmatrix} 0 & 0 & 1 \\ 0 & 0 & 0 \\ 0 & 0 & 0 \end{bmatrix}.$$

Thus,

$$\mathbf{u}^\infty = z\mathbf{i}_1. \tag{5-168}$$

The boundary condition at the sphere surface for the disturbance flow,

$$\mathbf{u}' \equiv \mathbf{u} - \mathbf{u}^\infty, \tag{5-169}$$

is thus

$$\mathbf{u}' = -\boldsymbol{\Gamma}\cdot\mathbf{x} \quad \text{for} \quad \mathbf{x}\epsilon\mathbf{x}_s \text{(sphere surface)}. \tag{5-170}$$

In this case, if the sphere does not rotate (that is, $\mathbf{u} = 0$ for $\mathbf{x}\epsilon\mathbf{x}_s$), there will be a torque in the direction of the 2 axis, and the solution must therefore involve a rotlet, as well as a stresslet and a potential quadrapole. Indeed, if we derive the expression for surface velocity as a superposition of these three singularity types located at the sphere center and compare it with the boundary condition (5–170) expressed in terms of spherical coordinates, it is not difficult to show that the solution can be expressed in the form

$$\mathbf{u}' = -\frac{5}{6}\mathbf{u}_{SS}(\mathbf{x};\mathbf{i}_x,\mathbf{i}_z) - \frac{1}{2}\mathbf{u}_R(\mathbf{x};\mathbf{i}_y) - \frac{1}{6}\mathbf{u}_{D4}(\mathbf{x};\mathbf{i}_x,\mathbf{i}_z), \tag{5-171}$$

with

$$\mathbf{u} = \mathbf{u}^\infty + \mathbf{u}'.$$

This solution is identical to (4–273).

Uniform Flow Past a Prolate Spheroid

We have seen in the preceding two subsections that solutions for creeping flow in the vicinity of a spherical body can be generated by means of a superposition of fundamental singularities at the center of the sphere. Although this technique, in some instances, may be simpler to apply or more convenient than the methods discussed earlier, the reader may nevertheless wonder at the need to have introduced yet another new technique. For solutions involving spheres only, this skepticism would be well founded. However, the use of internal distributions of singularities applies equally well to certain creeping flow problems involving nonspherical bodies. Although these problems could also be solved by generalization of the methods introduced earlier, these methods would require the use of elliptical or other nonspherical coordinates. This is not true of the use of internal distributions of singularities.

The solution of Stokes flow problems by internal distributions of singularities for nonspherical bodies has been discussed in a series of papers by Chwang and Wu,[19] and more recently by Kim.[20] Here, as an example to demonstrate basic principles, we consider the relatively simple problem of uniform flow past a prolate spheroid

$$\frac{x^2}{a^2} + \frac{r^2}{b^2} = 1 \quad (r^2 \equiv y^2 + z^2, \ a \geq b), \tag{5-172}$$

with the free-stream velocity given by

$$\mathbf{u}^\infty = U_1\mathbf{i}_x + U_2\mathbf{i}_y \quad \text{as} \quad |\mathbf{x}| \to \infty, \tag{5-173}$$

where i_x and i_y are unit vectors. In dimensionless terms, with a characteristic length scale a and a characteristic velocity scale, say, U_1, the prolate spheroid is specified as

$$x^2 + \left(\frac{a^2}{b^2}\right) r^2 = 1 \tag{5-174}$$

with the free-stream velocity given by

$$u^\infty = i_x + \frac{U_2}{U_1} i_y \quad \text{as} \quad |x| \to \infty. \tag{5-175}$$

For bodies of nonspherical shape, it is generally necessary to use a *distribution* of singularities over an interior *surface*, but this does not offer much advantage over a solution in terms of a distribution of Stokeslets on the body surface itself, via (5-131) and (5-132) However, for prolate spheroids, we shall see that it is sufficient to utilize a *line* distribution of singularities along the central axis of revolution, between the two foci. In this case, the original problem of determining a *surface* distribution of Stokeslets is reduced to the *one-dimensional* problem of determining the distribution of Stokeslets and higher-order singularities on a line.

Although there are no firm rules to determine the *type* of singularities that should be used, we attempt to use the qualitative insight of the preceding sections. Thus, since the net force on the body will be nonzero, we try to construct a solution using a line distribution of Stokeslets, between the foci of the ellipsoid, $x = -c$ and $+c$ (dimensionless), where

$$c = \left(1 - \frac{b^2}{a^2}\right)^{1/2} = e \tag{5-176}$$

and e is also known as the eccentricity $0 \le e < 1$. In addition, since we have found earlier that the Stokeslet is always accompanied by a potential dipole, we also include a line distribution of potential dipoles along the centerline from $x = -c$ to c. Hence, we try to construct a solution for the disturbance flow in the form

$$u' = -\int_{-c}^{c} [\alpha_1 u_s(x - \xi; i_x) + \alpha_2 u_s(x - \xi; i_y)] d\xi$$

$$+ \int_{-c}^{c} (c^2 - \xi^2)[\beta_1 u_D(x - \xi; i_x) + \beta_2 u_D(x - \xi; i_y)] d\xi \tag{5-177}$$

and

$$p' = -\int_{-c}^{c} [\alpha_1 p_s(x - \xi; i_x) + \alpha_2 p_s(x - \xi; i_y)] d\xi, \tag{5-178}$$

where $\xi = \xi i_x$.

The proposed form (5-177) and (5-178) clearly satisfies the creeping flow Equations (4-6) and also satisfies the free-stream boundary condition, namely,

$$u' \to 0 \quad \text{as} \quad |x| \to \infty. \tag{5-179}$$

The question is whether the no-slip condition at the spheroid surface can be satisfied, that is,

$$\mathbf{u}' = -\mathbf{i}_x - \frac{U_2}{U_1}\mathbf{i}_y \quad \text{for} \quad \mathbf{x}\epsilon\mathbf{x}_s(\text{spheroid surface}), \tag{5-180}$$

through the choice of constants α_1, α_2, β_1, and β_2. If the suggested solution form is correct, then this should be possible. However, it is also possible that we may need additional singularities or that the singularities should be distributed with different weighting functions or over some other portion of the centerline, and in this case no choice of α_1, α_2, β_1, and β_2 will work. To see whether we can satisfy the no-slip conditions, we make use of the integrated form of $\mathbf{u}'$, which can be written as

$$\mathbf{u}' = -(2\alpha_1\mathbf{i}_x+\alpha_2\mathbf{i}_y)B_{1,0} - (\alpha_1 r\mathbf{i}_r+\alpha_2 y\mathbf{i}_x)\left(\frac{1}{R_2}-\frac{1}{R_1}\right)$$

$$+ (\alpha_1 r\mathbf{i}_x-\alpha_2 y\mathbf{i}_r)rB_{3,0} + \nabla\left\{-2\beta_1 B_{1,1}+\beta_2 y\left[\frac{x-c}{r^2}R_1 - \frac{x+c}{r^2}R_2+B_{1,0}\right]\right\},$$

$$\tag{5-181}$$

where $\mathbf{e}_r = (y\mathbf{e}_y+z\mathbf{e}_z)/r$ is the radial unit vector in the yz plane and

$$R_1 = [(x+c)^2+r^2]^{1/2},$$

$$R_2 = [(x-c)^2+r^2]^{1/2},$$

$$B_{1,0} = \log\frac{R_2-(x-c)}{R_1-(x+c)},$$

$$B_{1,1} = R_2-R_1+xB_{1,0},$$

$$B_{3,0} = \frac{1}{r^2}\left(\frac{x+c}{R_1}-\frac{x-c}{R_2}\right),$$

$$B_{3,1} = \left(\frac{1}{R_1}-\frac{1}{R_2}\right)+xB_{3,0}.$$

On the surface of the spheroid,

$$r^2 = (1-e^2)(1-x^2), \quad R_1 = 1+ex, \quad R_2 = 1-ex,$$

and

$$B_{1,0} = \log\frac{1+e}{1-e} = L_e, \quad B_{3,0} = \frac{2e}{(1-e^2)(1-e^2x^2)}.$$

Hence, after some manipulation, the velocity *at the spheroid surface* can be expressed in the form

$$
\mathbf{u}_s' = \left[\frac{2\alpha_1}{e} - 2(\alpha_1 + \beta_1)L_e \right] \mathbf{i}_x - \frac{2}{e}\left(\alpha_1 - \frac{2e^2\beta_1}{1-e^2} \right) \frac{b^2\mathbf{i}_x + e^2 x r \mathbf{i}_r}{1 - e^2 x^2}
$$

$$
- \left[\frac{2e^2\beta_2}{1-e^2} + (\alpha_2 - \beta_2)L_e \right] \mathbf{i}_y - 2e\left(\alpha_2 - \frac{2e^2\beta_2}{1-e^2} \right) y \frac{\dfrac{b^2}{a^2} x \mathbf{i}_x + r \mathbf{i}_r}{\dfrac{b^2}{a^2}(1 - e^2 x^2)}. \qquad (5\text{-}182)
$$

Thus, the no-slip condition (5–159) is satisfied if

$$
\alpha_1 = \frac{2\beta_1 e^2}{(1-e^2)} = e^2 \left[-2e + (1+e^2)L_e \right]^{-1} \qquad (5\text{-}183)
$$

and

$$
\alpha_2 = \frac{2\beta_2 e^2}{(1-e^2)} = 2\frac{U_2}{U_1} e^2 \left[2e + (3e^2 - 1)L_e \right]^{-1}. \qquad (5\text{-}184)
$$

Although this completes the formal solution and demonstrates that the distribution of Stokeslets and potential dipoles proposed in (5–177) is sufficient to solve the problem, it is of interest to calculate the force on the body.

   In view of the properties of the Stokeslet and potential dipole solutions, the force acting on the spheroid is simply represented by the accumulative strength of the Stokeslet distribution along the centerline of the spheroid. Thus, in dimensionless terms,

$$
\mathbf{F} = -8\pi \int_{-c}^{c} (-\alpha_1 \mathbf{i}_x - \alpha_2 \mathbf{i}_y)\, dx. \qquad (5\text{-}185)
$$

Hence, evaluating the integral using the expressions (5–183) and (5–184) for α_1 and α_2, we obtain

$$
\mathbf{F} = 6\pi \left[C_{F1}\mathbf{i}_x + \frac{U_2}{U_1} C_{F2}\mathbf{i}_y \right], \qquad (5\text{-}186)
$$

where

$$
C_{F1} = \frac{8}{3}e^3 \left[-2e + (1+e^2)\log\frac{1+e}{1-e} \right]^{-1} \qquad (5\text{-}187a)
$$

and

$$
C_{F2} = \frac{16}{3}e^3 \left[2e + (3e^2 - 1)\log\frac{1+e}{1-e} \right]^{-1}. \qquad (5\text{-}187b)
$$

In dimensional terms, then, the total force is

$$
\mathbf{F}_D = 6\pi\mu a \left[U_1 C_{F1}\mathbf{i}_x + U_2 C_{F2}\mathbf{i}_y \right]. \qquad (5\text{-}188)
$$

For the limiting case of a sphere, $e \to 0$, and

$$
C_{F1} = C_{F2} = 1,
$$

as expected. On the other hand, for very slender spheroids, with $b/a = (1 - e^2)^{1/2} \ll 1$, the force coefficients take the limiting form

$$C_{F1} \approx \frac{2}{3} \frac{1}{\log(2a/b) - 1/2}, \quad C_{F2} \approx \frac{4}{3} \frac{1}{\log(2a/b) + 1/2},$$

(5-189)

so that

$$\frac{C_{F1}}{C_{F2}} \longrightarrow \frac{1}{2} \quad \text{as} \quad \frac{b}{a} \longrightarrow 0.$$

(5-190)

This latter result also can be obtained via the approximate *slender-body* procedure outlined in the next section. The fact that the ratio of the force coefficients C_{F1}/C_{F2} approaches 1/2 only for an extremely elongated, needlelike shape is of qualitative interest because it reflects a general property of Stokes flows; namely, the resistance to motion of a body is remarkably insensitive to the geometric configuration. In particular, the ratio 1/2 means that an elongated, needlelike body will translate at a speed when moving parallel to its symmetry axis that is only a factor of 2 faster than the same body would translate under the action of the same force when moving perpendicular to its symmetry axis. The insensitivity of the resistance to the particle geometry is also evident in the fact that the force law (5-188) for a long slender body differs from that for a sphere of radius a primarily in the logarithmic factor $\log(2a/b)$, which varies only *slowly* with the axis ratio. In particular, the drag on a slender rod aligned parallel to its direction of motion is only $2/3\log(2a/b)$ times the drag on a sphere having the same diameter as the length of the rod. Although this ratio vanishes as $b/a \to 0$, it is only 1/12 for $b/2a \sim 10^{-4}$.

Approximate Solutions of Stokes's Equations via Slender-Body Theory

In the preceding subsection, we saw that exact solutions for creeping flow past a solid body can be obtained via internal distributions of fundamental singularities and that this can lead to significant simplification when the geometry of the body is either spherical or a prolate, axisymmetric ellipsoid (other body shapes that have been studied include oblate spheroids[21]). In the prolate ellipsoid case, instead of solving for the strength of a surface distribution of Stokeslets, as suggested by the integral formulation (5-127) or (5-128), we need determine only the distribution of Stokeslets and higher-order singularities on the centerline of the body (and this reduces to a single point at the center for a sphere). Hence, as noted earlier, the problem of solving Stokes flow problems is reduced to the solution of one-dimensional integral equations for the singularity distribution functions. Unfortunately, however, for complicated flows it is not clear what particular singularities are needed, nor what the functional forms of the density functions should be, and this currently reduces the practical usefulness of the singularity superposition technique.

In one case, however, the problem is greatly simplified. This is the creeping motion of a fluid relative to a *very slender body*—namely, a body in which the typical

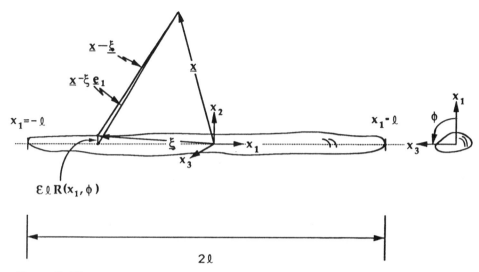

Figure 5–12 A schematic sketch of the geometrical configuration for the slender-body analysis of an elongated body of arbitrary cross-sectional shape [derivation of Equation (5-191)].

cross-sectional dimension is very small compared with the body's length. In this case, an approximate solution can be obtained for bodies of *arbitrary* cross-sectional shape using only a Stokeslet distribution along the centerline of the body. Intuitively, it is clear that this must be the case. An exact solution can always be obtained, at least in principle, via a *surface* distribution of Stokeslets, and this solution, in turn, can be expressed in terms of an equivalent *internal* distribution of Stokeslets, Stokes dipoles, and higher-order singularities via the generalized Taylor series approximation that was initially given as equation (5-134). However, for a slender body, the magnitude of the position vector which connects points at the body surface and a point on the centerline is very small (for the same x_1), and thus the Stokeslet distribution at the body surface can be accurately approximated using only a Stokeslet distribution on the centerline. This qualitative discussion can be formalized in a straightforward manner.

Let us then consider the elongated solid particle that is sketched in Figure 5-12. The length of the particle is $2l$, and the surface of the particle can be denoted in the form

$$\sqrt{x_2^2 + x_3^2} = \epsilon l R(x_1, \phi),$$

where the x_1 axis is assumed to be parallel to the particle centerline, ϕ is the azimuthal angle about the x_1 axis, R is an $0(1)$ quantity, and ϵ is the slenderless ratio. In dimensionless terms, the equation for the body surface becomes

$$\sqrt{x_2^2 + x_3^2} = \epsilon R(x_1, \phi),$$

where we have taken the half-length of the body l as a characteristic length scale, and the body now lies in the range $-1 \le x_1 \le 1$. For the slender-body approximation that we shall seek here, it is assumed that

$$\epsilon \ll 1.$$

In the case of an axisymmetric body, the cross-sectional shape at any point is a circle (R independent of ϕ), and for the special case of a spheroid, $\epsilon \equiv a/b$, where a and b are the semiminor and semimajor axes lengths.

We have already seen in Equation (5-132) that a general solution of the creeping flow equations for flow past a solid body is

$$\mathbf{u}(\mathbf{x}) = \mathbf{u}^\infty(\mathbf{x}) - \frac{1}{8\pi} \int_{\partial D} \left[\frac{\mathbf{I}}{R} + \frac{(\mathbf{x} - \boldsymbol{\xi})(\mathbf{x} - \boldsymbol{\xi})}{R^3} \right] \cdot \mathbf{t}(\boldsymbol{\xi}) dA_\xi,$$

where $\mathbf{t}(\boldsymbol{\xi}) = \mathbf{n} \cdot \mathbf{T}$. Now, $\mathbf{x}$ is the position vector for a fixed point in the flow domain, and $\boldsymbol{\xi}$ is the position vector for a variable point of integration on the body surface, ∂D. But, for a slender body, the vector $\mathbf{x} - \boldsymbol{\xi}$ can be approximated by the vector $\mathbf{x} - \xi \mathbf{e}_1$ as suggested by the sketch in Figure 5-11, where $-1 \le \xi \le 1$. Thus, the general solution (5-132) can be approximated in the form of a line integral along the centerline of the particle,

$$u_i(\mathbf{x}) = u_i^\infty(\mathbf{x}) - \frac{1}{8\pi} \int_{-1}^{1} \left\{ \frac{\delta_{ij}}{r} + \frac{(x_i - \xi\delta_{i1})(x_j - \xi\delta_{j1})}{r^3} \right\} f_j(\xi) d\xi, \qquad (5\text{-}191)$$

where

$$r \equiv \sqrt{(x_1 - \xi)^2 + \rho^2}, \rho^2 \equiv x_2^2 + x_3^2$$

and $f_i(\xi)$ is the integral of the local surface stress vector $t_j(\xi)$ around the closed perimeter of the particle at each point $x_1 = \xi$. For convenience, we have reverted to a component form for (5-191). The error in approximating the exact solution (5-132) by the approximate form (5-191) generally depends on the geometry of the ends of the body. For a particle with rounded ends, the error is of order $\epsilon^2 \ln \epsilon$.[22]

To apply (5-191) to obtain the solution for a specific problem, we must determine the weighting function $f_j(\xi)$ by applying the boundary condition $\mathbf{u} = 0$ at ∂D. Thus, proceeding formally,

$$u_i^\infty(\mathbf{x}) = \frac{1}{8\pi} \int_{-1}^{1} \left[\frac{\delta_{ij}}{r} + \frac{(x_i - \xi\delta_{i1})(x_j - \xi\delta_{j1})}{r^3} \right] f_j(\xi) d\xi \qquad (5\text{-}192)$$

for $\mathbf{x} \in \partial D$. Although the reader may question whether the approximation (5-192) remains valid for $\mathbf{x} \in \partial D$, since the approximation $\mathbf{x} - \boldsymbol{\xi} \approx \mathbf{x} - \xi \mathbf{e}_1$ is clearly in error when $\boldsymbol{\xi}$ is in the neighborhood of $\mathbf{x}$, it can be shown rigorously that (5-191) remains valid (basically because the approximation referred to above fails over a portion of ∂D that is very small). Thus, the solution of creeping flow problems for slender bodies is reduced to solving an integral equation of the first kind, (5-192), for the Stokeslet density function $f_j(\xi)$.

To this end, let us consider the integral terms in (5-192), beginning with the first term. With $\rho = \epsilon R(\mathbf{x})$, the integral is

$$\int_{-1}^{1} \frac{f_i(\xi)d\xi}{\sqrt{(x-\xi)^2 + \epsilon^2 R^2}}. \tag{5-193}$$

Since this integral is singular at $x = \xi$ in the limit $\epsilon \to 0$, we rewrite it in the form

$$\int_{-1}^{1} \frac{f_i(\xi)d\xi}{\sqrt{(x-\xi)^2 + \epsilon^2 R^2}} \equiv f_i(x) \int_{-1}^{1} \frac{d\xi}{\sqrt{(x-\xi)^2 + \epsilon^2 R^2}} + \int_{-1}^{1} \frac{f_i(\xi) - f_i(x)}{\sqrt{(x-\xi)^2 + \epsilon^2 R^2}} d\xi. \tag{5-194}$$

In the limit $\epsilon \to 0$, the right-hand side of (5–194) is dominated by the first term. Specifically,

$$I \equiv \int_{-1}^{1} \frac{d\xi}{\sqrt{(x-\xi)^2 + \epsilon^2 R^2}} = \log\left[(x+1) + [(x+1)^2 + \epsilon^2 R^2]^{1/2}\right]$$

$$-\log\left[(x-1) + [(x-1)^2 + \epsilon^2 R^2]\right]$$

$$\cong -2\ln\epsilon + \log\frac{4(1-x^2)}{R^2} + \ldots = -2\ln\epsilon + 0(1). \tag{5-195}$$

The symbol $0(1)$ means that the error in approximating the integral as $-2\ln\epsilon$ is independent of ϵ as $\epsilon \to 0$, and thus is much smaller than $-2\ln\epsilon$. The second integral in (5–194) also represents an error of $0(1)$. Similarly,

$$\int_{-1}^{1} \frac{(x-\xi)^2 d\xi}{\{(x-\xi)^2 + \epsilon^2 R^2\}^{3/2}} \longrightarrow I - 2 + 0(\epsilon^2) = -2\ln\epsilon + 0(1). \tag{5-196}$$

Hence, utilizing (5–195) and (5–196), the integral equation (5–192) can be approximated in the form

$$\boxed{u_i^\infty(x_1, 0, 0) = \frac{\ln 1/\epsilon}{4\pi}\{f_i(x_1) + \delta_{i1} f_1(x_1)\} + 0(1).}\tag{5-197}$$

With $\mathbf{u}^\infty(\mathbf{x})$ specified, the algebraic equation (5–197) can thus be solved to determine a first approximation to the Stokeslet density distribution in the slender-body limit, $\epsilon \to 0$.

In general, it can be seen from (5–197) that

$$\boxed{f_1(x_1) \cong \frac{2\pi u_1^\infty(x_1, 0, 0)}{\ln 1/\epsilon} \quad \text{and} \quad f_i \approx \frac{4\pi u_i^\infty(x_1, 0, 0)}{\ln 1/\epsilon} \quad \text{for } i = 1,2.}\tag{5-198}$$

Thus, for $\mathbf{u}^\infty = \mathbf{e}_1$ (that is, uniform flow parallel to the slender-body axis, with $\mathbf{u}^\infty$ nondimensionalized by the magnitude of the velocity, U),

$$f_1 \approx \frac{2\pi}{\ln 1/\epsilon}, \quad f_2 = f_3 = 0, \tag{5-199}$$

and the dimensionless drag force on the body is therefore

$$F_1 = -\int_{-1}^{1} f_1 \, dx_1 = \frac{4\pi}{\ln 1/\epsilon} .$$

(5-200)

The corresponding dimensional drag force is thus

$$(F_D)_1 = \frac{4\pi\mu lU}{\ln 1/\epsilon} .$$

(5-201)

For $\mathbf{u}^{\infty} = \mathbf{e}_2$, on the other hand,

$$f_2 \cong \frac{4\pi}{\ln \epsilon}, \quad f_1 = f_3 = 0,$$

(5-202)

and the force

$$(F_D)_2 = -\mu Ul \int_{-1}^{1} f_2 \, dx_1 = \frac{8\pi\mu lU}{\ln 1/\epsilon} .$$

(5-203)

Comparing the slender-body results (5-201) and (5-203) with the limiting form of the exact solution for prolate ellipsoids (spheroids) [Equations (5-186) and (5-187)], we see that the two are identical. Again, the force on a very elongated body is seen to be approximately proportional to its length. Furthermore, the ratio of the force for motion parallel and perpendicular to the particle axis, respectively, is just $(F_D)_2/(F_D)_1 = 2$. This result provides further evidence of the relative insensitivity of the drag to the body geometry in the creeping flow limit.

C. Further Topics in Creeping Flow Theory

The preceding sections have been concerned primarily with direct solution techniques for problems in creeping flow theory. Here, we discuss several general topics that evolve directly from these developments. The first two involve application of the so-called *reciprocal theorem* of low Reynolds number hydrodynamics.

The Reciprocal Theorem

The reciprocal theorem is derived directly from the general integral formula (5-121). For this purpose, we identify $\mathbf{u}$ and $\hat{\mathbf{u}}$, as well as $\mathbf{T}$ and $\hat{\mathbf{T}}$, as the solutions of two creeping flow problems for flow past the same body but with different boundary conditions on the body surface ∂D.

Then, since

$$\nabla \cdot \mathbf{T} = 0, \quad \nabla \cdot \hat{\mathbf{T}} = 0,$$

(5-121) can be written in the form

$$\int_{\partial D} \hat{\mathbf{u}} \cdot \mathbf{f} \, dA = \int_{\partial D} \mathbf{u} \cdot \hat{\mathbf{f}} \, dA,$$

(5-204)

where

$$\mathbf{f} = \mathbf{T} \cdot \mathbf{n} \quad \text{and} \quad \hat{\mathbf{f}} = \hat{\mathbf{T}} \cdot \mathbf{n}.$$

The relationship (5–204) is the famous *reciprocal theorem* of Lorentz.[23] As we shall see shortly, it is an extremely useful result that often leads to the impression of "getting something for nothing."

Faxen's Law for a Body in an Unbounded Fluid

As a first application of the reciprocal theorem, we consider its use in calculating the hydrodynamic force on a body in an undisturbed flow, $u^\infty(x)$.

In particular, let us suppose that we have obtained the solution of the creeping motion equations for uniform flow U past a stationary body ∂D. Equivalently, we may consider the case of a particle that translates with velocity $-U$ in a fluid at rest at infinity. We denote the solution of this problem as u and the corresponding surface force function on ∂D as f. Then, on applying the reciprocal theorem, we find that

$$U \cdot \int_{\partial D} \hat{f} dA = U \cdot \hat{F} = \int_{\partial D} \hat{u} \cdot f dA. \tag{5–205}$$

But this simple formula provides an example of the idea of something for nothing. For if we have actually solved the uniform flow problem, we can immediately deduce the force on the same body held fixed in any undisturbed flow $u^\infty(x)$ that satisfies the creeping flow equations. In particular,

$$U \cdot \hat{F} = \int_{\partial D} u^\infty \cdot f dA. \tag{5–206}$$

Hence, we can obtain the force $\hat{F}$ for the undisturbed flow $u^\infty(x)$ by means of a simple integration of the right-hand side of (5–206), with f known from the solution of the uniform flow problem but without any need to actually solve the creeping flow problem involving $u^\infty(x)$. To obtain all three components of $\hat{F}$, we require a solution for uniform flow past the same body along any three mutually perpendicular directions.

An especially powerful result is obtained if we apply the formula (5–206) to a stationary solid sphere of radius a in the undisturbed flow $u^\infty(x)$. In this case, the solution for uniform flow past a stationary solid sphere yields the general result

$$f = \frac{3}{2a} \mu U. \tag{5–207}$$

Hence, from (5–206),

$$U \cdot \hat{F} = \int_{\partial D} u^\infty(x) \cdot \left(\frac{3}{2a} \mu U \right) dA.$$

Since U is arbitrary,

$$\hat{F} = \frac{3}{2a} \mu \int_{\partial D} u^\infty(x) dA. \tag{5–208}$$

Finally, supposing the origin to be at the center of the sphere, we can expand $u^{(\infty)}$ at ∂D in a multipole representation in terms of the values of $u^{(\infty)}$ and higher-order derivatives of $u^{(\infty)}$ at the position occupied by the center of the sphere,

$$\mathbf{u}^{(\infty)}(\mathbf{x}) = \mathbf{u}^{(\infty)}(0) + \mathbf{x} \cdot (\nabla \mathbf{u}^{(\infty)})_0 + \frac{\mathbf{x}\mathbf{x}}{2!} : (\nabla(\nabla \mathbf{u}^{(\infty)}))_0 + \cdots. \qquad (5\text{-}209)$$

Inserting (5-209) into (5-208), we can show that

$$\hat{\mathbf{F}} = 6\pi\mu a \left[\mathbf{u}^{(\infty)}(0) + \frac{a^2}{6} (\nabla^2 \mathbf{u}^{(\infty)})_0 + \text{const} \, (\nabla^4 \mathbf{u}^{(\infty)})_0 + \cdots \right]. \qquad (5\text{-}210)$$

In deriving (5-210), it may be noted that all the odd number terms in (5-209) vanish since ∂D is a sphere. However, since

$$\nabla^2 \mathbf{u} - \frac{1}{\mu} \nabla p, = 0$$

we see that $\nabla^4 \mathbf{u} = 0$, as well as all other terms, $\nabla^{2n} \mathbf{u}$, for $n \geq 2$, so that

$$\boxed{\hat{\mathbf{F}} = 6\pi\mu a \left[\mathbf{u}^{(\infty)}(0) + \frac{a^2}{6} (\nabla^2 \mathbf{u}^{(\infty)})_0 \right].} \qquad (5\text{-}211)$$

This important result is known as *Faxen's* law.[24] According to this law, if we specify the undisturbed velocity $\mathbf{u}^{(\infty)}(\mathbf{x})$, then the force on a sphere can be calculated directly from the formula (5-211), without any need to actually solve the flow problem corresponding to the free-stream velocity $\mathbf{u}^{(\infty)}(\mathbf{x})$.

A number of interesting results can be obtained from the Faxen formula, (5-211). The first is to notice that the force on a sphere in an arbitrary linear flow

$$\mathbf{u}^\infty = \mathbf{U}^\infty + \mathbf{\Gamma} \cdot \mathbf{x}, \qquad (5\text{-}212)$$

where $\mathbf{U}_\infty$ and $\mathbf{\Gamma}$ are independent of $\mathbf{x}$, depends only on the *relative translational velocity* between the body and the undisturbed fluid evaluated at the body center and is otherwise independent of the velocity gradient tensor $\mathbf{\Gamma}$, that is,

$$\boxed{\hat{\mathbf{F}} = 6\pi\mu a \left[\mathbf{U}^\infty + \mathbf{\Gamma} \cdot \mathbf{x}_p - \mathbf{u}_p \right].} \qquad (5\text{-}213)$$

Here, $\mathbf{x}_p$ denotes the position of the center of the sphere, and $\mathbf{u}_p$ is the velocity of the sphere. Thus, in the absence of an external force on the sphere (in the presence of gravity, we assume $\rho_p = \rho_{\text{fluid}}$), it must translate at steady state with the undisturbed velocity of the fluid evaluated at the position occupied by its center point, that is,

$$\boxed{u_p = \mathbf{U}^\infty + \mathbf{\Gamma} \cdot \mathbf{x}_p.} \qquad (5\text{-}214)$$

It follows that a neutrally buoyant sphere in any linear flow can be treated as though the origin of coordinates for the undisturbed flow is coincident with the center of the sphere, namely,

$$\mathbf{u}^\infty = \mathbf{\Gamma} \cdot (\mathbf{x} - \mathbf{x}_p). \qquad (5\text{-}215)$$

In a flow with *quadratic* dependence upon spatial position, on the other hand, the hydrodynamic force on a sphere will not simply equal Stokes drag. For example, let us suppose that we have a sphere in a two-dimensional, Poiseuille flow

$$\mathbf{u}^{\infty} = \frac{G}{2\mu}(dy - y^2)\mathbf{i}_x, \tag{5-216}$$

where G is the negative of the pressure gradient and d is the distance between the two walls. Then, according to Faxen's law, the hydrodynamic force is

$$\hat{\mathbf{F}} = 6\pi\mu a \left[\frac{G}{2\mu}(dy_p - y_p^2)\mathbf{i}_x - \mathbf{u}_p - \frac{Ga^2}{6\mu}\mathbf{i}_x \right]. \tag{5-217}$$

Again, suppose that there is no external force acting on the sphere (suppose it is neutrally buoyant). Then, according to (5-217),

$$\mathbf{u}_p = \frac{Gd^2}{2\mu}\left[\bar{y}_p - \bar{y}_p^2 - \frac{1}{3}\left(\frac{a}{d}\right)^2 \right]\mathbf{i}_x, \tag{5-218}$$

where $\bar{y}_p = y_p/d$. Thus, the particle translates slower by an amount $(1/3)(a/d)^2$ than it would do if the center of the particle simply translated with the undisturbed velocity of the fluid, evaluated at the position occupied by the particle center. It should be noted, however, that the results (5-217) and (5-218) were obtained without taking any account of the hydrodynamic interaction between the sphere and the channel walls; that is, the force on the sphere is calculated by implicitly assuming that the only effect of the channel walls is to create the parabolic profile (5-216), via the no-slip condition. This should be a reasonable approximation for $a/d \ll 1$, provided the particle is not too near either of the walls, that is, $\bar{y}_p \neq 0,1$. In this case, however, the correction in (5-218) will be very small. This suggests that it would be valuable to try to include the direct hydrodynamic effect of the walls, but this is beyond our present scope (notice, however, that we will briefly consider the interaction with a single wall later in this section).

Inertial Corrections to the Force on a Body

Another consequence of the integral theorem (5-121) is that we can calculate *inertial corrections* to the force on a body directly from the creeping flow solution. In particular, let us recall that the creeping flow equations are an approximation to the full Navier-Stokes equations obtained by taking the limit Re → 0. Thus, if we start with the full equations of motion for a steady flow in the form

$$\nabla\cdot\mathbf{T} = \text{Re}(\mathbf{u}\cdot\nabla\mathbf{u}), \tag{5-219}$$

then the creeping flow equation is obtained as the limit Re → 0. Alternatively, however, it is intuitively obvious that the creeping flow solution also can be obtained as the first term in a series approximation of the form

$$\mathbf{u} = \mathbf{u}_0 + \text{Re}\,\mathbf{u}_1 + \cdots,$$
$$\mathbf{T} = \mathbf{T}_0 + \text{Re}\,\mathbf{T}_1 + \cdots \tag{5-220}$$

for Re ≪ 1. We shall have more to say about this type of approximation scheme in

Chapter 6. For now, we simply note that substitution of (5–220) into (5–219) yields the approximation

$$\nabla \cdot \mathbf{T}_0 + \mathrm{Re}(\nabla \cdot \mathbf{T}_1 - \mathbf{u}_0 \cdot \nabla \mathbf{u}_0) + \cdots = 0. \tag{5-221}$$

Hence, if this approximation is to hold for small, but arbitrary, values of the Reynolds number, it is obvious that each term in (5–221) must individually be equal to zero, that is,

$$\boxed{\nabla \cdot \mathbf{T}_0 = 0,} \tag{5-222}$$

$$\boxed{\nabla \cdot \mathbf{T}_1 - \mathbf{u}_0 \cdot \nabla \mathbf{u}_0 = 0,} \tag{5-223}$$

and so on. The first of these equations is just the creeping flow equation. Thus, as suggested above, the creeping flow equation is the first approximation to the full equations of motion for small values of the Reynolds number, $\mathrm{Re} \ll 1$. The advantage of deriving it in the manner outlined above is that we also obtain a governing equation for the second and higher-order approximations in the series (5–220). It should be noted that the derivation of (5–222) and (5–223) does not guarantee that an approximate solution of the form (5–220) actually exists. This can only be established by showing that solutions to (5–222), (5–223), etc. exist that satisfy the boundary conditions at each order of approximation.

For present purposes, we assume that the approximation (5–220) is valid. Then, if we wished to obtain a correction to the solution of Stokes Equation (5–222) to account for the first influence of the acceleration or inertial contributions to the equations of motion, we would solve (5–223) subject to appropriate boundary conditions with $\mathbf{u}_0$ being Stokes solution. Hence, to calculate a first correction to the force on ∂D, we would expect to solve (5–223) for $\mathbf{u}_1$ and only then to calculate $\mathbf{F}_1$:

$$\mathbf{F} = \mathbf{F}_0 + \mathrm{Re}\,\mathbf{F}_1 + \cdots, \tag{5-224}$$

where

$$\boxed{\mathbf{F}_1 = \int_{\partial D} \mathbf{T}_1 \cdot \mathbf{n} \, dA.} \tag{5-225}$$

Surprisingly, however, it is not actually necessary to solve for the velocity and pressure fields, $\mathbf{u}_1$ and p_1, in order to determine the first correction to the force $\mathbf{F}_1$. Instead, by manipulating the integral theorem (5–121), we can determine $\mathbf{F}_1$ based on the solution of creeping flow problems only.

In order to see that this is true, let us identify the velocity field $\mathbf{u}$ in (5–121) as the creeping flow solution for translation of the body ∂D with unit velocity in the $\mathbf{e}$ direction. Then,

$$\int_V \mathbf{u} \cdot (\nabla \cdot \hat{\mathbf{T}}) \, dV = \int_{\partial D} \hat{\mathbf{u}} \cdot (\mathbf{T} \cdot \mathbf{n}) \, dA - \mathbf{e} \cdot \int_{\partial D} \hat{\mathbf{T}} \cdot \mathbf{n} \, dA. \tag{5-226}$$

Now let $\hat{\mathbf{u}}, \hat{\mathbf{T}}$ stand for the solution $(\mathbf{u}_1, \mathbf{T}_1)$ of (5–223), subject to the boundary conditions,

$$\mathbf{u}_1 = 0 \quad \text{on} \quad \partial D. \tag{5-227}$$

This condition derives from the fact that the boundary condition for the disturbance flow is

$$\hat{\mathbf{u}} = -\mathbf{u}^{(\infty)}(\mathbf{x}) \quad \text{on} \quad \partial D.$$

Hence, for arbitrary Re,

$$\mathbf{u}_0 = -\mathbf{u}^{(\infty)}(\mathbf{x}) \quad \text{on} \quad \partial D$$

and

$$\mathbf{u}_1 = 0 \quad \text{on} \quad \partial D.$$

Now, if we carefully examine (5–226), the first integral on the right-hand side is zero because of (5–227), while the second integral is just $\mathbf{e} \cdot \mathbf{F}_1$. Hence,

$$\boxed{\mathbf{e} \cdot \mathbf{F}_1 = -\int_V \mathbf{u} \cdot (\mathbf{u}_0 \cdot \nabla \mathbf{u}_0) dV.} \tag{5–228}$$

To calculate the component of $\mathbf{F}_1$ in the $\mathbf{e}$ direction, we therefore require the solution of the original Stokes flow problem to obtain $\mathbf{u}_0$, plus the solution of a second Stokes flow problem for translation through a quiescent fluid to obtain $\mathbf{u}$. However, we do *not* have to determine $\mathbf{u}_1, p_1$ to calculate $\mathbf{F}_1$, and this represents a substantial simplification of the problem.

The formulation leading to (5–228) has been used to great advantage in calculating inertial corrections to the motion of rigid bodies in flows with small, but nonzero, Reynolds number.[25] It must be remembered that the force $\mathbf{F}_1$ represents only a small correction to the actual force on the body. Nevertheless, the configurations of particles or bodies in Stokes flows are often indeterminate in the sense that they depend solely on the initial configuration, and thus corrections to the particle trajectory that are very small at any instant may have a major accumulative effect on the system. One example of this is the radial position of a solid spherical particle that is being carried through a circular tube in a Newtonian, incompressible fluid. In this case, as we have already seen in Section **B** of Chapter 4, the particle in creeping flow remains precisely at the radial position that was set by the initial configuration—that is, lateral or cross-stream motions (migrations) do not occur. However, if we take fluid inertia into account, there is no reason why the particle may not translate sideways, and the formula (5–228) provides a means to evaluate the lateral force based only on *creeping* flow solutions, for the original problem and for translation normal to the channel walls. Indeed, when the formula for the inertial contribution $\mathbf{F}_1$ to the force is evaluated, it is found that a lateral force does exist which drives particles away from the walls and away from the centerline, toward an intermediate, equilibrium position which is 60 percent of the way from the centerline to the walls. This same result has been observed experimentally, beginning with the famous papers of Segre and Silberberg (1962).[26]

The reader may note that the preceding developments actually provide a basis for calculating corrections to the force on bodies, ∂D, due to a number of different kinds of weak departures from the creeping flow limit for Newtonian fluids. We cite one additional example here, namely, the effects of weak non-Newtonian contributions to the motion of a body. In this case, for vanishingly small Reynolds numbers, we can symbolically write the equation of motion in the form

$$\boxed{\nabla \cdot \mathbf{T} = \mathrm{We}\,\mathbf{f(u)},} \qquad (5\text{--}229)$$

where the left-hand side is the basic Stokes operator and the right-hand side is the *non-linear* correction due to the non-Newtonian fluid properties. The parameter We is known as the *Weissenberg number* and provides a measure of the relative magnitudes of the Newtonian and non-Newtonian contributions to the fluid's behavior. Clearly, a development that mirrors the preceding analysis can be used to obtain an equation from which the first correction, $\mathrm{We}\hat{\mathbf{F}}_1$, to the force on a body can be calculated using the solution of the original creeping flow problem only, plus one additional creeping flow solution for translation through a quiescent fluid. The result, analogous to equation (5–228), is

$$\boxed{\mathbf{e} \cdot \hat{\mathbf{F}}_1 = -\int_V \mathbf{u} \cdot \mathbf{f(u_0)}\,dV.} \qquad (5\text{--}230)$$

As before, $\mathbf{u}_0$ is the creeping flow solution of the same problem, while $\mathbf{u}$ is the solution for translation with velocity $\mathbf{e}$ through a quiescent fluid. The interested reader may find examples of the application of the formula (5–230) in published papers.[27]

Hydrodynamic Interactions in Low Reynolds Number Flows

The presence of a rigid body in a Stokes flow produces a disturbance in the velocity field that decays only algebraically with increasing distance from the body. The rate of decay depends upon the type of disturbance: If there is a net hydrodynamic force on the body, the far-field disturbance flow is dominated by the Stokeslet velocity field (5–133) with a strength α that depends upon the net force, and the velocity disturbance decays as $|r|^{-1}$; if there is no net force on the body, but the undisturbed flow is linear so that the dominant disturbance mode is a rotlet or stresslet, the disturbance decays as $|r|^{-2}$, and so forth. As a consequence of the long-range perturbation to the velocity field in such a case, we may expect significant hydrodynamic interactions when other bodies or boundaries are present, even when the separation distance is relatively large. Thus, for example, to achieve a dilute suspension where each particle is hydrodynamically isolated from the others, we require an extremely small volume fraction of particles ($< 1\%$). Further, the sedimentation velocity of a solid particle is still significantly influenced by walls or other particles when separation distance is more than 10 particle radii away.

Clearly, the analysis of particle-particle and particle-wall interactions is very important. Unfortunately, analytical techniques are often very difficult to apply. Most analytical methods rely upon the existence of a coordinate system in which the particle and boundary surfaces are all coincident with coordinate surfaces, and there are very few particle-particle or particle-wall geometries that fall into this niche. The exceptions are two spheres or a sphere and a plane wall, where bispherical coordinates may be employed to obtain exact eigenfunction expansions for solution of the creeping flow equations. In fact, this coordinate system has been used to solve for the translation of two spheres along their line of centers and for a sphere near an infinite plane wall or interface. However, the bispherical coordinate system specific to these particular geometries is somewhat complicated to use, and the results for such entities as the force or torque are generally obtained in the form of an infinite series that converges very slowly when the separation distance between the surfaces are small. Thus, it is essential to have another solution procedure that does not rely upon a specific coordinate system.

The best available analytic technique, known as the *method of reflections,* is an iterative scheme that applies when the separation distance between the particles or between the particle and wall is large.[28] The basic idea is to approximate the solution as a series of terms that satisfy the creeping flow equations at each term, but only alternatively satisfy the boundary conditions at the solid surfaces. In order to illustrate the idea, let us suppose that we have two particles, A and B, that move in an unbounded fluid with velocities $\mathbf{U}_A$ and $\mathbf{U}_B$. We denote a length scale characteristic of the particles as a and denote the distance between the particles as d. The leading-order approximation to the velocity field in the vicinity of particle A is thus the creeping flow solution, $\mathbf{u}_1^A$, which vanishes at infinity and satisfies the boundary condition

$$\mathbf{u}_1^A = \mathbf{U}_A \quad \text{on} \quad \partial D_A. \tag{5-231}$$

Similarly, the leading-order approximation in the vicinity of B is the Stokes velocity field that satisfies the boundary condition

$$\mathbf{u}_1^B = \mathbf{U}_B \quad \text{on} \quad \partial D_B. \tag{5-232}$$

Now, neither $\mathbf{u}_1^A$ or $\mathbf{u}_1^B$ is exact, because neither satisfies the no-slip condition on the surface of the other particle. For example, the motion of particle A produces a disturbance velocity in the vicinity of B of order $|\mathbf{U}_A| \cdot a/d$ and the motion of B similarly produces a disturbance velocity of the fluid in the vicinity of A. Thus, to improve the approximation of the velocity field near particle A, we add to $\mathbf{u}_1^A$ a correction $\mathbf{u}_2^A$ that is a solution of the creeping flow equations that vanishes at infinity and satisfies the condition

$$\mathbf{u}_2^A = -\mathbf{u}_1^B \quad \text{on} \quad \partial D_A. \tag{5-233}$$

Near particle B, we add a correction $\mathbf{u}_2^B$ that satisfies the creeping flow equations, vanishes at infinity, and satisfies the condition

$$\mathbf{u}_2^B = -\mathbf{u}_1^A \quad \text{on} \quad \partial D_B. \tag{5-234}$$

Now, however, each of the correction fields $\mathbf{u}_2^A$ and $\mathbf{u}_2^B$ produces an additional disturbance motion near the *other* particle that is smaller again by an amount proportional to (a/d). Thus, to improve the approximation near each particle, we add velocity fields $\mathbf{u}_3^A$ and $\mathbf{u}_3^B$, which satisfy the boundary conditions

$$\mathbf{u}_3^A = -\mathbf{u}_2^B \text{ on } \partial D_A \quad \text{and} \quad \mathbf{u}_3^B = -\mathbf{u}_2^A \text{ on } \partial D_B. \tag{5-235}$$

Clearly, this procedure can be continued indefinitely. At each successive level of approximation, the magnitude of the correction decreases in proportion to (a/d). In the vicinity of ∂D_A the velocity field is approximated as

$$\mathbf{u}^A = \mathbf{u}_1^A + \mathbf{u}_2^A + \mathbf{u}_3^A + \cdots, \tag{5-236}$$

and the hydrodynamic force is similarly approximated as

$$\mathbf{F}^A = \mathbf{F}_1^A + \mathbf{F}_2^A + \mathbf{F}_3^A. \tag{5-237}$$

Similar expressions hold for the particle B.

We can now apply the preceding general analysis to the specific case of two *identical* spheres of radius a, the centers of which are separated by a distance d, as illustrated in Figure 5-13. The line connecting the particle centers is assumed to make an angle θ with the horizontal. Thus, at $\theta = 0$, the line of centers is horizontal, while for $\theta = \pi/2$ the line of centers is vertical. In the *absence* of any particle interaction, we assume that the spheres would translate vertically. At the leading order, the solution for both spheres is thus Stokes solution

$$\mathbf{u}_1^A = \mathbf{u}_S\left(\mathbf{x}; \frac{3}{4}\mathbf{U}_A^0\right) + \mathbf{u}_D\left(\mathbf{x}; -\frac{1}{4}\mathbf{U}_A^0\right) \tag{5-238a}$$

and

$$\mathbf{u}_1^B = \mathbf{u}_S\left(\mathbf{x}; \frac{3}{4}\mathbf{U}_B^0\right) + \mathbf{u}_D\left(\mathbf{x}; -\frac{1}{4}\mathbf{U}_B^0\right) \tag{5-238b}$$

Here, the superscript zero signifies that $\mathbf{U}_A^0$ and $\mathbf{U}_B^0$ are the velocities that the spheres would have in the absence of any hydrodynamic interaction between them. Since the particles are *identical*,

$$\mathbf{U}_A^0 = \mathbf{U}_B^0 = -U^0\mathbf{i}_z. \tag{5-239}$$

Hence, the velocity induced at particle A due to the motion of B and the velocity at particle B due to the motion of A are identical, that is,

$$\mathbf{u}_1^A(B) = \mathbf{u}_1^B(A) = -\frac{3}{4}\left(\frac{a}{d}\right)U^0(1 + \sin^2\theta)\mathbf{i}_z + \frac{3}{4}\left(\frac{a}{d}\right)U^0\sin\theta\cos\theta\mathbf{i}_x + 0\left(\left(\frac{a}{d}\right)^3 U^0\right). \tag{5-240}$$

But each of these disturbance velocities is just the superposition of a pair of uniform streaming flows.

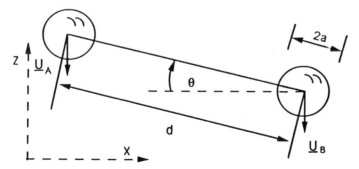

Figure 5-13 Two identical spheres of radius a, separated by a distance d, that are translating through a viscous fluid with velocity $\mathbf{U}_A = \mathbf{U}_B$. The line of centers makes an angle θ with the horizontal. For $0 < \theta < \pi/2$, the pair of spheres drifts toward the right. For $\pi/2 < \theta < \pi$, it drifts to the left.

Now, one can consider two possible consequences of the disturbance flow that is induced in the vicinity of each particle by the motion of the other particle. If the particle *velocity* is assumed to be *fixed*, in the form of equation (5–239), there will be a change in the hydrodynamic force acting on the body [that is, the external force required to produce the specified motion, (5–234), will be changed]. If, on the other hand, the *external* (and thus the hydrodynamic) *force* is held *constant*, the velocity of the particles will change, and (5–234) is then just the first approximation to the particle velocity.

In the analysis outlined earlier in this section, we essentially assumed that the velocities of the particles were fixed, and calculated the resulting hydrodynamic forces. Thus, we apply the boundary conditions (5–233) and (5–234) to determine $\mathbf{u}_2^A$ and $\mathbf{u}_2^B$, and we see that $\mathbf{u}_2^A$ and $\mathbf{u}_2^B$ are just the (creeping flow) disturbance velocities for simple translation of spheres A and B with velocities $-\mathbf{u}_1^B(A)$ and $-\mathbf{u}_1^A(B)$ through an unbounded fluid. The solutions for $\mathbf{u}_2^A$ and $\mathbf{u}_2^B$ therefore can be expressed in the same form as (5–238a) and (5–238b), namely,

$$\mathbf{u}_2^A = \mathbf{u}_S\left(\mathbf{x}; -\frac{3}{4}\mathbf{u}_1^B(A)\right) + \mathbf{u}_D\left(\mathbf{x}; \frac{1}{4}\mathbf{u}_1^B(A)\right) \tag{5–241a}$$

and

$$\mathbf{u}_2^B = \mathbf{u}_S\left(\mathbf{x}; -\frac{3}{4}\mathbf{u}_1^A(B)\right) + \mathbf{u}_D\left(\mathbf{x}; \frac{1}{4}\mathbf{u}_1^A(B)\right) \tag{5–241b}$$

The corresponding hydrodynamic force on the particles is

$$\mathbf{F} = 6\pi\mu a U^0\left(1 - \frac{3}{4}\left(\frac{a}{d}\right)(1+\sin^2\theta)\right)\mathbf{i}_z + 6\pi\mu a U^0\left(\frac{a}{d}\right)\cdot\frac{3}{4}\sin\theta\cos\theta\,\mathbf{i}_x + 0\left(\frac{a}{d}\right)^3.$$

$$\tag{5–242}$$

Hence, in order to maintain the constant velocity $-U^0\mathbf{i}_z$, it would be necessary to apply extra forces to both spheres, one with a vertical component of magnitude

$$6\pi\mu a U_0\left(\frac{3}{4}\right)\left(\frac{a}{d}\right)(1+\sin^2\theta)$$

and a horizontal component, from right to left, of magnitude

$$6\pi\mu a U_0\left(\frac{3}{4}\right)\left(\frac{a}{d}\right)\sin\theta\cos\theta.$$

The other approach to analysis, as indicated earlier, is to hold the force on the body constant and allow its velocity to vary. In this case, we assume

$$U_A = U_A^0 + \left(\frac{a}{d}\right)U_A^1 + \cdots,$$

$$U_B = U_B^0 + \left(\frac{a}{d}\right)U_B^1 + \cdots, \tag{5–243}$$

and the boundary conditions (5–233) and (5–234) are replaced by

$$\mathbf{u}_2^B = -\mathbf{u}_1^A(B) + \mathbf{U}_B^1\left(\frac{a}{d}\right) \quad \text{on } \partial D_B, \tag{5-244}$$

$$\mathbf{u}_2^A = -\mathbf{u}_1^B(A) + \mathbf{U}_A^1\left(\frac{a}{d}\right) \quad \text{on } \partial D_A. \tag{5-245}$$

Then, in place of the solutions (5–241), we find

$$\mathbf{u}_2^A = \mathbf{u}_S\left(\mathbf{x}; -\frac{3}{4}\left(\mathbf{u}_1^B(A) - \mathbf{U}_A^1\left(\frac{a}{d}\right)\right)\right) + \mathbf{u}_D\left(\mathbf{x}; \frac{1}{4}\left(\mathbf{u}_1^B(A) - \mathbf{U}_A^1\left(\frac{a}{d}\right)\right)\right)$$

and

$$\mathbf{u}_2^B = \mathbf{u}_S\left(\mathbf{x}; -\frac{3}{4}\left(\mathbf{u}_1^A(B) - \mathbf{U}_B^1\left(\frac{a}{d}\right)\right)\right) + \mathbf{u}_D\left(\mathbf{x}; \frac{1}{4}\left(\mathbf{u}_1^A(B) - \mathbf{U}_B^1\left(\frac{a}{d}\right)\right)\right), \tag{5-246}$$

and the hydrodynamic force on the spheres is

$$\mathbf{F}_B = \left[6\pi\mu a U^0\left(1 - \frac{3}{4}\left(\frac{a}{d}\right)(1 + \sin^2\theta)\right)\right]\mathbf{i}_z + 6\pi\mu a \mathbf{U}_B^1\left(\frac{a}{d}\right)$$

$$- 6\pi\mu a U^0 \cdot \frac{3}{4}\cdot\left(\frac{a}{d}\right)\sin\theta\cos\theta\,\mathbf{i}_x, \tag{5-247}$$

$$\mathbf{F}_A = -\left[6\pi\mu a U^0\left(1 - \frac{3}{4}\left(\frac{a}{d}\right)(1 + \sin^2\theta)\right)\right]\mathbf{i}_z + 6\pi\mu a \mathbf{U}_A^1\left(\frac{a}{d}\right)$$

$$- 6\pi\mu a U^0 \cdot \frac{3}{4}\cdot\left(\frac{a}{d}\right)\sin\theta\cos\theta\,\mathbf{i}_x.$$

Now, if the net force on the spheres is assumed to be fixed (thus equal to the hydrodynamic force in the absence of interactions $-6\pi\mu a U^0\mathbf{i}_z$), it follows that

$$\mathbf{U}_A^1 = \mathbf{U}_B^1 = U^0\left[-\frac{3}{4}(1 + \sin^2\theta)\mathbf{i}_z + \frac{3}{4}\sin\theta\cos\theta\,\mathbf{i}_x\right]. \tag{5-248}$$

Thus, the overall motion is

$$\boxed{\mathbf{U}_A = \mathbf{U}_B = U^0\left(-1 - \frac{3}{4}\left(\frac{a}{d}\right)(1 + \sin^2\theta)\right)\mathbf{i}_z + \frac{3}{4}\left(\frac{a}{d}\right)U^0\sin\theta\cos\theta\,\mathbf{i}_x.} \tag{5-249}$$

The two spheres move at the *same* velocity to this order of approximation, but the configuration shown in Figure 5-13 drifts toward the *right* as it sediments vertically downward. The exceptional cases are $\theta = 0, \pi/2$. For $\theta = 0$, the two spheres lie in the horizontal plane and sediment vertically downward with a speed

$$U^0\left(1 + \frac{3}{4}\left(\frac{a}{d}\right)\right). \tag{5-250}$$

On the other hand, for $\theta = \pi/2$, the line of centers is vertical and the two spheres again sediment vertically downward with a speed

$$U^0\left(1 + \frac{3}{2}\left(\frac{a}{d}\right)\right). \tag{5-251}$$

The reader may note that the far-field hydrodynamic interactions cause the vertical configuration to sediment slightly faster than the horizontal configuration. This is qualitatively similar to the fact that a vertical rod falls faster than the same rod in a horizontal orientation [see equations (5–201) and (5–203)].

Homework Problems

1. Consider the case of a motionless liquid layer meeting a plane vertical rigid wall as sketched in the figure below. The liquid makes an angle θ with the solid wall, known as the static contact angle, which is a property of the materials involved—the solid, the liquid, and the gas. Derive the relationship between the curvature of the gas-liquid interface and the height of the interface $\zeta(y)$. What are the appropriate boundary conditions for the unknown function $\zeta(y)$. Determine the height h to which the liquid climbs at the rigid wall.

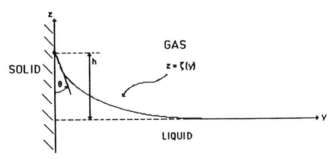

2. A rigid sphere of radius a rests on a flat surface, and a small amount of liquid surrounds the point of contact making a concave-planar lens whose diameter is small compared with a. The angle of contact θ_c with each of the solid surfaces is zero (see Problem 1), and the tension in the air-liquid surface is α. Show that there is an adhesive force of magnitude $4\pi a\alpha$ acting on the sphere. (The fact that this adhesive force is independent of the volume of liquid is noteworthy. Note also that the force is repulsive when $\theta_c = \pi$.)

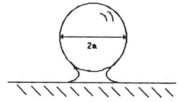

3. A soda straw of inside diameter d is placed in a pan of liquid of density ρ.

The surface tension between the water and the air is α, and the water makes a contact angle θ at the water-straw interface.

a. Show that the water rises in the straw to an equilibrium height H_0,

$$H_0 = \frac{2\alpha\cos\theta}{\rho g a}.$$

State any assumptions that may be necessary to obtain this result.

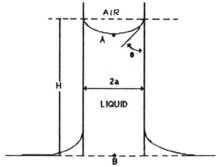

b. Consider the motion of capillary rise from an initial state $H = 0$ to the final equilibrium condition $H = H_0$. Show that the rate at which the capillary will rise is

$$\frac{dH}{dt} = \frac{a^2}{8\mu}\frac{(-\Delta p)}{H},$$

where

$$\Delta p \equiv P_B - P_A$$

assuming that the velocity profile at any instant of time is given by the *steady* Poiseuille profile. Under what conditions is this assumption valid?

c. Express the pressure drop, $-\Delta p$, in terms of α, θ, a, ρ, g, and H, and then show that the dimensionless rate of capillary rise can be expressed in the form

$$\frac{d\bar{H}}{d\bar{t}} = \frac{1-\bar{H}}{\bar{H}},$$

where $\bar{H} \equiv H/H_0$ and $\bar{t} = t/t_c$, where t_c is the characteristic time to attain the equilibrium height H_0. Obtain t_c in terms of the dimensional parameters of the problem.

d. Obtain an exact solution for $\bar{H}$ as a function of $\bar{t}$.

e. Show that when $\bar{H} \ll 1$ (that is, for short times),

$$\bar{H} = \sqrt{2}\, \bar{t}^{1/2}.$$

From this result, demonstrate that the dimensional height of the liquid column, H, increases as the radius of the tube increases, that is, $H \sim a^{1/2}$. Explain the reason for this result in physical terms.

f. At large times, when $\bar{H} \approx 1$, show that

$$\bar{H} = 1 - e^{-\bar{t}}.$$

This implies that

$$H \sim \frac{1}{a},$$

that is, H decreases as a increases. How can you explain the apparent discrepancy between (e) and (f)?

4 A spherical drop of viscosity $\hat{\mu}$, density $\hat{\rho}$ and radius a translates under the action of gravity with a velocity U through a stationary liquid of viscosity μ and density ρ. Assume that the creeping flow approximation is valid for the fluid's motion, both inside and outside the drop. Assume, in addition, that there are no surfactants present and the system is isothermal so that the interfacial tension is uniform on the drop surface.

a. Use the eigenfunction expansion approach outlined in Chapter 4 to obtain the solution (5–30) and (5–31). What is the drop velocity U in terms of the other parameters?

b. Solve the same problem using the general method of Section F in Chapter 4, in which the solutions are constructed via the superposition of vector harmonic functions.

5. A viscous drop of viscosity $\hat{\mu}$ and density $\hat{\rho}$ is carried along in the unidirectional motion of an incompressible, Newtonian fluid of viscosity μ and density $\rho \equiv \hat{\rho}$ between two infinite plane walls. The radius of the undeformed drop is denoted as a, and the distance between the walls is d. We assume that the capillary number, $\delta \equiv a\mu G/\sigma$, is small so that the drop deformation is also small. Here, σ is the interfacial tension, and G is the mean shear rate of the undisturbed flow.

The flow configuration is sketched below. The coordinate direction normal to the plane walls is specified as x_3, and the velocity of the center of mass is denoted as $\mathbf{U}_S$. Further, the undisturbed velocity, pressure, and stress are denoted, respectively, as $(\mathbf{V}, Q, \mathbf{T})$. In the most general case

$$\mathbf{V} = (\alpha + \beta x_3 + \gamma x_3^2)\mathbf{e}_1 - \mathbf{U}_S,$$

where we have adopted a coordinate frame that moves with the drop velocity, $\mathbf{U}_S$.

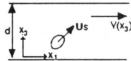

Now, if the Reynolds number of the flow is sufficiently small for the creeping motion approximation to apply, it can be shown by the arguments of Section **B** in Chapter 4, that no lateral motion of the drop is possible unless the drop deforms. In other words, $U_S = U_S e_1$ in this case, though, of course, U_S is not generally equal to the undisturbed velocity of the fluid evaluated at the x_3 position of the drop center. The drop may either lag or lead (in principle) due to a combination of interaction with the walls and the hydrodynamic effect of the quadratic form of the undisturbed velocity profile—see Faxen's law. Because the drop deforms, however, lateral migration can occur even in the complete absence of inertia (or non-Newtonian) effects. In this problem, our goal is to formulate two possible ways to evaluate the lateral velocity of the drop: first, via a full solution of the problem to $0(\delta)$, and second, via the reciprocal theorem that involves only velocity and stress fields for motion of the *undeformed* drop.

a. Let us denote the velocity, pressure, and stress fields outside and inside the drop as (U, P, S) and $(\tilde{U}, \tilde{P}, \tilde{S})$, respectively. Further, let us denote the drop surface in terms of F,

$$F \equiv r - 1 - f = 0.$$

Suppose that $\delta \ll 1$, so that all of the velocity, pressure, and stress fields can be expressed in terms of an asymptotic expansion of the form

$$U = U_0 + \delta U_1 + \ldots ,$$

$$\tilde{U} = \tilde{U}_0 + \delta \tilde{U}_1 + \ldots ,$$

and

$$F \equiv r - 1 - \delta f_1 + \ldots = 0.$$

Derive dimensionless equations and boundary conditions whose solution would be sufficient to determine the drop velocity (and shape) to $0(\delta)$. Use the method of domain perturbations to express all boundary conditions at the deformed drop interface in terms of equivalent conditions at the spherical surface of the undeformed drop.

b. Instead of solving part (a) directly to terms of $0(\delta)$, the option is to apply the reciprocal theorem. In order to apply the reciprocal theorem to obtain $U_S^{(1)}$, it is only necessary to solve the $0(1)$ problem from part (a), and the "complementary" creeping flow problem of a drop translating perpendicular to the walls through a quiescent Newtonian fluid. Write the governing equations and boundary conditions for translation of a *spherical* drop with velocity e_3 normal to the two plane walls. The velocity, pressure, and stress fields for this complementary problem should be denoted as (u, q, t) and $(\tilde{u}, \tilde{q}, \tilde{t})$ for the fluid outside and inside the drop, respectively.

Note: To apply the reciprocal theorem, the shape of the drop for the complementary problem generally would have to be exactly the *same* as the shape in the problem of interest. However, since the original problem can be reduced via domain perturbations to an equivalent problem with the boundary conditions applied at the spherical surface, $r = 1$, we may also conveniently choose the drop to be spherical for this complementary problem.

c. Beginning with the reciprocal theorem for the disturbance velocity fields in the fluid volume outside the undeformed drop surface,

$$\int_{V_f} [(\nabla \cdot \mathbf{S} - \nabla \cdot \mathbf{T}) \cdot (\mathbf{u} + \mathbf{e}_s) - \nabla \cdot \mathbf{t} \cdot (\mathbf{U} - \mathbf{V})] dV = 0,$$

and the corresponding expression for the fluid volume inside the under-deformed drop show that

$$\int_{A_d} [(\mathbf{S} - \kappa \tilde{\mathbf{S}}) \cdot \mathbf{u} - (\mathbf{t} - \kappa \tilde{\mathbf{t}}) \cdot \mathbf{U} - \kappa \tilde{\mathbf{t}} \cdot (\mathbf{U} - \tilde{\mathbf{U}}) - \mathbf{T} \cdot \mathbf{u} + \mathbf{t} \cdot \mathbf{V}] \cdot \mathbf{n} dA,$$

where A_d is the spherical surface of the undeformed drop and κ is the viscosity ratio $\kappa \equiv \hat{\mu}/\mu$.

Derive an expression for the drop velocity normal to the plane walls at $0(\delta)$, and express this formula entirely in terms of the stress and velocity fields $\mathbf{S}, \tilde{\mathbf{S}}, \mathbf{U}, \tilde{\mathbf{U}}$ at $0(1)$, plus the shape function $f^{(1)}$ at $0(\delta)$. Show that $f^{(1)}$ can be calculated from the velocity and stress fields at $0(1)$.

[(Reference: Chan and Leal, *JFM* 92, 131–170 (1979).]

6. a. A long solid cylinder of length $2L$ and radius a is held stationary in a pure straining flow, the undisturbed velocity of which is given by $u_1 = Gx_1$, $u_2 = -Gx_2$, and $u_3 = 0$. The center of the cylinder is at $x_1 = L/4$, $x_2 = 0$, $x_3 = 0$. The slenderness ratio, $\epsilon \equiv a/L$, is asymptotically small, $\epsilon \ll 1$. Obtain an expression for the force on the cylinder.

b. Determine the hydrodynamic torque [up to $O(\ln \epsilon)^{-2}$] exerted on a slender particle that is rotating with a constant angular velocity $\mathbf{\Omega}$ through a quiescent Newtonian fluid.

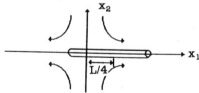

c. We consider the motion of a slender torus through a quiescent Newtonian fluid. The radius of the torus is R, and its cross-sectional radius is a. We assume $a/R \ll 1$. Apply slender-body analysis to show that the ratio in force for motion parallel to the torus (say, x_1) and broadside (say, x_3) is

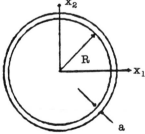

$$\frac{F_3}{F_1} = \frac{4}{3} + O(\ln \epsilon)^{-1}.$$

7. Consider a solid spherical particle that is held stationary in a parabolic shear flow. In dimensionless terms

$$\mathbf{u}_\infty = (x^2 + y^2)\mathbf{i}_z,$$

where the spatial variables x and y are nondimensionalized using the sphere radius. Solve for the velocity field in the creeping flow limit by means of a superposition of a Stokeslet, a potential dipole, a Stokes quadrapole, and a potential octupole, all located at the center of the sphere. What is the force on the sphere? Is this consistent with the solution of Problem 6 in Chapter 4?

8. We consider an axisymmetric, prolate ellipsoid whose surface is given by

$$\frac{x^2}{a^2} + \frac{r^2}{b^2} = 1, \text{ where } r^2 = y^2 + z^2 \text{ and } a \geq b.$$

The foci of the ellipsoid are located at $x = \pm c$, where

$$c = (a^2 - b^2)^{1/2}.$$

Suppose that the spheroid is held fixed in a simple shear flow,

$$\mathbf{u}_\infty = Gy\mathbf{e}_x.$$

Show that the disturbance velocity field necessary to satisfy no-slip boundary conditions at the surface of the ellipsoid can be expressed in terms of a line distribution of stresset, rotlet, and potential quadrapole singularities of the form

$$\mathbf{u} = \int_{-c}^c (c^2 - \xi^2) \left[\alpha \mathbf{U}_{SS}(\mathbf{x} - \boldsymbol{\xi}; \mathbf{e}_x, \mathbf{e}_y) + \gamma \mathbf{U}_R(\mathbf{x} - \boldsymbol{\xi}; \mathbf{e}_x; \mathbf{e}_a) \right] d\xi$$

$$+ \beta \int_{-c}^c (c^2 - \xi^2) \frac{\partial}{\partial y} \mathbf{U}_D(\mathbf{x} - \boldsymbol{\xi}; \mathbf{e}_x) d\xi,$$

where α, γ, and β are constants. What is the torque on the ellipsoid?

   Note: The integrals can all be expressed in terms of the function $B_{m,n}(\mathbf{x})$, defined by

$$B_{m,n}(\mathbf{x}) \equiv \int_{-c}^c \frac{\xi^n d\xi}{|\mathbf{x} - \boldsymbol{\xi}|^m} \quad (n = 0, 1, 2, \ldots; \ m = -1, 1, 3, 5, \ldots).$$

A recurrence formula that proves useful is

$$B_{m,n} = -\frac{C^{n-1}}{m-2} \left(\frac{1}{R_2^{m-2}} + \frac{(-1)^n}{R_1^{m-2}} \right) \quad (n \geq 2)$$

$$+ \frac{n-1}{m-2} B_{m-2,n-2} + x B_{m,n-1},$$

where

$$R_1 = \left[(x+c)^2 + r^2 \right]^{1/2}, \ R_2 = \left[(x-c)^2 + r^2 \right]^{1/2}$$

and

$$B_{1,0} = \log \frac{R_2 - (x-c)}{R_1 - (x+c)}, \quad B_{1,1} = R_2 - R_1 + xB_{1,0},$$

$$B_{3,0} = \frac{1}{r^2}\left(\frac{x+c}{R_1} - \frac{x-c}{R_2}\right), \quad B_{3,1} = \left(\frac{1}{R_1} - \frac{1}{R_2}\right) + xB_{3,0}.$$

[Chwang and Wu, *JFM* 67:787–815 (1975).]

9. We wish to consider the effect of interface contamination by surfactant on the deformation of a drop in a steady linear flow at very low Reynolds number, that is, $\mathbf{u}_\infty = G[(1/2)\boldsymbol{\omega}\wedge\mathbf{x} + \mathbf{E}\cdot\mathbf{x}]$. The local interface concentration of surfactant is denoted as Γ^*, specified in units of mass per unit of interfacial area. We assume that Γ^* is small so that a linear relationship exists between Γ^* and interfacial tension, σ,

$$\sigma_S - \sigma = \Gamma^* RT,$$

where R is the gas constant, T is the absolute temperature, and σ_S is the interfacial tension of the clean interface without surfactant ($\Gamma^* = 0$). In the absence of motion, the surfactant concentration at the interface is uniform, Γ_0.

a. Derive *dimensionless* forms of the governing creeping flow equations and boundary conditions for the problem, assuming that the surfactant can be approximated as *insoluble* in the bulk fluids, so that the distribution of surfactant at the interface is governed by a balance between advection with the surface velocity $\mathbf{u}_S'$, diffusion with interface diffusivity D_S, and changes in the local concentration due to stretching and distortion of the interface; namely, in dimensional terms,

$$\frac{\partial \Gamma^*}{\partial t'} + \nabla_S' \cdot [\Gamma^* \mathbf{u}_S' - D_s \nabla_S' \Gamma^*]$$

$$+ \Gamma^* (\nabla_S' \cdot \mathbf{n})(\mathbf{u}' \cdot \mathbf{n}) = 0.$$

To define a dimensionless surface concentration, use $\Gamma = \Gamma^*/\Gamma_0$. The dimensionless parameters which should appear upon nondimensionalization are

$$\lambda = \frac{\hat{\mu}}{\mu}, \quad \mathrm{Ca}_s = \frac{\mu Ga}{\sigma_2}, \quad \beta = \frac{\Gamma_0 RT}{\sigma_s},$$

and

$$\mathrm{Pe}_s = \frac{Ga^2}{D_s}.$$

The latter is known as the interface Peclet number and provides a measure of the relative importance of convection and diffusion in determining the surfactant distribution on the interface.

Note: In dimensionless terms, the uniform equilibrium concentration of surfactant is $\Gamma = 1$, and the corresponding interfacial tension is

$$\sigma^* = \sigma_s(1 - \beta).$$

For some purposes, it is more convenient to compare results obtained for the full problem with results for the uniformly contaminated surface, σ^*, rather than the clean interface value σ_s. Thus, express all results in terms of the capillary number Ca*, based on σ^*, by noting Ca* $= \text{Ca}_s/(1 - \beta)$.

b. Let us now consider the changes in drop shape produced by increases of the local shear rate G. For this purpose, it is convenient to define

$$\gamma \equiv \frac{\sigma^* a}{\mu D_s}$$

so that $\text{Pe}_s = \gamma \text{Ca}^*$, where γ is a parameter that depends only on material properties. We consider the limit

$$\text{Ca}^* \ll 1 \text{ and } \gamma = 0(1) \text{ (i.e. Pe}_s \ll 1)$$

so that both the drop shape and the surfactant concentration distribution are only slightly perturbed from their equilibrium values. For the general linear flow considered here, the first *corrections* to the description of the surface shape and the surfactant distribution are expected to take the forms

$$r = 1 + \text{Ca}^* b_r(t) \frac{\mathbf{x} \cdot \mathbf{E} \cdot \mathbf{x}}{r^2}$$

and

$$\Gamma = 1 + \gamma \text{Ca}^* b_\Gamma(t) \frac{\mathbf{x} \cdot \mathbf{E} \cdot \mathbf{x}}{r^2}.$$

Use the general representation of solutions for creeping flows in terms of vector harmonic functions to solve for the velocity and pressure fields in the two fluids, as well as the deformation and surfactant concentration distribution functions *at steady state*. You should find

$$b_r = \frac{5}{4} \frac{(16 + 19\lambda) + 4\beta\gamma/(1 - \beta)}{10(1 + \lambda) + 2\beta\gamma/(1 - \beta)},$$

plus a comparable expression for b_Γ.

c. Discuss your solution in physical terms. The classical limit of a drop in the absence of surfactant, Equation (5–93), is obtained for both $\beta \to 0$ and $\gamma \to 0$. Explain this. Is the deformation increased or decreased in the presence of surfactant, at a fixed dimensionless shear rate Ca*? Explain.

10. Consider the flow in the long rectangular two-dimensional, cavity-shaped region as sketched below. The upper surface is moved at constant velocity **U**, thereby generating a flow in the cavity. Notice that the cavity is closed so that no mass may enter or leave the chamber. Clearly, near the end walls,

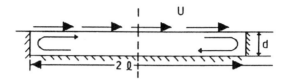

the fluid motion may be rather complicated. However, away from the end walls, near the middle of the chamber, it appears to be a good approximation to treat the flow as unidirectional.

a. Nondimensionalize, and demonstrate that the governing equations reduce to unidirectional form in the limit $A = d/l \ll 1$. Solve for the velocity distribution and the pressure for the central region in this limit. You should find that the pressure gradient is nonzero, even though it is zero for the unbounded domain problem of the motion induced by tangential motion of an infinite solid wall. Does this make sense? Explain.

b. A very similar problem is the flow induced by motion of a lower plane surface, in a shallow cavity whose upper surface is a gas-liquid interface. The interfacial tension is a constant, equal to γ. Assume that $A = d/l \ll 1$, and nondimensionalize the governing equations and boundary conditions. Calculate the velocity and pressure fields for this case in the central region assuming that the capillary number, $Ca = \mu U/\gamma$, is small, $Ca \ll 1$. Use this solution to calculate the interface shape to $0(Ca)$ assuming that the liquid sticks to the sharp edge at the end walls so that

$$h(\pm l) = d.$$

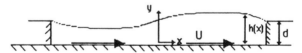

In calculating the interface shape, you may assume that the pressure distribution is valid over the whole liquid layer, all the way to the end walls.

c. Now, consider the related problem of the flow induced in a shallow, two-dimensional cavity that is differentially heated as shown.

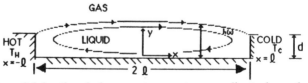

The differential heating induces a temperature gradient along the gas-liquid interface, which in turn induces a surface-tension gradient along the interface. Thermocapillary flows of this general type are known to be important in the containerless processing of single crystals. Consider, for example, the configuration shown below in which a cylindrical solid passes through a heating coil, melts, and then resolidifies into a single crystal of high quality.

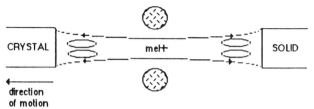

CRYSTAL   melt   SOLID

direction
of motion

In view of this connection, and its possible application to materials processing in space, we wish to consider thermocapillary flow in the thin two-dimensional cavity in the *absence* of gravity (that is, no buoyancy effects to drive motion in the fluid layer).

i. Formulate and nondimensionalize the problem of thermocapillary flow in a shallow two-dimensional cavity, employing the following assumptions where necessary.

a) At steady state, the heat transfer process as the gas-liquid interface reduces to the balance

$$k_1(\mathbf{n}\cdot\nabla T) + k_g(T - T_g) = 0,$$

where k_1 is the thermal conductivity of the liquid, k_g is the heat transfer coefficient in the gas, $\mathbf{n}$ is the normal to the interface, and T_g is the temperature of the gas phase.

b) The temperature distribution in the gas is known to be

$$T_g = \frac{T_H + T_c}{2} - \frac{T_H - T_c}{2l}x.$$

c) The bottom of the cavity is insulated so that

$$\mathbf{n}\cdot\nabla T = 0$$

(that is, the heat flux out the bottom of the cavity is zero).

d) The interfacial tension is prescribed in terms of the temperature as

$$\gamma(T) = \gamma_0 - \beta\left[T - \frac{1}{2}(T_H + T_c)\right],$$

where γ_0 is the interfacial tension of the liquid at the reference temperature $(1/2)(T_H + T_c)$.

e) The thermal energy equation in the liquid—namely, equation (2–50), with $\mathbf{q}$ given by (2–70)—can be approximated as

$$\rho C_p\left(u\frac{\partial T}{\partial x} + v\frac{\partial T}{\partial y}\right) = k_1\left(\frac{\partial^2 T}{\partial x^2} + \frac{\partial^2 T}{\partial y^2}\right).$$

f) The interface will be slightly deformed relative to the flat surface

$$h(x) = d.$$

g) The characteristic velocity is determined by the balance between the viscous stress and the tangential stress produced by the surface tension gradient at the interface.

Dimensionless parameters that should appear in your analysis are

$$A = \frac{d}{l}, \qquad \text{Re} = \frac{\beta A (T_H - T_c) d}{\mu \nu} \quad \text{(Reynolds number)},$$

$$M = \frac{\beta A (T_H - T_c) d}{\mu \kappa} \quad \text{(Marangoni number)},$$

$$\text{Ca} = \frac{\beta A (T_H - T_c)}{\gamma_0} \quad \text{(Capillary number)},$$

$$L = \frac{k_g d}{k_1} \quad \text{(Biot number)}.$$

Based upon your nondimensionalized version of the governing equations and boundary conditions, discuss the physical significance of these dimensionless groups.

ii. Determine the velocity, temperature and pressure distributions, together with the shape of the interface, in the central region away from the end walls for the asymptotic limit $A \to 0$. Assume that the interface sticks to the sharp edge at the end walls. Furthermore, consider the following asymptotic limits of the other dimensional parameters:

$$R = \bar{R} A, \quad M = \bar{M} A, \quad C = \bar{C} A^4, \quad L = 0(1),$$

with $\bar{R}$, $\bar{M}$, and $\bar{C}$ all $0(1)$. In this case, to a leading level of approximation, the solution consists of a parallel flow in the core with a flat interface. It may be noted that the leading order approximations for temperature distribution and the interface shape both satisfy the natural boundary conditions to $x = \pm l$. Determine your solution to $0(A)$ in all unknowns, including the interface shape.

References/Notes

1. An excellent review of work through 1972 is Harper, J.F., "The Motion of Bubbles and Drops Through Liquids," *Adv. in Applied Mech.* 12:59–129 (1972).

2. Scriven, L.E., "Dynamics of a Fluid Interface," *Chem. Eng. Sci.* 12:98–108 (1960).

3. Taylor, T.D., and Acrivos, A., "On the Deformation and Drag of a Falling Viscous Drop at Low Reynolds Number," *J. Fluid Mech.* 18:466–476 (1964).

4. Kojima, M., Hinch, E.J., and Acrivos, A., "The Formation and Expansion of a Toroidal Drop Moving in a Viscous Fluid," *Physics of Fluids* 27:19–32 (1984).

5. Koh, C.J., and Leal, L.G., "The Stability of Drop Shapes for Translation at Zero Reynolds Number Through a Quiescent Fluid," *Phys. Fluids A.* 1:1309–1313 (1989).

6. Frumkin, A.N., and Levich, V.G., *Zhur Fiz. Khim.* 21:1183 (1947) (in Russian). This work, as well as related research on the motion of drops and bubbles in fluids, is summarized

in the textbook (translated from the Russian) V.G. Levich, *Physicochemical Hydrodynamics.* Prentice-Hall: Englewood Cliffs, NJ (1962).

7. Davis, R.E., and Acrivos, A., "The Influence of Surfactants on the Creeping Motion of Small Bubbles," *Chem. Eng. Sci.* 21:681–685 (1966).

Holbrook, J.A., and LeVan, M.D., "Retardation of Droplet Motion by Surfactant," Part 1, "Theoretical Development and Asymptotic Solutions," *Chem. Eng. Commun.* 20:191–207 (1983); Part 2, "Numerical Solutions for Exterior Diffusion, Surface Diffusion, and Adsorption Kinetics, *Chem. Eng. Commun.* 20:273–290 (1983).

8. The textbook by Levich is referenced in 6 above.

9. A useful compilation of research on bubble and drop dynamics in the context of space experiments and space processing of materials can be found in the proceedings of a series of colloquia sponsored by NASA.

a. *Proceedings of the International Colloquium on Drops and Bubbles*, D.J. Collins, M.S. Plesset, and M.M. Saffren (eds.). NASA: JPL (1976).

b. *Proceedings of the Second International Colloquium on Drops and Bubbles*, D.H. Le Croissette (ed.). JPL Publication 82–87, March 1, 1982.

c. *Proceedings of the Third International Colloquium on Drops and Bubbles*, ed. Taylor G. Wang, AIP Conference Proceedings, *197* (1989).

Additional papers that contain reference to materials processing applications in low-gravity environments (space) include

d. Barton, K.D., and Subramanian, R.S., "The Migration of Liquid Drops in a Vertical Temperature Gradient," *J. Coll. Interface Sci.* 133:211–222 (1989).

e. Merritt, R.M., and Subramanian, R.S., "Migration of a Gas Bubble Normal to a Plane Horizontal Surface in a Vertical Temperature Gradient," *J. Coll. Interface Sci.* 131: 514–525 (1989).

f. Subramanian, R.S., "The Motion of Bubbles and Drops in Reduced Gravity," in *Transport Processes in Bubbles, Drops and Particles.* R.P. Chhabra and D. DeKee (eds.), Hemisphere: New York (1990).

10. This problem was first considered by Young, N.O., Goldstein, J.S., and Block, M.J., "The Motion of Bubbles in a Vertical Temperature Gradient," *J. Fluid Mech.* 6:350–356 (1959).

11. See Reference 10.

12. See Reference 2 of Chapter 3.

13. A large number of papers have been written on the boundary-integral technique. Among them, good general references are

a. Youngren, G.K., and Acrivos, A., "Stokes Flow Past a Particle of Arbitrary Shape: A Numerical Method of Solution," *J. Fluid Mech.* 69:377–403 (1975).

b. Weinbaum, S., Ganatos, P., and Yan, Z.Y., "Numerical Multipole and Boundary Integral Equation Techniques in Stokes Flow," *Ann. Rev. of Fluid Mech.* 22:275–316 (1990).

14. Some representative examples of this type of application of the boundary-integral method are

a. Rallison, J.M., "A Numerical Study of the Deformation and Burst of a Viscous Drop in General Shear Flows," *J. Fluid Mech.* 109:465–482 (1981).

b. Stone, H.A., and Leal, L.G., "Relaxation and Breakup of an Initially Extended Drop in an Otherwise Quiescent Fluid," *J. Fluid Mech.* 198:399–427 (1989).

c. Stone, H.A., and Leal, L.G. "The Influence of Initial Deformation on Drop Breakup in Subcritical Time-Dependent Flow at Low Reynolds Numbers," *J. Fluid Mech.* 206:223–263 (1989).

d. Stone, H.A., and Leal, L.G., "Breakup of Concentric Double Emulsion Droplets in Linear Flows," *J. Fluid Mech.* 211:123–156 (1990).

15. Hancock, G.J., "The Self-Propulsion of Microscopic Organisms Through Liquids," *Proc. Royal Soc.* A217:96–121 (1953).

16. Ladyzhenskaya, O.A., *The Mathematical Theory of Viscous Incompressible Flow.* Gordon & Breach: New York (1963).

17. Chwang, A.T., and Wu, T.Y., "Hydromechanics of Low-Reynolds Number Flow," Part 1, "Rotation of Axisymmetric Prolate Bodies," *J. Fluid Mech.* 63:607–622 (1974).
Part 2, "Singularity Method for Stokes Flows," *J. Fluid Mech.* 67:787–815 (1975).
Part 4, "Translation of Spheroids," *J. Fluid Mech.* 75:677–689 (1975).
  Chwang, A.T., "Hydromechanics of Low-Reynolds Number Flow," Part 3, "Motion of a Spheroidal Particle in Quadratic Flows," *J. Fluid Mech.* 72:17–34 (1975).
18. a. Batchelor, G.K., "Slender-Body Theory for Particles of Arbitrary Cross-Section in Stokes Flow," *J. Fluid Mech.* 44:419–440 (1970).
  b. Cox, R.G., "The Motion of Long Slender Bodies in a Viscous Fluid," Part 1, "General Theory," *J. Fluid Mech.* 44:791–810 (1970); Part 2, "Shear Flow," *J. Fluid Mech.* 45:625–657 (1971).
  c. Keller, J.B., and Rubinow, S.T., "Slender-Body Theory for Slow Viscous Flow," *J. Fluid Mech.* 75:705–714 (1976).
  d. Johnson, R., "An Improved Slender-Body Theory for Stokes Flow," *J. Fluid Mech.* 99:411–431 (1980).
  e. Yang, S.M., and Leal, L.G., "Particle Motion in Stokes Flow Near a Plane Fluid-Fluid Interface," Part 1, "Slender Body in a Quiescent Fluid," *J. Fluid Mech.* 136:393–421 (1983).
19. See Reference 17.
20. Kim, S., "Singularity Solutions for Ellipsoids in Low-Reynolds Number Flows: With Applications to the Calculation of Hydrodynamic Interactions in Suspensions of Ellipsoids," *Int. J. Multiphase Flow* 12:469–491 (1986).
21. Chapter 3 of the book: Kim, S., and Karrila, S.J., *Microdynamics: Principles and Selected Applications,* Butterworth-Heinemann, Boston (1991).
22. See Cox, Reference 18.
23. Lorentz, H.A., "Ein Allgemeiner Satz, die Bewegung einer Reibendon Flüssigkeit Betreffend, Nebst Einigen Anwendungen Desselben" (a general theorem concerning the motion of a viscous fluid and a few consequences derived from it), *Versl. Kon. Akad. Wetensch. Amsterdam* 5:168–174 (1896).
24. The original reference to this work was Faxen's Ph.D. thesis, Uppsala Univ., Uppsala, Sweden (1921).
25. A review of applications of this idea is given in Reference 5 of Chapter 4.
26. See Reference 3 of Chapter 4.
27. See Reference 25 above and the references cited therein.
28. An alternative description of the application of the method of reflections can be found in the textbook by Happel and Brenner (see Reference 2 of Chapter 3).

Asymptotic Approximations for Unidirectional, One-Dimensional, and Nearly Unidirectional Flows

Although the full Navier-Stokes equations are nonlinear, we have studied a number of problems in the preceding chapters in which the flow was either geometrically constrained so that the nonlinear terms $\mathbf{u} \cdot \nabla \mathbf{u}$ were identically equal to zero or characterized by very small Reynolds numbers so that the nonlinear terms could be approximated as zero compared to viscous and pressure-gradient terms. In either case, the governing equations were linearized, and this allowed a wide range of different analytical techniques to be used to obtain solutions.

In general, however, neither the constraint to unidirectional flow nor the creeping flow approximation will be applicable, and we must return to the original, nonlinear equations of motion. In such cases, the analytic methods of the preceding chapters do not apply because they rely either explicitly or inherently upon the so-called *superposition principle,* according to which a sum of solutions of the governing equations is still a solution. In fact, no generally applicable analytic method exists for the solution of nonlinear partial differential equations. Nevertheless, it is often possible to obtain approximate solutions for such problems, applicable for very large or very small values of the dimensionless parameters that characterize a particular problem.

For example, if we consider the motion of an unbounded incompressible, Newtonian fluid past a solid body, such as a sphere or cylinder, then nondimensionalization was shown previously to lead to the conclusion that the problem is completely characterized by a single dimensionless parameter, the Reynolds number. Solutions cannot be obtained to such a problem for arbitrary values of the Reynolds number. However, it is often possible to achieve *approximate* solutions when the Reynolds number is either very small or very large.

In fact, we have already seen that the limit of very *small* Reynolds numbers will generally allow solutions via the creeping flow equations. However, our view of the creeping flow equations to this point has been essentially as an ad hoc approximation, where the nonlinear terms in the Navier-Stokes equations are neglected completely. Later, we shall see that the creeping flow solution actually represents a *first* approximation, in a systematic approximation scheme that applies for small, but nonzero, values of the Reynolds number. Such approximation schemes are known as *asymptotic* methods, in recognition of the fact that they apply for limiting (or asymptotic) values of dimensionless parameters. The generalization from creeping flow to a systematic asymptotic approximation scheme for small, but nonzero, Reynolds numbers will be discussed in detail in Chapter 8. It is, perhaps, not surprising that the limit of very small Reynolds numbers should lead to an approximate analytic solution scheme, since the first approximation of the Navier-Stokes equations in this limit completely neglects the nonlinear terms. What may be more surprising is that the same methods that allow asymptotic approximations at low Reynolds numbers also provide a basis for approximate solutions in the limit of very large Reynolds numbers. Readers with a prior knowledge of fluid mechanics or transport phenomena may recognize this limit as the domain of boundary-layer theory. However, we are getting ahead of our story.

In the next two chapters, we begin the transition toward more general, approximate solution methods. Rather than starting directly with the problem areas just described, however, we begin with the application of asymptotic methods to unidirectional flows, and to some close cousins involving one-dimensional (but not unidirectional) flows and flows that deviate only slightly from unidirectional geometries. This class of problems is interesting and quite important in its own right, especially the near-unidirectional flow geometries, because these lead to a general discussion of lubrication phenomena. However, it also provides a convenient framework for the introduction of asymptotic approximation methods.

We shall begin with a problem that was analyzed in Chapter 3, namely, pulsatile flow in a circular tube with a periodic pressure gradient. In particular, we begin by showing how asymptotic methods can be applied to obtain both the high and low frequency approximations, $R_\omega \ll 1$ and $R_\omega \gg 1$, for this problem. Although the exact solution and its low frequency approximation were given in Chapter 3, we shall see here that approximate solutions for both the high and low frequency limits can be achieved much more easily by *direct approximation of the governing equations* (the basis of asymptotic methods), rather than first solving the exact equation and then trying to find approximate forms of the exact solution. Following this, in the second section, we briefly consider the general features of asymptotic approximations. Finally, in the remainder of this chapter and in Chapter 7, we apply the same methods to nearly unidirectional and one-dimensional flows.

A. Pulsatile Flow in a Circular Tube Revisited— Asymptotic Solutions for High and Low Frequencies

The analysis leading from the governing equation (3–134) for pulsatile flow in a circular tube to the exact solution for large times, (3–159), was straightforward, requiring only the recognition of Bessel's equation of order zero (3–156) and its general solutions $J_0(z)$ and $Y_0(z)$. However, the evaluation of this solution for any arbitrary R_ω requires considerable effort, and we thus considered only the limiting case $R_\omega \ll 1$. The asymptotic form for $R_\omega \gg 1$ is very difficult to obtain from the exact solution.

In this section we show that asymptotic solutions for $R_\omega \ll 1$ or $R_\omega \gg 1$ at large times can be obtained much more easily if we look for them by directly approximating (3–149), rather than first deriving the exact solution and then trying to deduce its limiting forms. The basis of this direct, asymptotic approximation of the governing equations for $R_\omega \ll 1$ (or $R_\omega \gg 1$) is to neglect terms that become asymptotically small (compared with other terms in equations) in these limiting cases. For this purpose, it is essential that the problem is formulated in correct dimensionless form so that we can determine the relative magnitudes of terms by their dependence (or lack of dependence) on R_ω.

In the present case, where the basic governing equation is linear, the asymptotic analysis serves only to simplify the solution procedure, for example, by avoiding the need to deal with Bessel's equation when $R_\omega \ll 1$. Later, however, we shall see that the same basic methods may often allow approximate analytic solutions to be obtained for nonlinear problems, even when no exact solution is possible.

Asymptotic Solution for $R_\omega \ll 1$.

Let us begin by considering the case $R_\omega \ll 1$. It is sufficient to consider the problem (3–149) and (3–150) for $G(\bar{r})$, that is,

$$\frac{d^2G}{d\bar{r}^2} + \frac{1}{\bar{r}}\frac{dG}{d\bar{r}} - iGR_\omega = -1, \tag{6–1}$$

$$G(\bar{r}) = 0 \quad \text{at} \quad \bar{r} = 1,$$

$$G \text{ bounded at } \bar{r} = 0. \tag{6–2}$$

In Chapter 3, we derived a general *exact* solution of this problem in terms of Bessel functions J_0 for arbitrary R_ω and then obtained an approximate form for $R_\omega \ll 1$ by approximating this solution. Instead, in the present section, let us suppose from the outset that $R_\omega \ll 1$ and try to seek an approximate solution directly by approximating (6–1) and (6–2).

Assuming that the scaling in (6–1) is correct, terms that do not contain R_ω explicitly are independent of R_ω. Thus, in the limit as $R_\omega \to 0$, equation (6–1) reduces to the approximate form

$$\frac{d^2G}{d\bar{r}^2} + \frac{1}{\bar{r}}\frac{dG}{d\bar{r}} = -1. \tag{6–3}$$

The solution of this equation subject to the boundary conditions (6–2) is just

$$G = \frac{1}{4}(1 - \bar{r}^2), \tag{6–4}$$

and if we express this solution in terms of $u_z^{(1)}$ using (3–148), we obtain

$$u_z^{(1)} = \frac{1}{4}\sin\bar{t}(1 - \bar{r}^2). \tag{6–5}$$

Thus, in the limit $R_\omega \to 0$, the problem reduces to a quasi-steady Poiseuille flow with an instantaneous pressure gradient $\sin \bar{t}$. In view of the analysis in Chapter 3, this result is not surprising, but we do note that the solution (6-4) was easier to obtain in this case where we *directly approximate the differential equation* rather than first solving the exact problem and then approximating the solution.

Of course, the limiting case (6-3) contains no influence of inertia. In order to determine the effects of inertia for very small, but nonzero, values of R_ω, we look for an approximate solution of (6-1) in the form of an asymptotic expansion in which successive terms are proportional to $R_\omega{}^n$ for $n = 0, 1, 2, \dots$,

$$\boxed{G = G_0(\bar{r}) + R_\omega G_1(\bar{r}) + R_\omega^2 G_2(\bar{r}) + 0(R_\omega^3).} \tag{6-6}$$

Evidently if such a solution exists, the first term, $G_0(\bar{r})$, must be just (6-4). Although it is already known from the exact solution in Chapter 3 that the approximation for small R_ω should take this form, we shall see later that we can determine both the form of asymptotic expansions and the specific functions G_0, G_1, G_2, and so on without any prior knowledge of the exact solution. The form proposed in (6-6) is called a *regular perturbation* (or *asymptotic*) *expansion* of $G(\bar{r})$ for $R_\omega \ll 1$. We shall discuss the properties of such expansions in more detail in the next section. Here we simply note that the expansion is called *regular* because it is assumed that the same form holds throughout the domain, $0 \le \bar{r} \le 1$. We also remark that the convergence of the right-hand side of (6-6) to $G(\bar{r})$ is strictly a consequence of the limit $R_\omega \to 0$ so that each successive term can be made arbitrarily small compared with the terms before it. This includes the error of $0(R_\omega^3)$, which can thus be made arbitrarily small compared with the terms that we have retained in (6-6). *Asymptotic convergence* of this type does not imply convergence to G for some fixed R_ω as the number of terms retained increases, as is true, for example, in power-series approximations of a function. Indeed, for fixed R_ω, we would not necessarily achieve any better approximation to G by the addition of more terms in (6-6), but we shall discuss these questions later.

If we now substitute (6-6) into the governing equation (6-1) for G, we obtain

$$\frac{d^2 G_0}{d\bar{r}^2} + R_\omega \frac{d^2 G_1}{d\bar{r}^2} + R_\omega^2 \frac{d^2 G_2}{d\bar{r}^2} + \frac{1}{\bar{r}} \frac{dG_0}{d\bar{r}} +$$

$$R_\omega \frac{1}{\bar{r}} \frac{dG_1}{d\bar{r}} + R_\omega^2 \frac{1}{\bar{r}} \frac{dG_2}{d\bar{r}} - iG_0 R_\omega - iG_1 R_\omega^2 + 0(R_\omega^3) = -1. \tag{6-7}$$

The symbol $0(R_\omega^3)$ represents all terms proportional to R_ω^3 and higher-order terms that have been neglected in writing (6-7) and is known as the *error* term since it represents the largest of the terms neglected in (6-7) by truncating (6-6) after only three terms. Substituting (6-6) into the boundary conditions (6-2), we obtain

$$G_0(1) + R_\omega G_1(1) + R_\omega^2 G_2(1) + 0(R_\omega^3) = 0. \tag{6-8}$$

Now, the parameter R_ω is assumed to be asymptotically small but is otherwise arbitrary; that is, the equalities in (6-7) and (6-8) must be satisfied for any small, but arbitrary, value of R_ω. Thus, if we rewrite (6-7) in the form

$$\left[1 + \frac{1}{\bar{r}}\frac{d}{d\bar{r}}\left(\bar{r}\frac{dG_0}{d\bar{r}}\right)\right] + R_\omega\left(\frac{1}{\bar{r}}\frac{d}{d\bar{r}}\left(\bar{r}\frac{dG_1}{d\bar{r}}\right) - iG_0\right) +$$

$$R_\omega^2\left(\frac{1}{\bar{r}}\frac{d}{d\bar{r}}\left(\bar{r}\frac{dG_2}{d\bar{r}}\right) - iG_1\right) + 0(R_\omega^3) = 0, \quad (6\text{-}9)$$

it is obvious that the terms at each level in R_ω must individually be equal to zero. Similarly, if (6-8) is to hold for arbitrary R_ω, it is evident that $G_0(1) = G_1(1) = G_2(1) = 0$. Thus, as we have already noted, the function G_0 satisfies (6-3) plus the condition $G_0(1) = 0$ and is given by (6-4). The functions $G_1(\bar{r})$ and $G_2(\bar{r})$ can be seen from (6-8) and (6-9) to satisfy

$$\frac{1}{\bar{r}}\frac{d}{d\bar{r}}\left(\bar{r}\frac{dG_1}{d\bar{r}}\right) = iG_0 \text{ with } G_1(1) = 0 \text{ and } G_1(0) \text{ bounded} \quad (6\text{-}10)$$

and

$$\frac{1}{\bar{r}}\frac{d}{d\bar{r}}\left(\bar{r}\frac{dG_2}{d\bar{r}}\right) = iG_1 \text{ with } G_2(1) = 0 \text{ and } G_2(0) \text{ bounded.} \quad (6\text{-}11)$$

With $G_0(\bar{r})$ given by (6-4), the solution to (6-10) is obtained easily by integrating twice with respect to $\bar{r}$ and applying boundary conditions. The result is

$$G_1 = \frac{i}{16}\left(\bar{r}^2 - \frac{\bar{r}^4}{4} - \frac{3}{4}\right). \quad (6\text{-}12)$$

Similarly, with G_1 given by (6-12), the equation and boundary conditions (6-11) can be solved easily to give

$$G_2 = -\frac{1}{196}\left(\bar{r}^4 - \frac{\bar{r}^6}{9} - 3\bar{r}^2 + \frac{19}{9}\right). \quad (6\text{-}13)$$

Evidently, the same procedure could be used to obtain as many terms as we like in the expansion (6-6). In the present development we stop with G_2. Substituting (6-4) for G_0, plus (6-12) and (6-13) into the asymptotic expansion (6-6), we obtain

$$G = \frac{1}{4}(1 - \bar{r}^2) + R_\omega\frac{i}{16}\left[\bar{r}^2 - \frac{\bar{r}^4}{4} - \frac{3}{4}\right]$$

$$+ R_\omega^2\frac{1}{196}\left(\frac{\bar{r}^6}{9} - \bar{r}^4 + 3\bar{r}^2 - \frac{19}{9}\right) + 0(R_\omega^3). \quad (6\text{-}14)$$

Now, according to the connection between $\hat{u}_z$ and $u_z^{(1)}$ from Chapter 3,

$$u_z^{(1)} = -\mathcal{Re}\left[ie^{i\bar{t}}G(\bar{r})\right].$$

Substituting for $G(\bar{r})$ from (6–14), we obtain

$$
u_z^{(1)} = \frac{1}{4}\sin\bar{t}(1 - \bar{r}^2) - \frac{1}{16}\cos\bar{t}(R_\omega)\left\{\frac{3}{4} - \bar{r}^2 + \frac{\bar{r}^4}{4}\right\}
$$
$$
+ \frac{1}{196}\sin\bar{t}(R_\omega^2)\left(\frac{\bar{r}^6}{9} - \bar{r}^4 + 3\bar{r}^2 - \frac{19}{9}\right) + 0(R_\omega^3). \tag{6–15}
$$

The first two terms in this solution are identical to (3–124), which was obtained from the exact solution by asymptotic expansion of $J_0[(R_\omega/i)^{1/2}\bar{r}]$ for $R_\omega \ll 1$.

There are several important remarks to make concerning the analysis leading to (6–15). First, by introducing the asymptotic form (6–6) at the beginning and solving the approximate form of the governing equation (6–9), the analysis has been simplified significantly. For example, at each level of approximation in the asymptotic approach, the ordinary differential equations can be solved directly with two integrations with respect to $\bar{r}$, and one requires no knowledge of the Bessel functions $J_0(z)$ or $Y_0(z)$. Second, the solution that was generated (and, indeed, the whole analysis) is precisely equivalent to a solution of (6–1) and (6–2) by the method of successive approximations in which $G_0(\bar{r})$ is obtained by neglecting altogether the term proportional to R_ω in (6–1). This equivalence between successive approximation and asymptotic expansion methods always exists when the asymptotic expansion is regular (that is, valid at all points in the solution domain). It is thus evident that a regular asymptotic expansion will always proceed in increasing powers of the small parameter that appears either in the differential equations or the boundary conditions, and this fact allows us to anticipate the form (6–6) without the necessity of having an exact solution available for comparison. Finally, it is important to reiterate the role played by nondimensionalization and the choice of characteristic scales in the asymptotic solution procedure that we have outlined. Specifically, if we return to (6–1), we have implicitly assumed that the term iGR_ω is arbitrarily small in the limit $R_\omega \to 0$ compared with the other terms in the equation for all $\bar{r}$ in the flow domain, $0 \le \bar{r} \le 1$. This will clearly be true provided the characteristic scales that are inherent in (6–1)—namely $t_c = 1/\omega$, plus u_c and l_c from (3–133)—provide correct measures of velocity, length, and time scales at all points in the domain. For, in this case, the magnitude of the terms $d^2G/d\bar{r}^2$ and $1/\bar{r}(dG/d\bar{r})$ will be independent of R_ω. Thus, as $R_\omega \to 0$, we can be sure that iGR_ω will become arbitrarily small compared with these terms. If, on the other hand, the characteristic scales listed above were not relevant everywhere in the domain, the magnitude of terms like $d^2G/d\bar{r}^2$ would no longer necessarily be independent of R_ω, and we could make a mistake if we neglected iGR_ω relative to $d^2G/d\bar{r}^2$ even in the limit $R_\omega \to 0$. Fortunately in this case, the scaling and solutions that we have generated are uniformly valid throughout the domain.

Of course, we can never change a physical (or mathematical) problem by simply nondimensionalizing variables, no matter what the choice of scale factors. It is only when we attempt to simplify a problem by neglecting some terms compared to others on the basis of nondimensionalization that the correct choice of characteristic scales becomes essential. Fortunately, as we shall see, an incorrect choice of characteristic scales resulting in incorrect approximations of the equations or boundary conditions will always become apparent by the appearance of some inconsistency in the asymptotic solution scheme. The main cost of incorrect scaling is thus lost labor (depending on how far we must go to expose the inconsistency for a particular problem), rather than errors in the solution.

Asymptotic Solution for $R_\omega \gg 1$.

The next question that we explore is whether the same asymptotic methods that were outlined in the preceding subsection for $R_\omega \ll 1$ could prove equally useful in obtaining explicit analytical results for the asymptotic limit, $R_\omega \to \infty$.

Thus, we again consider (6–1) and (6–2) but now for the limit $R_\omega \gg 1$. In this case, we might expect the acceleration term to be larger than the viscous term, which is just the opposite of the limit $R_\omega \ll 1$. However, it appears as though the acceleration term, iGR_ω, also becomes larger than the pressure gradient term in (6–1). But this is clearly impossible! The motion of the fluid exists only because of the pressure gradient, and the acceleration *cannot* exceed the pressure gradient. Instead, we should expect that the acceleration and pressure gradient terms remain in balance even as $R_\omega \to \infty$. Examining (6–1), we see that this can only occur if the magnitude of G decreases as R_ω increases according to

$$G(\bar{r}) = \frac{1}{R_\omega} \bar{G}(\bar{r}).$$

In this case, (6–1) takes the form

$$\boxed{i\bar{G}(\bar{r}) - 1 = \frac{1}{R_\omega}\left[\frac{1}{\bar{r}}\frac{d}{d\bar{r}}\left(\bar{r}\frac{d\bar{G}}{d\bar{r}}\right)\right]} \tag{6-16}$$

for $R_\omega \gg 1$. Then, in the limit $R_\omega \to \infty$,

$$i\bar{G} = 1 \quad \text{or} \quad \bar{G} = -i, \tag{6-17}$$

and this is consistent with the expectation of a balance between acceleration and pressure gradient terms.

We saw in the previous example for $R_\omega \ll 1$ that the approximate solution (6–4), obtained by taking the limit $R_\omega \to 0$ in the exact equation (6–1), was just the first term in an asymptotic solution for $R_\omega \ll 1$. Here, the approximate equation and solution (6–17) that we obtain by taking the limit $R_\omega \to \infty$ in (6–16) should also be recognized as the first term in a formal asymptotic expansion of $\bar{G}$ for $R_\omega \gg 1$, that is,

$$\bar{G} = \bar{G}_0(\bar{r}) + \frac{1}{R_\omega}\bar{G}_1(\bar{r}) + 0(R_\omega^{-2}). \tag{6-18}$$

Now, since $\bar{G}_0 = -i$, it follows that

$$\boxed{G = -\frac{i}{R_\omega} + 0(R_\omega^{-2}),} \tag{6-19}$$

where the error estimate corresponds to the second term in the expansion (6–18). Thus, referring to (6–17) and (6–18), it follows that

$$
u_z^{(1)} = -\frac{1}{R_\omega} \cos\bar{t} + 0(R_\omega^{-2}). \tag{6-20}
$$

Thus, equation (6–19) yields a uniform (plug) flow that is periodic in time but lags $\pi/2$ radians behind the imposed pressure gradient ($\sim \sin\bar{t}$) as a first approximation to the velocity field for $R_\omega \gg 1$. The fact that the velocity should lag behind the pressure gradient as $R_\omega \to \infty$ is not surprising, as noted in the dicussion in Chapter 3, because the balance in (6–1) is increasingly between the pressure gradient and acceleration terms for large R_ω.

In spite of the fact that some features of the result (6–20) seem reasonable, however, one feature is definitely wrong. The solution (6–19) or (6–20) does *not* satisfy the no-slip boundary conditions $u_z^{(1)} = 0$ at $\bar{r} = 1$. This means that it cannot represent a uniformly valid approximate solution of the original problem, either (3–143) with $u_z^{(1)} = 0$ at $\bar{r} = 1$ or (6–14) and (6–20). What has gone wrong?

The source of the difficulty is seen easily by reexamining the governing differential equation (6–16). In this form, the limiting process, $R_\omega \to \infty$, seems to imply that the viscous terms should be negligible everywhere compared with the acceleration and pressure gradient terms that appear on the left-hand side. But the neglect of viscous terms reduces (6–16) from a second-order ordinary differential equation whose solution can be expected to satisfy both the boundary and boundedness conditions on G to an algebraic equation whose solution cannot satisfy these boundary conditions. From a physical point of view, the problem is that the limiting form of (6–16) as $R_\omega \to \infty$ neglects viscous terms, and it is momentum transfer associated with these terms that is responsible for the fact that the fluid does not slip at the boundary (the no-slip condition). It is therefore understandable that the resulting solution does not satisfy the no-slip boundary condition at the tube wall, and this failure indicates that there is something wrong with the limiting process $R_\omega \to \infty$ when applied to (6–16).

The nature of the problem is illustrated in Figure 6–1, where we have plotted velocity profiles from the exact solution, (3–159), for various values of R_ω. We see that as R_ω increases, the velocity profile does become increasingly blunt in the center of the tube, and the change in $u_z^{(1)}$ with respect to $\bar{r}$ occurs over an increasingly short distance near the tube walls. It thus appears that the problem with (6–16) is that the scaling inherent in its nondimensionalization, namely,

$$
l_c = R,
$$

is not characteristic of the velocity gradients in the vicinity of the tube walls for $R_\omega \gg 1$. The same conclusion can be reached without recourse to examination of the exact solution for R_ω. For, if the scaling were correct, then the terms $d^2\bar{G}/d\bar{r}^2$ and $1/\bar{r}\,(d\bar{G}/d\bar{r})$ in (6–16) would be $0(1)$ as actually assumed in the limiting process that led to the solution (6–19) [or (6–20)]. But this scaling cannot be correct, at least near the walls, because the solution obtained does not satisfy the no-slip condition. Apparently, viscous terms must remain important there even as $R_\omega \to \infty$ since otherwise the

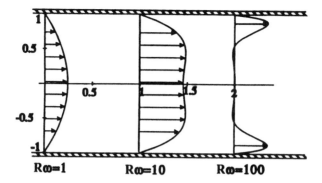

Figure 6–1 Velocity profiles for pulsatile flow in a circular tube for three different values of R_ω, all plotted at $\bar{t} = \pi/2$.

no-slip condition cannot be satisfied. The characteristic scale $l_c = R$ and the resulting solution (6–19) is perfectly reasonable for most values of $\bar{r}$ in the tube, but neither the solution nor the scaling can be correct near the tube wall.

Let us then reconsider the near-wall region, beginning again with equation (6–16). Before proceeding, it is convenient to introduce a change of variables

$$y = 1 - \bar{r} \quad \text{or} \quad \bar{r} = 1 - y$$

so that

$$\frac{1}{R_\omega}\left(\frac{d^2\bar{G}}{dy^2} - \frac{1}{(1-y)}\frac{d\bar{G}}{dy}\right) - i\bar{G} = -1 \tag{6-21}$$

and $y = 0$ at the tube wall. Now, all of the preceding discussion indicates that viscous terms must remain important near $y = 0$ even as $R_\omega \to \infty$. Thus,

$$\frac{1}{R_\omega}\left(\frac{d^2\bar{G}}{dy^2}\right) \quad \text{and/or} \quad \frac{1}{R_\omega}\frac{d\bar{G}}{dy}$$

cannot be small compared to unity as suggested by (6–21); that is, derivatives of $\bar{G}$ with respect to y must become large in this region as $R_\omega \to \infty$. But, if derivatives of $\bar{G}$ do become large, it can only mean that $\bar{G}$ must vary over a much smaller length scale than $l_c = R$, and, thus, a different nondimensionalization should be introduced into the governing equations in this near-wall region.

Rather than starting with the original dimensional equations and searching for an appropriate characteristic length scale for the near-wall region, the correct form can be determined by simply *rescaling* the previously nondimensionalized equation (6–21). To do this, let us introduce a new independent spatial variable,

$$\boxed{Y = yR_\omega^\alpha,} \tag{6-22}$$

where α is a constant that we will determine later. The motivation for this rescaling can be explained in at least two ways. The simplest idea is that R_ω represents a ratio of the two "natural" length scales of the problem, $l_c \equiv R$ and $L \equiv \nu/\omega R$, so that rescaling according to (6–22) simply redefines the radial variable scaled with respect to a new length scale, say, l^*. To see this, we can rewrite (6–22) in the form

$$Y = \left(\frac{y'}{l_c}\right)\left(\frac{l_c}{L}\right)^\alpha,$$

where y' is the dimensional radial variable. Evidently, if $\alpha = 1$, then

$$Y = \frac{y'}{L},$$

and $L \equiv \nu/\omega R$ would be the new length scale l^* for nondimensionalizing y'. On the other hand, for some other α,

$$Y = \frac{y'}{l^*},$$

where

$$l^* = (l_c)^{1-\alpha} L^\alpha.$$

Thus, rescaling in the form (6–22) is precisely equivalent to introducing a new characteristic length scale l^* in the near-wall region. Of course, to completely specify l^* we need to determine α, and we shall see shortly how this can be done. A second way to motivate (6–22) is to simply note that we are *stretching* the variable normal to the wall in such a way that $\bar{G}$ changes from 0 to its mainstream value in an increment of the *rescaled* variable $\Delta Y = 0(1)$, whereas in the original nondimensionalized form, $\bar{G}$ apparently changes over a dimensionless distance $\Delta y = 0(R_\omega^{-\alpha})$. Thus, in the rescaled system, all spatial derivatives, say, $d^2\bar{G}/dY^2$, will be $0(1)$ and independent of R_ω in the near-wall region, and this will simplify the problem of deciding which terms in the governing equations are to be retained in that part of the domain as $R_\omega \to \infty$.

The coefficient α and the relevant form of the governing equations in the near-wall region are determined by substitution of (6–22) into (6–21), which thus becomes

$$\frac{1}{R_\omega}\left[R_\omega^{2\alpha}\frac{d^2\bar{G}}{dY^2} - R_\omega^\alpha\left[\frac{1}{1 - YR_\omega^{-1}}\right]\frac{d\bar{G}}{dY}\right] - i\bar{G} = -1. \tag{6–23}$$

It can be seen that the largest viscous term is $0(R_\omega^{2\alpha-1})$, and thus it follows from (6–23) that

$$2\alpha - 1 = 0$$

if viscous effects are to be equally important as acceleration and pressure gradient effects in the near-wall region for $R_\omega \to \infty$. Thus

$$\boxed{\alpha = \frac{1}{2}} \tag{6–24}$$

and

$$\boxed{\frac{d^2\bar{G}}{dY^2} - i\bar{G} + 1 = R_\omega^{-1/2}\frac{d\bar{G}}{dY} + 0(R_\omega^{-3/2}).} \tag{6–25}$$

If we go back to the definition of the rescaling (6–22), we see that the region $\Delta Y = 0(1)$

is actually very thin compared with the radius of the tube. Indeed, in terms of y (which is scaled with respect to R), the near-wall region is only $0(R_\omega^{-1/2})$ in dimension for $R_\omega \gg 1$. This very thin region near the wall where viscous effects are important is called a *boundary layer*. We shall see many other examples of boundary layers in later chapters of this book. Since $d^2\bar{G}/dY^2$ and $d\bar{G}/dY$ are $0(1)$ in the near-wall region, we see that a first approximation to the governing equation in the boundary layer can be obtained by letting $R_\omega \to \infty$ in (6–25).

Before going further with this analysis, it will probably be helpful to recapitulate what we have shown, starting with the original nondimensionalized version of the problem for G, equation (6–16) plus the boundary condition $G(1) = 0$. We have seen that the limiting form of (6–16) for $R_\omega \gg 1$ represents a balance between the pressure gradient and acceleration terms with the viscous terms neglected altogether and has a leading-order solution (6–19) or (6–20) that does not satisfy boundary conditions at the walls. We surmise from this that the scaling, $l_c = R$, inherent in (6–16) is not valid in the vicinity of the tube walls. Thus, we search for a new scaling that is consistent with the fact that viscous terms cannot be negligible close to the walls if the no-slip condition is to be satisfied. This new length scale is inherent in the rescaling (6–22) with $\alpha = 1/2$, and we see from this that the region where viscous effects are important is only

$$\Delta y \sim 0(R_\omega^{-1/2}) \quad \text{for} \quad R_\omega \to \infty.$$

In this region, the relevant approximation to the governing equation for G for $R_\omega \gg 1$ is (6–25). Thus, as illustrated in Figure 6–2, the solution domain, $0 \leq \bar{r} \leq 1$, splits into two parts for $R_\omega \gg 1$: One is the interior region away from the walls where $l_c = R$ and the appropriate nondimensional form of the governing equations is (6–16); the other

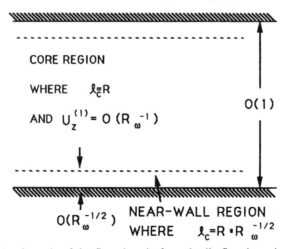

Figure 6–2 A schematic of the flow domain for pulsatile flow in a circular tube at very large values of R_ω. In the core region, the velocity field is characterized by a length scale $l_c = R$ and the velocity field is dominated by inertia (acceleration) effects due to the time-dependent pressure gradient. In the near-wall region, on the other hand, the characteristic length scale for changes in velocity is much shorter, $0(R \cdot R_\omega^{-1/2})$, and viscous effects remain important even for very large values of R_ω.

is a narrow region near the wall of dimension $0(R_\omega^{-1/2})$ relative to the tube radius where the relevant nondimensionalized form of the governing equation is (6–25). If we develop an asymptotic solution of (6–16) for $R_\omega \gg 1$, such as (6–17), we cannot expect it to satisfy boundary conditions at the tube walls because the limiting form of (6–16) is not valid within the region of $0(R_\omega^{-1/2})$ near the tube walls. On the other hand, an asymptotic solution of (6–25) for $R_\omega \gg 1$ will satisfy boundary conditions at the tube wall but cannot ordinarily be expected to give the correct solution outside the narrow, boundary-layer region. Problems in which an asymptotic approximation to the solution requires two (or more) distinct expansions, each valid in different parts of the domain, are called *singular,* and the solutions are known as *singular* or *matched asymptotic expansions.* The word *matched* comes from the observation that two (or more) approximations of the solution to a problem, each valid in different parts of the domain, must have the same functional form in any region of space where their individual domains of validity overlap. We shall see later that this concept of *matching* plays a crucial role in determining solutions of the singular perturbation type.

Let us now return to the solution of our problem for $R_\omega \gg 1$. Although the arguments leading to (6–25) were complex, the resulting equation itself is simple compared with the original Bessel's equation. Our objective here is an asymptotic approximation of the solution for the boundary-layer region. In general, we may expect an asymptotic expansion of the form

$$\bar{G}(Y) = \bar{G}_0(Y) + \frac{1}{R_\omega^m}\bar{G}_1(Y),$$

where m is a positive coefficient that must be determined as part of the solution procedure. Here, we seek only the first term in this approximate solution form, corresponding to the fact that we have also obtained only a leading-order approximation to the solution in the interior region away from the walls. Since $d^2\bar{G}/dY^2$ and $d\bar{G}/dY$ are $0(1)$ in the near-wall region, the governing equation for the leading-order term in the asymptotic expansion can be obtained by letting $R_\omega \to \infty$ in (6–25). The result is

$$\frac{d^2\bar{G}_0}{dY^2} - i\bar{G}_0 + 1 = 0,$$

and the general solution of this equation is

$$\bar{G}_0 = -i + A'\exp\left(\frac{\sqrt{2}}{2}Y\right)\exp\left(\frac{\sqrt{2}}{2}iY\right) + B\exp\left(-\frac{\sqrt{2}}{2}Y\right)\exp\left(-\frac{\sqrt{2}}{2}iY\right). \quad (6\text{–}26)$$

The boundary condition $\bar{G}_0(0) = 0$, then requires that

$$B = i - A$$

or

$$\bar{G}_0 = i\left(-1 + \exp\left(-\frac{\sqrt{2}}{2}Y\right)\exp\left(-\frac{\sqrt{2}}{2}iY\right)\right)$$
$$+ A\left[\exp\left(\frac{\sqrt{2}}{2}Y\right)\exp\left(\frac{\sqrt{2}}{2}iY\right) - \exp\left(-\frac{\sqrt{2}}{2}Y\right)\exp\left(-\frac{\sqrt{2}}{2}iY\right)\right]. \quad (6\text{–}27)$$

The constant A cannot be determined from the boundary condition at the wall but must be obtained from the *matching* requirement that (6-27) reduce to the form of the core solution (6-17) in the region of overlap between the boundary layer and the interior region. Now, any arbitrarily large, but finite, value of Y will fall within the boundary-layer domain; on the other hand, the corresponding value of y can be made arbitrarily small in the asymptotic limit $R_\omega \to \infty$. Thus, the condition of matching is often expressed in the form

$$[\text{B.L.SOLN}]_{Y \gg 1} \Longleftrightarrow [\text{INTERIOR SOLN}]_{y \ll 1} \text{ for } R_\omega \to \infty.$$

Although this notation might appear confusing at first, the limiting formulas $Y \gg 1$ and $y \ll 1$, respectively, are intended to serve as a reminder that the region of overlapping validity for $R_\omega \to \infty$ corresponds to large (but finite) values of Y and small (but non-zero) values of y. The implication of the matching formula is that the functional forms of the solution in the different parts of the domain must be the same, to within an error that is asymptotically small in the limit $R_\omega \to \infty$. The magnitude of the error in matching, expressed in the form of an estimate proportional to R_ω^{-n}, depends on the number of terms that have been evaluated in the asymptotic expansions for the solution in the different parts of the domain. In the present case, we have calculated only a single term in both the boundary-layer and interior solutions, and the matching condition can be expressed in the form

$$\frac{1}{R_\omega} \bar{G}_0 |_{Y \gg 1} = -\frac{i}{R_\omega} \text{ as } R_\omega \to \infty, \tag{6-28}$$

where the right-hand side is the leading-order approximation for G in the interior of the tube and the left-hand side is the boundary-layer solution (6-27), including the R_ω^{-1} scaling from (6-16). It can be seen from the condition (6-28) that

$$A = 0.$$

It may be noted that the boundary-layer solution with this value for A does not match perfectly with the core solution, $-i/R_\omega$. There is a mismatch in the term

$$\frac{i}{R_\omega} \left[\exp\left(-\frac{\sqrt{2}}{2} Y \right) \exp\left(-\frac{\sqrt{2}}{2} iY \right) \right].$$

However, this mismatch is asymptotically small in the limit $R_\omega \to \infty$, as can be seen by expressing it in terms of the original radial variable r using (6-22), that is,

$$\frac{i}{R_\omega} \left[\exp\left(-\frac{\sqrt{2}}{2} R_\omega r \right) \exp\left(-i\frac{\sqrt{2}}{2} R_\omega r \right) \right] \text{ for } R_\omega \to \infty.$$

This small mismatch need not concern us at this stage in the solution scheme since we have so far considered only the first, leading-order approximation for $\bar{G}$ in the boundary layer and in the core [see equation (6-18)]. With $A = 0$, the final solution for $\bar{G}$ can be expressed in the form

$$\bar{G} = i \left[-1 + \exp\left(-\frac{\sqrt{2}}{2} Y \right) \left(\cos\frac{\sqrt{2}}{2} Y - i\sin\frac{\sqrt{2}}{2} Y \right) \right]. \tag{6-29}$$

Thus, in the boundary-layer region, it follows from (6–1) and (6–2) that

$$
\begin{aligned}
u_z^{(1)} = \frac{1}{R_\omega}\Bigg[&-\cos\bar{t}\left\{1-\exp\left(-\frac{\sqrt{2}}{2}Y\right)\cos\frac{\sqrt{2}}{2}Y\right\} \\
&+\sin\bar{t}\left\{\exp\left(-\frac{\sqrt{2}}{2}Y\right)\sin\frac{\sqrt{2}}{2}Y\right\}\Bigg],
\end{aligned}
\tag{6–30}
$$

where

$$
Y = yR_\omega^{1/2} = R_\omega^{1/2}(1-\bar{r}).
$$

In the rest of the tube, we have [equation (6–20)]

$$
u_z^{(1)} = -\frac{1}{R_\omega}\cos\bar{t}.
$$

Thus, near the wall, there is both an *in-phase* and *out-of-phase* contribution to the velocity field, whereas the approximation to the velocity field in the interior has no in-phase component. Both the in-phase and out-of-phase contributions to (6–30) are modestly complicated in form. We plot their spatial variation in Figure 6–3 for a single, fixed $\bar{t}$. In the region away from the walls, viscous effects are negligible to the leading order of approximation, and the acceleration of the fluid is in-phase with the pressure gradient (that is, the velocity lags behind the pressure gradient by 180°). No simple explanation for the complicated wavelike character of the solution in the boundary-layer region is possible.

In any case, our point in presenting this high frequency asymptotic solution is primarily as an introduction to the concepts of singular (or matched) asymptotic expansions rather than as a basis for a physical explanation of the flow itself. The reader should not be discouraged if there are detailed questions about the asymptotic techniques that we have discussed in this section that he or she does not yet fully understand. The

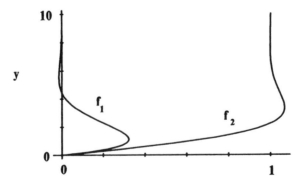

Figure 6–3 The magnitudes of the in-phase and out-of-phase components of the velocity profile, equation (6–30), for the axial velocity $u_z^{(1)}$. Note that we have plotted $u_z^{(1)}R_\omega$ in order to eliminate the R_ω^{-1} dependence in Equation (6–30).

main ideas of this section will be repeated in the next several sections, when we present a more general discussion of asymptotic approximation methods.

B. Asymptotic Expansions— General Considerations

In the previous section we demonstrated the application of asymptotic expansion techniques to obtain the high and low frequency limits of the velocity field for flow in a circular tube driven by an oscillatory pressure gradient. In the process, we introduced such fundamental notions as the difference between a *regular* and *singular* asymptotic expansion and, in the latter case, the concept of *matching* of the asymptotic approximations that are valid in different parts of the domain. However, all of the presentation was ad hoc, without the benefit of any formal introduction to the properties of asymptotic expansions. The present section is intended to provide at least a partial remedy for that shortcoming. We note, however, that the presentation will still be largely ad hoc. It is not our intention to provide a complete mathematical statement of the current status of asymptotic approximation methods; there are a number of comprehensive texts which do a good job of performing that task.[1] In this textbook, we strive only to give the student sufficient information so that he or she can *use* the methods and understand both their advantages and shortcomings. To a large extent, this goal will be pursued via the many examples of approximate solutions, obtained via asymptotic methods, that will be found throughout the following chapters. In this section, we present a number of necessary and useful general facts about asymptotic expansions.

To do this, let us consider a function, say, $T(\mathbf{x};\epsilon)$, which depends on spatial position and upon a dimensionless parameter ϵ that we may assume to be arbitrarily small. In the context of problems to be considered in this text, this function will usually be defined by a differential equation and boundary conditions, and the parameter ϵ then appears as a dimensionless parameter either in the equation or boundary conditions. We suppose, for purpose of discussion, that T has an asymptotic expansion for small ϵ. The general form of such an expansion is

$$T(\mathbf{x};\epsilon) = f_0(\epsilon)T_0(\mathbf{x}) + f_1(\epsilon)T_1(\mathbf{x}) + f_2(\epsilon)T_2(\mathbf{x}) + 0(f_3(\epsilon)). \qquad (6\text{--}31)$$

The functions T_0, T_1, and T_2 are independent of ϵ, and the magnitude of each term is therefore given by the gauge functions $f_n(\epsilon)$. We indicate the size of the error made by truncating (6–31) after three terms with the so-called order symbol, 0( ). Actually, there are two such symbols that could have been used in (6–31). The one known as big 0, which appears in (6–31), means that the largest neglected term varies in the asymptotic limit $\epsilon \to 0$ as $f_3(\epsilon)$; the use of the order of magnitude symbol implies that the magnitude of the next term in the expansion is known. This may not always be the case. Then, a second symbol known as little o, designated as $o(\ )$, is used. In the expansion (6–31) of $T(\mathbf{x};\epsilon)$, we would write $o(f_2(\epsilon))$, meaning that the largest neglected term is *smaller* than $f_2(\epsilon)$ in the asymptotic limit, $\epsilon \to 0$. As implied above, this second symbol is usually used only when we do not know the magnitude of the next largest term in an asymptotic series.

A very important property of an asymptotic expansion is the manner in which it converges to the function that it is intended to represent. Two facts can be stated that relate intimately to the nature of the convergence of an asymptotic expansion. First, if a function such as $T(\mathbf{x};\epsilon)$ has an asymptotic expansion for small ϵ (either for all $\mathbf{x}$ or at least in some subdomain of $\mathbf{x}$), then this expansion is unique (at least in the subdomain). But, second, more than one function T may have the same asymptotic representation through any finite number of terms. The second of these statements implies that one cannot sum an asymptotic expansion to find a unique function $T(\mathbf{x};\epsilon)$ as would be possible (in the domain of convergence) with a normal power-series representation of a function. This distinction between an asymptotic and infinite series representation is reflected in a more formal statement of the convergence properties of both an infinite series and an asymptotic expansion. In the case of an infinite-series representation of some function $T(\mathbf{x};\epsilon)$, namely,

$$T(\mathbf{x};\epsilon) = \sum_{n=1}^{\infty} a_n T^{(n)}(\mathbf{x};\epsilon), \tag{6-32}$$

convergence requires that the difference between T and its series representation can be made arbitrarily small by taking a sufficiently large number of terms in the series, that is,

$$\lim_{N \to \infty} \frac{T - \sum_{n=1}^{N} a_n T^{(n)}}{T} \to 0 \tag{6-33}$$

for fixed ϵ and $\mathbf{x}$ in the *domain of convergence*. An asymptotic expansion is not necessarily convergent at all in this sense. Instead,

$$\boxed{\lim_{\epsilon \to 0} \frac{T - \sum_{n=1}^{N} f_n(\epsilon) T_n(\mathbf{x})}{T} \to 0} \tag{6-34}$$

for fixed N. A necessary condition for asymptotic convergence is

$$\boxed{\lim_{\epsilon \to 0} \frac{f_{n+1}(\epsilon)}{f_n(\epsilon)} \to 0 \quad \text{for all } n.} \tag{6-35}$$

Thus, according to (6–34), the difference between T and its asymptotic expansion can be made arbitrarily small for any fixed N by taking the limit $\epsilon \to 0$. It is very important to recognize that asymptotic convergence does not imply that a better approximation will be achieved by taking more terms for any fixed ϵ, even if ϵ is small. Indeed, it is possible that the difference between T and its asymptotic expansion may actually diverge as we add more terms while holding ϵ fixed.

We have already noted that there are two types of asymptotic expansions: regular and singular. If the asymptotic expansion of a function $T(\mathbf{x};\epsilon)$ is *regular,* the convergence criterion (6–34) must be satisfied for all $\mathbf{x}$ in the domain of interest. The likelihood of a regular asymptotic expansion is improved if the domain is finite. If the asymptotic expansion is *singular,* on the other hand, the convergence criterion (6–34) is still satisfied but only in some subdomain of the region of interest. For this subdomain, the expansion

$$\sum_{n=1}^{N} f_n(\epsilon) T_n$$

provides a perfectly adequate representation, but for the rest of the domain one or more additional asymptotic expansions are necessary, each satisfying (6–34) in their subdomain of convergence. Let us suppose, for some particular example, that only one additional expansion is necessary, which we may denote as

$$T = \sum_{n=1}^{N} F_n(\epsilon) \hat{T}_n(\hat{\mathbf{x}}). \qquad (6\text{–}36)$$

In this second formula, neither the gauge functions nor the spatially dependent coefficients are normally the same as those appearing in the first representation of T. Furthermore, the spatial variable $\hat{\mathbf{x}}$ will frequently be scaled differently from its nondimensionalization in the portion of the domain where

$$\sum_{n=1}^{N} f_n(\epsilon) T_n(\mathbf{x})$$

is relevant. However, the two asymptotic expansions

$$T = \sum_{n=1}^{N} f_n(\epsilon) T_n(\mathbf{x})$$

and

$$T = \sum_{n=1}^{N} F_n(\epsilon) \hat{T}_n(\hat{\mathbf{x}})$$

are both approximations of the same function T, simply valid in different parts of the domain. To provide a uniformly valid asymptotic representation of T for all $\mathbf{x}$, it is necessary that the subdomains overlap. In this overlap region, which is a region of common validity for the expansions, the individual expansions must take the same functional form. This is known as the *matching principle*. For two representations of a function to match, it is not sufficient that they have the same numerical value at some fixed $\mathbf{x}$ (indeed, this property is sometimes associated with the name *patching* to distinguish it from *matching*); rather, they must adopt the same functional form to within an error that can be made arbitrarily small in the limit $\epsilon \to 0$.

Often the function $T(\mathbf{x};\epsilon)$ is defined as the solution of a differential equation and boundary conditions. In this case, the asymptotic expansion for T will be generated from an asymptotic approximation of the governing equation and boundary conditions for T. We have seen an example of this idea in the preceding section. When the asymptotic expansion for T is singular, different approximations of the equations (and thus of T) will be relevant in the different subdomains of the full solution domain. In this case, the original boundaries of the domain will always be split among the various subdomains, and thus not all of the original boundary conditions can be used for the asymptotic expansion in any given subdomain. This means that *too few* boundary conditions usually will be available to uniquely determine the asymptotic solution of the governing equation in any subdomain. In this case, the concept of matching plays a critical role; the approximate (asymptotic) solutions must match in the regions of overlap between the subdomains, and this provides sufficient additional conditions to obtain unique solutions. Later we shall see many examples of the principles described in this paragraph. Matching was used to determine the high frequency, singular perturbation solution for pulsatile flow in a circular tube that was discussed in Section **A**.

A question that is asked frequently is how one can determine, for a particular problem, whether the asymptotic expansion will be *regular* or *singular*. It is, in fact, not always easy to tell a priori. One telltale sign of a *singular limit*, corresponding to the necessity for singular or matched asymptotic expansions, occurs when *the limiting process, say, $\epsilon \to 0$, causes the governing equation to be reduced in order.* An example is the high frequency limit of the pulsatile flow problem considered in Section **A**. In this case, if we take the limit $R_\omega^{-1} \to 0$ in the governing equation (6–16), the problem is reduced from a second-order ordinary differential equation to an algebraic equation. Whenever the order of the governing equation is reduced, the solution of the resulting equation cannot satisfy all the boundary conditions of the original problem. Hence, as in the pulsatile flow problem, a different approximation of the problem must be required in the vicinity of the boundaries at least, and the asymptotic approximation of the solution for $\epsilon \ll 1$ will be singular. However, reduction of the order of the governing equation is only one obvious sign of singular asymptotic behavior, and there are many problems that are singular without exhibiting this particular characteristic. Generally speaking, a singular limit will be characterized by the necessity for multiple characteristic scales (especially length scales), each valid in different parts of the solution domain.

When in doubt about whether a particular asymptotic limit is regular or singular, it is frequently necessary simply to proceed under the assumption that the asymptotic solution will be *regular,* that is, that the problem can be characterized by a single set of characteristic scales, valid for all **x** in the solution domain. If the limit is, in fact, regular, the asymptotic solution scheme based on a single nondimensionalization of the governing equations and boundary conditions will proceed successfully at all levels of approximation. Otherwise, an obvious sign of singularity will appear; for example, the solution at some level of approximation either may not satisfy boundary conditions or simply may not exist. In this case, the problem can then be reformulated in a form appropriate to the singular limit.

C. The Motion of a Fluid Through a Slightly Curved Tube

We noted in the introduction to Chapter 3 that unidirectional flows only occur when the direction of motion is strictly in a straight-line path. For example, in flow through a circular tube, we assume that the tube axis is straight. An obvious question is whether the analysis and the predicted relationship between volumetric flow rate and the applied pressure gradient will apply if the tube is slightly bent. In this section, we explore this question, closely following an asymptotic analysis that was first published by W.R. Dean in 1927.[2] Let us denote the tube radius as a, while the coil radius is denoted as R. The small parameter in this analysis is the ratio of the tube radius to the radius of curvature of the tube axis, which is assumed to be coiled into a circle, that is, $\epsilon \equiv a/R \ll 1$. The limiting case of a perfectly straight tube, for which the unidirectional Poiseuille flow solution applies, corresponds to $\epsilon = 0$ (that is, $R = \infty$). Of course, a very long tube that is coiled in a circular arc must exhibit a slight pitch orthogonal to the main coil in order to avoid intersecting itself after 360° of coil, but we ignore this pitch in the present analysis. The motion of the fluid is due to an imposed pressure gradient along the tube, which we denote as G.

In order to analyze this problem, a system of coordinates must be specified. We adopt the coordinate system used originally by Dean (1927) and sketched in Figure 6-4. The cross section of the tube is circular, with the centerline denoted as C. As stated previously, the centerline traces a circular arc of radius R, and the section of the tube that is shown lies in a plane that makes an angle θ with a fixed axial plane. Within the cross section of the tube, the position of a point P is specified by the polar cylindrical coordinates (r, ψ), defined relative to the centerline of the tube. In this system, the polar angle ψ is measured from the vertical axis that is orthogonal to the OC plane traced by the circular arc of the centerline of the tube. Thus, in general, the point P is specified by the orthogonal coordinates (r, ψ, θ).* The components of

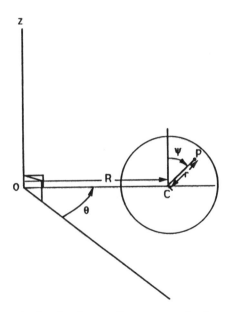

Figure 6–4 A schematic showing the coordinate system for the motion of a fluid through a slightly curved tube (after Dean [1927]).

---

*An alternative would be a toroidal coordinate system, as described by Happel and Brenner (1965), appendix, A-20.[3]

velocity corresponding to these coordinates will be designated as (u, v, w), respectively. It may be noted that distance along the tube, corresponding to the axial variable z in a straight circular tube geometry, is measured by $R\theta$.

In this analysis, we neglect entry and exit effects and concentrate upon the *fully developed* flow regime where the motion (u, v, w) is independent of θ. In a straight circular tube, we have already seen that the velocity field takes the form $(0, 0, w(r))$. However, in the case of a coiled tube, the tube geometry is no longer unidirectional, and there is no reason to suppose that the velocity field will be so simple. The equations of motion and continuity, specified in dimensional form, for a fully developed, three-dimensional flow are

$$\rho\left(u'\frac{\partial u'}{\partial r'} + \frac{v'}{r'}\frac{\partial u'}{\partial \psi} - \frac{v'^2}{r} - \frac{w'^2\sin\psi}{R + r\sin\psi}\right) =$$
$$-\frac{\partial p'}{\partial r'} + \mu\left[\left(\frac{1}{r'}\frac{\partial}{\partial \psi} + \frac{\cos\psi}{R' + r'\sin\psi}\right)\left(\frac{\partial v'}{\partial r'} + \frac{v'}{r'} - \frac{1}{r'}\frac{\partial u'}{\partial \psi}\right)\right], \qquad (6\text{-}37)$$

$$\rho\left(u'\frac{\partial v'}{\partial r'} + \frac{v'}{r'}\frac{\partial v'}{\partial \psi} + \frac{u'v'}{r'} - \frac{w'^2\cos\psi}{R' + r'\sin\psi}\right) =$$
$$-\frac{1}{r'}\frac{\partial p'}{\partial \psi} + \mu\left[\left(\frac{\partial}{\partial r'} + \frac{\sin\psi}{R' + r'\sin\psi}\right)\left(\frac{\partial v'}{\partial r'} + \frac{v'}{r'} - \frac{1}{r'}\frac{\partial u'}{\partial \psi}\right)\right], \qquad (6\text{-}38)$$

$$\rho\left(u'\frac{\partial w'}{\partial r'} + \frac{v'}{r'}\frac{\partial w'}{\partial \psi} + \frac{u'w'\sin\psi}{R' + r'\sin\psi} + \frac{v'w'\cos\psi}{R' + r'\sin\psi}\right) = -\frac{1}{R' + r'\sin\psi}\frac{\partial p'}{\partial \theta}$$
$$+ \mu\left[\left(\frac{\partial}{\partial r'} + \frac{1}{r'}\right)\left(\frac{\partial w'}{\partial r'} + \frac{w'\sin\psi}{R' + r'\sin\psi}\right) + \frac{1}{r'}\frac{\partial}{\partial \psi}\left(\frac{1}{r'}\frac{\partial w'}{\partial \psi} + \frac{w'\cos\psi}{R' + r'\sin\psi}\right)\right],$$
$$(6\text{-}39)$$

and

$$\frac{\partial u'}{\partial r'} + \frac{u'}{r'} + \frac{u'\sin\psi}{R' + r'\sin\psi} + \frac{1}{r'}\frac{\partial v'}{\partial \psi} + \frac{v'\cos\psi}{R' + r'\sin\psi} = 0. \qquad (6\text{-}40)$$

These equations reduce to the equations of motion in cylindrical coordinates if $R^{-1} = 0$ and $1/R(\partial/\partial\theta) = \partial/\partial z$. In the form shown, they apply to a tube of arbitrary radius a and arbitrary radius of curvature R.

To proceed, we nondimensionalize equations (6-37)–(6-40) using the radius of the tube as a characteristic length scale,

$$l_c = a,$$

the axial velocity at the centerline of a *straight* tube with the same axial pressure gradient G as a characteristic velocity scale,

$$u_c = \frac{Ga^2}{4\mu},$$

and a viscous pressure scale

$$p_c = \frac{\mu u_c}{a}.$$

The result of nondimensionalization for the radial velocity component, (6–37), is

$$\mathrm{Re}\left[u\frac{\partial u}{\partial r} + \frac{v}{r}\frac{\partial u}{\partial \psi} - \frac{v^2}{r} - \frac{w^2\sin\psi}{(R/a)+r\sin\psi}\right] =$$

$$-\frac{\partial p}{\partial r} + \left(\frac{1}{r}\frac{\partial}{\partial \psi} + \frac{\cos\psi}{(R/a)+r\sin\psi}\right)\left(\frac{\partial v}{\partial r} + \frac{v}{r} - \frac{1}{r}\frac{\partial u}{\partial \psi}\right).$$

If we consider the case of a slightly bent tube, where $a/R \ll 1$, this equation can be written in the alternative form

$$\mathrm{Re}\left[u\frac{\partial u}{\partial r} + \frac{v}{r}\frac{\partial u}{\partial \psi} - \frac{v^2}{r} - w^2\sin\psi\left(\frac{a}{R}\right)\left(1 - \frac{a}{R}r\sin\psi + 0\left(\frac{a}{R}\right)^2\right)\right] =$$

$$-\frac{\partial p}{\partial r} + \frac{1}{r}\frac{\partial}{\partial \psi}\left(\frac{\partial v}{\partial r} + \frac{v}{r} - \frac{1}{r}\frac{\partial u}{\partial \psi}\right)$$

$$+ \cos\psi\left(\frac{a}{R}\right)\left(1 - \frac{a}{R}r\sin\psi + 0\left(\frac{a}{R}\right)^2\right)\left(\frac{\partial v}{\partial r} + \frac{v}{r} - \frac{1}{r}\frac{\partial u}{\partial \psi}\right). \qquad (6\text{–}41)$$

Similarly, the nondimensionalized equations for the other velocity components can be written in the form appropriate for small values of a/R.

$$\mathrm{Re}\left[u\frac{\partial v}{\partial r} + \frac{v}{r}\frac{\partial v}{\partial \psi} + \frac{uv}{r} - w^2\cos\psi\left(\frac{a}{R}\right)\left(1 - \frac{a}{R}r\sin\psi + 0\left(\frac{a}{R}\right)^2\right)\right] =$$

$$-\frac{1}{r}\frac{\partial p}{\partial \psi} + \left(\frac{\partial}{\partial r} + \sin\psi\left(\frac{a}{R}\right)\left(1 - \frac{a}{R}r\sin\psi + 0\left(\frac{a}{R}\right)^2\right)\right)\left(\frac{\partial v}{\partial r} + \frac{v}{r} - \frac{1}{r}\frac{\partial u}{\partial \psi}\right)$$

and

$$\mathrm{Re}\left[u\frac{\partial w}{\partial r} + \frac{v}{r}\frac{\partial w}{\partial \psi} + (uw\sin\psi + wv\cos\psi)\left(\frac{a}{R}\right)\left(1 - \frac{a}{R}r\sin\psi + 0\left(\frac{a}{R}\right)^2\right)\right] =$$

$$-\left(1 - \frac{a}{R}r\sin\psi + 0\left(\frac{a}{R}\right)^2\right)\frac{1}{R}\frac{\partial p}{\partial \theta}$$

$$+ \left(\frac{\partial}{\partial r} + \frac{1}{r}\right)\left(\frac{\partial w}{\partial r} + \frac{a}{R}w\sin\psi\left(1 - \frac{a}{R}r\sin\psi + 0\left(\frac{a}{R}\right)^2\right)\right)$$

$$+ \frac{1}{r}\frac{\partial}{\partial \psi}\left(\frac{1}{r}\frac{\partial w}{\partial \psi} + \frac{a}{R}w\cos\psi\left(1 - \frac{a}{R}r\sin\psi + 0\left(\frac{a}{R}\right)^2\right)\right)$$

$$(6\text{–}42) \text{ and } (6\text{–}43)$$

Finally, the continuity equation is

$$\frac{\partial u}{\partial r} + \frac{u}{r} + \frac{1}{r}\frac{\partial v}{\partial \psi} + (u\sin\psi + v\cos\psi)\left(\frac{a}{R}\right)\left(1 - \frac{a}{R}r\sin\psi + 0\left(\frac{a}{R}\right)^2\right) = 0.$$

(6–44)

The general problem, then, is to solve the system of (6–41)–(6–44), subject to the boundary conditions

$$u = v = w = 0 \quad \text{at} \quad r = 1$$

(6–45)

and the requirement that u, v, and w are bounded at the origin, $r = 0$. It can be seen, by examination of these equations, that the problem for nonzero values of (a/R) and Re is highly *nonlinear*. Here, however, we seek only an asymptotic approximation to the solution in the limiting case $a/R \ll 1$.

As a starting point, we recall that the limit $a/R \equiv 0$ corresponds to a straight circular tube, with the flow described by the Poiseuille flow solution $w = (1 - r^2)$, $u = v = 0$. In the present context, we consider small, but *nonzero*, values of a/R, and recognize the Poiseuille flow solution as a first approximation in an asymptotic approximation scheme. In particular, if we assume that a solution exists for **u** in the form of a *regular asymptotic expansion*,

$$\mathbf{u} = \mathbf{u}_0 + \left(\frac{a}{R}\right)\mathbf{u}_1 + \left(\frac{a}{R}\right)^2\mathbf{u}_2 + \cdots,$$

(6–46)

the Poiseuille flow solution corresponds to $\mathbf{u}_0$, and our goal is to obtain additional terms in the expansion, namely, $\mathbf{u}_1$, $\mathbf{u}_2$, and so on. It is, of course, not self-evident that a regular perturbation expansion actually exists for this problem. However, instead of trying to establish this fact in an a priori manner, it is common practice initially to seek a solution in the form of a regular expansion, with the proof of existence for such an expansion then residing in the ability to satisfy the resulting differential equations and boundary conditions to arbitrary order in a/R. If this can be done, it implies that the scaling inherent in the governing equations is valid throughout the flow domain, and a regular expansion exists.

We assume then, that an expansion of the form (6–46) exists for each of the components of **u** and for p, namely,

$$w = w_0 + \left(\frac{a}{R}\right)w_1 + \left(\frac{a}{R}\right)^2 w_2 + \cdots,$$

$$u = u_0 + \left(\frac{a}{R}\right)u_1 + \left(\frac{a}{R}\right)^2 u_2 + \cdots,$$

$$v = v_0 + \left(\frac{a}{R}\right)v_1 + \left(\frac{a}{R}\right)^2 v_2 + \cdots,$$

and

$$p = p_0 + \left(\frac{a}{R}\right)p_1 + \left(\frac{a}{R}\right)^2 p_2 + \cdots.$$

(6–47)

The coefficients in these expansions—w_i, u_i, v_i, and p_i— can be functions of (r, ψ) but must be independent of (a/R). In order to obtain governing equations and boundary conditions for these coefficients, we substitute the expansions (6–47) into the governing equations and boundary conditions (6–41)–(6–45) and collect all terms at each order in a/R. The result is a set of four equations plus boundary conditions, each of which takes the form

$$(\text{fnc}_0 \text{ of } r, \psi) + \left(\frac{a}{R}\right)(\text{fnc}_1 \text{ of } r, \psi) + \left(\frac{a}{R}\right)^2 (\text{fnc}_2 \text{ of } r, \psi) + \cdots = 0. \quad (6\text{--}48)$$

Since a/R is arbitrary, each of the coefficients labeled (fnc of r, ψ) must equal zero, and this yields governing equations for the functions w_i, u_i, v_i, and p_i in equation (6–47).

At $O(1)$, we obtain a set of coupled equations for the first terms in (6–47), namely,

$$\text{Re}\left[u_0 \frac{\partial u_0}{\partial r} + \frac{v_0}{r}\frac{\partial u_0}{\partial \psi} - \frac{v_0^2}{r}\right] = -\frac{\partial p_0}{\partial r} + \frac{1}{r}\frac{\partial}{\partial \psi}\left(\frac{\partial v_0}{\partial r} + \frac{v_0}{r} - \frac{1}{r}\frac{\partial u_0}{\partial \psi}\right), \quad (6\text{--}49\text{a})$$

$$\text{Re}\left[u_0 \frac{\partial v_0}{\partial r} + \frac{v_0}{r}\frac{\partial v_0}{\partial \psi} + \frac{u_0 v_0}{r}\right] = -\frac{1}{r}\frac{\partial p_0}{\partial \psi} + \frac{\partial}{\partial r}\left(\frac{\partial v_0}{\partial r} + \frac{v_0}{r} - \frac{1}{r}\frac{\partial u_0}{\partial \psi}\right), \quad (6\text{--}49\text{b})$$

$$\text{Re}\left[u_0 \frac{\partial w_0}{\partial r} + \frac{v_0}{r}\frac{\partial w_0}{\partial \psi}\right] = -\frac{\partial p_0}{\partial z} + \frac{\partial^2 w_0}{\partial r^2} + \frac{1}{r}\frac{\partial w_0}{\partial r} + \frac{1}{r}\frac{\partial^2 w_0}{\partial \psi^2}, \quad (6\text{--}49\text{c})$$

and

$$\frac{\partial u_0}{\partial r} + \frac{u_0}{r} + \frac{1}{r}\frac{\partial v_0}{\partial \psi} = 0, \quad (6\text{--}49\text{d})$$

with boundary conditions

$$u_0 = v_0 = w_0 = 0 \quad \text{at} \quad r = 1. \quad (6\text{--}49\text{e})$$

Here, we have expressed the axial pressure gradient in the form $\partial p_0/\partial z$, where $\partial/\partial z \sim R^{-1} (\partial/\partial \theta)$, to emphasize the connection between this problem and the steady motion in a straight tube, which was considered in Chapter 3. We previously indicated that the dimensional magnitude of the axial pressure gradient is G. Hence, referring to the definitions of characteristic scales following Equation (6–40), it follows that

$$-\frac{\partial p_0}{\partial z} = 4,$$

and it is evident from (6–49c) that $w_0 \neq 0$. On the other hand, the equations and boundary conditions for $u_0 = v_0$ are completely homogeneous and independent of w_0, so that an obvious solution is $u_0 = v_0 = 0$, with p_0 being a linear function of z only. In this case, if we assume axisymmetry for w_0, the governing Equation (6–49c) reduces to

$$\frac{1}{r}\frac{d}{dr}\left(r\frac{dw_0}{dr}\right) = -4, \quad (6\text{--}50)$$

and the solution, subject to (6–49e), is

$$w_0 = 1 - r^2.$$

(6–51)

Of course, this is just the familiar Poiseuille solution as expected.

The first influence of the curvature of the tube is felt in the terms of $0(a/R)$ in the expansions (6–47). The governing equations at $0(a/R)$ are obtained by substituting the expansions (6–47) into the dimensionless equations (6–42)–(6–44), and collecting all terms of $0(a/R)$,

$$-\mathrm{Re}\,w_0^2\sin\psi = -\frac{\partial p_1}{\partial r} + \frac{1}{r}\frac{\partial}{\partial\psi}\left(\frac{\partial v_1}{\partial r} + \frac{v_1}{r} - \frac{1}{r}\frac{\partial u_1}{\partial\psi}\right),$$

(6–52a)

$$-\mathrm{Re}\,w_0^2\cos\psi = -\frac{1}{r}\frac{\partial p_1}{\partial\psi} + \frac{\partial}{\partial r}\left(\frac{\partial v_1}{\partial r} + \frac{v_1}{r} - \frac{1}{r}\frac{\partial u_1}{\partial\psi}\right),$$

(6–52b)

$$\mathrm{Re}\left[u_1\frac{\partial w_0}{\partial r}\right] = \frac{\partial p_0}{\partial z}r\sin\psi - \frac{\partial p_1}{\partial z} + \left[\frac{\partial^2 w_1}{\partial r^2} + \frac{1}{r}\frac{\partial w_1}{\partial r} + \frac{1}{r^2}\frac{\partial^2 w_1}{\partial\psi^2}\right]$$
$$+ \left(\frac{\partial}{\partial r} + \frac{1}{r}\right)w_0\sin\psi + \frac{1}{r}\frac{\partial}{\partial\psi}(w_0\cos\psi),$$

(6–52c)

and

$$\frac{\partial u_1}{\partial r} + \frac{u_1}{r} + \frac{1}{r}\frac{\partial v_1}{\partial\psi} = 0.$$

(6–52d)

The boundary conditions are

$$u_1 = v_1 = w_1 = 0 \quad \text{at} \quad r = 1.$$

(6–52e)

Although the original governing equations are *nonlinear* for arbitrary Re and (a/R), we see that (6–52) (and all higher-order equations) are *linear*. This is typical of regular perturbation solutions of nonlinear partial differential equations. Equation (6–50) for the leading-order term, w_0, contains only terms that were linear in the original equations. At the next level of approximation, represented by (6–52), nonlinear terms appear, but these are evaluated using the leading-order solution that is *now known*. Further examination of (6–52) clearly indicates that the flow at this level of approximation $0(a/R)$, will no longer be one-dimensional. All three velocity components, u_1, v_1, and w_1, must be nonzero, since (6–52a–c) are nonhomogeneous. Hence, we see that a slight departure from the straight tube geometry, $a/R \ll 1$, causes an important qualitative change in the nature of the flow.

To ascertain the details of the changes represented by u_1, v_1, w_1, and p_1, we must solve the system of equations (6–52a–e). We can begin with the pressure. It follows from (6–52c) that the most general possible form for p_1 is

$$p_1 = H(r,\psi)z + B(r,\psi).$$

However, according to (6–52a) or (6–52b), $\partial p_1/\partial r$ and $\partial p_1/\partial\psi$ are both functions, at most, of r and ψ. Thus, we see that

$$H(r,\psi) \equiv 0$$

and

$$\boxed{p_1 = B(r,\psi) \text{ only.}} \tag{6-53}$$

Hence, the pressure contribution at $0(a/R)$ enters, at most, in the pressure gradient terms in (6–52a) and (6–52b).

To solve (6–52a) and (6–52b), it is convenient to introduce a function $f_1(r,\psi)$, defined such that

$$\boxed{ru_1 \equiv -\frac{\partial f_1}{\partial \psi} \quad \text{and} \quad v_1 \equiv \frac{\partial f_1}{\partial r}.} \tag{6-54}$$

Then, substituting for w_0 from (6–51), equations (6–52a,b) can be written in the forms

$$-\mathrm{Re}(1 - r^2)^2 \sin\psi = -\frac{\partial p_1}{\partial r} - \frac{1}{r}\frac{\partial}{\partial \psi}(\nabla_1^2 f_1) \tag{6-55a}$$

and

$$-\mathrm{Re}(1 - r^2)^2 \cos\psi = -\frac{1}{r}\frac{\partial p_1}{\partial \psi} + \frac{\partial}{\partial r}(\nabla_1^2 f_1), \tag{6-55b}$$

where

$$\nabla_1^2 \equiv \frac{\partial^2}{\partial r^2} + \frac{1}{r}\frac{\partial}{\partial r} + \frac{1}{r^2}\frac{\partial^2}{\partial \psi^2}. \tag{6-56}$$

Hence, eliminating p_1 by cross-differentiating and combining (6–55a) and (6–55b), we obtain a single equation for f_1,

$$\boxed{r\nabla_1^2(\nabla_1^2 f_1) = 4\mathrm{Re}\, r^2(1 - r^2)\cos\psi.} \tag{6-57}$$

It may be noted from the definitions (6–54) that f_1 plays a role in the present problem at $0(a/R)$ that is analogous (in the plane) to the streamfunction for a general two-dimensional flow. Hence, with u_1 and v_1 defined in terms of f_1, according to (6–54), the continuity equation (6–52d) is satisfied exactly.

The governing equation for w_1 is obtained by substituting for w_0 as indicated in (6–52c) and noting that $\partial p_0/\partial z \equiv -4$. Since $\partial p_1/\partial z \equiv 0$,* the resulting equation takes the form

$$\nabla_1^2 w_1 = 6r\sin\psi - (2ru_1)\mathrm{Re}. \tag{6-58}$$

Hence, w_1 can be obtained only after first solving for u_1 from equations (6–57) and (6–54).

---

*In this problem, the pressure gradient is *fixed* at a dimensionless value of -4, and any changes in the flow may only produce changes in the volumetric flow rate. Hence, $\partial p_n/\partial z = 0$ for all $n \geq 1$. A closely related problem would have the volumetric flow rate fixed, and, in this case, changes in the flow would produce changes in the required pressure gradient so that $\partial P_n/\partial z \neq 0$ for $n \geq 1$.

The solution of (6–57) is straightforward. The boundary conditions (6–52e) expressed in terms of f_1 are

$$\frac{\partial f_1}{\partial r} = \frac{\partial f_1}{\partial \psi} = 0 \quad \text{at} \quad r = 1.$$

We leave the details of solving (6–57) to the reader. The solution, satisfying the boundary conditions, is

$$f_1 = \frac{\text{Re}}{288} r(1 - r^2)^2 (4 - r^2) \cos \psi. \tag{6–59}$$

The corresponding velocity components are

$$u_1 = \frac{\text{Re}}{288} (1 - r^2)^2 (4 - r^2) \sin \psi \tag{6–60}$$

and

$$v_1 = -\frac{\text{Re}}{288} (7r^4 - 23r^2 + 4)(1 - r^2) \cos \psi. \tag{6–61}$$

With u_1 known, (6–58) can now be solved for w_1. Again, the solution is straightforward. The result, after applying the boundary condition (6–52e), is

$$w_1 = -\frac{3}{4} \sin \psi \, r(1 - r^2) + \text{Re}^2 \frac{\sin \psi}{11520} r(1 - r^2)(19 - 21r^2 + 9r^4 - r^6). \tag{6–62}$$

Thus, the presence of a small amount of curvature of the tube axis is seen to produce a significant departure in the nature of the flow even at the leading-order correction, $0(a/R)$. The motion is no longer simply in the axial direction, but there is also motion in the cross-sectional plane. This motion, superposed upon the primary axial flow, is known as a *secondary* flow. A sketch of the streamlines for the secondary flow, given by contours of constant f_1, is shown in Figure 6–5. The motion at the center of the tube is from the inside to the outside, driven essentially by the centrifugal force exerted on the fluid as it traverses the curved path followed by the tube axis. The return flow nearest the tube walls at the top and bottom is necessary to satisfy continuity.

The path of a typical fluid element is a superposition of the recirculating secondary flow and the primary flow in the axial direction. The result is a helical motion as the fluid element moves through the tube. In addition to the secondary flow, however, the curvature of the tube axis also contributes the modification (6–62) in the axial profile. Combining (6–62) and (6–51) in the asymptotic form (6–47), the axial velocity component takes the form

$$w = (1 - r^2) + \left(\frac{a}{R}\right)\left[-\frac{3}{4} r(1 - r^2) \right.$$

$$\left. + \frac{\text{Re}^2}{11520} r(1 - r^2)(19 - 21r^2 + 9r^4 - r^6) \right] \sin \psi + 0\left(\frac{a}{R}\right)^2. \tag{6–63}$$

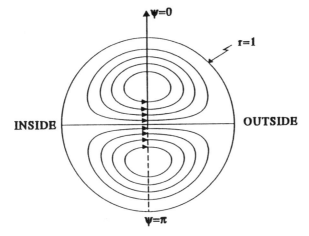

Figure 6–5 The streamlines for the *secondary* flow of a fluid that is moving through a slightly curved tube. Contour values are plotted in increments of 0.2222.

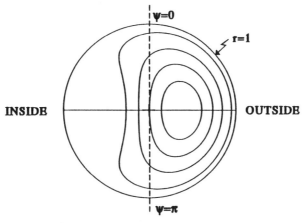

Figure 6–6 Contours of constant axial velocity for pressure-driven flow through a slightly curved tube. Values are plotted in increments of 0.3819.

Not only is the radial dependence of w changed, but also the axial velocity profile is no longer independent of the polar angle ψ, as shown in Figure 6–6. Indeed, the contribution to w at $O(a/R)$ consists of an *asymmetric* profile, with a slightly increased axial flow in the range $0 < \psi < \pi$, which lies on the outer half of the tube, and a slightly decreased flow for $\pi < \psi < 2\pi$, which lies on the inside half of the tube. It should be noted, however, that there is no change at this level of approximation in the total volumetric flow rate through the tube.

Experimentally, it is found that the volumetric flow rate through a curved tube actually decreases relative to the flow rate in a straight tube if all other factors are held constant (namely, the tube length and radius, the pressure gradient, and the fluid viscosity). However, this effect does not appear in the first correction to the Poiseuille flow solution for small a/R. To predict this effect, we must proceed to higher-order corrections in the asymptotic expansions (6–47). The decrease in volumetric flow rate first appears, in fact, at the next level of approximation, $0(a/R)^2$. However, the algebraic effort necessary to calculate the next terms (u_2, v_2, w_2, p_2) is large, and we do not pursue the detailed calculations here. The physical explanation for decreased flow rates is simply that some of the energy imparted to the fluid through the applied pressure gradient is diverted to driving the secondary flow, which is an added source of dissipation from kinetic energy to heat.

The main conclusions to be drawn from the preceding analysis are two-fold: First, the departure from a straight tube to a slightly curved tube destroys the unidirectional nature of the Poiseuille flow, and we may generally expect small departures from strict adherence to a unidirectional geometry to produce similar transitions in the flow for other problems; second, for small a/R, the influence of curvature of the tube is described by a regular perturbation expansion, based upon the governing equations in the nondimensionalized forms (6–41)–(6–44). We should keep in mind that the solution form (6–47) is valid only in the asymptotic limit $a/R \rightarrow 0$. If the tube is more tightly coiled so that the ratio a/R is not small, some other solution technique must be used. Although analytic approximation is possible for some cases, the vast majority of solutions for finite a/R have been carried out numerically. The interested reader may find a summary of additional work on this problem in a recent review article written by Berger, Talbot, and Yao (1983).[4]

D. Bubble Growth in a Quiescent Fluid

The critical defining property of a *unidirectional* flow is that the fluid moves strictly in a single direction, and we saw in the preceding section that even slight deviations from strict adherence to the unidirectional geometry can lead to important, qualitative changes in the nature of the flow. Another important distinction is the difference between a *unidirectional* flow and a *one-dimensional* flow. In general, a one-dimensional flow consists of a single velocity component in one specific coordinate direction. Of course, a unidirectional flow is an example of a one-dimensional flow, in which the coordinate direction is Cartesian and the fluid follows a straight-line path. However, there are many one-dimensional flows that are not unidirectional.

One important example is the motion induced by changes in the volume of a single spherical bubble that is suspended in an incompressible fluid that is at rest at infinity. The motion of the fluid, in this case, is strictly radial, that is, $u_r \neq 0$, $u_\theta = u_\phi = 0$; hence, the flow is one-dimensional but not unidirectional. The problem has a long history, beginning with important contributions of Lord Rayleigh.[5] In general, contemporary interest in the problem is still motivated largely by the sound produced by bubbles that undergo a time-dependent change of volume, particularly in the context of cavitation bubbles that are produced in hydromachinery or in the propulsion systems of submarines and other underwater vehicles.[6] The objective of theoretical analyses for single bubble motions is usually to predict the bubble radius as a function of time in a prescribed ambient pressure field. Once the bubble radius is known as a function of time, it is a simple matter to determine the velocity and pressure fields throughout the liquid.

In this section, we consider this problem in some detail, with a particular focus on the application of asymptotic analyses. As we shall see, one important difference between unidirectional flows and this particular class of one-dimensional flows is that the governing equations are nonlinear and, thus, most features can only be exposed by either numerical or asymptotic techniques. We begin by deriving the well-known *Rayleigh-Plesset* equation, which governs the time dependence of the bubble radius in a prescribed ambient pressure field, subject to the assumption that the bubble remains strictly spherical.

The Rayleigh-Plesset Equation

We consider a single bubble that undergoes a time-dependent change of volume in an incompressible, Newtonian fluid that is at rest at infinity. The bubble may either be a vapor bubble (that is, contain the vapor of the liquid) or it may contain a ''contaminant'' gas that is insoluble in the liquid (or at least only dissolves very slowly compared with the time scales associated with changes in the bubble volume) or a combination of vapor and contaminant gases. We assume, at the outset, that the bubble remains strictly sperical, and thus that the bubble surface moves only in the radial direction. It follows that the motion induced in the liquid must also be radial, so that

$$u_r = u_r(r,t); \quad u_\theta = u_\phi = 0 \tag{6-64}$$

when expressed in terms of velocity components in a spherical coordinate system. Thus, by the assumption of a spherical shape, we implicitly assume that the fluid motion will be one-dimensional.

The assumption of a one-dimensional, radial flow leads to a great simplification of the fluid mechanics problem. Indeed, the continuity equation, corresponding to (6-64), takes the very simple form

$$\frac{1}{r^2} \frac{\partial}{\partial r} (r^2 u_r) = 0, \tag{6-65}$$

and we see, by integrating, that u_r is completely determined to within a single unknown function of time by the requirement of mass conservation, that is,

$$\boxed{u_r(r,t) = \frac{F(t)}{r^2}.} \tag{6-66}$$

From a physical viewpoint, this form for u_r ensures that the total mass flux is the same across any closed spherical surface that encloses the bubble.

The time-dependent function $F(t)$ is determined by the rate of increase or decrease in the bubble volume. The governing equations and boundary conditions that remain to be satisfied are (1) the radial component of the Navier-Stokes equation; (2) the kinematic condition, in the form of equation (5-5), at the bubble surface; and (3) the normal stress balance (5-10) at the bubble surface with $\hat{\tau} \equiv 0$. Generally, for a gas bubble, the zero shear-stress condition also must be satisfied at the bubble surface, but $\tau_{r\theta} \equiv 0$ for a purely radial velocity field of the form (6–64), and this condition thus provides no useful information for the present problem.

The relationship between $F(t)$ and the bubble radius $R(t)$ is determined from the kinematic boundary condition. In particular, for a bubble containing only an insoluble gas, the kinematic condition takes the form

$$u_r(R,t) = \frac{dR}{dt}. \tag{6–67}$$

For the case of a vapor bubble containing a combination of vapor and insoluble gas, there will be a net mass flux across the interface as the liquid vaporizes or the vapor condenses, and this means that the condition (6–67) is not exactly true. In the case of a pure vapor bubble,

$$u_r(R,t) = \frac{dR}{dt}\left(1 - \frac{\rho_v}{\rho}\right),$$

where ρ_v is the saturation vapor density at the bubble temperature and ρ is the liquid density. Hence, since $\rho_v/\rho \ll 1$ in most cases, it follows that the relationship (6–67) can be used as a good approximation for both gas and vapor bubbles. Thus, combining (6–66) and (6–67), it follows that

$$\boxed{F(t) = R^2 \frac{dR}{dt}.} \tag{6–68}$$

Finally, to determine $F(t)$ or, equivalently, $R(t)$ as a function of time, we must consider the radial component of the Navier-Stokes equation, that is,

$$\frac{\partial u_r}{\partial t} + u_r \frac{\partial u_r}{\partial r} = -\frac{1}{\rho}\frac{\partial p}{\partial r} + \nu\left[\frac{1}{r^2}\frac{\partial}{\partial r}\left(r^2 \frac{\partial u_r}{\partial r}\right) - \frac{2u_r}{r^2}\right]. \tag{6–69}$$

We purposely maintain this equation in dimensional form for the time being. Now, substituting (6–66) into (6–69) and carrying out the indicated differentiation, we obtain a differential equation for $F(t)$,

$$\boxed{-\frac{1}{\rho}\frac{\partial p}{\partial r} = \frac{1}{r^2}\frac{\partial F}{\partial t} - \frac{2F^2}{r^5}.} \tag{6–70}$$

Thus, integrating with respect to r,

$$\boxed{\frac{1}{\rho}\left[p(r,t) - p_\infty(t)\right] = \frac{1}{r}\frac{\partial F}{\partial t} - \frac{F^2}{2r^4},} \tag{6–71}$$

where $p_\infty(t)$ is the ambient pressure in the fluid at a large distance from the bubble. Finally, if (6–71) is evaluated at the bubble surface $r = R(t)$ and expressed in terms of R using (6–68), we obtain

$$\boxed{R\ddot{R} + \frac{3}{2}(\dot{R})^2 = \frac{1}{\rho}\left[p(R,t) - p_\infty(t)\right],} \tag{6–72}$$

where $p(R, t)$ is the pressure at the bubble surface in the liquid. This gives a direct relationship between the pressure in the liquid and the rate of change of the bubble radius.

A more useful alternative is to express (6–72) in terms of the pressure $p_B(t)$ inside the bubble rather than $p(R, t)$. To do this, we must use the normal stress balance (5–10) at the bubble surface, which takes the form

$$-p(R, t) + 2\mu \frac{\partial u_r}{\partial r}\bigg|_{r=R} + p_B(t) - \frac{2\gamma}{R} = 0. \tag{6–73}$$

Here, we have assumed that the viscous stress in the gas is negligible and the pressure within the gas is uniform, p_B. The last term in (6–73) is the capillary pressure contribution due to surface tension, γ. Evaluating $\partial u_r / \partial r |_{r=R}$ using (6–66) and (6–68), we thus obtain

$$-p(R, t) - \frac{4\mu \dot{R}}{R} + p_B(t) - \frac{2\gamma}{R} = 0. \tag{6–74}$$

Thus, substituting (6–74) into (6–72), we obtain the famous Rayleigh-Plesset equation

$$\boxed{\frac{p_B(t) - p_\infty(t)}{\rho} - \frac{1}{\rho}\left(\frac{2\gamma}{R} + \frac{4\mu \dot{R}}{R}\right) = R\ddot{R} + \frac{3}{2}(\dot{R})^2.} \tag{6–75}$$

All that remains is a theory to calculate $p_B(t)$ in terms of $R(t)$. For this purpose, we assume

$$p_B = p_v + p_G, \tag{6–76}$$

where p_v is the vapor pressure at the bubble temperature and p_G is the pressure contribution for a contaminant gas, which we model here as an ideal gas, so that

$$p_G = G \frac{T}{R^3}. \tag{6–77}$$

In general, the temperatures in the gas and liquid both will change with time due primarily to the release (or use) of latent heat at the interface as vapor condenses (or liquid vaporizes). However, for present purposes, we neglect this effect and consider that $T = T_\infty = $ constant everywhere in the system. In terms of practical application, this is a severe assumption, and it is well known that thermal effects can have an important influence on the dynamics of gas/vapor bubbles. However, the isothermal problem retains many of the interesting qualitative features of the full problem and is sufficiently general for our present illustrative purposes. Thus, we assume that

$$p_B(t) = p_v(T_\infty) + \frac{\hat{G}}{R^3(t)}, \tag{6–78}$$

where $p_v(T_\infty)$ and $\hat{G}$ are both constants. The Rayleigh-Plesset equation takes the form

$$\boxed{\frac{p_v - p_\infty(t)}{\rho} + \frac{1}{\rho}\left[\frac{\hat{G}}{R^3} - \frac{2\gamma}{R} - \frac{4\mu \dot{R}}{R}\right] = R\ddot{R} + \frac{3}{2}(\dot{R})^2.} \tag{6–79}$$

We shall discuss the solution of the Rayleigh-Plesset equation (6–79) shortly.

First, however, it may be useful to consider the Rayleigh-Plesset equation in dimensionless form. For this purpose, we note that the driving force for change in the bubble volume is the difference between the pressure differential $[p_v - p_\infty(t)]/\rho$ and the value at equilibrium, $(p_v - \hat{p}_\infty)/\rho$, which corresponds to an equilibrium bubble radius, R_E. The equilibrium condition is

$$\frac{p_v - \hat{p}_\infty}{\rho} + \frac{1}{\rho}\left[\frac{\hat{G}}{R_E^3} - \frac{2\gamma}{R_E}\right] = 0. \tag{6-80}$$

Thus, to describe the dynamics of changes in bubble radius due to changes in the ambient pressure, it is convenient to introduce the change of variables

$$R = R_E(1 + r)$$

and

$$p_\infty = \hat{p}_\infty(1 + p_\infty^*) \tag{6-81}$$

so that

$$\frac{p_v - \hat{p}_\infty}{\rho} - \frac{\hat{p}_\infty}{\rho}p_\infty^* + \frac{1}{\rho}\left[\frac{\hat{G}}{R_E^3(1+r)^3} - \frac{2\gamma}{R_E(1+r)} - \frac{4\mu\dot{r}}{(1+r)}\right] = R_E^2\left[(1+r)\ddot{r} + \frac{3}{2}(\dot{r})^2\right].$$

Now, introducing $t = t_c\bar{t}$, where

$$t_c = R_E\left(\frac{\rho}{p_v - \hat{p}_\infty}\right)^{1/2}, \tag{6-82}$$

and denoting $p_v - \hat{p}_\infty$ as Δp, the Rayleigh-Plesset equation can be expressed in dimensionless form

$$(1+r)\ddot{r} + \frac{3}{2}(\dot{r})^2 = \left\{1 - \alpha p_\infty^* + \frac{\beta}{(1+r)^3}\right\} - \left[\frac{2}{\text{We}}\frac{1}{(1+r)} + \frac{4}{\text{Re}}\frac{\dot{r}}{(1+r)}\right], \tag{6-83}$$

where

$$\alpha \equiv \frac{\hat{p}_\infty}{(p_v - \hat{p}_\infty)},$$

$$\beta \equiv \frac{\hat{G}}{R_E^3\Delta p},$$

$$\text{We} \equiv \frac{R_E\Delta p}{\gamma} = \frac{\rho R_E u_c^2}{\gamma} \quad \text{(Weber number)},$$

$$\text{Re} \equiv \frac{\rho^{1/2}R_E(\Delta p)^{1/2}}{\mu} = \frac{\rho R_E u_c}{\mu} \quad \text{(Reynolds number)},$$

and

$$u_c \equiv \frac{R_E}{t_c} = \left(\frac{p_v - \hat{p}_\infty}{\rho}\right)^{1/2}.$$

Before discussing the solution of the Rayleigh-Plesset equation, in either of the forms (6–79) and (6–83), a few comments may be useful to put the derivation leading to (6–79) into the context of earlier sections of this book (and this chapter). From a general perspective, the most important point is the distinction between a *unidirectional* and *one-dimensional* (but not unidirectional) flow. In particular, one-dimensional flows are often characterized by a strong (or even dominant) contribution from the *inertial* (or nonlinear) terms in the equation of motion. For the special case considered here, the viscous terms in (6–69) are, in fact, identically zero, and the dynamics within the fluid are dominated by *Eulerian acceleration* of fluid elements associated with time-dependent changes in $R(t)$ and by *Lagrangian acceleration* due to the *divergence* of streamlines in the radial flow. Lagrangian accelerations are, of course, impossible in a unidirectional flow. In view of the importance of Lagrangian acceleration effects here, it is not surprising that the Rayleigh-Plesset equation is highly *nonlinear*.

As a consequence of this nonlinearity, it is impossible to obtain analytic solutions of the Rayleigh-Plesset equation for most problems of interest, where $p_\infty(t)$ is specified and the bubble radius $R(t)$ is to be calculated. Indeed, most comprehensive studies of (6–79) or (6–83) have been carried out numerically. These show a richness of dynamic behavior that lies beyond the capabilities of analytic approximation. For example, a typical case might have $p_\infty(t)$ first decrease below $p_\infty(0)$ and then recover its initial value, as illustrated in Figure 6–7. The bubble radius $R(t)$ first grows up to a maximum (which typically occurs *after* the minimum in p), but then undergoes collapse in a dramatic fashion to a minimum radius, followed by a succession of rebounds and collapses whose amplitude is attenuated by viscosity. It is this latter collapse sequence that is responsible for the sharp crackling sounds that can be heard in cavitating hydromachinery. The reader who is interested in a comprehensive review of the predicted behavior of gas and vapor bubbles in time-dependent pressure fields should consult the excellent paper of Plesset and Prosperetti (1977).[7] Here, we concentrate on results that can be obtained analytically by regular asymptotic approximations that are based on *domain perturbations*. In particular, we seek approximate, asymptotic solutions that correspond to small changes in either bubble *volume* or *shape* relative to some case where the solution is known (or obtained easily).

Before turning to these asymptotic theories, however, we conclude this subsection by considering one special case where the Rayleigh-Plesset equation can be integrated exactly. This is the case of a step change in p_∞ from some initial value $p_\infty(0)$ to a different constant value p_∞^*. In order to integrate (6–79), we first multiply by $2R^2\dot{R}$. Then every term in the resulting equation is an exact differential (and can be integrated directly) except for the viscous term, which cannot be integrated in this way. We thus consider the approximate case where $\mu \equiv 0$ [namely, Re $\equiv \infty$ in (6–83)]. In spite of the neglect of viscous effects, the solution provides useful qualitative insight into some of the behavior that is inherent in the Rayleigh-Plesset equation. The result is

$$(\dot{R})^2 = \frac{2}{3}\frac{p_v - p_\infty^*}{\rho}\left(1 - \frac{R_0^3}{R^3}\right) - \frac{2}{3}\frac{\hat{G}}{\rho}\frac{1}{R^3}\ln\left(\frac{R_0^3}{R^3}\right) - \frac{2\gamma}{\rho R}\left(1 - \frac{R_0^2}{R^2}\right), \quad (6\text{–}84)$$

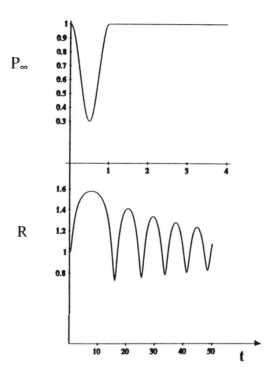

Figure 6–7 The response of a gas bubble to a perturbation of the ambient pressure $p_\infty(t)$. The trace of pressure $p_\infty(t)$ as a function of time is shown in the upper half of the figure, while the bubble radius is shown as a function of time in the lower half. Particularly noteworthy is the time scale for changes in radius, compared to that of the pressure change (note that the *scale* is different in the two parts of the figure).

where we have assumed that

$$R(0) = R_0$$

and, for simplicity,

$$\dot{R}(0) = 0.$$

Now we consider the two cases of bubble growth and collapse. For bubble growth, $p_\infty^* < p_\infty(0)$ and the bubble will either reach a new steady equilibrium radius or else continue to grow until $R \gg R_0$. In the latter case,

$$\dot{R} \sim \left[\frac{2}{3} \frac{(p_v - p_\infty^*)}{\rho} \right]^{1/2}, \tag{6–85}$$

and the increase in the bubble radius asymptotes to a constant rate. This corresponds to a smooth increase in the bubble volume at a rate dV/dt, proportional to t^2. For $p_\infty^* > p_\infty(0)$, on the other hand, the bubble collapses, and if $R \ll R_0$, the radius decreases at a rate

$$\dot{R} \sim \left(\frac{R_0}{R}\right)^{3/2} \left[\frac{2}{3}\frac{p_\infty^* - p_v}{\rho} + 2\frac{\gamma}{\rho R_0} - \frac{2\hat{G}}{\rho R_0^3}\ln\left(\frac{R_0}{R}\right)\right]^{1/2}. \tag{6-86}$$

If the bubble contains mostly insoluble gas, it is unlikely that the bubble will ever satisfy the inequality $R \ll R_0$. In other cases, however, (6–86) shows that the bubble radius will decrease at an ever-increasing rate until the last term finally balances the first two. This occurs at a minimum bubble size

$$R_{\min} = R_0 \exp\left\{-\frac{R_0^3}{3G}\left(p_\infty^* - p_v + \frac{3\gamma}{R_0}\right)\right\}. \tag{6-87}$$

The relatively mild growth and the violent collapse processes predicted by the asymptotic forms (6–85) and (6–86) are characteristic of the dynamics obtained by more general *numerical* studies of the Rayleigh-Plesset equation, but additional *results* for cases of large volume change are not possible by analytic solution. In the remainder of this section, we consider additional results that can be obtained by asymptotic methods.

Equilibrium Solutions and Their Stability

We have noted in the previous section that a bubble in equilibrium must have a radius R_E that satisfies the condition (6–80), namely,

$$R_E^3(p_v - \hat{p}_\infty) - 2\gamma R_E^2 + \hat{G} = 0.$$

Here, $\hat{p}_\infty$ is a constant, ambient pressure, and p_v is the vapor pressure at the ambient temperature.

Now, if we solve this cubic equation for R_E, we find that there are *three real roots* when $p_v - \hat{p}_\infty$ is less than a critical value

$$\boxed{(\Delta p)_{\text{crit}} = \left(\frac{32\gamma^3}{27G}\right)^{1/2}} \tag{6-88}$$

but no positive real roots for larger values of Δp. Two of the roots for R_E are positive, and one is negative (thus meaningless in this context). Thus, if we plot R_E versus $(p_v - \hat{p}_\infty)$, we obtain the result sketched in Figure 6–8. In the language of nonlinear systems, there is a limit point at Δp_{crit}, with two steady solutions for smaller values of Δp, and no solutions for $\Delta p > \Delta p_{\text{crit}}$. The bubble radius at the limit point, which is denoted in Figure 6–8 as R_{crit}, can be found directly from (6–80) by calculating the derivative

$$\frac{\partial R_E}{\partial(\Delta p)} = \frac{R_E^2}{(4\gamma - 3R_E\Delta p)}$$

and determining the point where $\partial R/\partial(\Delta p) \to \infty$. Thus,

$$R_{\text{crit}} = \frac{4\gamma}{3\Delta p_{\text{crit}}}. \tag{6-89}$$

Substituting this value for R_E into (6–80) leads to the critical value of Δp given by

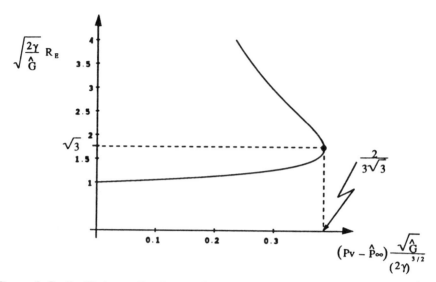

Figure 6–8 Equilibrium radius R_E as a function of the pressure driving force $(p_v - \hat{p}_\infty)$. For $(p_v - \hat{p}_\infty) < \Delta p_{crit}$ [Equation (6–88)], there are two equilibrium radii possible. However, for $(p_v - \hat{p}_\infty) > \Delta p_{crit}$, no equilibrium radius exists, and the bubble must undergo time-dependent growth.

(6–88). Finally, using the expression (6–88), we can express R_{crit} solely in terms of the material parameters $\hat{G}$ and γ, namely,

$$\boxed{R_{crit} = \sqrt{3\hat{G}/2\gamma}.} \tag{6–90}$$

Although two steady solutions exist for all $\Delta p \le \Delta p_{crit}$, an important question is whether either of these solutions is *stable*. Thus, if we were to consider an arbitrarily small change in the bubble volume from one of the predicted equilibrium values, we ask whether the bubble radius will return to the equilibrium value or continue to either grow or collapse.[8] In the former case, the corresponding equilibrium state is said to be *stable* to infinitesimal *perturbations* in the bubble volume, while it is said to be *unstable* in the latter case. A steady, equilibrium solution that is unstable to infinitesimal perturbations will not be realizable in any real physical system.

Now, the dynamics of changes in bubble radius with time, starting from some initial radius that differs slightly from an equilibrium value, is a problem that is ideally suited to solution via a regular asymptotic approximation. Of course, the governing equation is still the Rayleigh-Plesset equation. In the analysis of stability that follows, we seek an approximate, asymptotic solution of this equation in which the small parameter is the dimensionless *magnitude* of the initial departure from the equilibrium radius R_E. Hence, we seek a solution in the form

$$\boxed{R = R_E\left(1 + \sum_{n=1}^{\infty} \epsilon^n \delta_n(t)\right),} \tag{6–91}$$

where ϵ is a small parameter that provides a measure of the magnitude of the change in radius from R_E. Here, we consider the asymptotic limit $\epsilon \to 0$. The governing equations for the functions $\delta_n(t)$ can be obtained by substituting (6–91) into the Rayleigh-Plesset equation and collecting all terms of equal order in ϵ. Referring to (6–83), we obtain

$$(1 - \alpha p_\infty^*) + \beta - \frac{2}{\text{We}} + \epsilon\left[-\ddot{\delta}_1 - \frac{4}{\text{Re}}\dot{\delta}_1 - \left(3\beta - \frac{2}{\text{We}}\right)\delta_1 \right]$$

$$+ \epsilon^2\left[-\ddot{\delta}_2 - \frac{4}{\text{Re}}\dot{\delta}_2 - \left(3\beta - \frac{2}{\text{We}}\right)\delta_2 + \left(6\beta - \frac{2}{\text{We}}\right)\delta_1^2 + \left(\frac{4}{\text{Re}} - 1\right)\delta_1\dot{\delta}_1 - \frac{3}{2}\dot{\delta}_1^2 \right] + 0(\epsilon^3) = 0.$$

For arbitrary though small ϵ, the term at each level in ϵ is equal to zero, that is,

$$0(1) \quad 1 - \alpha p_\infty^* + \beta - \frac{2}{\text{We}} = 0,$$

$$0(\epsilon) \quad \ddot{\delta}_1 + \frac{4}{\text{Re}}\dot{\delta}_1 + \left(3\beta - \frac{2}{\text{We}}\right)\delta_1 = 0,$$

$$0(\epsilon^2) \quad \ddot{\delta}_2 + \frac{4}{\text{Re}}\dot{\delta}_2 + \left(3\beta - \frac{2}{\text{We}}\right)\delta_2 = \left(6\beta - \frac{2}{\text{We}}\right)\delta_1^2 + \left(\frac{4}{\text{Re}} - 1\right)\delta_1\dot{\delta}_1 - \frac{3}{2}\dot{\delta}_1^2.$$

For convenience, at $t = 0$, we assume that

$$\delta_1(0) = 1, \ \delta_2(0) = \delta_3(0) = \cdots = \delta_n(0) = 0.$$

The equation at $0(1)$ is just the equilibrium condition (6–80) now expressed in dimensionless terms [compare equation (6–83)].* The equation at $0(\epsilon)$ is

$$\boxed{\ddot{\delta}_1(t) + \frac{4}{\text{Re}}\dot{\delta}_1(t) + \left(3\beta - \frac{2}{\text{We}}\right)\delta_1(t) = 0.} \qquad (6\text{–}92)$$

To determine the *linear* stability of the equilibrium solution $R = R_E$, it is sufficient to determine whether the magnitude of the initial perturbation to the bubble radius, $\delta_1(0) = 1$, grows or decays with time. For this purpose, let us suppose that

$$\delta_1(t) = \delta_0 e^{st}. \qquad (6\text{–}93)$$

Then the steady solution is stable if the real part of s is negative, but unstable if it is positive. A characteristic equation for s is obtained by substituting (6–93) into (6–92). The result is

$$s^2 + \frac{4}{\text{Re}}s + \left(3\beta - \frac{2}{\text{We}}\right) = 0, \qquad (6\text{–}94)$$

and the two roots of this equation are

$$s = -\frac{2}{\text{Re}}\left[1 \pm \left(1 - \frac{\text{Re}^2}{4}\left(3\beta - \frac{2}{\text{We}}\right)\right)^{1/2}\right]. \qquad (6\text{–}95)$$

*Note that $p_\infty^* \equiv 0$ for this equilibrium solution.

Thus, the condition for *instability* is

$$\left(3\beta - \frac{2}{\text{We}}\right) < 0$$

or, upon substitution for β and We,

$$\boxed{R_E > \sqrt{3\hat{G}/2\gamma}.}$$
(6-96)

Referring back to Figure 6-8 and equation (6-90), we see that this limiting value for stability is precisely the same as the critical value R_{crit}. Thus, the lower part of the equilibrium solution curve (the lower branch of solutions) for

$$R_E \leq R_{\text{crit}}$$

is predicted to be *stable* to infinitesimal changes in the bubble radius, while the upper solution branch for

$$R_E > R_{\text{crit}}$$

is *unstable*.

Several comments are in order with regard to the preceding analysis. First, since the governing equation at $0(\epsilon)$ is linear, it pertains equally to the case $\delta_1(0) < 0$ or $\delta_1(0) > 0$. Second, the initial magnitude of $\epsilon\delta_1(0)$ is small but arbitrary. In particular, for $R_E > R_{\text{crit}}$, even an *infinitesimal* perturbation in the bubble volume leads to instability. Third, the governing equation (6-92) describes only the leading order (linearized) approximation to the departure from the steady equilibrium radius R_E. Thus, although the first term in the regular perturbation analysis can tell us whether an arbitrarily small change in the bubble radius will initially grow or decay, it cannot determine the ultimate fate of a growing disturbance because the underlying perturbation expansion will break down if the condition $\epsilon\delta_1(t) \ll 1$ is not maintained. For example, if $\delta_1(0) < 0$ and $R_E > R_{\text{crit}}$, the linearized analysis tells us that the bubble radius initially decreases exponentially with time, but it cannot tell us whether the radius ultimately goes to the equilibrium value on the lower solution branch. Similarly, for $\delta_1(0) > 0$, the bubble radius initially increases exponentially from R_E for $R_E > R_{\text{crit}}$, but the eventual fate of the bubble cannot be determined from this analysis. Finally, we should point out that the stability condition (6-96), expressed in terms of the bubble radius R, also provides a useful criteria for the critical pressure difference $(p_v - p_\infty)_{\text{crit}}$ that leads to instability. In particular, let us suppose that the bubble is subjected to an environment where the ambient pressure is changing slowly. Then, according to (6-80), as $p_v - \hat{p}_\infty$ increases, the radius of a stable bubble must also increase. But, according to (6-96), there is a maximum that the radius can achieve before the bubble becomes unstable. Thus, there is a corresponding maximum in $(p_v - \hat{p}_\infty)$, which is just the critical value $(\Delta p)_{\text{crit}}$, given in (6-88),

$$(p_v - \hat{p}_\infty)_{\text{max}} = \frac{4}{3}\gamma\left(\frac{2\gamma}{3G}\right)^{1/2}.$$

If $\hat{p}_\infty$ decreases below this value, we should expect continuous bubble growth.

For the reader who is knowledgeable about the dynamics of nonlinear systems, it will not be surprising that the stability analysis predicts an exchange of stability for steady solutions of (6–80) as we pass through the limit point from the lower to the upper solution branch. This behavior is expected on general grounds from the theories of nonlinear dynamical systems (see Iooss and Joseph [1980]).[9]

Bubble Oscillation for Periodic Pressure Oscillations

Another important characteristic of the gas bubble is its response to a periodic oscillation of the ambient pressure p_∞. For large amplitude oscillations of the pressure, or for an initial condition that is not near a stable equilibrium state for the bubble, the response can be very complicated, including the possibility of *chaotic* variations in the bubble radius.[10] However, such features are outside the realm of simple, analytical solutions of the governing equations, and we focus our attention here on the bubble response to asymptotically small oscillations of the ambient pressure, namely,

$$\boxed{p_\infty(t) = p_\infty^0(1 + \epsilon \sin \omega t),} \tag{6–97}$$

where $\epsilon \ll 1$ and the frequency ω is arbitrary. Furthermore, we restrict our attention to systems that deviate only slightly from a stable equilibrium radius, R_E. This implies that the initial state must lie within a small neighborhood of the equilibrium point.

The basis of our analysis is, again, a very simple variation of the regular perturbation technique in which we assume that the bubble radius can be expressed in terms of a regular asymptotic expansion of the form

$$\boxed{R = R_E\left[1 + \sum_n \epsilon^n g_n(t)\right].} \tag{6–98}$$

Inherent in this form for R is the assumption that the $0(\epsilon)$ forcing in the ambient pressure leads to an $0(\epsilon)$ change in the bubble radius, at least at the leading order of approximation.

The governing equations for the coefficient functions $g_n(t)$ are obtained by the standard perturbation procedure of substituting the expansion (6–98) into the Rayleigh-Plesset equation in the form (6–79) and then equating the sum of all terms at each order in ϵ to zero. Alternatively, we may use the nondimensionalized form (6–83). In either case, the result at $0(1)$, expressed in dimensional terms, is

$$p_v + \frac{\hat{G}}{R_E^3} - p_\infty^0 - \frac{2\gamma}{R_E} = 0.$$

At $0(\epsilon)$, we obtain the governing equation for g_1, that is,

$$\boxed{\ddot{g}_1 + \left(\frac{4\mu}{\rho R_E^2}\right)\dot{g}_1 + \frac{1}{\rho}\left(\frac{3\hat{G}}{R_E^5} - \frac{2\gamma}{R_E^3}\right)g_1 = -\frac{p_\infty^0}{\rho R_E^2}\sin \omega t.}$$

This can be expressed in the generic form

$$\boxed{\ddot{g}_1 + \sigma \dot{g}_1 + \omega_0^2 g_1 = \Delta \sin \omega t} \tag{6-99}$$

of a *forced harmonic oscillator* with damping factor

$$\sigma \equiv \frac{4\mu}{\rho R_E^2}$$

and a natural frequency

$$\omega_0^2 = \frac{1}{\rho}\left(\frac{3\hat{G}}{R_E^5} - \frac{2\gamma}{R_E^3}\right).$$

The response depends on the relationship between the forcing frequency ω and the natural frequency ω_0.

Prior to seeking solutions of (6–99), it is advantageous to rewrite it in dimensionless form. This is done by introducing ω_0^{-1} as a characteristic time scale (note that g_1 is dimensionless by definition). Then,

$$\boxed{\frac{d^2 g_1}{dt^2} + \frac{4}{\text{Re}_\omega}\frac{dg_1}{d\bar{t}} + g_1 = -\bar{p}_\infty \sin\left(\frac{\omega}{\omega_0}\bar{t}\right),} \tag{6-100}$$

where $\bar{t} = \omega_0 t$ and

$$\frac{4}{\text{Re}_\omega} \equiv \frac{4\mu}{\rho R_E^2 \omega_0} = \frac{\sigma}{\omega_0},$$

$$-\bar{p}_\infty = -\frac{p_\infty^0}{\rho(\omega_0 R_E)^2} = \frac{\Delta}{\omega_0^2}.$$

We begin by considering the solution of (6–100) for the case $\omega \neq \omega_0$. Let us first examine the limiting case of zero damping where $4/\text{Re}_\omega \equiv 0$, so that the governing equation is

$$\boxed{\frac{d^2 g_1}{dt^2} + g_1 = -\bar{p}_\infty \sin\left(\frac{\omega}{\omega_0}\bar{t}\right).} \tag{6-101}$$

We seek a solution of this equation subject to initial conditions

$$g_1(0) = \dot{g}_1(0) = 0, \tag{6-102}$$

which imply that

$$R(0) = R_E, \quad \dot{R}(0) = 0.$$

The general solution of (6–101) is

$$g_1(\bar{t}) = C_1 \sin \bar{t} + C_2 \cos \bar{t} - \frac{\bar{p}_\infty}{1 - \left(\dfrac{\omega}{\omega_0}\right)} \sin\left(\frac{\omega}{\omega_0}\bar{t}\right). \tag{6-103}$$

Then, applying the boundary conditions (6–102), we obtain

$$g_1(\bar{t}) = \frac{p_\infty^0}{1 - \left(\dfrac{\omega}{\omega_0}\right)^2} \left[\frac{\omega}{\omega_0}\sin\bar{t} - \sin\left(\frac{\omega}{\omega_0}\bar{t}\right)\right]. \qquad (6\text{–}104)$$

Thus, in the absence of viscous damping, we find that the bubble radius oscillates periodically with an amplitude of $0(\epsilon)$ in response to the oscillating pressure field, provided only that $\omega \neq \omega_0$ as we have assumed. A plot is given in Figure 6–9 showing the time dependence of g_1/p_∞^0 for several different values of (ω/ω_0). A key point to note about the solution (6–104), however, is that the magnitude of g_1 becomes *unbounded* in the limit $\omega \to \omega_0$. Indeed, in the limit $\omega = \omega_0$, no bounded solution of the asymptotic form (6–98) exists. This is a consequence of the *resonant interaction* that occurs when the forcing frequency ω is equal to the natural oscillation frequency of the bubble, ω_0.

A resonant effect is also evident even when the viscous damping term is included in equation (6–99). In this case, it is convenient to reformulate the problem in the form

$$\ddot{G}_1 + \frac{4}{\text{Re}_\omega}\dot{G}_1 + G_1 = -\bar{p}_\infty e^{i(\omega/\omega_0)\tau}, \qquad (6\text{–}105)$$

where $g_1 = -\mathcal{R}e(iG_1)$. In this case, we can seek a solution in the form

$$G_1 = Ce^{st}. \qquad (6\text{–}106)$$

Substituting into (6–105) thus yields

$$Ce^{st}\left(s^2 + \frac{4}{\text{Re}_\omega}s + 1\right) = -\bar{p}_\infty e^{i(\omega/\omega_0)\tau}. \qquad (6\text{–}107)$$

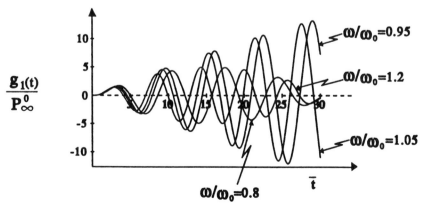

Figure 6–9 Amplitude function for the oscillation of bubble radius in an inviscid fluid due to a time-dependent oscillation in the ambient pressure.

The particular solution is thus obtained by letting $s = i\omega$, then

$$C = \frac{-\bar{p}_\infty}{-\left(\dfrac{\omega}{\omega_0}\right)^2 + \dfrac{4}{\text{Re}_\omega}\left(\dfrac{\omega}{\omega_0}\right)i + 1}. \tag{6-108}$$

Homogeneous solutions are obtained by letting s be the two roots of the characteristic equation

$$s^2 + \frac{4}{\text{Re}_\omega}s + 1 = 0. \tag{6-109}$$

These are

$$s = -\frac{2}{\text{Re}_\omega}\left[1 \pm \left(1 - \frac{\text{Re}_\omega^2}{4}\right)^{1/2}\right]. \tag{6-110}$$

Thus, the general solution of (6–105) is

$$G_1 = C_1 e^{s_1 t} + C_2 e^{s_2 t} + \frac{-\bar{p}_\infty}{-\left(\dfrac{\omega}{\omega_0}\right)^2 + \dfrac{4}{\text{Re}_\omega}\left(\dfrac{\omega}{\omega_0}\right)i + 1} e^{i(\omega/\omega_0)\tau} \tag{6-111}$$

with the boundary conditions

$$G_1 = \dot{G}_1 = 0 \quad \text{at} \quad t = 0. \tag{6-112}$$

There are two cases: When $\text{Re}_\omega^2/4 < 1$, both roots s_1 and s_2 are real and negative; when $\text{Re}_\omega^2/4 > 1$, s_1 and s_2 are complex conjugates with negative real parts. Hence, in both cases, the homogeneous terms in (6–111) vanish as $t \to \infty$ for all nonzero values of σ. For present purposes, we restrict our attention only to the large-time asymptote of (6–111), which thus takes the form

$$G_1 = \frac{-\bar{p}_\infty}{\left(1 - \left(\dfrac{\omega}{\omega_0}\right)^2\right) + i\dfrac{4}{\text{Re}_\omega}\left(\dfrac{\omega}{\omega_0}\right)} e^{i(\omega/\omega_0)\tau}. \tag{6-113}$$

The corresponding real part of $(-iG_1)$ is

$$g_1 = -\frac{\bar{p}_\infty}{\left[1 - \left(\dfrac{\omega}{\omega_0}\right)^2\right]^2 + \left(\dfrac{4}{\text{Re}_\omega} \cdot \dfrac{\omega}{\omega_0}\right)^2}\left[\left(1 - \left(\dfrac{\omega}{\omega_0}\right)^2\right)\sin\frac{\omega}{\omega_0}\bar{t} - \frac{4}{\text{Re}_\omega}\left(\dfrac{\omega}{\omega_0}\right)\cos\frac{\omega}{\omega_0}\bar{t}\right].$$

$$\tag{6-114}$$

The solution $g_1/\bar{p}_\infty$ is plotted in Figure 6–10 for several values of (ω/ω_0) and several values of Re_ω. It will be noted that (6–114) does not reduce to (6–104) in the limit as $\text{Re}_\omega^{-1} \to 0$. This is because the roots s_1 and s_2 have negative real parts for all $\text{Re}_\omega^{-1} \neq 0$, so that the homogeneous terms in (6–111) vanish for sufficiently large times but are purely imaginary for $\text{Re}_\omega^{-1} = 0$ ($s = \pm i$) and thus do not vanish even as $t \to \infty$. Indeed, the second of the two terms in (6–104) is a homogeneous solution,

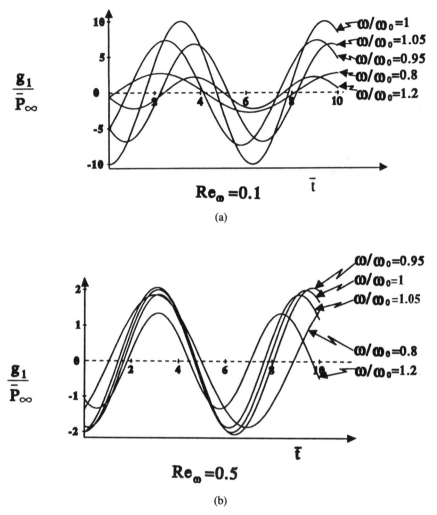

Figure 6–10 Amplitude function for the oscillation of bubble radius in a *viscous* fluid due to a time-dependent oscillation in the ambient pressure (a) $Re_\omega = 0.1$ and (b) $Re_\omega = 0.5$.

which has no counterpart in (6–114) for $Re_\omega^{-1} \to 0$. In the resonant limit $\omega \to \omega_0$, the solution (6–114) approaches a maximum magnitude for g_1,

$$(g_1)_{max} \sim \frac{\bar{p}_\infty Re_\omega}{4} \cos \bar{t}$$

though this remains bounded for finite $Re_\omega^{-1} \sim 0(1)$. If, on the other hand, $Re_\omega^{-1} \sim 0(\epsilon^p)$—that is, the viscous damping is small and proportional to the magnitude of the pressure variation to some positive power, p—the asymptotic scheme again breaks down in the resonant limit, and some different type of solution must be attempted.

The resonant case $\omega = \omega_0$ is, itself, quite interesting and also amenable to asymptotic analysis. We have seen, in this case, that an approximation of the form (6–98) is not possible for either $\mathrm{Re}_\omega^{-1} = 0$ or $\mathrm{Re}_\omega^{-1} = 0(\epsilon^p)$—that is, a response of $0(\epsilon)$ does not occur in spite of the fact that the pressure variation is $0(\epsilon)$ and $\epsilon \ll 1$. An obvious question is whether the amplitude of changes in bubble radius actually remains bounded in the resonant case $\omega = \omega_0$ when $\mathrm{Re}_\omega^{-1} \ll 1$. We explore this question briefly in the following pages for the undamped case $\mathrm{Re}_\omega^{-1} \equiv 0$. The problem for very small, but nonzero, damping is similar, but we shall not consider it here.

If we reexamine the governing equation (6–101) for $\mathrm{Re}_\omega^{-1} \equiv 0$, we see that the mathematical difficulty for $\omega = \omega_0$ is that the forcing function is actually a solution of the homogeneous equation. The forcing function in this case is known as a *secular* term. Obviously, when a secular forcing term is present, a particular solution cannot exist in the form $C \sin \bar{t}$ as is possible for $\omega \neq \omega_0$—see, for example (6–103)—but must instead take the form $C \bar{t} \cos \bar{t}$. Indeed, substituting this form into the governing equation (6–101) with $\omega = \omega_0$, we find

$$\boxed{(g_1)_{\text{particular}} = \frac{\bar{p}_\infty}{2} \bar{t} \cos \bar{t} \quad (\omega \equiv \omega_0).}$$

Although this solution is perfectly well behaved for $\bar{t} \ll 1$, we see that it grows *without bound* for $\bar{t} \to \infty$. It thus follows that the asymptotic expansion (6–98) cannot provide a valid asymptotic approximation scheme for $\epsilon \ll 1$ for all $\bar{t}$, when $\omega = \omega_0$. The fact is that if a bounded solution is to exist for the variation of bubble radius due to small pressure oscillations at the resonant frequency, the $0(\epsilon)$ pressure variation must eventually be balanced by one or more of the *nonlinear* terms that were neglected in deriving (6–101) from the limiting form of the Rayleigh-Plesset equation with $\mathrm{Re}_\omega^{-1} \equiv 0$. In the analysis that follows, we show that the resonant response from the Rayleigh-Plesset equation for small ϵ is that the bubble radius oscillates with an amplitude of $0(\epsilon^{1/3})$—that is, it is still asymptotically small but *much larger* than the $0(\epsilon)$ amplitude of the pressure forcing and much larger than assumed in the expansion (6–98).

The following analysis for the resonant case is based upon the observation that the response of a nonlinear oscillator to weak periodic forcing at a *resonant* frequency is usually characterized by *multiple time scales*, a slow, periodic oscillation superposed on a higher frequency oscillation that is at (or near) the forcing frequency. A special kind of asymptotic approximation scheme has been developed for this kind of problem, which is known as a *two-timing* or *two-variable expansion procedure*. A complete description and motivation for the two-timing technique is beyond the scope of this book.[11] Nevertheless, it is perhaps useful to illustrate its application to this specific example problem by proceeding in a purely *formal* manner. Thus, we simply note that the basis of the two-timing procedure is to consider the unknown function as depending on two *independent* variables: the regular time $\bar{t}$, which is of $0(1)$, and a second asymptotically longer time scale variable τ that is related to $\bar{t}$ via a rescaling of the form

$$\tau = \epsilon^m \bar{t}.$$

Hence, any time derivative that appears in the governing equation is now approximated as

$$\frac{\partial}{\partial \bar{t}} \rightarrow \frac{\partial}{\partial \bar{t}} + \epsilon^m \frac{\partial}{\partial \tau}.$$

With this general formalism, the solution for the present problem can be shown to exist in the specific asymptotic form

$$R = R_E\left(1 + \sum_n \epsilon^{n/3} g_n(t, \tau)\right), \qquad (6\text{-}115)$$

where

$$\tau = \epsilon^{2/3} t. \qquad (6\text{-}116)$$

The governing equations for the functions g_n are obtained from the Rayleigh-Plesset equation in the usual manner. We thus substitute (6–115) into the inviscid form of the Rayleigh-Plesset equation and collect terms of like powers in ϵ. The first several functions $g_n(t, \tau)$ are found to satisfy the dimensionless equations

$$\ddot{g}_1 + g_1 = 0, \qquad (6\text{-}117)$$

$$\ddot{g}_2 + g_2 = -g_1 \ddot{g}_1 - \frac{3}{2}(\dot{g}_1)^2 + \frac{1}{\omega_0^2}\left[\omega_0^2 + \frac{3G}{\rho R_E^5}\right] g_1^2, \qquad (6\text{-}118)$$

and

$$\ddot{g}_3 + g_3 = -2\frac{\partial^2 g_1}{\partial \bar{t}\partial \tau} - \bar{p}_\infty \sin \bar{t} - \left\{ g_1 \ddot{g}_2 + g_2 \ddot{g}_1 + 3\dot{g}_1 \dot{g}_2 \right\}$$

$$+ \left[2 + \frac{6\hat{G}}{\rho \omega_0^2 R_E^5} \right] g_1 g_2 - \left[1 + \frac{7\hat{G}}{\rho \omega_0^2 R_E^5} \right] g_1^3. \qquad (6\text{-}119)$$

Here, the derivatives with respect to $\bar{t}$ are represented as $\dot{g}_n$ and $\ddot{g}_n$.

Now, equation (6–117) is nothing more than the homogeneous equation obtained in our earlier analysis, and it might appear that nothing has been accomplished since the secular term will now appear in (6–119). However, it is at this point that the analysis departs from the preceding work. For in this case, we consider g_1 to depend not only on the independent variable $\bar{t}$, which appears explicitly in (6–117), but also on the slow time scale τ. Hence, a convenient form for the general solution of (6–117) is

$$g_1 = A(\tau) \sin(\bar{t} - \phi(\tau)), \qquad (6\text{-}120)$$

where the slowly varying functions $A(\tau)$ and $\phi(\tau)$ remain unspecified. But now $A(\tau)$ and $\phi(\tau)$ can be chosen in such a way that the secular terms that would otherwise appear on the right-hand side of (6–119) are canceled so that g_3 remains bounded for all t. To see that this is possible, we must proceed with the solution of equations (6–118) and (6–119).

Hence, using g_1 from (6–120), equation (6–118) for g_2 takes the form

$$\ddot{g}_2 + g_2 = \frac{A^2}{4}\left(1 + \frac{6G}{\omega_0^2 \rho R_E^5}\right) - \frac{A^2}{4}\left(7 + \frac{6G}{\omega_0^2 \rho R_E^5}\right)\cos 2(\bar{t} - \phi(\tau)), \qquad (6\text{-}121)$$

and a general solution is

$$
g_2 = C(\tau)\sin(\bar{t} - \delta(\tau)) + \frac{A^2}{4}\left(1 + \frac{6G}{\rho R_E^5 \omega_0^2}\right) + \frac{A^2}{4}\left(\frac{7}{3} + \frac{2G}{\rho R_E^5 \omega_0^2}\right)\cos 2(\bar{t} - \phi).
$$

(6–122)

Again, the long time scale functions $C(\tau)$ and $\delta(\tau)$ can be chosen to avoid secular behavior for one of the higher-order terms in the asymptotic expansion.

Finally, we consider the solution of (6–119) for the $0(\epsilon)$ term in the expansion (6–115). The first thing to notice is that the right-hand side of (6–119) contains a secular term corresponding to the direct pressure forcing at $0(\epsilon)$, and our immediate reaction may be that it will again be impossible to obtain a bounded solution for g_3. However, in this case, the other terms on the right-hand side of (6–119) involve the as yet undetermined functions $A(\tau)$, $\phi(\tau)$, $C(\tau)$, and $\delta(\tau)$. Clearly, if a bounded solution is to exist in the form (6–115), these functions must be chosen in such a way as to eliminate the secular behavior in the equation for g_3. To see that this is possible, we evaluate the right-hand side of (6–119) using the general solutions (6–120) and (6–122) for g_1 and g_2. The result is

$$
\ddot{g}_3 + g_3 = -2\left[\frac{dA}{d\tau}\cos(\bar{t} - \phi(\tau)) + A\frac{d\phi}{d\tau}\sin(\bar{t} - \phi(\tau))\right] - \bar{p}_\infty\sin\bar{t}
$$
$$
+ \left(\frac{A^3}{4}\right)\left[-\frac{7}{6} - 5\left(\frac{G}{\rho R_E^5 \omega_0^2}\right) + 30\left(\frac{G}{\rho R_E^5 \omega_0^2}\right)^2\right]\sin(\bar{t} - \phi(\tau))
$$
$$
+ \text{nonsecular terms.}
$$

(6–123)

Thus, to avoid secular terms and, therefore, to ensure bounded solutions for g_3, we must choose $A(\tau)$ and $\phi(\tau)$ so that the secular terms on the right-hand side of (6–123) are equal to zero. After some manipulation, this condition leads to the coupled pair of equations

$$
A\frac{d\phi}{d\tau} = -(\bar{p}_\infty)\frac{\cos\phi}{2} + \left(\frac{A^3}{8}\right)\left[-\frac{7}{6} - 5\left(\frac{G}{\rho R_E^5 \omega_0^2}\right) + 30\left(\frac{G}{\rho R_E^5 \omega_0^2}\right)^2\right]
$$

(6–124)

and

$$
\frac{dA}{d\tau} = -(\bar{p}_\infty)\frac{\sin\phi}{2}.
$$

(6–125)

Hence, provided solutions to these equations exist, we see that

$$
R = R_E\left[1 + \epsilon^{1/3}g_1 + 0(\epsilon^{2/3})\right],
$$

(6–126)

with

$$
g_1 = A(\tau)\sin(\omega_0 t - \phi(\tau))
$$

(6–127)

and

$$
\tau = \epsilon^{2/3}t,
$$

where $A(\tau)$ and $\phi(\tau)$ satisfy the differential equations (6–124) and (6–125). If we proceed to the next level of approximation to determine g_4, the functions $C(\tau)$ and $\delta(\tau)$ would be chosen to eliminate secular terms in that problem, and so on.

To complete the analysis, we must obtain solutions for the amplitude and phase functions $A(\tau)$ and $\phi(\tau)$ from (6–124) and (6–125). This is most easily accomplished by first rescaling time according to

$$t^* = \tau \left(\frac{\bar{p}_\infty}{2} \right)$$

so that (6–124) and (6–125) can be expressed in the simpler form

$$A \frac{d\phi}{dt^*} = -\cos\phi + C\, A^3, \tag{6–128}$$

$$\frac{dA}{dt^*} = -\sin\phi, \tag{6–129}$$

where

$$C = \frac{1}{8} \left(\frac{2}{\bar{p}_\infty} \right) \left(30 \left(\frac{G}{\rho R_E^5 \omega_0^2} \right)^2 - 5 \left(\frac{G}{\rho R_E^5 \omega_0^2} \right) - \frac{7}{6} \right).$$

Then, introducing $x \equiv A \cos \phi$ and $y = A \sin \phi$, it follows that the equations (6–128) and (6–129) can be replaced by the equivalent pair of dynamical equations

$$\frac{dy}{dt^*} = -1 + C(x^2 + y^2)x \tag{6–130}$$

and

$$\frac{dx}{dt^*} = -C(x^2 + y^2)y. \tag{6–131}$$

The advantage of this latter transformation is that (6–130) and (6–131) can be written in a classical Hamiltonian form,

$$\frac{dy}{dt^*} = \frac{\partial H}{\partial x} \quad ; \quad \frac{dx}{dt^*} = -\frac{\partial H}{\partial y}, \tag{6–132}$$

where the so-called Hamiltonian function H is defined as

$$H \equiv \frac{C}{4} (x^2 + y^2)^2 - x. \tag{6–133}$$

The solution trajectories $x(t^*)$ and $y(t^*)$ of a system of this type correspond to contours of constant values of H.

The qualitative significance of this solution of the dynamic system of equations (6–130) and (6–131) can be deduced by plotting the contours of constant H in the x-y plane, as shown in Figure 6–11 (for two different positive values of C). The relationships between a point on one of these contours and the original functions $A(\tau)$ and

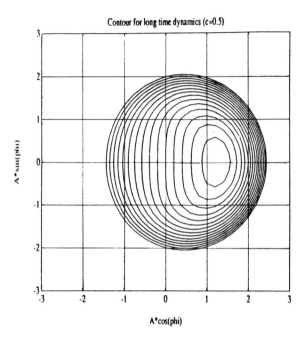

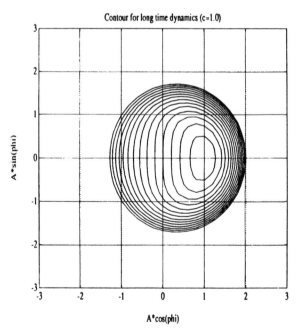

Figure 6–11 The solution trajectories $x(t^*)$ and $y(t^*)$ of the system (6–130) and (6–131), which correspond to contours of constant values of H (c = 0.5 and c = 1.0).

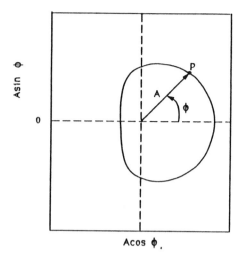

Figure 6–12 A schematic showing a single contour from Figure 6–11 in order to indicate the physical significance of a point on the contour in terms of $A(\tau)$ and $\phi(\tau)$.

$\phi(\tau)$ is indicated in Figure 6–12. The direction of motion *around* the contours of Figure 6–11 is counterclockwise, as we can easily verify by examining the sign of dy/dt^* at *various* points along the x axis (that is, $y = 0$), using (6–130). The qualitative nature of the variation of A and ϕ with time depends upon the initial conditions $A(0)$ and $\phi(0)$. For each C, there is a single *fixed* point at $\phi = 0$, $A = 1/\sqrt{C}$ where both ϕ and A are time independent. All other solution trajectories give *oscillatory* values for A. Those trajectories that lie wholly to the *right* of the origin also give oscillatory variations of the phase angle ϕ, while all other trajectories yield a monotonic *increase* of ϕ. Hence, the predicted response to a time-dependent oscillation of pressure at the *natural* oscillation frequency of the bubble *is* an oscillatory response in bubble volume at the same frequency, with a slow variation in the *amplitude* of this oscillation described by $A(\tau)$ and a slow variation in the phase angle described by $\phi(\tau)$. A sketch of g_1 as a function of t for $\epsilon = 0.1$ and $C = 1$ is shown in Figure 6–13 for two arbitrarily chosen sets of initial values for A and ϕ at $t = 0$.

We see that even in the resonant case $\omega = \omega_0$ with zero viscous damping, the changes in bubble volume remain small for small amplitude pressure forcing. Thus, although the analysis is more complicated, it is still possible to obtain approximate solutions for $\epsilon \ll 1$ by means of an asymptotic perturbation expansion for small changes in the bubble radius. However, it is significant in this case that the $0(\epsilon)$ forcing produces an $0(\epsilon^{1/3})$ response in the changes of bubble volume. The solutions of (6–124) and (6–125) describe the slow modulation of amplitude and phase shift in the solution as discussed above. The reader who is interested in more details of the two-timing asymptotic technique may wish to consult one of the several excellent books on applied mathematics and asymptotic methods, such as Bender and Orszag (1978) or Kevorkian and Cole (1981).[12]

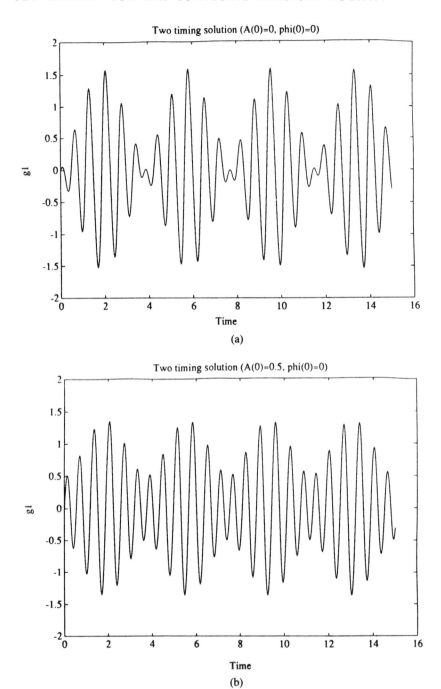

Figure 6–13 The function g_1, defined in Equation (6–127), for $A(0) = \phi(0) = 0$ and $A(0) = 0.5, \phi(0) = 0$. The qualitative difference in these plots is indicative of the sensitivity to initial conditions.

Stability to Nonspherical Disturbances

All of the preceding analysis in this section is predicated on the critical assumption that the bubble remains spherical, so that the flow is *radial*. An important question, which can also be answered by a regular perturbation analysis for small deformations in the bubble *shape*, is whether the spherical surface is stable or whether we should expect deviations to nonspherical shape as the bubble either expands or contracts. It is well known that the shape of a bubble near a wall, or in a mean flow such as simple shear, will be nonspherical.[13] Here, however, we are concerned with instabilities in the shape of an expanding or collapsing spherical bubble in an unbounded, quiescent fluid. The fact that the interface of a bubble may be unstable in these circumstances is suggested by the early work of Taylor (1950),[14] who showed that an infinite *plane* interface between two fluids of different density in accelerated motion is either stable or unstable depending on whether the acceleration is directed from the heavier to the lighter fluid, or conversely. The possible instability of an expanding or collapsing spherical interface may be viewed as a generalization of Taylor's result.[15] The critical difference is that the amplitude of a disturbance on a spherical surface must be *decreased* as the surface expands and *increased* as the surface contracts. Conversely, the wavelength of the disturbance must increase as the surface expands and decrease as it contracts.

In the present case, we consider a bubble whose interface is described in terms of a spherical coordinate system in the asymptotic form

$$r_s = R(t) + \sum_n \epsilon^n f_n(\theta, \phi, t). \qquad (6\text{--}134)$$

Here, ϵ is a small parameter that will form the basis of an asymptotic approximation for the dynamics of the bubble surface. The question here is whether a bubble with a nonspherical initial shape of small amplitude, $0(\epsilon)$, will return to a sphere—that is, $f_n(\theta, \phi, t) \to 0$ as $t \to \infty$—or whether the initial disturbance in shape will grow with increase of t.

To analyze this problem, we need to go back to a statement of the problem in general fluid dynamical terms. Thus, we begin by restating the governing equations and boundary conditions for an oscillating bubble in a quiescent, incompressible fluid. These are the Navier-Stokes and continuity equations; the three boundary conditions

$$\mathbf{u} \cdot \mathbf{n} = -\frac{1}{|\nabla F|} \frac{\partial F}{\partial t} \quad \text{(kinematic condition)}, \qquad (6\text{--}135)$$

$$(\boldsymbol{\tau} \cdot \mathbf{n}) \cdot \mathbf{t}_i = 0 \; (i = 1,2) \quad \text{(zero shear stress)}, \qquad (6\text{--}136)$$

and

$$p_B - p + (\boldsymbol{\tau} \cdot \mathbf{n}) \cdot \mathbf{n} = \gamma(\nabla \cdot \mathbf{n}) \quad \text{(normal stress condition)} \qquad (6\text{--}137)$$

that are applied at the surface of the deformed bubble

$$F \equiv r - \left(R(t) + \sum_n \epsilon^n f_n(\theta, \phi, t) \right) = 0;$$

and the condition

$$\boxed{\mathbf{u} \to 0 \quad \text{as} \quad |\mathbf{r}| \to \infty.} \tag{6-138}$$

Notice in the conditions above that p_B is the pressure inside the bubble, γ is the interfacial tension, and $\mathbf{t}_i$ represents one of the pair of orthogonal unit vectors that are tangent at any point to the bubble surface. The problem is to calculate velocity and pressure fields in the fluid, as well as to determine the functions $R(t)$ and $f_n(\theta, \phi, t)$ that describe the geometry of the bubble surface. In the present context, it is the latter part of the problem that is the focus of our interest.

One complication is that the boundary conditions (6–135)–(6–137) must be applied at the bubble surface, which is both unknown [that is, specified in terms of functions $R(t)$ and $f_n(\theta, \phi, t)$ that must be calculated as part of the solution] and *nonspherical*. Further, the normal and tangent unit vectors $\mathbf{n}$ and $\mathbf{t}_i$ that appear in the boundary conditions are also functions of the bubble shape. In this analysis, we use the small deformation limit $\epsilon \ll 1$ to simplify the problem. First, we note that the unit normal and tangent vectors can be approximated for small ϵ in the forms

$$\mathbf{n} = \frac{\nabla F}{|\nabla F|} = \mathbf{i}_r - \epsilon \left[\mathbf{i}_\theta \frac{1}{R} \frac{\partial f_1}{\partial \theta} + \mathbf{i}_\phi \frac{1}{R \sin\theta} \frac{\partial f_1}{\partial \phi} \right] + 0(\epsilon^2), \tag{6-139}$$

$$\mathbf{t}_i = \mathbf{i}_\theta + \epsilon \frac{1}{R} \frac{\partial f_1}{\partial \theta} \mathbf{i}_r + 0(\epsilon^2), \tag{6-140}$$

and

$$\mathbf{t}_2 = \mathbf{i}_\phi + \epsilon \frac{1}{R \sin\theta} \frac{\partial f_1}{\partial \phi} \mathbf{i}_r + 0(\epsilon^2). \tag{6-141}$$

Further, the curvature at the bubble surface becomes

$$\nabla \cdot \mathbf{n} = \frac{2}{R} - \epsilon \left[\frac{\cos\theta}{\sin\theta} \frac{\partial f_1}{\partial \theta} + \frac{\partial^2 f_1}{\partial \theta^2} + \frac{1}{\sin^2\theta} \frac{\partial^2 f_1}{\partial \phi^2} \right] \frac{1}{R^2} + 0(\epsilon^2). \tag{6-142}$$

Thus, at the bubble surface

$$\mathbf{u} \cdot \mathbf{n} = u_r - \epsilon \frac{1}{R} \frac{\partial f_1}{\partial \theta} u_\theta - \epsilon \frac{1}{R \sin\theta} \frac{\partial f_1}{\partial \phi} u_\phi + 0(\epsilon^2), \tag{6-143}$$

while

$$(\boldsymbol{\tau} \cdot \mathbf{n}) \cdot \mathbf{t}_1 = \tau_{r\theta} - \epsilon \left[\frac{1}{R} \frac{\partial f_1}{\partial \theta} \tau_{\theta\theta} + \frac{1}{R \sin\theta} \frac{\partial f_1}{\partial \phi} \tau_{\theta\phi} - \frac{1}{R} \frac{\partial f_1}{\partial \theta} \tau_{rr} \right] + 0(\epsilon^2), \tag{6-144}$$

$$(\boldsymbol{\tau} \cdot \mathbf{n}) \cdot \mathbf{t}_2 = \tau_{r\phi} - \epsilon \left[\frac{1}{R} \frac{\partial f_1}{\partial \theta} \tau_{\theta\phi} + \frac{1}{R \sin\theta} \frac{\partial f_1}{\partial \phi} \tau_{\phi\phi} - \frac{1}{R \sin\theta} \frac{\partial f_1}{\partial \theta} \tau_{rr} \right] + 0(\epsilon^2), \tag{6-145}$$

and

$$(\boldsymbol{\tau} \cdot \mathbf{n}) \cdot \mathbf{n} = \tau_{rr} - \epsilon \left[2 \frac{1}{R} \frac{\partial f_1}{\partial \theta} \tau_{r\theta} + 2 \frac{1}{R \sin\theta} \frac{\partial f_1}{\partial \phi} \tau_{r\phi} \right] + 0(\epsilon^2). \tag{6-146}$$

Finally, we can combine (6–143)–(6–146) with the boundary conditions (6–135)–(6–137).

It should be noted, however, that the application of these conditions is still complicated by the fact that they must be applied at the slightly deformed surface

$$r_s = R(t) + \epsilon f_1(\theta, \phi, t) + 0(\epsilon^2).$$

To avoid this, we use an idea known as *domain perturbation theory* to transform from the exact boundary conditions applied at $r_s = R + \epsilon f_1$ to asymptotically equivalent boundary conditions applied at the *spherical* surface $r_s = R(t)$. For example, instead of a condition on u_r at $r = R(t) + \epsilon f_1$ from the kinematic condition, we can obtain an asymptotically eqivalent condition at $r = R$ via the Taylor series approximation

$$u_r|_{r=r_s} = u_r|_{r=R} + \frac{\partial u_r}{\partial r}\bigg|_{r=R} \epsilon f_1 + 0(\epsilon^2) \qquad (6\text{–}147)$$

and similarly for $\tau_{r\theta}$, $\tau_{r\phi}$, and τ_{rr}. This simple idea of transforming from boundary conditions at a surface that is not a coordinate surface to asymptotically equivalent boundary conditions on a nearby surface that is a coordinate surface is the essence of the *domain perturbation* technique. Putting everything together, then, the boundary conditions (6–135)–(6–137) become

$$u_r|_{r=R} + \epsilon\left[\frac{\partial u_r}{\partial r}\bigg|_{r=R} f_1 - \frac{1}{R}\frac{\partial f_1}{\partial\theta}u_\theta\bigg|_{r=R} - \frac{1}{R\sin\theta}\frac{\partial f_1}{\partial\phi}u_\phi\bigg|_{r=R}\right] + 0(\epsilon^2) =$$

$$= \frac{dR}{dt} + \epsilon\frac{\partial f_1}{\partial t} + 0(\epsilon^2) \qquad (6\text{–}148)$$

$$\tau_{r\theta}|_{r=R} + \epsilon\left[\frac{\partial \tau_{r\theta}}{\partial r}\bigg|_{r=R} f_1 - \frac{1}{R}\frac{\partial f_1}{\partial\theta}\tau_{\theta\theta}\bigg|_R - \frac{1}{R\sin\theta}\frac{\partial f_1}{\partial\phi}\tau_{\theta\phi}\bigg|_R + \frac{1}{R}\frac{\partial f_1}{\partial\theta}\tau_{rr}|_R\right] + 0(\epsilon^2) = 0,$$

$$(6\text{–}149)$$

$$\tau_{r\phi}|_{r=R} + \epsilon\left[\frac{\partial \tau_{r\phi}}{\partial r}\bigg|_R f_1 - \frac{1}{R}\frac{\partial f_1}{\partial\theta}\tau_{\theta\phi}\bigg|_R - \frac{1}{R\sin\theta}\frac{\partial f_1}{\partial\phi}\tau_{\phi\phi} + \frac{1}{R\sin\theta}\frac{\partial f_1}{\partial\phi}\tau_{rr}|_R\right] + 0(\epsilon^2) = 0,$$

and

$$(p_B - p_{r=R}) + \frac{\partial}{\partial t}(-p)|_R \epsilon f_1 + \tau_{rr}|_{r=R}$$

$$+ \epsilon\left[\frac{\partial \tau_{rr}}{\partial r}\bigg|_{r=R} f_1 - \frac{2}{R}\frac{\partial f_1}{\partial\theta}\tau_{r\theta}\bigg|_{r=R} - \frac{2}{R\sin\theta}\frac{\partial f_1}{\partial\phi}\tau_{r\phi}\bigg|_{r=R}\right] + 0(\epsilon^2) =$$

$$\gamma\left[\frac{2}{R} - \epsilon\left[\frac{\cos\theta}{\sin\theta}\frac{\partial f_1}{\partial\theta} + \frac{\partial^2 f_1}{\partial\theta^2} + \frac{1}{\sin^2\theta}\frac{\partial^2 f_1}{\partial\phi^2}\right]\frac{1}{R^2} + 0(\epsilon^2)\right]. \qquad (6\text{–}150)$$

Now, the objective is an asymptotic solution for small ϵ. For simplicity, we consider only the limiting case in which we neglect viscous effects altogether. In the limit $\epsilon = 0$, considered earlier in deriving the Rayleigh-Plesset equations, the viscous terms in the Navier-Stokes equations were identically equal to zero, and it was only in application of the normal stress condition that viscosity entered the problem. Here, however, the exact problem is more complicated, and neglect of viscous effects is a significant simplification. It should be noted that the neglect of viscous effects as an approximation for large Reynolds number, $\text{Re} \to \infty$, is not as severe an approximation for fluid motion near a gas bubble as for motion near a no-slip surface; in fact, we shall see in Chapter 10 that the limiting form of the Navier-Stokes equation for $\mu = 0$ (that is, the so-called *inviscid equations of motion*) provides a uniformly valid first approximation for the velocity field in this case as $\text{Re} \to \infty$. Nevertheless, neglect of viscous effects is treated here strictly as an ad hoc approximation. The resulting problem retains much of the essential physics of the full problem, but the reader who wishes to apply the results in a practical context would be well advised to consult with one of the many texts or research papers that discuss viscous effects for this problem.[16]

The form of the asymptotic solution that we seek is

$$\boxed{\begin{aligned} \mathbf{u} &= \mathbf{u}^0 + \epsilon \mathbf{u}^1 + 0(\epsilon^2), \\ p &= p_0 + \epsilon p_1 + 0(\epsilon^2), \end{aligned}} \tag{6-151}$$

and the governing equations in the inviscid approximation are thus

$$\boxed{\rho\left(\frac{\partial \mathbf{u}^0}{\partial t} + \mathbf{u}^0 \cdot \nabla \mathbf{u}^0\right) = -\nabla p_0, \quad \nabla \cdot \mathbf{u}^0 = 0,} \tag{6-152}$$

$$\boxed{\rho\left(\frac{\partial \mathbf{u}^1}{\partial t} + \mathbf{u}^0 \cdot \nabla \mathbf{u}^1 + \mathbf{u}^1 \cdot \nabla \mathbf{u}^0\right) = -\nabla p_1, \quad \nabla \cdot \mathbf{u}^1 = 0.} \tag{6-153}$$

The boundary conditions, obtained by substituting the expansions (6–151) into the conditions (6–148) and (6–150) are at $0(1)$ and $0(\epsilon)$:

$$\boxed{u_r^0|_{r=R} = \frac{dR}{dt},} \tag{6-154}$$

and

$$\boxed{u_r^1|_{r=R} + \left(\frac{\partial u_r^0}{\partial r}\right)_{r=R} f_1 - \frac{1}{r}\frac{\partial f_1}{\partial \theta} u_\theta^0\bigg|_{r=R} - \frac{1}{r\sin\theta}\frac{\partial f_1}{\partial \phi} u_\phi^0\bigg|_{r=R} = \frac{\partial f_1}{\partial t},} \tag{6-155}$$

$$\boxed{p_B^0 - p^0|_{r=R} = \frac{2\gamma}{R},} \tag{6-156}$$

$$\boxed{-p^1|_{r=R} + \frac{\partial}{\partial r}(-p^0)|_{r=R} f_1 = -\frac{\gamma}{R^2}\left[\frac{\cos\theta}{\sin\theta}\frac{\partial f_1}{\partial \theta} + \frac{\partial^2 f_1}{\partial \theta^2} + \frac{1}{\sin^2\theta}\frac{\partial^2 f_1}{\partial \phi^2}\right].} \tag{6-157}$$

The tangential stress condition is satisfied automatically for an inviscid fluid ($\mu = 0$), since $\tau \equiv 0$ in this case, and there is no viscous contribution to the normal stress balance.

Thus, examining (6–152), (6–154), and (6–156), which is the full problem at $0(1)$, we see that it is identical to the differential equations and boundary conditions that led to the Rayleigh-Plesset equation, except for the neglect of the viscous stress term in (6–156). Thus, the solution at $0(1)$ is

$$u_\theta^0 = u_\phi^0 = 0, \quad u_r^0 = \frac{R^2 \dot{R}}{r^2}, \tag{6-158}$$

$$p_0(r,t) = p_\infty(t) + \rho \left[2R(\dot{R})^2 + R^2\ddot{R} - \frac{1}{2r^4}(R^4\dot{R}^2) \right], \tag{6-159}$$

where

$$R\ddot{R} + \frac{3}{2}(\dot{R})^2 = \frac{1}{\rho}\left[p_0(R,t) - p_\infty(t)\right] \tag{6-160}$$

and

$$p_0(R,t) = p_B^0(t) - \frac{2\gamma}{R}, \tag{6-161}$$

with

$$p_B^0 = p_v + \frac{\hat{G}}{R^3}. \tag{6-162}$$

The problem at $0(\epsilon)$ is then to solve (6–153), which can now be expressed in component notation as

$$\rho \frac{\partial u_r^1}{\partial t} + \rho u_r^0 \frac{\partial u_r^1}{\partial r} + \rho u_r^1 \frac{\partial u_r^0}{\partial r} = -\frac{\partial p_1}{\partial r},$$

$$\rho \frac{\partial u_\theta^1}{\partial t} + \rho u_r^0 \frac{\partial u_\theta^1}{\partial r} + \rho \frac{u_r^0 u_\theta^1}{r} = -\frac{1}{r}\frac{\partial p_1}{\partial \theta}, \tag{6-163}$$

$$\rho \frac{\partial u_\phi^1}{\partial t} + \rho u_r^0 \frac{\partial u_\phi^1}{\partial r} + \rho \frac{u_\phi^1 u_r^0}{r} = -\frac{1}{r\sin\theta}\frac{\partial p_1}{\partial \phi},$$

$$\frac{1}{r^2}\frac{\partial}{\partial r}(r^2 u_r^1) + \frac{1}{r\sin\theta}\frac{\partial}{\partial \theta}(\sin\theta u_\theta^1) + \frac{1}{r\sin\theta}\frac{\partial u_\phi^1}{\partial \phi} = 0, \tag{6-164}$$

with boundary conditions

$$u_r^1\big|_{r=R} + \left(\frac{\partial u_r^0}{\partial r}\right)_{r=R} f_1 = \frac{\partial f_1}{\partial t}, \tag{6-165}$$

$$-p^1\big|_{r=R} - \frac{\partial p_0}{\partial r}\bigg|_{r=R} f_1 = -\frac{\gamma}{R^2}\left[\frac{\cos\theta}{\sin\theta}\frac{\partial f_1}{\partial \theta} + \frac{\partial^2 f_1}{\partial \theta^2} + \frac{1}{\sin^2\theta}\frac{\partial^2 f_1}{\partial \phi^2}\right], \tag{6-166}$$

and

$$u_r^1 \to 0 \quad \text{as} \quad r \to \infty. \tag{6-167}$$

This problem is linear in $\mathbf{u}^1$, p^1, and f_1. Hence, to consider an arbitrary initial bubble shape, we can consider the function f_1 to be a superposition of spherical harmonics, that is,

$$f_1 = \sum_{k,l} a_{kl}(t) P_k^l(\cos\theta) e^{il\phi}. \tag{6-168}$$

To examine the stability of some arbitrary initial perturbation in the shape, we must examine the behavior of the time-dependent coefficients $a_{kl}(t)$. Since this problem at $0(\epsilon)$ is completely linear, the dynamics of the various modes are uncoupled, and it is sufficient to examine any arbitrary single mode,

$$f_1 = a_{kl}(t) P_k^l(\cos\theta) e^{il\phi}. \tag{6-169}$$

Then, the kinematic boundary condition (6–165) takes the form

$$u_r^1|_{r=R} = \left[\left(\frac{2\dot{R}}{R} \right) a_{kl} + \frac{da_{kl}}{dt} \right] P_k^l(\cos\theta) e^{il\phi}, \tag{6-170}$$

while the normal stress balance is

$$-p_1|_{r=R} = \left[2\rho \frac{\dot{R}^2}{R} a_{kl}(t) P_k^l(\cos\theta) e^{il\phi} - \frac{\gamma}{R^2} a_{kl}(t) D^2 \left(P_k^l(\cos\theta) e^{il\phi} \right) \right], \tag{6-171}$$

where

$$D^2 \equiv \frac{\cos\theta}{\sin\theta} \frac{\partial}{\partial\theta} + \frac{\partial^2}{\partial\theta^2} + \frac{1}{\sin^2\theta} \frac{\partial^2}{\partial\phi^2}.$$

To solve (6–163), (6–164), (6–167), (6–170), and (6–171), we note that the continuity equation for an inviscid, irrotational flow is satisfied by a velocity field that can be expressed as the gradient of a scalar function, namely,

$$u^1 = \nabla\phi, \tag{6-172}$$

provided ϕ is a harmonic function, so that

$$\nabla \cdot u^1 = \nabla \cdot (\nabla\phi) = \nabla^2 \phi = 0. \tag{6-173}$$

But solutions of Laplace's equation, expressed in terms of spherical coordinates, are just the spherical harmonics

$$\phi = \begin{cases} r^n P_n^m(\cos\theta) e^{im\phi} \\ r^{-(n+1)} P_n^m(\cos\theta) e^{im\phi}, \end{cases} \quad \text{where } n, m \geq 0. \tag{6-174}$$

In order that u^1 satisfies the far-field condition (6–167), ϕ must be limited to the negative harmonics, that is,

$$\phi = \sum_{n,m} A_{nm} r^{-(n+1)} P_n^m(\cos\theta) e^{im\phi}. \tag{6-175}$$

But according to the kinematic condition (6–170),

$$u_r^1|_{r=R} \equiv \frac{\partial\phi}{\partial r}\bigg|_{r=R} \cong P_k^l(\cos\theta) e^{il\phi}.$$

Hence, for the distortion mode (6-169), the potential function ϕ must be of the form

$$\phi = A_{kl} r^{-(k+1)} P_k^l (\cos\theta) e^{il\phi},$$

and substitution into the kinematic condition (6-170) shows that

$$\phi = -\frac{R^{k+2}}{k+1} \left[2\frac{\dot{R}}{R} a_{kl} + \frac{da_{kl}}{dt} \right] \frac{1}{r^{k+1}} P_k^l(\cos\theta) e^{il\phi}. \tag{6-176}$$

Taken together with (6-172), this result for ϕ is analogous to the result (6-158) and (6-160) for the 0(1) problem. All that remains is to obtain the dynamic equation for the coefficients $a_{kl}(t)$ i.e. For this purpose, we follow the example from the earlier derivation of the Rayleigh-Plesset equation. First, we calculate the pressure $p_1(r,t)$ by means of the equations of motion (6-163), and, second, apply the normal stress condition (6-171) to obtain the dynamic equation for the coefficients $a_{kl}(t)$. In the interest of brevity, we do not display all of the details here but simply report the result:

$$\ddot{a}_{kl} + 3\frac{\dot{R}}{R}\dot{a}_{kl} + (k-1)\left[-\frac{\ddot{R}}{R} + (k+1)(k+2)\frac{\gamma}{\rho R^3} \right] a_{kl} = 0. \tag{6-177}$$

It will be noted that this result is independent of the azimuthal index l, and it is thus convenient to drop this subscript so that a_{kl} is denoted simply as a_k. Further, the mode $k=1$ is a special case as can be seen by writing the expression for the bubble surface for a pure $k=1$ deformation, namely,

$$s = r - R(t) - a_1(t)P_1(\cos\theta).$$

Now, $P_1(\cos\theta) \equiv \cos\theta$. Thus, if $a_1(t) \neq 0$, the *center* of the bubble translates a distance $a_1(t)$ from the origin, but the bubble *shape* is unchanged. To discuss the stability of the spherical shape, we must determine whether the coefficients $a_k(t)$ for $k \geq 2$ increase or decrease with time.

Even without a detailed analysis, there are a few observations and general conclusions that can be made from (6-177). First, for a bubble of *constant volume*,

$$\ddot{a}_k + (k-1)(k+1)(k+2)\frac{\gamma}{\rho R^3} a_k = 0, \tag{6-178}$$

and we see that the *natural* oscillation frequency for the k mode is

$$\omega_{0,k}^2 = (k-1)(k+1)(k+2)\frac{\gamma}{\rho R^3}. \tag{6-179}$$

This result is well known. More interesting is the coupling between the radial and shape oscillations when $\dot{R} \neq 0$.

Let us recall the properties of a standard linear oscillator equation in the form

$$\ddot{F} + a\dot{F} + bF = 0. \tag{6-180}$$

If the second term is zero ($a = 0$), the solution is

$$F = e^{st}, \text{ where } s^2 + b = 0. \tag{6-181}$$

So, if $b > 0$, $s = \pm i\sqrt{b}$, and the solution is oscillatory. On the other hand, if $b < 0$, $s = \pm\sqrt{-b}$, and one of the two solutions grows exponentially with time. The second term in (6-180) plays the role of damping if $a > 0$ but is destabilizing (somewhat analogous to a negative viscosity) if $a < 0$. If we now examine (6-177), we see that surface tension is always stabilizing (in the sense that it contributes to $b > 0$), while the acceleration term $-\ddot{R}/R$ is stabilizing when $\ddot{R} < 0$ but destabilizing when $\ddot{R} > 0$. The instability that occurs for $\ddot{R}$ sufficiently large and positive is essentially the *Rayleigh-Taylor instability* for this problem. Maximum positive values for $\ddot{R}$ tend to occur particularly at the *beginning* of an expansion or near the *minimum* radius when the bubble is collapsing. The term $3(\dot{R}/R)\dot{a}_k$ corresponds to positive damping for $\dot{R} > 0$ but tends to be destabilizing for $\dot{R} < 0$. Indeed, as we shall see shortly, it is possible for the bubble shape to be unstable if $\dot{R} < 0$ even though $\ddot{R} < 0$. The physical origin of the damping or destabilization associated with this term is strictly kinematic in origin. Due to the diverging/converging nature of the streamlines for the $0(1)$ radial motion, the area of the bubble surface is either increased with time or decreased with time depending on whether $\dot{R} > 0$ or $\dot{R} < 0$. When the surface area is increased, the magnitude of any surface disturbance is decreased and the wavelength is increased. On the other hand, when the area is decreased, the amplitude of the disturbance is increased and its wavelength is decreased. The increase or decrease in amplitude corresponds to a destabilizing or stabilizing effect.

A more quantitative discussion of the stability is possible. A convenient first step is to introduce the transformation

$$a_k = \frac{1}{R^{3/2}} \alpha_k \tag{6-182}$$

into (6-177). This leads to the transformed equation

$$\boxed{\ddot{\alpha}_k - G_k(t)\alpha_k = 0,} \tag{6-183}$$

where

$$G_k(t) = \frac{3}{2}\frac{d}{dt}\left(\frac{\dot{R}}{R}\right) + \frac{9}{4}\left(\frac{\dot{R}}{R}\right)^2 + \left[(k-1)\left\{\frac{\ddot{R}}{R} - (k+1)(k+2)\frac{\gamma}{\rho R^3}\right\}\right]. \tag{6-184}$$

Now, the time dependence of $G_k(t)$ is determined by the time dependence of $R(t)$, which in turn is dependent upon the variation in ambient pressure $p_\infty(t)$. Although the solutions of (6-183) are generally quite complicated, we may hope to learn something that is at least qualitatively useful by examining cases in which the form for G is simple. For example, if $G_k = c$(const), then the solutions of (6-183) are periodic and stable provided $c < 0$.* On the other hand, at least one of the solutions of (6-183) is exponentially growing for $c > 0$, and thus the spherical bubble shape is *unstable*. Other forms for $G_k(t)$ are more difficult to analyze in detail, though one may hope that the general

---

*The amplitude of oscillation remains constant. Thus, any dissipation associated with weak viscous effects will lead to a damped oscillation.

"rule" of stability for $G_k < 0$ and instability for $G_k > 0$ might still carry over. In that case, some useful insight can be achieved by determining the factors that control the sign of G_k. For this purpose, it is convenient to rewrite G_k in the form

$$G_k(t) = \frac{3}{4}\left(\frac{\dot{R}}{R}\right)^2 + \left(k + \frac{1}{2}\right)\frac{\ddot{R}}{R} - (k-1)(k+1)(k+2)\frac{\gamma}{\rho R^3}. \tag{6-185}$$

Generally, the surface tension is stabilizing (in the sense that it contributes a negative term to G), and this tendency is enhanced with increase of k.† Let us consider the case $\gamma \equiv 0$. Then,

$$G_k(t) = \frac{3}{4}\left(\frac{\dot{R}}{R}\right)^2 + \left(k + \frac{1}{2}\right)\frac{\ddot{R}}{R}. \tag{6-186}$$

Clearly, $G_k > 0$ for $\ddot{R} \geq 0$. Instability corresponding to $\ddot{R} > 0$ is the analogue of Rayleigh-Taylor instability. For a flat interface, we could have instability *only* for $\ddot{R} \geq 0$. Here, however, we may still have instability even if $\ddot{R} < 0$, provided $\dot{R}/R$ is sufficiently large. To determine the condition for $G > 0$ in terms of controllable parameters, we can substitute for $\ddot{R}/R$ in (6-186) using the inviscid form of the Rayleigh-Plesset equation (6-75) with $\gamma \equiv 0$, that is,

$$\frac{\ddot{R}}{R} = -\frac{3}{2}\left(\frac{\dot{R}}{R}\right)^2 + \frac{p_B - p_\infty}{\rho R^2}. \tag{6-187}$$

Then,

$$G_k(t) = k\frac{\ddot{R}}{R} + \frac{p_B - p_\infty}{2\rho R^2}. \tag{6-188}$$

Hence, for $\ddot{R} < 0$, $G(t) > 0$ if

$$\frac{p_B - p_\infty}{\rho} > 2kR|\ddot{R}|. \tag{6-189}$$

Thus, all else being equal, a smaller pressure difference is necessary to induce instability for smaller values of k.

It is important to recognize that the criteria of *instability* for $G_k > 0$ and *stability* for $G_k < 0$ is only strictly valid for G_k independent of t. Indeed, if G_k depends on t, the differential equation (6-183) is *nonautonomous*, and it is well known that stability can generally be established only by integrating the equation exactly. If G_k is "slowly varying," a local analysis based upon an instantaneous value of G_k will be qualitatively correct for a finite time interval, but for more general time-dependent forms for G_k, we should not be surprised if the situation turns out to be more complex. To see an example of this, we can consider the special case when the bubble volume changes periodically with time about some mean value as may occur in response to an oscillatory variation in p_∞ at a frequency different from (6-100). Thus, we let

$$R = R_E(1 + \epsilon \cos \omega t) \tag{6-190}$$

---

†Note, however, that once G is negative, the effect of increased k, in a system with no viscous damping, is to produce bubble oscillation at higher frequency.

and evaluate $G_k(t)$. The result is

$$G_k(t) = -\left(k + \frac{1}{2}\right)\epsilon\omega^2\cos\omega t - (k-1)(k+1)(k+2)\frac{\gamma}{\rho R_E^3}(1 - 3\epsilon\cos\omega t) + 0(\epsilon^2),$$

(6–191)

and thus letting $\tau = \omega t$, we find that

$$\ddot{\alpha}_k + \left[\frac{\omega_{0,k}^2}{\omega^2} + \left[\left(k + \frac{1}{2}\right) - 3\frac{\omega_{0,k}^2}{\omega^2}\right]\epsilon\cos\tau\right]\alpha_k = 0,$$

(6–192)

where $\omega_{0,k}$ is given by equation (6–179).

Equation (6–192) is a special case of a differential equation that is known as *Mathieu's equation*, for which the standard form is

$$\ddot{y} + (a + 2b\cos t)y = 0.$$

(6–193)

In the present case,

$$a = a_k = \frac{\omega_{0,k}^2}{\omega^2}$$

(6–194)

and

$$b = b_k = \frac{\epsilon}{2}\left[\left(k + \frac{1}{2}\right) - 3\frac{\omega_{0,k}^2}{\omega^2}\right].$$

(6–195)

The general theory for linear differential equations with periodic coefficients is known as *Floquet theory*. This theory predicts that *unstable* solutions exist for some values of a and b (these solutions consist of modulated oscillations that grow exponentially with increasing t). A plot of the stability boundaries of solutions of Mathieu's equation is reproduced from the book by Bender of Orszag (1978) in Figure 6–14.[17] All solutions are stable in the white regions of the (a,b) plane, while unstable solutions exist in the crosshatched region. For much of the (a,b) plane, the existence of *unstable* solutions represents a violation of the general stability rule proposed previously, which suggested that solutions of equation (6–183) would tend to be stable for any combination of a and b for which $G_k(t)$ is negative for all t. The condition $G_k < 0$ requires that

$$a_k > 2|b_k|.$$

The straight line corresponding to this condition is superposed on Figure 6–14. Clearly, regions of instability exist in spite of the fact that a_k exceeds $2|b_k|$ so that $G_k < 0$.

Although the behavior of solutions of Mathieu's equation is generally quite complicated, analytic approximations can be obtained via an asymptotic expansion for $b \ll 1$. In this case, b is small provided the magnitude of changes in bubble volume is small, that is, $\epsilon \ll 1$ in (6–190). Therefore, for our problem, the limit $b \ll 1$ corresponds to modulations in bubble shape driven by weak oscillations in the bubble volume, that is, weak oscillations in the ambient pressure $p_\infty(t)$. In this regime, some features of the stability boundary shown in Figure 6–14 can be predicted analytically.

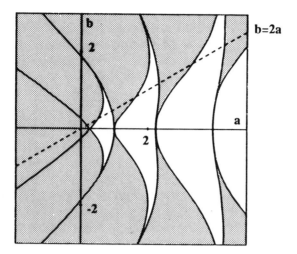

b=2a

Figure 6-14 Stability boundaries for solutions of Mathieu's equation. (Reproduced from Reference 1b.) The shaded regions correspond to unstable regions in the a-b plane, while the unshaded regions are stable regions.

An added bonus of carrying out the necessary analysis is that it exposes the *physical basis* for instability of a bubble in the regime where a_k is asymptotically large compared with b_k so that $G_k < 0$ for all t.

In the remainder of this section, we thus consider the solution of (6–183) with $G_k(t)$ in the form (6–191) and $b_k \ll 1$, namely,

$$\ddot{\alpha}_k + (a_k + 2\epsilon \bar{b}_k \cos t)\alpha_k = 0. \tag{6-196}$$

Let us suppose, initially, that a solution of (6–196) exists in the form of a regular asymptotic expansion for $\epsilon \ll 1$, namely,

$$\alpha_k(t) = \alpha_k^{(0)}(t) + \epsilon \alpha_k^{(1)}(t) + \epsilon^2 \alpha_k^{(2)}(t) + \cdots. \tag{6-197}$$

Then, the governing equations for the successive terms in the asymptotic expansion for α can be expressed in the form

$$0(1) \quad \ddot{\alpha}_k^{(0)} + a_k \alpha_k^{(0)} = 0, \tag{6-198a}$$

$$0(\epsilon) \quad \ddot{\alpha}_k^{(1)} + a_k \alpha_k^{(1)} = -2\bar{b}_k \cos t \, \alpha_k^{(0)}, \tag{6-198b}$$

$$0(\epsilon^2) \quad \ddot{\alpha}_k^{(2)} + a_k \alpha_k^{(2)} = -2\bar{b}_k \cos t \, \alpha_k^{(1)}, \tag{6-198c}$$

and so on. Now, the general solution of (6–198a) is

$$\alpha_k^{(0)}(t) = A\exp(i\sqrt{a_k}t) + B\exp(-i\sqrt{a_k}t). \tag{6-199}$$

Thus, equation (6–198b) for $\alpha_k^{(1)}$ becomes

$$\ddot{\alpha}_k^{(1)} + a_k \alpha_k^{(1)} = -2\bar{b}_k A e^{i(1+\sqrt{a_k})t} - 2\bar{b}_k B e^{i(1-\sqrt{a_k})t}. \tag{6-200}$$

Now, however, we can see that if

$$\sqrt{a_k} \pm 1 = \pm \sqrt{a_k}, \tag{6-201}$$

the terms on the right are *secular* terms, and $\alpha_k^{(1)}$ will grow without bound as $t \to \infty$. The condition (6-201) is satisfied only if $a_k = 1/4$. Thus, provided $a_k \neq 1/4$, the solutions for $\alpha_k^{(1)}$ are bounded for all time, but the existence of secular terms for even this one value of a_k in (6-200) already provides significant insight into the physics that underlies the possibility of *unstable* solutions to Mathieu's equation (even though the time-dependent oscillation of the coefficient $a_k + 2\epsilon \bar{b}_k \cos t$ is very weak for $\epsilon \ll 1$). The fact is that oscillations of this time-dependent coefficient at the natural frequency of the constant bubble volume limit (that is, $b_k \equiv 0$), produces a resonant interaction with the eigenmode for this case (namely, the solution $\alpha_k^{(0)}$ in the present context) and the possibility of growth in the amplitude with time. Subsequent equations in the series have no secular terms provided $a_k \neq n^2/4$. Thus, provided $a_k \neq n^2/4$, the solution of Mathieu's equation for $\epsilon \ll 1$ is bounded for all t. The condition

$$a_k \neq \frac{n^2}{4} \tag{6-202a}$$

also can be written as a condition on the frequency ω,

$$\omega \neq \frac{2}{n} \omega_{0,k}. \tag{6-202b}$$

Thus, for each deformation mode k, there are a discrete set of frequencies for which the bubble shape is unstable. This is consistent with the stability diagram in Figure 6-14 for the limit $\epsilon \bar{b} \to 0$.

Additional insight into the nature of the solutions of Mathieu's equation can be obtained by extending the asymptotic solution described previously to small, but nonzero, values of ϵ. Of particular interest is the behavior of solutions near the first resonant instability point $a_k = 1/4$. Referring to Figure 6-14, we see that there is a finite region around $a_k = 1/4$ where solutions for nonzero values of $\epsilon \bar{b}_k$ are predicted to be unstable. In this section, we seek an asymptotic expression in terms of ϵ for the critical values of $a_k \approx 1/4$ that separate the regions of stable and unstable solutions. Thus, we suppose that

$$a = \frac{1}{4} + \epsilon a_1 + \cdots \tag{6-203}$$

and again seek stable solutions of (6-196). As is typical of problems that exhibit secular behavior due to resonant forcing, the solution for $a \approx 1/4$ exhibits two time scales: a short scale comparable with the time scale of the forcing and a longer time scale representative of the slow modulation of the induced oscillation. Thus, we follow the usual two-timing asymptotic procedure and suppose that

$$\alpha_k = \alpha_k(t, \tau) \tag{6-204}$$

and

$$\tau = \epsilon t. \tag{6-205}$$

In this case, the governing Equation (6–196) takes the form

$$0 = \frac{d^2\alpha_k}{dt^2} + 2\epsilon\frac{d^2\alpha_k}{dt d\tau} + \epsilon^2\frac{d^2\alpha_k}{d\tau^2} + \left(\frac{1}{4} + \epsilon a_1^{(k)} + \cdots + 2\epsilon\bar{b}_k\cos t\right)\alpha_k, \quad (6\text{–}206)$$

and we seek an asymptotic expansion for α_k of the form

$$\alpha_k = \alpha_k^{(0)}(t,\tau) + \epsilon\alpha_k^{(1)}(t,\tau) + \cdots . \quad (6\text{–}207)$$

The governing equations at the first two levels of approximation are then

$$\frac{d^2\alpha_k^{(0)}}{dt^2} + \frac{1}{4}\alpha_k^{(0)} = 0 \quad (6\text{–}208)$$

and

$$\frac{d^2\alpha_k^{(1)}}{dt^2} + \frac{1}{4}\alpha_k^{(1)} = -2\frac{d^2\alpha_k^{(0)}}{dt d\tau} - a_1^{(k)}\alpha_k^{(0)} - 2\bar{b}_k\cos t\alpha_k^{(0)}. \quad (6\text{–}209)$$

The solution of (6–208) is simply

$$\alpha_k^{(0)} = A(\tau)e^{1/2(it)} + A^*(\tau)e^{-1/2(it)}, \quad (6\text{–}210)$$

where $A(\tau)$ and $A^*(\tau)$ are complex conjugates. Substituting into (6–209), we then obtain the governing equation for $\alpha_k^{(1)}$:

$$\frac{d^2\alpha_k^{(1)}}{dt^2} + \frac{1}{4}\alpha_k^{(1)} =$$

$$-\left[\left(i\frac{dA}{d\tau} + a_1^{(k)}A + \bar{b}_kA^*\right)e^{1/2(it)} - \left(i\frac{dA^*}{d\tau} - a_1^{(k)}A^* - \bar{b}_kA\right)e^{-1/2it} + \bar{b}_kAe^{3/2it} + \bar{b}_kA^*e^{-3/2it}\right].$$

$$(6\text{–}211)$$

Thus, to eliminate secular terms, we require that

$$i\frac{dA}{d\tau} + a_1^{(k)}A + b_kA^* = 0 \quad \text{and} \quad i\frac{dA^*}{d\tau} - a_1^{(k)}A^* - b_kA = 0, \quad (6\text{–}212)$$

and this provides dynamic equations for the slow time-scale modulation of (6–204). In order to solve these equations, it is convenient to express A in terms of its *real* and *imaginary* parts:

$$A = B(\tau) + iC(\tau), \quad A^* = B(\tau) - iC(\tau).$$

Then,

$$\frac{dB}{d\tau} + (a_1^{(k)} - \bar{b}_k)C = 0,$$

$$\frac{dC}{d\tau} - (a_1^{(k)} + \bar{b}_k)B = 0. \quad (6\text{–}213)$$

If we eliminate C, we thus obtain a second-order equation for B,

$$\frac{d^2B}{d\tau^2} = (\bar{b}_k^2 - a_1^{(k)2})B, \quad (6\text{–}214)$$

and the solution of this equation is

$$B = K \exp\left[\pm \sqrt{\bar{b}_k^2 - a_1^{(k)2}}\right] \tau. \tag{6-215}$$

Thus, if $(\bar{b}_k^2 - a_1^{(k)2}) > 0$, the solution (6-210) is unstable since the slowly varying amplitude functions grow exponentially in time. On the other hand, for $(\bar{b}_k^2 - a_1^{(k)2}) < 0$, the solution is stable with a two-time scale oscillation in the bubble shape. The stability boundary thus occurs at

$$a_1^{(k)} = \pm \bar{b}_k, \tag{6-216}$$

and so the overall stability boundary for solutions of (6-196) occurs at

$$a_k = \frac{1}{4} \pm \epsilon \bar{b}_k + 0(\epsilon^2). \tag{6-217}$$

This corresponds to the pair of crossing straight lines in Figure 6-14 that pass through the point $(a_k = 1/4, \epsilon \bar{b}_k = 0)$.

Hence, we can see from this example of a bubble with weak time-periodic oscillations of volume that the qualitative condition for stability $G_k < 0$ does *not* apply unless G_k is constant or only very slowly varying with time (and in the latter case, only strictly for an interval of time that is short compared with the characteristic time scale for variation of G_k). If the coefficient G_k varies at a rate that is comparable to the natural frequency of shape oscillation for a bubble of constant volume, then the bubble may undergo a resonant oscillation of shape to very large amplitudes even though $G_k < 0$ for all times.

Homework Problems

1. Find the total kinetic energy E_k of the liquid outside a spherical gas bubble that is undergoing time-dependent changes in volume in an unbounded, incompressible, Newtonian fluid. Show that the net rate of working by the pressure inside the bubble $\hat{p}$ at the inner side of the bubble boundary is

$$\dot{W} = 4\pi(\hat{p} - p_\infty)R^2\dot{R},$$

where $R(t)$ is the bubble radius. Hence, use the principle of energy conservation in the macroscopic form

$$\dot{W} = \dot{E}_k + \dot{E}_s + \Phi,$$

where E_s is the surface free energy and Φ is the rate of viscous dissipation in the fluid, to derive the Rayleigh-Plesset equation in the form of Equation (6-79).

2. We consider the flow of an incompressible, Newtonian fluid between two infinite, plane walls, with a time-dependent pressure gradient parallel to the walls

$$-\frac{\partial p}{\partial z} = G_0 \sin \omega t.$$

The walls are separated by a distance d, and the kinematic viscosity of the fluid is ν. This is a time-dependent unidirectional flow that can be solved exactly as in Chapter 3. We are concerned here with the long-time asymptotic behavior only.

    a. Assume ω to be asymptotically small, $\omega \ll 1$. Determine the first three terms in an asymptotic approximation of the exact solution for $(\omega a^2/\nu) \ll 1$, using the direct method of regular perturbation expansion.

    b. Determine the first approximation everywhere in the domain for $(\omega a^2/\nu) \gg 1$, again using asymptotic techniques.

3. Consider two infinite parallel disks that are separated by a gap H. The lower disk rotates about the central axis with angular velocity Ω_1, and the upper disk rotates about the same axis with angular velocity Ω_2. In Chapter 4, we considered the solution of the problem of a single rotating disk with $|\Omega_1| = \Omega$ and $\Omega_2 \equiv 0$, in the creeping flow limit,

$$\mathrm{Re} \equiv \frac{\rho H^2 \Omega}{\mu} \ll 1,$$

where inertia could be neglected and the problem reduced to a single nonzero velocity component $u_\theta(r,z)$. When the Reynolds number is *not* vanishing, inertia effects will induce secondary currents so that u_z and u_r are nonzero.

    a. Determine the inertia-induced changes in the flow, using a regular perturbation approximation scheme for arbitrarily small, but nonzero, Reynolds numbers, including terms to $O(\mathrm{Re}^2)$. Sketch the secondary flow patterns that appear at $O(\mathrm{Re})$ in the (r,z) plane, paying special attention to the changes in the flow patterns that occur as the magnitudes and signs of the angular velocities, Ω_1 and Ω_2, are varied. Calculate the changes in the shear stress distributions on the two disks, and determine the torque required to maintain angular velocities Ω_1 and Ω_2. Discuss the couplings between the shear stress distribution and secondary flow in terms of vorticity transport. Discuss the sign of the torque components—does it make sense?

    b. The problem of flow due to rotation of two infinite parallel disks is one of very few examples in which an exact solution of the full nonlinear Navier-Stokes equations is possible for arbitrary values of the Reynolds number, in the sense that the governing equations can be reduced to a coupled set of ordinary differential equations. The form for the velocity component $u_\theta(r,z)$ is dictated by the r dependence of u_θ at the boundaries of the rotating disks to be

$$\bar{u}_\theta = \bar{r} F(\bar{z}),$$

where the nondimensionalization is the same as that used in Chapter 4. The forms of the other velocity components $\bar{u}_r$ and $\bar{u}_z$ and the pressure $\bar{p}$ are then determined from $\bar{u}_\theta$, using the r and z components of the Navier-Stokes equations plus the continuity equation, to be

$$\bar{u}_r = \bar{r} H(\bar{z}), \quad \bar{u}_z = G(\bar{z}), \quad \bar{p} = P(\bar{z}).$$

    i. Derive the coupled set of ordinary differential equations for F, H, G, and P, and obtain corresponding boundary conditions.

    ii. Demonstrate that the *form* of the solution in the limiting case where the upper disk is removed ($H \to \infty$, so that we have a single disk rotating in a semi-infinite fluid) is independent of *all* dimensional parameters. Hint: To demonstrate this fact, you need to show that you can rescale all equations and boundary conditions so that none of the dimensional parameters appears explicitly.

    iii. In the case of two disks, with H finite, it is not possible to eliminate the Reynolds number

$$\text{Re} = \frac{\rho \Omega H^2}{\mu}.$$

Show that the *coupled* set of ordinary differential equations for H, G, F, and P can be solved in the limit $\text{Re} \ll 1$ by the method of regular perturbation expansions, and that the solution to $O(\text{Re}^2)$ is identical to that obtained directly in part (a).

4. The flow of a viscous fluid radially outward between two circular disks is a useful model problem for certain types of polymer mold-filling operations, as well as lubrication systems. We consider such a system, as sketched here,

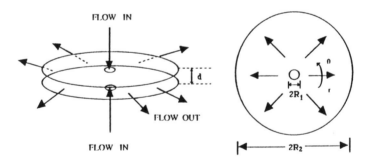

in which the flow takes place because of a pressure drop ΔP between the inner and outer radii R_1 and R_2. The two disks are separated by a gap width d.

    a. Assume that the flow is directed radially so that $\mathbf{u} = (u, 0, 0)$ in cylindrical coordinates (r, θ, z). Show that u must then take the form

$$u = \frac{F(z)}{r}.$$

Derive a governing equation for $F(z)$, and show that no solution exists for this equation unless the inertia term involving F^2 is neglected. What is the significance of this result?

    b. In most analyses of this problem, it is assumed, on the basis of the result from part (a), that the relevant Reynolds number is sufficiently small that the creeping flow approximation can be made, so that the velocity field is purely radial. Here, however, we will seek to understand the role of inertia in this problem by obtaining a solution for small, but nonzero, Reynolds numbers, using the techniques of a regular perturbation expansion.

i. Obtain a dimensionless form of the governing equations and boundary conditions by assuming that derivatives in the radial direction scale as R_2^{-1}, while derivatives with respect to z scale as d^{-1}. Note that the solution to the problem will depend upon whether the pressure difference ΔP or the volumetric flow rate Q is held constant (in the former case, the flow rate Q will depend upon the Reynolds number, while ΔP will depend on the Reynolds number if Q is held constant). The choice of a characteristic velocity scale will depend on which problem we consider. Eventually, we will consider both cases, so you may wish to consider both possibilities at this point. In any case, you should end up with three dimensionless parameters, $\epsilon = d/R_2$, $\delta = R_2/R_1$, and Reynolds number

$$\mathrm{Re} \equiv \frac{\rho u_c d^2}{\mu R_2}.$$

ii. Assume $\mathrm{Re} \ll 1$ and $\epsilon^2 \ll \mathrm{Re}$. Derive governing equations and boundary conditions for the first two terms in an asymptotic expansion of the form

$$\mathbf{u} = \mathbf{u}_0 + \mathrm{Re}\,\mathbf{u}_1 + \cdots,$$

$$p = p_0 + \mathrm{Re}\,p_1 + \cdots.$$

It is clear from part (a) that the fluid motion cannot be purely radial except, possibly, in the creeping flow limit, $\mathrm{Re} \to 0$. However, you should be able to demonstrate that the swirl velocity component in the θ direction is zero.

iii. Obtain the creeping flow solution for the leading-order terms $\mathbf{u}_0$ and p_0 in the expansion for $\mathrm{Re} \ll 1$. Neglect any rearrangement of the flow that occurs in the immediate vicinity of the entry and exit regions at $\bar{r} = \delta$ and $\bar{r} = 1$. You should find that $u_0 \neq 0$, $w_0 = 0$, and $p_0 = p_0(r)$. Derive the relation between volume flow rate Q and pressure drop ΔP, which corresponds to this solution, that is,

$$Q = \frac{4\pi d^3 \Delta P}{\ln (R_2/R_1)}.$$

iv. Determine the influence of inertia by solving for the second terms in the expansion, namely, $\mathbf{u}_1$ and p_1. As noted above, the solution will depend on what is held constant. If the pressure difference is maintained at ΔP, independent of Re, the volume flow rate must change. If, on the other hand, the volume flux is maintained at a constant value independent of Re, the required pressure drop ΔP must depend on Re. Consider both cases. In the first case, calculate the effect of inertia on Q. In the second, determine the change in ΔP that is necessary to maintain Q independent of Re.

References/Notes

1. Among the textbooks on this topic are

    a. Nayfeh, A.H., *Perturbation Methods*. John Wiley and Sons: New York (1973).

    b. Bender, C.M., and Orszag, S.A., *Advanced Mathematical Methods for Scientists and Engineers*. McGraw-Hill: New York (1978).

    c. Van Dyke, M., *Perturbation Methods in Fluid Mechanics,* annotated ed. Parabolic Press: Stanford (1975).

    d. Kevorkian, J., and Cole, J.D. *Perturbation Methods in Applied Mathematics*. Springer-Verlag (Applied Math. Sciences Series):34, New York (1981).

    e. Lagerstrom, P.A., *Matched Asymptotic Expansions, Ideas and Techniques*. Springer-Verlag (Applied Math. Sciences Series):76 New York (1988).

2. Dean, W.R., "Note on the Motion of Fluid in a Curved Pipe," *Phil. Magazine* (Series 7) 4:208–219 (1927).

    Dean, W.R., "The Stream-line Motion of Fluid in a Curved Pipe," *Phil. Magazine* (Series 7) 5:673–695 (1928).

3. See Reference 2 of Chapter 3.

4. Berger, S.A., Talbot, L., and Yao, L.S., "Flow in Curved Pipes," *Ann. Rev. of Fl. Mech.* 15:461–512 (1983).

5. Rayleigh L. (J.W. Strutt), "On the Pressure Developed in a Liquid During the Collapse of a Spherical Cavity," *Phil. Magazine* (Series 6) 34:94–98 (1917).

6. a. Hammitt, F.G., *Cavitation and Multiphase Flow Phenomena*. McGraw-Hill: New York (1980).

    b. Clancy, T. *The Hunt for Red October*. Naval Institute Press: Annapolis, Maryland (1984).

    c. Young, F.R., *Cavitation*. McGraw-Hill: New York (1989).

7. Plesset, M.S., and Prosperetti, A., "Bubble Dynamics and Cavitation," *Ann. Review of Fl. Mech.* 9:145–185 (1977).

8. An analysis of the dynamics of spherical bubbles that are perturbed away from an equilibrium volume by a finite (not arbitrarily small) amount has been reported for the case of a constant differential pressure Δp by Cheng, H.C., and Chen, L.H., "Growth of a Gas Bubble in a Viscous Fluid," *Phys. Fluids* 29:3530–3536 (1986).

9. Iooss, G., and Joseph, D.D., *Elementary Stability and Bifurcation Theory*. Springer-Verlag: New York (1980).

10. Two relatively recent papers that discuss the transition from regular oscillations of the bubble radius to chaotic oscillations are

    a. Kameth, V., and Prosperetti, A., "Numerical Integration Methods in Gas-Bubble Dynamics," *J. Acoust. Soc. Am.* 85:1538–1548 (1989).

    b. Szeri, A.J., and Leal, L.G., "The Onset of Chaotic Oscillations and Rapid Growth of a Spherical Bubble at Subcritical Conditions in an Incompressible Liquid," *Phys. of Fluids* A3, 551–555 (1991).

11. A description of multivariable expansion procedures, of which two-timing or two-variable expansions are a special case, can be found in Reference 1d.

12. See Reference 1.

13. A useful review of recent research on the dynamics of cavitation bubbles near boundaries is Blake, J.R., and Gibson, D.C., "Cavitation Bubbles Near Boundaries," *Ann. Rev. of Fl. Mech.* 19:99–123 (1987).

14. This type of instability is known as Rayleigh-Taylor instability. The stability of the interface between two fluids of different density that is accelerated in a direction perpendicular to the interface was studied by

    a. Taylor, G.I., "The Instability of Liquid Surfaces when Accelerated in a Direction Perpendicular to their Planes, I," *Proc. Royal Soc.* (London) A201: 192–196 (1950).

A much earlier analysis by Rayleigh of the stability of a heterogeneous fluid accelerated perpendicular to the plane of stratification is

b. Rayleigh, L., "Investigation of the Character of the Equilibrium of an Incompressible Heavy Fluid of Variable Density," *Scientific Papers* (Cambridge) 2:200–207 (1900).

A readable account summarizing this and related work is given in the book by Chandrasekhar, Reference 4a of Chapter 3.

15. The generalization of Taylor's result to the acceleration of a spherical interface was independently reported, shortly after publication of Taylor's paper, by two authors:

a. Plesset, M.S., "On the Stability of Fluid Flows with Spherical Symmetry," *J. Appl. Phys.* 25:96–98 (1954).

b. Birkhoff, G., "Note on Taylor Instability," *Quart. Appl. Math* 12:306–309 (1954); "Stability of Spherical Bubbles," *Quart. Appl. Math* 13:451–453 (1956).

16. See Reference 7 for a brief introduction with references.

17. Figure 6–14 is reproduced from figure 11–11 of Reference 1b.

CHAPTER 7

Thin Films, Lubrication, and Related Problems

In Chapter 6 we explored the consequences of a weak departure from strict adherence to the conditions for unidirectional flow, namely, the effect of slight curvature of the centerline in flow through a circular tube. For that case, the centripetal acceleration associated with the curved path of the primary flow was shown to produce a weak secondary motion in the plane orthogonal to the tube axis. In this chapter we consider the class of deviations from unidirectional flow that are caused by *nonparallel boundaries* (that is, variations in the cross-sectional area of the channel in the flow direction).

The fact that the boundaries of the flow domain are not parallel means that the magnitude of the primary velocity component must vary as a function of distance in the flow direction. In general, this means that the Navier-Stokes equations cannot be linearized, and exact analytical solutions are no longer possible. In this chapter, we thus consider only approximate analytical solutions obtained via asymptotic techniques for the *thin-film limit* in which the distance between the boundaries is small compared with their dimension in the flow direction and/or the limit in which the boundaries are *nearly parallel.* In the former case, the governing equations reduce, at leading order, to the famous *Reynolds lubrication equations*. The thin-film or lubrication limit is of great practical importance because the combination of a thin gap and nonparallel boundaries can lead to very high pressures that can support a considerable load without allowing the boundaries to touch.

Rather than beginning with a general formulation for lubrication flows, however, we first consider the *specific* problem of flow between two rotating cylinders whose axes are parallel but offset to produce the *eccentric cylinder* geometry sketched in Figure 7-1. Although an exact analytic solution is possible in the creeping flow limit for an arbitrary ratio of cylinder radii and an arbitrary degree of eccentricity, the analysis requires the use of a bicylindrical coordinate system.[1] On the other hand, *approximate* analytic solutions can be obtained using circular *cylindrical* coordinates, but only for two distinct limiting cases: first, the *slightly eccentric* limit when the offset in the cylinder axes is small compared with the gap width between the cylinders; and, second, the *thin-film* limit when the gap width is very small compared with the cylinder radius.

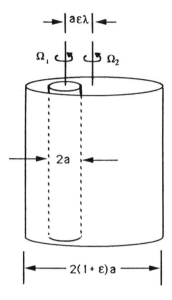

Figure 7-1 The eccentric cylinder geometry.

The eccentric cylinder problem in the latter limit is a special case of lubrication theory. Thus, once it is solved, we turn in Section B to a derivation of the governing lubrication equations for an arbitrary, thin-gap geometry. These general equations are then applied to two specific problems in Section C. Finally, in the last section, we consider inertia effects on the dynamics of fluid motions in a thin film.

A. The Eccentric Cylinder Problem

As stated, we begin with the problem of flow between two rotating cylinders whose axes are parallel but offset to the eccentric cylinder geometry shown in Figure 7-1. In the *concentric limit,* this is the famous *Couette flow* problem, which is analyzed in the first part of this section. In the *thin-film limit,* the eccentric cylinder flow is known as the *journal-bearing* problem. The journal bearing is one of the most important practical lubrication configurations. For example, the journal bearing is used in all standard internal combustion (automobile) engines to connect the piston shaft to both the piston and the crankshaft.

1. The Concentric Cylinder Limit—Couette Flow

As a preliminary to the *eccentric* rotating cylinder problem, we begin with an analysis of the flow between two rotating cylinders that are *concentric.* This is the famous Couette flow problem, which is very important from a practical standpoint because the rotating concentric cylinder geometry is the simplest way to produce a close approximation to a linear, simple shear flow (see Chapter 3). Hence, it is used as the basic flow device for many of the standard commercial rheometers.[2] It may appear intuitively evident to the reader, even without any detailed analysis, that the basic velocity profile between two rotating, concentric cylinders will approach the idealized linear shear flow in the limit as the gap width between the cylinders becomes small relative to their radii, and we shall see shortly that this is true. However, the flow is *not unidirectional* regardless how small the gap width becomes. Nevertheless, provided the rotation rate of the cylinders is not too large, the fluid will follow a circular path between the cylinders. This may at first seem surprising because we have shown in Chapter 6 that centripetal acceleration due to curvature of the streamlines of the primary flow through a curved tube led to a *secondary* motion in the plane orthogonal to the basic flow. However, in the Couette flow configuration, the centrifugal force that arises because of centripetal acceleration of fluid elements is balanced exactly by a radial pressure gradient so that the flow remains exactly circular.

To see why this is true, it is simplest to derive a solution for the concentric cylinder problem. For this purpose, we consider the Navier-Stokes equations expressed in terms of circular cylindrical coordinates (r, θ, z), with the z axis parallel to the central axis of the two cylinders. For steady flow, these equations are

$$\frac{1}{r}\frac{\partial}{\partial r}(ru_r) + \frac{1}{r}\frac{\partial u_\theta}{\partial \theta} + \frac{\partial u_z}{\partial z} = 0, \tag{7-1}$$

$$\rho\left[u_r\frac{\partial u_r}{\partial r} + \frac{u_\theta}{r}\left(\frac{\partial u_r}{\partial \theta} - u_\theta\right) + u_z\frac{\partial u_r}{\partial z}\right] =$$
$$-\frac{\partial p}{\partial r} + \mu\left[\frac{1}{r}\frac{\partial}{\partial r}\left(r\frac{\partial u_r}{\partial r}\right) + \frac{1}{r^2}\frac{\partial^2 u_r}{\partial \theta^2} + \frac{\partial^2 u_r}{\partial z^2} - \frac{2}{r^2}\frac{\partial u_\theta}{\partial \theta} - \frac{u_r}{r^2}\right], \tag{7-2}$$

$$\rho\left[u_r\frac{\partial u_\theta}{\partial r} + \frac{u_\theta}{r}\left(\frac{\partial u_\theta}{\partial \theta} + u_r\right) + u_z\frac{\partial u_\theta}{\partial z}\right] =$$
$$-\frac{1}{r}\frac{\partial p}{\partial \theta} + \mu\left[\frac{1}{r}\frac{\partial}{\partial r}\left(r\frac{\partial u_\theta}{\partial r}\right) + \frac{1}{r^2}\frac{\partial^2 u_\theta}{\partial \theta^2} + \frac{\partial^2 u_\theta}{\partial z^2} + \frac{2}{r^2}\frac{\partial u_r}{\partial \theta} - \frac{u_\theta}{r^2}\right], \tag{7-3}$$

and

$$\rho\left[u_r\frac{\partial u_z}{\partial r} + \frac{u_\theta}{r}\frac{\partial u_z}{\partial \theta} + u_z\frac{\partial u_z}{\partial z}\right] =$$
$$-\frac{\partial p}{\partial z} + \mu\left[\frac{1}{r}\frac{\partial}{\partial r}\left(r\frac{\partial u_z}{\partial r}\right) + \frac{1}{r^2}\frac{\partial^2 u_z}{\partial \theta^2} + \frac{\partial^2 u_z}{\partial z^2}\right], \tag{7-4}$$

with boundary conditions

$$u_z = u_r = 0 \quad \text{at } r = a \quad \text{and} \quad r = a(1+\epsilon), \tag{7-5}$$

$$u_\theta = a\Omega_1 \quad \text{at } r = a \quad \text{and} \quad u_\theta = a(1+\epsilon)\Omega_2 \quad \text{at } r = a(1+\epsilon). \tag{7-6}$$

In writing the boundary conditions (7–5) and (7–6), we have denoted the radius of the inner cylinder as $r = a$ and the radius of the outer cylinder as $r = a(1 + \epsilon)$. Further, we assume that the inner cylinder rotates with angular velocity Ω_1, while the outer cylinder rotates with a different angular velocity Ω_2.

Now, in general, there is no reason to suppose that the solution of $(7-1)$–$(7-4)$ will be anything but fully three-dimensional, with $u_r = u_r(r,\theta,z)$, $u_\theta = u_\theta(r,\theta,z)$, and $u_z = u_z(r,\theta,z)$. However, for the following analysis, we assume that the cylinders are infinitely long, thus neglecting any z dependence of the flow due to end effects that would occur for cylinders of finite length. Of course, in reality, we may make the cylinders very long in comparison with their radius, but not infinite, and the neglect of end effects introduces an approximation into the problem that we do not analyze here because it is extremely complex.[3] With this approximation, we shall assume that the solution takes the simpler form

$$u_r = u_r(r,\theta), \quad u_\theta = u_\theta(r,\theta), \quad \text{and} \quad u_z = u_z(r,\theta). \tag{7-7}$$

Then, reexamining the governing equations and boundary conditions, we see that the problem for $u_z(r,\theta)$ is completely *homogeneous*—that is, the boundary conditions are $u_z \equiv 0$ on both cylinder surfaces, and the governing equation is uncoupled from u_r and u_θ. Thus, an acceptable solution for u_z is

$$\boxed{u_z(r,\theta) \equiv 0 \quad \text{all } (r,\theta).} \tag{7-8}$$

On the other hand, u_θ is nonzero and *independent* of θ at the boundaries. Thus, u_θ must be nonzero within the flow domain, and a plausible assumption for u_θ is that

$$\boxed{u_\theta = G(r) \quad \text{only.}} \tag{7-9}$$

In this case, it can be seen from the continuity equation $(7-1)$ that the most general form for u_r is

$$u_r = \frac{F(\theta)}{r}.$$

But since $u_r = 0$ at both $r = a$ and $r = a(1 + \epsilon)$, it follows that $F(\theta) \equiv 0$ and thus

$$\boxed{u_r(r,\theta) \equiv 0.} \tag{7-10}$$

Finally, the pressure, corresponding to $(7-9)$, must be independent of θ.

In summary, then, we see that an acceptable form for the solution is

$$u_\theta = u_\theta(r), \quad p = p(r), \quad \text{and} \quad u_r = u_z = 0. \tag{7-11}$$

The governing equations $(7-2)$–$(7-4)$ in this case reduce to the simplified form

$$-\rho \frac{u_\theta^2}{r} = -\frac{\partial p}{\partial r} \tag{7-12}$$

and

$$0 = \mu \left(\frac{1}{r} \frac{\partial}{\partial r} \left(r \frac{\partial u_\theta}{\partial r} \right) - \frac{u_\theta}{r^2} \right). \tag{7-13}$$

The first of these equations is the balance between centrifugal forces and the radial pressure gradient that is responsible for the fact that $u_r = 0$. Thus, in the Couette flow problem *the acceleration associated with curved streamlines does not generally lead to secondary flow*, but is balanced exactly by a radial pressure gradient.

If we choose a characteristic length scale $l_c = a$, a velocity scale $u_c = a\Omega_1$, and a pressure scale $p_c = \rho a^2 \Omega_1^2$, (7–12) and (7–13) can be rewritten in dimensionless form as

$$\frac{\bar{u}_\theta^2}{\bar{r}} = \frac{\partial \bar{p}}{\partial \bar{r}} \tag{7–14}$$

and

$$0 = \frac{1}{\bar{r}} \frac{\partial}{\partial \bar{r}} \left(\bar{r} \frac{\partial \bar{u}_\theta}{\partial \bar{r}} \right) - \frac{\bar{u}_\theta}{\bar{r}^2}, \tag{7–15}$$

with

$$\bar{u}_\theta = 1 \quad \text{at} \quad \bar{r} = 1,$$

$$\bar{u}_\theta = (1 + \epsilon) \frac{\Omega_2}{\Omega_1} \quad \text{at} \quad \bar{r} = 1 + \epsilon. \tag{7–16}$$

It should be noticed that these equations are completely characterized by the two dimensionless parameters ϵ and Ω_2/Ω_1. In particular, the Reynolds number, $\mathrm{Re} \equiv a^2 \Omega_1 / \nu$, does not appear in spite of the fact that it would have appeared in the full equations (7–2)–(7–4) if these had been nondimensionalized in the same way. From a mathematical point of view, this is because the viscous terms are identically equal to zero in (7–2), while the inertia and pressure terms are zero in (7–3)—compare (7–12) and (7–13). Thus, the form of the velocity and pressure fields in the Couette flow problems is completely independent of Re. In this sense, the Couette flow problem is very similar to the unidirectional flows of Chapter 3.

Solving Equations (7–14) and (7–15) with the boundary conditions (7–16), we find that

$$\bar{u}_\theta = C_1 \bar{r} + \frac{C_2}{\bar{r}}, \quad \text{with } C_1 = \frac{\frac{\Omega_2}{\Omega_1}(1+\epsilon)^2 - 1}{(1+\epsilon)^2 - 1} \text{ and } C_2 = -\frac{\left(\frac{\Omega_2}{\Omega_1} - 1\right)(1+\epsilon)^2}{(1+\epsilon)^2 - 1}, \tag{7–17}$$

and

$$\bar{p} = C_1^2 \frac{\bar{r}^2}{2} + 2 C_1 C_2 \ln \bar{r} - \frac{C_2^2}{2} \frac{1}{\bar{r}^2} + \text{const}, \tag{7–18}$$

where the constant of integration in p can be chosen so that p is the pressure on the inner (or outer) cylinder when $\bar{r} = 1$ (or $\bar{r} = 1 + \epsilon$). The Couette flow solution in dimensional terms can be expressed in the form

$$u_\theta = C_1 r + \frac{C_2}{r}, \quad \text{with } C_1 = \frac{\Omega_2 R_2^2 - \Omega_1 R_1^2}{R_2^2 - R_1^2} \text{ and } C_2 = -\frac{(\Omega_2 - \Omega_1) R_2^2 R_1^2}{R_2^2 - R_1^2}, \tag{7–19}$$

where

$$R_1 = a \quad \text{and} \quad R_2 = a(1 + \epsilon).$$

Summarizing the results for *Couette* flow, we see that the fluid moves in circular paths and, thus, the fluid particles are being accelerated. As a consequence, the inertial terms in the equations of motion are nonzero—clearly, the flow is *not* unidirectional. Unlike the flow in a curved pipe, however, this acceleration does not produce a secondary flow, because the nonlinear acceleration terms are exactly balanced by a radial pressure gradient. The general tendency of acceleration due to changes in flow direction to produce secondary motions (as seen in Chapter 6, Section C) does, however, make the Couette flow prone to instabilities when it is subjected to small disturbances. In particular, when

$$\Omega_1 R_1^2 > \Omega_2 R_2^2, \tag{7-20}$$

there is a tendency for centrifugal forces acting on small disturbances of the velocity field to produce a nonzero radial velocity—that is, to break the precise balance between centrifugal forces and the forces due to the pressure gradient. This tendency is counteracted by viscous damping for sufficiently large μ, but detailed analysis as well as careful experiment show that the basic Couette flow becomes unstable at a critical value of the so-called Taylor number,

$$T_c \equiv \frac{4\Omega_1^2 R_1^4}{\nu^2} \left[\frac{\left(1 - \frac{\Omega_2}{\Omega_1}\right)\left(1 - \frac{R_1^2 \Omega_2}{R_2^2 \Omega_1}\right)}{\left(1 - \frac{R_2^2}{R_1^2}\right)^2} \right] = F\left(\frac{\Omega_2}{\Omega_1}\right),$$

where

$$F\left(\frac{\Omega_2}{\Omega_1}\right) \cong \frac{3416}{1 + \frac{\Omega_2}{\Omega_1}} \quad \text{for} \quad 0 \le \frac{\Omega_2}{\Omega_1} < 1. \tag{7-21}$$

For Taylor numbers exceeding T_c, the flow develops a secondary flow pattern in which u_r and u_z are both nonzero. A sketch of the stability criteria given by (7-21) is shown in Figure 7-2. The reader who is interested in a detailed description of the stability analysis that leads to the criteria (7-21) is encouraged to consult one of the standard textbooks on hydrodynamic stability theory (see Chandrashekhar [1961] for a particularly lucid discussion of the instability of Couette flows).[4]

2. The Eccentric Cylinder Problem

When the two cylinders are no longer concentric, the flow produced in the gap between them becomes much more complex. In particular, the velocity component u_θ must now depend on both θ and r since the distance between the cylinders is now a function of θ. Hence, it is clear from the continuity equation (7-1) that the radial velocity component u_r must also be nonzero. Apart from the creeping flow limit, which can be solved exactly using bicylindrical coordinates as stated above, the resulting fluid mechanics problem is sufficiently difficult, even for the case of infinitely long cylinders where $u_z = 0$, that it must be solved numerically.[5] However, there are two limiting cases where the problem can be simplified to allow approximate (asymptotic) analytical

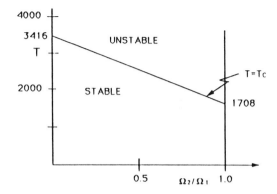

Figure 7–2 Stability diagram for Couette flow with $0 \leq \Omega_2/\Omega_1 < 1$, plotted as Taylor number T versus Ω_2/Ω_1. The flow is stable for $T < T_c$ and unstable for $T > T_c$.

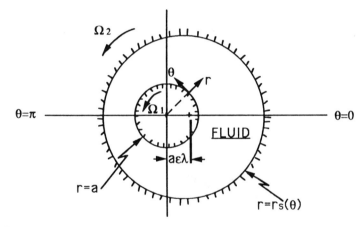

Figure 7–3 A top view of the eccentric cylinder geometry demonstrating the cylindrical coordinate system with origin at the center of the inner cylinder.

solutions: first is the *slightly eccentric* limit, where the offset of the cylinder axes is small compared to their radii; and second is the *narrow-gap* or *lubrication* approximation, where the cylinder radii are approximately equal. These two limiting cases are analyzed in the following subsections.

A starting point for *both* of these limiting problems is the governing equations and boundary conditions for the general eccentric cylinder geometry, as depicted schematically in Figure 7–3. It is convenient for the analyses that follow to adopt the circular cylindrical coordinate system that is also shown in Figure 7–3. In this system, the origin of the coordinate system is chosen to be coincident with the central axis of the inner cylinder, which is assumed to have a radius a and to be rotating with angular velocity Ω_1 in the direction shown. The radius of the outer cylinder is assumed to be

$a(1 + \epsilon)$, and its center is offset relative to the inner cylinder axis by the amount of $a\epsilon\lambda$. The surface of the inner cylinder is thus

$$r = a,$$

while the equation describing the surface of the outer eccentric cylinder is

$$r_s(\theta) = a\epsilon\lambda\cos\theta + \sqrt{(a\epsilon\lambda\cos\theta)^2 - a^2\left[\epsilon^2\lambda^2 - (1+\epsilon)^2\right]}. \tag{7-22}$$

The parameter λ determines the degree of eccentricity. If $\lambda = 1$, the two cylinders touch at $\theta = \pi$, whereas they are concentric if $\lambda = 0$. The range of allowable values for λ is thus

$$0 \le \lambda < 1. \tag{7-23}$$

The gap width between the cylinders varies as a function of θ, being a maximum at $\theta = 0$ and a minimum at $\theta = \pi$. The general expression for the gap width is

$$h = a(1 + \epsilon)\left[1 - \frac{\epsilon^2\lambda^2\sin^2\theta}{(1+\epsilon)^2}\right]^{1/2} + a\epsilon\lambda\cos\theta - a. \tag{7-24}$$

The governing equations for arbitrary λ and ϵ are still the Navier-Stokes and continuity equations (7-1)-(7-4), specialized to the case $u_z \equiv 0$ and with all derivatives with respect to z set equal to zero, that is,

$$\frac{1}{r}\frac{\partial}{\partial r}(ru_r) + \frac{1}{r}\frac{\partial u_\theta}{\partial \theta} = 0, \tag{7-25}$$

$$\rho\left[u_r\frac{\partial u_r}{\partial r} + \frac{u_\theta}{r}\left(\frac{\partial u_r}{\partial \theta} - u_\theta\right)\right] =$$
$$-\frac{\partial p}{\partial r} + \mu\left[\frac{1}{r}\frac{\partial}{\partial r}\left(r\frac{\partial u_r}{\partial r}\right) + \frac{1}{r^2}\frac{\partial^2 u_r}{\partial \theta^2} - \frac{2}{r^2}\frac{\partial u_\theta}{\partial \theta} - \frac{u_r}{r^2}\right], \tag{7-26}$$

and

$$\rho\left[u_r\frac{\partial u_\theta}{\partial r} + \frac{u_\theta}{r}\left(\frac{\partial u_\theta}{\partial \theta} + u_r\right)\right] =$$
$$-\frac{1}{r}\frac{\partial p}{\partial \theta} + \mu\left[\frac{1}{r}\frac{\partial}{\partial r}\left(r\frac{\partial u_\theta}{\partial r}\right) + \frac{1}{r^2}\frac{\partial^2 u_\theta}{\partial \theta^2} + \frac{2}{r^2}\frac{\partial u_r}{\partial \theta} - \frac{u_\theta}{r^2}\right]. \tag{7-27}$$

The boundary conditions at the cylinder surfaces are just the no-slip and kinematic conditions as before,

$$u_r = 0, \quad u_\theta = \Omega_1 a \quad \text{at} \quad r = a, \tag{7-28}$$

and

$$\mathbf{u}\cdot\mathbf{n} = 0, \quad \mathbf{u}\cdot\mathbf{t} = \Omega_2(1+\epsilon)a \quad \text{at} \quad r = r_s(\theta). \tag{7-29}$$

Now, however, the normal and tangential velocity components at the surface of the *outer* cylinder do not correspond to the radial and azimuthal velocity components u_r and u_θ, specified relative to cylindrical coordinates that are fixed on the inner cylinder. To complete the problem formulation, we must express the unit normal and tangent vectors at r_s in terms of cylindrical coordinate components. For this purpose, we use the well-known formula from differential geometry for the unit normal to a surface,

$$\mathbf{n} \equiv \frac{\nabla F}{|\nabla F|}. \tag{7-30}$$

Here, $F(\mathbf{x})$ is the function whose roots define the surface as the set of points $\mathbf{x}_s$, such that

$$F(\mathbf{x}_s) \equiv 0. \tag{7-31}$$

In the present case,

$$F \equiv r - r_s(\theta), \tag{7-32}$$

and thus

$$\boxed{\mathbf{n} \equiv \frac{\left(\mathbf{i}_r - \mathbf{i}_\theta\left(\dfrac{1}{r_s(\theta)}\dfrac{dr_s}{d\theta}\right)\right)}{\left(1 + \dfrac{1}{r_s^2(\theta)}\left(\dfrac{dr_s}{d\theta}\right)^2\right)^{1/2}}.} \tag{7-33}$$

The unit tangent $\mathbf{t}$ is

$$\boxed{\mathbf{t} \equiv \frac{\left(\mathbf{i}_\theta + \mathbf{i}_r\left(\dfrac{1}{r_s(\theta)}\dfrac{dr_s}{d\theta}\right)\right)}{\left(1 + \dfrac{1}{r_s^2(\theta)}\left(\dfrac{dr_s}{d\theta}\right)^2\right)^{1/2}}.} \tag{7-34}$$

The general problem defined by (7-25)–(7-34) is highly nonlinear and can generally only be solved numerically, as noted earlier. In this section, however, we consider the two limiting cases

$$\lambda \ll 1, \quad \epsilon = 0(1) \quad \text{(slight eccentricity)}$$

and

$$\epsilon \ll 1, \quad \lambda = 0(1) \quad \text{(narrow gap)},$$

and in these two cases, analytic approximations to the solution are possible. We begin with the slightly eccentric cylinder problem.

(a) Slight Eccentricity, $\lambda \ll 1$, $\epsilon = 0(1)$

We consider the case of a slight degree of eccentricity, $\lambda \ll 1$, for cylinders of arbitrary radius ratio, that is, $\epsilon = 0(1)$. In this case, the geometry of the system can be defined in simpler terms by considering the limit of equations (7-22) and (7-24) for $\lambda \ll 1$ with $\epsilon = 0(1)$.

The surface of the outer cylinder in this case then can be approximated in the form

$$\boxed{r_s(\theta) = a(1 + \epsilon) + a\epsilon\lambda\cos\theta + 0(\lambda^2),} \tag{7-35}$$

and a corresponding approximation for the gap width is

$$h(\theta) = a\epsilon \left[1 + \lambda\cos\theta + 0(\lambda^2)\right]. \tag{7-36}$$

In addition, the unit normal to the outer cylinder surface now becomes

$$\mathbf{n} \equiv \mathbf{i}_r + \mathbf{i}_\theta \left[\frac{\epsilon\lambda}{1+\epsilon} \sin\theta\right] + 0(\lambda^2), \tag{7-37}$$

and the unit tangent vector $\mathbf{t}$ is thus

$$\mathbf{t} \equiv \mathbf{i}_\theta - \mathbf{i}_r \left[\frac{\epsilon\lambda}{1+\epsilon} \sin\theta\right] + 0(\lambda^2). \tag{7-38}$$

To proceed with an approximate solution for small λ, we must nondimensionalize the governing equations and boundary conditions. In this way, if the characteristic length and velocity scales are chosen properly, we can identify terms that can be neglected in the limit $\lambda \ll 1$ by considering the dependence of all terms on λ. To nondimensionalize the governing equations (7-25)–(7-27), we assume that the velocities are proportional to the magnitude of the boundary velocity,*

$$u_c = \Omega_1 a. \tag{7-39}$$

In addition, we pick the inner cylinder radius as a characteristic length scale†

$$l_c = a. \tag{7-40}$$

Thus, nondimensionalizing velocities according to (7–39), the radial variable r according to (7–40), and the pressure using

$$p_c = \rho u_c^2 = \rho a^2 \Omega_1^2, \tag{7-41}$$

we obtain the nondimensionalized equations

$$\left[\bar{u}_r \frac{\partial \bar{u}_r}{\partial \bar{r}} + \frac{\bar{u}_\theta}{\bar{r}}\left(\frac{\partial \bar{u}_r}{\partial \theta}\right) - \frac{\bar{u}_\theta^2}{\bar{r}}\right] =$$
$$-\frac{\partial \bar{p}}{\partial \bar{r}} + \frac{1}{\text{Re}}\left[\frac{1}{\bar{r}}\frac{\partial}{\partial \bar{r}}\left(\bar{r}\frac{\partial \bar{u}_r}{\partial \bar{r}}\right) + \frac{1}{\bar{r}^2}\frac{\partial^2 \bar{u}_r}{\partial \theta^2} - \frac{2}{\bar{r}^2}\frac{\partial \bar{u}_\theta}{\partial \theta} - \frac{\bar{u}_r}{\bar{r}^2}\right], \tag{7-42}$$

$$\left[\bar{u}_r \frac{\partial \bar{u}_\theta}{\partial \bar{r}} + \frac{\bar{u}_\theta}{\bar{r}}\frac{\partial \bar{u}_\theta}{\partial \theta} + \frac{\bar{u}_\theta\bar{u}_r}{\bar{r}}\right] =$$
$$-\frac{1}{\bar{r}}\frac{\partial \bar{p}}{\partial \theta} + \frac{1}{\text{Re}}\left[\frac{1}{\bar{r}}\frac{\partial}{\partial \bar{r}}\left(\bar{r}\frac{\partial \bar{u}_\theta}{\partial \bar{r}}\right) + \frac{1}{\bar{r}^2}\frac{\partial^2 \bar{u}_\theta}{\partial \theta^2} + \frac{2}{\bar{r}^2}\frac{\partial \bar{u}_r}{\partial \theta} - \frac{\bar{u}_\theta}{\bar{r}^2}\right], \tag{7-43}$$

and

$$\frac{1}{\bar{r}}\frac{\partial}{\partial \bar{r}}(\bar{r}\bar{u}_r) + \frac{1}{\bar{r}}\frac{\partial \bar{u}_\theta}{\partial \theta} = 0. \tag{7-44}$$

*Note that we also could have chosen $u_c = \Omega_2(1 + \epsilon)a$, and this would be essential if $\Omega_1 \equiv 0$. Generally, if Ω_1 and Ω_2 are of approximately equal magnitude, either choice is satisfactory, and the ratio Ω_1/Ω_2 will then appear as one of the dimensionless parameters that characterizes the problem.

†Again, we could use the radius of the outer cylinder. Provided $\epsilon = 0(1)$, so that the gap width

The Reynolds number appears in these equations as the only dimensionless parameter and is defined as

$$\text{Re} \equiv \frac{\rho a^2 \Omega_1}{\mu}.$$

The boundary conditions at the outer cylinder are (7–29), with **n**, **t**, and $r_s(\theta)$ given, in dimensional form, by (7–35), (7–37), and (7–38). Thus,

$$\left. \begin{aligned} u_r - \frac{1}{r_s(\theta)} \frac{dr_s}{d\theta} u_\theta &= 0 \\[2mm] u_\theta + \frac{1}{r_s(\theta)} \frac{dr_s}{d\theta} u_r &= \Omega_2(1+\epsilon)a \end{aligned} \right\} \quad \text{at } r = r_s(\theta).$$

In dimensionless terms, these conditions are

$$\boxed{\left. \begin{aligned} \bar{u}_r - \frac{1}{\bar{r}_s(\theta)} \frac{d\bar{r}_s}{d\theta} \bar{u}_\theta &= 0 \\[2mm] \bar{u}_\theta + \frac{1}{\bar{r}_s(\theta)} \frac{d\bar{r}_s}{d\theta} \bar{u}_r &= \frac{\Omega_2}{\Omega_1}(1+\epsilon) \end{aligned} \right\} \quad \text{at } \bar{r} = \bar{r}_s(\theta),} \qquad (7\text{–}45)$$

where

$$\bar{r}_s(\theta) \equiv (1+\epsilon) + \epsilon\lambda\cos\theta + 0(\lambda^2). \qquad (7\text{–}46)$$

The boundary conditions on the inner cylinder, also in dimensionless form, are

$$\boxed{\bar{u}_r = 0 \quad \text{and} \quad \bar{u}_\theta = 1 \quad \text{at} \quad \bar{r} = 1.} \qquad (7\text{–}47)$$

Three additional dimensionless parameters thus appear in the boundary conditions, the two geometric parameters ϵ and λ, and the ratio of angular velocities Ω_2/Ω_1. In this section, we seek an asymptotic solution of (7–42)–(7–47) for ϵ, Re and Ω_2/Ω_1 all of $0(1)$, and $\lambda \ll 1$.

One difficulty in actually solving this problem analytically is that the surface of the outer cylinder does *not* correspond to a coordinate surface in the cylindrical coordinate system that we have used, except in the limit $\lambda \to 0$, when the problem reduces to the concentric cylinder geometry that was already considered in the previous section. Our interest here is for small, but *nonzero*, values of λ. Fortunately, the problem for $\lambda \ll 1$ can be recast in a form that is amenable to analytic solution using a circular cylindrical coordinate system. This is done using the so-called *domain perturbation* technique,[6] in which a Taylor series approximation is used to reexpress the *exact* boundary conditions (7–45) that are applied at the surface $\bar{r} = \bar{r}_s(\theta)$ in terms of asymptotically *equivalent* boundary conditions applied at the nearby surface, $\bar{r} = 1 + \epsilon$, which is a coordinate surface for the cylindrical coordinate system.

---

between the cylinders is not small compared to the cylinder radii, the radius of either cylinder provides an equally valid measure of the distance over which the velocity components vary by $0(u_c)$, and the radius ratio appears as a second dimensionless parameter, namely, $(1 + \epsilon)$.

For this purpose, we express $\bar{u}_r$ and $\bar{u}_\theta$ at $\bar{r} = \bar{r}_s(\theta)$ in terms of $\bar{u}_r$, $\bar{u}_\theta$, and their derivatives with respect to $\bar{r}$, all evaluated at $\bar{r} = 1 + \epsilon$, that is,

$$\bar{u}_r|_{\bar{r} = \bar{r}_s(\theta)} = \bar{u}_r|_{\bar{r} = 1 + \epsilon} + \left(\frac{\partial \bar{u}_r}{\partial \bar{r}}\right)_{\bar{r} = 1 + \epsilon} \epsilon \lambda \cos\theta + 0(\lambda^2) \tag{7-48a}$$

and

$$\bar{u}_\theta|_{\bar{r} = \bar{r}_s(\theta)} = \bar{u}_\theta|_{\bar{r} = 1 + \epsilon} + \left(\frac{\partial \bar{u}_\theta}{\partial \bar{r}}\right)_{\bar{r} = 1 + \epsilon} \epsilon \lambda \cos\theta + 0(\lambda^2). \tag{7-48b}$$

Thus, using these equations, the boundary conditions (7–45) can be expressed in the revised form

$$\bar{u}_r + \left[\left(\frac{\partial \bar{u}_r}{\partial \bar{r}}\right) \epsilon \cos\theta + \frac{1}{r}(\epsilon \sin\theta)\bar{u}_\theta\right]\lambda + 0(\lambda^2) = 0 \text{ at } \bar{r} = 1 + \epsilon \tag{7-49a}$$

and

$$\bar{u}_\theta + \left[\left(\frac{\partial \bar{u}_\theta}{\partial \bar{r}}\right) \epsilon \cos\theta - \frac{1}{r}\epsilon \sin\theta \bar{u}_r\right]\lambda + 0(\lambda^2) = \frac{\Omega_2}{\Omega_1}(1 + \epsilon) \text{ at } \bar{r} = 1 + \epsilon. \tag{7-49b}$$

We see that the exact boundary conditions (7–45) at $\bar{r} = \bar{r}_s(\theta)$ are replaced by the asymptotically equivalent conditions (7–49) applied at $\bar{r} = 1 + \epsilon$. This fundamental, but simple, idea of transforming from exact boundary conditions at a surface that is not a coordinate surface to asymptotically equivalent boundary conditions on a nearby surface that *is* a coordinate surface can be used for any problem involving boundaries that differ only slightly from a coordinate surface.[6]

The problem now is to solve the governing equations (7–42)–(7–44) subject to the boundary conditions (7–47) and (7–49). We seek an approximate solution in the form of a regular asymptotic expansion for $\lambda \ll 1$,

$$\bar{u}_r = \bar{u}_r^{(0)} + \lambda \bar{u}_r^{(1)} + \cdots,$$
$$\bar{u}_\theta = \bar{u}_\theta^{(0)} + \lambda \bar{u}_\theta^{(1)} + \cdots,$$

and

$$\bar{p} = \bar{p}^{(0)} + \lambda \bar{p}^{(1)} + \cdots. \tag{7-50}$$

Clearly, in the limit $\lambda \to 0$, the eccentric cylinder geometry reduces to the concentric geometry, and the leading-order terms in this expansion should be the Couette flow solution that was derived in the preceding section, that is,

$$\bar{u}_r^{(0)} = 0, \ \bar{u}_\theta^{(0)} = \bar{u}_\theta(\bar{r}), \text{ and } \bar{p}^{(0)} = p(\bar{r}), \tag{7-51}$$

with $\bar{u}_\theta(\bar{r})$ given by (7–17). To see that this is true, we derive governing equations and boundary conditions at each order in λ. The procedure, as outlined in Chapter 6, is to substitute the expansions (7–50) into the full equations and boundary conditions (7–42)–(7–44), (7–47), and (7–49) and then set the sum of the terms at each order in λ equal to zero.

At 0(1), the governing equations for $\bar{u}_r^{(0)}$, $\bar{u}_\theta^{(0)}$, and $\bar{p}^{(0)}$ are, in fact, identical to the full equations (7–42)–(7–44), and we do not repeat them here. On the other hand, the boundary conditions at 0(1) are just

$$\bar{u}_r^{(0)} = 0, \quad \bar{u}_\theta^{(0)} = 1 \quad \text{at} \quad \bar{r} = 1,$$

$$\bar{u}_r^{(0)} = 0, \quad \bar{u}_\theta^{(0)} = \frac{\Omega_2}{\Omega_1}(1 + \epsilon) \quad \text{at} \quad \bar{r} = 1 + \epsilon. \tag{7-52}$$

These are identical to the boundary conditions (7–16) for the concentric cylinder problem, and the solution is just the Couette flow solution (7–17) and (7–18), as stated earlier.

At $O(\lambda)$, the eccentricity of the geometry produces a change in the velocity and pressure fields relative to the Couette flow solution. Taking account of the fact that $\bar{u}_r^{(0)} = 0$, the governing equations at $O(\lambda)$ become

$$\frac{\bar{u}_\theta^{(0)}}{\bar{r}} \frac{\partial \bar{u}_r^{(1)}}{\partial \theta} - \frac{2\bar{u}_\theta^{(0)}\bar{u}_\theta^{(1)}}{\bar{r}} =$$
$$-\frac{\partial \bar{p}^{(1)}}{\partial \bar{r}} + \frac{1}{\text{Re}}\left[\frac{1}{\bar{r}}\frac{\partial}{\partial \bar{r}}\left(\bar{r}\frac{\partial \bar{u}_r^{(1)}}{\partial \bar{r}}\right) + \frac{1}{\bar{r}^2}\frac{\partial^2 \bar{u}_r^{(1)}}{\partial \theta^2} - \frac{2}{\bar{r}^2}\frac{\partial \bar{u}_\theta^{(1)}}{\partial \theta} - \frac{\bar{u}_r^{(1)}}{\bar{r}^2}\right], \tag{7-53}$$

$$\bar{u}_r^{(1)}\frac{\partial \bar{u}_\theta^{(0)}}{\partial \bar{r}} + \frac{\bar{u}_\theta^{(0)}}{\bar{r}}\frac{\partial \bar{u}_\theta^{(1)}}{\partial \theta} + \frac{\bar{u}_\theta^{(0)}\bar{u}_r^{(1)}}{\bar{r}} =$$
$$-\frac{1}{\bar{r}}\frac{\partial \bar{p}^{(1)}}{\partial \theta} + \frac{1}{\text{Re}}\left[\frac{1}{\bar{r}}\frac{\partial}{\partial \bar{r}}\left(\bar{r}\frac{\partial \bar{u}_\theta^{(1)}}{\partial \bar{r}}\right) + \frac{1}{\bar{r}^2}\frac{\partial^2 \bar{u}_\theta^{(1)}}{\partial \theta^2} + \frac{2}{\bar{r}^2}\frac{\partial \bar{u}_r^{(1)}}{\partial \theta} - \frac{\bar{u}_\theta^{(1)}}{\bar{r}^2}\right], \tag{7-54}$$

and

$$\frac{1}{\bar{r}}\frac{\partial}{\partial \bar{r}}(\bar{r}\bar{u}_r^{(1)}) + \frac{1}{\bar{r}}\frac{\partial \bar{u}_\theta^{(1)}}{\partial \theta} = 0. \tag{7-55}$$

The boundary conditions at $O(\lambda)$ are

$$\bar{u}_r^{(1)} = \bar{u}_\theta^{(1)} = 0 \quad \text{at} \quad \bar{r} = 1 \tag{7-56}$$

and

$$\bar{u}_r^{(1)} = -\frac{1}{1 + \epsilon}(\epsilon\sin\theta)\bar{u}_\theta^{(0)}$$
$$\bar{u}_\theta^{(1)} = -\frac{\partial \bar{u}_\theta^{(0)}}{\partial \bar{r}}\epsilon\cos\theta \quad\left.\right\} \quad \text{at} \quad \bar{r} = 1 + \epsilon.$$

With $\bar{u}_\theta^{(0)}$ and $\partial\bar{u}_\theta^{(0)}/\partial\bar{r}$ evaluated using the Couette flow solution at $\bar{r} = 1 + \epsilon$, the latter conditions can be expressed in the form

$$\bar{u}_r^{(1)} = -\frac{\epsilon\sin\theta}{(1+\epsilon)^2}\left[C_1(1+\epsilon)^2 + C_2\right] = -\epsilon\frac{\Omega_2}{\Omega_1}\sin\theta,$$

$$\bar{u}_\theta^{(1)} = -\frac{\epsilon\cos\theta}{(1+\epsilon)^2}\left[C_1(1+\epsilon)^2 - C_2\right] = \epsilon\cos\theta\left[\frac{2 - \frac{\Omega_2}{\Omega_1}\left[(1+\epsilon)^2 + 1\right]}{(1+\epsilon)^2 - 1}\right] \quad\left.\right\} \quad \text{at} \quad \bar{r} = 1 + \epsilon.$$

$$\tag{7-57}$$

We may note, from (7–53) and (7–54), that the form of the velocity and pressure fields for this leading-order departure from the Couette flow limit depends upon the Reynolds number in addition to Ω_2/Ω_1 and ϵ [note that the eccentricity parameter λ appears directly in the expansion (7–50)]. This dependence on the Reynolds number, for even a slight degree of eccentricity, is a fundamental change from the Couette solution, which depends only on ϵ and Ω_2/Ω_1.

Now since the problem at $0(\lambda)$ is *linear*, it is tempting to suppose that a solution could be found in which $\bar{u}_r^{(1)}$ and $\bar{u}_\theta^{(1)}$ show the same dependence on θ as the boundary data (7–57) at $\bar{r} = 1 + \epsilon$. In fact, this is not directly possible—notice, for example, that the left side of (7–53) involves $\partial\bar{u}_r^{(1)}/\partial\theta$, while the right side involves $\bar{u}_r^{(1)}$ and $\partial^2\bar{u}_r^{(1)}/\partial\theta^2$. However, a very similar procedure can be used if we introduce a complex velocity and pressure defined such that

$$\bar{u}^{(1)} = \mathrm{Re}(\hat{u}) \text{ and } p^{(1)} = \mathrm{Re}(\hat{p}). \qquad (7\text{–}58)$$

Let us suppose that $\hat{u}$ and $\hat{p}$ satisfy (7–53)–(7–55), but with $\hat{u}$ and $\hat{p}$ substituted for $u^{(1)}$ and $p^{(1)}$. Furthermore, let us assume that $\hat{u}$ and $\hat{p}$ satisfy the boundary conditions

$$\hat{u}_r = \hat{u}_\theta = 0 \quad \text{at} \quad \bar{r} = 1, \qquad (7\text{–}59)$$

$$\text{and} \quad \hat{u}_r = \frac{\epsilon}{(1+\epsilon)^2}\left[C_1(1+\epsilon)^2 + C_2\right]ie^{i\theta} = \epsilon\frac{\Omega_2}{\Omega_1}ie^{i\theta},$$

$$\hat{u}_{(\theta)} = -\frac{\epsilon}{(1+\epsilon)^2}\left[C_1(1+\epsilon)^2 - C_2\right]e^{i\theta} =$$

$$\left. \phantom{\rule{0pt}{3em}} \right\} \quad \text{at } \bar{r} = 1 + \epsilon.$$

$$\epsilon\left[\frac{2 - \dfrac{\Omega_2}{\Omega_1}\left[(1+\epsilon)^2 + 1\right]}{(1+\epsilon)^2 - 1}\right]e^{i\theta} \qquad (7\text{–}60)$$

Then, in view of the linearity of the basic equations and boundary conditions, it is evident that the real part of $\hat{u}$ and $\hat{p}$ represents a solution of the original problem (7–53)–(7–57) for $u^{(1)}$ and $p^{(1)}$.

Now, however, the form of the solution for $\hat{u}$ can be anticipated from the boundary data (7–60) at $\bar{r} = 1 + \epsilon$. In particular, let us suppose that

$$\hat{u}_r = F(\bar{r})e^{i\theta} \quad \text{and} \quad \hat{u}_\theta = H(\bar{r})e^{i\theta}, \qquad (7\text{–}61)$$

where $F(\bar{r})$ and $H(\bar{r})$ are both complex functions of $\bar{r}$. This choice is consistent with the forms of $\hat{u}_r$ and $\hat{u}_\theta$ at the boundary, $\bar{r} = 1 + \epsilon$, but since F and H can be complex, it does not force $u_r^{(1)}$ and $u_\theta^{(1)}$ to be proportional to $\sin\theta$ and $\cos\theta$, respectively, as one might have guessed from (7–57) if we had attempted to solve directly for $u^{(1)}$ and $p^{(1)}$. The relationship between $F(\bar{r})$ and $H(\bar{r})$ is determined by substituting (7–61) into the continuity equation,

$$\left(\frac{dF}{d\bar{r}} + \frac{F}{\bar{r}}\right)e^{i\theta} + \frac{1}{\bar{r}}iH(\bar{r})e^{i\theta} = 0. \qquad (7\text{–}62)$$

Thus,

$$-iH(\bar{r}) = \frac{d}{d\bar{r}}(\bar{r}F) \qquad (7\text{–}63)$$

or

$$F(\bar{r}) = -\frac{i}{\bar{r}}\int H(\bar{r})\,d\bar{r}. \qquad (7\text{-}64)$$

Let us suppose that

$$H(\bar{r}) \equiv \frac{dG}{d\bar{r}}$$

for convenience. Then

$$F(\bar{r}) = -\frac{iG}{\bar{r}}$$

and

$$\boxed{\hat{u}_r = -\frac{iG(\bar{r})}{\bar{r}}e^{i\theta} \quad \text{and} \quad \hat{u}_\theta = \frac{dG}{d\bar{r}}e^{i\theta}.} \qquad (7\text{-}65)$$

With this assumed form for $\hat{u}_r$ and $\hat{u}_\theta$, the equations of motion (7–53) and (7–54), with $\hat{\mathbf{u}}$ and $\hat{p}$ replacing $\mathbf{u}^{(1)}$ and $p^{(1)}$, can now be written in the form of a pair of coupled partial differential equations for $G(\bar{r})$ and $\hat{p}$, namely,

$$u_\theta^{(0)}\left[\frac{2}{\bar{r}}\frac{dG}{d\bar{r}} - \frac{G}{\bar{r}^2}\right]e^{i\theta} = \frac{\partial\hat{p}}{\partial\bar{r}} + \frac{i}{\mathrm{Re}}\left[\frac{1}{\bar{r}}\frac{d^2G}{d\bar{r}^2} + \frac{1}{\bar{r}^2}\frac{dG}{d\bar{r}} - \frac{G}{\bar{r}^3}\right]e^{i\theta} \qquad (7\text{-}66)$$

and

$$-i\left[G\frac{\partial u_\theta^{(0)}}{\partial\bar{r}} - u_\theta^{(0)}\left(\frac{dG}{d\bar{r}} - \frac{G}{\bar{r}}\right)\right]e^{i\theta} =$$

$$-\frac{\partial\hat{p}}{\partial\theta} + \frac{1}{\mathrm{Re}}\left[\bar{r}\frac{d^3G}{d\bar{r}^3} + \frac{d^2G}{d\bar{r}^2} - \frac{2}{\bar{r}}\frac{dG}{d\bar{r}} + \frac{2G}{\bar{r}^2}\right]e^{i\theta}. \qquad (7\text{-}67)$$

If we eliminate $\hat{p}$ from (7–66) and (7–67) by cross differentiating—that is, (7–66) by θ and (7–67) by $\bar{r}$—and substituting one equation into the other, we obtain a single fourth-order ordinary differential equation for G,

$$\boxed{\bar{r}^4 G'''' + 2\bar{r}^3 G''' - 3\bar{r}^2 G'' + 3\bar{r}G' - 3G = i\mathrm{Re}(C_1\bar{r}^2 + C_2)(\bar{r}^2 G'' + \bar{r}G' - G).}$$

$$(7\text{-}68)$$

The constants C_1 and C_2 come from the Couette flow solution for $u_\theta^{(0)}(\bar{r})$ and were given earlier in Equation (7–17).

Thus, at $0(\lambda)$, the slightly eccentric cylinder problem is reduced to solving a single, fourth-order linear, ordinary differential equation. In order to satisfy the boundary conditions (7–59) and (7–60), we require that

$$\boxed{G = \frac{dG}{d\bar{r}} = 0 \quad \text{at} \quad \bar{r} = 1} \qquad (7\text{-}69)$$

and

$$G = -\frac{\epsilon}{(1+\epsilon)}\left[C_1(1+\epsilon)^2 + C_2\right] = -\epsilon(1+\epsilon)\frac{\Omega_2}{\Omega_1}$$

plus

$$\frac{dG}{d\bar{r}} = -\frac{\epsilon}{(1+\epsilon)^2}\left[C_1(1+\epsilon)^2 - C_2\right] = \epsilon\left[\frac{2 - \frac{\Omega_2}{\Omega_1}\left[(1+\epsilon)^2 + 1\right]}{(1+\epsilon)^2 - 1}\right] \text{ at } \bar{r} = 1 + \epsilon.$$

(7-70)

An *exact* solution of (7-68)-(7-70) is possible. The simplest way to derive this solution is to note that the fourth-order equation for G can be expressed in terms of an equivalent pair of second-order equations, namely,

$$\bar{r}^2\frac{d^2H}{d\bar{r}^2} - 3\bar{r}\frac{dH}{d\bar{r}} + \left[3 - i\text{Re}(C_1\bar{r}^2 + C_2)\right]H = 0$$

(7-71)

and

$$r^2\frac{d^2G}{d\bar{r}^2} + \bar{r}\frac{dG}{d\bar{r}} - G = H.$$

(7-72)

Thus, to obtain a general solution for G, we first solve the homogeneous Equation (7-71) for H and then solve (7-72) as a second-order, nonhomogeneous equation for G.

Equation (7-71) is a form of the so-called transformed Bessel's equation with complex coefficients. A general solution for H can thus be expressed in the form

$$H(\bar{r}) = A_1\bar{r}^2J_q(\alpha\bar{r}) + A_2\bar{r}^2Y_q(\alpha\bar{r}),$$

(7-73)

where J_q and Y_q are Bessel's functions of the first and second kind, with

$$q = \sqrt{1 + i\text{Re}C_2}$$

and

$$\alpha = \frac{\sqrt{2}}{2}(1 - i)\sqrt{C_1\text{Re}}.$$

An exact solution of (7-68) can then be achieved by solving (7-72) with $H(\bar{r})$ given by (7-73). The homogeneous solution of (7-72) is obtained easily if we recognize (7-72) as an example of a second-order Euler's equation. The solutions of this type of equation are of the form $\bar{r}^s$, where s represents the roots of the characteristic equation that is obtained by simple substitution. In the present case, substituting $\bar{r}^s$ into the homogeneous form of (7-72), we obtain

$$\bar{r}^s\left[s(s-1) + s - 1\right] = 0.$$

Hence,

$$s = \pm 1,$$

(7-74)

and the homogeneous solution is thus

$$G_h = d_1 \bar{r} + \frac{d_2}{\bar{r}}.$$

(7-75)

Since the homogeneous solution is known, the particular solutions of (7-72) can be obtained by the method of *variation of parameters*. In this technique, we seek a particular solution of the form

$$G_p(\bar{r}) = u_1(\bar{r}) \cdot \bar{r} + \frac{u_2(\bar{r})}{\bar{r}}.$$

(7-76)

Hence, differentiating and substituting into (7-72), we find that

$$\bar{r}^2 \left[u_1'' \bar{r} + \frac{u_2''}{\bar{r}} + 2 \left(u_1' - \frac{u_2'}{\bar{r}^2} \right) \right] + \bar{r} \left[u_1' \bar{r} + \frac{u_2'}{\bar{r}} \right] = A_1 \bar{r}^2 J_q(\alpha \bar{r}) + A_2 \bar{r}^2 Y_q(\alpha \bar{r}).$$

(7-77)

Here, for simplicity, we have denoted derivatives with respect to $\bar{r}$ via a prime, that is, $u_1' \equiv du_1/d\bar{r}$, and so on. Now, it follows from (7-77) that a necessary condition for existence of a particular solution of the form (7-76) is that the coefficients $u_1(\bar{r})$ and $u_2(\bar{r})$ satisfy the pair of equations

$$u_1' \bar{r} + \frac{u_2'}{\bar{r}} = 0$$

(7-78a)

and

$$u_1' - \frac{u_2'}{\bar{r}^2} = A_1 J_q(\alpha \bar{r}) + A_2 Y_q(\alpha \bar{r}).$$

(7-78b)

But these equations are just a pair of linear algebraic equations for u_1' and u_2'. After some minor manipulation, these equations can be solved for u_1' and u_2', and the resulting expressions integrated to find u_1 and u_2. Combining the result with (7-76) and (7-75), we thus obtain a general solution for G:

$$G = d_1 \bar{r} + \frac{d_2}{\bar{r}} + A_1 \frac{\bar{r}}{2} F_1(\bar{r}) + A_2 \frac{1}{2\bar{r}} F_2(\bar{r}),$$

(7-79)

where

$$F_1(\bar{r}) \equiv \int_1^{\bar{r}} (1 - t^2) J_q(\alpha t) dt$$

and

$$F_2(\bar{r}) \equiv \int_1^{\bar{r}} (1 - t^2) Y_q(\alpha t) dt.$$

The boundary conditions (7-69) at $\bar{r} = 1$ require that

$$d_1 = d_2 = 0.$$

(7-80)

The conditions (7-70) then determine the constants A_1 and A_2, via the pair of equations

$$A_1 \frac{(1 + \epsilon)}{2} F_1(1 + \epsilon) + \frac{A_2}{2(1 + \epsilon)} F_2(1 + \epsilon) = -\frac{\epsilon}{1 + \epsilon} \left[C_1(1 + \epsilon)^2 + C_2 \right] \quad (7-81)$$

and

$$A_1 \left[\frac{F_1(1+\epsilon)}{2} + \frac{(1+\epsilon)}{2}\left[1-(1+\epsilon)^2\right]J_q(\alpha(1+\epsilon)) \right]$$
$$+ A_2 \left[-\frac{F_2(1+\epsilon)}{2(1+\epsilon)^2} + \frac{\left[1-(1+\epsilon)^2\right]}{2(1+\epsilon)}Y_q(\alpha(1+\epsilon)) \right] = -\frac{\epsilon}{(1+\epsilon)^2}\left[C_1(1+\epsilon)^2 - C_2\right]. \quad (7\text{--}82)$$

Now, the solution (7–78), with the coupled algebraic equations for A_1 and A_2 is an *exact* solution of the original problem (7–68)–(7–70) for *arbitrary* values of the Reynolds number. Nevertheless, it is quite complex, and it would be highly advantageous if we could achieve approximate solutions that were simpler in form. We shall consider this possibility shortly. First, however, we briefly call attention to one special case where the *exact* solution is somewhat simpler.

This is the case in which C_1 from the 0(1) Couette flow solution is *zero*. Referring back to the definition of C_1 in Equation (7–17), we see that the case

$$C_1 = 0$$

corresponds to the special condition in which

$$\frac{\Omega_2}{\Omega_1} = \frac{1}{(1+\epsilon)^2}. \quad (7\text{--}83)$$

In this case,

$$u_\theta^{(0)} = \frac{C_2}{r}, \quad (7\text{--}84)$$

and the Couette flow solution is *irrotational* in the sense that

$$\nabla \wedge \mathbf{u}^{(0)} \equiv 0.$$

The governing Equation (7–71) for $H(\bar{r})$ now becomes a Euler equation,

$$\bar{r}^2\frac{d^2H}{d\bar{r}^2} - 3\bar{r}\frac{dH}{d\bar{r}} + (3 - i\,\mathrm{Re}\,C_2)H = 0, \quad (7\text{--}85)$$

which has a pair of solutions of the form

$$H = r^s, \quad (7\text{--}86)$$

where s represents the roots of the characteristic equation

$$s^2 - 4s + (3 - i\,\mathrm{Re}\,C_2) = 0. \quad (7\text{--}87)$$

These are

$$\boxed{s_1 = 2 + (1 + i\,\mathrm{Re}\,C_2)^{1/2}}$$

and

$$\boxed{s_2 = 2 - (1 + i\,\mathrm{Re}\,C_2)^{1/2}.} \quad (7\text{--}88)$$

It then follows from (7–72) that the general form of the solution for G is

$$G = d_1\bar{r} + \frac{d_2}{\bar{r}} + d_3 r^{s_1} + d_4 r^{s_2}.$$

(7–89)

Applying the boundary conditions (7–69) and (7–70) leads to a set of four algebraic relationships for the coefficients d_i:

$$d_1 + d_2 + d_3 + d_4 = 0,$$

$$d_1 - d_2 + s_1 d_3 + s_2 d_4 = 0,$$

$$d_1(1 + \epsilon) + \frac{d_2}{(1 + \epsilon)} + d_3(1 + \epsilon)^{s_1} + d_4(1 + \epsilon)^{s_2} = -\frac{C_2\epsilon}{(1 + \epsilon)},$$

$$d_1 - \frac{d_2}{(1 + \epsilon)^2} + s_1 d_3(1 + \epsilon)^{s_1-1} + s_2 d_4(1 + \epsilon)^{s_2-1} = +\frac{C_2\epsilon}{(1 + \epsilon)^2}.$$

(7–90)

Although these equations are tedious to solve, the solution (7–89) is much simpler in form than the general solution (7–79). Nevertheless, even this solution is sufficiently complicated so that there would be a major advantage if we could obtain approximate solutions of the general problem (arbitrary C_1) that were simpler in form.

We shall see that much simpler results can be obtained in the asymptotic limits Re $\ll 1$ or Re $\gg 1$. In these cases, instead of starting with the exact solution, (7–79) or (7–89), and attempting to simplify this solution for the two limits, it is much easier to start again with the full problem, (7–68)–(7–70), and seek approximate asymptotic solutions *directly* by approximating the governing equations and boundary conditions. In this limited sense, the current problem is reminiscent of the flow through a tube with a periodic pressure gradient that was considered at the beginning of Chapter 6.

Let us begin with the simpler of the two limiting cases, namely, Re $\ll 1$.

i.) Approximate Solution at $0(\lambda)$ for the Asymptotic Limit, Re $\ll 1$ In the case Re $\ll 1$, it is natural to seek a solution of (7–68)–(7–70) in the general asymptotic form

$$G = G_0(\bar{r}) + \text{Re}\, G_1(\bar{r}) + \cdots.$$

(7–91)

The first term in this expansion is the "creeping flow" approximation, and the second represents the first inertial contributions in this limiting case.

Governing equations and boundary conditions for G_0 and G_1 are obtained in the usual manner of substituting the expansion (7–91) into the full problem (7–68)–(7–70) and setting the sum of all terms at each order in Reynolds number equal to zero. At $0(1)$, we obtain

$$\bar{r}^4 G_0'''' + 2\bar{r}^3 G_0''' - 3\bar{r}^2 G_0'' + 3\bar{r} G_0' - 3G_0 = 0,$$

(7–92)

with boundary conditions

$$G_0 = \frac{dG_0}{dr} = 0 \quad \text{at} \quad \bar{r} = 1$$

(7–93)

and

$$G_0 = -\epsilon(1+\epsilon)\frac{\Omega_2}{\Omega_1},$$

$$\left.\frac{dG_0}{dr} = \epsilon\left[\frac{2 - \dfrac{\Omega_2}{\Omega_1}\left[(1+\epsilon)^2 + 1\right]}{(1+\epsilon)^2 - 1}\right]\right\} \quad \text{at } \bar{r} = 1+\epsilon. \qquad (7\text{-}94)$$

The governing Equation (7–92) is a fourth-order version of Euler's equation, and we therefore seek solutions in the general form

$$G_0 \cong r^s. \qquad (7\text{-}95)$$

In this case, the roots of the characteristic equation obtained by substituting (7–95) into (7–92) are

$$s = 1 \,(\text{repeated}), \; 3, \; -1,$$

and the most general solution of Equation (7–92) is thus

$$\boxed{G_0(\bar{r}) = A\bar{r} + B\bar{r}\ln\bar{r} + C\bar{r}^3 + D\bar{r}^{-1}.} \qquad (7\text{-}96)$$

Here, A, B, C, and D are constants that are determined by the boundary conditions (7–93) and (7–94).

The former require that

$$0 = A + C + D \qquad (7\text{-}97)$$

and

$$0 = A + B + 3C - D, \qquad (7\text{-}98)$$

whereas the latter take the form

$$A(1+\epsilon) + B(1+\epsilon)\ln(1+\epsilon) + C(1+\epsilon)^3 - D(1+\epsilon)^{-1} = -\epsilon(1+\epsilon)\frac{\Omega_2}{\Omega_1} \qquad (7\text{-}99)$$

and

$$A + B\left[1 + \ln(1+\epsilon)\right] + 3C(1+\epsilon)^2 - D(1+\epsilon)^{-2} = \epsilon\left[\frac{2 - \dfrac{\Omega_2}{\Omega_1}\left[(1+\epsilon)^2 + 1\right]}{(1+\epsilon)^2 - 1}\right]. \qquad (7\text{-}100)$$

Solving these algebraic equations for the coefficients A, B, C, and D is straightforward. However, we do not concern ourselves with the details here. The key point, for this bounded domain problem, is that we can obtain a uniformly valid first approximation in the expansion (7–91) that satisfies all four boundary conditions (7–93) and (7–94).

This suggests that the small Reynolds number limit is *regular*. Hence a second term in the expansion (7–91) should be obtained, in principle, by solving

$$\boxed{\bar{r}^4 G_1'''' + 2\bar{r}^3 G_1''' - 3\bar{r}^2 G_1'' + 3\bar{r}G_1' - 3G_1 = i(C_1\bar{r}^2 + C_2)\left[\bar{r}^2\bar{G}_0'' + \bar{r}G_0' - G_0\right],}$$

$$(7\text{-}101)$$

with boundary conditions

$$G_1 = \frac{dG_1}{d\bar{r}} = 0 \quad \text{at} \quad \bar{r} = 1, 1 + \epsilon.$$

(7–102)

Evaluating the right-hand side of (7–101) using the solution (7–96), the governing equation becomes

$$\bar{r}^4 G_1'''' + 2\bar{r}^3 G_1''' - 3\bar{r}^2 G_1'' + 3\bar{r} G_1' - 3G_1 = i(C_1\bar{r}^2 + C_2)\left[8C\bar{r}^3 + 2B\bar{r}\right].$$

(7–103)

The homogeneous solution of (7–103) is identical in form with G_0, namely, (7–96). After some manipulation, the complete solution, including the homogeneous and particular parts, may be shown to take the form

$$G_1 = \alpha\bar{r} + \frac{\beta}{\bar{r}} + \delta\bar{r}\ln\bar{r} + \gamma\bar{r}^3 + i\frac{CC_1}{24}\bar{r}^5 +$$
$$i\left(\frac{4C_2C + BC_1}{8}\right)\bar{r}^3\ln\bar{r} - i\left(\frac{BC_2}{4}\right)\bar{r}(\ln\bar{r})^2.$$

(7–104)

Again, the four unknown coefficients α, β, δ, and γ can be obtained by applying the boundary conditions in equation (7–102), but the algebra is tedious, and we shall not present the details here. It is sufficient for present purposes to note that the *regular* perturbation solutions (7–91) exists, presumably to all orders in Re.

Although we have not carried through the detailed algebra in the text that is necessary to evaluate the constants A, B, C, and D in (7–96) and α, β, δ, and γ in (7–104), this can be done if quantitative results are required. However, several general features of the solution can be discerned without obtaining quantitative values for these constants, and these are worth exploring briefly. First, we may note from (7–97)–(7–100) that the constants A, B, C, and D are all real. Thus, the leading-order term in the expansion in (7–91) is real. It follows from this, and the definitions in (7–58) and (7–65), that

$$u_r^{(1)} = \frac{G_0(\bar{r})}{\bar{r}}\sin\theta + 0(\text{Re}),$$

$$u_\theta^{(1)} = \frac{dG_0}{d\bar{r}}\cos\theta + 0(\text{Re}).$$

(7–105)

Hence,

$$u_r = \lambda\left[\frac{G_0(\bar{r})}{\bar{r}}\sin\theta + 0(\text{Re})\right] + 0(\lambda^2)$$

(7–106)

and

$$u_{(\theta)} = \left(C_1\bar{r} + \frac{C_2}{\bar{r}}\right) + \lambda\left[\frac{dG_0}{d\bar{r}}\cos\theta + 0(\text{Re})\right] + 0(\lambda^2).$$

Thus, in the *absence* of inertia—that is, neglecting the terms of 0(Re)—the velocity field maintains the same θ symmetry as is imposed by the boundary conditions (7–52) and (7–57). Thus, based upon the values at $\bar{r} = 1 + \epsilon$, the azimuthal velocity is a minimum at $\theta = 0$, where the gap is widest, and a maximum at $\theta = \pi$, where the gap is narrowest. The magnitude of the radial velocity, on the other hand, is a maximum at $\theta = \pi/2$ and $3\pi/2$. When we add weak inertia by inclusion of the 0(Re) correction to $u_r^{(1)}$ and $u_\theta^{(1)}$, we find that

$$u_r = \lambda\left[\frac{G_0(\bar{r})}{\bar{r}}\sin\theta + \mathrm{Re}\left\{\frac{G_1^R(\bar{r})}{\bar{r}}\sin\theta + \frac{G_1^I(\bar{r})}{\bar{r}}\cos\theta\right\} + 0(\mathrm{Re}^2)\right] + 0(\lambda^2) \qquad (7\text{–}107a)$$

and

$$u_\theta = \left(C_1\bar{r} + \frac{C_2}{\bar{r}}\right) + \lambda\left[\frac{dG_0}{d\bar{r}}\cos\theta + \mathrm{Re}\left\{\frac{dG_1^R}{d\bar{r}}\cos\theta - \frac{dG_1^I}{d\bar{r}}\sin\theta\right\} + 0(\mathrm{Re}^2)\right] + 0(\lambda^2),$$

$$(7\text{–}107b)$$

where G_1^R and G_1^I are the real and imaginary parts of G_1. Thus, at 0(Re), the θ symmetry imposed by the boundary geometry is broken, and the positions of the maximum and minimum azimuthal velocity component are no longer coincident with the points where the gap width is an extremum, as before. The presence of inertia in the flow thus induces a type of *phase shift* in the velocity field.

Two other features of the low Reynolds number approximation that are easily obtained are general formulas for the torque required to rotate the inner cylinder and for the magnitude of any *force* induced on the inner cylinder by the pressure or viscous stress distribution in the slightly eccentric cylinder geometry. To calculate the torque per unit length on the inner cylinder (made dimensionless with respect to $\mu\Omega_1 a^2$), we must evaluate

$$T = \int_0^{2\pi} (-\bar{\tau}_{r\theta})\bar{r}^2 d\theta \quad \text{at} \quad \bar{r} = 1, \qquad (7\text{–}108)$$

where

$$\bar{\tau}_{r\theta} = \bar{r}\frac{\partial}{\partial\bar{r}}\left(\frac{u_\theta}{\bar{r}}\right) + \frac{1}{\bar{r}}\frac{\partial u_r}{\partial\theta} \qquad (7\text{–}109)$$

in dimensionless terms. Using the formulas for u_r and u_θ above, we see (symbolically) that

$$\bar{\tau}_{r\theta} = -\frac{2C_2}{\bar{r}^2} + \lambda\left[(H_0(\bar{r}) + \mathrm{Re}H_1(\bar{r}))\cos\theta + \mathrm{Re}H_2(\bar{r})\sin\theta + 0(\mathrm{Re}^2)\right] + 0(\lambda^2),$$

$$(7\text{–}110)$$

where

$$(H_0, H_1, \text{ and } H_2) \equiv \left(\bar{r}\frac{d}{d\bar{r}}\left(\frac{1}{\bar{r}}\frac{d}{d\bar{r}}\right) + \frac{1}{\bar{r}^2}\right)[G_0, G_1^R, -G_1^I].$$

Hence,

$$\boxed{T = 4\pi C_2,} \qquad (7\text{–}111)$$

and the torque is seen to be *unchanged* from the concentric cylinder case, at least to $0(\lambda\mathrm{Re}^2)$ and $0(\lambda^2)$.

The *force* on the inner cylinder can be calculated as well. In particular, the only possible nonzero component is in the vertical direction (see Figure 7–3). The formula for this force component (nondimensionalized with respect to $\mu \Omega_1 a$) is

$$F = \int_0^{2\pi} \left[(-\operatorname{Re}\bar{p} + \bar{\tau}_{rr})\sin\theta + \bar{\tau}_{r\theta}\cos\theta \right]_{\bar{r}=1} d\theta, \qquad (7\text{–}112)$$

where

$$\bar{\tau}_{rr} = 2 \frac{\partial u_r}{\partial r} \qquad (7\text{–}113)$$

and $\bar{p}$ can be calculated from $\bar{u}_r$ and $\bar{u}_\theta$ using Equations (7–42) and (7–43).

Clearly, for the Couette case $\bar{\tau}_{r\theta}|_{\bar{r}=1} = \text{const}$, $\bar{\tau}_{rr}|_{\bar{r}=1} = 0$, and $\bar{p}|_{\bar{r}=1} = \text{const}$, so that $F = 0$. This result is obvious in any case, due to the symmetry of the concentric cylinder geometry. On the other hand, including contributions at 0 (λ), we find that

$$\bar{\tau}_{rr}|_{\bar{r}=1} = 2\lambda \left(\frac{1}{\bar{r}}\frac{d}{d\bar{r}} - \frac{1}{\bar{r}^2} \right) \left[G_0 \sin\theta + \operatorname{Re}(G_1^R \sin\theta + G_1^I \cos\theta) + 0(\operatorname{Re}^2) \right]|_{\bar{r}=1} \qquad (7\text{–}114)$$

and

$$\bar{p}|_{\bar{r}=1} = \frac{\lambda}{\operatorname{Re}} \int \left(\frac{1}{\bar{r}}\frac{d^2 G_0}{d\bar{r}^2} + \frac{1}{\bar{r}^2}\frac{dG_0}{d\bar{r}} - \frac{G_0}{\bar{r}^3} \right) \sin\theta \, d\bar{r}|_{\bar{r}=1} + 0(\lambda, \lambda\operatorname{Re}) + \text{const.} \qquad (7\text{–}115)$$

Thus, the contributions of $\bar{\tau}_{rr}$ and $\bar{\tau}_{r\theta}$ to (7–112) are $0(\lambda\operatorname{Re})$. On the other hand, the contribution from the pressure $\bar{p}$ is $0(\lambda)$. This is the dominant term for $\operatorname{Re} \ll 1$. If we evaluate $\bar{p}_{r=1}$ using (7–115), we obtain

$$p|_{r=1} = \frac{\lambda}{\operatorname{Re}}\left[8C - 2B \right]\sin\theta + 0(\lambda, \lambda\operatorname{Re}) + \text{const.}, \qquad (7\text{–}116)$$

and thus

$$f = \lambda(4B)\pi. \qquad (7\text{–}117)$$

The right-hand side is *not* zero. The sign at $0(\lambda)$ can be shown to depend upon the directions and relative rates of rotation of the two cylinders. Although we do not present detailed results, since we have not evaluated the various coefficients in the solution for $\mathbf{u}^{(1)}$, it is worth noting that the force is vertically upward when the rotation of the inner cylinder is clockwise (see Figure 7–3) and $\Omega_2 \equiv 0$. The most important conclusion for this slightly eccentric, low Reynolds number case is that there is, in fact, an *induced lift* on the inner cylinder. Of course, the magnitude is small since $\lambda \ll 1$ and $\epsilon = 0(1)$. One interesting note is that the asymmetry of the flow due to *inertia* does not contribute to the induced lift at this order of approximation. Rather, it is the viscous contribution to the pressure that produces the most important effect.

ii.) Approximate Solution at $0(\lambda)$ for the Asymptotic Limit, $\operatorname{Re} \gg 1$ The other case that we can attack analytically for this slightly eccentric cylinder problem is the limit $\operatorname{Re} \gg 1$. In this case we seek an asymptotic approximation for G with $\operatorname{Re}^{-1} \ll 1$ as the small parameter, that is,

$$G(\bar{r}; \operatorname{Re}) = G_0(\bar{r}) + F_1(\operatorname{Re}^{-1})G_1(\bar{r}) + \cdots . \qquad (7\text{–}118)$$

Here, we anticipate that the magnitude of the leading term is $0(1)$ and denote the gauge function of the second term as F_1. Governing equations for G_0, G_1, and so forth can be obtained by substituting this expansion into the governing Equation (7–68) in the usual manner (see the first section in Chapter 6). However, in the present analysis we restrict our attention to the leading-order approximation G_0 only. The governing equations and boundary conditions for the leading-order term in an asymptotic expansion can always be obtained directly by taking the asymptotic limit, in this case $Re^{-1} \to 0$, in the full governing equations. We follow this simpler procedure.

For this purpose, it is convenient to consider the problem in terms of the pair of second-order equations (7–71) and (7–72). In particular, the limiting form of the former equation is just

$$\boxed{H \equiv 0} \qquad (7\text{–}119a)$$

so that G_0 satisfies the homogeneous equation

$$\boxed{\bar{r}^2 G_0'' + \bar{r} G_0' - G_0 = 0.} \qquad (7\text{–}119b)$$

In order to facilitate the later discussion, let us denote the solution of this equation as G_0^*. The general solution for G_0^* is

$$\boxed{G_0^* = A_1 \bar{r} + A_2 \frac{1}{\bar{r}}.} \qquad (7\text{–}120)$$

Clearly, this solution *cannot* be made to satisfy all four of the original boundary conditions (7–69) and (7–70). Although the solution (7–120) can satisfy the kinematic conditions on u_r, at $\bar{r} = 1$ and $\bar{r} = 1 + \epsilon$, it cannot simultaneously satisfy the no-slip conditions on u_θ. In particular, if we apply the kinematic conditions

$$G_0^* = 0 \quad \text{at} \quad \bar{r} = 1$$

and

$$G_0^* = -\frac{\epsilon}{1+\epsilon} \left[C_1 (1+\epsilon)^2 + C_2 \right] \quad \text{at} \quad \bar{r} = 1 + \epsilon,$$

we find that

$$\boxed{G_0^* = -\frac{\epsilon (C_1 (1+\epsilon)^2 + C_2)}{(1+\epsilon)^2 - 1} \left[\bar{r} - \frac{1}{\bar{r}} \right].} \qquad (7\text{–}121)$$

However, as suggested above, this solution does not satisfy the no-slip conditions. Specifically,

$$\frac{dG_0^*}{d\bar{r}} \Big|_{\bar{r}=1} = -\frac{2\epsilon \left[C_1 (1+\epsilon)^2 + C_2 \right]}{(1+\epsilon)^2 - 1}$$

and

$$\frac{dG_0^*}{d\bar{r}} \Big|_{\bar{r}=1+\epsilon} = -\frac{\epsilon \left[C_1 (1+\epsilon)^2 + C_2 \right]}{(1+\epsilon)^2 - 1} \left(\frac{(1+\epsilon)^2 + 1}{(1+\epsilon)^2} \right),$$

and we see that there is a $0(1)$ difference between these values and the no-slip values

$$\frac{dG}{d\bar{r}} = 0 \text{ at } \bar{r} = 1 \quad \text{and} \quad \frac{dG}{d\bar{r}} = -\frac{\epsilon}{(1+\epsilon)^2}\left[C_1(1+\epsilon)^2 - C_2\right] \text{ at } \bar{r} = 1 + \epsilon.$$

This shows that the asymptotic approximation (7–118), with (7–121) as the first term, is *not* uniformly valid but presumably breaks down near the boundaries at $\bar{r} = 1$ and $\bar{r} = 1 + \epsilon$. This is the classic signature of a singular perturbation problem. The asymptotic expansion of G, based upon (7–68), is not uniformly valid over the whole domain and must be supplemented by an alternate asymptotic approximation that is valid in the vicinity of the cylinders that satisfies the no-slip conditions.

It is worth reviewing the formulation of the solution for Re $\gg 1$ in some detail, for the problem defined by (7–68)–(7–70) is relatively simple and can thus serve as a useful model for later problems that are more complex from a mathematical point of view. The starting point is to ask ourselves why the solution (7–120) breaks down in the vicinity of the cylinders. From a mathematical point of view, the answer is that the first approximation of (7–68) for Re $\gg 1$ leads to a governing equation for G_0 that is only second order in $\bar{r}$, whereas the original exact equation was fourth order. It is this neglect of the highest-order derivatives that is responsible for the fact that the leading term in the expansion (7–118) cannot satisfy all of the boundary conditions of the original problem. This is, in fact, one of the most common signals of a singular asymptotic limit. If the order of the differential equations is reduced at any level of approximation, the corresponding terms in the asymptotic expansion will not be able to satisfy all of the boundary conditions of the original problem, and the expansion therefore cannot be uniformly valid (that is, it must be singular, rather than regular).

Clearly, in the present problem, if we are to satisfy all of the boundary conditions, we must retain the highest-order derivatives in the approximating equation, at least in the vicinity of the cylinders. On the other hand, the exact equation in the nondimensional form (7–68) seems to suggest that these terms should be negligible in the limit Re $\gg 1$, and this is what led to the solution (7–120). The problem is that the length scale a that was used for nondimensionalization of (7–68) is not uniformly valid throughout the domain. Wherever a is the appropriate characteristic length scale, all of the derivatives of G in (7–68) are 0(1)—that is, independent of Re—and the relative magnitude of the various terms for Re $\gg 1$ is determined solely by the appearance of Re on the right-hand side, but not on the left. Thus, in these parts of the domain, the expansion (7–118) and it's leading-order term (7–120) are guaranteed to be correct. Near the cylinders, however, G actually varies over a much shorter length scale (which actually decreases in size relative to a as the Reynolds number is increased). Thus, the magnitude of derivatives of G with respect to $\bar{r}$ will generally depend on Re in this region, and it is *not* necessarily true that the terms on the left-hand side of (7–68) are negligible compared with those on the right in the limit Re $\rightarrow \infty$.

To obtain a uniformly valid approximation to the solution of (7–68) for Re $\gg 1$, we need to use the method of matched asymptotic expansions as outlined in Chapter 6. For the region away from the cylinder walls where the characteristic length scale is a, the correct nondimensionalized version of the governing equation is (7–68), and the asymptotic form of the solution for Re $\rightarrow \infty$ is (7–118), with (7–119) as the governing equation for the leading term in this expansion and (7–120) as its solution. For the region in the immediate vicinity of the cylinder walls, known as the *boundary layer*, the characteristic length scale is not a, and we require a different approximation to the solution of the problem. To obtain this so-called boundary-layer approximation, we must determine the correct characteristic length scale and the correct nondimensionalized form of the governing equation.

In the remainder of this section, we pursue the approximate solution of (7–68) for $\mathrm{Re} \gg 1$ via the method of matched asymptotic expansions. In doing this, it is advantageous to reformulate the problem by introducing a new dependent function $\hat{G}$ as the unknown, which is related to G in the form

$$\boxed{\hat{G} \equiv G - G^*.} \tag{7–122}$$

Here, the function G^* is precisely the inviscid solution given by Equation (7–121). However, rather than emphasizing this point, it is perhaps better simply to think of G^* as a useful function that transforms the problem for G into a more convenient and exactly equivalent problem for $\hat{G}$. This problem is

$$\boxed{\bar{r}^4\hat{G}'''' + 2\bar{r}^3\hat{G}''' - 3\bar{r}^2\hat{G}'' + 3r\hat{G} - 3\hat{G} = i\,\mathrm{Re}(C_1\bar{r}^2 + C_2)(\bar{r}^2\hat{G}'' + \bar{r}\hat{G}' - \hat{G}),} \tag{7–123}$$

with boundary conditions

$$\boxed{\hat{G} = 0 \quad \text{at} \quad \bar{r} = 1,\ 1 + \epsilon} \tag{7–124}$$

and

$$\boxed{\begin{aligned} \frac{d\hat{G}}{d\bar{r}} &= \frac{2\epsilon\left[C_1(1+\epsilon)^2 + C_2\right]}{(1+\epsilon)^2 - 1} \quad \text{at} \quad \bar{r} = 1, \\[2mm] \frac{d\hat{G}}{d\bar{r}} &= \frac{2\epsilon(C_1 + C_2)}{(1+\epsilon)^2 - 1} \quad \text{at} \quad \bar{r} = 1 + \epsilon, \end{aligned}} \tag{7–125}$$

where $C_1 + C_2 = 1$. It is important to recognize that no approximation is inherent in the transformation (7–122). The resulting problem (7–123)–(7–125) is *exactly* equivalent to (7–68)–(7–70) for *arbitrary* Reynolds number. The main advantage of the transformed problem is that the natural symmetry of the boundary conditions near the two cylinders is exposed; for comparison, see the boundary conditions (7–69) and (7–70) in terms of G.

We now focus on obtaining an asymptotic solution of our problem in the form (7–123)–(7–125) for $\mathrm{Re} \gg 1$. Of course, these equations and boundary conditions are still nondimensionalized using a as the characteristic length scale. Thus, there exists an asymptotic expansion of $\hat{G}$,

$$\hat{G} = \hat{G}_0 + F_1(\mathrm{Re}^{-1})\hat{G}_1 + \cdots, \tag{7–126}$$

that is precisely equivalent to the expansion (7–118) for G. The governing equation for $\hat{G}_0$ is

$$\bar{r}^2\hat{G}_0'' + \bar{r}\hat{G}_0 - \hat{G}_0 = 0, \tag{7–127}$$

and the general solution is

$$\hat{G}_0 = \alpha \bar{r} + \frac{\beta}{\bar{r}}. \qquad (7\text{--}128)$$

We already have pointed out that this expansion is valid only in the region of the domain that is not too near the cylinders. Thus, the constants α and β cannot strictly be determined except by *matching* after we have considered the asymptotic approximation for $\hat{G}$ in the boundary-layer region near the cylinder walls. Here, however, in the interest of brevity we short-circuit the solution process and simply assume that $\alpha = \beta = 0$. In effect, this is simply saying that the inviscid solution (7–121) is the first approximation to G in the region away from the walls. We shall verify this assumption a posteriori by showing that the solution $\hat{G}_0 = 0$ matches at leading order with the solutions from the boundary-layer regions at $\bar{r} = 1$ and $\bar{r} = 1 + \epsilon$.

The deficiency with the formulation (7–123)–(7–125) for the region near $\bar{r} = 1$ and $\bar{r} = 1 + \epsilon$ is that it is based upon the assumption that G varies over a characteristic distance that is comparable to the cylinder radius a, whereas the velocity components near the two cylinders actually vary over a *much shorter* scale for large Re, such that the viscous terms in (7–123) continue to play a role in this region, even in the leading-order approximation for Re $\to \infty$. To obtain the correct leading-order approximation to (7–123) in the vicinity of the cylinder surfaces for Re $\to \infty$, we must nondimensionalize again to reflect the much shorter length scale that is characteristic of changes in velocity in this region. This is a classic example of a *singular* perturbation problem, since different characteristic length scales and different approximations to the exact solution will be found in different parts of the domain.

To obtain a first approximation in the vicinity of the cylinder surfaces for large Re, we must therefore rescale the equations in these regions and match the resultant solutions with the interior solution (7–128). Let us begin with the inner cylinder, $\bar{r} = 1$. It is convenient to change variables from $\bar{r}$ to $y \equiv \bar{r} - 1$, so that the inner cylinder is $y = 0$ and the outer cylinder $y = \epsilon$. With this change, the governing Equation (7–123) takes the form

$$(1+y)^4 \frac{d^4\hat{G}}{dy^4} + 2(1+y)^3 \frac{d^3\hat{G}}{dy^3} - 3(1+y)^2 \frac{d^2\hat{G}}{dy^2} + 3(1+y)\frac{d\hat{G}}{dy} - 3\hat{G} =$$
$$i\text{Re}\left[C_1(1+y)^2 + C_2\right]\left((1+y)^2 \frac{d^2\hat{G}}{dy^2} + (1+y)\frac{d\hat{G}}{dy} - \hat{G}\right). \qquad (7\text{--}129)$$

Now, the simplest way to introduce a change of length scale for nondimensionalization into (7–129) is to *rescale* by changing variables from y to Re^{-m}Y, where m is a constant that must be determined as part of the solution, that is,

$$y = \text{Re}^{-m}Y. \qquad (7\text{--}130)$$

As noted in conjunction with the analysis of the time-dependent flow in a circular tube, this change of variables is precisely equivalent to a change of length scale in the non-dimensionalization. The new nondimensionalized variable Y is just

$$Y = \frac{y'}{a}\left(\frac{\Omega_1 a^2}{\nu}\right)^m, \qquad (7\text{--}131)$$

where y' is the *dimensional* variable. But this is just a definition of a new nondimensionalized radial variable with characteristic length scale

$$l_c = a\left(\frac{\Omega_1 a^2}{\nu}\right)^{-m}. \tag{7-132}$$

From a physical point of view, the change of variable (7–130) conditions the governing equation to the fact that changes in velocity occur over a distance near the inner cylinder that is vanishingly thin for $\text{Re} \gg 1$. In particular, a change in the rescaled variable Y of $0(1)$ corresponds to a region of vanishing thickness, $0(\text{Re}^{-m})$, when expressed in terms of y.

Although it may be tempting to substitute the rescaling (7–130) directly into (7–129), it is important to note first that any rescaling of y must be accompanied by a corresponding change in the nondimensionalization for $\hat{G}$. To see this, we need only note that the derivative of $\hat{G}$ with respect to y is the tangential velocity in the vicinity of the inner cylinder, and this is clearly going to remain $0(1)$ for any value of Re, as we can see from the boundary condition (7–125a) (now expressed in terms of y),

$$\frac{d\hat{G}}{dy} = \frac{2\epsilon\left[C_1(1+\epsilon)^2 + C_2\right]}{(1+\epsilon)^2 - 1} \quad \text{at} \quad y = 0. \tag{7-133}$$

Thus, if the rescaling (7–130) is applied to y, it is clear that $\hat{G}$ must also be rescaled in the form

$$\hat{G} = \frac{1}{\text{Re}^m} \bar{G}. \tag{7-134}$$

Then, substituting (7–130) and (7–134) into (7–129), we find that

$$\text{Re}^{3m}\frac{d^4\bar{G}}{dY^4} + 0(\text{Re}^{2m}) = i\,\text{Re}(C_1 + C_2)\left[\text{Re}^m\frac{d^2\bar{G}}{dY^2} + 0(1)\right] \tag{7-135}$$

in the limit $\text{Re} \gg 1$. Hence, if viscous effects are to remain as important as inertia effects in the region near the cylinder surface where (7–135) applies, it follows that

$$m = \frac{1}{2}. \tag{7-136}$$

Thus, the governing equation for the *leading-order term* in an asymptotic expansion for this region is

$$\frac{d^4\bar{G}}{dY^4} - i\frac{d^2\bar{G}}{dY^2} + 0(\text{Re}^{-1/2}) = 0, \tag{7-137}$$

where we have made use of the fact that $(C_1 + C_2) \equiv 1$. The boundary conditions at $Y = 0$, taken from (7–124) and (7–125) are

$$\bar{G} = 0, \quad Y = 0,$$

$$\frac{d\bar{G}}{dY} = 2\alpha, \quad Y = 0, \tag{7-138}$$

where we have represented the right-hand side of (7–125a) as

$$\alpha \equiv \frac{\epsilon\left[C_1(1+\epsilon)^2 + C_2\right]}{(1+\epsilon)^2 - 1}.$$

In addition, the solution must match the leading-order approximation in the interior region (7–128). With the assumption $\alpha = \beta = 0$ as indicated earlier, the matching conditions are

$$\left(\frac{1}{\sqrt{\text{Re}}}\bar{G}\right)_{Y \gg 1} \to 0 \quad \text{as} \quad \text{Re} \to \infty \tag{7-139}$$

and

$$\left(\frac{d\bar{G}}{dY}\right)_{Y \gg 1} \to 0 \quad \text{as} \quad \text{Re} \to \infty. \tag{7-140}$$

The $\text{Re}^{-1/2}$ factor that appears in (7–139) is due to the scaling (7–134).

It is a simple matter to solve this problem for $\bar{G}$. The general solution of (7–137) is

$$\bar{G} = aY + b + \frac{1}{i}\left(ce^{+\sqrt{i}Y} + de^{-\sqrt{i}Y}\right). \tag{7-141}$$

The matching condition (7–139) is satisfied with an error that is no larger than $0(\text{Re}^{-1/2})$ for any choice of the constants a, b, c, and d. However, the second matching condition (7–140) requires

$$a = c = 0.$$

It then remains to satisfy the two boundary conditions (7–138) at $Y = 0$. To do this, we require that

$$b = \frac{2\alpha}{\sqrt{i}} \quad \text{and} \quad d = -2\alpha\sqrt{i},$$

so that

$$\boxed{\bar{G} = \frac{2\alpha}{\sqrt{i}}\left(1 - e^{-\sqrt{i}Y}\right).} \tag{7-142}$$

This yields a complete solution for G of the form

$$G = G^* + \frac{1}{\sqrt{\text{Re}}}\left[\frac{2\alpha}{\sqrt{i}}\left(1 - e^{-\sqrt{i}Y}\right)\right] + o(\text{Re}^{-1/2}), \tag{7-143}$$

where

$$G^* = -\alpha\left(\bar{r} - \frac{1}{\bar{r}}\right) \quad \text{and} \quad Y = \sqrt{\text{Re}}(\bar{r} - 1).$$

The corresponding result for the velocity components $\bar{u}_r^{(1)}$ and $\bar{u}_\theta^{(1)}$ in the boundary-layer is

$$\bar{u}_r^{(1)} = -\alpha\left(1 - \frac{1}{\bar{r}^2}\right)\sin\theta +$$

$$\sqrt{\frac{2}{\mathrm{Re}}}\,\alpha\left[\frac{1}{\bar{r}}(1 - e^{-\beta}\cos\beta - e^{-\beta}\sin\beta)\sin\theta - \right.$$

$$\left.\frac{1}{\bar{r}}(1 - e^{-\beta}\cos\beta + e^{-\beta}\sin\beta)\cos\theta\right] + \cdots, \qquad (7\text{–}144)$$

and

$$\bar{u}_\theta^{(1)} = -\alpha\left(1 + \frac{1}{\bar{r}^2}\right)\cos\theta + \alpha\left(2e^{-\beta}\cos\beta\cos\theta + 2e^{-\beta}\sin\beta\sin\theta\right), \qquad (7\text{–}145)$$

where

$$\beta \equiv \sqrt{\frac{\mathrm{Re}}{2}}(\bar{r} - 1).$$

At the cylinder surface, $\bar{r} = 1$, both components $\bar{u}_r^{(1)}$ and $\bar{u}_\theta^{(1)}$ are zero. For any non-zero value of $\bar{r} - 1$, the maximum departure from the inviscid solution is $0(\mathrm{Re}^{-1/2})$, other terms being exponentially small in the limit $\mathrm{Re} \to \infty$. One interesting feature is that the presence of viscosity introduces an oscillatory dependence of both $\bar{u}_r^{(1)}$ and $\bar{u}_\theta^{(1)}$ on the radial variable $\bar{r}$. We shall return to this point in a moment. First, however, let us consider the nature of the solution in the immediate vicinity of the outer cylinder.

At the outer cylinder, a very similar analysis can be applied. In this case, the most convenient choice of independent variable is

$$y^* = (1 + \epsilon) - \bar{r} \qquad (7\text{–}146)$$

so that $y^* = 0$ at $\bar{r} = 1 + \epsilon$ and increases for $\bar{r}$ moving inward into the gap between the two cylinders. We leave it to the reader to provide the details that are necessary to analyze the velocity field in the thin boundary-layer region near $\bar{r} = 1 + \epsilon$. In this case, it can be shown easily that the rescaled variables are

$$\bar{G} = \sqrt{\mathrm{Re}}\,\hat{G} \qquad (7\text{–}147)$$

and

$$\bar{Y} = \sqrt{\mathrm{Re}}\,y^*,$$

and the governing equation for the leading-order approximation to $\bar{G}$ now takes the form

$$\frac{d^4\bar{G}}{d\bar{Y}^4} - i\frac{\Omega_2}{\Omega_1}\frac{d^2\bar{G}}{d\bar{Y}^2} = 0, \qquad (7\text{–}148)$$

with matching conditions

$$\left(\frac{\bar{G}}{\sqrt{\mathrm{Re}}}\right)_{\bar{Y}\gg 1} \to 0 \quad \text{as} \quad \mathrm{Re} \to \infty, \qquad (7\text{–}149)$$

and

$$\left(\frac{d\bar{G}}{d\bar{Y}}\right)_{\bar{Y}\gg} \to 0 \quad \text{as} \quad \mathrm{Re} \to \infty, \qquad (7\text{–}150)$$

and boundary conditions

$$\bar{G} = 0 \quad \text{at} \quad \bar{Y} = 0 \tag{7-151}$$

and

$$\frac{d\bar{G}}{d\bar{Y}} = -\gamma \quad \text{at} \quad \bar{Y} = 0. \tag{7-152}$$

The solution is

$$\boxed{\bar{G} = -\frac{\gamma}{\sqrt{\Omega_2/\Omega_1}} \frac{1}{\sqrt{i}} \left(1 - e^{\sqrt{i}\sqrt{\Omega_1/\Omega_2}\,\bar{Y}}\right).} \tag{7-153}$$

The existence of the approximate solutions (7–142) and (7–153), which match with the interior solution in the form $\hat{G} = 0$, provides a posteriori evidence that the coefficients α and β in the general solution (7–128) are, in fact, zero as assumed immediately following (7–128). The form of the component profiles for $\bar{u}_r^{(1)}$ and $\bar{u}_\theta^{(1)}$, corresponding to (7–153), is very similar to (7–144) and (7–145). The only difference is that the characteristic scale for both the exponential decay and the oscillatory dependence on $\bar{r}$ is now

$$0\left(\sqrt{\text{Re}\frac{\Omega_2}{\Omega_1}}\right) = 0\left(\frac{\Omega_2 a^2}{\nu}\right)^{1/2}.$$

It is interesting to note that the characteristic dimension of the boundary-layer structure near each of the cylinders is determined by the angular velocity of that cylinder.

From a qualitative point of view, one important observation is that the form of the steady-state solution for the slightly eccentric cylinder geometry is *dependent* on the Reynolds number, at least in the vicinity of the cylinder surfaces, whereas the Couette solution for the *concentric* cylinder geometry is *independent* of the Reynolds number. In the latter case, there is no acceleration or deceleration in the streamwise direction, and the motion of all fluid elements is absolutely steady. Thus, the *rate* of vorticity diffusion in the radial direction has no influence on the form of the steady-state velocity profile, and the Reynolds number does not appear in the solution. When the geometry becomes eccentric, on the other hand, there *is* a time scale associated with changes in the streamwise velocity of a fluid element, and the form of the velocity profile depends upon this time scale compared to the time scale for diffusion of vorticity across the gap between cylinders. This is evident both in the form of the exact solution (7–79) for arbitrary Reynolds numbers, and in the forms of the boundary-layer solutions near the cylinder surfaces for Re $\gg$ 1. In the latter case, the Reynolds number dependence is confined to the regions very close to the cylinders at the leading order of approximation, and the Lagrangian accelerations and decelerations experienced by fluid elements lead to an oscillatory steady-state flow structure [see (7–144) and (7–145)]. Interestingly, in the latter case, the Reynolds number dependence of the flow structure in the boundary-layers is determined by the relative magnitudes of the diffusion scale, a^2/ν, and the *local* flow-induced time scale for accelerations and decelerations in the azimuthal direction, that is, Ω_1^{-1} near the inner cylinder and Ω_2^{-1} near the outer cylinder.

In summary, then, we have considered the effects of a very small degree of eccentricity on the flow structure between two rotating circular cylinders whose axes are parallel. Although the effects are small and thus not of immediate practical utility, the results are *qualitatively* interesting—particularly, the existence of *lift* on the inner cylinder due to eccentricity at low Reynolds numbers and the complex, oscillatory boundary-layer structure found for $\mathrm{Re} \gg 1$.

(b) Narrow-Gap–Lubrication Limit $\epsilon \ll 1$, $\lambda = 0(1)$

Let us now turn our attention to the other limiting case that is amenable to asymptotic approximation, namely, the narrow-gap limit $\epsilon \ll 1$ with arbitrary eccentricity, $\lambda = 0(1)$. As suggested in the introduction to this section, this problem is a specific example of the general class of so-called *lubrication* problems in which very large pressures are generated by a relative translational motion of solid boundaries due to the fact that the boundaries are not parallel and the gap between them is very thin. These large pressures lead to a significant force that tends to maintain the gap even when the boundaries are supporting a significant load. The narrow-gap, eccentric cylinder geometry is known in lubrication theory as the *journal bearing* and derives its practical significance from the fact that the pressure forces for the geometric configuration sketched in Figure 7–4 are capable of supporting a significant load in the y direction at a relatively low cost in terms of the torque that is required to maintain rotation of the inner cylinder at an angular velocity Ω.[7]

The geometry of the eccentric cylinder configuration in the narrow-gap limit is described by the limiting form as $\epsilon \rightarrow 0$ of the general equations (7–22), (7–24), (7–33), and (7–34). Thus,

$$r_s(\theta) = a(1 + \epsilon) + a\epsilon\lambda\cos\theta + 0(\epsilon^2), \tag{7-154}$$

$$\mathrm{h}(\theta) = a\epsilon\left[1 + \lambda\cos\theta\right] + 0(\epsilon^2) \tag{7-155}$$

and

$$\mathbf{n} = \mathbf{i}_r + \mathbf{i}_\theta(\epsilon\lambda\sin\theta) + 0(\epsilon^2), \tag{7-156}$$

$$\mathbf{t} = \mathbf{i}_\theta - \mathbf{i}_r(\epsilon\lambda\sin\theta) + 0(\epsilon^2). \tag{7-157}$$

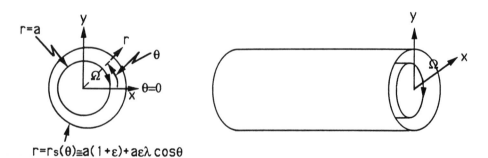

Figure 7–4 The journal-bearing configuration.

It will be convenient for comparisons between this and later sections of this chapter if we introduce a modified radial variable,

$$y \equiv r - a, \qquad (7\text{-}158)$$

in place of r, so that y varies from 0 at the surface of the inner cylinder to $y = h(\theta)$ on the surface of the outer cylinder. In this case, the governing differential equations and boundary conditions are again (7-25)-(7-29). When these differential equations are expressed in terms of y instead of r, they take the forms

$$\frac{\partial}{\partial y}\left[(y+a)u_r\right] + \frac{\partial u_\theta}{\partial \theta} = 0, \qquad (7\text{-}159)$$

$$\rho\left[u_r\frac{\partial u_r}{\partial y} + \frac{u_\theta}{y+a}\left(\frac{\partial u_r}{\partial \theta} - u_\theta\right)\right] = -\frac{\partial p}{\partial y} +$$

$$\mu\left[\frac{1}{y+a}\frac{\partial}{\partial y}\left((y+a)\frac{\partial u_r}{\partial y}\right) + \frac{1}{(y+a)^2}\frac{\partial^2 u_r}{\partial \theta^2} - \frac{u_r}{(y+a)^2} - \frac{2}{(y+a)^2}\frac{\partial u_\theta}{\partial \theta}\right], \qquad (7\text{-}160)$$

and

$$\rho\left[u_r\frac{\partial u_\theta}{\partial y} + \frac{u_\theta}{y+a}\left(\frac{\partial u_\theta}{\partial \theta} + u_r\right)\right] = -\frac{1}{(y+a)}\frac{\partial p}{\partial \theta} +$$

$$\mu\left[\frac{1}{y+a}\frac{\partial}{\partial y}\left((y+a)\frac{\partial u_\theta}{\partial y}\right) + \frac{1}{(y+a)^2}\frac{\partial^2 u_\theta}{\partial \theta^2} + \frac{2}{(y+a)^2}\frac{\partial u_r}{\partial \theta} - \frac{u_\theta}{(y+a)^2}\right]. \qquad (7\text{-}161)$$

We consider only the *journal-bearing* configuration in which $\Omega_1 = \Omega$ and $\Omega_2 = 0$. Then, the boundary conditions are

$$u_r = 0, \ u_\theta = -\Omega a \quad \text{at} \quad y = 0, \qquad (7\text{-}162)$$

and

$$u_r = u_\theta = 0 \quad \text{at} \quad y = h(\theta). \qquad (7\text{-}163)$$

In order to see how the *thin-gap* approximation $\epsilon \ll 1$ simplifies this problem, it is again necessary to nondimensionalize the governing equations and boundary conditions. The characteristic velocity for the azimuthal velocity component is the same as in the previous section,

$$u_c = \Omega a.$$

However, the length scale characteristic of velocity gradients in the thin gap is not the cylinder radius a, as in the concentric and slightly eccentric cylinder problems, but the characteristic distance across the gap,

$$l_c = \epsilon a.$$

Thus,

$$\boxed{u_\theta = a\Omega\bar{u}_\theta \quad \text{and} \quad y = \epsilon a\bar{y},} \tag{7-164}$$

and we should expect that the dimensionless derivative $\partial\bar{u}_\theta/\partial\bar{y}$ obtained in this way is $0(1)$. Gradients along the gap (in the θ direction), on the other hand, are characterized by a characteristic length scale $a\theta$, and this is inherent in the governing equations (7-159)–(7-161). The large difference between the length scale characteristic of changes in velocity *along* the gap and *across* the gap—that is, a and $a\epsilon$, respectively—is typical of thin-gap, lubrication-type geometries. The only dimensional quantity that remains to be nondimensionalized is the radial velocity component u_r. In order to determine a characteristic scale for u_r, we utilize the continuity equation (7-159) along with the dimensionless forms (7-164) for u_θ and y. For convenience, let us denote the characteristic magnitude of u_r as V. Then, substituting $u_r = V\bar{u}_r$, plus (7-164), into (7-159), we obtain

$$\frac{1}{\epsilon a}\frac{\partial}{\partial\bar{y}}\left[(\epsilon a\bar{y} + a)V\bar{u}_r\right] + a\Omega\frac{\partial\bar{u}_\theta}{\partial\theta} = 0.$$

Solving for V, this can be expressed in the alternate form

$$V = -\epsilon\Omega a\left[\frac{\dfrac{\partial\bar{u}_\theta}{\partial\theta}}{\dfrac{\partial}{\partial\bar{y}}((1 + \epsilon\bar{y})\bar{u}_r)}\right]. \tag{7-165}$$

Since the quantity in square brackets is $0(1)$, it follows that

$$V = 0(\epsilon a\Omega)$$

and

$$u_r = \epsilon a\Omega\bar{u}_r. \tag{7-166}$$

Evidently, if ϵ were $0(1)$, the scales characteristic of u_r and u_θ would be comparable. Here, however, we are concerned with the limit $\epsilon \ll 1$, and in this case, we see that the characteristic magnitude of the radial velocity component is very much smaller than the azimuthal component, u_θ. Although it is somewhat unusual in fluid mechanics for the velocity components in different directions to exhibit different scales, we see (from continuity) that this is a consequence of the extreme geometry of the thin-gap limit in the present problem.

Finally, in order to nondimensionalize the equations of motion, we require a characteristic pressure scale. However, it is not immediately obvious what choice is appropriate. We temporarily postpone the decision by introducing the formal non-dimensionalization

$$p = \pi\bar{p}, \tag{7-167}$$

with a characteristic pressure π yet to be determined. With the dimensionless variables, $\bar{u}_r$, $\bar{u}_\theta$, $\bar{y}$, and $\bar{p}$ defined by (7-164), (7-166), and (7-167), the continuity and Navier-Stokes equations take the forms

$$\frac{\partial}{\partial \bar{y}}\left((1+\epsilon \bar{y})\bar{u}_r\right) + \frac{\partial \bar{u}_\theta}{\partial \theta} = 0, \tag{7-168}$$

$$\left(\frac{\rho a^2 \Omega}{\mu}\right)\epsilon \left[\epsilon \left(\bar{u}_r \frac{\partial \bar{u}_r}{\partial \bar{y}} + \frac{\bar{u}_\theta}{1+\epsilon \bar{y}}\frac{\partial \bar{u}_r}{\partial \theta}\right) - \frac{\bar{u}_\theta^2}{(1+\epsilon \bar{y})}\right] = -\frac{\pi}{\mu \Omega}\frac{\partial \bar{p}}{\partial \bar{y}} +$$

$$\left[\frac{1}{1+\epsilon \bar{y}}\frac{\partial}{\partial \bar{y}}\left((1+\epsilon \bar{y})\frac{\partial \bar{u}_r}{\partial \bar{y}}\right) + \frac{\epsilon^2}{(1+\epsilon \bar{y})^2}\frac{\partial^2 \bar{u}_r}{\partial \theta^2} - \frac{\epsilon^2 \bar{u}_r}{(1+\epsilon \bar{y})^2} - \frac{2\epsilon}{(1+\epsilon \bar{y})^2}\frac{\partial \bar{u}_\theta}{\partial \theta}\right], \tag{7-169}$$

and

$$\left(\frac{\rho a^2 \Omega}{\mu}\right)\epsilon^2 \left[\bar{u}_r \frac{\partial \bar{u}_\theta}{\partial \bar{y}} + \frac{\bar{u}_\theta}{1+\epsilon \bar{y}}\frac{\partial \bar{u}_\theta}{\partial \theta} + \epsilon \frac{\bar{u}_\theta \bar{u}_r}{(1+\epsilon \bar{y})}\right] = -\frac{\pi \epsilon^2}{\mu \Omega}\frac{1}{(1+\epsilon \bar{y})}\frac{\partial \bar{p}}{\partial \theta} +$$

$$\left[\frac{1}{1+\epsilon \bar{y}}\frac{\partial}{\partial \bar{y}}\left((1+\epsilon \bar{y})\frac{\partial \bar{u}_\theta}{\partial \bar{y}}\right) + \frac{\epsilon^2}{(1+\epsilon \bar{y})^2}\left(\frac{\partial^2 \bar{u}_\theta}{\partial \theta^2} - \bar{u}_\theta\right) + \epsilon^3 \frac{2}{(1+\epsilon \bar{y})^2}\frac{\partial \bar{u}_r}{\partial \theta}\right]. \tag{7-170}$$

The boundary conditions become

$$\bar{u}_r = 0, \ \bar{u}_\theta = -1 \quad \text{at} \quad \bar{y} = 0,$$
$$\bar{u}_r = \bar{u}_\theta = 0 \quad \text{at} \quad \bar{y} = \bar{h}(\theta), \tag{7-171}$$

where

$$\bar{h}(\theta) = 1 + \lambda \cos \theta \tag{7-172}$$

is the dimensionless gap width.

Upon examining (7–169) and (7–170), we can see that the nonlinear, inertial terms are at least

$$0\left(\frac{\rho a^2 \Omega}{\mu} \cdot \epsilon\right)$$

or smaller compared with the largest of the viscous terms. Thus, in the limit $\epsilon \to 0$, the nonlinear terms in the equations of motion become negligible compared with the viscous terms. Indeed, if we seek an asymptotic solution of (7–168)–(7–170) in the form

$$\bar{\mathbf{u}} = \bar{\mathbf{u}}^{(0)} + \epsilon \bar{\mathbf{u}}^{(1)} + \cdots,$$
$$\bar{p} = \bar{p}^{(0)} + \epsilon \bar{p}^{(1)} + \cdots, \tag{7-173}$$

the governing equations for the first term ($\bar{\mathbf{u}}^{(0)}$, $\bar{p}^{(0)}$) can be obtained by taking the limit $\epsilon \to 0$ in the full equations. The result is

$$\frac{\partial \bar{u}_r^{(0)}}{\partial \bar{y}} + \frac{\partial \bar{u}_\theta^{(0)}}{\partial \theta} = 0, \tag{7-174}$$

$$\frac{\partial^2 \bar{u}_r^{(0)}}{\partial \bar{y}^2} - \frac{\pi}{\mu \Omega}\frac{\partial \bar{p}^{(0)}}{\partial \bar{y}} = 0, \tag{7-175}$$

and

$$\frac{\partial^2 \bar{u}_\theta^{(0)}}{\partial \bar{y}^2} - \frac{\pi \epsilon^2}{\mu \Omega} \frac{\partial \bar{p}^{(0)}}{\partial \theta} = 0 \tag{7-176}$$

In contrast to the full equations (7-169) and (7-170), these equations are *linear*. Although the parameter ϵ^2 appears in the pressure gradient term of (7-176), making it seem at first that this term should have been dropped in the limit $\epsilon \to 0$, we must remember that the characteristic pressure π has not been specified yet, and it is possible that $\pi \epsilon^2$ does not become small in the limit $\epsilon \to 0$. The reader may note that the normal component (7-175) differs fundamentally from the balance (7-14) for the concentric cylinder problem. The acceleration term u_θ^2/r, which is the dominant contribution to the radial pressure gradient in the Couette flow problem, is smaller by $O(\epsilon)$ than the viscous term shown in (7-175). This is a consequence of the fact that the pressure gradients that are induced in the thin fluid layer because of the nonuniform gap width are *much larger* than those associated with the centripetal acceleration for $\epsilon \ll 1$ and $\lambda = 0(1)$.

It is important to notice that (7-174)–(7-176) are identical in form to the equations that would be obtained if the problem had been formulated from the beginning in terms of a local Cartesian coordinate system, with $\bar{y}$ normal to the surface of the inner cylinder, and

$$\bar{x} \approx \theta$$

tangent to that surface, instead of beginning with the cylindrical coordinate system that was actually used. Thus, we see that the *curvature* of the boundaries, which occurs on the scale a, can be neglected in the thin-gap limit when formulating the leading-order (lubrication) equations of motion and continuity. All of the terms in (7-168)–(7-170) that existed because the cylindrical coordinate system was curvilinear instead of Cartesian, drop out at this leading order of approximation. Although we have demonstrated the reduction to a local Cartesian form only for the specific case of a circular cylindrical boundary geometry, the same result is true for curved boundaries of any shape provided that the thin-film approximation is applicable. We shall use this fact later in this chapter without any added proof.

To complete our derivation of the leading-order equations for flow in the thin gap, we must specify the characteristic pressure π. It is evident in examining (7-174)–(7-176) that one of the pressure gradient terms must be retained in the limit $\epsilon \to 0$, since the system of equations would otherwise be overdetermined. Thus, either $\pi = \mu \Omega$ or $\pi = (1/\epsilon^2)\mu \Omega$. Either choice gives a set of three equations for the three unknowns $\bar{u}_r^{(0)}$, $\bar{u}_\theta^{(0)}$, and $\bar{p}^{(0)}$. However, only the equations corresponding to $\pi = \mu \Omega/\epsilon^2$ yield solutions that can satisfy all of the boundary conditions (7-171). In particular, if $\pi = \mu \Omega$, it can be shown that the condition $\bar{u}_r^{(0)} = 0$ at $\bar{y} = \bar{h}$ can be satisfied only if $\partial \bar{h}/\partial \theta = 0$ so that the cylinders are concentric and $\bar{u}_r^{(0)} \equiv 0$ everywhere. Thus,

$$\pi = \frac{\mu \Omega}{\epsilon^2}, \tag{7-177}$$

and the governing equations reduce to the form

$$\frac{\partial \bar{u}_r^{(0)}}{\partial \bar{y}} + \frac{\partial \bar{u}_\theta^{(0)}}{\partial \theta} = 0, \tag{7-178}$$

$$\frac{\partial^2 \bar{u}_\theta^{(0)}}{\partial \bar{y}^2} - \frac{\partial \bar{p}^{(0)}}{\partial \theta} = 0, \tag{7-179}$$

and

$$\frac{\partial \bar{p}^{(0)}}{\partial \bar{y}} = 0(\epsilon^2). \tag{7-180}$$

According to (7–180), the pressure $\bar{p}^{(0)}$ depends only on θ, and the problem reduces to the solution of (7–178) and (7–179) subject to the boundary conditions (7–171). Equations (7–178) and (7–179) are known as the *lubrication equations*. We see that they resemble the equations for unidirectional flow. However, in this case, the boundaries are not required to be parallel. Thus, $\bar{u}_\theta^{(0)}$ can be a function of the streamwise variable θ, and $\bar{u}_r^{(0)}$ will not be zero in general. Furthermore, since $\bar{u}_\theta^{(0)}$ is a function of θ, so is $\partial \bar{p}^{(0)}/\partial \theta$. Finally, whereas the unidirectional flow equations are exact, the lubrication equations are only the leading order approximation to the exact equations of motion and continuity in the asymptotic, thin-gap limit, $\epsilon \to 0$.

The solution of (7–178) and (7–179) is straightforward. We integrate (7–179) twice with respect to $\bar{y}$, keeping in mind that $\partial p^{(0)}/\partial \theta$ is independent of $\bar{y}$ according to (7–180). Thus,

$$\bar{u}_\theta^{(0)} = \frac{d\bar{p}^{(0)}}{d\theta} \frac{\bar{y}^2}{2} + a\bar{y} + b. \tag{7-181}$$

Applying the boundary conditions on $\bar{u}_\theta^{(0)}$ from (7–171), this becomes

$$\bar{u}_\theta^{(0)} = -\left(1 - \frac{\bar{y}}{h}\right) + \frac{d\bar{p}^{(0)}}{d\theta}\left(\frac{\bar{y}^2}{2} - \frac{\bar{y}h}{2}\right). \tag{7-182}$$

It is evident that $\bar{u}_\theta^{(0)}$ depends both on $\bar{y}$ and θ, though the velocity field has the same form locally (that is, for fixed θ) as for unidirectional flow between parallel plane boundaries. In this "journal-bearing" problem, the motion is due to the motion of the boundary only. The pressure gradient is a *consequence* of this motion and the fact that the gap width is not constant.

To determine the pressure gradient, we turn to the continuity equation (7–178) and the two boundary conditions on $\bar{u}_r^{(0)}$ from (7–171). Substituting for $\partial \bar{u}_\theta^{(0)}/\partial \theta$ from (7–182), the continuity equation becomes a first-order differential equation for $\bar{u}_r^{(0)}$,

$$\frac{\partial \bar{u}_r^{(0)}}{\partial \bar{y}} = +\frac{\bar{y}}{h^2}\frac{dh}{d\theta} - \frac{d^2\bar{p}^{(0)}}{d\theta^2}\left(\frac{\bar{y}^2}{2} - \frac{\bar{y}h}{2}\right) + \frac{d\bar{p}^{(0)}}{d\theta}\left(\frac{\bar{y}}{2}\frac{\partial h}{\partial \theta}\right). \tag{7-183}$$

Integrating with respect to $\bar{y}$, we find that

$$\bar{u}_r^{(0)} = +\frac{\bar{y}^2}{2h^2}\frac{dh}{d\theta} - \frac{d^2\bar{p}^{(0)}}{d\theta^2}\left(\frac{\bar{y}^3}{6} - \frac{\bar{y}^2h}{4}\right) + \frac{d\bar{p}^{(0)}}{d\theta}\left(\frac{\bar{y}^2}{4}\frac{dh}{d\theta}\right), \tag{7-184}$$

where the constant of integration has been set equal to zero in order that $\bar{u}_r^{(0)}$ satisfies the boundary condition

$$\bar{u}_r^{(0)} = 0 \quad \text{at} \quad \bar{y} = 0. \tag{7-185}$$

The second boundary condition on $\bar{u}_r^{(0)}$ from (7-171) then yields a differential equation for the unknown pressure $\bar{p}^{(0)}$. Thus, applying

$$\bar{u}_r^{(0)} = 0 \quad \text{at} \quad \bar{y} = \bar{h} \tag{7-186}$$

to (7-184), we obtain

$$\boxed{\frac{d}{d\theta}\left(\bar{h}^3 \frac{d\bar{p}^{(0)}}{d\theta}\right) = -6\frac{d\bar{h}}{d\theta}.} \tag{7-187}$$

This equation is a special case of the famous Reynolds equation from general lubrication theory, which will be discussed in the next section.[8]

If we integrate (7-187) once with respect to θ, we find that

$$\frac{d\bar{p}^{(0)}}{d\theta} = -\frac{6}{h^2} + \frac{c}{h^3}, \tag{7-188}$$

where c is a constant of integration. Since the pressure $p^{(0)}$ must be periodic in θ,

$$\int_0^{2\pi} \frac{d\bar{p}^{(0)}}{d\theta}\, d\theta \equiv 0, \tag{7-189}$$

and, thus, integrating (7-188) we find that

$$c = +6\frac{\int_0^{2\pi} d\theta/h^2}{\int_0^{2\pi} d\theta/h^3}. \tag{7-190}$$

The pressure gradient $d\bar{p}^{(0)}/d\theta$ can now be expressed in the form

$$\boxed{\frac{d\bar{p}^{(0)}}{d\theta} = -\frac{6}{h^2} + \frac{6}{h^3}\frac{I_2}{I_3},} \tag{7-191}$$

where

$$I_n(\lambda) \equiv \int_0^{2\pi} \frac{d\theta}{(1 + \lambda\cos\theta)^n}.$$

Finally, substituting for $d\bar{p}^{(0)}/d\theta$ in (7-182), we have

$$\boxed{\bar{u}_\theta^{(0)} = \left\{\frac{-3}{(1+\lambda\cos\theta)^2}\right\}\left(1 - \frac{I_2}{I_3(1+\lambda\cos\theta)}\right)(\bar{y}^2 - (1+\lambda\cos\theta)\bar{y}) - 1 + \frac{\bar{y}}{(1+\lambda\cos\theta)}.}$$
$$\tag{7-192}$$

The analysis of this section is typical of lubrication problems. First, the equations of motion are solved to obtain a profile for the tangential velocity component, which is always locally similar in form to the profile for unidirectional flow between parallel plane boundaries, but with the streamwise pressure gradient unknown. The continuity equation is then integrated to obtain the normal velocity component, but this requires only one of the two boundary conditions for the normal velocity. The second condition then yields a differential equation (known as the *Reynolds equation*) that can be used to determine the pressure distribution.

The main results of this subsection are equations (7–191) and (7–192) for $d\bar{p}_0/d\theta$ and $\bar{u}_\theta^{(0)}$, plus the corresponding result for $\bar{u}_r^{(0)}$, which can be calculated from (7–184). It will be convenient for discussion (and also for comparison with the general results that are presented in the next section) to express these results in dimensional form. Thus, reincorporating dimensional variables from (7–164), (7–166), and (7–167), we obtain

$$\frac{dp^{(0)}}{d\theta} = \frac{6\mu\Omega}{\epsilon^2}\left[-\frac{1}{(1 + \lambda\cos\theta)^2} + \frac{1}{(1 + \lambda\cos\theta)^3}\frac{I_2}{I_3} \right], \qquad (7\text{–}193)$$

$$u_\theta^{(0)} = \Omega a\left[\left(\frac{-3}{1 + \lambda\cos\theta)^2} \right)\left(1 - \frac{I_2}{I_3(1 + \lambda\cos\theta)} \right) \right.$$
$$\left[\left(\frac{y}{\epsilon a}\right)^2 - (1 + \lambda\cos\theta)\left(\frac{y}{\epsilon a}\right) \right] - 1 + \frac{(y/\epsilon a)}{(1 + \lambda\cos\theta)} \right]. \quad (7\text{–}194)$$

and

$$u_r^{(0)} = \epsilon\Omega a\bar{u}_r^{(0)}. \qquad (7\text{–}195)$$

Finally, incorporating the value of I_2/I_3 from the previous page

$$\frac{dp^{(0)}}{d\theta} = \frac{6\mu\Omega}{\epsilon^2(2 + \lambda^2)}\left[\frac{-3\lambda^2 - (2 + \lambda^2)\lambda\cos\theta}{(1 + \lambda\cos\theta)^3} \right].$$

and this can be integrated to give an explicit expression for the pressure

$$p^{(0)}(\theta) - p^{(0)}(0) = -\frac{6\mu\Omega}{\epsilon^2}\left[\frac{\lambda\sin\theta(2 + \lambda\cos\theta)}{(2 + \lambda^2)(1 + \lambda\cos\theta)^2} \right].$$

We noted at the beginning of this section that the practical significance of journal-bearing geometry with $\epsilon \ll 1$ is that large pressure variations are generated in the gap, which can be used to support a load that is "attached" to the inner cylinder. Now that the pressure and the velocity components $u_r^{(0)}$ and $u_\theta^{(0)}$ are known, we can actually calculate the total force on the inner cylinder.

It is instructive to first consider the problem from a qualitative point of view, beginning with the pressure distribution in the gap as a function of θ.

We begin by recalling that $h = \text{constant}$ if $\lambda = 0$, and, in this case of two concentric cylinders, $\partial p^{(0)}/\partial\theta \equiv 0$. It is therefore obvious that a *net* pressure force on the inner cylinder can be generated only if $\lambda \neq 0$. We have plotted $p^{(0)}(\theta) - p^{(0)}(0)$ as a function of θ for several values of λ, $0 < \lambda < 1$, in Figure 7–5. It is evident that the variation in $p^{(0)}(\theta)$ increases as λ increases, as expected on qualitative grounds. Further, referring to Figure 7–4, we see that there is a strong minimum on the upper half of the cylinder in the quadrant between $\pi/2 < \theta < \pi$, and a corresponding maximum on the bottom half in the quadrant $\pi < \theta < 3\pi/2$.

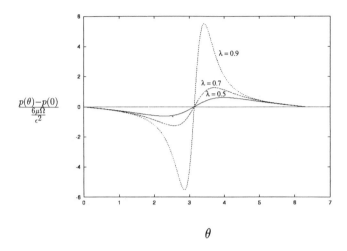

Figure 7–5 The pressure distribution in a journal-bearing flow.

Thus, a net pressure *lift* is exerted upon the inner cylinder as a consequence of its rotation. Note that if the inner cylinder were rotated in the opposite direction, the sign of the pressure gradient would change, and the force on the inner cylinder would be directed down instead of up. In any case, since $dp^{(0)}/d\theta \sim 0(\epsilon^{-2})$, it is evident that the pressure variations in the gap, and hence the lift on the inner cylinder, can be *very large* in the limit $\epsilon \to 0$. In order to produce this lift, we must rotate the inner cylinder. The torque required to do this will be proportional to $\partial u_\theta^{(0)}/\partial y|_{y=0} = 0(\Omega/\epsilon)$. Thus, by applying a torque of $0(\epsilon^{-1})$, we achieve a lift of $0(\epsilon^{-2})$. The generation of a very large normal force, $0(\epsilon^{-2})$, at a smaller $0(\epsilon^{-1})$ cost in mechanical energy to maintain tangential motion between the two surfaces is the basic principle of hydrodynamic lubrication. Indeed, the eccentric cylinder configuration, also known as the *journal bearing*, is one of the classic lubrication geometries.

The discussion of the preceding paragraph was mainly qualitative. However, explicit results for the total force acting in the x and y directions (see Figure 7–4) can be calculated from the solutions that we have obtained. To do this, we recall from Chapter 2 that the stress vector (the force per unit area) at any surface in the fluid is given in terms of the stress tensor, **T**, as

$$\mathbf{t}(\mathbf{n}) = \mathbf{T} \cdot \mathbf{n}.$$

Thus, the total force per unit length on the inner cylinder is

$$F_i = \int_S T_{ij} n_j dS = - \int_S p n_i dS + 2\mu \int_S e_{ij} n_j dS, \qquad (7\text{--}196)$$

where we have used the general expression (2–88) for the stress in a Newtonian fluid. Let us denote the pressure and viscous contributions to the force as $F_i^{(p)}$ and $F_i^{(v)}$, respectively. The outer unit normal vector **n** is given in terms of the

Cartesian components as

$$\mathbf{n} = \cos\theta\,\mathbf{e}_1 + \sin\theta\,\mathbf{e}_2. \tag{7–197}$$

Let us first consider $F_i^{(p)}$, corresponding to the lubrication approximation to the pressure distribution, $p^{(0)}$. The component $F_x^{(p)}$ can be evaluated immediately. In particular, since $p^{(0)}$ is an odd functon of θ

$$F_x^{(p)} = \int_0^{2\pi} -p^{(0)}\cos\theta\,ad\theta = 0. \tag{7–198}$$

On the other hand, the contribution $F_y^{(p)}$ is not zero:

$$F_y^{(p)} = \int_0^{2\pi} -p^{(0)}\sin\theta\,ad\theta. \tag{7–199}$$

If we integrate (7–199) by parts, the right-hand side becomes

$$F_y^{(p)} = a[\,p^{(0)}\cos\theta\,]\Big|_0^{2\pi} - \int_0^{2\pi}\cos\theta\left(\frac{dp^{(0)}}{d\theta}\right)ad\theta. \tag{7–200}$$

The first term on the right is zero. The second term can be evaluated by substituting for $(dp^{(0)}/d\theta)$ from (7–193). The result is

$$F_y^{(p)} = +\frac{6\mu\Omega a}{\epsilon^2}\left[K_2(\lambda) - \frac{I_2(\lambda)}{I_3(\lambda)}K_3(\lambda)\right], \tag{7–201}$$

where

$$K_n(\lambda) \equiv \int_0^{2\pi}\frac{\cos\theta}{(1+\lambda\cos\theta)^n}\,d\theta.$$

Since

$$K_n = -\frac{1}{\lambda}(I_n - I_{n-1}),$$

the result (7–201) also can be expressed in the explicit form

$$\boxed{F_y^{(p)} = \frac{12\pi\mu\Omega a}{\epsilon^2}\left[\frac{\lambda}{(2+\lambda^2)(1-\lambda^2)^{1/2}}\right].} \tag{7–202}$$

Now let us consider the viscous contributions to the force on the cylinder corresponding to the velocity distributions (7–194) and (7–195). These are

$$F_i^{(v)} = 2\mu\int_0^{2\pi} e_{ij}^{(0)}n_j\,ad\theta. \tag{7–203}$$

To evaluate the integrals in (7–203), we would need to determine the components of the rate of strain tensor in the Cartesian $(1,2)$ coordinate system that is shown in Figure 7–5 from the velocity components $u_r^{(0)}$ and $u_\theta^{(u)}$. This is not a difficult task. However, in the present case, we will be content to show that $F_i^{(v)}$ is $0(\epsilon^{-1})$ and is thus asymptotically small compared with $F_2^{(p)}$. To see this, we note that

$$e^{(0)} = \frac{1}{2}(\nabla\mathbf{u}^{(0)} + \nabla\mathbf{u}^{(0)T}).$$

and the maximum contribution to $\nabla\mathbf{u}^{(0)}$ comes from $\partial u_\theta^{(0)}/\partial r$. But

$$\frac{\partial u_\theta^{(0)}}{\partial r} = 0\left(\frac{\Omega}{\epsilon}\right),$$

and thus, from (7–203),

$$|F_i^{(v)}| = 0\left(\frac{\mu\Omega a}{\epsilon}\right). \tag{7–204}$$

In fact, it can be shown by symmetry considerations that $F_1^{(v)} \equiv 0$, but this is not essential in the present context. The most important conclusion is that the dominant contribution to the force on the inner cylinder is a lift generated by the strong pressure asymmetry in the gap and that viscous contributions to the force and the torque on the inner cylinder are smaller by $0(\epsilon)$.

It is perhaps worthwhile to dwell briefly on the use of the journal-bearing configuration in practical lubrication applications. In such circumstances, the rotating inner cylinder is normally allowed to "float" to seek a position in which the hydrodynamic force precisely balances the load. In the "standard" configuration, sketched in Figure 7-4, there is a net vertical force but no horizontal force. Moreover, examination of (7–202) shows that the magnitude of the vertical force for a given pair of cylinders (so that a and ϵ are fixed) is determined by Ω and λ. The λ dependence is contained in the factor

$$\alpha = \frac{\lambda}{(2+\lambda^2)(1-\lambda^2)^{1/2}}.$$

The dependence of α on λ is shown in Figure 7-6 for $0 \le \lambda < 1$. Clearly, as $\lambda \to 1$, so that the geometry becomes increasingly eccentric, the vertically directed force increases. Hence, if we increase the load on an established journal-bearing system, the configuration will change to a more eccentric geometry (that is, an increased value for λ). The load limit is determined when the value of λ necessary to support the load reaches the maximum possible value, $\lambda = 1$.

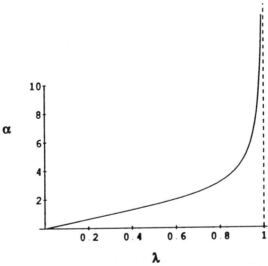

Figure 7–6 The dependence of the magnitude of the hydrodynamic lubrication force on the degree of eccentricity of the journal bearing, $\alpha \equiv F/(12\pi\mu\Omega a/\epsilon^2)$.

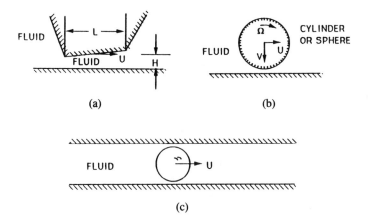

Figure 7–7 Some common lubrication configurations: (a) the slider block; (b) a sphere or cylinder moving in the vicinity of plane wall; (c) a sphere translating axially in a circular tube.

B. General Equations for Lubrication Problems

The journal-bearing problem that we have just considered is one example of a large class of so-called lubrication problems, in which the thin-film asymptotic approximation can be used to analyze the flow in all, or at least part, of the flow domain. However, the journal-bearing problem is somewhat special in that the whole of the flow domain actually satisfies the basic assumptions of the thin-film approximation. As a consequence, the solution can be obtained in terms of a *regular* perturbation expansion for $\epsilon \ll 1$, and the pressure distribution in the gap is specified completely by the condition of periodicity. Finally, in the journal-bearing problem, the boundary motion is completely tangential.

Typically, these special features of the journal-bearing problem will not occur in other lubrication applications. In particular, the thin-film region will normally constitute only a portion of the flow domain. The best known example of this class of problems is the so-called *slider block* configuration that is depicted in Figure 7–7a. In this problem, a relative tangential motion is produced between two plane boundaries that are nearly parallel and separated by a thin gap. As a consequence, large pressures are produced in the thin gap that act to support the verical load that is associated with the slider block. Although the thin film approximation may be applied in the gap if the ratio $\epsilon \equiv H/L$ is small (see Figure 7–7a), there are clearly large regions of the flow domain where this approximation cannot be used. Then, because the pressure must be specified at the ends of the lubrication region in order to determine the interior pressure distribution, it is not evident that it will be possible to consider the motion in the thin gap independently of the flow in the rest of the domain. This is in spite of the fact that the *form* of the velocity profile in the thin film depends only on the *local* values of the pressure gradient and boundary velocity.

A second example, closely related to the slider block, is the motion of a body, such as a cylinder or sphere, near a plane boundary. Here, as illustrated in Figure 7–7b, the body could be translating either parallel or normal to the wall (other directions could be considered by superposition of the results for these two fundamental cases in the thin-film or lubrication approximation), or it could be rotating (again, combinations of rotation and translation can be considered by superposition in the lubrication limit). In all of these cases, the thin-film approximation can be applied locally if the cylinder or sphere is sufficiently close to the wall, but there are again large areas of the flow domain where the approximation cannot be used. A third, closely related example is the motion of a tightly fitting sphere inside a circular tube, shown as Figure 7–7c. Many other lubrication-type problems could be cited. Some of these will be included as exercises at the end of this chapter.

Each of these problems is sufficiently complicated geometrically that it would be impossible to obtain an analytic solution for the flow field in the absence of some type of approximate analysis. At the same time, each is characterized by a local region of width, $O(H)$, that is very much smaller than its length, $O(L)$, where one can apply the thin-film or lubrication approximation. In this section, we will generalize the analysis described in the preceding section to include a more general class of prescribed boundary motions, including, specifically, the possibility of normal as well as tangential motions and the class of lubrication problems in which the thin film constitutes only a local part of the flow domain. For the latter problems, we will show that

1. The dominant hydrodynamic contributions to the force or torque on a body occur in the thin gap where the lubrication equations are relevant;
2. The magnitude of the dynamic pressure outside the thin gap is asymptotically small compared with those within the lubrication region.

The first of these results leads to the immediate conclusion that it is the velocity and pressure fields in the thin gap that we should try to determine. The second result implies that the motion outside the thin lubrication layer need not be known in order to obtain the complete *leading-order* approximation to the pressure and velocity distributions within the gap. This decoupling of the leading-order problem within the lubrication layer from the fluid mechanics problem in the rest of the domain is a particularly powerful result (in view of the dominance of the hydrodynamic force and torque contributions from the thin film) because it means that systems with extremely complicated overall geometries can be analyzed to a first level of approximation by considering only the simpler, ''universal'' flow structure within the lubrication layer. Let us now carry out the detailed analysis that is necessary to show that the results cited above are true.

Evidently, an asymptotic theory for any of the problems depicted in Figure 7–7, or, indeed, any other problems of the same general class, for $\epsilon \to 0$ must involve a *singular* perturbation analysis. It was noted earlier that such an analysis is inevitable when there are different characteristic length scales in different parts of the flow domain, and this is definitely the case in the present class of problems. Here, outside the gap, the scale

$$l_c = L \qquad (7\text{–}205)$$

is relevant to velocity gradients in all coordinate directions. On the other hand, within the gap, gradients across the flow are characterized by the length scale $0(H)$ of the gap width, whereas gradients along the lubrication layer are characterized by the length scale $0(L)$. The fact that different scales are relevant in different parts of the domain means that different approximations to the full governing equations will be relevant in different parts of the domain for $\epsilon \ll 1$, and, thus, different asymptotic forms for the solutions will also appear. This is the classic signature of a singular asymptotic limit, where we must use the method of matched asymptotic expansions. For purposes of discussion, we will denote the region within the thin gap as the *inner* region, and the rest of the flow domain as the *outer* region. In principle, our objective is an asymptotic expansion for the solution in both the inner and outer regions, with the two solutions connected in the common region near the ends of the thin gap by *matching* conditions. However, we shall see, as we have already noted, that the first term within the inner region actually can be obtained without an explicit calculation of the first term in the outer region, and we will ultimately limit our detailed calculations to that first inner solution. This theory, based only on the first term of the asymptotic solution in the thin gap, is what is normally called *lubrication theory*. We begin by considering the complete asymptotic theory for $\epsilon \to 0$ within the inner region (the thin gap).

Insofar as the inner region is concerned, it is sufficient for present purposes to consider the two-dimensional configuration that is depicted in Figure 7–8. One bounding surface is assumed to be planar and stationary. The other surface is separated from it by a distance $h(x,t)$ that depends upon position and time and is assumed to be moving with velocity components $(U(x,t), V(x,t),0)$. Although the corresponding three-dimensional problem with $h=h(x,z,t)$ and $U_{y=h}=(U,V,W)$ could be considered without substantially more difficulty, we limit our considerations to the simpler two-dimensional case as indicated earlier. It should be noted also that U, V, and h cannot all be specified independently, because there is a simple relationship between them:

$$\boxed{V = \frac{\partial h}{\partial t} + U \frac{\partial h}{\partial x}.} \tag{7–206}$$

Nevertheless, it is convenient, for the moment, to treat U, V, and h as though they were independent, and we shall continue to do so. In spite of the apparent limitations in the assumed geometry, as depicted in Figure 7–8, the problem that we have posed is sufficiently general to encompass any two-dimensional thin-film configuration in which

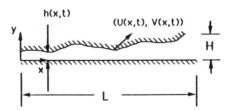

Figure 7–8 A schematic illustration of the two-dimensional, thin-film (lubrication) configuration.

variations in h of $O(H)$ occur only over distances of $O(L)$. In this case, the boundary curvature can be neglected, though not the variation in the film thickness h with position, and the governing equations can be expressed in terms of a local Cartesian coordinate system, denoted in Figure 7–8 as (x, y, z). This fact was demonstrated in detail for the eccentric cylinder problem in the preceding subsection, and we shall adopt it without further proof here.

To obtain governing equations for the general, two-dimensional lubrication layer, depicted in Figure 7–8, we follow the usual procedure for asymptotic approximations: First, we nondimensionalize the full equations of motion and continuity; second, we introduce asymptotic expansions for the unknown, independent variables $(\mathbf{u}, p)$; and, third, we obtain approximate equations for the various terms in this expansion by collecting terms with the same dependence on the small parameter (in this case $\epsilon \equiv H/L$) and requiring that the equations and boundary conditions be satisfied independently at each level in ϵ. Alternatively, the governing equations and boundary conditions for the *leading* term (only) in the asymptotic expansion can be obtained by taking the limit $\epsilon \to 0$ in the full, nondimensionalized equations and boundary conditions.

The scaling relevant for the (inner) lubrication layer in this general formulation follows closely the scaling that was used earlier in the journal-bearing problem. Thus, we assume that there exists a length scale H that is characteristic of the distance across the gap and a length scale L that is characteristic of the distance over which the velocity varies along the gap. The choice of a characteristic velocity scale depends somewhat on the details of the particular problem. In the journal-bearing problem, the normal velocity component at the boundary was zero, and the tangential component was constant, $U = a\Omega$. In the present, more general situation, both U and V may be nonzero. Furthermore, a consequence of the extreme ratio, H/L, of length scales in the lubrication layer is that the velocity components in the x and y (that is, tangent and normal) directions within this region will always differ by $O(\epsilon)$. To see this, let us denote the characteristic magnitude of u as u_c and the characteristic magnitude of v as v_c. Then, if we nondimensionalize the continuity equation, we find that

$$\left(\frac{u_c}{L}\right)\frac{\partial \bar{u}}{\partial \bar{x}} + \left(\frac{v_c}{H}\right)\frac{\partial \bar{v}}{\partial \bar{y}} = 0, \tag{7–207}$$

where $u = u_c \bar{u}$, $v = v_c \bar{v}$, $L\partial/\partial x = \partial/\partial \bar{x}$, and $H\partial/\partial y = \partial/\partial \bar{y}$. Since $(\partial \bar{u}/\partial \bar{x})$ and $(\partial \bar{v}/\partial \bar{y})$ are $O(1)$, it follows that

$$v_c = \epsilon u_c. \tag{7–208}$$

Thus, if motion in the lubrication layer is a result of relative tangential motion of the boundaries, so that an appropriate choice for the magnitude u_c is*

$$u_c = U, \tag{7–209}$$

then it follows from the relationship (7–129) that

$$v_c = \epsilon U. \tag{7–210}$$

---

*Note that if $U = U(x, t)$, as would be the case for a deformable boundary (or boundaries), the u_c should be chosen as an average of $U(\mathbf{x}, t)$ over the surface in question, that is $u_c = <U>$.

If, on the other hand, the motion in the lubrication layer results from relative motions of the boundaries toward (or away from) one another so that

$$v_c = V, \tag{7-211}$$

it follows from (7–208) that

$$u_c = \frac{V}{\epsilon}. \tag{7-212}$$

Thus, independent of whether the motion in the lubrication layer is a consequence of tangential or normal motions of the boundaries, the dominant fluid velocity component will be in the tangential direction. We can choose either (7–209) and (7–210) or (7–211) and (7–212) to nondimensionalize velocities depending upon the relative magnitudes of U and V/ϵ. Finally, the appropriate characteristic pressure is

$$p_c = \frac{\mu u_c}{\epsilon H}, \tag{7-213}$$

which is the same choice that was used for the journal-bearing problem (for journal bearing, $u_c = a\Omega$ where $a = L$ in the present nomenclature, and so $p_c = \mu\Omega/\epsilon^2$). It is adopted here for precisely the same reason.

With the characteristic scales (7–208), (7–213), and either (7–209) or (7–212), plus H and L as defined in connection with (7–207), the limiting form of the full Navier-Stokes and continuity equations for the lubrication layer (that is, the inner region in the asymptotic solution scheme for $\epsilon \to 0$) is identical to that obtained for the journal-bearing problem, equations (7–178)–(7–180), namely,

$$\frac{\partial \bar{u}^{(0)}}{\partial \bar{x}} + \frac{\partial \bar{v}^{(0)}}{\partial \bar{y}} = 0, \tag{7-214}$$

$$\frac{\partial^2 \bar{u}^{(0)}}{\partial \bar{y}^2} = \frac{\partial \bar{p}^{(0)}}{\partial \bar{x}}, \tag{7-215}$$

and

$$\frac{\partial \bar{p}^{(0)}}{\partial \bar{y}} = 0(\epsilon^2). \tag{7-216}$$

The superscript (0) is added to remind us that these equations govern the first term only in an asymptotic expansion of $\bar{u}$, $\bar{v}$, and $\bar{p}$ for small ϵ, that is,

$$\bar{u} = \bar{u}^{(0)} + \epsilon^n \bar{u}^{(1)} + \cdots,$$

where it can be shown that $n = 1$ or 2 depending on whether the boundaries are curved or flat on the scale, $0(L)$. It is most important that these governing equations for the first term in the asymptotic approximation are *linear*, in spite of the fact that the complete equations for arbitrary values of ϵ are highly nonlinear.

Thus, we see that the so-called *lubrication equations* [(7–214)–(7–216)] have a universal form independent of the overall boundary geometry, provided only that the thin-gap (or lubrication) approximation $\epsilon \equiv H/L \ll 1$ is valid. All that really changes from problem to problem in this inner (thin-film) part of the flow domain is the relevant boundary conditions and the dependence of the gap width on position and time. In the general case considered here, the boundary conditions for $y = 0$ and $y = h$, expressed in dimensionless form, are

$$\bar{u}^{(0)} = \bar{U}(x,t) = \frac{U(x,t)}{u_c}, \quad \bar{v}^{(0)} = \bar{V}(x,t) = \frac{V(x,t)}{\epsilon u_c} \quad \text{at} \quad \bar{y} = \bar{h}(x,t) \equiv \frac{h(x,t)}{H},$$

and

$$\bar{u}^{(0)} = 0, \ \bar{v}^{(0)} = 0 \quad \text{at} \quad \bar{y} = 0.$$

(7–217)

It may be noted that the governing differential equation (7–214)–(7–216) involve derivatives with respect to $\bar{x}$ as well as $\bar{y}$. Thus, a unique solution will generally be possible only with the imposition of additional boundary conditions at the ends of the lubrication layer. This is the point where the inner and outer domains merge, and we should expect these conditions to derive from the requirement of *matching* between the approximate solutions in the thin gap and in the rest of the flow domain. Before being concerned with the details of matching, however, it is useful to proceed a bit further with the analysis in the inner region.

In particular, let us try to integrate (7–214)–(7–216) along with the known boundary conditions (7–217). Starting with (7–215) and noting that $\partial \bar{p}^{(0)}/\partial \bar{x}$ will be independent of $\bar{y}$, according to (7–216), we have

$$\bar{u}^{(0)} = \frac{1}{2}\left(\frac{d\bar{p}^{(0)}}{d\bar{x}}\right)\bar{y}^2 + c_1(x)\bar{y} + c_2(x).$$ (7–218)

Thus, applying the boundary conditions (7–217) on $\bar{u}^{(0)}$ at $\bar{y} = 0$ and $\bar{h}$, we find

$$\bar{u}^{(0)} = \frac{1}{2}\left(\frac{d\bar{p}^{(0)}}{d\bar{x}}\right)\bar{y}(\bar{y} - \bar{h}) + \bar{U}\frac{\bar{y}}{\bar{h}}.$$ (7–219)

This solution is essentially identical to the solution (7–182) for the journal-bearing problem. We may also note that the form $\bar{u}^{(0)}$ is identical to that found in Chapter 3 for motion between parallel plane walls with boundary motion and an imposed pressure gradient. In the present case, however, $\bar{h} \neq$ constant, and the pressure gradient in the gap is usually due to the fact that U (or V) is nonzero at $\bar{y} = \bar{h}$, rather than being a consequence of an applied pressure gradient. Indeed, although a pressure difference can be imposed between the two ends of a lubrication layer, the resulting contribution to the streamwise pressure gradient is usually a small fraction of the total pressure gradient. This should be contrasted with unidirectional flow between parallel plane boundaries where the pressure gradient is completely determined by the imposed pressure difference between two points. The pressure distribution within the thin lubrication layer is *dominated* by the relative motion of the boundaries in conjunction with the nonuniform gap width, and the imposed pressures at the ends of the gaps thus play a secondary role unless the imposed pressure difference becomes extremely large, $0(1/\epsilon^2)$. For example, $d\bar{p}^0/d\bar{x}$ will be strongly nonzero within a lubrication layer, even if the pressures at the two ends of the gap are equal.

The solution (7–219) appears to be formally complete, but it is not, because the pressure gradient has not yet been determined. We shall see shortly that the necessary condition to determine $d\bar{p}^{(0)}/d\bar{x}$ comes naturally from an attempt to calculate the normal velocity component $\bar{v}^{(0)}$.

The only remaining equation from which to obtain $\bar{v}^{(0)}$ is the continuity equation. Substituting for $\partial \bar{u}^{(0)}/\partial \bar{x}$ from (7–219), the continuity equation can be written in the form of a first-order differential equation for $\bar{v}^{(0)}$, namely,

$$\frac{\partial \bar{v}^{(0)}}{\partial \bar{y}} = -\frac{1}{2}\frac{d^2\bar{p}^{(0)}}{d\bar{x}^2}\bar{y}(\bar{y}-\bar{h}) + \frac{1}{2}\frac{d\bar{p}^{(0)}}{d\bar{x}}\bar{y}\frac{d\bar{h}}{d\bar{x}} - \left(\frac{d\bar{U}}{d\bar{x}}\frac{\bar{y}}{\bar{h}} - \bar{U}\frac{\bar{y}}{\bar{h}^2}\frac{d\bar{h}}{d\bar{x}}\right).$$

The right-hand side is just $-\partial \bar{u}^{(0)}/\partial \bar{x}$. Integrating,

$$\bar{v}^{(0)} = -\frac{1}{2}\frac{d^2\bar{p}^{(0)}}{d\bar{x}^2}\left(\frac{\bar{y}^3}{3} - \frac{\bar{y}^2\bar{h}}{2}\right) + \frac{1}{2}\frac{d\bar{p}^{(0)}}{d\bar{x}}\frac{d\bar{h}}{d\bar{x}}\frac{\bar{y}^2}{2} - \left(\frac{d\bar{U}}{d\bar{x}}\frac{1}{\bar{h}} - \bar{U}\frac{1}{\bar{h}^2}\frac{d\bar{h}}{d\bar{x}}\right)\frac{\bar{y}^2}{2} + C_3(\bar{x}).$$

$$(7\text{–}220)$$

Now, $\bar{v}^{(0)}$ is supposed to satisfy two boundary conditions according to (7–217), but there is only a single undetermined integration constant available in (7–220). It must be remembered, however, that the pressure gradient remains to be determined also. The condition $\bar{v}^{(0)} = 0$ at $\bar{y} = 0$ implies that

$$C_3(\bar{x}) = 0.$$

When we apply the second condition on $\bar{v}^{(0)}$ at $\bar{y} = \bar{h}$, we obtain

$$\bar{V} = \frac{1}{12}\frac{d^2\bar{p}^{(0)}}{d\bar{x}^2}\bar{h}^3 + \frac{1}{4}\frac{d\bar{p}^{(0)}}{d\bar{x}}\bar{h}^2\frac{d\bar{h}}{d\bar{x}} - \frac{1}{2}\frac{d\bar{U}}{d\bar{x}}\bar{h} + \frac{1}{2}\bar{U}\frac{d\bar{h}}{d\bar{x}}.$$

But upon rearrangement, this equation can be written as

$$\boxed{\frac{d}{d\bar{x}}\left(\bar{h}^3\frac{d\bar{p}^{(0)}}{d\bar{x}}\right) = 6\left[\bar{h}\frac{d\bar{U}}{d\bar{x}} - \bar{U}\frac{d\bar{h}}{d\bar{x}} + 2\bar{V}\right].}$$

$$(7\text{–}221)$$

This is the famous *Reynolds equation* of lubrication theory. It is a second-order ordinary differential equation for $\bar{p}^{(0)}$, which can be solved, in general, to determine the pressure distribution in the lubrication layer. The coefficient $\bar{h}^3$ and the nonhomogeneous term on the right-hand side of (7–221) are both known, provided that we know the gap geometry $\bar{h}$ and the boundary velocity components U and V. In practice, of course, two of $\bar{h}$, U, and V will be specified and the third determined by the relationship (7–206). To obtain a unique solution of (7–221), the values of $\bar{p}^{(0)}$ must be specified at the ends of the lubrication layer.* It is at this point that the solution within the lubrication layer is seen to depend upon boundary conditions at the ends of the lubrication layer. The necessary boundary values for the pressure come from the requirement that the solution within the lubrication layer must match the solution in

---

*Note that for an incompressible fluid, the pressure is determinate only to within an arbitrary reference pressure. Thus, $\bar{p}^{(0)}$ can be determined to within this arbitrary constant value by specifying the difference in pressure across the lubrication layer rather than values for the pressure itself at the two ends.

the outer domain, at the ends of the lubrication layer. Note, however, that this would seem to suggest that the solution within the lubrication layer cannot be determined completely without simultaneously obtaining the leading-order solution in the outer region, including particularly the dynamic pressure. We will turn to this outer problem in a moment.

Before doing that, there are several comments that we wish to make in regard to the analysis so far. First, the Reynolds equation is most often expressed in dimensional, rather than dimensionless, form as in (7–221), and we shall sometimes find it convenient to use the dimensional form. If we reintroduce dimensional variables into (7–221), we obtain

$$\frac{d}{dx}\left(h^3 \frac{dp^{(0)}}{dx}\right) = 6\mu\left[h\frac{dU}{dx} - U\frac{dh}{dx} + 2V\right]. \tag{7-222}$$

Second, the Reynolds equation (7–187) that was derived in conjunction with the journal-bearing problem is a special case of (7–221), with $\bar{x} = \theta$ and U given by $-a\Omega$, so that

$$V = \frac{U}{a}\frac{\partial h}{\partial \theta} = -\Omega\frac{\partial h}{\partial \theta}.$$

Third, and finally, the Reynolds equation (and the rest of our analysis) is easily generalized to the three-dimensional case in which the upper surface is at $y = h(x, z, t)$ and $\mathbf{u} = (U, V, W)$ on this surface. Similar arguments to those above give

$$\frac{\partial}{\partial x}\left(h^3 \frac{\partial p^{(0)}}{\partial x}\right) + \frac{\partial}{\partial z}\left(h^3 \frac{\partial p^{(0)}}{\partial z}\right) = 6\mu\left[h\left(\frac{\partial U}{\partial x} + \frac{\partial W}{\partial z}\right) - U\frac{\partial h}{\partial x} - W\frac{\partial h}{\partial z} + 2V\right]. \tag{7-223}$$

Let us now return to the problem of determining a value for the dynamic pressure $\bar{p}^{(0)}$ at the ends of the lubrication layer. We have seen that the first term in the asymptotic solution within the inner region is determined completely by a local analysis except for $d\bar{p}^{(0)}/d\bar{x}$, which is determined from the Reynolds equation, and depends on the gap geometry, the boundary motion, and the values of $\bar{p}^{(0)}$ at the ends of the gap. The necessary values of $\bar{p}^{(0)}$ at the ends of the gap must come from the first approximation to the pressure distribution in the outer region.

The asymptotic analysis for small ϵ in the outer domain proceeds via the same steps that we outlined for the inner region. Specifically, we nondimensionalize the full Navier-Stokes and continuity equations using characteristic scales relevant to the outer region (exterior to the thin gap). We then obtain approximate equations for the various terms in an asymptotic expansion of the solution by substituting the expansion into the nondimensionalized equations and requiring that they be satisfied at each order of magnitude in ϵ. As always, governing equations for the first term in the asymptotic series can be obtained by simply taking the limit $\epsilon \to 0$ in the full nondimensionalized equations for the outer region. The appropriate characteristic scales for this outer region were anticipated at the beginning of this section. Specifically, the characteristic length scale, $l_c = L$, equation (7–205), is now relevant to velocity gradients in all coordinate directions. Thus, the same characteristic velocity scale, $u_c = U$ (or V), is also relevant to both components of the velocity field in the outer domain. If we introduce nondimensional variables

$$\bar{u} = \frac{u}{u_c}, \quad \hat{v} = \frac{v}{u_c}, \quad \bar{x} = \frac{x}{L}, \quad \hat{y} = \frac{y}{L}, \quad \text{and} \quad \bar{p} = \frac{p}{p_c}$$

into the full Navier-Stokes equations, we obtain

$$\bar{u}\frac{\partial \bar{u}}{\partial \bar{x}} + \hat{v}\frac{\partial \bar{u}}{\partial \hat{y}} = -\frac{p_c}{\rho u_c^2}\frac{\partial \bar{p}}{\partial \bar{x}} + \frac{\mu}{\rho u_c L}\left[\frac{\partial^2 \bar{u}}{\partial \bar{x}^2} + \frac{\partial^2 \bar{u}}{\partial \hat{y}^2}\right],$$

$$\bar{u}\frac{\partial \hat{v}}{\partial \bar{x}} + \hat{v}\frac{\partial \hat{v}}{\partial \hat{y}} = -\frac{p_c}{\rho u_c^2}\frac{\partial \bar{p}}{\partial \hat{y}} + \frac{\mu}{\rho u_c L}\left[\frac{\partial^2 \hat{v}}{\partial \bar{x}^2} + \frac{\partial^2 \hat{v}}{\partial \hat{y}^2}\right], \qquad (7\text{--}224)$$

Although we will not use it here, it is interesting to note that the dimensional equation for the gap

$$y = H\bar{h}(x)$$

becomes

$$\hat{y} = \epsilon \bar{h}(x)$$

when expressed in terms of the outer scaling. Thus, as $\epsilon \to 0$, the geometric configuration, when seen from the outer point of view, reduces to that of the body "touching" the moving plane boundary with a line source and sink for mass of equal strength located at the two ends of the gap.

As yet unspecified in (7–224) is the characteristic pressure p_c. This is determined from the characteristic pressure in the inner region. The basic idea is that $\bar{p}$ from the inner and outer domains must match at the entry and exit to and from the gap. It follows from this reasoning that

$$p_c = \frac{\mu u_c L}{H^2}.$$

Hence, equation (7–224) becomes

$$\nabla \bar{p} = \epsilon^2\left[\nabla^2 \mathbf{u} + \left(\frac{\rho u_c L}{\mu}\right)\mathbf{u}\cdot\nabla\mathbf{u}\right], \qquad (7\text{--}225)$$

where $\mathbf{u} = (\bar{u}, \hat{v})$. This shows that the effect of the fluid motion on the pressure in the outer region is negligible relative to the pressure variations that are possible in the inner region as $\epsilon \to 0$. Thus, insofar as the first approximation in the thin gap is concerned, it is only the values of dynamic pressure for a *stationary* fluid that are needed at the ends of the gap. Frequently, the values of the dynamic pressure for a *stationary* fluid will be equal at the two ends of the gap. For example, in the case of a body of finite extent moving in a large body of fluid near a plane wall, the dynamic pressure difference between the two ends of the gap will be zero—i.e., $p^{(0)}(0) = p^{(0)}(L)$—independent of the orientation of the plane wall. On the other hand, the dynamic pressure difference will not always be zero. Consider, for example, the configuration

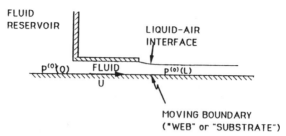

Figure 7-9 A sketch of a film-coating process.

depicted in Figure 7-9, which is typical of film-coating applications. Here, the pressure on the upstream side of the thin gap (the lubrication layer) can be controlled by pressurizing or depressurizing the fluid reservoir, and the static difference in dynamic pressure will not be zero across the gap, that is, $p^{(0)}(0) = p_{\text{reservoir}}$ and $p^{(0)}(0) \neq p^{(0)}(L)$ in general.

In any case, the very important conclusion from (7-225) is that the leading-order approximation to the solution in the thin-film region always can be determined completely without the need to determine anything about the velocity field in the rest of the domain where the geometry is much more complicated. This constitutes a very considerable simplification! When the lubrication approximation can be made, we need focus our attention on only the lubrication equations within the thin gap, and in this region the general solutions (7-219)-(7-221) have been worked out already. We shall see in the next section that the dominant contributions to the forces or torques acting on a body near a second boundary always occur in the lubrication layer when $\epsilon \ll 1$.

C. Applications of the General Equations of Lubrication Theory

In this section we consider the detailed analysis for two applications of lubrication theory: the classic slider block problem that was depicted in the previous section and the motion of a circular cylinder toward an infinite plane wall when the cylinder is very close to the wall. It is the usual practice in lubrication theory to focus directly on the motion in the thin gap using (7-214)-(7-216), or their solutions (7-219)-(7-221), without any mention of the asymptotic nature of the problem or of the fact that these equations (and their solutions) represent only a first approximation to the full solution in the lubrication layer. We adopt the same approach here but with the formal justification of the preceding section.

1. The Slider Block Problem

We begin by considering the classic slider block problem that is sketched in Figure 7-10. Here, a two-dimensional cylindrical body of arbitrary cross section, but with one face that is flat, moves with a velocity U parallel to an infinite plane stationary

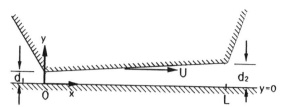

Figure 7-10   The slider-block configuration.

boundary.* The plane boundary and the smooth face of the slider block form a lubrication layer that is wider at the front and narrower toward the rear of the body but is everywhere thin in the sense that $d_2/L \ll 1$, where d_2 is the maximum gap width and L is the characteristic length of the lubrication layer. The goals of analysis for this problem are to determine the magnitude of the lift force (normal to the boundary) that is generated by the large pressure and viscous stresses in the lubrication layer and the magnitude of the tangential force that is necessary to maintain the constant lateral velocity U. The slider block problem illustrates the key features of the most common class of lubrication configurations—namely, the generation of very large forces that tend to maintain the separation between two solid boundaries due to a relative *tangential* motion of the boundaries that is maintained by a tangential force of smaller magnitude.

The analysis of the preceding section and the solutions (7–219)–(7–221) can be applied directly to this problem. We need only specify the gap width as a function of position, $h(x)$, and the boundary velocities, U and V. The geometric configuration for the slider block problem is independent of time, and the lubrication approximation is intrinsically a quasi-steady theory. Thus, we can formulate the problem using a Cartesian coordinate system with origin fixed on the lower wall, which is thus $y = 0$, with $x = 0$ and $x = L$ corresponding, at the instant of solution, to the trailing and leading edges of the slider block. Relative to this coordinate system, the moving upper surface of the lubrication layer is at

$$y = h(x) = \frac{d_2 - d_1}{L}x + d_1. \qquad (7\text{-}226)$$

If we adopt d_1 as the characteristic gap width H and L as the characteristic streamwise length scale, the dimensionless gap width function is

$$\boxed{\bar{h}(\bar{x}) = 1 + \left(\frac{d_2}{d_1} - 1\right)\bar{x}.} \qquad (7\text{-}227)$$

The velocity of the slider block is parallel to the plane wall with a magnitude U, and this is an appropriate choice for the characteristic velocity u_c, which appears in

---

*Alternatively, we could consider the slider block to be stationary and the plane boundary to be moving in its own plane with velocity $-U$.

(7-219)-(7-221). Thus, in dimensionless terms, $\bar{U} = 1$. Now, let us apply the solutions (7-219)-(7-221) to the slider block problem.

The dimensionless tangential velocity distribution in the lubrication layer is given by

$$\bar{u}^{(0)} = \frac{1}{2}\left(\frac{d\bar{p}^{(0)}}{d\bar{x}}\right)\bar{y}(\bar{y} - \bar{h}) + \frac{\bar{y}}{\bar{h}} \tag{7-228}$$

since $u_c = U$. Then, the unknown pressure distribution in the lubrication layer can be obtained from the Reynolds Equation (7-221), which takes the form

$$\frac{d}{d\bar{x}}\left(\bar{h}^3\frac{d\bar{p}^{(0)}}{d\bar{x}}\right) = 6\left(1 - \frac{d_2}{d_1}\right). \tag{7-229}$$

Integrating once with respect to $\bar{x}$, we find that

$$\frac{d\bar{p}^{(0)}}{d\bar{x}} = 6\left(1 - \frac{d_2}{d_1}\right)\frac{\bar{x}}{\bar{h}^3} + \frac{c_1}{\bar{h}^3}. \tag{7-230}$$

Here, c_1 is a constant of integration. Integrating again,

$$\bar{p}^{(0)} = 6\left(1 - \frac{d_2}{d_1}\right)\int_0^{\bar{x}}\frac{\bar{s}d\bar{s}}{\bar{h}^3(\bar{s})} + c_1\int_0^{\bar{x}}\frac{d\bar{s}}{\bar{h}^3(\bar{s})} + c_2. \tag{7-231}$$

The integration constants c_1 and c_2 are given by the boundary values for $\bar{p}^{(0)}$ at $\bar{x} = 0$ and $\bar{x} = 1$, and these come from the dynamic pressure values for a stationary fluid in the region outside the lubrication layer. We assume that the slider block is moving through a large body of fluid that completely surrounds it. In this case, it is obvious that

$$p^{(0)}(0) = p^{(0)}(1) = p_0, \tag{7-232}$$

where p_0 is an arbitrary reference pressure. The fact that p_0 is arbitrary is a consequence of the assumed incompressibility of the fluid. Applying this condition (7-232) to (7-231), we find that

$$c_2 = p_0$$

and

$$c_1 = \frac{6\left(\dfrac{d_2}{d_1} - 1\right)\displaystyle\int_0^1\frac{\bar{s}d\bar{s}}{\bar{h}^3(\bar{s})}}{\displaystyle\int_0^1\frac{d\bar{s}}{\bar{h}^3(\bar{s})}} = 6\left(\frac{d_2 - d_1}{d_1 + d_2}\right). \tag{7-233}$$

Hence,

$$\bar{p}^{(0)}(\bar{x}) - p_0 = 6\left(\frac{d_1 - d_2}{d_1 + d_2}\right)\frac{(\bar{x}^2 - \bar{x})}{(1 + b\bar{x})^2} \quad \text{with} \quad b \equiv \frac{d_2 - d_1}{d_1}. \tag{7-234}$$

Hence, in dimensional terms,

$$p^{(0)}(x) - p_0 = \frac{6\mu\, UL}{d_1^2}\left(\frac{d_1 - d_2}{d_1 + d_2}\right)\frac{(x^2 - Lx)}{\left(L + \left(\frac{d_2 - d_1}{d_2}\right)x\right)^2}, \tag{7-235}$$

and the pressure gradient is

$$\frac{dp^{(0)}}{dx} = \frac{6\mu\, UL}{d_1^2}\left(\frac{d_1 - d_2}{d_1 + d_2}\right)\left[\frac{\left(\frac{d_1 + d_2}{d_1}\right)xL - L^2}{\left(\frac{d_2 - d_1}{d_1}x + L\right)^3}\right]. \tag{7-236}$$

With $\partial p/\partial x$ known, the velocity distribution (7-228) can be specified completely, and the solution for the velocity and pressure fields in the lubrication layer is complete.

Let us now calculate the hydrodynamic forces on the slider block. In general, for any body in a viscous fluid

$$\mathbf{F} = \int_A \mathbf{t}\, dA = \int_A \mathbf{n}\cdot\mathbf{T}\, dA = -\int_A p^{(0)}\mathbf{n}\, dA + \int_A \mathbf{n}\cdot\boldsymbol{\tau}^{(0)}\, dA, \tag{7-237}$$

where A represents the complete exterior surface of the slider block, including both the surface inside and outside the lubrication layer, and p is the dynamic pressure relative to any arbitrary reference pressure (note that a constant reference pressure produces no contribution to the integral involving p). In (7-235), $p^{(0)}(0)$ was used as the background reference pressure. Thus, to a first approximation,

$$\mathbf{F} = \int_A -\mathbf{n}(p^{(0)} - p(0))\, dA + \int_A \mathbf{n}\cdot\boldsymbol{\tau}^{(0)}\, dA. \tag{7-238}$$

Let us begin by calculating the forces on the flat surface of the slider block that forms the upper boundary of the lubrication layer. To do this we require an expression for the outer normal $\mathbf{n}$ to this surface. In general, for a surface $y = g(x)$, the unit normal is

$$\mathbf{n} = \pm \frac{\nabla(y - g)}{|\nabla(y - g)|}. \tag{7-239}$$

Here, $g = mx + d_1$, thus $\nabla(y - g) = \mathbf{e}_y - m\mathbf{e}_x$ and $|\nabla(y - g)| = \sqrt{1 + m^2}$, where $m \equiv (d_2 - d_1)/L$. It follows from the definition (7-239) that

$$\mathbf{n} = \pm\left(\frac{-m}{\sqrt{1 + m^2}}\mathbf{e}_x + \frac{1}{\sqrt{1 + m^2}}\mathbf{e}_y\right). \tag{7-240}$$

The choice of the plus or minus sign in (7-240) determines whether the unit normal points into the slider block or points into the gap and is the outer normal as required. It is easy to see which sign to choose by considering the limiting case $m = 0$ where the surface of the slider block is parallel to the plane wall. In this case

$$\mathbf{n} = \pm\mathbf{e}_y,$$

from which it is evident that the outer normal to the slider plate corresponds to the minus sign in (7-240) and

$$\mathbf{n} = \frac{m}{\sqrt{1+m^2}}\mathbf{e}_x - \frac{1}{\sqrt{1+m^2}}\mathbf{e}_y. \tag{7-241}$$

Thus,

$$\mathbf{n} \cdot \boldsymbol{\tau}^{(0)} = \frac{m}{\sqrt{1+m^2}}(\tau_{xx}^{(0)}\mathbf{e}_x + \tau_{xy}^{(0)}\mathbf{e}_y) - \frac{1}{\sqrt{1+m^2}}(\tau_{yx}^{(0)}\mathbf{e}_x + \tau_{yy}^{(0)}\mathbf{e}_y). \tag{7-242}$$

Finally, to calculate the force per unit width on the slider block we must integrate along s. For this purpose it is convenient to express a differential change along the slider block, ds, in terms of a differential change in x using

$$ds = \sqrt{1+(dy/dx)^2}\,dx = \sqrt{1+m^2}\,dx. \tag{7-243}$$

Thus, the x component of $\mathbf{F}$ on the bottom surface of the slider block is

$$F_x = \int_0^L -\left[p^{(0)}(x) - p^{(0)}(0)\right]m\,dx + \int_0^L (m\tau_{xx}^{(0)} - \tau_{yx}^{(0)})\,dx. \tag{7-244}$$

Substituting for $\left[p^{(0)}(x) - p^{(0)}(0)\right]$ and $(m\tau_{xx}^{(0)} - \tau_{yx}^{(0)})$ using (7–235) and the corresponding expressions for $\tau_{xx}^{(0)}$ and $\tau_{yx}^{(0)}$ calculated from the known velocity field in the lubrication layer, we find that

$$F_x = -\frac{18\mu UL}{d_2+d_1}\left[1 - \frac{4}{9}\frac{d_2+d_1}{d_2-d_1}\ln\frac{d_2}{d_1} + 0\left(\frac{d_2-d_1}{L}\right)^2\right]. \tag{7-245}$$

Thus, in the limit $\epsilon \to 0$,

$$F_x \sim -\frac{18\mu UL}{d_1+d_2}. \tag{7-246}$$

Thus, we see that the force necessary to maintain the constant slider block velocity U is proportional to ϵ^{-1} for $\epsilon \ll 1$, where $\epsilon \equiv d_2/L$. The normal (or y) component of force on the slider block can be seen from (7–235) and (7–241)–(7–243) to be

$$F_y = \int_0^L \left[p^{(0)}(x) - p^{(0)}(0)\right]dx + \int_0^L (m\tau_{xy}^{(0)} - \tau_{yy}^{(0)})\,dx. \tag{7-247}$$

Noting that $m = 0(\epsilon)$, it can be shown that the second term in this expression is $0(\epsilon^{-1})$ but the pressure contribution is $0(\epsilon^{-2})$. Hence, to a first approximation for small ϵ,

$$F_y = \int_0^L \left[p^{(0)}(x) - p^{(0)}(0)\right]dx. \tag{7-248}$$

Again, substituting for $\left[p^{(0)} - p^{(0)}(0)\right]$ and integrating, we find that

$$F_y = -\frac{6\mu UL}{d_2^2-d_1^2}\left[2 - \frac{d_1+d_2}{d_2-d_1}\ln\left(\frac{d_2}{d_1}\right)\right] \tag{7-249}$$

or, in the limit $\epsilon \to 0$,

$$F_y \sim \frac{12\mu UL^2}{d_2^2-d_1^2}. \tag{7-250}$$

The normal force is thus $0(\epsilon^{-2})$ for $\epsilon \ll 1$.

These results for the slider block are typical of this general class of lubrication problems. The largest force in the lubrication layer is the normal lift force, and this is dominated, for small ϵ, by the pressure contribution. This force is always $0(\epsilon^{-2})$ in magnitude, where ϵ is the dimensionless ratio of gap width to gap length. The tangential force necessary to maintain the relative motion between the boundaries is also large $0(\epsilon^{-1})$ for small ϵ, but is smaller than the normal force by $0(\epsilon)$. Again, this is typical of lubrication problems of this general class.

The only remaining point is the contribution to F_x and F_y due to pressure and viscous stresses acting on the portions of the slider block that are not included in the lubrication region. At most, however, these contributions are $0(1)$ compared to the $0(\epsilon^{-1})$ and $0(\epsilon^{-2})$ contributions from the lubrication layer, and thus they do not need to be calculated explicitly for $\epsilon \ll 1$. To see this, we note (according to the scaling that we have already done) that the largest viscous stresses in the lubrication layer are proportional to $\partial u^{(0)}/\partial y$ and are thus of order $0\left[\mu(U/H)\right]$. The largest contribution to the pressure in the lubrication layer [relative to $p(0)$] is $0\left[\mu(UL/H^2)\right]$. Outside the lubrication layer, however, the largest viscous stresses are $0(\mu U/L)$, and this is the same magnitude as the largest pressures in this region. The ratio of viscous stresses inside the lubrication layer compared to those outside is thus $0(\epsilon^{-1})$, while the ratio of dynamic pressures is $0(\epsilon^{-2})$. Since the surface area of the slider block is the same order of magnitude inside and outside the lubrication layer, the estimate we made of the relative magnitude of the contributions to the force follows. We thus see that the calculations for F_x and F_y from the lubrication layer, (7–246) and (7–250), are, in fact, the correct leading-order approximations (for small ϵ) to the *total* drag and lift on the slider block. We again note that the dominance of contributions from the lubrication region is typical of problems of this general class. One need not calculate velocity and pressure fields anywhere but in the thin-gap region, for $\epsilon \rightarrow 0$, if the objective is to determine the leading-order approximation to the force and/or torque on the body.

2. The Motion of a Circular Cylinder Toward a Solid, Plane Boundary

A second example of the application of lubrication theory is its use in analyzing the motion of a circular cylinder toward a solid plane boundary when the axis of the cylinder is parallel to the wall and the gap between the cylinder and the wall at the point of closest approach is small compared with the radius of the cylinder.[9] Although this problem is geometrically similar to the slider block in the sense that a cylindrically shaped body is moving in the vicinity of an infinite plane boundary, the problem differs in that the motion is normal to the boundary rather than parallel to it.

The geometry of the problem is sketched in Figure 7–11. We consider a cylinder of radius a, whose center is a distance $a + b$ from an infinite plane solid boundary when measured along the normal to the wall that passes through the center of the cylinder. The cylinder moves toward the wall under the action of a constant applied force, F, and the objective is to calculate its velocity or its position relative to the wall as a function of time. In the present development, we consider the asymptotic solution for $b/a \ll 1$. The flow domain then divides into two parts, the region between the cylinder

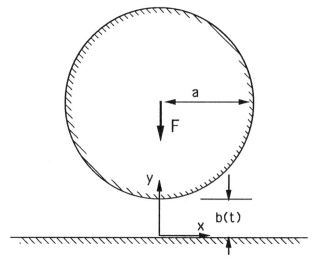

Figure 7–11 A sketch of the geometry and coordinate system for analysis of the motion of a circular cylinder toward a plane wall under the action of an applied force, F.

and the wall where the lubrication equations (and solutions) pertain, and the rest of the domain where the lubrication equations cannot be used. As usual, the dominant contributions to the hydrodynamic force on the body come from the thin lubrication layer, and we thus focus our attention only on obtaining an approximate solution in that regime, using the general equations from Section **B**. One question may occur immediately to the reader: If lubrication theory can be applied in this case, where does the lubrication layer end, since it is evident that the gap between the cylinder and wall varies from a minimum value of b to a maximum equal to $b + a$ as one traverses parallel to the wall. We shall see that it does not really matter what choice we make—the asymptotic results for $\epsilon \equiv b/a \to 0$ will turn out to be independent of this choice.

Again, we employ the general solutions obtained in Section **B** to analyze the problem. In this case, however, the applied force is held constant, the cylinder velocity will vary with time, and it is not entirely obvious what choice to make for a characteristic velocity. Rather than being concerned with this point, we simply solve the problem in terms of dimensional variables. In order to utilize the solutions and analysis from Section **B**, we must first specify the gap width function, $h(x)$. For this purpose, we use a Cartesian coordinate system with its origin on the solid wall at the point where the gap width is the minimum. Then, from geometry, an exact expression for $h(x)$ is

$$h(x) = a + b(t) - \sqrt{a^2 - x^2}, \qquad (7\text{–}251)$$

where $-a \leq x \leq a$. However, we expect the thin-film approximation to apply only when $x/a \ll 1$. In this case,

$$h(x) \sim b(t) + \frac{a}{2}\left(\frac{x}{a}\right)^2 + \cdots. \qquad (7\text{–}252)$$

Now, the dimensional form for the velocity profile parallel to the flat bottom wall is, from (7–218),

$$u^{(0)} = \frac{1}{2\mu} \left(\frac{dp^{(0)}}{dx} \right) y^2 + c_1(x) y + c_2(x). \tag{7-253}$$

But in this case, the motion of the cylinder is strictly normal to the wall, and so

$$u^{(0)} = 0 \quad \text{at} \quad y = 0,$$

$$u^{(0)} = 0 \quad \text{at} \quad y = h. \tag{7-254}$$

Thus it follows that

$$u^{(0)} = \frac{1}{2\mu} \left(\frac{dp^{(0)}}{dx} \right) (y^2 - yh(x)), \tag{7-255}$$

which is simply the dimensional form of (7–219) with $U \equiv 0$. To determine the pressure gradient, $dp^{(0)}/dx$, we use the Reynolds equation (7–222) with $U \equiv 0$ and $V = V(t)$, so that

$$\frac{d}{dx} \left(h^3 \frac{dp^{(0)}}{dx} \right) = 12\mu V(t). \tag{7-256}$$

Thus, integrating once with respect to x, we find that

$$\frac{dp^{(0)}}{dx} = 12\mu V(t) \frac{x}{h^3(x)} + \frac{c_1}{h^3(x)}.$$

The integration constant c_1 can be determined using the condition, evident from the symmetry of the problem, that $p^{(0)}(x)$ is an even function of x and, thus, $dp^{(0)}/dx$ is odd. Since $1/h^3(x)$ is an even function of x, it is evident that $c_1 = 0$, and

$$\frac{dp^{(0)}}{dx} = 12\mu V(t) \frac{x}{h^3(x)}. \tag{7-257}$$

Integrating again with respect to x,

$$p^{(0)}(x) - p^{(0)}(-X) = 12\mu V(t) \int_{-X}^{x} \frac{s}{h^3(s)} \, ds, \tag{7-258}$$

where $-X$ symbolizes the left-hand "edge" of the lubrication layer. Although X has not been specified explicitly, we shall see eventually that it drops out of the limiting form for the solution as $\epsilon \to 0$. Substituting the approximate form for the gap width function $h(s)$ from (7–252) and then carrying out the integration yields

$$p^{(0)}(x) - p^{(0)}(-X) = -6\mu a V(t) \left[\frac{1}{\left(b + \frac{x^2}{2a} \right)^2} - \frac{1}{\left(b + \frac{X^2}{2a} \right)^2} \right]. \tag{7-259}$$

Now, given the pressure field, we can examine the vertical force exerted by the fluid on the cylinder. As usual in lubrication analyses, the force on the cylinder is dominated by the forces from within the lubrication layer, and in this region, pressure forces are $0(X/b) \gg 1$ larger than forces arising due to viscous stresses in the fluid. Thus, it follows that the vertical force per unit length exerted by the fluid on the cylinder is

$$\mathbf{e}_y \cdot \mathbf{F} = \mathbf{e}_y \cdot \int_1 -p^{(0)} \mathbf{n} dl = -\int_1 p^{(0)} (\mathbf{n} \cdot \mathbf{e}_y) dl, \tag{7-260}$$

where l represents the perimeter of the cylinder. Now,

$$\mathbf{n} = \frac{x}{a} \mathbf{e}_x - \sqrt{1-(x/a)^2} \mathbf{e}_y, \tag{7-261}$$

and thus

$$\mathbf{n} \cdot \mathbf{e}_y = -\sqrt{1-(x/a)^2}. \tag{7-262}$$

Consequently,

$$\mathbf{e}_y \cdot \mathbf{F} = \int_1 p^{(0)} \sqrt{1-(x/a)^2} dl. \tag{7-263}$$

Furthermore, the differential element of length dl along the surface of the cylinder in the thin gap can be expressed in terms of a differential element dx, in the form

$$dl = \sqrt{1+(dy/dx)^2} dx \quad \text{on} \quad y = h(x), \tag{7-264}$$

where

$$\frac{dy}{dx} = \frac{dh}{dx} = \frac{x}{\sqrt{a^2-x^2}}.$$

Thus,

$$dl = \frac{dx}{\sqrt{1-(x/a)^2}}, \tag{7-265}$$

and therefore

$$\mathbf{e}_y \cdot \mathbf{F} = \int_{-X}^{X} \left[p^{(0)}(x) - p^{(0)}(-X) \right] dx. \tag{7-266}$$

Since $p = p(-X)$ everywhere outside the lubrication layer, to a first approximation, it is the pressure difference $p^{(0)}(x) - p^{(0)}(-X)$ within the lubrication layer that is responsible for the net lift force on the cylinder.

Substituting for $p^{(0)}(x) - p^{(0)}(-X)$ for (7-259), the net hydrodynamic force on the cylinder can be evaluated explicitly as

$$\mathbf{e}_y \cdot \mathbf{F} = -\frac{\mu a V(t)}{b^2} \left\{ \frac{6X}{\left(1+\frac{X^2}{2ab}\right)} + 6\sqrt{2ab} \ \tan^{-1}\left(\frac{X}{\sqrt{2ab}}\right) - \frac{12X}{\left(1+\frac{X^2}{2ab}\right)^2} \right\} \tag{7-267}$$

But within the lubrication layer, $X^2/2ab \gg 1$, and therefore

$$\mathbf{e}_y \cdot \mathbf{F} \sim -\frac{\mu a V(t)}{b^2} \left\{ \frac{12ab}{X} + 3\pi\sqrt{2ab} - \frac{48a^2}{X}\left(\frac{b}{X}\right)^2 \right\}. \qquad (7\text{-}268)$$

Since $X = 0(a)$, the dominate term here is the one in the middle, and so

$$\boxed{\mathbf{e}_y \cdot \mathbf{F} \sim 3\sqrt{2}\pi\mu|V(t)|\left(\frac{a}{b}\right)^{3/2}.} \qquad (7\text{-}269)$$

This is the final asymptotic result for the hydrodynamic force on the cylinder, caused by its motion toward an infinite plane wall. It will be noted, however, that the right-hand side contains both $V(t)$ and b, which will also vary with time. Thus, though it appears that the force should increase dramatically as $a/b \to \infty$, this may not be true since $|V(t)|$ could decrease even more rapidly as $b/a \to 0$.

To determine exactly what will happen, we need to proceed a few more steps with the analysis. First, note that the cylinder is moving under the action of a *constant* applied force F. Thus, neglecting the inertia of the cylinder,

$$\boxed{F = -3\sqrt{2}\pi\mu V(t)\frac{a^{3/2}}{b^{3/2}},} \qquad (7\text{-}270)$$

and we see that $b = b(t)$ must change in such a way that

$$\frac{V(t)}{b^{3/2}} = \text{const.} \qquad (7\text{-}271)$$

Now, however,

$$V = \frac{db}{dt}, \qquad (7\text{-}272)$$

and therefore

$$\frac{1}{b^{3/2}}\frac{db}{dt} = -\frac{F}{3\sqrt{2}\pi\mu a^{3/2}} \qquad (7\text{-}273)$$

It follows that

$$\boxed{\frac{1}{\sqrt{b(t)}} - \frac{1}{\sqrt{b(0)}} = \frac{F}{6\sqrt{2}\pi\mu a^{3/2}}t,} \qquad (7\text{-}274)$$

where $b(0)$ represents the gap width at some initial moment $t = 0$. It can be seen that $b(t)$ approaches zero asymptotically as $t \to \infty$. We therefore conclude that *the cylinder will not contact the wall in a finite amount of time* under the action of any finite force.

The result (7-274) is an apparent disagreement with experimental observation, because it is generally agreed that a real cylinder will actually contact the wall in a finite time interval when acted upon by a finite force that is pushing it toward the wall. It should be remembered, however, that the lubrication analysis just presented assumes that the cylinder and the wall are *absolutely smooth*. In reality, neither surface will exactly satisfy this requirement. In normal circumstances, the characteristic length scales of a flow are sufficiently large that the small, microscale imperfections in a

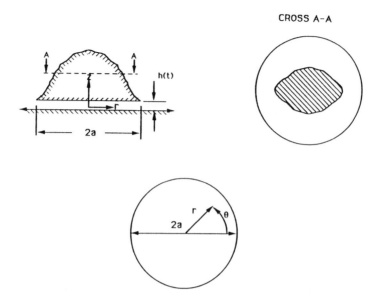

Figure 7–12 A sketch of a body with a flat, circular base moving toward a parallel plane wall when the inertia of the body and of the fluid in the gap are not negligible.

solid surface are imperceptible. In lubrication theory, however, we are concerned with the motion of the cylinder when it is very close to the wall, and in this case, such imperfections may become significant if they are not vanishingly small compared with the already small gap width. Presumably, any microscopic surface roughness will eventually become important as $b \to 0$, and this fact may account for the apparent difference between theory and experiment that was mentioned at the beginning of this paragraph.

D. Thin Films with Inertia

In all of the preceding sections that have dealt with flow in thin films, we have limited our considerations to circumstances where the Reynolds number based on gap height is vanishingly small so that all inertial effects are negligible. Here, we discuss an example of a thin-film problem when inertial effects may be important. In particular, we consider the case of a solid body with a *flat, circular lower surface* moving under the action of a constant applied force, such as gravity, toward a parallel plane wall when the inertia of the body and the inertia of the fluid in the gap are not negligible. The analysis presented here is based upon the original work of Weinbaum and co-workers (1985),[10] who showed that the problem for flow in the thin gap can be reduced to the solution of a coupled set of ordinary differential equations that are amenable to asymptotic analysis, even for large Reynolds numbers.

A schematic representation of the basic problem is given in Figure 7–12. We denote the radius of the circular lower surface as a, the gap height as $h(t)$, and the net force acting on the body when the system is at rest as W. This net force is comprised, in general, of a net gravitational contribution plus any additional external force f, namely,

$$\mathbf{W} = -\left((m - m_b)g + f\right)\mathbf{i}_z, \tag{7-275}$$

where m is the mass of the body, m_b is the mass of the fluid displaced by the body, and $\mathbf{i}_z$ is a unit vector in the direction perpendicular to the plane boundaries. The velocity of the body relative to the fixed plane boundary is denoted as $V(t)$. The initial value of h is denoted as h_o. In this case, we again consider the thin-film limit

$$\frac{h_o}{a} \ll 1 \tag{7-276}$$

but *without* assuming that the inertia of the body or of the fluid in the thin film is negligible; that is, the thin-film Reynolds number,

$$\mathrm{Re} \equiv \frac{\rho V_c h_o}{\mu}, \tag{7-277}$$

is *not necessarily small*, where V_c is an appropriate characteristic velocity, as discussed later.

In view of the axisymmetry of the geometry of the thin-film region, we assume that the velocity field generated in the gap by the motion of the object will be axisymmetric. Thus, with respect to cylindrical coordinates (r, θ, z), centered on the axis of symmetry, as shown in Figure 7–12, the velocity field will take the form

$$\mathbf{u} = \left(u_r(r,z,t), \, 0, \, u_z(r,z,t)\right), \tag{7-278}$$

and the pressure will also be axisymmetric, that is,

$$p = p(r,z,t). \tag{7-279}$$

Thus, the full, dimensional Navier-Stokes and continuity equations for the fluid in the thin film are

$$\frac{1}{r}\frac{\partial}{\partial r}(ru_r) + \frac{\partial u_z}{\partial z} = 0, \tag{7-280}$$

$$\rho\left[\frac{\partial u_r}{\partial t} + u_r\frac{\partial u_r}{\partial r} + u_z\frac{\partial u_r}{\partial z}\right] = -\frac{\partial p}{\partial r} + \mu\left[\frac{1}{r}\frac{\partial}{\partial r}\left(r\frac{\partial u_r}{\partial r}\right) + \frac{\partial^2 u_r}{\partial z^2} - \frac{u_r}{r^2}\right], \tag{7-281}$$

and

$$\rho\left[\frac{\partial u_z}{\partial t} + u_r\frac{\partial u_z}{\partial r} + u_z\frac{\partial u_z}{\partial z}\right] = -\frac{\partial p}{\partial z} + \mu\left[\frac{1}{r}\frac{\partial}{\partial r}\left(r\frac{\partial u_z}{\partial r}\right) + \frac{\partial^2 u_z}{\partial z^2}\right]. \tag{7-282}$$

The boundary conditions, also in dimensional terms, are

$$u_r = u_z = 0 \quad \text{at} \quad z = 0, \text{ all } r,$$

(7–283a)

$$u_r = 0, \; u_z = -V(t) \quad \text{at} \quad z = h(t) \text{ for } r \le a.$$

(7–283b)

As indicated earlier, we assume that the body moves under the action of a constant net force, $-W\mathbf{i}_z$. Hence, in addition to the equations and boundary conditions for the fluid, we must add an equation of motion for the body. In dimensional terms, this is

$$m\frac{d^2h}{dt^2} = -W + D(t),$$

(7–284)

where D is the hydrodynamic force on the body. It may be noted that

$$V(t) \equiv -\frac{dh}{dt}$$

(7–285)

so that h and V are not independent. For simplicity, we assume that both the body and fluid are initially motionless so that

$$h = h_o, \; u_r = u_z = 0 \quad \text{at} \quad t = 0.$$

(7–286)

It is important to remember that we have *not* assumed that the whole body or the external flow domain is necessarily axisymmetric. It is only the thin-film region that is assumed to be axisymmetric, in accord with the forms (7–278) and (7–279) for the velocity and pressure fields, and we should not expect equations (7–280)–(7–282) to apply anywhere else. In particular, the flow outside the thin film may be three-dimensional, and we would then require the full three-dimensional Navier-Stokes and continuity equations. Thus, in defining the problem in terms of the equations and boundary conditions (7–280)–(7–286), we anticipate that the full solution of the problem in the thin-film limit, $h_o/a \ll 1$, can still be expressed in the form of a singular or matched asymptotic expansion, with changes in the velocity and pressure fields *outside* the thin film characterized by the length scale, $L_c = a$, of the body, while the velocity and pressure fields *within* the thin film vary over a characteristic dimension, $l_c = h_o$, in the z direction. Based upon the earlier analysis of thin-film flows in the *usual lubrication limit*, $h_o/a \ll 1$ and $Re \ll 1$, we may anticipate that the leading-order approximation for the velocity and pressure fields in the thin film can be obtained independently of the form of the solution in the outer part of the domain. In particular, we shall see that the leading-order thin-film approximation does not require an explicit match with the leading-order outer solution. Furthermore, the first approximation to the hydrodynamic force on the body $D(t)$ is dominated by contributions from the thin-film region so that the equation of motion for the body, equation (7–284), can be applied to obtain h as a function of t using only the thin-film approximation for the fluid mechanics. We leave a formal proof of these claims to the reader, based upon the arguments from Section **B**, and proceed directly to a discussion of a first approximation of the equations and boundary conditions (7–280)–(7–286) for the thin-film region.

1. Governing Equations in the Thin-Film Region

To begin, we nondimensionalize the problem (7–280)–(7–286) using the usual approach for a thin-film geometry in which gradients with respect to z are assumed to be characterized by the length scale $l_c = h_o$, that is,

$$\frac{\partial}{\partial z} \equiv \frac{1}{h_o}\frac{\partial}{\partial \bar{z}}, \tag{7–287}$$

while gradients in the r direction are smaller by $O(h_o/a)$ and are thus nondimensionalized using the disk radius,

$$\frac{\partial}{\partial r} = \frac{1}{a}\frac{\partial}{\partial \bar{r}}. \tag{7–288}$$

The only slight complication in the present problem is that it is the force W that is specified rather than the velocity of the body, and it is not immediately obvious how to specify a characteristic velocity. In fact, we shall see that the most appropriate choice for a characteristic velocity depends upon the conditions of the problem. For the moment, we simply denote the characteristic velocity for motion in the z direction as V, that is,

$$u_z = V\bar{u}_z. \tag{7–289}$$

Thus, by means of the continuity equation,

$$u_r = V\left(\frac{a}{h_o}\right)\bar{u}_r \tag{7–290}$$

so that

$$\boxed{\frac{1}{\bar{r}}\frac{\partial}{\partial \bar{r}}(\bar{r}\bar{u}_r) + \frac{\partial \bar{u}_z}{\partial \bar{z}} = 0.} \tag{7–291}$$

The corresponding dimensionless forms for the Navier-Stokes equations require, in addition, a characteristic time scale t_c and a characteristic pressure scale that we denote, for the moment, as Π, that is,

$$t = t_c\bar{t} \quad \text{and} \quad p = \Pi\bar{p}. \tag{7–292}$$

Both t_c and Π will be specified later. Introducing (7–287)–(7–290) and (7–292) into (7–281) and (7–282), we obtain the dimensionless forms

$$\boxed{\begin{aligned}
\text{Re}&\left[\alpha\frac{\partial \bar{u}_r}{\partial \bar{t}} + \bar{u}_r\frac{\partial \bar{u}_r}{\partial \bar{r}} + \bar{u}_z\frac{\partial \bar{u}_r}{\partial \bar{z}}\right] = \\
&-\frac{\Pi h_o}{\mu V}\left(\frac{h_o}{a}\right)^2\frac{\partial \bar{p}}{\partial \bar{r}} + \left(\frac{h_o}{a}\right)^2\left(\frac{\partial}{\partial \bar{r}}\left(\frac{1}{\bar{r}}\frac{\partial}{\partial \bar{r}}(\bar{r}\bar{u}_r)\right)\right) + \frac{\partial^2 \bar{u}_r}{\partial \bar{z}^2},
\end{aligned}} \tag{7–293}$$

where

$$\text{Re} \equiv \frac{\rho V h_o}{\mu}, \quad \alpha \equiv \frac{h_o}{t_c}\frac{1}{V}, \tag{7-294}$$

and

$$\boxed{\begin{aligned} \text{Re}\left[\alpha\frac{\partial \bar{u}_z}{\partial \bar{t}} + \bar{u}_r\frac{\partial \bar{u}_z}{\partial \bar{r}} + \bar{u}_z\frac{\partial \bar{u}_z}{\partial \bar{z}}\right] = \\ -\frac{\Pi h_o}{\mu V}\frac{\partial \bar{p}}{\partial \bar{z}} + \frac{\partial^2 \bar{u}_z}{\partial \bar{z}^2} + \left(\frac{h_o}{a}\right)^2\frac{1}{\bar{r}}\frac{\partial}{\partial \bar{r}}\left(\bar{r}\frac{\partial \bar{u}_z}{\partial \bar{r}}\right). \end{aligned}} \tag{7-295}$$

Note that if $t_c = h_o/V$, which is the time for the body to move with characteristic velocity V over a distance equal to the initial gap width, then $\alpha \equiv 1$, and $\text{Re} \equiv \rho V h_o/\mu$ appears in all of the inertial/acceleration terms on the left-hand side of (7–293) and (7–295). This is the most obvious choice to t_c, but Weinbaum and co-workers (1985) show that interesting behavior also occurs on other time scales, and we temporarily leave α as an independent dimensionless parameter. The boundary and initial conditions, in dimensionless form, are

$$\boxed{\bar{u}_r = \bar{u}_z = 0 \quad \text{at} \quad \bar{z} = 0, \text{ all } \bar{r},} \tag{7-296a}$$

$$\boxed{\bar{u}_r = 0, \ \bar{u}_z = \alpha\left(\frac{d\bar{h}}{d\bar{t}}\right) \quad \text{at} \quad \bar{z} = \bar{h}(\bar{t}), \ \bar{r} \le 1,} \tag{7-296b}$$

and

$$\boxed{\bar{h} = 1, \ \bar{u}_r = \bar{u}_z = 0 \quad \text{at} \quad \bar{t} = 0.} \tag{7-297}$$

We temporarily defer nondimensionalization of the equation of motion, (7–294), of the body.

Now, we are concerned with the limiting case of a thin film. The governing equations and boundary conditions for the leading-order approximation in an asymptotic expansion for

$$\frac{h_o}{a} \ll 1$$

are (7–293) and (7–295)–(7–297), with the two terms of $0(h_o/a)^2$ on the right-hand side of (7–293) and (7–295) neglected. We assume that the dimensionless parameters Re and α are *independent* of (h_o/a). Furthermore, we retain the pressure gradient terms in both (7–293) and (7–295) since we have not yet specified a characteristic pressure Π.

Now, the key to solving these limiting equations for the thin-film region is to note, from the axial symmetry of the flow and the form of the boundary conditions at $\bar{z} = 0$ and $\bar{z} = \bar{h}(\bar{t})$, that a solution can be obtained in terms of a streamfunction of the form*

$$\boxed{\psi = \bar{r}^2 F(\bar{z}, \bar{t}).} \tag{7-298}$$

---

*Here, $\bar{u}_r \equiv (1/r)(\partial \psi/\partial z)$ and $\bar{u}_z \equiv -(1/r)(\partial \psi/\partial \bar{r})$.

This corresponds to the assumption that the z component of velocity at this leading level of approximation is independent of $\bar{r}$ and thus, by continuity, that $\bar{u}_r$ varies only linearly with $\bar{r}$, that is,

$$\bar{u}_r = \bar{r}\frac{\partial F}{\partial \bar{z}} \quad \text{and} \quad \bar{u}_z = -2F(\bar{z},\bar{t}). \tag{7-299}$$

A solution of this form not only satisfies the continuity equation exactly but is also consistent with the boundary conditions at $\bar{z}=0$ and $\bar{h}$ provided that

$$F = \frac{\partial F}{\partial \bar{z}} = 0 \quad \text{at} \quad \bar{z}=0, \tag{7-300a}$$

$$-2F = \alpha\frac{d\bar{h}}{d\bar{t}}, \quad \text{and} \quad \frac{\partial F}{\partial \bar{z}} = 0 \quad \text{at} \quad \bar{z}=\bar{h}. \tag{7-300b}$$

The initial conditions (7–297) are also satisfied provided that

$$\bar{h}=1, \; F\equiv 0 \quad \text{at} \quad \bar{t}=0. \tag{7-300c}$$

The proposed solution form cannot, however, be forced to satisfy any boundary conditions at $\bar{r}=1$ without overspecifying F. However, since the thin-film approximation neglects the second-derivatives with respect to $\bar{r}$ in (7–293) and (7–295), a solution at leading order in h_o/a no longer requires boundary conditions at both $\bar{r}=0$ and 1. The condition at $\bar{r}=0$ is that the solution remain bounded, and this condition is retained. However, at $\bar{r}=1$ we require only that the thin-film solution match with the solution in the region outside the thin film, and this matching condition does not influence the form of the leading-order solution in the thin-film region, as we have already seen.

To obtain the governing equation for $F(z,t)$, we substitute (7–299) into the nondimensionalized Navier-Stokes equations (7–293) and (7–295) with the two viscous terms of $0(h_o/a)^2$ neglected. This results in the pair of equations

$$\bar{r}\left[\text{Re}\left(\alpha\frac{\partial^2 F}{\partial \bar{z}\partial \bar{t}} + \left(\frac{\partial F}{\partial \bar{z}}\right)^2 - 2F\frac{\partial^2 F}{\partial \bar{z}^2}\right) - \frac{\partial^3 F}{\partial \bar{z}^3}\right] = -\frac{\Pi h_o}{\mu V}\left(\frac{h_o}{a}\right)^2\frac{\partial \bar{p}}{\partial \bar{r}} \tag{7-301}$$

and

$$\text{Re}\left(-2\alpha\frac{\partial F}{\partial \bar{t}} + 4F\frac{\partial F}{\partial \bar{z}}\right) + 2\frac{\partial^2 F}{\partial \bar{z}^2} = -\frac{\Pi h_o}{\mu V}\frac{\partial \bar{p}}{\partial \bar{z}}. \tag{7-302}$$

But equation (7–301) can be rewritten in shorthand notation as a differential equation for $\bar{p}$, namely,

$$\bar{r}A(\bar{z},\bar{t}) = \frac{\Pi h_o}{\mu V}\left(\frac{h_o}{a}\right)^2\frac{\partial \bar{p}}{\partial \bar{r}}. \tag{7-303}$$

Hence, integrating with respect to $\bar{r}$,

$$\left(\frac{\Pi h_o}{\mu V}\right)\left(\frac{h_o}{a}\right)^2\bar{p} = \frac{\bar{r}^2}{2}A(\bar{z},\bar{t}) + B(\bar{z},\bar{t}). \tag{7-304}$$

The function $B(\bar{z}, \bar{t})$ is evaluated by substituting (7–304) into (7–302). The result is

$$\text{Re}\left(-2\alpha\frac{\partial F}{\partial \bar{t}} + 4F\frac{\partial F}{\partial \bar{z}}\right) + 2\frac{\partial^2 F}{\partial \bar{z}^2} = \left(\frac{a}{h_o}\right)^2\left(-\frac{\bar{r}^2}{2}\frac{\partial A}{\partial \bar{z}} - \frac{\partial B}{\partial \bar{z}}\right).$$

Now, since the left-hand side of this equation is independent of $\bar{r}$, it follows that

$$A = A(\bar{t}) \quad \text{only}, \tag{7–305}$$

and hence

$$-\frac{\partial B}{\partial \bar{z}} = \left(\frac{h_o}{a}\right)^2\left[\text{Re}\left(-2\alpha\frac{\partial F}{\partial \bar{t}} + 4F\frac{\partial F}{\partial \bar{z}}\right) + 2\frac{\partial^2 F}{\partial \bar{z}^2}\right]. \tag{7–306}$$

If we now integrate from $\bar{z}$ to $\bar{h}$ and use the boundary conditions for F at $\bar{z} = \bar{h}$, it can be shown easily that

$$B(\bar{z}, \bar{t}) = B(\bar{h}, \bar{t}) + \left(\frac{h_o}{a}\right)^2.$$

$$\left[-2\text{Re}\int_{\bar{z}}^{\bar{h}}\alpha\frac{\partial F}{\partial \bar{t}}\,d\bar{z} + \left(\frac{\text{Re}}{2}\right)\left(\alpha\frac{d\bar{h}}{d\bar{t}}\right)^2 - 2\text{Re}F^2(\bar{z}, \bar{t}) - 2\frac{\partial F}{\partial \bar{z}}(\bar{z}, \bar{t})\right]. \tag{7–307}$$

Thus, finally,

$$\left(\frac{\Pi h_o}{\mu V}\right)\left(\frac{h_o}{a}\right)^2\bar{p} = \frac{\bar{r}^2}{2}A(\bar{t}) + B(\bar{h}, \bar{t}) + 0\left(\frac{h_o}{a}\right)^2. \tag{7–308}$$

Since the right-hand side is $0(1)$, it follows that

$$\Pi = \frac{\mu V}{h_o}\left(\frac{a}{h_o}\right)^2 \tag{7–309}$$

so that

$$\boxed{\bar{p} = \frac{\bar{r}^2}{2}A(\bar{t}) + B(\bar{h}, \bar{t}) + 0\left(\frac{h_o}{a}\right)^2.} \tag{7–310}$$

It may be noted that if $\Pi = \mu V/h_o$, then $\bar{p}$ would be $0(a/h_o)^2$, and this is not self-consistent [that is, $\bar{p}$ is not $0(1)$ as it should be if we have made the correct choice for Π].

Now, $\bar{p}$ is determinate up to an arbitrary constant only. Hence, if we adopt the pressure at $\bar{r} = 1$ and $\bar{z} = \bar{h}$ as a reference pressure, then

$$\bar{p}(1, \bar{h}, \bar{t}) \equiv 0, \tag{7–311}$$

and it follows that

$$B(\bar{h}, \bar{t}) = -\frac{1}{2}A(\bar{t}).$$

Thus, finally,

$$\boxed{\bar{p}(\bar{r}, \bar{z}, \bar{t}) = \left(\frac{\bar{r}^2}{2} - \frac{1}{2}\right)A(\bar{t}) + 0\left(\frac{h_o}{a}\right)^2.} \tag{7–312}$$

With this result for $\bar{p}$, the equation and boundary conditions that remain for F are

$$\text{Re}\left[\alpha\frac{\partial^2 F}{\partial \bar{z}\partial \bar{t}}+\left(\frac{\partial F}{\partial \bar{z}}\right)^2-2F\frac{\partial^2 F}{\partial \bar{z}^2}\right]-\frac{\partial^3 F}{\partial \bar{z}^3}=-A(\bar{t}),\qquad(7\text{-}313)$$

with

$$F=\frac{\partial F}{\partial \bar{z}}=0\quad\text{at}\quad \bar{z}=0,\qquad(7\text{-}314a)$$

$$F=-\frac{1}{2}\alpha\frac{d\bar{h}}{d\bar{t}},\ \frac{\partial F}{\partial \bar{z}}=0\quad\text{at}\quad \bar{z}=\bar{h},\qquad(7\text{-}314b)$$

and

$$\bar{h}=1,\ F=0\quad\text{at}\quad \bar{t}=0.\qquad(7\text{-}314c)$$

To complete the problem formulation, we need to obtain an equation for $A(\bar{t})$. For this, we must now return to the equation of motion (7-284) for the body, that is,

$$\frac{mh_o}{t_c^2}\frac{d^2\bar{h}}{d\bar{t}^2}=-W+D(t).$$

We saw previously, when lubrication ideas were discussed, that the dominant contribution to the force on the body was due to pressure forces in the thin film. In particular, in this case

$$D(t)\approx-\int_{\text{disk}}T_{zz}|_{z=h}dS,\qquad(7\text{-}315)$$

where

$$T_{zz}=\frac{\mu V}{h_o}\left(\frac{a}{h_o}\right)^2\left[-\bar{p}+0\left(\frac{h_o}{a}\right)^2\right].\qquad(7\text{-}316)$$

Hence,

$$D(t)=\frac{\mu V}{h_o}\left(\frac{a}{h_o}\right)^2 2\pi a^2\left[\int_0^1\bar{p}(\bar{r},\bar{h},\bar{t})\bar{r}d\bar{r}\right].\qquad(7\text{-}317)$$

Now, since

$$\bar{p}(\bar{r},\bar{h},\bar{t})=\left(\frac{\bar{r}^2}{2}-\frac{1}{2}\right)A(\bar{t})+0\left(\frac{h_o}{a}\right)^2,$$

it follows that

$$D(t)=-\frac{\mu V}{h_o}\left(\frac{a}{h_o}\right)^2 2\pi a^2\frac{A(\bar{t})}{8},\qquad(7\text{-}318)$$

and (7-284) can be rewritten as an equation for $A(\bar{t})$ as a function of $\bar{h}$, namely,

$$A(\bar{t})=-\frac{4h_o^3}{\pi\mu Va^4}W-\frac{4mh_o^4}{\pi\mu Va^4}\frac{1}{t_c^2}\frac{d^2\bar{h}}{d\bar{t}^2},\qquad(7\text{-}319)$$

where W is the net body force on the body

$$W = (m - m_b)g + f$$

according to (7–285).

Thus, finally, (7–313) can be written in the alternative form

$$\alpha \frac{\partial^2 F}{\partial \bar{z} \partial \bar{t}} + \left(\frac{\partial F}{\partial \bar{z}}\right)^2 - 2F \frac{\partial^2 F}{\partial \bar{z}^2} - \frac{1}{\text{Re}} \frac{\partial^3 F}{\partial \bar{z}^3} = \frac{4h_o^3 m}{\pi \rho a^4 V^2 t_c^2} \frac{d^2 \bar{h}}{d\bar{t}^2} + \frac{4h_o^2 W}{\pi \rho a^4 V^2}. \tag{7-320}$$

Together with the first of the conditions, (7–314b),

$$F(\bar{z}, \bar{t}) = -\frac{\alpha}{2} \frac{d\bar{h}}{d\bar{t}} \quad \text{at} \quad \bar{z} = \bar{h}(\bar{t}). \tag{7-321}$$

We now have *two equations* relating the two unknown functions $F(\bar{z}, \bar{t})$ and $\bar{h}(\bar{t})$, together with the boundary and initial conditions (7–314). The equations and boundary conditions (7–314), (7–320), and (7–321) contain four dimensionless groups,

$$\alpha, \text{ Re}, \frac{4h_o^3 m}{\pi \rho a^4 V^2 t_c^2}, \text{ and } \frac{4h_o^2 W}{\pi \rho a^4 V^2}.$$

However, the characteristic time and velocity scales remain to be specified, and this will reduce the dimensionless parameters from four to two. The choice of a characteristic time scale depends on the aspect of the problem that we wish to consider. Here, we initially examine the evolution of the flow over the period of significant changes in the gap width $h(\bar{t})$. Then the natural characteristic time scale to choose is

$$t_c = \frac{h_o}{V}, \tag{7-322}$$

independent of whether the Reynolds number Re is large or small. In this case,

$$\alpha = 1, \tag{7-323}$$

and the coefficient of $\bar{h}_{\bar{t}\bar{t}}$ in (7–320) becomes

$$\beta \equiv \frac{4}{\pi}\left(\frac{h_o m}{\rho a^4}\right). $$

It remains only to determine an appropriate estimate for the characteristic velocity V, and this turns out to depend on the magnitude of the Reynolds number Re.

Of course, the ordinary differential equation (7–320) for F is *nonlinear*, and an exact solution is *not* possible for arbitrary values of the dimensionless parameters. However, the reduction of the full problem to the simpler form (7–320), which is the leading-order approximation for small h_o/a, does allow approximate solutions to be obtained for the two limiting cases of *weak* inertia, Re $\ll$ 1 (where the problem should reduce to the usual lubrication form at leading order), and strong inertia, Re $\gg$ 1.

When Re $\ll$ 1, we assume that inertia of both the fluid in the thin gap and of the body can be neglected to a first approximation (m finite). Hence, the velocity scale is determined by a balance between viscous pressure forces and the net body force. Now, $p = 0(\mu V a^2 / h_o^3)$ according to (7–309), and thus the force $D = 0(\mu V a^4 / h_o^3)$, which must balance W. This implies that

$$V = 0\left(\frac{Wh_o^3}{\mu a^4}\right).$$

For convenience, we choose

$$V = \frac{4}{\pi} \frac{Wh_o^3}{\mu a^4}, \tag{7–324}$$

and, thus,

$$t_c = \frac{h_o}{V} = \frac{\pi}{4} \frac{\mu a^4}{Wh_o^2}. \tag{7–325}$$

In this case, the governing equations and boundary conditions become

$$\frac{\partial^2 F}{\partial \bar{z} \partial \bar{t}} + \left(\frac{\partial F}{\partial \bar{z}}\right)^2 - 2F \frac{\partial^2 F}{\partial \bar{z}^2} - \frac{1}{\text{Re}} \frac{\partial^3 F}{\partial \bar{z}^3} = \frac{1}{\text{Re}} + \beta \frac{d^2 \bar{h}}{d\bar{t}^2} \tag{7–326}$$

and

$$F(\bar{z}, \bar{t}) = -\frac{1}{2} \frac{d\bar{h}}{d\bar{t}} \quad \text{at} \quad \bar{z} = \bar{h}, \tag{7–327}$$

with

$$F = \frac{\partial F}{\partial \bar{z}} = 0 \quad \text{at} \quad \bar{z} = 0, \tag{7–328a}$$

$$\frac{\partial F}{\partial \bar{z}} = 0 \quad \text{at} \quad \bar{z} = \bar{h}, \tag{7–328b}$$

and

$$F = 0, \quad \bar{h} = 1 \quad \text{at} \quad \bar{t} = 0. \tag{7–328c}$$

In this case, the leading-order thin-film approximation is specified completely in terms of the two dimensionless parameters β and Reynolds number,

$$\text{Re} = \frac{4\rho W h_o^4}{\pi \mu^2 a^4} \ll 1. \tag{7–329}$$

When $\text{Re} \gg 1$, on the other hand, inertial effects are important, and this suggests that the dominant balance in (7–320) should be between the body force term involving W and the inertial terms on the left-hand side. In this case,

$$\frac{h_o^2 W}{\rho a^4 V^2} = 0(1)$$

so that

$$V = 0\left(\frac{h_o}{a^2} \frac{W^{1/2}}{\rho^{1/2}}\right).$$

For convenience, we choose

$$V = \frac{2h_o}{a^2} \sqrt{\frac{W}{\pi\rho}}, \tag{7-330}$$

and thus

$$t_c = \frac{h_o}{V} = \frac{a^2}{2} \sqrt{\frac{\pi\rho}{W}}. \tag{7-331}$$

Alternatively, we can note that the characteristic magnitude of the pressure in an inertia-dominated flow is $p = 0\left[\rho(a^2/h_o^2)V^2\right]$, so that $F = 0\left[\rho(a^4/h_o^2)V^2\right]$, and if this is to be of the same magnitude as the net body force W, we again require (7-330). In this case, the governing Equation (7-320) takes the form

$$\frac{\partial^2 F}{\partial\bar{z}\partial\bar{t}} + \left(\frac{\partial F}{\partial\bar{z}}\right)^2 - 2F\frac{\partial^2 F}{\partial\bar{z}^2} - \frac{1}{\text{Re}}\frac{\partial^3 F}{\partial\bar{z}^3} = \beta\frac{d^2\bar{h}}{d\bar{t}^2} + 1, \tag{7-332}$$

while the other equations and boundary conditions remain as in (7-327) and (7-328). The Reynolds number Re is now

$$\text{Re} \equiv \frac{2h_o^2}{\mu a^2} \sqrt{\frac{W\rho}{\pi}} \gg 1.$$

2. Asymptotic Results for Re ≪ 1

Let us first consider the case of *small, but nonzero,* inertial effects in which the governing equations and boundary conditions are (7-326)–(7-328). The limit Re → 0 is just the normal *lubrication* problem for parallel surfaces. For small, but nonzero, values of Re, on the other hand, we seek a solution in the form of a regular perturbation expansion in Re, namely,

and

$$F = F_0(\bar{z},\bar{t}) + \text{Re}F_1(\bar{z},\bar{t}) + \cdots$$
$$\bar{h} = \bar{h}_0(\bar{t}) + \text{Re}\bar{h}_1(\bar{t}) + \cdots. \tag{7-333}$$

In this section, we explicitly calculate only the first two terms in these expansions. This gives the lubrication limit, plus the first inertial correction to that limit.

The governing equations at each order in (7-333) are obtained by substituting these expansions into the governing equations and boundary/initial conditions and collecting terms of equal order in Re as before. The governing equations for the first terms, on the other hand, can be obtained by simply taking the limit Re → 0 in these same equations and boundary conditions. The result of this process for values of $\beta = 0(1)$ is

$$\frac{\partial^3 F_0}{\partial\bar{z}^3} + 1 = 0 \tag{7-334}$$

and

$$F_0 = -\frac{1}{2}\frac{d\bar{h}_0}{d\bar{t}} \quad \text{at} \quad \bar{z} = \bar{h}_0, \tag{7-335}$$

with

$$F_0 = \frac{\partial F_0}{\partial \bar{z}} = 0 \quad \text{at} \quad \bar{z} = 0, \tag{7-336a}$$

$$\frac{\partial F_0}{\partial \bar{z}} = 0 \quad \text{at} \quad \bar{z} = \bar{h}_0, \tag{7-336b}$$

and

$$F_0 = 0, \, \bar{h}_0 = 1 \quad \text{at} \quad \bar{t} = 0. \tag{7-336c}$$

Examination of the problem posed by (7-334)-(7-336) immediately indicates that a uniformly valid asymptotic solution of the form (7-333) will not be possible. Since the governing equation for F_0 neglects the derivatives of F and $\bar{h}$ with respect to time, the solution (7-334) and (7-335) cannot be expected to satisfy the initial conditions at $\bar{t} = 0$.

Indeed, this is the case. The most general solution of (7-334) for F_0 is

$$F_0 = -\frac{1}{6}\bar{z}^3 + C_1\bar{z}^2 + C_2\bar{z} + C_3, \tag{7-337}$$

and the coefficients in this solution are completely determined from the boundary conditions (7-336a) and (7-336b):

$$C_2 = C_3 = 0, \tag{7-338}$$

$$C_1 = \frac{\bar{h}_0}{4}.$$

Thus,

$$F_0 = \frac{1}{4}\left[\bar{h}_0\bar{z}^2 - \frac{2}{3}\bar{z}^3\right], \tag{7-339}$$

and it is immediately evident that this solution does *not* satisfy the initial condition $F_0 = 0$ at $\bar{t} = 0$. The corresponding solution for $\bar{h}_0$ is obtained from (7-335), which now takes the form

$$\frac{d\bar{h}_0}{d\bar{t}} = -\frac{\bar{h}_0^3}{6}. \tag{7-340}$$

Thus, integrating,

$$\bar{h}_0 = \left(\frac{3}{\bar{t} + C_0}\right)^{1/2}, \tag{7-341}$$

where C_0 is a constant of integration. Although this solution can be made to satisfy the condition $\bar{h}_0 = 1$ at $\bar{t} = 0$ by choosing

$$C_0 = 3, \tag{7-342}$$

we have already seen that the solution for F_0 is not valid at (or near) $\bar{t} = 0$, and we should not expect necessarily to obtain the correct value for C_0 by requiring $\bar{h}_0$ (which is derived from F_0) to satisfy a condition at $\bar{t} = 0$.

Instead, we may recognize the solution of (7–326)–(7–328) for Re $\ll$ 1 as a singular perturbation problem, which differs from earlier examples primarily in the fact that the expansion (7–333), in conjunction with the governing equations (7–326)–(7–328), is *not uniformly valid in time*, rather than spatial position. The problem is that the choice of characteristic time scale inherent in the nondimensionalization of (7–326) leads to neglect of the local acceleration terms in the limit Re $\rightarrow$ 0. At very short times, the initial transient phase in the development of the flow is dominated by diffusion of vorticity from the walls into the thin-film region, in a manner that is highly analogous to the transient start-up of pressure-driven unidirectional flow, as discussed in Chapter 3. However, this process is characterized by the viscous diffusion time scale

$$t_{\text{diff}} = \frac{h_o^2}{\nu},\tag{7–343}$$

which is very much smaller than the time scale h_o/V that was used to nondimensionalize Equations (7–326)–(7–328). Indeed, the ratio of this diffusion time scale to the original characteristic scale h_o/V is just the Reynolds number Re that we have assumed to be asymptotically small.

In deriving the approximate equations (7–334) and (7–335) from (7–326), we have assumed that the characteristic time scale used for nondimensionalization is correct so that the relative magnitudes of the terms depend only on Re. Thus, in the limit Re $\rightarrow$ 0, the time-derivative term involving F in (7–326) was assumed to be negligible compared with the terms retained in (7–334). However, this is *not* correct at short times where the flow evolves on the much shorter time scale t_{diff}. In fact, time derivatives with respect to $\bar{t}$ in this regime actually depend on Re, and it is not true that the time derivative terms in (7–326) are necessarily small in the limit Re $\rightarrow$ 0.

Thus, again we see the classic signature of a singular perturbation, in which the solution of a problem is characterized by different characteristic scales in different parts of the solution domain. In the present case, we may anticipate that the asymptotic solution for Re $\ll$ 1 will take the following general form. For short times, which we shall refer to as the *inner* region, the governing Equations (7–326) and (7–327) must be nondimensionalized using t_{diff} as a characteristic time scale, and we should expect to obtain an asymptotic expansion that will satisfy the initial conditions at $t = 0$ but break down for time scales that greatly exceed t_{diff}. On the other hand, for larger times, which we shall refer to as the *outer* region, we expect the characteristic time scale to be h_o/V, as was assumed in deriving (7–326)–(7–328); the asymptotic expansion (7–333) is thus valid in this part of the time domain, with F_0 and h_0 given by (7–339) and (7–341). The critical difference is that we now recognize that these solutions do not apply for very small times, and we *cannot* apply the original initial conditions at $t = 0$. Instead, we require that the solutions for the inner and outer regions of the time domain *match*, and we shall see that this provides a (correct) basis to obtain C_0 (and the similar constants for higher-order terms of the expansion).

The Short-Time Solution ($\hat{t} = 0(1)$)—To obtain solutions for Re $\ll$ 1 in the inner (short time) regime via the asymptotic approach outlined previously, the governing equations must be nondimensionalized using t_{diff} instead of the time scale h_o/V. If we notice that

$$t_{\text{diff}} = t_c \cdot \text{Re}, \tag{7-344}$$

then the governing equation appropriate for small Re and short times can be obtained directly from (7-326) by transforming from $\bar{t}$ to $\hat{t}$ according to

$$\text{Re}\hat{t} = \bar{t}. \tag{7-345}$$

This change of variables, often called *rescaling*, is equivalent to rederiving the governing equation with time nondimensionalized using t_{diff} as a characteristic time scale, rather than t_c. The result of rescaling (7-326) according to (7-345) is

$$\frac{\partial^2 F}{\partial \bar{z} \partial \hat{t}} - \frac{\partial^3 F}{\partial \bar{z}^3} - 1 = \text{Re}\left[-\left(\frac{\partial F}{\partial \bar{z}}\right)^2 + 2F\frac{\partial^2 F}{\partial \bar{z}^2} \right] + \frac{\beta}{\text{Re}}\frac{d^2 h}{d\hat{t}^2}. \tag{7-346}$$

We seek to solve (7-346) for small $\text{Re} \ll 1$ subject to the boundary and initial conditions

$$F = 0, \ \bar{h} = 1 \quad \text{at} \quad t = 0, \tag{7-347a}$$

$$F = \frac{\partial F}{\partial \bar{z}} = 0 \quad \text{at} \quad \bar{z} = 0, \tag{7-347b}$$

$$\frac{\partial F}{\partial \bar{z}} = 0 \quad \text{at} \quad \bar{z} = \bar{h}, \tag{7-347c}$$

and

$$F = -\frac{1}{2}\frac{1}{\text{Re}}\frac{d\bar{h}}{d\hat{t}} \quad \text{at} \quad \bar{z} = \bar{h}. \tag{7-347d}$$

For this purpose, we assume that an asymptotic expansion for F exists in the form

$$F = F_0 + \text{Re}F_1 + \cdots, \tag{7-348}$$

and it thus follows from (7-347d) that

$$\bar{h} = 1 + \text{Re}\bar{h}_1 + \cdots. \tag{7-349}$$

The governing equations and boundary conditions for the leading-order terms F_0 and h_1 are then

$$\frac{\partial^2 F_0}{\partial \bar{z} \partial \hat{t}} - \frac{\partial^3 F_0}{\partial \bar{z}^3} = 1 + \beta\frac{d^2\bar{h}_1}{d\hat{t}^2}, \tag{7-350}$$

with

$$F_0 = \frac{\partial F_0}{\partial \bar{z}} = 0 \quad \text{at} \quad \bar{z} = 0, \tag{7-351a}$$

$$\frac{\partial F_0}{\partial \bar{z}} = 0, \ F_0 = -\frac{1}{2}\frac{d\bar{h}_1}{d\hat{t}} \quad \text{at} \quad \bar{z} = 1, \tag{7-351b}$$

and

$$F_0 = 0, \ \bar{h}_1 = 0 \quad \text{at} \quad \hat{t} = 0. \tag{7-351c}$$

It is worth noting that a solution to (7–350) and (7–351) will yield the $0(1)$ contribution to F but the $0(\text{Re})$ contribution to $\bar{h}$ for this inner region [i.e., for $\hat{\imath} = 0(1)$].

The particular solution for F_0 can be seen, by inspecting Equation (7–350), to be

$$F_0^{\text{particular}} = -\frac{1}{6}\bar{z}^3 + \beta\frac{d\bar{h}_1}{d\hat{\imath}}\bar{z}.$$

Hence, a general solution of (7–350) can be obtained via separation of variables,

$$F_0 = -\frac{1}{6}\bar{z}^3 + \beta\frac{d\bar{h}_1}{d\hat{\imath}}\bar{z} + \sum_\lambda\left[A_\lambda\sin\lambda z + B_\lambda\cos\lambda z + C_\lambda\right]e^{-\lambda^2\hat{\imath}} + d_1\frac{\bar{z}^2}{2} + d_2\bar{z} + d_3.$$

$$(7\text{--}352)$$

The constants A_λ, B_λ, C_λ, d_1, d_2, and d_3, plus the eigenvalue λ, are obtained by applying the boundary and initial conditions (7–351).

Specifically, the boundary conditions (7–351a) at $\bar{z}=0$ and (7–351b) at $\bar{z}=1$ yield the four simultaneous equations

$$0 = \sum_\lambda\left[B_\lambda + C_\lambda\right]e^{-\lambda^2\hat{\imath}} + d_3, \qquad (7\text{--}353\text{a})$$

$$0 = \beta\frac{d\bar{h}_1}{d\hat{\imath}} + \sum_\lambda\lambda A_\lambda e^{-\lambda^2\hat{\imath}} + d_2, \qquad (7\text{--}353\text{b})$$

$$0 = -\frac{1}{2} + \beta\frac{d\bar{h}_1}{d\hat{\imath}} + \sum_\lambda\left[\lambda A_\lambda\cos\lambda - \lambda B_\lambda\sin\lambda\right]e^{-\lambda^2\hat{\imath}} + d_1 + d_2, \qquad (7\text{--}353\text{c})$$

and

$$-\frac{1}{2}\frac{d\bar{h}_1}{d\hat{\imath}} = -\frac{1}{6} + \beta\frac{d\bar{h}_1}{d\hat{\imath}} + \sum_\lambda\left[A_\lambda\sin\lambda + B_\lambda\cos\lambda + C_\lambda\right]e^{-\lambda^2\hat{\imath}} + \frac{d_1}{2} + d_2 + d_3,$$

$$(7\text{--}353\text{d})$$

which are required to hold for all times $\hat{\imath}$. Equation (7–353a) yields

$$\boxed{d_3 = 0, \ B_\lambda = -C_\lambda.} \qquad (7\text{--}354)$$

Then, combining (7–353b) and (7–353c), we find that

$$\boxed{d_1 = \frac{1}{2}, \ \lambda A_\lambda(1-\cos\lambda) = -\lambda B_\lambda\sin\lambda.} \qquad (7\text{--}355)$$

Finally, comparing the condition (7–353d) in the form

$$\beta\frac{d\bar{h}_1}{d\hat{\imath}} = \frac{\beta}{\beta+\dfrac{1}{2}}\left[-d_2 - \frac{1}{12} - \sum_\lambda\left\{A_\lambda\sin\lambda + B_\lambda(\cos\lambda - 1)\right\}e^{-\lambda^2\hat{\imath}}\right]$$

with (7–353b), or (7–353c), we see that

$$\frac{\beta}{\beta + \frac{1}{2}}\left(-d_2 - \frac{1}{12}\right) = -d_2 \tag{7–356a}$$

and

$$\lambda A_\lambda = \frac{\beta}{\beta + \frac{1}{2}}\left(A_\lambda \sin\lambda + B_\lambda(\cos\lambda - 1)\right). \tag{7–356b}$$

The equation (7–356a) can be solved for d_2:

$$d_2 = \frac{\beta}{6}. \tag{7–357}$$

The equation (7–356b) can be combined with (7–355) to obtain

$$A_\lambda\left(\beta + \frac{1}{2}\right)\lambda\sin\lambda = A_\lambda 2\beta(1 - \cos\lambda).$$

Since we seek a nontrivial solution (that is, $A_\lambda \neq 0$), this gives the eigenvalue condition for λ

$$\left(\beta + \frac{1}{2}\right)\lambda\sin\lambda = 2\beta(1 - \cos\lambda). \tag{7–358}$$

Clearly, the eigenvalues λ are functions of β. For $\beta = 0$ (that is, the inertia of the body is neglected),

$$\lambda_n = n\pi, \quad n = 0, 1, 2 \ldots . \tag{7–359}$$

For $\beta \neq 0$, we must solve the transcendental equation (7–358) numerically for each value of β. We denote the discrete set of eigenvalues as λ_n. Numerical values for $\lambda_1 \to \lambda_6$ are tabulated for several values of β in Table 7–1.

Table 7–1 Eigenvalues Versus β

$\lambda_\mathbf{n}$	$(2n-1)\pi$	$\beta = 1/10$	$\beta = 1/4$	$\beta = 1/2$	$\beta = 3/4$	$\beta = 1$	$\beta = 10$
λ_1	3.1416	2.9138	2.6484	2.3312	2.1056	1.9348	0.7502
λ_2	9.4248	9.3534	9.2814	9.2084	9.1644	9.1348	9.0080
λ_3	15.7080	15.6654	15.6226	15.5798	15.5540	15.5368	15.4628
λ_4	21.9911	21.9608	21.9304	21.8998	21.8816	21.8694	21.8170
λ_5	28.2743	28.2508	28.2270	28.2034	28.1892	28.1798	28.1390
λ_6	34.5575	34.5382	34.5190	34.4996	34.4878	34.4802	34.4470

Combining all of the results above, the function F_0 can now be expressed in the form

$$F_0 = -\frac{\bar{z}^3}{6} + \frac{\bar{z}^2}{4} + \frac{\beta\bar{z}}{6} + \beta\frac{dh_1}{d\hat{t}}\bar{z} + \sum_n A_n \left[\sin\lambda_n z + \frac{(\cos\lambda_n - 1)}{\sin\lambda_n}(\cos\lambda_n z - 1) \right] e^{-\lambda_n^2 \hat{t}}.$$

(7–360)

All that remains is to apply the initial condition on F_0 to determine A_n. This requires

$$\sum_n A_n \left[\sin\lambda_n z + \frac{(\cos\lambda_n - 1)}{\sin\lambda_n}(\cos\lambda_n z - 1) \right] = \frac{\bar{z}^3}{6} - \frac{\bar{z}^2}{4} - \frac{\beta\bar{z}}{6},$$

(7–361)

where we have taken account of the fact that $dh_1/d\hat{t} = 0$ at $\hat{t} = 0$, according to (7–351b) and (7–351c). The key to evaluating A_n is the so-called orthogonality property for the eigensolutions F_0. Let us define the function

$$U_n(z) \equiv \sin\lambda_n \sin\lambda_n z + (\cos\lambda_n - 1)\cos\lambda_n z.$$

Then the orthogonality condition can be expressed in the form

$$\int_0^1 U_m(z)\left[U_n(z) + 1 - \cos\lambda_n\right] dz = \begin{cases} 0 & \text{if } n \neq m \\ (1 - \cos\lambda_n)\left(1 - \frac{\sin\lambda_n}{\lambda_n}\right) & \text{if } n = m. \end{cases}$$

(7–362)

It follows from (7–361) and the orthogonality property (7–362) that

$$A_n \equiv \frac{-\sin\lambda_n}{\left(1 - \frac{\sin\lambda_n}{\lambda_n}\right)\left(\beta + \frac{1}{2}\right)\lambda_n^4}.$$

(7–363)

Hence, the complete solution for F_0 is

$$F_0 = -\frac{\bar{z}^3}{6} + \frac{\bar{z}^2}{4} + \frac{\beta\bar{z}}{6} + \beta\frac{dh_1}{d\hat{t}}\bar{z} + \sum_{n=1}^{\infty} \frac{-1}{\left(1 - \frac{\sin\lambda_n}{\lambda_n}\right)\left(\beta + \frac{1}{2}\right)\lambda_n^4}$$

$$\left[\sin\lambda_n \sin\lambda_n z + (\cos\lambda_n - 1)(\cos\lambda_n z - 1)\right] e^{-\lambda_n^2 \hat{t}}.$$

(7–364)

For the limiting case $\beta = 0$, this becomes

$$F_0 = -\frac{\bar{z}^3}{6} + \frac{\bar{z}^2}{4} - \sum_{n=1}^{\infty} \frac{2}{n^4 \pi^4}\left\{1 - (-1)^n\right\}(1 - \cos n\pi z) e^{-n^2\pi^2 \hat{t}}.$$

(7–365)

The final step in completing the solution for the inner (short-time) region is to determine $h_1(\hat{t})$. This can be done by integrating any of the conditions (7–353b, c, or d). Taking account of the values of the coefficients obtained above, (7–353d) can be expressed in the form

$$\left(\beta + \frac{1}{2}\right)\frac{d\bar{h}_1}{d\hat{t}} = -\frac{1}{6}\left(\beta + \frac{1}{2}\right) + \sum_n \frac{2(1 - \cos\lambda_n)}{\left(1 - \frac{\sin\lambda_n}{\lambda_n}\right)\left(\beta + \frac{1}{2}\right)\lambda_n^4}e^{-\lambda_n^2\hat{t}}. \quad (7\text{-}366)$$

Thus,

$$\bar{h}_1 + C = -\frac{\hat{t}}{6} - \sum_n \frac{2(1 - \cos\lambda_n)}{\left(1 - \frac{\sin\lambda_n}{\lambda_n}\right)\left(\beta + \frac{1}{2}\right)^2\lambda_n^6}e^{-\lambda_n^2\hat{t}},$$

and applying the initial condition $h_1 = 0$ at $\hat{t} = 0$, this becomes

$$\bar{h}_1 = -\frac{\hat{t}}{6} + \sum_{n=1}^{\infty} \frac{2(1 - \cos\lambda_n)}{\left(1 - \frac{\sin\lambda_n}{\lambda_n}\right)\left(\beta + \frac{1}{2}\right)^2\lambda_n^6}\left(1 - e^{-\lambda_n^2\hat{t}}\right). \quad (7\text{-}367)$$

Again, for $\beta = 0$ where $\lambda_n = n\pi$, this becomes

$$\bar{h}_1 = -\frac{\hat{t}}{6} + \sum_{n=1}^{\infty} \frac{8(1 - \cos n\pi)}{n^6\pi^6}\left(1 - e^{-n^2\pi^2\hat{t}}\right). \quad (7\text{-}368)$$

This completes our formal solution for F_0 and $\bar{h}_1$ in the inner (short-time) region. To the level of approximation considered here, we note that this solution is completely determined from the initial conditions, the boundary conditions at $\bar{z} = 0$, 1, and the governing equation (7-350) without any need to refer to the solution in the outer (long-time) region $\bar{t} = 0(1)$. It is also interesting that the leading order problem in the inner regime is sufficient to determine the leading *inertial* contributions at $0(\text{Re})$ to the motion of the body (that is, the time-dependence of the gap width $\bar{h}$) for $\hat{t} = 0(1)$. We shall return shortly to discuss the physics of the influence of inertia as predicted by equation (7-367) or (7-368). First, however, it is advantageous to complete our picture of inertia effects by calculating the $0(\text{Re})$ contributions in the outer (long-time) regime where $\bar{t} = 0(1)$.

The Long-Time-Solution ($\bar{t} = 0(1)$)—To this point, we have only obtained the first, $0(1)$, contributions to F and $\bar{h}$, and even this approximation retains one constant, C_0, that has not yet been determined, except on the ad hoc basis of setting $\bar{h}_0 = 1$ at $\bar{t} = 0$. Here, we first demonstrate that the leading-order terms, (7-339) and (7-341), actually match with the short-time approximation (7-364) and (7-367) at $0(1)$, and then we consider the second, $0(\text{Re})$, contributions to F and $\bar{h}$ in the outer (long-time) part of the solution domain.

Let us begin by considering the matching at $0(1)$. To do this, we compare the solutions (7-364) and (7-367) for $\hat{t} \gg 1$ with the solutions (7-339) and (7-341) for $\bar{t} \ll 1$, all in the asymptotic limit $\text{Re} \to 0$; that is, for $\bar{h}$, we require that

$$\lim_{\hat{t} \gg 1} 1 + \text{Re}\bar{h}_1 + \cdots \Longleftrightarrow \lim_{\bar{t} \ll 1} \bar{h}_0 + \cdots \quad \text{as Re} \to 0. \quad (7\text{-}369)$$

To carry out the matching, it is convenient to express the inner solution in terms of the outer variable $\bar{t}$.* This is done by simply substituting $\bar{t}/\text{Re}$ for $\hat{t}$. In terms of the *long-time* variable, the left-hand side of (7–369) thus becomes

$$1 + \text{Re} \left[-\frac{1}{6} \frac{\bar{t}}{\text{Re}} + \sum_{n=1}^{\infty} \frac{2(1-\cos\lambda_n)}{\left(1 - \dfrac{\sin\lambda_n}{\lambda_n}\right)\left(\beta + \dfrac{1}{2}\right)^2 \lambda_n^6} \left(1 - e^{\lambda_n^2(\bar{t}/\text{Re})}\right) \right] + \cdots.$$

Hence, in the limit $\text{Re} \to 0$, we obtain

$$1 - \frac{\bar{t}}{6} + \text{Re} \sum_{n=1}^{\infty} \frac{2(1-\cos\lambda_n)}{\left(1 - \dfrac{\sin\lambda_n}{\lambda_n}\right)\left(\beta + \dfrac{1}{2}\right)^2 \lambda_n^6} \tag{7–370}$$

through terms of 0(Re). On the right-hand side of (7–369), we have only the 0(1) approximation to $\bar{h}$, namely $\bar{h}_0$. Thus, the matching condition can be applied only to 0(1) at this point in the analysis, and we temporarily neglect the term of 0(Re) in (7–370). At 0(1), the matching condition (7–369) thus becomes

$$1 - \frac{\bar{t}}{6} + 0(\text{Re}) = \left(\frac{3}{\bar{t}+C_0}\right)^{1/2} \Bigg|_{\bar{t} \ll 1} + 0(\text{Re}) \qquad \text{as } \text{Re} \to 0. \tag{7–371}$$

For small $\bar{t}$, the right-hand side can be expressed as

$$\left(\frac{3}{\bar{t}+C_0}\right)^{1/2} \approx \left(\frac{3}{C_0}\right)^{1/2}\left(1 - \frac{\bar{t}}{2C_o} + 0(\bar{t}^2)\right). \tag{7–372}$$

Thus, comparing the left and right sides of the matching condition (7–371), we see

$$\boxed{C_0 = 3,} \tag{7–373}$$

and the match at 0(1) in Re is *exact* through terms of $0(\bar{t})$ on the right.

The reader may well ask, at this point, why the expansion (7–372) was truncated at $0(\bar{t})$ since terms of $0(\bar{t}^2)$ are thus neglected in the matching procedure. There is, in fact, a *mismatch* of $0(\bar{t}^2)$ in (7–372) with $C_0 = 3$. The justification for neglecting the $0(\bar{t}^2)$ contribution at this point in the analysis is that it represents a mismatch that is asymptotically smaller than the terms that we have calculated so far in the inner approximation to the solution. To see this, we note that to match a term proportional to $\bar{t}^2$ from the right-hand side of (7–371), we would require a term from the left-hand side that is proportional to $\hat{t}^2\text{Re}^2$. But such a term can only occur in the inner solution at $0(\text{Re}^2)$, and this is asymptotically smaller than the terms through 0(Re) that have been calculated so far. It should be noted, however, that all of the terms on the two sides of the matching condition must eventually be matched, and we would need to include the $0(\bar{t}^2)$ term if the approximation on the left side were carried to $0(\text{Re}^2)$.

The choice $C_0 = 3$ produces a match between the inner and outer approximations to the solution for the function $\bar{h}(t)$. Ironically, the value of C_0 obtained by matching is identical to the value calculated earlier by the incorrect procedure of applying the

---

*Since the matching condition is applied in a region of overlap, where both the short- and long-time solutions are valid, we could equally well express the long-time solution in terms of the short-time variable $\hat{t}$.

initial condition directly to the outer (long-time) approximation. The reader should not be misled by this occurrence. If we apply the same procedure to the second term in the expansion (7–333)—that is, apply $\bar{h}_1 = 0$ at $\bar{t} = 0$—we get the wrong result, as we shall see shortly. More commonly, we will find that matched asymptotic approximations will not satisfy any boundary or initial condition that is applied outside their domain of validity.

In the present problem, this is illustrated by the function $F(\bar{z}, \bar{t})$. We have already seen that the first approximation to this function in the outer (long-time) domain, (7–339), will not satisfy the initial condition $F_0 = 0$ at $\bar{t} = 0$. Of course, the inner approximation to this function, equation (7–364), does satisfy the initial condition. For completeness, we now show that this inner (short-time) solution also *matches* the outer approximation (7–339) at the current $0(1)$ level of approximation. Thus, in the limit Re $\rightarrow 0$, we compare

$$
\lim_{\bar{t} \ll 1} \left[\frac{1}{4} \left(\bar{h}_0 \bar{z}^2 - \frac{2}{3} \bar{z}^3 \right) + \cdots \right] \Longleftrightarrow
$$

$$
\lim_{\hat{t} \gg 1} \left[-\frac{\bar{z}^3}{6} + \frac{\bar{z}^2}{4} + \frac{\beta \bar{z}}{6} + \beta \frac{dh_1}{d\hat{t}} \bar{z} + \sum_{n=1}^{\infty} \frac{-1}{\left(1 - \frac{\sin \lambda_n}{\lambda_n} \right) \left(\beta + \frac{1}{2} \right) \lambda_n^4} \right.
$$

$$
\left. \left[\sin \lambda_n \sin \lambda_n z + (\cos \lambda_n - 1)(\cos \lambda_n z - 1) \right] e^{-\lambda_n^2 \hat{t}} \right]. \quad (7\text{–}374)
$$

Substituting for $\bar{h}_0$ on the left from (7–341) with $C_0 = 3$ and expressing the right-hand side in terms of $\bar{t}$, as above, we see that the two sides match

$$
\frac{1}{4} \left[\left(1 + 0(\bar{t}) \right) \bar{z}^2 - \frac{2}{3} \bar{z}^3 \right] + 0(\text{Re}) \Longleftrightarrow
$$

$$
-\left(\frac{\bar{z}^3}{6} - \frac{\bar{z}^2}{4} \right) + \frac{\beta \bar{z}}{6} + \beta \bar{z} \left(-\frac{1}{6} \right) + 0(\text{Re}) \quad \text{as Re} \rightarrow 0. \quad (7\text{–}375)
$$

In taking the limit Re $\rightarrow 0$ on the right, we have neglected exponentially small terms in Reynolds number. The symbol $0(\text{Re})$ is intended to serve as a reminder that we have not yet calculated the second term in the expansion (7–350). We note also that there is an unmatched term proportional to $\bar{t}$ from the small $\bar{t}$ approximation of the left-hand side. However, this need not concern us yet, because to obtain a term proportional to $\bar{t}$ from the short-time approximation, we would need to start with a term proportional to Re $\hat{t}$, and we have so far only considered terms of $0(1)$ for F in the expansion for the short-time region.

We have already obtained the $0(\text{Re})$ inertial contribution to the function $\bar{h}(t)$ in the inner domain, and it is of interest to continue for one more step in the solution procedure to determine the corresponding inertial correction, at $0(\text{Re})$, to the function $\bar{h}(\bar{t})$ in the outer region. This means calculating the second terms in the asymptotic expansion (7–333) in this regime. To obtain the governing equations at $0(\text{Re})$, we must substitute the full expansion (7–333) into (7–326)–(7–328) and collect terms of equal order in Re, following the formal procedure outlined in Chapter 6.

The governing equation at $0(\text{Re})$ and the boundary conditions at $\bar{z}=0$ are straightforward,

$$\frac{\partial^3 F_1}{\partial \bar{z}^3} = \frac{\partial^2 F_0}{\partial \bar{z} \partial \bar{t}} + \left(\frac{\partial F_0}{\partial \bar{z}}\right)^2 - 2F_0 \frac{\partial^2 F_0}{\partial \bar{z}^2} - \beta \frac{\partial^2 \bar{h}_0}{\partial \bar{t}^2} \tag{7-376}$$

and

$$F_1 = \frac{\partial F_1}{\partial \bar{z}} = 0 \quad \text{at } \bar{z}=0. \tag{7-377}$$

However, the boundary conditions at $\bar{z}=\bar{h}$ in (7–327) and (7–328b) require somewhat more care because $\bar{h}$ is itself expressed as an asymptotic expansion.

The simplest approach to applying these conditions in the asymptotic limit $\text{Re} \ll 1$ is to use the method of *domain perturbations*, first introduced in Chapter 6, to express the boundary conditions at $\bar{z}=\bar{h}$ in terms of equivalent conditions at $\bar{z}=\bar{h}_o$. For the condition (7–328b), this is nothing more than a Taylor series approximation in the form

$$\frac{\partial F}{\partial \bar{z}}\bigg|_{\bar{z}=\bar{h}_0} + \frac{\partial}{\partial \bar{z}}\left(\frac{\partial F}{\partial \bar{z}}\right)_{\bar{z}=\bar{h}_0} (\bar{h}-\bar{h}_0) + \cdots = 0.$$

Thus, at $0(1)$, we recover the condition (7–336b),

$$\frac{\partial F_0}{\partial \bar{z}} = 0 \quad \text{at } \bar{z}=\bar{h}_o, \tag{7-378a}$$

and at $0(\text{Re})$, the boundary condition becomes

$$\frac{\partial F_1}{\partial \bar{z}} = -\frac{\partial^2 F_0}{\partial \bar{z}^2}\bar{h}_1 \quad \text{at } \bar{z}=\bar{h}_0. \tag{7-378b}$$

Similarly, by using a Taylor series approximation for the two sides of (7–327), we obtain, at $0(1)$, the condition, (7–335),

$$F_0(\bar{z},\bar{t}) = -\frac{1}{2}\frac{d\bar{h}_0}{d\bar{t}} \quad \text{at } \bar{z}=\bar{h}_0 \tag{7-379a}$$

and, at $0(\text{Re})$, a condition relating F_1 and h_1,

$$F_1(\bar{z},\bar{t}) + \frac{\partial F_0}{\partial \bar{z}}\bar{h}_1 = -\frac{1}{2}\frac{d\bar{h}_1}{d\bar{t}} \quad \text{at } \bar{z}=\bar{h}_0. \tag{7-379b}$$

Finally, the *long-time* solution is required to match at each level of approximation with the *short-time* solution, that is,

$$\lim_{\bar{t} \ll 1} \bar{h}_0(\bar{t}) + \text{Re}\,\bar{h}_1(\bar{t}) + \cdots \iff \lim_{\hat{t} \gg 1} 1 + \text{Re}\,\bar{h}_1(\hat{t}) + \cdots \quad \text{as } \text{Re} \to 0 \tag{7-380}$$

and

$$\lim_{\bar{t} \ll 1} F_0(\bar{z},\bar{t}) + \text{Re}\, F_1(\bar{z},\bar{t}) + \cdots \Longleftrightarrow \lim_{\hat{t} \gg 1} F_0(\bar{z},\hat{t}) + \text{Re}\, F_1(\bar{z},\hat{t}) + \cdots \quad \text{as Re} \to 0,$$

(7–381)

so that the matching conditions play the role of initial conditions for the outer solutions at 0(Re)—that is, for $\bar{h}_1$ and $F_1(\bar{z},\bar{t})$ at $\bar{t} \ll 1$.

Thus, at 0(Re) in the outer region, the problem is to solve (7–376), subject to the conditions (7–377), (7–378b), (7–379b), as well as the matching conditions (7–380) and (7–381). We begin with the general solution of (7–376).

$$F_1 = \frac{1}{48} \frac{d\bar{h}_0}{d\bar{t}} \bar{z}^4 + \frac{1}{720}\left[\bar{h}_0 \bar{z}^6 - \frac{2}{7}\bar{z}^7\right] - \beta \frac{d^2\bar{h}_0}{d\bar{t}^2}\frac{\bar{z}^3}{6} + d_1\bar{z}^2 + d_2\bar{z} + d_3.$$

(7–382)

Applying the boundary conditions (7–377) at $\bar{z}=0$, we see immediately that

$$d_2 = d_3 = 0.$$

Then, the condition (7–378b) yields

$$d_1 = -\frac{\bar{h}_0^2}{24}\frac{d\bar{h}_0}{d\bar{t}} - \frac{2\bar{h}_0^5}{720} + \frac{\beta\bar{h}_0}{4}\frac{d^2\bar{h}_0}{d\bar{t}^2} + \frac{\bar{h}_1}{4},$$

and thus

$$F_1 = \frac{1}{48}\frac{d\bar{h}_0}{d\bar{t}}(\bar{z}^4 - 2\bar{h}_0^2\bar{z}^2) + \frac{1}{720}\left(\bar{h}_0\bar{z}^6 - \frac{2\bar{z}^7}{7} - 2\bar{h}_0^5\bar{z}^2\right) -$$
$$\beta\frac{d^2\bar{h}_0}{d\bar{t}^2}\left(\frac{\bar{z}^3}{6} - \frac{\bar{h}_0\bar{z}^2}{4}\right) + \frac{\bar{h}_1}{4}\bar{z}^2.$$

(7–383)

Then, finally, from the condition (7–379b), we find that

$$-\frac{1}{2}\frac{d\bar{h}_1}{d\bar{t}} = -\frac{1}{48}\bar{h}_0^4\frac{d\bar{h}_0}{d\bar{t}} - \frac{\bar{h}_0^7}{560} + \frac{\beta}{12}\frac{d^2\bar{h}_0}{d\bar{t}^2}\bar{h}_0^3 + \frac{\bar{h}_0^2}{4}\bar{h}_1,$$

(7–384)

and substituting for $\bar{h}_0$, using (7–341) and (7–373), this becomes

$$-\frac{1}{2}\frac{d\bar{h}_1}{d\bar{t}} = \frac{1}{4}\left(\frac{3}{\bar{t}+3}\right)h_1 + \frac{17}{10080}\left(\frac{3}{\bar{t}+3}\right)^{7/2} + \frac{\beta}{144}\left(\frac{3}{\bar{t}+3}\right)^4.$$

(7–385)

The general solution of this equation for $\bar{h}_1$ is

$$\bar{h}_1 = C_1\left(\frac{3}{\bar{t}+3}\right)^{3/2} + \frac{17}{1680}\left(\frac{3}{\bar{t}+3}\right)^{5/2} + \frac{\beta}{36}\left(\frac{3}{\bar{t}+3}\right)^3.$$

(7–386)

To determine the constant C_1 in this solution for $\bar{h}_1$, we must apply the matching condition (7–380). On the left-hand side, we have

$$\bar{h}_0(\bar{t}) + \mathrm{Re}\,\bar{h}_1(\bar{t}) + 0(\mathrm{Re}^2) =$$

$$\left(\frac{3}{\bar{t}+3}\right)^{1/2} + \mathrm{Re}\left[C_1\left(\frac{3}{\bar{t}+3}\right)^{3/2} + \frac{17}{1680}\left(\frac{3}{\bar{t}+3}\right)^{5/2} + \frac{\beta}{36}\left(\frac{3}{\bar{t}+3}\right)^3\right] + 0(\mathrm{Re}^2), \quad (7\text{--}387)$$

which is to be evaluated for $\bar{t} \ll 1$. On the right-hand side we have the inner (short-time) solution (7–349), where h_1 is given by (7–367). For purposes of matching, it is convenient to express this inner solution in terms of the outer variable $\bar{t}$. We have previously shown that the short-time solution to $0(\mathrm{Re})$ reduces to the form (7–370). Hence, expanding (7–387) for $\bar{t} \ll 1$, the matching condition becomes

$$\left(1 - \frac{\bar{t}}{6} + 0(\bar{t}^2)\right) + \mathrm{Re}\left[C_1(1 - 0(\bar{t})) + \frac{17}{1680}(1 - 0(\bar{t})) + \frac{\beta}{36}(1 - 0(\bar{t}))\right] + \cdots \iff$$

$$1 - \frac{\bar{t}}{6} + \mathrm{Re}\sum_{n=1}^{\infty}\frac{2(1 - \cos\lambda_n)}{\left(1 - \dfrac{\sin\lambda_n}{\lambda_n}\right)\left(\beta + \dfrac{1}{2}\right)^2\lambda_n^6} + \cdots \quad \text{for } \mathrm{Re} \to 0. \quad (7\text{--}388)$$

Terms of $0(\bar{t}^2)$ and $0(\mathrm{Re}\,\bar{t})$ on the left are not explicitly considered at this level of approximation because they could only appear in the inner solution at $0(\mathrm{Re}^2)$, and we have not yet obtained terms beyond $0(\mathrm{Re})$ in the inner expansion. Comparing the two sides of the matching condition (7–388), we see that the terms at $0(1)$ match exactly to the present level of approximation—indeed, C_0 was chosen earlier based upon this condition. The terms at $0(\mathrm{Re})$ also can be made to match provided C_1 is chosen to satisfy the condition

$$C_1 + \frac{17}{1680} + \frac{\beta}{36} = \sum_{n=1}^{\infty}\frac{2(1 - \lambda_n)}{\left(1 - \dfrac{\sin\lambda_n}{\lambda_n}\right)\left(\beta + \dfrac{1}{2}\right)^2\lambda_n^6}. \quad (7\text{--}389)$$

We denote the constant on the right as K_β. Then,

$$\boxed{C_1 = K_\beta - \frac{17}{1680} - \frac{\beta}{36}.} \quad (7\text{--}390)$$

Values of K_β and C_1 for various values of β are given in Table 7–2. In the limit $\beta = 0$, where $\lambda_n = n\pi$, the constant $K_\beta = 1/60$ and

$$\boxed{C_1 = \frac{11}{1680} \quad (\beta \equiv 0).} \quad (7\text{--}391)$$

It is, perhaps, worth reemphasizing that the coefficient C_1 could not have been evaluated correctly without first obtaining the short-time approximation to $0(\mathrm{Re})$ and then matching. In particular, when we considered the leading-order approximation for $\bar{h}_0$, Equation (7–341), we found that the value of the coefficient C_0 obtained from matching turned out to be identical to the value obtained from the ad hoc (and incorrect) procedure of simply applying the condition $\bar{h}_0 = 1$ at $\bar{t} = 0$, in spite of the fact that $\bar{t} = 0$ is outside the domain of validity of the expansion (7–333). If we would follow

Table 7–2 Values of K_β and C_1

	$\beta = 0$	$\beta = 0.1$	$\beta = 0.25$	$\beta = 0.5$	$\beta = 0.75$	$\beta = 1.0$	$\beta = 10$
K_β	0.01667	0.01944	0.02361	0.03055	0.03750	0.04445	0.30033
$C_1(\times 10^3)$	6.5476	6.5432	6.5465	6.5421	6.5476	0.65532	12.4332

the same ad hoc procedure at $0(\text{Re})$, however, the apparent condition on $\bar{h}_1$ would be $\bar{h}_1 = 0$ at $\bar{t} = 0$, and this would yield a value of C_1 from Equation (7–386) of

$$C_1 = -\frac{17}{1680} - \frac{\beta}{36}.$$

Clearly, this is incorrect, as we can see by comparison with (7–390).

The expression (7–390) for C_1 completes the solution for $\bar{h}$ through terms of $0(\text{Re})$. Collecting all of our results, the solution for the inner (short-time) regime, with $t_c = 0(h_0^2/\nu)$, is

$$\bar{h} = 1 - \text{Re}\left[\frac{\hat{t}}{6} - \sum_{n=1}^{\infty} \frac{2(1 - \cos\lambda_n)}{\left(1 - \dfrac{\sin\lambda_n}{\lambda_n}\right)\left(\beta + \dfrac{1}{2}\right)^2 \lambda_n^6}\left(1 - e^{-\lambda_n^2 \hat{t}}\right)\right] + 0(\text{Re}^2), \qquad (7\text{–}392)$$

For the outer (long-time) regime, with $t_c = h_0/V$, it is

$$\bar{h} = \left(\frac{3}{\bar{t}+3}\right)^{1/2} + \text{Re}\left[K_\beta\left(\frac{3}{\bar{t}+3}\right)^{3/2} - \frac{17}{1680}\left\{\left(\frac{3}{\bar{t}+3}\right)^{3/2} - \left(\frac{3}{\bar{t}+3}\right)^{5/2}\right\} - \right.$$
$$\left. \frac{\beta}{36}\left\{\left(\frac{3}{\bar{t}+3}\right)^{3/2} - \left(\frac{3}{\bar{t}+3}\right)^{3}\right\}\right] + 0(\text{Re}^2). \qquad (7\text{–}393)$$

Together, these provide a uniformly valid approximation for $\bar{h}$ for all times.

There is one feature of the solution (7–392) that is slightly misleading and therefore worth a brief discussion. This is the origin and significance of the term $\text{Re}\ \hat{t}/6$. Since this term does not appear until $0(\text{Re})$, it might be thought at first to arise from inertia effects. However, this is not true. It is nothing more than the linear approximation to the first term in (7–393), which is completely viscous in origin. The only reason that this term appears at $0(\text{Re})$ in the inner solution is because the characteristic time in this regime is so short. In particular, the actual contributions of the inertia of the fluid and the body for short times are all contained in the second term of the $0(\text{Re})$ contribution to (7–392).

The functions $\bar{h}(\hat{t})$ from (7–392) and $\bar{h}(\bar{t})$ from (7–393) are plotted in Figures 7–13 and 7–14, respectively, for $\beta = 0$ and for $\beta = 10$, both with $\text{Re} = 0.1$. In addition, we also show the corresponding result without any inertia in both cases. In the inner (short-time) region, shown in Figure 7–13, the inertial contribution to the function h_1 is positive, and this means that the thin film thins slower in the presence of inertia than it does without inertia. The parameter $\beta = 0$ corresponds to a case in which the inertia

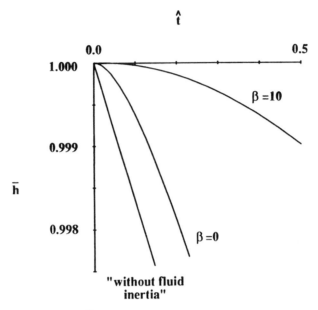

Figure 7–13 The gap width, $\bar{h}(\hat{t})$, in the inner (short-time) regime for the limit Re $\ll 1$, evaluated at Re $= 0.1$ for $\beta = 0$ and $\beta = 10$ from Equation (7–392). Also shown is the first term of Equation (7–392), which represents the gap width in the absence of inertia.

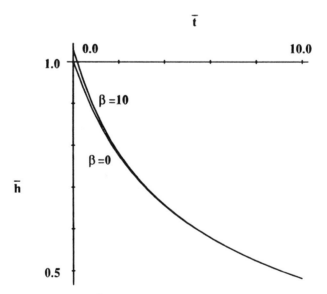

Figure 7–14 The gap width, $\bar{h}(\bar{t})$, in the outer (long-time) regime for Re $\ll 1$, evaluated from (7–393) for Re $= 0.1$ and $\beta = 0$ and $\beta = 10$. Also shown is the gap width calculated from (7–393) in the absence of inertia (that is, for Re $\equiv 0$).

of the solid body is negligible. On the other hand, $\beta = 10$ is representative of a case in which the mass (or inertia) of the body is finite. Thus, according to Figure 7-13, a more massive body will cause the film to thin more slowly. Similarly, an increased inertia effect in the fluid also causes the film to thin more slowly. These results may, at first, seem counterintuitive. However, in this early part of the time domain, the body is being *accelerated* from its initial rest state, and this acceleration process is *inhibited* by the inertia of both the fluid in the thin film and of the body. Indeed, in the limit $\beta \to \infty$ of an infinitely massive body, the film does not thin at all, because the finite force on the body is unable to cause it to move.

The film thickness in the *outer* part of the time domain in Figure 7-14 is much less sensitive to the mass of the body (that is, to the parameter β). This is a reflection of the fact that for Re = 0.1, the whole effect of inertia on the film-thinning process is weak. In fact, the lower of the two curves in Figure 7-14 is actually a superposition of two cases: Re = 0.1, $\beta = 0$; and Re $\equiv 0$. Of course, the magnitude of inertia effects will be increased for larger Reynolds numbers, but the present analysis is only valid asymptotically in the limit Re $\ll 1$. The only effect of inertia that is obvious in Figure 7-14 is the displacement of the two curves for small values of $\bar{t}$. This is, primarily, a reflection of the inertia effect at short times, which is felt in this outer part of the time domain via matching with the inner solution. It should be noted that the displacement of the trajectory for $\beta = 0$ to a value for short times that is greater than the initial film thickness does *not* mean that the film is actually ever increased above h_0. Rather, it is a reflection of the fact that the outer approximation breaks down for sufficiently small values of $\bar{t}$.

It is, perhaps, worthwhile to plot only the inertia contribution to $\bar{h}$ in (7-393)—namely, $\bar{h}_1(\bar{t})$—since otherwise the qualitative behavior of the inertial contribution to $\bar{h}$ is masked by the small value of Re in the asymptotic limit Re $\to 0$. Hence, in Figure 7-15, we plot $\bar{h}_1(\bar{t})$ for $\beta = 0$ and $\beta = 10$. Since $K_\beta > 0$, we see initially that $\bar{h}_1$ is positive—that is, initially, the film is thicker than in the absence of inertia effects—and since K_β increases as β increases, this effect is enhanced by the inertia of the body. But this behavior is a consequence of the effects of inertia in the *inner* part of the time domain. The other two terms in $\bar{h}_1$ are, in fact, *negative* for all $\bar{t} > 0$. Indeed, if we examine $d\bar{h}_1/d\bar{t}$, we find that it is positive, meaning that the film is actually thinning faster with inertia than without. This is a consequence of the fact that in the outer (long-time) region, the body is actually *decelerating* due (primarily) to viscous interaction with the walls. But in this case, the effect of inertia is to inhibit (or slow) the rate of deceleration, and the film actually thins faster than it does without any inertia present. Thus, in the outer region, we have the ironic situation in which the film thickness is *larger* with inertia because of the effect of inertia on the inner solution, but the rate of decrease of the film thickness is actually greater. Of course, for long times, $\bar{t} \gg 1$, all inertial effects are swamped by the viscous interactions between the body and the wall, and these effects are hardly discernable in Figure 7-14. This is reflected in the fact that the viscous term is proportional to $(3/\bar{t} + 3)^{1/2}$, while the inertia term for large times is proportional to $(3/\bar{t} + 3)^{3/2}$. Obviously, for any Re, the ratio of inertia to viscous contributions to $\bar{h}$ goes to *zero* as $\bar{t} \to \infty$.

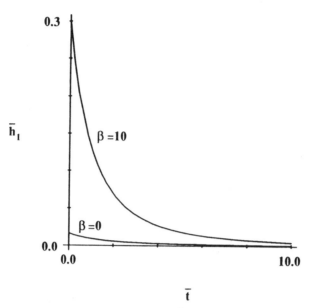

Figure 7–15 The 0(Re) inertia contribution, $\bar{h}_1(\bar{t})$, calculated from (7-393) for $\beta = 0$ and $\beta = 10$.

Finally, in Figure 7–16a, we show a composite of the inner and outer approximation for $\bar{h}$ at Re $= 0.1$ and $\beta = 10$, plotted versus the short time variable $\hat{t}$. This illustrates the nature of the matching between inner and outer solutions and demonstrates the sense in which the two approximations, each valid in different parts of the time domain, provide a composite approximation that is valid over the whole time domain. As we have seen earlier, if we extrapolate the outer solution back to $t = 0$, we find that we do not recover $\bar{h} = 1$, but we do see that this outer solution matches smoothly for $t \sim 4$–5. Clearly, the inner solution also breaks down for very large values of $\hat{t}$, though this is not visually evident over the limited range for $\hat{t}$ that is encompassed in Figure 7–16. If we examine (7–392), we see that $\bar{h}$ is predicted to become negative for sufficiently large $\hat{t}$ at any fixed value of Re. For illustrative purposes, we also show the same composite plot in Figure 7–16b, but for Re $= 0.5$ and $\beta = 10$. Clearly, the match between inner and outer approximations is less satisfactory than in Figure 7–16a for the smaller value of Re. This comparison is included to emphasize the fact that the solution is really valid only in the limit Re $\rightarrow 0$. This is the reason why we are always careful to express the matching conditions as being applicable only in the limit Re $\rightarrow 0$. At any given level of approximation in the asymptotic expansions for the two parts of the time domain, there is a mismatch between the inner and outer solutions, which is only guaranteed to vanish in the limit Re $\rightarrow 0$. For small, but finite, values of Re, there always will be a small, but finite, mismatch as is evident in Figure 7–16b, but less so in Figure 7–16a.

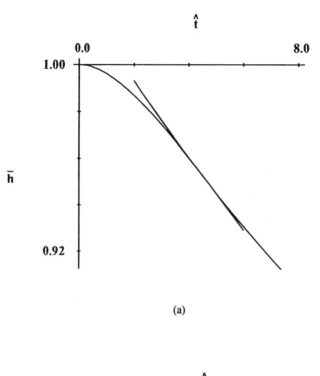

(a)

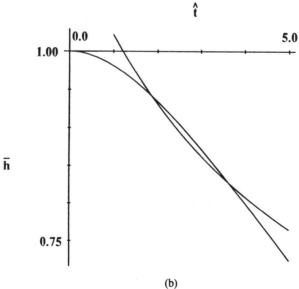

(b)

Figure 7–16 A composite of the inner and outer approximations to the gap width function: (a) Re = 0.1 and $\beta = 10$; (b) Re = 0.5 and $\beta = 10$.

3. Asymptotic Results for Re ≫ 1

Let us now turn to the case of very strong inertia, Re ≫ 1. The nondimensional-ized version of the governing equation for $F(\bar{z}, \bar{t})$, when $t_c = h_o/V$, is given by (7-332), namely,

$$\frac{\partial^2 F}{\partial \bar{z} \partial \bar{t}} + \left(\frac{\partial F}{\partial \bar{z}}\right)^2 - 2F\frac{\partial^2 F}{\partial \bar{z}^2} - \frac{1}{\text{Re}}\frac{\partial^3 F}{\partial \bar{z}^3} = \beta\frac{d^2\bar{h}}{d\bar{t}^2} + 1,$$

with the other equations and boundary conditions given in (7-327) and (7-328). Following the precedent of earlier sections in this chapter, we may anticipate that an asymptotic expression should exist for F (and h) in which the small parameter is $\text{Re}^{-1} \ll 1$. Thus, if we propose an expansion for F of the form

$$F = F_0 + \frac{1}{\text{Re}}F_1 + \cdots,$$

$$\bar{h}(\bar{t}) = \bar{h}_0(\bar{t}) + \frac{1}{\text{Re}}\bar{h}_1(\bar{t}) + \cdots,$$

(7-394)

the governing equation for the first term, F_0, is just the limit of (7-332) for $\text{Re} \to \infty$, that is,

$$\frac{\partial^2 F_0}{\partial \bar{z} \partial \bar{t}} + \left(\frac{\partial F_0}{\partial \bar{z}}\right)^2 - 2F_0\frac{\partial^2 F_0}{\partial \bar{z}^2} = \beta\frac{d^2\bar{h}_o}{d\bar{t}^2} + 1,$$

(7-395)

and the corresponding boundary and initial conditions, from (7-327) and (7-328), are

$$F_0 = -\frac{1}{2}\frac{d\bar{h}_0}{d\bar{t}} \quad \text{at} \quad \bar{z} = \bar{h}_0,$$

(7-396)

$$F_0 = \frac{\partial F_0}{\partial \bar{z}} = 0 \quad \text{at} \quad \bar{z} = 0,$$

(7-397a)

$$\frac{\partial F_0}{\partial \bar{z}} = 0 \quad \text{at} \quad \bar{z} = \bar{h}_0,$$

(7-397b)

and

$$F_0 = 0, \bar{h}_0 = 1 \quad \text{at} \quad \bar{t} = 0.$$

(7-397c)

Examination of (7-395) shows that the highest-order derivative with respect to $\bar{z}$ that appeared in the exact equation (7-332) is neglected in the limit $\text{Re} \to \infty$, and this term corresponds to the viscous contribution to the thin-film dynamics. The fact that the resulting approximate equation (7-395) is lower order in $\bar{z}$ than the full equation (7-332) means that the solution to (7-395) cannot satisfy all the boundary conditions of the original problem. Furthermore, since it is the viscous term that is neglected, we may anticipate that the solution will not satisfy the no-slip conditions at $\bar{z} = 0$ and $\bar{h}$. This means that the solution [and (7-395)] cannot be a uniformly valid first approximation to the original problem, since it does not satisfy all of the boundary conditions. In this sense, the Re $\gg 1$ limit of the present thin-film problem is very similar to the high-frequency limit, $R_\omega \gg 1$, for flow in a tube with a sinusoidal variation of the pressure gradient, discussed in Section A of Chapter 6, and to the high Reynolds limit of the slightly eccentric cylinder problem discussed earlier in this chapter. Based upon these analyses, it is clear that the limit Re $\gg 1$ requires a different boundary-layer approximation in the vicinity of the plane surfaces at $\bar{z} = 0$ and $\bar{h}$. However, we are getting ahead of the story. Let us begin by examining the solutions of (7–395).

Not surprisingly, the governing equation (7–395) is nonlinear. However, considerable simplification can be achieved if we can be clever enough to guess the general form of the solution. For this purpose, we may note that the neglect of viscous terms means that the radial velocity component u_r will not satisfy no-slip conditions at $\bar{z} = 0$, $\bar{h}$, and thus may be expected to be independent of $\bar{z}$ (that is, the velocity "profile" across the thin film is a flat, plug-flow form). The most general expression for F_0 that is consistent with this assumption is

$$\boxed{F_0 = \bar{z}\phi(\bar{t}).}$$ (7–398)

The velocity field corresponding to this form for F_0 is known as an *inviscid, stagnation point flow,* and is sketched in Figure 7–17. Assuming this *form* to be correct, the only remaining unknown is the time dependent function $\phi(t)$. It should be noted that the proposed solution (7–398) satisfies the kinematic boundary condition on F at $\bar{z} = 0$, and with proper choice for $\bar{h}_0$ can also satisfy the kinematic boundary condition (7–396) at $\bar{z} = \bar{h}_0$. It does *not* satisfy the no-slip conditions $\partial F_0/\partial \bar{z} = 0$ at $\bar{z} = 0$, $\bar{h}_0$. The initial condition (7–397c) can be satisfied with the condition $\phi(0) = 0$.

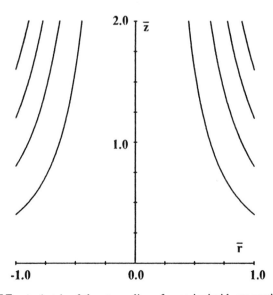

Figure 7–17 A sketch of the streamlines for an inviscid, stagnation point flow.

If we substitute (7–398) into (7–395), we find

$$\frac{d\phi}{d\bar{t}} + \phi^2 = \beta \frac{d^2\bar{h}_0}{d\bar{t}^2} + 1, \tag{7–399}$$

In addition, from (7–396), we obtain

$$\frac{d\bar{h}_0}{d\bar{t}} = -2\bar{h}_0\phi. \tag{7–400}$$

The pair of equations (7–399) and (7–400) is to be solved subject to the initial conditions

$$\begin{aligned} \bar{h}_0 &= 1 \quad \text{at} \quad \bar{t} = 0, \\ \phi &= 0 \quad \text{at} \quad \bar{t} = 0. \end{aligned} \tag{7–401}$$

Alternatively, we can eliminate ϕ by substitution of (7–400) into (7–399) to obtain a single second-order equation for $\bar{h}_0(\bar{t})$. The result is

$$\left[\beta + \frac{1}{2\bar{h}_0}\right] \frac{d^2\bar{h}_0}{d\bar{t}^2} - \frac{3}{4\bar{h}_0^2}\left(\frac{d\bar{h}_0}{d\bar{t}}\right)^2 + 1 = 0. \tag{7–402}$$

The initial condition (7–401) on ϕ now becomes a second initial condition on $\bar{h}_0$, namely,

$$\frac{d\bar{h}_o}{d\bar{t}} = 0 \quad \text{at} \quad \bar{t} = 0, \tag{7–403}$$

Equation (7–402), or the pair of equations (7–399) and (7–400), is nonlinear and can only be solved analytically for certain limiting values of the parameter β. We restrict our considerations here to the limit $\beta \ll 1$. A first approximation for this case is obtained by taking the limit $\beta \to 0$ in (7–399),

$$\frac{d\phi}{d\bar{t}} + \phi^2 = 1. \tag{7–404}$$

This is the governing equation for the first term in an asymptotic expansion of ϕ for $\beta \ll 1$.

The solution of (7–404) that satisfies the condition $\phi = 0$ at $\bar{t} = 0$ is

$$\phi(\bar{t}) = \tanh \bar{t}. \tag{7–405}$$

Therefore,

$$\frac{d\bar{h}_0}{d\bar{t}} = -2\bar{h}_0(\tanh \bar{t}), \tag{7–406}$$

which is to be solved subject to the condition (7–401). The solution of (7–406) is

$$\bar{h}_0 = \frac{1}{\cosh^2 \bar{t}}. \tag{7–407}$$

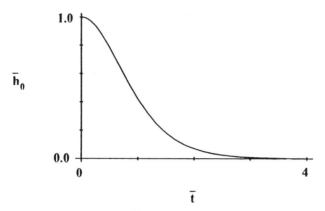

Figure 7–18   The gap width $\bar{h}_0$ as a function of $\bar{t}$, calculated from (7–409).

A plot showing $\bar{h}_0$ versus $\bar{t}$ is shown in Figure 7-18.

The solution (7–405) and (7–407) incorporates much of the qualitatively correct physics for the case of a body with negligible inertia ($\beta \ll 1$) moving under the action of a constant force toward a parallel plane wall at high Re, where the force on the body is balanced by dynamic pressure forces in the thin film. Thus, for example, the predicted behavior for $\bar{h}_0$ is qualitatively correct. However, the solution form (7–398) is still incomplete in the sense that it does not satisfy the no-slip conditions at the solid surfaces at $\bar{z} = 0$ and $\bar{h}$. In particular, it is clear that (7–395) is not a uniformly valid approximation of the exact equation (7–332), for if it were, the solution F_0 would satisfy all of the boundary and initial conditions of the original problem. By now, the explanation of the failure of (7–395) in the vicinity of the boundaries should be familiar. We have obtained (7–395) by taking the limit Re $\rightarrow \infty$ in the exact equation (7–332). The only change in this equation from the exact, *dimensional* equations is that it has been nondimensionalized as described in the first part of this section. We assume, in deriving (7–395), that the nondimensionalization is correct so that all of the terms in (7–332), except for the one involving Re *explicitly,* are O(1) and thus independent of Re for Re $\rightarrow \infty$. This implies that the viscous term is asymptotically small compared with any other term in (7–332) as Re $\rightarrow \infty$. But this cannot be correct everywhere in the thin-film region because the neglect of viscous terms means that the no-slip conditions cannot be satisfied. In particular, it is evident, based upon the preceding analyses of high Reynolds number flows that the assumption $l_c = h_o$ that is inherent in (7–332) is *incorrect* in the vicinity of the solid surfaces $\bar{z} = 0$ and $\bar{h}$.

If the nondimensionalized form of the governing Equation (7–332) is not valid everywhere in the domain, as suggested previously, then the expansion (7–394) based on this equation is also valid in only part of the domain. Thus, the asymptotic limit for Re $\gg 1$ is *singular,* and we require different forms for both the governing equation and the asymptotic expansion for large Re in different parts of the thin-film region. Equation (7–332) and the expansion (7–394), based upon the characteristic length scale

$l_c = h_o$, are relevant in the whole central region between the upper and lower boundaries—that is, for $0 < \bar{z} < \bar{h}$—with the exception of *very thin boundary-layer regions* in the immediate vicinity of the boundaries. In the boundary-layers, u_r goes from its value $u_r = \bar{r}\phi(\bar{t})$ in the interior to the no-slip value $u_r = 0$ at the walls. The distance over which this change occurs is very short compared with the gap width for Re $\gg 1$. It is this fact that accounts for the observation that the equations nondimensionalized with respect to h_o give the wrong limiting form in the vicinity of the walls. In fact, the velocity changes by an $0(1)$ amount over a characteristic distance that is *much smaller* than h_o, and the nondimensionalization using $l_c = h_o$ is wrong in this part of the thin-gap domain. Since the nondimensionalization is wrong, terms like $F_{\bar{z}\bar{z}\bar{z}}$ are not independent of Re, and the assumption cannot be made that $(1/\text{Re})F_{\bar{z}\bar{z}}$ is necessarily small for Re $\rightarrow \infty$ relative to the other terms in Equation (7–332). It should be noted, however, that the *radial* velocity outside the boundary-layers is determined solely by the radial pressure gradient in this limit, Re $\gg 1$, and is simply

$$u_r = \bar{r}\tanh \bar{t}.$$

This form for u_r is unaffected by the presence of thin boundary-layers, and the analysis presented here will be valid until the boundary-layers on the upper and lower boundaries *overlap*.

In order to determine the correct form for the governing equations in the vicinity of the boundaries $\bar{z} = 0$ and $\bar{z} = \bar{h}$, we must nondimensionalize again with the correct choice for the characteristic length scale l^*, so that gradients of F in the new nondimensionalized variable $\hat{z} = z'/l^*$ are $0(1)$ for Re $\gg 1$. This means that l^* must be representative of the actual distance over which the velocity changes from its value in the core of the thin-gap region to zero at the boundaries. In order to determine l^*, we again use the method that is known as *rescaling*. Instead of simply trying to "guess" l^* based upon some combination of dimensional parameters that has the dimensions of length, we introduce a rescaling of the nondimensionalized independent variable $\bar{z}$ in terms of the dimensionless parameter Re, namely,

$$\hat{z} = \bar{z}\text{Re}^{\delta}. \tag{7–408}$$

Since $\bar{z}$ is nondimensionalized with h_o, and Re$^{\delta}$ is dimensionless, this rescaling is exactly equivalent to nondimensionalizing by a new independent length scale

$$l^* = h_o\left(\frac{2h_0^2}{\mu a^2}\sqrt{\frac{W\rho}{\pi}}\right)^{-\delta}, \tag{7–409}$$

that is,

$$\hat{z} = \frac{z'}{l^*}, \tag{7–410}$$

where l^* is given by (7–409). The coefficient δ remains to be determined, and it is this factor that determines the physical significance of l^*. It may be noted, however, that for any $\delta > 0$, the length scale l^* characteristic of the boundary-layers at $\bar{z} = 0$ and $\bar{h}$ is extremely small compared with h_o for the limit Re $\gg 1$.

As we have seen in previous examples, the advantage of introducing a new non-dimensionalization through rescaling is that δ can be determined in a systematic way from the governing equation. Specifically, δ should have a value so that the viscous term in the newly nondimensionalized equation is of equal importance to the other largest terms for Re $\gg 1$. Thus, when the limit Re $\to \infty$ is applied to this rescaled equation to obtain the governing equation for the leading-order approximation in the vicinity of the boundaries, the viscous term must be retained. Before introducing (7–408) into (7–332), however, it is useful to apply (7–410) to the definitions of the velocity components $\bar{u}_r$ and $\bar{u}_z$ in terms of F, namely,

$$\bar{u}_r = \bar{r}\,\mathrm{Re}^\delta \frac{\partial F}{\partial \hat{z}} \quad \text{and} \quad \bar{u}_z = -2F. \tag{7–411}$$

Thus, if $\partial F/\partial \hat{z} = 0(1)$ as should be true if the nondimensionalization is correct, then the implication of (7–411) is that $\bar{u}_r$ in the boundary-layer is $0(\mathrm{Re}^\delta)$ for Re $\gg 1$. But, this cannot be true. The velocity $\bar{u}_r$ in the inviscid core of the thin gap is $0(1)$—that is, independent of Re—and this must match with $\bar{u}_r$ in the boundary-layer. Thus, we conclude that F in the boundary-layer must be $0(\mathrm{Re}^{-\delta})$, namely,

$$f = \mathrm{Re}^\delta F, \tag{7–412}$$

where $f(\hat{z}, \bar{t})$ is $0(1)$. This implies that

$$\bar{u}_r = \bar{r}\frac{\partial f}{\partial \hat{z}} = 0(1) \quad \text{and} \quad \bar{u}_z = -\frac{2}{\mathrm{Re}^\delta}f(\hat{z}, \bar{t}) = 0\left(\frac{1}{\mathrm{Re}^\delta}\right). \tag{7–413}$$

The latter means that $\bar{u}_z$ is arbitrarily small (for Re $\to \infty$) in the boundary-layer.

Introducing the rescaling (7–408) and (7–412) into the governing equation (7–332), we find that

$$\frac{\partial^2 f}{\partial \hat{z}\partial \bar{t}} + \left(\frac{\partial f}{\partial \hat{z}}\right)^2 - 2f\frac{d^2 f}{d\hat{z}^2} - \mathrm{Re}^{2\delta-1}\frac{d^3 f}{\partial \hat{z}^3} = \beta\frac{d^2 h}{d\bar{t}^2} + 1. \tag{7–414}$$

Thus, if the viscous term is to be retained in the boundary-layer for Re $\to \infty$, it must be of the same magnitude as other terms in (7–414). This implies that

$$2\delta - 1 = 0$$

or

$$\delta = \frac{1}{2}. \tag{7–415}$$

Thus, the rescalings (7–412) and (7–408) become

$$\boxed{z = \bar{z}\,\mathrm{Re}^{1/2} \quad \text{and} \quad f = F\cdot\mathrm{Re}^{1/2},} \tag{7–416}$$

and the characteristic length scale for the boundary-layer is

$$\boxed{\frac{l^*}{h_o} = \mathrm{Re}^{-1/2} \quad \text{for Re} \gg 1.} \tag{7–417}$$

This means that the boundary-layer becomes increasingly thin relative to h_o for increasingly large values of the Reynolds number. Of course, the width of the thin gap decreases monotonically with time, and eventually, the boundary-layers will merge no matter how large Re may be.

It may be noted that the governing equation in the boundary-layer region for this particular problem is just the complete original equation, now expressed in terms of boundary-layer variables, namely,

$$\frac{\partial^2 f}{\partial \hat{z} \partial \bar{t}} + \left(\frac{\partial f}{\partial \hat{z}}\right)^2 - 2f\frac{\partial^2 f}{\partial \hat{z}^2} - \frac{\partial^3 f}{\partial \hat{z}^3} = \beta \frac{d^2 h}{d\bar{t}^2} + 1. \tag{7-418}$$

However, this is unusual. Most often the governing equation in a boundary-layer will be simpler than the full equations. In the present problem, to determine the solution in the boundary-layer on the boundary $\hat{z} = 0$, we must solve (7–418) subject to the boundary conditions

$$f = \frac{\partial f}{\partial \hat{z}} = 0 \quad \text{at} \quad \hat{z} = 0 \tag{7-419}$$

plus the initial condition

$$f = 0 \quad \text{at} \quad \bar{t} = 0. \tag{7-420}$$

The boundary conditions at the other wall are not relevant to the solution of (7–418). In the dimensionless form shown, this equation applies only in the thin region within a distance $0(h_o \text{Re}^{-1/2})$ of the boundary at $\hat{z} = 0$. However, the equation is third order in $\hat{z}$, and we clearly need one more boundary condition in addition (7–421). This condition comes from the requirement that the solution in the boundary-layer matches the inviscid core solution. In this case, this is equivalent to the condition

$$f \cong \hat{z} \tanh \bar{t} \quad \text{for} \quad \hat{z} \gg 1. \tag{7-421}$$

We do not pursue the solution of this problem here. In view of the nonlinear nature of (7–418), an exact solution would likely require the use of numerical methods.

It is clear that the boundary-layer on the moving boundary at $\bar{z} = \bar{h}$ is the mirror image of that described previously for the wall at $\bar{z} = 0$. To obtain the equations and boundary conditions for this case, it is convenient to introduce the independent variable

$$\bar{y} = \bar{h}(\bar{t}) - \bar{z} \tag{7-422}$$

and to introduce the change of sign

$$F^* = -F. \tag{7-423}$$

Then (7–332) becomes

$$\frac{\partial^2 F^*}{\partial \bar{y} \partial \bar{t}} + \left(\frac{\partial F^*}{\partial \bar{y}}\right)^2 - 2F^*\frac{\partial^2 F^*}{\partial \bar{y}^2} - \frac{1}{\text{Re}}\frac{\partial^3 F^*}{\partial \bar{y}^3} = \beta \frac{d^2 h}{d\bar{t}^2} + 1.$$

The boundary-layer rescaling then takes the form

$$\hat{y} = \bar{y}\mathrm{Re}^{1/2}, \quad f = \mathrm{Re}^{1/2}F^*, \tag{7-424}$$

so that

$$\frac{\partial^2 f}{\partial \hat{y} \partial \bar{t}} + \left(\frac{\partial f}{\partial \hat{y}}\right)^2 - 2f\frac{\partial^2 f}{\partial \hat{y}^2} - \frac{\partial^3 f}{\partial \hat{y}^3} = \beta \frac{d^2 h}{d\bar{t}^2} + 1, \tag{7-425}$$

with boundary conditions

$$\left.\begin{array}{c} 2f = \dfrac{dh}{d\bar{t}}\sqrt{\mathrm{Re}} = \dfrac{d\hat{h}}{d\bar{t}} \\[2mm] \dfrac{\partial f}{\partial \hat{y}} = 0 \end{array}\right\} \hat{y} = 0, \tag{7-426}$$

$$f = 0 \quad \text{at} \quad \bar{t} = 0, \tag{7-427}$$

and matching

$$f \cong (\hat{y} - \hat{h})\tanh \bar{t} \quad \text{for} \quad \hat{y} \gg 1. \tag{7-428}$$

Again, a *numerical* method is necessary to solve (7-425)–(7-428).

We shall have much more to say later about *boundary-layers* and *singular* asymptotic expansions. At the present time, we simply note that the solution structure exhibited above is fairly typical when the asymptotic limit, based upon the obvious physical dimension as a characteristic length scale, yields an approximate form of the governing equations where the highest-order derivative is neglected. The resulting solution cannot possibly satisfy all of the boundary conditions of the original problem, and a different approximate form of the governing equations must then apply in the immediate vicinity of the boundaries. This implies that the characteristic length scale for this region must generally be different from the physical dimensions of the boundaries (and typically much smaller).

Homework Problems

1. Two parallel plane, circular disks of radius R lie one above the other a distance H apart. The space between them is filled with an incompressible Newtonian fluid. One disk approaches the other at constant velocity V, displacing the fluid. The pressure at the edge of the upper disk is atmospheric.

   a. Derive a dimensionless version of the above problem from the full Navier-Stokes equations and boundary conditions. What is the Reynolds number for this problem?

   b. Under what conditions can we apply the lubrication approximation to this problem? What is the appropriate choice for the characteristic pressure in the lubrication approximation?

c. Find the *dynamic* pressure distribution and show that the hydrodynamic force, F, resisting the motion in the lubrication limit is given by

$$F = - \frac{3\pi}{2} \frac{\mu R^4}{H^3} \frac{dH}{dt}.$$

d. Show that a constant force $-F$ applied to the upper disk will pull it well away from the bottom disk in a time $(3/4)\pi\mu R^4/(H_0^2 F)$, where H_0 is an initial spacing. Note that this time is large when H_0 is small. Give an example for the application of this phenomena.

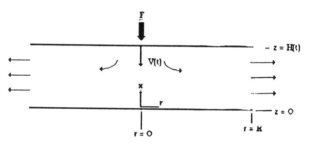

2. Two similar, rigid, plane rectangular plates are joined along an edge of length b by a smooth, oil-tight hinge and make a small angle $\alpha(t)$ with one another.

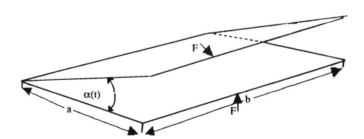

The space between them is filled with oil of viscosity μ, which is squeezed out as the plates are pushed together by a force, F, perpendicular to each plate, applied at the outer (free) edge. The outer edge is a distance $a \ll b$ from the hinge. Show that approximately

$$F = - \frac{3\mu ab}{2\alpha(t)^3} \frac{d\alpha}{dt}.$$

Note: A solution of this problem for arbitrary α was obtained in the creeping flow limit in Chapter 4. Here you are to use the fact $\alpha \ll 1$ to obtain an approximate solution via a lubrication type analysis.

3. Oil is forced by a constant, small pressure difference Δp through the narrow gap between two cylinders of radius a, with axes parallel and separated by a distance $2a + b$.

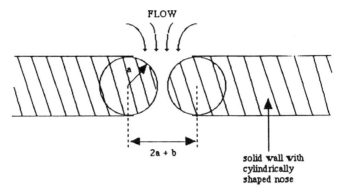

Show that if (b/a) and $(\rho\Delta p/\mu^2)\, b^3/a$ are sufficiently small, then the volume flux per unit width is approximately

$$\frac{2}{9\pi}\frac{\Delta p}{\mu}\left(\frac{b^5}{a}\right)^{1/2}$$

4. a. Two equal cylinders, radius a, with parallel axes are separated by a distance $2a + b$. The cylinders are rotated with angular velocity Ω and $-\Omega$, respectively. What is the volume flux of fluid/unit length pulled through the gap, what torque is required to maintain the angular velocity, and how much force must be exerted to maintain the gap width at a constant value?

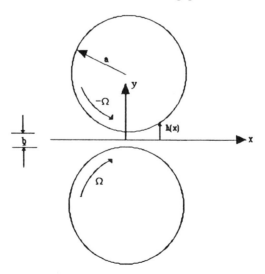

b. A cylinder of radius a is located with its axis parallel to an infinite, plane interface on which $\partial u/\partial y = 0$ and is separated from it by a distance $a + b$, where $b \ll a$. This cylinder rotates with angular velocity Ω and is attached at its ends to a piece of machinery above it that exerts a net downward force F on the cylinder. Can this load be supported by the high pressure generated in the lubrication layer, and if so, how fast must the cylinder be rotated? Does the answer change if the plane interface is replaced by an infinite, plane, no-slip boundary?

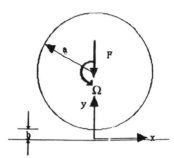

5. A sphere of radius a moves toward an infinite, plane, no-slip boundary under the action of a force F that is independent of time. The center of the sphere is located at a distance $b + a$ from the boundary, and we assume $b \ll a$.
   a. Under what conditions is the lubrication approximation valid?
   b. Determine the pressure distribution in the gap between the sphere and the wall using lubrication theory.
   c. Calculate the velocity of the sphere as a function of time.
   d. Determine how long it takes the sphere to contact the boundary, starting from some initial position $b(0)$.
   Repeat the above, assuming that the no-slip boundary is replaced by an interface that remains flat and nondeforming, and on which the shear stress vanishes. Explain any qualitative differences between the two cases, and discuss their implications for particle capture processes in qualitative terms. Can the results of this second calculation be applied to determine the motion of two spheres of equal radius that are acted upon by equal and oppositive forces along their line of centers? Explain.

6. A sphere is moving along the centerline of a circular cylinder. The sphere radius is a, and the cylinder radius is $a + b$, where $b \ll a$.
   a. Suppose that the cylinder is infinitely long and open at both ends to a large reservoir of the same fluid that is inside the cylinder. If the sphere is to move with a velocity U, what force must be applied? Is there any net volume flux of fluid through the tube? Calculate the relationship between volume flux and U.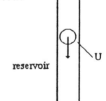
   b. It is proposed to use a cylinder and sphere arrangement like that described in (a) to estimate the viscosity of the fluid. Explain how this could be done. Suppose you have a cylinder with a *closed* end. How will this influence the relationship between the force applied and the sphere velocity? (First describe qualitatively what is going on, and then obtain a quantitative relationship between U and F.)

7. Consider converging flow (that is, flow inward toward the vertex) between two infinite plane walls with an included angle 2α, as shown in the sketch. The flow is best described in cylindrical coordinates (r, θ, z), with z being normal to the plane of the flow so that no derivatives exist with respect to z and $v_z = 0$. Also assume that the flow is entirely radial in the sense that fluid moves along rays $\theta = $ const toward the vertex, that is,

$$v_\theta = 0$$

(obviously, a solution of this form will only apply for large r, away from the slit at the vertex).

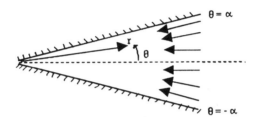

a. Show that $v_r = f(\theta)/r$.
b. Obtain the governing differential equation for $f(\theta)$,

$$\left(\frac{2\rho}{\mu}\right) f \frac{df}{d\theta} + \frac{d^3f}{d\theta^3} + 4\frac{df}{d\theta} = 0.$$

c. Assume no-slip conditions apply at the walls, and obtain a sufficient number of boundary conditions or other constraints on f to have a well-posed problem for $f(\theta)$.
d. Introduce a normalized angle ϕ and a normalized flow variable F, such that

$$\phi = \frac{\theta}{\alpha}, \quad F = \frac{\alpha f}{q},$$

where q is the volume flow rate per unit width in the z direction to obtain the dimensionless equation

$$\text{Re}\, F \frac{dF}{d\phi} + \frac{d^3F}{d\phi^3} + 4\alpha^2 \frac{dF}{d\phi} = 0$$

plus boundary conditions and auxiliary conditions on F where

$$\text{Re} \equiv \frac{2\rho q\alpha}{\mu} \quad \text{(Reynolds number)}.$$

e. Obtain an approximate solution for the creeping flow limit $\text{Re} \rightarrow 0$.
f. Obtain an approximate solution for $\alpha \rightarrow 0$.

g. Consider the solution for $Re \to \infty$; show that the solution reduces in this case to

$$F(\phi) = \frac{1}{2}.$$

This solution obviously does not satisfy no-slip conditions on the side walls, $\phi = \pm 1$. Explain why. Outline how a solution could be obtained for $Re \to \infty$ using the method of matched asymptotic expansions, which does satisfy boundary conditions on the walls. Be as detailed and explicit as possible, including actually setting up the equations and boundary conditions for the solution in the regions near the walls.

References/Notes

1. Kamal, M.M., "Separation in the Flow Between Eccentric Rotating Cylinders," *J. Basic Eng. ASME* 88:717–724 (1966).

Ballal, B.Y., and Rivlin, R.S., "Flow of a Newtonian Fluid Between Eccentric Rotating Cylinders: Inertial Effects," *Arch. Rational Mech. Anal.* 62:237–294 (1976).

2. A discussion of various rheometers, including the concentric cylinder configuration, can be found in a number of relatively recent books, for example, see Reference 6b and 6c in Chapter 2.

A more obscure reference specifically describing the concentric cylinder rheometer is Gabrysh, A.F., "An Automatic Concentric-Cylinder Rotational Viscometer for Rheological Measurements of non-Newtonian Flow," *Bulletin* (Univ. of Utah), Salt Lake City 51 (1960).

3. A very interesting series of studies of the influence of end effects in the rotating concentric cylinder problem has been published recently by Mullin and co-workers:

a. Mullin, T., "Mutations of Steady Cellular Flows in the Taylor Experiment," *J. Fluid Mech.* 121:207–218 (1982).

b. Benjamin, T.B., and Mullin, T., "Notes on the Multiplicity of Flows in the Taylor Experiment," *J. Fluid Mech.* 121:219–230 (1982).

c. Cliffe, K.A., and Mullin, T., "A Numerical and Experimental Study of Anomalous Modes in the Taylor Experiment," *J. Fluid Mech.* 153:243–258 (1985).

d. Pfister, G., Schmidt, H., Cliffe, K.A., and Mullin, T., "Bifurcation Phenomena in Taylor-Couette Flow in a Very Short Annulus," *J. Fluid Mech.* 191:1–18 (1988).

4. See Reference 4a, Chapter 3.

5. The following papers contain examples of numerical solutions for Newtonian fluids:

a. Beris, A.N., Armstrong, R.C., and Brown, R.A., "Finite Element Calculation of Viscoelastic Flow in a Journal Bearing, "I. Small Eccentricities," *J. Non-Newt. Fl. Mech.* 16:141–172 (1984); "II. Moderate Eccentricity," *J. Non-Newt. Fl. Mech.* 19:323–347 (1986).

b. Beris A.N., Armstrong, R.C., and Brown, R.A., "Spectral/Finite Element Calculations of the Flow of a Maxwell Fluid Between Eccentric Rotating Cylinders," *J. Non-Newt. Fl. Mech.* 22:129–167 (1987).

6. The method of domain perturbations was used for many years prior to its formal rationalization by D.D. Joseph:

a. Joseph, D.D., "Parameter and Domain Dependence of Eigenvalues of Elliptic Partial Differential Equations," *Arch. Rational Mech. Anal.* 24:325–351 (1967).

The method has been used for analysis of a number of different problems in fluid mechanics:

a. Beris, A., Armstrong, R.C., and Brown, R.A., "Perturbation Theory for Viscoelastic Fluids Between Eccentric Rotating Cylinders," *J. Non-Newt. Fl. Mech.* 13:109–148 (1983).

b. Cox, R.G., "The Deformation of a Drop in a General Time-Dependent Fluid Flow," *J. Fluid Mech.* 37:601–623 (1969).

c. Joseph, D.D., "Domain Perturbations: The Higher Order Theory of Infinitesimal Water Waves," *Arch. Rational Mech. Anal.* 51:295–303 (1975).

d. Joseph, D.D., and Fosdick, R., "The Free Surface on a Liquid Between Cylinders Rotating at Different Speeds," *Arch. Rational Mech. Anal.* 49:321–381 (1973).

7. The relevance of the rotating eccentric cylinder geometry to the journal-bearing lubrication system is discussed in many technical papers. As a starting point, the interested reader may wish to refer to

a. Booker, J.F., "Design of Dynamically Loaded Journal Bearings," *Fundamentals of the Design of Fluid Film Bearings.* ASME (1979).

b. Goenka, P.K., "Dynamically Loaded Journal Bearings: Finite Element Method Analysis," *Trans. ASME, J. of Tribology* 106:429–439 (1984).

c. Lodge, A.S., "A New Method of Measuring Multigrade Oil Shear Elasticity and Viscosity at High Shear Rates," SAE Tech. Paper 872043 (1987).

d. Walters, K., "Polymers as Additives in Lubricating Oils," *Colloques Internationaux de CNRS,* No. 233 *Polymeres et Lubrification* (Paris): 27–36 (1975).

8. A number of standard textbooks present general descriptions of lubrication theory, including the Reynolds equation and other aspects. The interested reader may be curious to examine the original paper of Reynolds:

a. Reynolds, O., "On the Theory of Lubrication and its Application to Mr. Beauchamp Tower's Experiments Including an Experimental Determination of the Viscosity of Olive Oil," *Phil. Trans. Roy. Soc.* A177:157 (1886).

A textbook description of lubrication phenomena can be found in

b. Cameron, A., *Principles of Lubrication.* Wiley: New York (1966).

c. Mitchell, A.G.M., *Lubrication: Its Principles and Practice.* Blackie: London (1950).

d. Pinkus, O., and Sternlicht, B., *Lubrication Theory.* McGraw-Hill: New York (1961).

9. Many papers have been written that discuss lubrication analyses of the interaction of a body and a plane wall or two bodies when the separation gap is vanishingly small compared with the length-scale of the body. The theoretical problem of a parallel cylinder and a wall considered here has been discussed by

a. Jeffrey, D.J., and Onishi, Y., "The Slow Motion of a Cylinder Next to a Plane Wall," *Quart. J. Mech. Appl. Meth.* 34:129–137 (1981).

A corresponding experimental investigation was reported by

b. Trahan, J.F., and Hussey, R.G., "The Stokes Drag on a Horizontal Cylinder Falling Toward a Horizontal Plane," *Physics of Fluids* 28:2961–2967 (1985).

The motion of a sphere near a plane wall in the lubrication limit was considered by

c. Cox, R.G., and Brenner, H., "The Slow Motion of a Sphere Through a Viscous Fluid Towards a Plane Surface, Part II, Small Gap Widths, Including Inertial Effects," *Chem. Eng. Sci.* 22:1753–1777 (1967).

d. Goldman, A.J., Cox, R.G., and Brenner, H., "Slow Viscous Motion of a Sphere Parallel to a Plane Wall, I. Motion Through a Quiescent Fluid," *Chem. Eng. Sci.* 22:637–651 (1967).

e. O'Neill, M.E., and Stewartson, K., "On the Slow Motion of a Sphere Parallel to a Nearby Wall," *J. Fluid Mech.* 27:705–724 (1967).

The motion of two drops was also considered recently by

f. Davis, R.H., Schonberg, J.A., and Rallison, J.M., "The Lubrication Force Between Two Viscous Drops," *Phys. Fluids* A1:77–81 (1989).

10. Weinbaum, S., Lawrence, C.J., and Kuang, Y., "The Inertial Draining of a Thin Fluid Layer Between Parallel Plates with a Constant Normal Force, Part 1, Analytic Solution; Inviscid and Small-But-Finite Reynolds Number Limits," *J. Fluid Mech.* 156:463–477 (1985).

Lawrence, C.J., Kuang, Y., and Weinbaum, S., "The Inertial Draining of a Thin Fluid Layer Between Parallel Plates with a Constant Normal Force, Part 2, Boundary-Layer and Exact Numerical Solutions," *J. Fluid Mech.* 156:479–494 (1985).

CHAPTER **8**

Weak Convection Effects

In Chapters 3, 4, and 5 we considered unidirectional and creeping flow problems in which the Navier-Stokes equations reduce, either exactly or approximately, to a linear form that is amenable to exact analytic solution. Then, in Chapters 6 and 7 we studied the consequences of small departures from unidirectional flows (as well as one-dimensional flows) and showed how asymptotic methods could be used to obtain approximate solutions even though the governing equations are nonlinear. In this chapter we turn back to fluid motions in the asymptotic limit $\text{Re} \to 0$ and extend our analysis to higher-order terms in the small Reynolds number solution, as is relevant for small, but nonzero, Reynolds number. This requires that we consider the first (weak) effects of momentum transport by convection. One motivation for this extension is to determine the magnitude and sign of inertial (or convective) corrections to the creeping flow predictions for the force and/or torque on a body that is moving in a viscous liquid. However, we shall see that an equally important outcome will be a better understanding of the limitations of the creeping flow approximation and the results obtained from it in Chapters 4 and 5.

An analogous problem to fluid motion at very low Reynolds numbers is the transfer of heat from a heated body to a fluid in the limit of "slow" motion where the transport process is dominated by conduction. In this case, the leading approximation is heat transfer by pure conduction, and we may again wish to extend the analysis to consider weak, but nonzero contributions due to convection. Although the two problems of fluid motion dominated by viscous effects (that is, diffusion of momentum or vorticity) and heat transfer dominated by conduction (that is, diffusion of heat) are very similar in structure, the heat transfer problem is simpler in detail and thus provides a convenient starting point for our discussion of *weak convection* effects.

Our organization in this chapter will be as follows. First, we briefly consider heat transfer from a heated three-dimensional body in the limiting case where conduction dominates convection; this requires, as a preliminary, some general discussion of convective heat transfer problems. Second, we attempt the addition of higher-order corrections via a regular perturbation scheme to account for weak, but nonzero, contributions due to convection in the heat transfer problem; this results in apparent fundamental difficulties. Similar, but more severe, problems are then shown to arise for an unbounded two-dimensional domain—for example, heat transfer from a heated circular cylinder. Third, we consider the source of these problems and show how they can be resolved via the method of matched asymptotic expansions that was first introduced in Chapter 6. Finally, following this completed discussion of the heat transfer case, we return to analyze the problem of streaming flow at small, but finite, Reynolds numbers.

A. Forced Convection Heat Transfer— General Considerations

We begin by considering the general problem of heat transfer from a hot or cold body to an adjacent flowing fluid. The starting point for any analysis of heat transfer effects in a flowing fluid is the thermal energy equation (2–117). The goal is to solve this equation to determine the temperature distribution in the fluid, given appropriate boundary conditions and the velocity field $\mathbf{u}$. In the problems of fluid motion considered in preceding sections of this book, we have assumed the fluid to be isothermal. One resultant simplification was that the fluid properties, such as density and viscosity, that depend upon temperature were treated as known constants, independent of position in the flow domain. When the fluid temperature is not uniform, however, all of the material properties, μ, ρ, C_p, and k, will vary from point to point throughout the domain. As a consequence, the problem of obtaining exact solutions for the velocity and temperature fields becomes highly coupled; in order to ascertain the values of μ, ρ, C_p, and k at any point, we must know the temperature T at that point. However, to determine T, we must know not only ρ, C_p, and k, which appear directly in the thermal energy equation, but also the velocity field $\mathbf{u}$. To obtain $\mathbf{u}$, we must know μ and ρ, and for this we must have T. Evidently, an exact solution would require that the equations of continuity, motion, and thermal energy be solved simultaneously.

As a consequence of the complexity of the full nonisothermal problem, as outlined in the preceding paragraph, it is necessary to introduce some rather severe approximations to make any progress. The most common approach is to completely neglect the spatial dependence of the material properties, approximating them as constant at the value appropriate to the ambient temperature of the fluid. The reader should recognize that this is a very severe approximation. At minimum, it requires that the actual variations in temperature be asymptotically small. Even then, the treatment of all material properties as constant is too harsh an assumption in some cases. For example, it is well known that a fluid near a heated (or cooled) body may move, even if there is no motion imposed on the fluid at large distances from the body. Such motions, called *natural convection,* are driven by the differences in density of the fluid that occur when the temperature of one part is different from that of another. Evidently, if we wish to analyze natural convection motions, we must retain at least the variation of density ρ with position in the nonisothermal fluid. Even if the fluid moves relative to a heated body for other reasons (for example, it may be "pumped"), there will always be some superposed *natural convection* present when the fluid is nonisothermal. However, if the density variations in the fluid are not too large, natural convection may

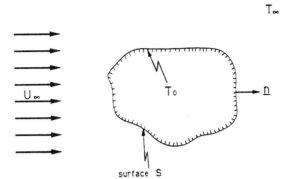

Figure 8–1 A schematic representation of a body of arbitrary shape and constant surface temperature, T_o, immersed in a fluid of constant properties, ρ, μ, C_p, and k, and ambient temperatue, T_∞, that undergoes a uniform motion of magnitude U_∞ relative to the body. The body surface is denoted as S, and the unit outer normal to the surface is **n**.

represent only a small correction to the motion that would be present in the absence of temperature variations, and it may be acceptable, as an approximation, to neglect natural convection altogether. This is tantamount to approximating the density as constant.

The approximation of all fluid properties as constant, including the complete neglect of any natural convection, is known as the *forced convection* approximation. We shall adopt it for this chapter, as well as Chapters 9 and 11. Unlike most approximations that we introduce in this book, the forced convection approximation will be adopted initially on an ad hoc basis, without a rigorous asymptotic justification. However, this is strictly a matter of convenience, and we shall return in Chapter 12 to the full problem, in which the temperature dependence of the density and other material properties is considered. At that point, we will be able to state explicitly the asymptotic limit inherent in the forced convection approximation and better understand the limitations inherent in results calculated on the basis that it is valid. For now, we simply accept it, being reassured that the resultant analysis is relevant to many conditions of practical interest by the fact that it has been adopted almost universally for analysis of heat transfer problems, in which the fluid motion is not due *solely* to natural convection.

In order to focus our subsequent discussion, let us begin by considering the problem of heat removal from a body of arbitrary shape that is immersed in a fluid that is undergoing a uniform motion of magnitude U_∞ relative to the body. The situation is illustrated in Figure 8-1. We shall suppose that the fluid has an ambient temperature T_∞ far from the body. Furthermore, we assume, for simplicity, that the body is heated in such a way that its surface is maintained at a constant, uniform temperature T_o. Finally, we adopt the forced convection approximation and treat the fluid properties ρ, μ, C_p, and k as constants with values corresponding to the ambient temperature T_∞. With all of these approximations, the thermal energy equation (2–117) becomes

$$\rho C_p \left(\frac{\partial T}{\partial t} + \mathbf{u} \cdot \nabla T \right) = k \nabla^2 T + 2\mu (\mathbf{E}:\mathbf{E}), \tag{8-1}$$

which is to be solved subject to boundary conditions

$$T = T_o \quad \text{on} \quad S, \tag{8-2a}$$

where S represents the body surface and

$$T \rightarrow T_\infty \quad \text{at} \quad |\mathbf{r}| \rightarrow \infty, \tag{8-2b}$$

where $\mathbf{r}$ is a position vector with origin inside S. The last term in (8–1) represents the irreversible dissipation of kinetic energy (associated with $\mathbf{u}$) to heat. In the forced convection approximation, with ρ and μ fixed, the Navier-Stokes and continuity equations can be solved (at least in principle) to determine the velocity field $\mathbf{u}$, and this solution is completely independent of the temperature distribution in the fluid. Once the velocity $\mathbf{u}$ is known, the thermal problem, represented by (8–1) and (8–2), can then be solved (again, in principle) to determine the temperature field, T. Since the boundary values of T are assumed to be constant, we may anticipate that the temperature distribution throughout the fluid will be independent of time (with the exception of some initial period after the heated body is first introduced into the moving fluid). It is the steady-state temperature field that is most often our goal. For this reason, the time derivative in (8–1) will be dropped in subsequent developments.

Certain features of the thermal problem can be ascertained via nondimensionalization alone. An appropriate characteristic velocity is U_∞. Furthermore, it is convenient to introduce a dimensionless temperature

$$\theta \equiv \frac{T - T_\infty}{T_0 - T_\infty}, \tag{8-3}$$

which varies between one at the body surface and zero at a large distance from the body. Finally, we suppose that variations in θ of $O(1)$ occur over a characteristic length scale that is proportional to a typical linear dimension of the body l. Then, following the examples of previous chapters, the steady-state nondimensionalized version of (8–1) becomes

$$\text{Pe}(\mathbf{u} \cdot \nabla \theta) = \nabla^2 \theta + 2\text{Br}(\mathbf{E}:\mathbf{E}). \tag{8-4}$$

Here, $\mathbf{u}$ and $\mathbf{E}$ represent the dimensionless velocity and rate-of-strain fields, and

$$\text{Pe} \equiv \frac{U_\infty l}{\kappa}, \quad \kappa = \frac{k}{\rho C_p}, \quad \text{and Br} \equiv \frac{\mu U_\infty^2}{k(T_0 - T_\infty)}.$$

The dimensionless parameter Pe is known as the *Peclet number* and provides a measure of the relative magnitude of the convection terms in (8–4) compared with conduction. It is sometimes useful to note that

$$\text{Pe} \equiv \text{RePr}, \tag{8-5}$$

where Re is the Reynolds number and

$$\text{Pr} \equiv \frac{\mu C_p}{k} = \frac{\nu}{\kappa} \tag{8-6}$$

is the *Prandtl number*. The Prandtl number depends on fluid properties only and is the ratio of the diffusivity for momentum (the kinematic viscosity) compared with the thermal diffusivity κ.

The relative magnitude of heat generation via viscous dissipation is determined by the dimensionless parameter Br, known as the *Brinkman number*. Although viscous dissipation can be significant for highly viscous fluids, the Brinkman number is often small, and we will consider only the limiting case Br $\rightarrow 0$, where the effects of viscous dissipation can be neglected completely.[1] In this case, (8-4) takes the simpler limiting form

$$\boxed{\text{Pe}(\mathbf{u} \cdot \nabla \theta) = \nabla^2 \theta.} \tag{8-7}$$

The dimensionless boundary conditions corresponding to equations (8-2a) and (8-2b) are

$$\boxed{\begin{aligned} \theta &= 1 \quad \text{on} \quad S, \\ \theta &\rightarrow 0 \quad \text{at} \quad |r| \rightarrow \infty. \end{aligned}} \tag{8-8}$$

It is clear from (8-7) and (8-8) that the form of the temperature distribution will depend upon the geometry of the heated body (represented by S) and the magnitude of the Peclet number, Pe. In addition, since the form of the forced convection velocity field depends on the Reynolds number, that is,

$$\mathbf{u} = \mathbf{u}(\mathbf{r}; \text{Re}),$$

we see from (8-7) and (8-8) that the temperature distribution for a body of specified shape also will depend upon Re, that is,

$$\theta = \theta(\mathbf{r}; \text{Re}, \text{Pe}) \tag{8-9a}$$

or, equivalently,

$$\theta = \theta(\mathbf{r}; \text{Re}, \text{Pr}). \tag{8-9b}$$

A primary objective of most calculations of the temperature field near a heated or cooled body is the overall rate of heat transfer between the body and the fluid. For the case of a body of fixed shape, the heat flux at the surface is due solely to heat conduction. Thus, the total rate of heat transfer is simply

$$\boxed{Q = \int_A -k(\nabla T \cdot \mathbf{n})\big|_s dA.} \tag{8-10}$$

Here, k is the thermal conductivity, while S stands for the body surface and $\mathbf{n}$ is the outer unit normal to the surface, as shown in Figure 8-1. The integration is carried out over the complete body surface, A.

The *dimensionless* total rate of heat transfer is known as the *Nusselt number*, Nu, and is normally defined as

$$\text{Nu} \equiv \frac{2Q}{(\text{surface area})k(T_o - T_\infty)/l}, \tag{8-11}$$

where l is the characteristic linear dimension of the body. Combining the definitions

(8–10) and (8–11) and expressing the integral in terms of dimensionless variables $(dA = l^2 d\bar{A})$,

$$\text{Nu} = \frac{-2l^2}{(\text{area})} \int_{\bar{A}} (\nabla\theta\cdot\mathbf{n})_{\bar{s}}\, d\bar{A}, \tag{8–12}$$

and thus, in view of (8–9), we see that

or

$$\boxed{\begin{aligned} \text{Nu} &= \text{Nu}(\text{Re}, \text{Pe}), \\ \text{Nu} &= \text{Nu}(\text{Re}, \text{Pr}). \end{aligned}} \tag{8–13}$$

Hence, from dimensional analysis alone, we see that a correlation should be expected relating Nu to the independent dimensionless parameters Re and Pe (or Re and Pr).

It was noted earlier that the dependence of θ and/or Nu on the Reynolds number is a consequence of the dependence of the velocity field on Re, as indicated just prior to (8–9). In some circumstances, however, the form of $\mathbf{u}$ is independent of Re. Two examples are the velocity profiles in unidirectional flows (Chapter 3) and the creeping flow approximation to the velocity fields for Re $\ll$ 1. In such cases

$$\mathbf{u} = \mathbf{u}(\mathbf{r})\ \text{only}$$

and

$$\boxed{\theta = \theta(\text{Pe}),\ \text{Nu} = \text{Nu}(\text{Pe}).} \tag{8–14}$$

In order to evaluate heat transfer rates via the definition (8–12) or (8–10), it is necessary to solve (8–7) and (8–8). Although this problem is linear in θ, the coefficient $\mathbf{u}$ in (8–7) is almost always too complex to allow exact analytic solutions for θ. The best that we can do, analytically, is to solve (8–7) approximately for limiting values of Re and Pe. We begin in this chapter with the limit Pe $\ll$ 1, in which convection effects are weak.

B. Heat Transfer by Conduction (Pe → 0)

When the dimensionless form of the thermal energy equation is compared with the dimensionless Navier-Stokes equation, it is clear that the Pectlet number plays a role for heat transfer that is analogous to the Reynolds number for fluid motion. Thus, it is natural to seek approximate solutions for asymptotically small values of the Peclet number, analogous to the low Reynolds number approximation of Chapters 4 and 5.

Let us suppose, then, that an asymptotic solution exists for Pe $\ll$ 1 in the form of a regular perturbation expansion

$$\boxed{\theta = \theta_0(\mathbf{x};\text{Re}) + \text{Pe}\,\theta_1(\mathbf{x};\text{Re}) + \cdots.} \tag{8–15}$$

The proposed asymptotic form (8–15) is completely analogous to the small Re approximation (5–220), where the first term was the creeping flow solution. If we substitute (8–15) into (8–7) and (8–8) and remember that $\mathbf{u}$ is independent of Pe, we see that the leading term satisfies

$$\boxed{\nabla^2\theta_0 = 0,} \tag{8–16}$$

with boundary conditions

$$\theta_0 = 1 \quad \text{on} \quad S,$$
$$\theta_0 \to 0 \quad \text{at} \quad |r| \to \infty. \tag{8-17}$$

Equation (8-16) is known as the steady-state *heat conduction equation* and is completely analogous to the creeping motion equation of Chapters 4 and 5. It can be seen that convection plays no role in the heat transfer process described by (8-16) and (8-17). Thus, the form of the velocity field is not relevant, and in spite of the initial assumption (8-15), there is no dependence of θ_0 on the Reynolds number of the flow. The solution of (8-16) and (8-17) depends only on the geometry of the body surface, represented in (8-17) by S.

A simple example where the problem (8-16) and (8-17) can be solved easily is the case of a heated sphere. In this case, we may choose the sphere radius as the characteristic length scale for nondimensionalization, and the problem is to solve (8-16) subject to the boundary condition $\theta = 1$ at $r = 1$. This can be done easily. We may first note that a general solution of Laplace's equation in spherical coordinates is

$$\theta_0 = \sum_{n=0}^{\infty} \left[A_n r^n + B_n r^{-(n+1)} \right] P_n(\eta). \tag{8-18}$$

Here $\eta \equiv \cos\theta$, and A_n and B_n are coefficients to be determined from the boundary conditions. The second of the conditions (8-17) requires that

$$A_n = 0 \quad \text{all } n.$$

The boundary condition at $r = 1$ is then satisfied provided that

$$B_n = 0, \ n \geq 1, \ \text{and} \ B_0 = 1.$$

Thus, the solution of (8-16) and (8-17) for a heated sphere is simply

$$\theta_0 = \frac{1}{r}. \tag{8-19}$$

According to this solution, which is plotted in Figure 8-2, the temperature falls off inversely with distance from the sphere in all directions. This very simple solution

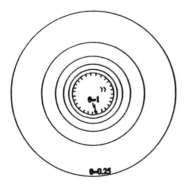

Figure 8-2 Contours of constant temperature for pure conduction heat transfer from a heated (or cooled) sphere. For constant surface temperature, $\theta = 1$, the isotherms are spherically symmetric, as indicated by Equation (8-19).

is the equivalent of Stokes's solution for creeping motion past a sphere. Indeed, it may be recalled from Chapter 4 that the disturbance to the velocity field due to a sphere in Stokes's flow also decreased as r^{-1} for large r.

With the approximate result (8–19) for the temperature field, we can easily evaluate the overall heat transfer rate, that is, the Nusselt number for this pure conduction limit. For a sphere, the general expression (8–12) for the Nusselt number becomes

$$ \text{Nu} = -\frac{1}{2\pi} \int_0^{2\pi} \int_{-1}^1 \left(\frac{\partial \theta}{\partial r} \right)_{r=1} d\eta d\phi, \tag{8–20} $$

where we have used $l = a$. Hence, substituting for $(\partial \theta / \partial r)_{r=1}$ from (8–19), we obtain

$$ \boxed{\text{Nu}_0 = 2.} \tag{8–21} $$

The subscript 0 in this case is intended to serve as a reminder that this result is only a first approximation to the dimensionless rate of heat transfer for $\text{Pe} \ll 1$.

C. Heat Transfer from a Solid Sphere in a Uniform Streaming Flow at Small, but Nonzero, Peclet Numbers

A natural question is how (or whether) the overall rate of heat transfer is modified by convection for small, but nonzero, values of the Peclet number. To obtain a more accurate estimate of Nu for this case, it would appear, from what we have said thus far, that we must calculate added terms in the regular asymptotic expansion (8–15) for θ. To attempt this, we must substitute (8–15) into (8–7) and (8–8) to obtain governing equations and boundary conditions for the subsequent terms θ_n. In this section, we consider only the second approximation, θ_1. The governing equation and boundary conditions derived from (8–7), (8–8), and (8–15) are

$$ \nabla^2 \theta_1 = \mathbf{u} \cdot \nabla \theta_0, \tag{8–22} $$

$$ \theta_1 = 0 \quad \text{on } S, \tag{8–23a} $$

and

$$ \theta_1 \to 0 \quad \text{at} \quad |\mathbf{r}| \to \infty. \tag{8–23b} $$

In order to complete the specification of the problem for θ_1, we must specify a particular velocity field $\mathbf{u}$. In case $\text{Re} \ll 1$, we can use the creeping flow solutions of Chapters 4 and 5, and it is again convenient to focus our attention on the case of a sphere in a uniform streaming flow, where a first approximation to the velocity field is given by the Stokes solution Equations (4–50), (4–64), and (4–65).

In this case, the full thermal energy Equation (8–7) takes the form

$$ \frac{1}{r^2} \left[\frac{\partial}{\partial r} \left(r^2 \frac{\partial \theta}{\partial r} \right) + \frac{\partial}{\partial \eta} \left((1 - \eta^2) \frac{\partial \theta}{\partial \eta} \right) \right] = $$
$$ \text{Pe} \left\{ \left(1 - \frac{3}{2r} + \frac{1}{2r^3} \right) \eta \frac{\partial \theta}{\partial r} + \frac{(1 - \eta^2)}{r} \left(1 - \frac{3}{4r} - \frac{1}{4r^3} \right) \frac{\partial \theta}{\partial \eta} \right\}. \tag{8–24} $$

Hence, the expansion (8–15) apparently can be written in the slightly simplified form [see equation (8–14)]

$$\theta = \theta_0(\mathbf{x}) + Pe\,\theta_1(\mathbf{x}) + \cdots, \tag{8–25}$$

and the governing Equation (8–22) for the second term, θ_1, becomes

$$\frac{1}{r^2}\left[\frac{\partial}{\partial r}\left(r^2\frac{\partial\theta_1}{\partial r}\right) + \frac{\partial}{\partial\eta}\left((1-\eta^2)\frac{\partial\theta_1}{\partial\eta}\right)\right] = -\frac{1}{r^2}\left(1 - \frac{3}{2r} + \frac{1}{2r^3}\right)P_1(\eta). \tag{8–26}$$

The homogeneous solution of this equation is just (8–18). A particular solution, corresponding to each of the terms on the right-hand side of (8–26), can be obtained in the form $r^s P_1(\eta)$. Combining the homogeneous and particular solutions, the general solution of (8–26) can thus be expressed as

$$\theta_1 = \sum_{n=0}^{\infty}\left[A_n r^n + B_n r^{-(n+1)}\right]P_n(\eta) + \left(\frac{1}{2} - \frac{3}{4r} - \frac{1}{8r^3}\right)P_1(\eta). \tag{8–27}$$

To this point, the regular perturbation scheme (8–15) [or (8–25)] has seemed to proceed in a completely straightforward manner. Now, however, when we attempt to determine values for A_n and B_n by application of the boundary conditions (8–23), we surprisingly find that no combination is satisfactory. The term $1/2\,P_1(\eta)$, which appears in the particular solution, does not vanish as $r\to\infty$, and there is no corresponding term in the homogeneous solution to cancel it so as to satisfy the condition (8–23b). The unavoidable, but seemingly paradoxical, conclusion is that no solution to (8–26) exists that will satisfy the boundary condition (8–23b) at infinity. Thus, an asymptotic solution of (8–24) with the boundary conditions (8–23) apparently does not exist in the form (8–25) or (8–15) for Pe $\ll 1$. This result is sometimes called *Whitehead's paradox* for convective heat transfer.

Although the conclusion of the preceding paragraph is clear, the reader may well be puzzled for an explanation of what has gone wrong. To ascertain the difficulty, we must carefully retrace our steps. In particular, the only two approximations that were made in the analysis leading to (8–26) were the neglect of viscous dissipation and the assumption inherent in (8–24) and (8–25) that convection terms in the thermal energy equation are negligible compared with conduction terms everywhere in the domain for Pe $\ll 1$, so that a uniformly valid *first* approximation to the temperature field in the limit Pe $\to 0$ can be obtained from the pure conduction equation (8–16). If both of these assumptions were correct, a uniformly valid asymptotic solution of the form (8–25) should be possible. The neglect of viscous dissipation is not a critical assumption; even if viscous dissipation were retained, the same difficulty would occur in attempting to obtain a solution of the form (8–25). The fact that such a solution does *not* exist can therefore indicate *only* that the neglect of convection everywhere in the fluid domain compared with conduction is incorrect, even as a first approximation for Pe $\to 0$. But this assumption is inherent in the nondimensionalized form of the thermal energy equation (8–7) or (8–24). In particular, at any point where the nondimensionalization is correct, the terms $\mathbf{u}\cdot\nabla\theta$ and $\nabla^2\theta$ are 0(1)—that is, *independent* of Pe—and the convection terms can be made arbitrarily small compared with conduction by simply letting Pe → 0. The unavoidable conclusion is that the characteristic scales $l_c = l$ and $u_c = U_\infty$ used to obtain equations (8–7) or (8–24) cannot be relevant everywhere in the domain. For purposes of simply writing the thermal energy equation in nondimensional form, this is not a serious problem. Indeed, in the transition between the exact dimensional and nondimensional forms—(8–1)–(8–4)—we are simply making a formal change of variables. If the original exact dimensional equation is correct, then the exact nondimensionalized equation is correct, independent of the length or velocity scales used in nondimensionalization. It is only when we attempt to approximate the nondimensionalized equation by estimating the magnitude of the various terms through the magnitude of dimensionless parameters such as Pe that we will make mistakes unless we have used the physically relevant characteristic scales in the nondimensionalization process. Since the problem in solving (8–26) arises in attempting to satisfy boundary conditions at infinity, we may surmise that Whitehead's paradox is a consequence of the fact that the radius of the sphere is not an appropriate characteristic length scale far from the sphere. Hence, the asymptotic approximation (8–25), based on (8–24), with (8–19) as the first term, is *not* valid in this region.

To corroborate this suggestion, it is instructive to consider the magnitudes of a typical conduction term in (8–25), $\partial^2\theta/\partial r^2$, and the largest convection term $\mathrm{Pe}\,\eta(\partial\theta/\partial r)$. If we evaluate the order of magnitude of these terms using the pure conduction approximation for θ, we find that

$$\frac{\partial^2\theta_0}{\partial r^2} \sim 0\left(\frac{1}{r^3}\right) \text{ and } \mathrm{Pe}\,\eta\,\frac{\partial\theta_0}{\partial r} \sim 0\left(\frac{\mathrm{Pe}}{r^2}\right) \text{ for } r \gg 1. \qquad (8\text{–}28)$$

The nondimensionalization leading to (8–7) and/or (8–24) has led to the conclusion that convection terms should be negligible compared with conduction terms anywhere in the domain if Pe is sufficiently small. However, the estimates (8–28) show clearly that this is not true. Specifically, for any arbitrarily small Pe, we can always find a value of $r \sim 0(\mathrm{Pe}^{-1})$ such that the conduction and convection terms are of equal importance. We conclude that the nondimensionalization leading to (8–7) can be valid only in the part of the domain within a distance $r < 0(\mathrm{Pe}^{-1})$ from the sphere. It follows that a regular (uniformly valid) asymptotic expansion cannot exist for Pe ≪ 1.

We may recall that the scenario described above bears a considerable resemblance to the problem of pulsatile flow at high frequencies, which was described in Chapter 6. We recommend that the reader review our discussion of the physical significance of nondimensionalization and the manner in which it may lead to incorrect approximations of governing equations when limits are applied to the dimensionless parameters that appear. In the pulsatile flow problem, we found that the obvious physical length scale, the tube radius, was not characteristic of velocity gradients in the region very near the tube wall for $\mathrm{Re}_\omega \gg 1$. Hence, a regular perturbation solution could not be found, and it was necessary to use the method of matched asymptotic expansions with different approximations to the governing equations valid in different parts of the domain, corresponding to the regions where the solution was characterized by different length scales. It may seem at first that a fundamental difference exists between the present problem and the pulsatile flow problem, since no solution satisfying boundary conditions could be obtained in the latter case, even for the first term in the asymptotic expansion, whereas in our problem we seemingly obtain a perfectly good solution for θ_0 but cannot find a solution for the second term, θ_1. However, this distinction is not fundamentally significant.

To illustrate this fact, we may consider the two-dimensional heat transfer problem of streaming flow past a heated circular cylinder with uniform surface temperature. In this case, if we look for a solution in the asymptotic form (8–15) for low Peclet numbers, the nondimensional governing equation and boundary conditions for θ_0 are again (8–16) and (8–17), but this time are expressed in cylindrical coordinates, namely,

$$\frac{1}{r}\frac{\partial}{\partial r}\left(r\frac{\partial\theta_0}{\partial r}\right) = 0, \tag{8–29}$$

$$\theta_0 = 1 \quad \text{at} \quad r = 1, \tag{8–30a}$$

and

$$\theta_0 \to 0 \quad \text{at} \quad r \to \infty. \tag{8–30b}$$

A general solution of (8–29) that is independent of position on the cylinder surface and at infinity, in accord with the boundary conditions (8–30), is

$$\theta_0 = C_1 \ln r + C_2, \tag{8–31}$$

where C_1 and C_2 are coefficients that must be determined from the boundary conditions. Upon examination, however, it is evident that no choice for these coefficients will actually satisfy the boundary conditions. In particular, to satisfy (8–30a), we require that $C_2 = 1$. However, the remaining term cannot be made to vanish at infinity for any choice of C_1. Hence, for this two-dimensional heat transfer problem, *no solution exists even for the first term* in a regular perturbation expansion for Pe $\ll 1$. This is analogous to the behavior found earlier for the pulsatile flow problem. In fact, it is the heated sphere problem that is unusual in the sense that a solution of (8–16) exists satisfying both (8–17a) and (8–17b) even though (8–16) is not a uniformly valid first approximation to (8–7), as we have already seen.

To obtain a valid approximate solution for heat transfer from a sphere in a uniform streaming flow at small, but nonzero, Peclet numbers, we must resort to the method of matched (or singular) asymptotic expansions.[2] In this method, as we have already seen in Chapters 6 and 7, two (or more) asymptotic approximations are proposed for the temperature field at Pe $\ll 1$, each valid in different portions of the domain but linked in a so-called overlap or matching region where it is required that the two approximations reduce to the same functional form. The approximate forms of (8–1), from which these matched expansions are derived, can be obtained by nondimensionalization using characteristic length scales that are *appropriate to each subdomain*.

Expansion in the Inner Region

Let us suppose, based upon the estimates of (8–28), that the dimensionless form (8–24) of the thermal energy equation is valid within the region $1 \le r \le 0(\text{Pe}^{-1})$. In other words, we suppose that within this so-called *inner* region, the sphere radius is an appropriate characteristic length scale as we have assumed in the nondimensionalization leading up to (8–24). Hence, within this region, the dimensionless temperature field can be represented in the form (8–25) with $\theta_0 = 1/r$, that is,

$$\theta(r, \eta, \text{Pe}) = \frac{1}{r} + \sum_{n=1}^{N} f_n(\text{Pe})\theta_n(r, \eta), \qquad (8\text{–}32)$$

where the gauge functions $f_n(\text{Pe})$ are, for the moment, unspecified except for the requirement that

$$f_1 \to 0 \quad \text{and} \quad \frac{f_{n+1}}{f_n} \to 0 \quad \text{as} \quad \text{Pe} \to 0. \qquad (8\text{–}33)$$

If $f_1(\text{Pe}) = \text{Pe}$, as assumed in the regular perturbation expansion (8–25), then, of course, the governing equation for θ_1 is still (8–26), and the general solution for θ_1 is still (8–27). However, we do not know a priori what form f_1 and the other $f_n(\text{Pe})$ in (8–32) should take; this must be determined as part of the solution of the problem.

The fact that the first term in (8–32) is still $1/r$ is a consequence of the fact that θ_0 must satisfy Laplace's equation, the boundary condition $\theta_0 = 1$ at $r = 1$, and also be a decreasing function of r in order that the total heat flux is conserved for increasing values of r. The only solution of $\nabla^2\theta_0 = 0$ with the latter two properties is $\theta_0 = 1/r$.

It should be noted, however, that the inner region does not extend to $r \to \infty$. Hence, boundary conditions cannot be imposed on any of the terms of (8–32) in the limit $r \to \infty$, and it is more or less accidental that the first term of (8–32) is consistent with the original boundary condition (8–8b) for $r \to \infty$. The most that we can require of the approximate solution form (8–32) is that it matches the corresponding small Pe approximation that is valid in the so-called outer region where $r \geq 0(\text{Pe}^{-1})$.

We shall return shortly to attempt to obtain a second approximation for θ, corresponding to the θ_1 term in (8–32). Before doing this, however, it is necessary to consider a first approximation to θ for the outer region.

Expansion in the Outer Region

In the outer part of the domain, the preceding discussion indicates that conduction and convection terms in the thermal energy equation will remain the same order of magnitude, even as Pe $\to 0$. Obviously, in the nondimensionalized form (8–7), this does not appear to be the case since both $(\mathbf{u} \cdot \nabla\theta)$ and $\nabla^2\theta$ should be $0(1)$ (that is, independent of Pe), provided that the choice of characteristic scales is correct. Thus, we conclude, as already stated, that the choice $l_c = a$ for a characteristic length scale must break down if we are too far from the sphere. Certainly, on intuitive grounds, this would seem to make sense. One would be surprised if any detailed feature of the temperature field at very large distances from a heated body were to remain sensitive to the body geometry—either its shape or size—except insofar as these factors control the total heat flux from the body to the fluid. Indeed, if we are sufficiently far from the sphere, we should anticipate that it will simply appear as a point source of heat in the uniform streaming flow. The question is, if the sphere radius is not appropriate as a characteristic length scale in this outer part of the domain, then what choice is appropriate and how shall we find it?

One approach is simply to guess until we (hopefully) hit the correct choice. An obvious possibility in the present problem is κ/U_∞, since this is another combination of independent dimensional parameters, independent of a, which has dimensions of length. However, this choice is by no means unique. There are, in fact, an infinite number of other possibilities, for example, $a^n(\kappa/U_\infty)^{1-n}$ for any n, $0 \le n < 1$. In some singular perturbation problems, the "obvious" guess turns out to be correct, but in others it does not. It is therefore imperative to develop a *systematic* strategy for determining the characteristic length scale for those parts of the domain where the physical dimension of the boundaries does not provide a characteristic length scale.

The most effective approach is to *rescale* the original nondimensionalized radial variable r in the very general form

$$\boxed{\rho = \mathrm{Pe}^m r.}$$

(8–34)

For any nonzero m, this newly rescaled radial variable ρ is just the original *dimensional* radial variable, r', nondimensionalized with respect to the new length scale

$$l^* = a^{1-m}\left(\frac{\kappa}{U_\infty}\right)^m.$$

In the present problem, the rescaling parameter m must be chosen in such a way that both conduction and convection terms are retained in the limit $\mathrm{Pe} \to 0$ when the thermal energy equation is expressed in terms of the outer variable ρ. Before actually attempting to determine m, however, the reader may wish to know why the radial variable is rescaled but not η or θ. In the case of θ, the governing equation is linear, and thus rescaling θ serves no purpose in spite of the fact that comparison with the leading term in the inner solution, expressed in terms of the outer variable ρ (as is appropriate in the overlap or matching region between the two solutions) shows clearly that the first approximation to θ in the outer region will be $0(\mathrm{Pe}^m)$. On the other hand, the independent variable η is bounded, $-1 \le \eta \le 1$, and there is simply no plausible basis to suggest that thermal gradients with respect to η will depend on Pe in some part of the domain. Thus, the only rescaling is (8–34). If (8–34) is now substituted into the dimensionless thermal energy equation (8–24) scaled with respect to a, we obtain the same equation scaled with respect to l^*,

$$\frac{\mathrm{Pe}^{2m}}{\rho^2}\left[\frac{\partial}{\partial\rho}\left(\rho^2\frac{\partial\oplus}{\partial\rho}\right) + \frac{\partial}{\partial\eta}\left((1-\eta^2)\frac{\partial\oplus}{\partial\eta}\right)\right] =$$
$$\mathrm{Pe}^{1+m}\left\{\left(1 - \frac{3\mathrm{Pe}^m}{2\rho} + \frac{\mathrm{Pe}^{3m}}{2\rho^3}\right)\eta\frac{\partial\oplus}{\partial\rho} + \frac{(1-\eta^2)}{\rho}\left(1 - \frac{3\mathrm{Pe}^m}{4\rho} - \frac{\mathrm{Pe}^{3m}}{4\rho^3}\right)\frac{\partial\oplus}{\partial\eta}\right\},$$

(8–35)

where we have denoted the dimensionless temperature as $\oplus$ in order to be able to distinguish it from the temperature approximation in the inner region, which we have denoted previously as θ. All that remains to completely specify the form of the governing equation in this outer region is to determine the parameter m and thus specify the characteristic length scale l^* for this region. To do this, we recall that conduction and convection terms are expected to remain in balance in this outer portion of the domain as Pe $\rightarrow 0$. We see from (8–35) that this implies

$$2m = 1 + m$$

or, simply,

$$m = 1. \tag{8–36}$$

In this case,

$$\rho = \mathrm{Pe}\, r = \left(\frac{U_\infty a}{\kappa}\right)\left(\frac{r'}{a}\right).$$

Thus, the appropriate characteristic length scale l^* in this outer portion of the domain is just κ/U_∞. Although this choice was recognized immediately in this problem as an obvious possibility (and thus could have been guessed a priori), we shall see eventually that the rescaling of variables is not usually so obvious.

With $m = 1$, the thermal energy Equation (8–35) can now be rewritten in the dimensionless form appropriate to the outer region of the domain:

$$\nabla_\rho^2 \oplus = \left[\eta\, \frac{\partial \oplus}{\partial \rho} + \frac{(1 - \eta^2)}{\rho}\frac{\partial \oplus}{\partial \eta}\right] +$$
$$\mathrm{Pe}\left[-\frac{3\eta}{2\rho}\frac{\partial \oplus}{\partial \rho} - \frac{3}{4\rho^2}(1 - \eta^2)\frac{\partial \oplus}{\partial \eta}\right] + \mathrm{Pe}^3\left[\frac{\eta}{2\rho^3}\frac{\partial \oplus}{\partial \rho} + \frac{(1 - \eta^2)}{4\rho^4}\frac{\partial \oplus}{\partial \eta}\right]. \tag{8–37}$$

If we examine the terms on the right-hand side, we see that the largest convection contribution for Pe $\rightarrow 0$ is that due to the undisturbed, uniform streaming flow. The corrections to the rate of convection due to the modifications of the velocity field by the sphere are $0(\mathrm{Pe})$ and $0(\mathrm{Pe}^3)$, respectively. According to (8–8b), the solution of (8–37) is required to vanish as $\rho \rightarrow \infty$. The boundary condition (8–8a) does not apply to the outer solution since the sphere surface is not a part of the outer region. Instead, the solution of (8–37) is required to match in the limit Pe $\rightarrow 0$ with the inner approximation to the solution of the full problem, (8–32), for $\rho \ll 1$. A schematic of the solution domain, summarizing the scaling, governing equations and boundary conditions is shown in Figure 8–3.

We now suppose that the outer solution for Pe $\ll 1$ can be represented by an asymptotic expansion of the form

$$\oplus(\rho, \eta, \mathrm{Pe}) = \sum_{n=0}^{N} F_n(\mathrm{Pe})\oplus_n(\rho, \eta), \tag{8–38}$$

where the gauge functions $F_n(\mathrm{Pe})$ are unknown but must again satisfy the requirement that

$$\frac{F_{n+1}}{F_n} \rightarrow 0 \quad \text{as} \quad \mathrm{Pe} \rightarrow 0.$$

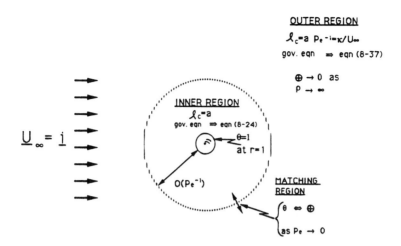

Figure 8-3 A schematic representation of the solution domain for heat transfer from a sphere in a uniform flow in the limit of small Peclet numbers, that is, Pe → 0.

On substituting (8–38) into (8–37), we find that, to leading order,

$$\nabla_\rho^2 \oplus_0 = \eta \frac{\partial \oplus_0}{\partial \rho} + \frac{(1 - \eta^2)}{\rho} \frac{\partial \oplus_0}{\partial \eta},$$
(8–39)

which, as we have already noted, is nothing more than the thermal energy equation with the velocity **u** replaced by the nondimensionalized, free stream velocity, $\mathbf{i}_\infty$. To solve (8–39), it is convenient to introduce the transformation, first discovered by Goldstein (1929),[3]

$$\oplus_0 = e^{\rho\eta/2}\Phi,$$
(8–40)

with Φ now considered as the independent unknown. Substituting (8–40) into (8–39), we obtain

$$\nabla_\rho^2 \Phi - \frac{1}{4}\Phi = 0.$$
(8–41)

A general solution of this equation, which is bounded at $\eta = \pm 1$, can be obtained easily by separation of variables in terms of modified Bessel functions of the first and second kind, $I_{n+1/2}(\rho/2)$ and $K_{n+1/2}(\rho/2)$. However, $I_{n+1/2}(\rho/2) \to \infty$ *exponentially* as $\rho \to \infty$. Thus, in view of the condition $\oplus \to 0$ as $\rho \to \infty$, the most general allowable solution for Φ is

$$\Phi = \sum_0^\infty A_n \sqrt{\frac{\pi}{\rho}} K_{n+1/2}\left(\frac{\rho}{2}\right) P_n(\eta).$$
(8–42)

After substituting back into (8–40), we have

$$\oplus_0 = \sqrt{\frac{\pi}{\rho}}\, e^{\rho\eta/2} \sum_{k=0}^{\infty} A_k K_{k+1/2}\left(\frac{\rho}{2}\right) P_k(\eta), \qquad (8\text{–}43)$$

where $K_{k+1/2}$ is the Bessel function of the second kind and half order, and $P_k(\eta)$ is the Legendre polynomial of order k.

The coefficients A_k must be determined so that the solution (8–43) matches, at leading order, the first (pure conduction) approximation for the inner solution θ, that is,

$$F_0(\text{Pe})\oplus_0|_{\rho \ll 1} \Longleftrightarrow \frac{1}{r} \quad \text{as} \quad \text{Pe} \to 0. \qquad (8\text{–}44)$$

In order to carry out this matching, we first note that

$$K_{k+1/2}\left(\frac{\rho}{2}\right) = \sqrt{\frac{\pi}{\rho}}\, e^{-\rho/2} \sum_{m=0}^{k} \frac{(k+m)!}{(k-m)!\,m!} \frac{1}{\rho^m} \qquad (8\text{–}45)$$

and then express the inner approximation $1/r$ in terms of outer variables so that the two sides of (8–44) can be compared directly, that is,

$$F_0(\text{Pe})\oplus_0(\rho,\eta)|_{\rho \ll 1} \Longleftrightarrow \frac{\text{Pe}}{\rho} \quad \text{as} \quad \text{Pe} \to 0. \qquad (8\text{–}46)$$

It is immediately obvious that the first approximation in the inner and outer regions can match only if

$$F_0(\text{Pe}) = \text{Pe}. \qquad (8\text{–}47)$$

In order to complete the matching, we may note that the matching region between inner and outer expansions corresponds to $\rho \ll 1$ and/or $r \gg 1$. Indeed, for any arbitrarily large, but finite, value of r, the corresponding value of ρ can be made arbitrarily small, though nonzero, by taking the limit $\text{Pe} \to 0$. Thus, for purposes of matching, we can approximate the left-hand side of (8–46) by expanding the exponential function as a power series in ρ. At leading order, $\exp(\rho(\eta-1)/2) \sim 1 + 0(\rho)$. Thus, the matching condition (8–46) can be expressed in the form

$$\text{Pe}\left(\pi \sum_{k=0}^{\infty} A_k P_k(\eta)\left[1 + 0(\rho)\right] \sum_{m=0}^{k} \frac{(k+m)!}{(k-m)!\,m!} \frac{1}{\rho^{m+1}}\right) \Longleftrightarrow \frac{\text{Pe}}{\rho} \qquad (8\text{–}48)$$

as $\text{Pe} \to 0$. Comparing the two sides of (8–48), it is evident that the inner and outer approximations can match only if

$$A_0 = \frac{1}{\pi}, \quad A_k = 0 \ (k \geq 1). \qquad (8\text{–}49)$$

With $A_0 = 1/\pi$, it is obvious that there is a term, $1/\rho$, on the left-hand side that matches precisely with the inner solution. The fact that all other $A_k = 0$ may require some additional explanation.

Let us suppose, for example, that $A_1 \neq 0$. Then there would be an additional term in (8–48) from the outer solution of the form

$$\text{Pe}\left[\pi A_1 P_1(\eta)\left[1 + 0(\rho)\right]\left(\frac{1}{\rho} + \frac{2}{\rho^2}\right)\right].$$

But such a term could not be matched with the solution in the inner region, either in the present form or even after we have added additional terms to the inner expansion. For example, to obtain a term from the inner solution that would equal Pe/ρ^2 when expressed in terms of outer variables, it would be necessary to start with a term in the inner solution of the form

$$\frac{1}{\text{Pe}}\left(\frac{1}{r^2}\right).$$

But this would imply the existence of a term $0(\text{Pe}^{-1})$ in the inner approximation, and such a term would be asymptotically large compared with the largest term in the inner expansion (8–32), which we have already proven to be $0(1)$, due to the boundary value $\theta = 1$ at $r = 1$. It is therefore evident that no such term can exist and $A_1 = 0$. Similarly, it can be shown that the other $A_k \equiv 0$, as these would require terms of $0(\text{Pe}^{-k})$ in the inner expansion for matching.

Finally, let us examine the matching condition (8–48) once more with the coefficients A_k evaluated according to (8–49). The result is

$$(\text{Pe})\frac{1}{\rho}\left[1 + 0(\rho)\right] \Longleftrightarrow \frac{1}{\rho}(\text{Pe}) \quad \text{for Pe} \to 0. \tag{8–50}$$

The $0(\rho)$ term on the left-hand side is just the second term of the power-series approximation of $\exp(\rho(\eta - 1)/2)$. We see that the first term on the left matches precisely with the first approximation in the inner region. However, there is a *mismatch* due to the second term from the exponential, which has no counterpart in the first approximation for the inner solution. However, this term is $0(\text{Pe})$ and independent of ρ. A term of this form could only be generated in the inner region from a term that is $0(\text{Pe})$ and independent of r. But, we have so far considered only the first term in the inner region, which is $0(1)$. The mismatch evident in (8–50) at $0(\text{Pe})$ can be made arbitrarily small by taking the asymptotic limit $\text{Pe} \to 0$ and can only be removed by considering additional terms in the expansion (8–32) for the inner region. We will return to this task shortly.

First, however, it is worthwhile to consider briefly the qualitative features of the first approximation to the temperature field in the outer region, which we have just shown to be

$$\boxed{\oplus(\rho, \eta, \text{Pe}) = \frac{\text{Pe}}{\rho}\, e^{-\rho(1-\eta)/2} + 0(F_1(\text{Pe})),} \tag{8–51}$$

where $F_1/\text{Pe} \to 0$ as $\text{Pe} \to 0$. This is, in fact, nothing more than the fundamental solution for a point source of heat at $\rho = 0$ in a uniform streaming flow. Contours of constant temperature, corresponding to this solution, are shown in Figure 8–4. In contrast

HEATED SPHERE

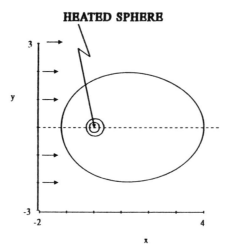

Figure 8–4 Contours of constant temperature (isotherms) for heat transfer from a sphere in a uniform flow at low Peclet numbers according to Equation (8–51). Note that the sphere appears in this representation, in which $l_c = \kappa/U_\infty$, as a *point* source (sink) of heat (the radius of the sphere, a, is vanishingly small compared with κ/U_∞ in the limit as Pe $\to 0$). In the inner region, near to the sphere, the isotherms at leading order of approximation are still spherical as illustrated in Figure 8-2. The three contours plotted are $\oplus = 0.25$, 2.75, and 5.25.

to the temperature field for pure conduction, which we showed to decrease linearly with distance from the sphere, we see from (8–51) that the presence of convection in the outer region causes $\oplus$ to diminish exponentially with distance from the sphere, except in a region arbitrarily near $\eta = 1$, where it continues to decrease algebraically as $1/\rho$. Thus, in spite of the fact that the pure conduction solution appeared to satisfy the original boundary condition $\theta = 0$ as $r \to \infty$, and thus to provide a uniformly valid first approximation to the temperature field for Pe $\ll 1$, we see now that it did not actually have the correct functional dependence on r for $r \to \infty$. The symmetry axis $\eta = 1$, where $\oplus$ continues to decrease algebraically, lies downstrem of the sphere, and the solution (8–51) is thus seen to exhibit a *thermal wake* structure. Heat released from the sphere is, at very large distances, mainly concentrated very near the symmetry axis $\eta = 1$, where it is transported by convection with the uniform streaming motion of the fluid. Unlike the first approximation in the inner region, which is radially symmetric, this first approximation in the outer region contains a contribution due to convection, and the fore-aft symmetry of the inner (conduction-dominated) temperature field is lost as the motion of the fluid carries heat in the downstream direction.

A Second Approximation in the Inner Region

The first approximation to the temperature field in the outer region, Equation (8–51), has been constructed in such a way that it asymptotically matches the first approximation in the inner region, $\theta = 1/r$, in the region of overlap. However, we have already noted that the match between these two leading-order approximations is not perfect. Rather, there is a $0(\mathrm{Pe})$ mismatch, which can be made arbitrarily small in the limit $\mathrm{Pe} \to 0$ but is nonetheless nonzero. To determine the precise form of this mismatch, we investigate the approximate form of (8–51) for $\rho \ll 1$. In particular, if we expand the exponential and retain the first two terms, we find that

$$\oplus(\rho, \eta, \mathrm{Pe}) \to \frac{\mathrm{Pe}}{\rho} + \frac{\mathrm{Pe}}{2}(P_1 - P_0) + 0(\mathrm{Pe}\rho) + 0(F_1(\mathrm{Pe})).$$

When written in terms of the inner variable $r = \rho/\mathrm{Pe}$, this becomes

$$\oplus(\rho, \eta, \mathrm{Pe}) = \frac{1}{r} + \frac{\mathrm{Pe}}{2}(P_1 - P_0) + 0(\mathrm{Pe}^2 r) + 0(F_1(\mathrm{Pe})). \qquad (8\text{--}52)$$

The first term matches identically with the pure conduction, leading-order approximation in the inner region, $1/r$, but the second term clearly represents an $0(\mathrm{Pe})$ mismatch between the first approximations in the two regions. The third term from the power-series approximation of the exponential is $0(\mathrm{Pe}\rho)$, but we see from (8–52) that this leads to a mismatch that is $0(\mathrm{Pe})^2$ when expressed in terms of inner variables. A term of $0(\mathrm{Pe}^2)$ cannot appear until at least the third approximation in the inner region.

We conclude from (8–52) that the largest mismatch between the first terms in the asymptotic expansions for the inner and outer regions is $0(\mathrm{Pe})$. In addition, we may note that the terms that were neglected in the governing equation for the inner region at the leading order of approximation are also $0(\mathrm{Pe})$. Hence, the magnitude of the second term in the inner expansion must be $0(\mathrm{Pe})$—that is, $f_1(\mathrm{Pe}) \equiv \mathrm{Pe}$—and the solution for θ_1 will both include the first $0(\mathrm{Pe})$ convection contributions in the governing equation (8–24) for the inner region and also match the $0(\mathrm{Pe})$ term in (8–52) in order to remove the largest mismatch between the inner and outer approximations. With $f_1(\mathrm{Pe}) \equiv \mathrm{Pe}$, the governing equation for θ_1 is precisely the same, Equation (8–26), that led to Whitehead's paradox. Hence, the general solution is exactly (8–27). Now, however, we recognize θ_1 as the second term in an expansion that holds only in the inner portion of the domain, rather than the second term in a regular expansion that was assumed valid throughout the domain. Thus, unlike the analysis that led to Whitehead's paradox, we do not attempt to apply boundary conditions on θ_1 for $r \to \infty$, but instead require that this solution match at $0(\mathrm{Pe})$ with the first approximation in the outer region.

Hence, to determine the correct form for θ_1 in the inner region, we first apply the boundary condition

$$\theta_1 = 0 \quad \text{at} \quad r = 1 \qquad (8\text{--}53)$$

to (8–27). The result is

$$A_n = -B_n, \; n = 0, 2, 3, \cdots,$$

and

$$A_1 = \frac{3}{8} - B_1. \qquad (8\text{--}54)$$

Thus,

$$\theta_1 = \sum_{n=0}^{\infty} A_n \left[r^n - r^{-(n+1)} \right] P_n(\eta) + \left(\frac{1}{2} - \frac{3}{4r} + \frac{3}{8r^2} - \frac{1}{8r^3} \right) P_1(\eta), \qquad (8\text{-}55)$$

and we are left with the task of determining the coefficients A_n. To do this, let us consider the matching between the outer solution for small ρ, equation (8-52), and the first two terms of the inner solution, namely,

$$\theta_0 + \text{Pe}\,\theta_1 + 0(f_2(\text{Pe}))\big|_{r \gg 1} \Longleftrightarrow \text{Pe}\oplus_0 + 0(F_1(\text{Pe}))\big|_{\rho \ll 1} \quad \text{as} \quad \text{Pe} \to 0. \qquad (8\text{-}56)$$

Using (8-55) for large r and (8-52), this matching condition reduces to

$$\frac{1}{r} + \text{Pe} \left[\sum_{n=0}^{\infty} A_n r^n P_n(\eta) + \frac{1}{2} P_1 + 0\left(\frac{1}{r} \right) \right] + 0(f_2(\text{Pe})) \Longleftrightarrow$$

$$\frac{1}{r} + \frac{\text{Pe}}{2} (P_1 - P_0) + 0(\text{Pe}^2 r) + 0(F_1(\text{Pe})) \quad \text{as} \quad \text{Pe} \to 0. \qquad (8\text{-}57)$$

Hence, we see that the two expressions are identical up to and including all terms of $0(\text{Pe})$ if

$$A_o = -\frac{1}{2} \quad \text{and} \quad A_n = 0 \quad \text{for} \quad n \geq 1. \qquad (8\text{-}58)$$

We also see that the term $\text{Pe}\,P_1/2$, which led to Whitehead's paradox, matches automatically with a term in the outer solution. As we suggested earlier, it is now evident that Whitehead's paradox resulted from an attempt to apply boundary conditions on the inner solution for $r \to \infty$, where it is not a valid approximation, instead of matching asymptotically with the outer solution for large r in the limit $\text{Pe} \to 0$.

Before proceeding, it is perhaps useful to return momentarily to the matching condition (8-57) to ensure that we understand why terms like $0(\text{Pe}/r)$ in the inner solution and $0(\text{Pe}^2 r)$ in the outer solution can be neglected in determining A_n. The latter is neglected at the present level of approximation because we have so far considered only the $0(1)$ and $0(\text{Pe})$ terms in the inner expansion. If we proceed to a term of $0(\text{Pe}^2)$ in the inner expansion, however, the $\text{Pe}^2 r$ term from $F_0 \oplus_0$ would clearly have to be included in the matching. In the inner expansion on the left-hand side of (8-57), we have neglected $0\left[(1/r)\text{Pe} \right]$, as well as smaller terms $0\left[(1/r^n)\text{Pe} \right]$. This is because such terms can be matched only to higher-order terms in the outer expansion. To see this, we can express the term Pe/r in outer variables

$$0\left(\frac{1}{r} \text{Pe} \right) \to 0\left(\frac{1}{\rho} \text{Pe}^2 \right).$$

Hence, we see that it corresponds to a contribution of $0(\text{Pe}^2)$ in the outer expansion and thus again need not be considered in the matching at the present level of approximation where we have included only the first $0(\text{Pe})$ term in the outer region.

With the coefficients A_n determined, according to (8-58), the complete inner solution to $0(\text{Pe})$ is

$$\theta = \frac{1}{r} + \text{Pe}\left\{\frac{1}{2}\left(\frac{1}{r} - 1\right) + \eta\left(\frac{1}{2} - \frac{3}{4r} + \frac{3}{8r^2} - \frac{1}{8r^3}\right)\right\} + 0(f_2(\text{Pe})). \qquad (8\text{-}59)$$

Although terms like $(1/r)\text{Pe}$ can be neglected in the matching process, they are not generally negligible in the inner region and thus must be included in (8-59).

We noted at the beginning of this chapter that a primary objective in heat transfer calculations is the overall rate of heat transfer. In dimensionless terms this is the Nusselt number [see equation (8-20)]. With the first two terms of the inner solution in hand, we are now in a position to extend our previous result for pure conduction from a sphere, (8-21), to include the first $0(\text{Pe})$ contribution of convection for small Pe. In particular, substituting the approximation (8-59) into (8-20) and carrying out the integration, we find that

$$\text{Nu} = 2\left(1 + \frac{1}{2}\text{Pe} + 0(f_2(\text{Pe}))\right). \qquad (8\text{-}60)$$

The influence of convection at small Pe is a small increase in the overall rate of heat transfer above its value for pure conduction.

Higher-Order Approximations

A great potential advantage of the matched asymptotic expansion procedure is that we can generate as many terms as we wish in the asymptotic expansion by simply extending the procedure described above. Of course, the effort required to do this increases as we proceed, due to the increased algebraic complexity of the problem at each level of approximation. One way around this difficulty is to use so-called symbolic manipulation packages to carry out the algebra on a computer, and this has actually been done in a few instances. However, we will not discuss this approach here. Part of the reason is that the addition of large numbers of additional terms to a result like (8-60) often does not significantly extend the range of the small parameter where the asymptotic expression gives accurate results.

In spite of the limited practical value of calculating many terms in the asymptotic expansion for small Pe, it is nevertheless instructive to consider briefly the addition of the next term, for it illustrates a fundamental difference between regular and singular asymptotic expansions. When there is a regular asymptotic solution for some differential equation and boundary conditions, the expansion must always proceed sequentially in increasing powers of the small parameter that appears in the problem. Thus, if the parameter is ϵ, the expansion will be $(1, \epsilon, \epsilon^2, \epsilon^3, \dots)$. The fact that this is true can be seen easily if we note that the sequence of equations generated by a regular perturbation scheme is identical to that obtained in a solution by the method of successive approximations. Returning to the heated sphere problem, it is tempting to assume that the third term in the inner expansion (8-32) should be $0(\text{Pe}^2)$, that is,

$$\theta = \frac{1}{r} + \text{Pe}\,\theta_1 + \text{Pe}^2\,\theta_2 + \cdots . \qquad (8\text{-}61)$$

However, Acrivos and Taylor (1962)[4] have shown that the particular solution to the equation for θ_2 contains a term $-1/2 \ln r$ [see their equation (31)] that when multiplied by Pe^2 and expressed in the outer variable, $\rho = r Pe$, generates a term $-(Pe^2/2) \ln Pe$ that must be matched with a corresponding term in the outer expansion. But this is not possible since the outer solution does not contain a term $0(Pe^2 \ln Pe)$. Consequently, we must backtrack and assume that $f_2(Pe)$ in (8–32) is $(Pe^2 \ln Pe)$ instead of Pe^2. In this case

$$\boxed{\nabla_r^2 \theta_2 = 0} \tag{8–62}$$

and

$$\boxed{\theta_2 = \frac{1}{2}\left(\frac{1}{r} - 1\right).} \tag{8–63}$$

Acrivos and Taylor actually evaluated two additional terms in the inner expansion. Their expression for the Nusselt number is

$$\boxed{Nu = 2\left(1 + \frac{1}{2}Pe + \frac{1}{2}Pe^2 \ln Pe + 0.41465 Pe^2 + \frac{1}{4}Pe^3 \ln Pe + 0(Pe^3)\right).} \tag{8–64}$$

Unlike the regular perturbation expansion discussed earlier, the method of matched asymptotic expansions often leads to a sequence of gauge functions that contain terms like $Pe^2 \ln Pe$ or $Pe^3 \ln Pe$ that are intermediate to simple powers of Pe. Thus, unlike the regular perturbation case, where the form of the sequence of gauge functions can be anticipated in advance, this is not generally possible when the asymptotic limit is singular. In the latter case, the sequence of gauge functions must be determined as a part of the matched asymptotic solution procedure.

D. Heat Transfer from a Body of Arbitrary Shape in a Uniform Streaming Flow at Small, but Nonzero, Peclet Numbers

We have invested considerable effort to analyze the rate of heat transfer from a heated sphere in a uniform streaming flow at low Peclet number. Now that we understand the asymptotic structure of the low Peclet number limit, however, we can very easily extend our result for the first two terms in the expression for Nu, (8–60), for the same flow to bodies of *arbitrary* shape, which may be either *solid* or *fluid,* and to *arbitrary values of the Reynolds number* provided only that Pe $\ll$ 1. This extension was first demonstrated by Brenner (1963),[5] and our discussion largely follows his original analysis.

Let us first consider the general formulation of the problem. If we adopt the linear dimension of the body as a characteristic length scale, we have seen that the governing equations and boundary conditions for forced convection heat transfer can be written in the forms (8–7) and (8–8), namely,

$$Pe(\mathbf{u} \cdot \nabla\theta) = \nabla^2\theta, \tag{8-65}$$

$$\theta = 1 \quad \text{on} \quad S, \tag{8-66a}$$

and

$$\theta \to 0 \quad \text{as} \quad |\mathbf{r}| \to \infty, \tag{8-66b}$$

and this is true independent of the body geometry. Furthermore, for small values of Pe, the solution of (8-65) for a body of arbitrary shape must take the form of a matched asymptotic expansion, with the solution in the previous section for a solid sphere being a special case. Thus, there exists an inner region, within a dimensionless distance $|\mathbf{r}| < 0(\text{Pe}^{-1})$ from the body, where

$$\theta = \theta_0 + Pe\theta_1 + o(Pe); \tag{8-67}$$

and the governing equations and boundary conditions are

$$\nabla^2\theta_0 = 0,$$
$$\theta_0 = 1 \quad \text{on} \quad S, \tag{8-68}$$

and

$$\nabla^2\theta_1 = \mathbf{u} \cdot \nabla\theta_0,$$
$$\theta_1 = 0 \quad \text{on} \quad S, \tag{8-69}$$

There must also be an outer region with

$$\tilde{x} = Pe\,x, \quad \tilde{y} = Pe\,y, \quad \tilde{z} = Pe\,z, \tag{8-70}$$

and a governing equation and boundary condition

$$\bar{\nabla}^2 \oplus = \mathbf{V} \cdot \bar{\nabla} \oplus, \quad \text{with}$$
$$\oplus \to 0 \quad \text{as} \quad \tilde{r} = (\tilde{x}^2 + \tilde{y}^2 + \tilde{z}^2)^{1/2} \to \infty. \tag{8-71}$$

Here

$$\mathbf{V} = \mathbf{u}\left(x = \frac{\tilde{x}}{Pe}, \ y = \frac{\tilde{y}}{Pe}, \ z = \frac{\tilde{z}}{Pe}\right), \tag{8-72}$$

and thus, in the limit $Pe \to 0$, with $\tilde{x}$, $\tilde{y}$, and $\tilde{z}$ fixed, which is applicable to the outer region,

$$\mathbf{V} = \mathbf{i}, \tag{8-73}$$

where it is assumed that the undisturbed streaming flow is parallel to the x axis. Based upon the analysis for the special case of a sphere, we assume that there exists an asymptotic expansion in this outer region of the form

$$\oplus = Pe \oplus_0 + Pe^2 \oplus_1 + o(Pe^2), \tag{8-74}$$

and thus, substituting (8-74) into (8-71), we have

$$\bar{\nabla}^2 \oplus_0 = \mathbf{i} \cdot \bar{\nabla} \oplus_0 = \frac{\partial \oplus_0}{\partial \tilde{x}}, \text{ with}$$

$$\oplus_0 \rightarrow 0 \quad \text{as} \quad \tilde{r} \rightarrow \infty. \tag{8-75}$$

Finally, we note that the inner and outer expansions must match in the region of overlap between the two regions.

So far, we have done no more than restate the governing equations and boundary conditions that were derived in Section C for the heated sphere in a uniform streaming flow, but without specifying the geometry of the body that we wish to consider. Clearly, if we wish to determine the details of the temperature distribution in the fluid, it will be necessary to restrict ourselves to a specific body geometry, both because of the need to apply boundary conditions on the body surface in (8-68) and (8-69) and because the velocity field that appears in (8-69) (and higher order approximations in both the inner and outer regions) depends on the body geometry. However, if our objective is a prediction of the correlation between Nu and Pe for small Pe, we can determine the first convective contribution in a form similar to equation (8-60) without the need to restrict ourselves to a particular body geometry.

To be more specific, if we use the definition of the Nusselt number

$$\text{Nu} = \frac{Q'}{2\pi k l_c (T_o - T_\infty)}, \text{ with } Q' = \int \int_S \mathbf{q}' \cdot \mathbf{n} \, dS, \tag{8-76}$$

where Q' is the overall dimensional rate of heat transfer, l_c is a characteristic length scale chosen so that the surface area is $4\pi l_c^2$, and $\mathbf{q}'$ is the local dimensional heat flux vector

$$\mathbf{q}' = -k\nabla T, \tag{8-77}$$

then corresponding to (8-67) we have

$$\boxed{\text{Nu} = \text{Nu}_0 + Pe\text{Nu}_1 + o(\text{Pe}),} \tag{8-78}$$

where

$$\text{Nu}_0 = \frac{1}{2\pi} \int \int_S \mathbf{q}_0 \cdot \mathbf{n} \, dS; \quad \mathbf{q}_0 = -\nabla\theta_0, \tag{8-79}$$

and

$$\text{Nu}_1 \equiv \frac{1}{2\pi} \int \int_S \mathbf{q}_1 \cdot \mathbf{n} \, dS; \quad \mathbf{q}_1 = -\nabla\theta_1. \tag{8-80}$$

The key result that we shall demonstrate in this section is that

$$\boxed{\text{Nu}_1 = \frac{1}{4}(\text{Nu}_0)^2.} \tag{8-81}$$

It is clear that Nu_0, which is the dimensionless overall heat flux in the pure conduction limit, will depend upon the geometry of the body. However, once Nu_0 is known (either by theoretical calculation or, perhaps, by experiment), we can calculate the first convective contribution in (8–78) by means of (8–81), with no extra work, so that

$$\boxed{\text{Nu} = \text{Nu}_0 + \frac{1}{4}(\text{Nu}_0)^2 \text{Pe} + o(\text{Pe}),} \qquad (8\text{–}82)$$

and we shall demonstrate that this is true independent of the body shape, the Reynolds number of the flow, or whether the dynamic boundary condition at the body surface is the no-slip condition (as is appropriate for a solid body) or continuity of velocity and stress (as is appropriate for a drop). The only requirements are that the flow far from the body be a uniform streaming motion

$$\mathbf{u} \to \mathbf{i} \quad \text{as} \quad r \to \infty \qquad (8\text{–}83)$$

[as already assumed in (8–73)] and that

$$\mathbf{u} \cdot \mathbf{n} = 0 \quad \text{on} \quad S, \qquad (8\text{–}84)$$

which implies that the body shape is fixed with respect to axes fixed in the undisturbed flow and that the boundary S is impermeable.

The key to obtaining the result (8–82) is the fact that Nu_0 and Nu_1, which would seem to require a specific body geometry in view of the definitions (8–79) and (8–80), can both be calculated via equivalent integral expressions over a spherical surface at some arbitrarily large distance from the body. The reader may anticipate that the form of the temperature field at a large distance from the body is not strongly dependent on the body shape. However, the more important fact is that the details of the *far-field* temperature distribution that are necessary to calculate Nu_0 and Nu_1 can be determined completely from a knowledge of the asymptotic structure of the problem without the need to specify a particular body geometry.

To demonstrate these facts, and hence prove (8–81), let us begin by deriving formulas from which Nu_0 and Nu_1 can be calculated on the basis of the far-field temperature distribution. First, for Nu_0, the desired result is obtained easily from the governing equation (8–68a) for θ_0 in the inner region, rewritten in the form

$$\nabla \cdot \mathbf{q}_0 = 0. \qquad (8\text{–}85)$$

Now, let σ denote the surface of a sphere that circumscribes the arbitrarily shaped body S and is constrained to lie within the inner region where (8–85) holds, but is otherwise assumed to have an arbitrarily large radius (that is, the radius $r_\sigma \gg 1$ but subject to $r_\sigma < 0(\text{Pe}^{-1})$ as $\text{Pe} \to 0$). Then,

$$\int_V \nabla \cdot \mathbf{q}_0 \, dV \equiv 0, \qquad (8\text{–}86)$$

where V denotes the portion of the fluid domain between the body surface S and the spherical surface σ. Now, applying the divergence theorem to (8–86), we find that

$$-\int_S \mathbf{q}_0 \cdot \mathbf{n} \, dS + \int_\sigma \mathbf{q}_0 \cdot \mathbf{n}_\sigma \, dS = 0. \qquad (8\text{–}87)$$

Here, $\mathbf{n}$ is the outer normal for the arbitrarily shaped body, and $\mathbf{n}_\sigma$ is the outer normal for the circumscribed sphere at $r_\sigma (\equiv (x^2 + y^2 + z^2)^{1/2})$. But according to the definition (8–79), the first term in (8–87) is just $2\pi \text{Nu}_0$, and it therefore follows that

$$\text{Nu}_0 = -\frac{1}{2\pi}\int_\sigma \mathbf{q}_0 \cdot \mathbf{n}_\sigma dS. \tag{8-88}$$

Hence, instead of calculating Nu_0 from (8–79) applied at the body surface, S, we see that it can be obtained by evaluating the integral above, taken over the surface of a circumscribed sphere at a large distance from the body.

A similar expression to (8–88) also can be obtained for Nu_1. As before, we begin with the governing equation for θ_1 in the form

$$\nabla \cdot (\mathbf{q}_1 + \mathbf{u}\theta_0) = 0. \tag{8-89}$$

Now, this equation can be integrated over V, and the divergence theorem again applied to obtain

$$\text{Nu}_1 = -\frac{1}{2\pi}\int_\sigma \mathbf{q}_1 \cdot \mathbf{n}_\sigma \partial S. \tag{8-90}$$

However, this expression does not prove to be convenient for calculating Nu_1, as we shall see shortly. Instead, we derive an alternative formula for Nu_1, obtained by multiplying (8–89) by θ_0 and rewriting the result in the form

$$\nabla \cdot \left(\theta_0 \mathbf{q}_1 - \theta_1 \mathbf{q}_0 + \frac{1}{2}\theta_0^2 \mathbf{u} \right) = 0. \tag{8-91}$$

Here we have used the fact that

$$\mathbf{q}_1 \cdot \nabla \theta_0 = \mathbf{q}_0 \cdot \nabla \theta_1 = \nabla \cdot (\theta_1 \mathbf{q}_0)$$

since $\nabla \cdot \mathbf{q}_0 = 0$. Now we can apply the divergence theorem after integrating (8–91) over the domain V between the body and the large sphere of radius r_σ. Taking account of the facts that $\mathbf{u} \cdot \mathbf{n} = 0$, $\theta_1 = 0$, and $\theta_0 = 1$ on the body surface S, we obtain

$$-\int_S \mathbf{q}_1 \cdot \mathbf{n} dS + \int_\sigma \theta_0 \mathbf{q}_1 \cdot \mathbf{n}_\sigma dS - \int_\sigma \theta_1 \mathbf{q}_0 \cdot \mathbf{n}_\sigma dS + \frac{1}{2}\int_\sigma \theta_0^2 \mathbf{u} \cdot \mathbf{n}_\sigma dS = 0. \tag{8-92}$$

But, the first term in (8–92) is just

$$2\pi \text{Nu}_1 \equiv \int_S \mathbf{q}_1 \cdot \mathbf{n} dS,$$

so that Nu_1 can be calculated, at least in principle, by evaluating the remaining terms in (8–92), all of which are integrals over the surface of the circumscribed sphere at r_σ. Of course, it is not evident at this stage why it should be advantageous to evaluate these three integrals instead of the single integral in (8–90). We shall return to this point shortly.

Let us now show that the desired results (8–81) and (8–82) can be obtained from (8–88) and (8–92) under the very general conditions listed just after (8–82). To do this, let us begin with (8–81) and the general observation that Nu_0, and thus the integral in (8–88), cannot depend upon the radius of the circumscribed sphere (r_σ), since it is arbitrary. Furthermore, Nu_0 must be bounded. The only contribution to $\mathbf{q}_0$ that satisfies these requirements is the one that varies as $0(r^{-2})$ for $r \gg 1$. Thus, to calculate Nu_0 from (8–88), it is obvious that we need only determine the $1/r^2$ dependence of q_0. Given the definition (8–79) for q_0 in terms of θ_0, it is clear that we need only evaluate terms of θ_0 that vary like $0(r^{-1})$, for large $r \gg 1$.

If we now examine solutions of Laplace's equation (8–68) for θ_0, we find that there is only one term that exhibits a r^{-1} dependence, namely, C/r, where C is an arbitrary constant. Furthermore, Brenner (1963) has shown that the next term in an expression for θ_0 for large $r \gg 1$ will be a solid spherical harmonic of $0(r^{-3})$, provided only that the origin of coordinates is chosen at the proper point inside the body, and this is always true independent of the body shape. Thus, in general,

$$\theta_0 = \frac{C}{r} + 0(r^{-3}) \tag{8–93}$$

for large values, $r \gg 1$. If we combine (8–93) with (8–88), the constant C can be evaluated in terms of Nu_0, and the result is

$$\boxed{\theta_0 = \frac{Nu_0}{2r} + 0(r^{-3}).} \tag{8–94}$$

Again, we emphasize that this far-field *form* for θ_0 at large r is independent of the body shape. Only the constant Nu_0 and the term at $0(r^{-3})$ will depend on the geometry of the body.

Now, we may note, by analogy to the analysis of Section C for the specific case of a solid sphere, that the asymptotic form (8–94) of the first term in the inner expansion for $r \gg 1$ is all that we need to obtain a *complete* solution for the first term, $\oplus_0$, in the expansion for the outer region. Indeed, if we examine the governing equation and boundary condition (8–75) and the matching condition derived from (8–94),

$$\lim_{\tilde{r} \ll 1} \oplus_0 \Longleftrightarrow \frac{Nu_0}{2\tilde{r}} + 0\left(\frac{Pe^2}{\tilde{r}^3}\right), \tag{8–95}$$

we see that the leading-order outer problem is *identical* to that for the solid sphere, and this is true independent of dynamic boundary conditions on the body surface, the Reynolds number of the flow, or the body shape. The solution is therefore

$$\boxed{\oplus_0 = \frac{Nu_0}{2\tilde{r}} \exp\left[-\frac{1}{2}(\tilde{r} - \tilde{x})\right].} \tag{8–96}$$

But now we can turn to the second term, $0(Pe)$, in the asymptotic expansion for the inner region. Matching conditions for this $0(Pe)$ inner problem can be obtained by expanding (8–96) for small $\tilde{r}$ and expressing the result in terms of inner variables. The result is

$$\theta_1|_{r \gg 1} \Longleftrightarrow -\frac{Nu_0}{4} + \frac{Nu_0 x}{4r}. \tag{8–97}$$

Together with the governing equation and boundary condition, (8–69), the matching condition (8–97) yields a well-posed problem that could be solved, in principle, to obtain θ_1. Note, however, that the solution will depend on the body geometry—both through the boundary condition on S and through the form of the velocity field $\mathbf{u}$ near the body, which now enters the problem for the first time through the governing equation for θ_1. Thus, if we really require a knowledge of θ_1 in the vicinity of the body, we cannot avoid an explicit dependence upon body geometry. If, on the other hand, we only require the contribution of θ_1 to the overall heat transfer rate (Nu_1), this can be determined directly without the need to solve for θ_1 or specify the boundary geometry.

To do this, we return to (8–92). We have seen already that the first term is just $2\pi Nu_1$. The other terms can be evaluated knowing only the far-field form (8–97) for θ_1 obtained from matching. We note, first, that all three terms must be independent of the radius of the spherical surface, r_σ, for $r_\sigma \gg 1$. This means that the integrands must be $0(r^{-2})$ or smaller for $r \gg 1$, and in the limit $r_\sigma \gg 1$, we need only evaluate the $0(r^{-2})$ contributions to the integrands to evaluate Nu_1. Let us begin with the second term. Since $\theta_0 \sim 0(r^{-1})$ for large r, we require $\mathbf{q}_1$ only to $0(r^{-1})$, or, equivalently, θ_1 to $0(1)$ for $r \gg 1$. But the asymptotic form for θ_1 at $0(1)$ is known from the matching condition (8–97),

$$\theta_1 = -\frac{Nu_0}{4} + \frac{Nu_0 x}{4r} + 0(r^{-1}).$$

However, when we substitute this expression for $\nabla\theta_1$ (or $\mathbf{q}_1$) in the second integral of (8–92), we find that

$$\int_\sigma \theta_0 \mathbf{q}_1 \cdot \mathbf{n}_\sigma dS \equiv 0.$$

Similarly, if we examine the last term in (8–92), we know that $\theta_0^2 \sim 0(r^{-2})$, and thus we require $\mathbf{u}$ only to $0(1)$ to evaluate the integral. But again, this is known from the far-field boundary condition on $\mathbf{u}$, namely,

$$\mathbf{u} \sim \mathbf{i}.$$

Substituting this form for $\mathbf{u}$ and the expression (8–94) for θ_0 into the last integral in (8–92), we again find that

$$\int_\sigma \theta_0^2 \mathbf{u} \cdot \mathbf{n}_\sigma dS \equiv 0.$$

Finally, let us consider the third term in (8–92). Here, we require θ_1 to $0(1)$—the asymptotic form (8–97)—and $\mathbf{q}_0$ to $0(r^{-2})$—that is, $-\nabla\theta_0$ with θ_0 evaluated from (8–94). In this case, however, when we evaluate the integral, we do not get zero, but

$$-\int_\sigma \theta_1 \mathbf{q}_0 \cdot \mathbf{n}_\sigma dS \equiv \pi Nu_0^2/2.$$

Hence, we find from (8–92) that

$$Nu_1 = \frac{Nu_0^2}{4},$$

as already given in equation (8–81). The intriguing feature of the preceding analysis is that the result (8–81) for *bodies of arbitrary shape* is obtained for "free," once the asymptotic analysis has been carried out for one specific case (the sphere) so that the solutions (8–94) and (8–96) are known, and the asymptotic structure of the problem is understood.

One comment before going further is that Nu_1 could not have been evaluated via a similar analysis starting from (8–90). The problem with (8–90) is that it requires q_1 to $0(r^{-2})$ and thus θ_1 to $0(r^{-1})$. However, from the far-field matching condition (8–97), we only know θ_1 to $0(1)$. Hence, to evaluate Nu_1 through (8–90), we would have to actually solve the inner problem to get θ_1 throughout the inner region. Anticipation of this difficulty with (8–90) was the motivation for derivation of (8–92).

In reviewing our analysis of (8–92) leading to (8–81), we may note that we have only used the conditions (8–83) and (8–84) on the velocity field. Thus, as stated earlier, the result (8–81) or (8–82) is valid in streaming flow for any heated body with a uniform surface temperature provided that these conditions are satisfied and that Pe $\ll 1$. Higher order terms in (8–82) will depend upon the details of the flow, and thus upon the Reynolds number Re, as well as the shape and orientation of the body relative to the free stream. However, in the creeping flow limit, Brenner (1963) was able to extend (8–82) to one additional term for a particle of *arbitrary* shape,

$$\frac{Nu}{Nu_0} = 1 + \frac{Nu_0}{4} Pe + \frac{Nu_0}{4} f Pe^2 \ln Pe + 0(Pe^2), \qquad (8\text{–}98)$$

where f is the magnitude of the force exerted by the fluid on the particle, divided by $6\pi\mu l_c U$.

E. Heat Transfer from a Sphere in Simple Shear Flow at Low Peclet Numbers

In previous sections of this chapter, we have considered forced convection heat transfer at low Peclet number from particles in a uniform streaming flow. The results are applicable if the density of the particle is different from that of the fluid, so that the particle is subject to gravitational and/or inertial forces that give it a translational motion relative to the fluid. In this case, we have seen that the relationship between Nu and Pe takes a common form,

$$Nu = Nu_0 + PeNu_1 + 0(Pe^2 \ln Pe), \qquad (8\text{–}99)$$

for bodies of arbitrary fixed shape, which may be either fluid or solid, and the first two terms hold even if the Reynolds number is unrestricted provided the flow remains laminar and steady.

In many physical processes, however, the particle has a density very nearly equal to that of the fluid, and the particle then has no (or negligible) translational velocity relative to the fluid. Provided that the linear dimensions of the particle are small compared with distances over which the velocity gradient in the ambient flow field changes significantly, the flow near the particle is then effectively that due to a force-free particle in an ambient velocity field that can be approximated as varying linearly with position. An important question is whether the rate of heat transfer from a heated sphere in such a flow is still given by a correlation of the form (8–99) for Pe $\ll 1$. More generally, of course, we may ask whether the type of flow that exists far from a heated particle has any effect on the rate of heat transfer from the particle for arbitrary Pe. Although an analysis for Pe $\ll 1$ cannot provide a complete answer to this more general question, it can serve as a useful initial test case.

Thus, in this section, we consider the special case of heat transfer from a rigid sphere when the undisturbed fluid motion, relative to axes that translate with the sphere, is a simple, linear shear flow,

$$\mathbf{u}_\infty = \gamma y \mathbf{i}. \tag{8-100}$$

The sphere is heated, as before, in such a way that its surface temperature is constant. We assume that the sphere is free of any external couples so that it can rotate with the fluid. Thus, the angular velocity of the sphere is just $\Omega = \gamma/2$. The situation is sketched in Figure 8–5. It is convenient to solve the problem using a spherical coordinate system (r, θ, ϕ), where the azimuthal angle ϕ is measured from the yz plane around the z axis and the polar angle θ is measured from the z axis. We have illustrated the relationship between spherical and Cartesian coordinates in Figure 8–5.

The governing differential equation and boundary conditions for the temperature field are again (8–1) and (8–2), while the dimensionless equation and boundary conditions are (8–7) and (8–8), with the same definition for the dimensionless temperature (8–3) and the sphere radius as a characteristic length scale. Only the form of the velocity field, $\mathbf{u}$, and the choice of characteristic velocity are different for this case of a linear shear flow far from the sphere. The appropriate choice for the characteristic velocity is

$$u_c = \gamma a. \tag{8-101}$$

Hence, the Peclet number that appears in (8–7) is defined as

$$\boxed{\mathrm{Pe} \equiv \frac{a^2 \gamma \rho C_p}{k}.} \tag{8-102}$$

In order to specify the velocity field $\mathbf{u}$, we must solve the Navier-Stokes equations subject to the boundary condition (8–100) at infinity. For present purposes, we follow the example of Section C and assume that the Reynolds number, defined here as $\mathrm{Re} \equiv a^2 \gamma \rho/\mu$, is very small so that the creeping flow solution for a sphere in shear flow (obtained in Chapter 4 Section F) can be applied throughout the domain in which θ differs significantly from unity. Hence, from Chapter 4, Section F we have

$$\boxed{u_r = \frac{1}{2}\left(r - \frac{5}{2}r^{-2} + \frac{3}{2}r^{-4}\right)\sin^2\theta \sin 2\phi,} \tag{8-103a}$$

$$u_\theta = \frac{1}{2}(r - r^{-4})\sin\theta\cos\theta\sin 2\phi, \tag{8-103b}$$

and

$$u_\phi = -\frac{1}{2}r\sin\theta + \frac{1}{2}(r - r^{-4})\sin\theta\cos 2\phi, \tag{8-103c}$$

where these velocity components also have been nondimensionalized with respect to the characteristic velocity scale (8–101).

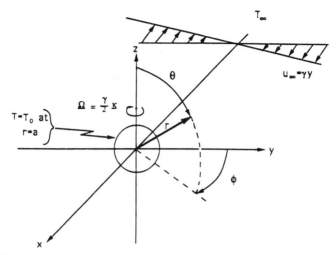

Figure 8–5 A schematic representation of a sphere of radius a and a constant surface temperature T_o in a fluid of ambient temperature T_∞ that is undergoing a simple shear flow, $\mathbf{u}_\infty = \gamma y \mathbf{i}$. Since the sphere is assumed to rotate freely in the ambient flow, it rotates with an angular velocity $\mathbf{\Omega} = (\gamma/2)\mathbf{\kappa}$ about the z axis.

We now seek a solution of (8–7) and (8–8) for small values of the Peclet number, Pe $\ll 1$, using the matched asymptotic expansion procedure that were detailed for streaming flow past a sphere in Section C. Although the reader may not immediately see that the derivation of an asymptotic solution for this new problem necessitates use of the matched asymptotic expansion technique, an attempt to develop a regular expansion for θ for Pe $\ll 1$ leads to a Whitehead-type paradox similar to that encountered for the streaming flow problem.

As in the streaming flow problem, the solution domain divides into two parts. In the so-called inner region, the sphere diameter provides an appropriate characteristic length scale, and the dimensionless form of (8–7) is applicable, along with the boundary condition (8–8a) at the sphere surface. We assume, in this region, that an asymptotic expansion for θ exists in the form

$$\theta = \theta_0 + \sum_n f_n(\text{Pe})\theta_n, \tag{8–104}$$

with boundary conditions on θ_n being

$$\theta_0 = 1, \ \theta_n = 0 \ (n \geq 1) \quad \text{at} \quad r = 1. \tag{8–105}$$

Obviously, for Pe $\ll 1$, the governing equation for θ_0 is again Laplace's equation for pure conduction,

$$\nabla^2 \theta_0 = 0, \tag{8–106}$$

and we have already seen that the solution of this equation, satisfying the condition (8–105) on the sphere surface is

$$\boxed{\theta_0 = \frac{1}{r}.}$$

(8–107)

The governing equation at the next level in the expansion (8–104) depends, of course, on the gauge function f_1(Pe). Although we might be tempted to guess that f_1(Pe) will be equal to Pe, analogous to the streaming flow problem, we have seen that the gauge functions do not necessarily follow in the simple sequence (1, Pe, Pe², ...), which would appear if the asymptotic solution were regular. Thus, it is prudent to be cautious at this stage and solve for the first approximation in the outer domain before we attempt to anticipate the form for the second approximation in the inner region.

To obtain the governing equation for the outer domain, we introduce the general rescaling

$$\rho = r\mathrm{Pe}^m$$

(8–108)

for the radial variable, following the analysis of Section C and recall that the rescaling parameter m must be chosen in such a way that both conduction and convection terms are retained when the thermal energy equation is expressed in terms of ρ and the limit Pe → 0 is applied with $\rho = 0(1)$. In order to distinguish the outer temperature field from the inner temperature field, it is also convenient to change symbols for the temperature from θ to $\oplus$.

The rescaled governing equation, expressed in terms of ρ and $\oplus$, then takes the form

$$\mathrm{Pe}^{2m}\left\{\frac{1}{\rho^2}\left[\frac{\partial}{\partial\rho}\left(\rho^2\frac{\partial\oplus}{\partial\rho}\right) + \frac{1}{\sin\theta}\frac{\partial}{\partial\theta}\left(\sin\theta\frac{\partial\oplus}{\partial\theta}\right) + \frac{1}{\sin^2\theta}\frac{\partial^2\oplus}{\partial\phi^2}\right]\right\} =$$

$$\mathrm{Pe}\left\{\left[\left(\frac{1}{2}\frac{\rho}{\mathrm{Pe}^m} + 0(\mathrm{Pe}^{2m})\right)\sin^2\theta\sin2\phi\right]\mathrm{Pe}^m\frac{\partial\oplus}{\partial\rho}\right.$$

$$+ \left[\frac{1}{2}\left(\frac{\rho}{\mathrm{Pe}^m} - 0(\mathrm{Pe}^{4m})\right)\sin\theta\cos\theta\sin2\phi\right]\frac{\mathrm{Pe}^m}{\rho}\frac{\partial\oplus}{\partial\theta}$$

$$+ \left.\left[-\frac{1}{2}\frac{\rho}{\mathrm{Pe}^m}\sin\theta + \frac{1}{2}\left(\frac{\rho}{\mathrm{Pe}^m} - 0(\mathrm{Pe}^{4m})\right)\sin\theta\cos2\phi\right]\frac{\mathrm{Pe}^m}{\rho\sin\theta}\frac{\partial\oplus}{\partial\phi}\right\}. \quad (8\text{–}109)$$

Examination of the right-hand side shows that in this outer region, the largest convection terms derive from the linear $0(r)$ contributions to the velocity field (8–103). But this is not at all surprising. These terms are just the dimensionless undisturbed linear shear flow $\mathbf{u}_\infty = y\mathbf{i}$ expressed in spherical coordinates, and the rescaled equation is simply reflecting the obvious result that the dominant convection contribution, sufficiently far from the sphere, is just that associated with the undisturbed linear shear flow. A very similar result was also obtained in the streaming flow problem where it was shown that the leading-order approximation to the convection terms in the outer region was due to the undisturbed *uniform* velocity, $\mathbf{u}_\infty = \mathbf{i}$.

If we retain only the largest terms on the right-hand side of (8–109), the equation takes the form

$$Pe^{2m}\left[\nabla_\rho^2\oplus\right] = Pe\left[\frac{1}{2}\rho\sin^2\theta\sin2\phi\,\frac{\partial\oplus}{\partial\rho} + \frac{1}{2}\sin\theta\cos\theta\sin2\phi\,\frac{\partial\oplus}{\partial\theta}\right.$$

$$\left.+\left(-\frac{1}{2}\sin\theta + \frac{1}{2}\sin\theta\cos2\phi\right)\frac{1}{\sin\theta}\,\frac{\partial\oplus}{\partial\phi}\right] + 0(Pe^{3m+1}). \quad (8\text{-}110)$$

Following the reasoning outlined in Section **C**, the rescaling parameter m is now chosen by requiring that Equation (8-110) preserve a balance between conduction and convection terms in the limit as $Pe\to 0$. Thus,

$$2m = 1 \quad\text{or}\quad m = \frac{1}{2}, \quad (8\text{-}111)$$

and the rescaling equation (8-108) becomes

$$\boxed{\rho = r Pe^{1/2}.} \quad (8\text{-}112)$$

Further, with $m = 1/2$, the governing equation (8-110) can be rewritten as

$$\boxed{\nabla_\rho^2\oplus - \hat{y}\,\frac{\partial\oplus}{\partial\hat{x}} = 0(Pe^{3/2}).} \quad (8\text{-}113)$$

Here, $\hat{x} = x Pe^{1/2}$, $\hat{y} = y Pe^{1/2}$, and $\rho = (\hat{x}^2 + \hat{y}^2 + \hat{z}^2)^{1/2}$.

Before discussing the solution of (8-113), it is perhaps useful to review once more the rationale for rescaling and the choice $m = 1/2$ for the rescaling parameter. The change of variables (8-108) is just one way of changing the nondimensionalization of the radial variable r' from the sphere radius a to the "new" length scale

$$l = a\left(\frac{a^2\gamma\rho C_p}{k}\right)^{-m}$$

that is appropriate to the region far from the sphere. If l is chosen properly, it should reflect the distance over which $\oplus$ changes in this region, and then derivatives of $\oplus$ with respect to ρ will be $0(1)$. The only choice possible for m that is consistent with the idea that convection should be as important as conduction in the outer region is (8-111).

Let us now turn to the solution for the temperature distribution in the outer region, which we assume has an asymptotic expansion for small Pe of the form

$$\boxed{\oplus = \sum_n F_n(Pe)\oplus_n,} \quad (8\text{-}114)$$

where, as usual,

$$\lim_{Pe\to 0}\frac{F_{n+1}}{F_n}\to 0.$$

The governing equation for the first term in this expansion is the limit of (8-113) for $Pe\to 0$,

$$\boxed{\nabla_\rho^2\oplus_0 - \hat{y}\,\frac{\partial\oplus_0}{\partial\hat{x}} = 0.} \quad (8\text{-}115)$$

This equation is to be solved subject to the boundary condition

$$\oplus_0 \to 0 \quad \text{as} \quad \rho \to \infty. \tag{8–116}$$

Furthermore, in the limit $Pe \to 0$, the solution should match the first approximation for θ in the inner domain, that is,

$$F_0(Pe)\oplus_0(\rho,\theta,\phi)|_{\rho \ll 1} \Longleftrightarrow Pe^{1/2}\frac{1}{\rho} \quad \text{as} \quad Pe \to 0, \tag{8–117}$$

where the right-hand side is just the inner solution, $1/r$, expressed in terms of the outer variable. It is evident from (8–117) that

$$\boxed{F_0(Pe) = Pe^{1/2}.} \tag{8–118}$$

A general solution to equation (8–115) that satisfies the boundary condition (8–116) was obtained some years ago by Elrick (1962) using Fourier transforms.[6] Elrick was seeking temperature distribution produced by a point source of heat in a simple shear flow. It is not surprising that his solution should be relevant to the present problem. When viewed with the scale of resolution

$$l = a\left[\frac{a^2\gamma\rho C_p}{k}\right]^{-1/2},$$

which is characteristic of the outer region for $Pe \to 0$, the sphere of radius a appears only as a point source of heat. Elrick's solution has the form

$$\oplus_0 = \frac{A}{2\sqrt{\pi}}\int_0^\infty \frac{ds}{\left(1+\frac{1}{12}s^2\right)^{1/2}s^{3/2}}\exp\left[-\frac{\left(\hat{x}-\frac{1}{2}\hat{y}s\right)^2}{4s\left(1+\frac{1}{12}s^2\right)}-\frac{\hat{y}^2+\hat{z}^2}{4s}\right], \tag{8–119}$$

in which A is an arbitrary constant related to the strength of the point heat source and s is a dummy variable of integration. In the present application, the constant A must be determined from the matching condition (8–117). In order to apply this condition, we require the limiting form of (8–119) for $\rho \ll 1$. It is straightforward to show that

$$\lim_{\rho \to 0}\oplus_0 \sim A\left(\frac{1}{\rho}+\frac{\Gamma}{2\sqrt{\pi}}+0(\rho)\right), \tag{8–120}$$

where

$$\Gamma \equiv \int_0^\infty \frac{ds}{s^{3/2}}\left[\left(1+\frac{1}{12}s^2\right)^{-1/2}-1\right] = -0.9104.$$

Now, applying the matching condition (8–117),

$$Pe^{1/2}A\left[\frac{1}{\rho}+\frac{\Gamma}{2\sqrt{\pi}}+0(\rho)\right] \Longleftrightarrow Pe^{1/2}\frac{1}{\rho} \quad \text{as} \quad Pe \to 0. \tag{8–121}$$

It is evident that $A = 1$. There is a mismatch in ρ of $0(1)$ for $\rho \ll 1$. However, this mismatch need not concern us at the present level of approximation. A corresponding term, independent of r (since $\Gamma/2\sqrt{\pi}$ is independent of ρ), will be generated only by a term in the inner expansion (8–104) that has a gauge function $0(\mathrm{Pe}^{1/2})$. At the present leading order of approximation, such terms have not yet been considered.

Now, with A determined, our solution for the first term in the outer expansion is completed, and we can turn to the problem of obtaining a second term in the asymptotic expansion for θ in the inner region. In view of the fact that the mismatch between the first terms in the inner and outer expansions has been shown in the previous paragraph to be $0(\mathrm{Pe}^{1/2})$, it is clear that the gauge function for the second term in the inner expansion (8–104) must be

$$f_1(\mathrm{Pe}) = \mathrm{Pe}^{1/2}. \tag{8–122}$$

With (8–122), the inner expansion has the form

$$\theta = \frac{1}{r} + \mathrm{Pe}^{1/2}\theta_1 + \sum_{n=2} f_n(\mathrm{Pe})\theta_n,$$

and the governing equation for θ_1 is easily seen from (8–7) to be

$$\boxed{\nabla^2\theta_1 = 0,} \tag{8–123}$$

with the boundary condition (8–105)

$$\boxed{\theta_1 = 0 \quad \text{at} \quad r = 1.} \tag{8–124}$$

In addition to (8–123) and (8–124), θ_1 must satisfy the matching condition

$$\mathrm{Pe}^{1/2}\left[\frac{1}{\rho} + \frac{\Gamma}{2\sqrt{\pi}} + 0(\rho)\right]_{\rho \ll 1} \iff \left(\frac{\mathrm{Pe}^{1/2}}{\rho} + \mathrm{Pe}^{1/2}\theta_1\right)_{r \gg 1} \quad \text{as} \quad \mathrm{Pe} \to 0, \tag{8–125}$$

that is,

$$\theta_1 \to \frac{\Gamma}{2\sqrt{\pi}} \quad \text{for} \quad r \gg 1. \tag{8–126}$$

The reader may have noted that the governing differential equation at this level of approximation is still the steady-state conduction equation. Thus, even at this second level of approximation, convection plays no direct role in the heat transfer process in the inner region. Indeed, the governing equation (8–123) and the boundary condition (8–124) are both homogeneous. Thus, the temperature distribution in the inner region changes at $0(\mathrm{Pe}^{1/2})$ only because of the effect of convection on the solution in the outer region and the resultant mismatch (8–125) at $0(\mathrm{Pe}^{1/2})$, which leads to the nonhomogeneous matching condition (8–126). This is quite different from the uniform streaming problem, analyzed in Section **C**, where the first correction to the pure conduction solution in the inner region occurred at $0(\mathrm{Pe})$ and contained a direct contribution due to convection.

The general solution of (8–123) for θ_1, which satisfies (8–124) and is a function of r only [as suggested by (8–126)], is

$$\boxed{\theta_1 = C\left(1 - \frac{1}{r}\right).} \tag{8–127}$$

The matching condition (8–126) then requires

$$C = \frac{\Gamma}{2\sqrt{\pi}}. \tag{8–128}$$

As usual, even with the condition (8–126) satisfied, there is still a mismatch between the solutions in the two parts of the domain. To see that this mismatch can be neglected at this level of approximation, let us reexamine the full matching condition (8–125) with θ_1 incorporated. The condition is

$$Pe^{1/2}\left[\frac{1}{\rho} + \frac{\Gamma}{2\sqrt{\pi}} + 0(\rho)\right]_{\rho \ll 1} + 0(F_1(Pe)) \Longleftrightarrow$$

$$\left(\frac{Pe^{1/2}}{\rho} + Pe^{1/2}\left[\frac{\Gamma}{2\sqrt{\pi}}\left(1 - \frac{Pe^{1/2}}{\rho}\right)\right]\right)_{\rho \gg 1} + 0(f_2(Pe)) \quad \text{as} \quad Pe \to 0. \tag{8–129}$$

The right-hand side is the first two terms of the inner solution $\theta_0 + Pe^{1/2}\theta_1$, expressed in terms of outer variables. The first two terms on the two sides of (8–129) match exactly. However, there is a mismatch at $0(Pe)$. In particular, on the right-hand side, the term $\Gamma/2\sqrt{\pi}(-Pe/\rho)$ is not yet matched with a corresponding term in the outer solution. Of course, this is to be expected since we have so far considered only the leading-order solution of $0(Pe^{1/2})$ in the outer region. Likewise, the term $0(\rho Pe^{1/2})$ in the outer solution has not yet been matched with any term in the inner expansion, but this is, again, to be expected since a term of this form could derive only from a term $0(rPe)$ in the inner solution, and we have not yet considered the $0(Pe)$ approximation in the inner expansion (8–104).

Evidently, it would be a relatively straightforward matter to proceed to higher order approximations in both the inner and outer portions of the domain. The next step would be to seek a second term in the outer solution, and the form of the mismatch in (8–129) suggests that this term should be $0(Pe)$, that is, $F_1(Pe) = Pe$. However, we shall not pursue further correction terms here but rather content ourselves with calculating the first two terms in the approximation to the Nusselt number for small Pe, corresponding to the first two terms in the inner expansion for the temperature distribution, that is,

$$\theta \sim \frac{1}{r} + Pe^{1/2}\left[\frac{\Gamma}{2\sqrt{\pi}}\left(1 - \frac{1}{r}\right)\right] + o(Pe^{1/2}). \tag{8–130}$$

Substituting (8–130) into the definition (8–20) for the Nusselt number,

$$Nu = -\frac{1}{2\pi}\int_0^{2\pi}\int_{-1}^1 \left(\frac{\partial\theta}{\partial r}\right)_{r=1} d\eta d\phi,$$

we obtain

$$Nu = 2\left(1 + \frac{0.9104}{2\sqrt{\pi}}Pe^{1/2} + o(Pe^{1/2})\right). \tag{8–131}$$

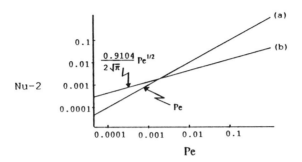

Figure 8–6 Correlation between the Nusselt number and Peclet number for heat transfer at low Peclet number from a sphere with constant surface temperature in (a) uniform streaming flow and (b) simple shear flow.

The first term in this expression is, of course, the familiar result for pure conduction from a heated sphere and is the same for all flows. The second term represents the first contribution of convection and should be compared to the second term in Equation (8–60), which is the Nusselt number for forced convection heat transfer at low Pe when the flow is uniform streaming. The most important observation is that the dependence on Pe is different in the two cases, being $O(Pe)$ in the uniform streaming flow and $O(Pe^{1/2})$ in simple shear flow. The two results, (8–60) and (8–131), are plotted in Figure 8–6. Evidently, the Nusselt number for simple shear flow exceeds the value for uniform streaming flow for Pe $\ll$ 1 where the two asymptotic predictions are valid. Although the numerical difference between the two results is small, the most important conclusion from the analysis is not the numerical magnitude of corrections to the pure conduction heat flux but rather the fact that the asymptotic form of the convection contribution clearly depends on flow type. In general, *heat transfer correlations developed for one type of flow will not carry over to some other type of flow.*

Before leaving the present problem, we wish to call the reader's attention to several generalizations of the predicted relationship (8–131) between Nu and Pe for Pe $\ll$ 1. These generalizations clearly illustrate the power of the asymptotic method to provide insight into the form of correlations between dimensionless parameters, with a minimum of detailed analysis. The first is due to Batchelor (1979)[7] and Acrivos (1980),[8] who showed that the correlation (8–131), first derived for a sphere in linear shear flow, could be generalized easily and extended to the much more general case of a rigid, heated sphere in an arbitrary linear flow

$$\mathbf{u} = \Gamma \cdot \mathbf{x}. \tag{8–132}$$

Here, Γ is a constant second-order tensor (the velocity gradient tensor) that is arbitrary except for the requirement $\mathrm{tr}\,\Gamma = 0$ so that $\nabla \cdot \mathbf{u} \equiv 0$.

The class of general linear flows represented by (8–132) includes simple shear flow as a special case,

$$\Gamma = \begin{pmatrix} 0 & \gamma & 0 \\ 0 & 0 & 0 \\ 0 & 0 & 0 \end{pmatrix}, \tag{8–133}$$

and a number of commonly studied *extensional flows* such as

$$\text{uniaxial extension} \quad \Gamma = \begin{pmatrix} 2 & 0 & 0 \\ 0 & -1 & 0 \\ 0 & 0 & -1 \end{pmatrix} E, \tag{8–134}$$

$$\text{biaxial extension} \quad \Gamma = \begin{pmatrix} -2 & 0 & 0 \\ 0 & 1 & 0 \\ 0 & 0 & 1 \end{pmatrix} E, \tag{8–135}$$

and

$$\text{hyperbolic flow} \quad \Gamma = \begin{pmatrix} 1 & 0 & 0 \\ 0 & -1 & 0 \\ 0 & 0 & 0 \end{pmatrix} E. \tag{8–136}$$

These flows are sketched in Figure 8–7. Another interesting subset of the general case, (8–132), is two-dimensional linear flows. In this case, the complete set of possible motions can be represented as a one parameter family of the form (8–132) with

$$\Gamma = \begin{pmatrix} 1+\lambda & 1-\lambda & 0 \\ -(1-\lambda) & -(1+\lambda) & 0 \\ 0 & 0 & 0 \end{pmatrix} \frac{E}{2} \tag{8–137}$$

and $-1 \leq \lambda \leq 1$. The case $\lambda = 1$ is just the hyperbolic flow listed as equation (8–136); $\lambda = -1$ is a purely rotational flow with streamlines that are circles; and $\lambda = 0$ is simple shear flow, expressed in terms of axes that are rotated $45°$ from the direction of motion. The parameter E that appears in the various expressions for Γ is usually expressed in units of $\sec^{-1}$ and is often called the shear rate.

In spite of the apparent generality of the class of motions considered, Batchelor was able to show that

$$\boxed{\frac{\text{Nu}}{\text{Nu}_0} = 1 + \alpha \, \text{Nu}_0 \text{Pe}^{1/2} + o(\text{Pe}^{1/2})} \tag{8–138a}$$

for a rigid spherical particle in any of the class of linear flows represented by (8–132). Shortly thereafter, Acrivos showed that (8–138a) could actually be extended for spherical particles to include two more terms with a minimum of additional effort. Acrivos's result is

$$\boxed{\frac{\text{Nu}}{\text{Nu}_0} = 1 + \alpha \, \text{Nu}_0 \text{Pe}^{1/2} + \alpha^2 \text{Nu}_0^2 \text{Pe} + \alpha^3 \text{Nu}_0^3 \text{Pe}^{3/2} + o(\text{Pe}^{3/2}).} \tag{8–138b}$$

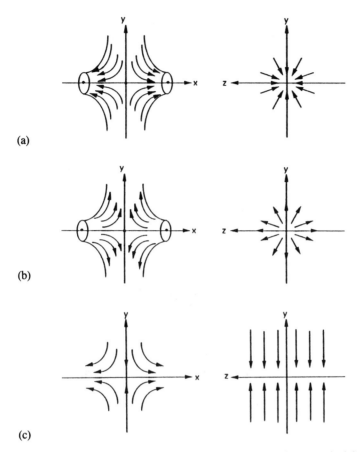

(a)

(b)

(c)

Figure 8–7 A schematic sketch showing the flow patterns for (a) uniaxial extensional flow, (b) biaxial extensional flow, and (c) hyperbolic (or two-dimensional) extensional flow. Each part of the figure shows the flow from two perspectives: one along the z axis toward the xy plane and the other along the x axis toward the yz plane. In the first two cases, the flow is axisymmetric, with the x axis being the symmetry axis. In the two-dimensional case, the flow is invariant in the z direction.

In both (8–138a) and (8–138b), $Nu_0 = 2$, and α is a constant coefficient that depends on the type of linear flow that we consider, that is, on the form of Γ. A general formula for computing α from the specified form for Γ is contained in Batchelor's original paper.

The essence of Batchelor's generalization can be summarized as follows. First, the leading approximation in the outer region for all flows of the general class, (8–132), satisfies an equation of the same form,

$$\nabla_\rho^2 \oplus_0 - (G_{ij}\hat{x}_j) \cdot \frac{\partial \oplus_0}{\partial \hat{x}_i} = 0, \qquad (8\text{–}139)$$

representing a balance between conduction and convection with the undisturbed velocity field. Second, in view of the common form (8-139), it is clear that the rescaling (8-112), originally derived for the special case of simple shear flow, actually applies to all problems in which the ambient flow is a linear function of spatial position. Third, the fundamental solution of (8-139) will always reduce to the asymptotic form

$$\oplus_0 \sim \frac{A_1}{\rho} - \alpha \quad \text{for} \quad \rho \ll 1 \tag{8-140}$$

(that is, for all G_{ij}), in which A_1 and α are constants of $0(1)$ that depend on the form of G_{ij}. Hence, the mismatch between the pure conduction solution in the inner region and the solution of (8-139) in the outer region will be $0(\text{Pe}^{1/2})$, and the second term in the inner expansion will thus satisfy (8-123) and (8-124) and be of the general form (8-127). Only the flow dependent constant $C = \alpha$ will vary from case to case for the general class of linear flows, and Batchelor's result (8-138) thus follows directly from the analysis for simple shear flow. Acrivos's generalization is based on the fact that (8-139) and (8-140) actually hold in the outer region up to the terms of $0(\text{Pe}^{3/2})$, when the first neglected convection terms in (8-113) come into play.

Acrivos (1980) also employed arguments very similar to those of Brenner (1963) and Batchelor (1979), outlined above, to show that (8-138a) was actually valid for particles of *arbitrary shape* in an arbitrary linear flow. For such particles, the only change in the arguments of the preceding section is a generalization of Laplace's equation for conduction to the general form (8-99) and the observation that the coefficients A_1 and α in (8-140) will depend on particle geometry as well as flow type.

F. Uniform Flow Past a Solid Sphere at Small, but Nonzero, Reynolds Numbers

Now, let us return to the fluid mechanics problem of streaming flow past a solid sphere at small, but nonzero, Reynolds numbers. As we shall see, this problem and its solution are very similar to the problem of convective heat transfer at low Pe from a sphere in a uniform streaming flow, which we studied in Section C. The problem of obtaining an inertial correction to Stokes solution was first investigated in 1889 by the (then) young English mathematician Whitehead,[9] who was attempting to obtain corrections to Stokes's formula for the drag force on a sphere since the latter had been found, experimentally, to provide accurate predictions *only* for very small Reynolds numbers, Re < 0(0.1). Our experience from earlier sections in this chapter should alert us to the fact that the creeping flow solution of Stokes is only the first approximation in a *singular* asymptotic expansion for Re → 0. However, Whitehead did not have the benefit of any knowledge of asymptotic methods, which were not developed until much later, and he attempted to obtain a solution for small, but nonzero, Reynolds numbers by the method of successive approximation. As we have seen, this is equivalent for small Re, to a regular perturbation scheme, and we should anticipate that it will fail. Nevertheless, it is instructive to consider Whitehead's original analysis, expressed in the language of asymptotic analysis.

Thus, we begin by considering the full Navier-Stokes equation expressed in terms of the streamfunction ψ and nondimensionalized using the streaming velocity of the fluid relative to the sphere U_∞ as a characteristic velocity and the sphere radius a as a characteristic length scale. Using spherical coordinates, with $\eta = \cos\theta$, this equation is

$$E^4\psi = \text{Re}\left[\frac{1}{r^2}\left\{\frac{\partial\psi}{\partial r}\frac{\partial}{\partial\eta}(E^2\psi) - \frac{\partial\psi}{\partial\eta}\frac{\partial}{\partial r}(E^2\psi)\right\} + \frac{2}{r^2}E^2\psi\left\{\frac{\partial\psi}{\partial r}\frac{\eta}{1-\eta^2} + \frac{1}{r}\frac{\partial\psi}{\partial\eta}\right\}\right].$$

$$(8\text{-}141)$$

The boundary conditions are (see Chapter 4 Section E)

$$\frac{\partial\psi}{\partial r} = \psi = 0 \quad \text{at} \quad r = 1,$$

$$\psi = 0 \qquad \eta = \pm 1,$$

$$\psi \to -r^2 Q_1(\eta) \qquad \text{as} \quad r \to \infty. \qquad (8\text{-}142)$$

Following Whitehead's approach, we seek a solution of (8-141) and (8-142) via a regular perturbation expansion of the form

$$\psi(r,\eta;\text{Re}) = \psi_0(r,\eta) + \text{Re}\,\psi_1(r,\eta) + \cdots. \qquad (8\text{-}143)$$

Substituting (8-143) into (8-141) and (8-142), we obtain

$$E^4\psi_0 + \text{Re}\,E^4\psi_1 + 0(\text{Re}^2) = \text{Re}\left[\frac{1}{r^2}\left\{\frac{\partial\psi_0}{\partial r}\frac{\partial}{\partial\eta}(E^2\psi_0) - \frac{\partial\psi_0}{\partial\eta}\frac{\partial}{\partial r}(E^2\psi_0)\right\}\right.$$

$$\left. + \frac{2}{r^2}E^2\psi_0\left\{\frac{\partial\psi_0}{\partial r}\frac{\eta}{1-\eta^2} + \frac{1}{r}\frac{\partial\psi_0}{\partial\eta}\right\}\right] + 0(\text{Re}^2), \quad (8\text{-}144)$$

with

$$\frac{\partial\psi_0}{\partial r} + \text{Re}\frac{\partial\psi_1}{\partial r} + 0(\text{Re}^2) = \psi_0 + \text{Re}\,\psi_1 + 0(\text{Re}^2) = 0, \quad r = 1,$$

$$\psi_0 + \text{Re}\,\psi_1 + 0(\text{Re}^2) = 0, \, \eta = \pm 1,$$

$$\psi_0 + \text{Re}\,\psi_1 + 0(\text{Re}^2) \to -r^2 Q_1(\eta) \quad \text{as} \quad r \to \infty. \qquad (8\text{-}145)$$

Thus, since Re is an arbitrary small parameter,

$$E^4\psi_0 = 0, \qquad (8\text{-}146)$$

$$\frac{\partial\psi_0}{\partial r} = \psi_0 = 0 \quad \text{at} \quad r = 1,$$

$$\psi_0 = 0 \quad \text{at} \quad \eta = \pm 1,$$

$$\psi_0 \to -r^2 Q_1(\eta) \quad \text{as} \quad r \to \infty, \qquad (8\text{-}147)$$

and, at $0(\text{Re})$,

$$E^4\psi_1 = \frac{1}{r^2}\left\{\frac{\partial\psi_0}{\partial r}\frac{\partial}{\partial\eta}(E^2\psi_0) - \frac{\partial\psi_0}{\partial\eta}\frac{\partial}{\partial r}(E^2\psi_0)\right\} + $$

$$\frac{2}{r^2}E^2\psi_0\left\{\frac{\partial\psi_0}{\partial r}\frac{\eta}{1-\eta^2} + \frac{1}{r}\frac{\partial\psi_0}{\partial\eta}\right\}, \qquad (8\text{-}148)$$

with

$$\frac{\partial\psi_1}{\partial r} = \psi_1 = 0 \quad \text{at} \quad r = 1,$$

$$\psi_1 = 0 \quad \text{at} \quad \eta = \pm 1, \qquad (8\text{-}149)$$

$$\frac{1}{r}\frac{\partial\psi_1}{\partial r}, \frac{1}{r^2}\frac{\partial\psi_1}{\partial\eta} \to 0 \quad \text{as} \quad r \to \infty.$$

The last condition follows from the fact that the first approximation, ψ_0, satisfies the nonhomogeneous boundary condition $\psi \to -r^2 Q_1(\eta)$ as $r \to \infty$ (that is, $\mathbf{u}_0 \to \mathbf{i}$ as $r \to \infty$), so that $\mathbf{u}_1 \to 0$ as $r \to \infty$.

The solution of (8-146) and (8-147) is Stokes's solution, equation (4-206), derived in Chapter 4. In view of our experience with the analogous heat transfer problem at low Peclet numbers, it should come as no surprise that a solution for the next term, ψ_1, satisfying (8-148) and (8-149), is not possible. In particular, the governing equation for ψ_1 is obtained by substituting the known solution for ψ_0 into the right-hand side of (8-148). This gives

$$E^4\psi_1 = \frac{9}{2}\left(\frac{2}{r^2} - \frac{3}{r^3} + \frac{1}{r^5}\right)Q_2(\eta), \qquad (8\text{-}150)$$

and a general solution of this equation is

$$\psi_1 = \sum_1^\infty \left[A_n r^{n+3} + B_n r^{n+1} + C_n r^{2-n} + D_n r^{-n}\right]Q_n(\eta) + \frac{3}{16}\left(2r^2 - 3r - \frac{1}{r}\right)Q_2(\eta),$$

$$(8\text{-}151)$$

with A_n, B_n, C_n, and D_n being arbitrary constants that should, in principle, be determined from the boundary conditions (8-149). However, if we examine (8-151), we find that there is no choice of the constants that will satisfy the boundary condition on ψ_1 as $r \to \infty$. In particular, there is no term in the homogeneous part of the solution that can be used to cancel the term $(3/8)r^2 Q_2(\eta)$ that appears in the particular solution and clearly violates the requirement that $\psi \sim 0(r^n)$ with $n < 2$. We conclude that *no solution exists* at $0(\text{Re})$ in the regular expansion (8-143) that satisfies the differential equation and boundary conditions in equation (8-148) and (8-149). This is known as *Whitehead's paradox*. The inability to generate a second approximation for small Reynolds numbers, based on Stokes's solution ψ_0 as the first approximation, so discouraged Whitehead that he is said to have given up his career as a mathematician and turned to philosophy.

Of course, our experience with the analogous thermal problem at low Peclet numbers suggest that it is the assumption of a *regular* perturbation expansion, valid for all r, that leads to Whitehead's paradox. However, as we have noted, Whitehead did not have the benefit of any knowledge of singular perturbations or the method of matched asymptotic expansions, which only came into being some eigthy years later, with the pioneering publications of Kaplun (1957)[10] and Proudman and Pearson (1957).[11] Indeed, even an explanation for the Whitehead paradox did not appear until 1911, when the German mathematician C.W. Oseen published a landmark paper that not only explained the difficulty but also provided a method of circumventing the problem.[12] Oseen's explanation was based on a comparison of the magnitude of viscous terms and inertial terms in (8-141), using Stokes's solution. In deriving Stokes's solution, it is assumed via the expansion (8-143) that viscous terms are dominant over inertia terms *everywhere* in the solution domain ($1 \le r \le \infty$) provided only that the Reynolds number is small enough. However, as $r \to \infty$, the magnitude of the largest term on the right-hand side of (8-141) can be estimated using Stokes's solution to be

$$\text{Re}\left(u_r \frac{\partial u_r}{\partial r}\right) \sim 0\left(\frac{\text{Re}}{r^2}\right). \tag{8-152}$$

A typical viscous term, on the other hand, is

$$\frac{\partial^2 u_r}{\partial r^2} \sim 0\left(\frac{1}{r^3}\right). \tag{8-153}$$

Comparing (8-152) and (8-153), it is clear that inertia terms are small compared with viscous terms only for

$$r < 0\left(\frac{1}{\text{Re}}\right). \tag{8-154}$$

Hence, as Oseen noted, we cannot expect the Stokes solution to provide a uniformly valid first approximation to the solution of (8-141), but we should expect that it will break down for large values of $r \ge 0(\text{Re}^{-1})$. Thus, Whitehead's attempt to evaluate the second term in the expansion (8-143) was unsuccessful for large r. Indeed, as noted earlier in conjunction with the thermal problem, it is not so much a surprise that we cannot obtain a solution for ψ_1 as it is a surprise to be able to solve for ψ_0 using the boundary condition (8-147) for $r \to \infty$, in spite of the fact that the governing equation (8-146) is not a valid first approximation to the full Navier-Stokes equation except for $r < 0(1/\text{Re})$.

If we examine the corresponding problem of streaming flow past a circular cylinder for $\text{Re} \to 0$, we find, in fact, that a solution of Stokes's equation cannot be obtained to satisfy boundary conditions at infinity. In particular, for a two-dimensional flow, the creeping motion approximation to the Navier-Stokes equation can be expressed in the form

$$\boxed{\nabla^4 \psi_0 = 0,} \tag{8-155}$$

where ∇^4 is the biharmonic operator ($\nabla^2 \cdot \nabla^2$). If we consider streaming flow past a

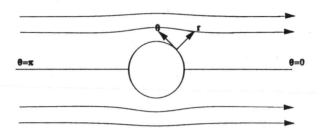

Figure 8–8 A sketch showing the cylindrical coordinate system used for analysis of streaming flow past a circular cylinder.

circular cylinder, the uniform velocity condition at infinity is most conveniently expressed in terms of a circular cylindrical coordinate system

$$\psi_0 \to r\sin\theta \quad \text{as} \quad r \to \infty, \tag{8-156}$$

where $\theta = 0$ and π are the downstream and upstream symmetry axes, respectively. A sketch of the problem, including the cylindrical coordinate system is shown in Figure 8-8. In addition to (8-156), the solution must satisfy boundary conditions $u_r = u_\theta = 0$ at the cylinder surface, plus the symmetry condition $u_\theta = 0$ at $\theta = 0, \pi$. In terms of the streamfunction ψ_0, these conditions take the form

$$\psi_0 = \frac{\partial \psi_0}{\partial r} = 0 \quad \text{at} \quad r = 1,$$

$$\psi_0 = 0 \quad \text{at} \quad \theta = 0, \pi. \tag{8-157}$$

We leave it to the reader to supply the details of solving (8-155). However, the most general solution that also satisfies the boundary conditions (8-157) is

$$\psi_0 = C\left(\frac{1}{r} - r + r\ln r\right)\sin\theta. \tag{8-158}$$

The coefficient C in (8-158) is arbitrary, and the condition (8-156) for $r \to \infty$ has not been applied in deriving (8-158). Thus, it would seem that C should be chosen to satisfy this condition. However, it is obvious in examining (8-158) that there is no value of C that will work. For any nonzero value of C, the right-hand side of (8-158) increases for large r in proportion to $r\ln r$, and this term is arbitrarily large compared with the required form (8-156). Thus, in the case of streaming flow past a circular cylinder, we must conclude that *no solution exists* for the creeping flow equations that satisfies boundary conditions for $r \to \infty$. In fact, the same conclusion is true for streaming flow past a two-dimensional cylinder of arbitrary cross-sectional shape. The fact that no creeping flow solution exists for streaming flow in an unbounded two-dimensional flow was actually discovered by Stokes and is known as *Stokes's paradox*.

A very similar result was found in Section C for heat conduction from a heated cylinder in an unbounded two-dimensional domain.

Of course, neither the Stokes' nor Whitehead's paradox is a true paradox, for we now understand both why they occur and how they can be avoided by use of the method of matched asymptotic expansions. In the remainder of this section, we will show how this technique can be applied to obtain higher order approximations to the solution for streaming flow past a rigid sphere at small Reynolds number. Our analysis will be very similar to that presented earlier for heat transfer from a sphere at low Peclet number, though the details are more complicated. The reader may wish to refer to the original paper of Proudman and Pearson (1957) for a somewhat more complete account of the problem for a sphere, as well as a discussion of the very similar analysis for low Reynolds number streaming flow past a two-dimensional, circular cylinder. Before considering the analysis for a sphere in an unbounded domain, it is important to point out that the singular nature of the low Reynolds number approximation (necessitating use of the matched asymptotic expansion technique) is a consequence of the assumption of an unbounded domain. Specifically, as long as r remains smaller than $0(\mathrm{Re}^{-1})$, Stokes' Equation (8–146), and hence the solution of the equation, will provide a valid first approximation of the complete problem (8–141) and (8–142) for small Reynolds numbers. In a *bounded* domain, all points can be held within the region of validity of Stokes' solution provided that we consider sufficiently small Reynolds numbers so that $r_{\mathrm{max}} < 0(\mathrm{Re}^{-1})$. For an *unbounded* domain, on the other hand, there *always* will be a region where the Stokes' approximation breaks down.

Let us now return to the problem of streaming flow past a sphere in an unbounded domain at small Reynolds numbers. It is presumably obvious, by analogy to the low Peclet number heat transfer problem, that the source of Whitehead's difficulty in obtaining a regular perturbation solution of (8–141) is the fact that the sphere radius used in nondimensionalizing (8–141) is appropriate as a characteristic length scale only over the region $r < 0(\mathrm{Re}^{-1})$. Far from the sphere, there must exist a second characteristic length scale, hence a second nondimensional form of the Navier-Stokes equation [replacing (8–141)] and a second asymptotic expansion [replacing (8–143)]. We have seen this situation now several times: When two different length scales are relevant in different parts of the domain, two different approximations to the governing equations will result in the limit $\mathrm{Re} \to 0$, and this requires two different asymptotic approximations to the solution. The asymptotic limit is *singular*, and we can attack the problem using the method of matched asymptotic expansions. The situation is illustrated schematically in terms of the solution domain for $\mathrm{Re} \ll 1$ in Figure 8–9.

In the so-called inner region nearest the sphere, the sphere radius is appropriate as the characteristic length scale, and thus equation (8–141) is applicable. To solve this equation for $\mathrm{Re} \ll 1$, we again assume the existence of an asymptotic expansion, but unlike (8–143), we do not presuppose any knowledge of the form of the gauge functions, so that

$$\psi = \sum_{n=0}^{N} f_n(\mathrm{Re}) \psi_n(r, \eta), \qquad (8\text{–}159)$$

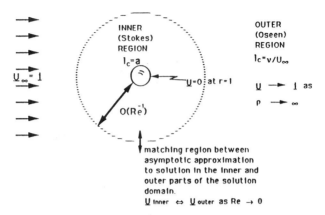

Figure 8–9 A schematic of the asymptotic solution domain for *low* Reynolds number streaming flow past a sphere, that is, Re ≪ 0.

subject only to the convergence condition

$$\lim_{\text{Re} \to 0} \frac{f_{n+1}(\text{Re})}{f_n(\text{Re})} \to 0. \tag{8-160}$$

The terms in (8–159) are expected to satisfy the symmetry conditions at $\eta = \pm 1$ and boundary conditions at the sphere surface, namely,

$$\psi_n(r, \eta) = 0, \quad \eta = \pm 1,$$

$$\psi_n(r, \eta) = \frac{\partial \psi_n}{\partial r} = 0 \quad \text{at } r = 1. \tag{8-161}$$

However, the boundary condition at infinity is no longer applied to the inner expansion since $r \to \infty$ lies outside the inner region. Instead, we will require that the inner expansion match the outer expansion in the region of overlap between the two approximations.

In the outer region, we must first determine the appropriate characteristic length scale. To do this, we follow the example of the small Peclet number heat transfer problem from Chapter 6, Section C, and introduce a rescaling,

$$\boxed{\rho = r\text{Re}^m.} \tag{8-162}$$

As before, we recognize this rescaling as defining a new nondimensionalized radial variable, with a new characteristic length scale

$$l = a(\text{Re})^{-m} \tag{8-163}$$

depending on m. To determine m, we must substitute the rescaled variable into (8–141). First, however, we note that rescaling (8–162) in this case requires a corresponding rescaling of the streamfunction ψ. This is because velocities, nondimensionalized with respect to U_∞, are expected to be $0(1)$ in both the inner and outer regions. But in terms of the streamfunction,

$$u_r = -\frac{1}{r^2}\frac{\partial \psi}{\partial \eta}, \quad u_\eta = -\frac{1}{r\sqrt{1-\eta^2}}\frac{\partial \psi}{\partial r}.$$ (8-164)

Thus, if we introduce the rescaling (8-162) into these definitions, it is evident that ψ should be rescaled in the outer region according to

$$\Psi = \text{Re}^{2m}\psi.$$ (8-165)

Now the scaling constant m is determined by substituting (8-162) and (8-165) into the Navier-Stokes equation (8-141) and requiring that the limiting form of the resulting equation for $\text{Re} \to 0$ contain viscous terms and at least one inertia term, as suggested by Oseen's argument. Upon substitution of outer variables, (8-141) becomes

$$\text{Re}^{2m}E_\rho^4\Psi = \text{Re}^{m+1}\left[\frac{1}{\rho^2}\left\{\frac{\partial \Psi}{\partial \rho}\frac{\partial}{\partial \eta}(E_\rho^2\Psi) - \frac{\partial \Psi}{\partial \eta}\frac{\partial}{\partial \rho}(E_\rho^2\Psi)\right\} + \right.$$
$$\left. \frac{2}{\rho^2}E_\rho^2\Psi\left\{\frac{\partial \Psi}{\partial \rho}\frac{\eta}{1-\eta^2} + \frac{1}{\rho}\frac{\partial \Psi}{\partial \eta}\right\}\right],$$ (8-166)

where

$$E_\rho^2 \equiv \frac{\partial^2}{\partial \rho^2} + \frac{1-\eta^2}{\rho^2}\frac{\partial^2}{\partial \eta^2}.$$ (8-167)

Thus, since inertial terms and viscous terms are of equal importance in the outer region for $\text{Re} \to 0$, it follows that

$$2m = m+1 \quad \text{or} \quad m = 1.$$ (8-168)

Thus,

$$\rho = r\text{Re}, \quad \Psi = \text{Re}^2\psi,$$ (8-169)

and the characteristic length scale in this outer region has been shown to be

$$l = a(\text{Re})^{-1} = \frac{\nu}{U_\infty}.$$ (8-170)

The governing equation in the outer region is just (8-166) with $m = 1$.

We assume, following our previous analyses, that there exists an asymptotic expansion for Ψ in the form

$$\Psi = \sum_{n=0}^{N}F_n(\text{Re})\Psi_n(\rho, \eta),$$ (8-171)

with

$$\lim_{\text{Re} \to 0}\frac{F_{n+1}}{F_n} \to 0.$$ (8-172)

In addition to the governing equation (8-166), the solution in this outer region must satisfy

$$\Psi(\rho, \pm 1) = 0 \qquad (8\text{-}173a)$$

and

$$\Psi \rightarrow -\rho^2 Q_1 \quad \text{as} \quad \rho \rightarrow \infty. \qquad (8\text{-}173b)$$

The latter condition is the uniform velocity at infinity, expressed in terms of outer variables. Of course, the outer region solution is not required to satisfy boundary conditions at the sphere surface since this is not part of the outer domain. Instead, we require matching between the solutions in the two parts of the domain, that is,

$$\text{Re}^2 \psi|_{r \gg 1} \Longleftrightarrow \Psi_{\rho \ll 1} \quad \text{as} \quad \text{Re} \rightarrow 0. \qquad (8\text{-}174)$$

In the limit as $\text{Re} \rightarrow 0$, the outer portion of the inner region, $r \gg 1$, corresponds to arbitrarily small values of the outer variable, $\rho \equiv \text{Re}\, r \ll 1$.

In the present problem, our analysis begins with a first approximation to the solution in the outer region. This is obtained by simply noticing that the asymptotic form (8-173b) for large ρ is actually an exact solution of the full governing equation (8-166). Thus, we find that

$$\boxed{\Psi_0 = -\rho^2 Q_1, \; F_0(\text{Re}) = 1.} \qquad (8\text{-}175)$$

Since the boundary condition at infinity provides a valid first approximation to the solution in the whole of the outer domain, we see why it was possible to obtain Stokes' solution by applying the free stream boundary condition for $r \rightarrow \infty$.

However, rather than simply accepting Stokes' solution as the first approximation for $\text{Re} \ll 1$ in the inner region, we will show how it is obtained in the present framework of matched asymptotic expansions. To do this, we note that the governing equation for ψ_0 in the expansion (8-159) is simply the Stokes equation for any choice of the gauge function $f_o(\text{Re})$, namely,

$$E^4 \psi_0 = 0, \qquad (8\text{-}176)$$

which has the general solution (see Chapter 4, Section **E**), satisfying the symmetry condition (8-161a),

$$\psi_0 = \sum_{n=1}^{\infty} \left[A_n r^{n+3} + B_n r^{n+1} + C_n r^{2-n} + D_n r^{-n} \right] Q_n(\eta). \qquad (8\text{-}177)$$

In our earlier derivation of Stokes's solution, (4-206), we applied the boundary conditions (8-161b), plus the free stream condition for $r \rightarrow \infty$. Here, however, the boundary condition at infinity is formally replaced by the matching condition (8-174) using the first approximation (8-175) to the solution in the outer region. To carry out the matching, we express ψ_0 in terms of outer variables by multiplying by Re^2, as indicated in (8-174), and converting from r to ρ. Then the matching condition becomes

$$-\rho^2 Q_1 \Longleftrightarrow f_0(\text{Re}) \left[\sum_{n=1}^{\infty} \left[A_n \rho^{n+3} \frac{1}{\text{Re}^{n+1}} + B_n \rho^{n+1} \frac{1}{\text{Re}^{n-1}} + \right.\right.$$
$$\left.\left. C_n \rho^{2-n} \text{Re}^n + + D_n \rho^{-n} \text{Re}^{n+2} \right] Q_n(\eta) \right]. \quad (8\text{-}178)$$

Here, the left-hand side is just the first approximation (8-175) in the outer region. Let us examine the various terms on the right. There is only one term that exhibits a ρ^2 dependence, namely,

$$f_0(\text{Re}) B_1 \rho^2 Q_1(\eta).$$

Hence, if the two sides of (8-178) are to match, it is evident that

$$\boxed{f_0(\text{Re}) = 1}$$

and

$$B_1 = -1. \quad (8\text{-}179)$$

The terms involving coefficients A_n are all $0(\text{Re}^{-2})$ or larger. Since the biggest term in the outer region is $0(1)$, it is clear that matching requires that

$$A_n \equiv 0, \quad \text{all} \quad n \geq 1. \quad (8\text{-}180)$$

Similarly, all of the terms involving B_n for $n \geq 2$ are $0(\text{Re}^{-1})$ or larger, and hence

$$B_n \equiv 0, \quad \text{all} \quad n \geq 2. \quad (8\text{-}181)$$

Thus, we obtain values for all the coefficients A_n and B_n in the first inner approximation (8-177) from the matching condition (8-178). Since the first approximation in the outer region is just the free stream boundary condition, expressed in outer variables, it should not be surprising that the coefficients A_n and B_n via matching are found to have the same values that were originally found by application of the boundary condition (8-147) for $r \to \infty$. The other coefficients, C_n and D_n, cannot be determined by matching. For example, the terms in (8-178) involving C_n are all $0(\text{Re})$ or smaller, when expressed in outer variables, and therefore cannot be required to match in (8-178) at the present level of approximation, since we have considered only the leading $0(1)$ term so far in the outer region. Similarly, the terms involving D_n are $0(\text{Re}^3)$ or smaller and cannot be required to match at this leading level of approximation. Instead, to determine C_n and D_n, we must apply the boundary conditions (8-161b) at $r = 1$. The values we obtain are just those of Stokes' solution, namely,

$$C_n = D_n = 0,$$
$$C_1 = \frac{3}{2}, \quad D_1 = -\frac{1}{2}, \quad (8\text{-}182)$$

and thus

$$\boxed{\psi_0 = -\left(r^2 - \frac{3}{2} r + \frac{1}{2r} \right) Q_1(\eta).} \quad (8\text{-}183)$$

So far, we have just reproduced Stokes' solution, though we now recognize it as the first approximation in the inner region only, rather than a universally valid first approximation everywhere in the domain. It is perhaps worth reiterating that Stokes' solution matches the first approximation in the outer region to 0(1) only. The terms involving C_1 and D_1 cause a mismatch between the leading-order approximate solutions in the inner and outer regions of 0(Re) and 0(Re^3), respectively. To see this, we simply reexpress (8-183) in terms of outer variables,

$$\text{Re}^2\psi_0 = -\left(\rho^2 - \frac{3}{2}\rho\,\text{Re} + \frac{1}{2\rho}\,\text{Re}^3\right)Q_1(\eta), \tag{8-184}$$

and compare with the left-hand side of (8-178).

To obtain higher order approximations to (8-175) and (8-183), we begin with the second term in the outer expansion, (8-171), which now takes the form

$$\boxed{\Psi = -\rho^2 Q_1 + F_1(\text{Re})\,\Psi_1(\rho,\eta) + \cdots.} \tag{8-185}$$

The condition (8-172) requires that

$$\lim_{\text{Re}\to 0} F_1(\text{Re}) \to 0.$$

If we substitute (8-185) into the governing equation in the outer region, (8-166), and retain only terms of 0(F_1(Re)), we obtain

$$\boxed{E_\rho^4\Psi_1 = \left\{\eta\frac{\partial}{\partial\rho}(E_\rho^2\Psi_1) + \frac{1-\eta^2}{\rho}\frac{\partial}{\partial\eta}(E_\rho^2\Psi_1)\right\},} \tag{8-186}$$

which is to be solved subject to the boundary conditions

$$\frac{1}{\rho}\frac{\partial\Psi_1}{\partial\rho}\to 0, \quad \frac{1}{\rho^2}\frac{\partial\Psi_1}{\partial\eta}\to 0 \quad\text{as}\quad \rho\to\infty, \tag{8-187}$$

plus the matching condition (8-174) with the inner solution. The boundary conditions follow immediately from the fact that the leading term in the expression for Ψ already satisfies the uniform velocity condition as $\rho\to\infty$. The governing equation (8-186) is known as the *Oseen equation*. In view of the fact that the Oseen equation and the boundary conditions (8-187) are all homogeneous, it should be evident that a nonzero second term in (8-185) arises only because of the mismatch, at 0(Re), between $-\rho^2 Q_1$ and the right-hand side of (8-184). Thus, we can anticipate that

$$\boxed{F_1(\text{Re}) = \text{Re}.} \tag{8-188}$$

To solve (8-186), it is convenient to introduce the change of variable, first discovered by Goldstein (1929),

$$E_\rho^2\Psi_1 = e^{\rho\eta/2}\Phi(\rho,\eta). \tag{8-189}$$

If we substitute (8-189) into (8-186), we obtain a simple governing equation for the function Φ, namely,

$$\left(E_\rho^2 - \frac{1}{4}\right)\Phi = 0. \tag{8-190}$$

When expressed in terms of $E_\rho^2 \Psi_1$, the most general solution of this equation that is bounded as $\rho \to \infty$ and at $\eta = \pm 1$ is

$$\boxed{E_\rho^2 \Psi_1 = e^{\rho\eta/2} \sum_1^\infty A_n \sqrt{\frac{\rho}{2}} K_{n+1/2}\left(\frac{\rho}{2}\right) Q_n(\eta).} \tag{8-191}$$

To solve for Ψ_1, we must integrate (8-191). Before doing this, however, it is convenient to match

$$E^2\psi_0|_{r \gg 1} \iff E_\rho^2(-\rho^2 Q_1 + \mathrm{Re}\,\Psi_1)|_{\rho \ll 1} \quad \text{as} \quad \mathrm{Re} \to 0, \tag{8-192}$$

since this will allow us to determine the constants A_n in (8-191) prior to trying to solve for Ψ_1. The reader may, at first, question the validity of (8-192) since it is the streamfunction itself that must ultimately be matched according to (8-174). However, the implication of matching is that the two approximations from the inner and outer regions must have the same functional form in the matching zone, and this means that the solutions *and all their derivatives* must match to the same order of approximation. In the present case, the left-hand side of (8-192) is evaluated using (8-183):

$$E^2\psi_0 = -\frac{3}{r} Q_1(\eta), \tag{8-193}$$

and when this is expressed in terms of outer variables, it becomes

$$= -\frac{3}{\rho} \mathrm{Re}\, Q_1(\eta). \tag{8-194}$$

The right-hand side of (8-192), on the other hand, is just

$$E_\rho^2(-\rho^2 Q_1 + \mathrm{Re}\,\Psi_1) = \mathrm{Re}\, E_\rho^2 \Psi_1, \tag{8-195}$$

and the matching condition becomes

$$-\frac{3}{\rho}\mathrm{Re}\, Q_1(\eta) \iff \mathrm{Re}\left[e^{\rho\eta/2}\sum_1^\infty A_n \sqrt{\rho/2}\, K_{n+1/2}\left(\frac{\pi}{2}\right) Q_n(\eta)\right]\Bigg|_{\rho \ll 1} \quad \text{as} \quad \mathrm{Re} \to 0. \tag{8-196}$$

In order to carry out the indicated matching, it is convenient to introduce a series representation for the modified Bessel function,

$$\sqrt{\frac{\rho}{2}}\, K_{n+1/2}\left(\frac{\rho}{2}\right) = \sqrt{\frac{\pi}{2}}\, e^{-\rho/2} \sum_{s=0}^n \frac{(n+s)!}{(n-s)!\, s!}\frac{1}{\rho^s}. \tag{8-197}$$

Thus, we obtain

$$-\mathrm{Re}\,\frac{3}{\rho} Q_1(\eta) \iff \mathrm{Re}\left[\sqrt{\pi/2}\, e^{\rho(\eta-1)/2} \sum_{n=1}^\infty A_n Q_n(\eta) \sum_{s=0}^\infty \frac{(n+s)!}{(n-s)!\, s!}\frac{1}{\rho^s}\right]_{\rho \ll 1}. \tag{8-198}$$

Now, the coefficients A_n must be chosen so that all terms $0(1/\rho^2)$ or larger ($\rho \ll 1$) are excluded from the right-hand side of (8-198). The reason is that such terms cannot be matched with the left-hand side of (8-198), nor with any terms that will arise at higher order approximations in the inner region. Suppose, for example, that a term like

$$\frac{\text{Re}}{\rho^2}$$

is nonzero in the outer solution. If this term is expressed in terms of inner variables, we find that

$$\frac{1}{\text{Re}}\left(\frac{1}{r^2}\right),$$

meaning that it could be matched only by a term of $0(\text{Re}^{-1})$ in the inner expansion. However, we have already seen that the largest and leading-order term in the inner expansion is $0(1)$. Hence, the matching condition (8-198) can be satisfied only if

$$A_n = 0, \ n \geq 2. \tag{8-199}$$

With these values, the matching condition becomes

$$-\frac{3}{\rho} Q_1 \text{Re} \Longleftrightarrow \sqrt{\pi/2}\, e^{\rho(\eta-1)/2} \text{Re} A_1 Q_1 \left(1 + \frac{2}{\rho}\right)\Bigg|_{\rho \ll 1}. \tag{8-200}$$

For small ρ,

$$e^{\rho(\eta-1)/2} \sim 1 + \frac{1}{2}\rho(\eta - 1) + 0(\rho^2). \tag{8-201}$$

Thus, (8-200) can be written in the form

$$-\frac{3}{\rho} Q_1 \text{Re} \Longleftrightarrow \sqrt{\pi/2}\, A_1 Q_1 \text{Re}\left[\frac{2}{\rho} + \eta + 0(\rho)\right], \tag{8-202}$$

and the coefficient A_1 is chosen so that the first term on the right-hand side of (8-202) matches the term on the left,

$$A_1 = -\frac{3}{2}\left(\sqrt{\pi/2}\right)^{-1}. \tag{8-203}$$

The term

$$\sqrt{\pi/2}\, A_1 Q_1 \text{Re}\,\eta$$

in (8-202) need not be matched at this level of approximation because it is independent of ρ and could thus be obtained only in the inner solution at $0(\text{Re})$. At the present level of approximation, however, we have considered only terms of $0(1)$ in the inner region.

It remains to solve (8-192) for Ψ_1. With the constants A_n chosen according to (8-199) and (8-203), we see that

$$E_\rho^2 \Psi_1 = -\frac{3}{2} e^{-\rho(1-\eta)/2}\left\{1 + \frac{2}{\rho}\right\} Q_1(\eta). \tag{8-204}$$

We leave the details of solving this equation to the reader. However, upon solving (8-204), with the boundary conditions (8-187) and matching according to (8-174), with ψ_0 from the inner solution, the result for Ψ_1 is found to be

$$\Psi_1 = -\frac{3}{2}(1+\eta)\left\{1 - e^{-\rho(1-\eta)/2}\right\}. \tag{8-205}$$

The first two terms in the asymptotic expansion for Ψ in the outer region are thus

$$\Psi \sim -\rho^2 Q_1 - \mathrm{Re}\left[\frac{3}{2}(1+\eta)\left(1 - e^{-\rho(1-\eta)/2}\right)\right] + 0(F_2(\mathrm{Re})). \tag{8-206}$$

We may note that Ψ behaves quite differently for large ρ than the Stokes solution ψ_0 for large r. In particular, ψ_0 is symmetric about $\eta = 0$, whereas Ψ is asymmetric at $0(\mathrm{Re})$. It is at this level of approximation that the first effects of convection appear in the asymptotic solution for $\mathrm{Re} \to 0$.

The term at $0(\mathrm{Re})$ in (8-206) represents a wakelike solution for the vorticity, which is analogous to the thermal wake in Equation (8-51). To see this, we can calculate the vorticity associated with (8-206) via the definition

$$\omega \equiv E_\rho^2 \Psi = \frac{3}{2}\mathrm{Re}\, Q_1 e^{-\rho(1-\eta)/2}. \tag{8-207}$$

Except for a small neighborhood $\eta = 1$ (the downstream axis of symmetry), the vorticity, which is generated by the no-slip condition at the sphere surface, decreases exponentially with increase of ρ. Very near $\eta = 1$, on the other hand, the vorticity decreases only algebraically with ρ. This is a consequence of the fact that the vorticity generated at the sphere surface is swept downstream by the flow into a wakelike region. For comparison, it may be noted that the vorticity in the inner region, which is wholly transported by diffusion at $0(1)$, only decreases as $1/r$. A difference between (8-207) and the thermal wake solution for $\oplus_0$ is that $\omega \equiv 0$ at $\eta \equiv \pm 1$, whereas $\oplus_0$ was maximum at $\eta = 1$. The vorticity generated on opposite sides of the sphere has opposite signs, and thus cancels at $\eta = 1$.

The $0(\mathrm{Re})$ term in (8-206) also can be associated with the existence of a "defect" in the flux of momentum downstream of the sphere, at $\eta = 1$, relative to that approaching the sphere from upstream. This *momentum defect* is proportional in magnitude to the drag force on the sphere.

To obtain higher order corrections to Stokes's law, we must obtain additional terms in the inner (and outer) expansions. So far we have the first two terms in the expansion (8-171) for the outer part of the domain, given by (8-206), but only the first term in the inner region, which turned out to be Stokes's solution. There is a mismatch of $0(\mathrm{Re}\rho^2)$ between these solutions, from which it can be shown that the second term in the inner expansion must be $0(\mathrm{Re})$,

$$f_1(\mathrm{Re}) = \mathrm{Re}. \tag{8-208}$$

To see this, we can compare the existing (first) term in the inner expansion expressed in terms of outer variables, namely,

$$\text{Re}^2 \psi_0 = -Q_1 \left(\rho^2 - \frac{3}{2} \rho \text{Re} + \frac{1}{2\rho} \text{Re}^3 \right), \tag{8-209}$$

with the first two terms in the outer expansion for small ρ,

$$-\rho^2 Q_1 + \frac{3}{2} \rho \text{Re} Q_1 + \frac{3}{16} \text{Re}(1+\eta)(1-\eta)^2 \rho^2 + 0(\rho^3 \text{Re}). \tag{8-210}$$

With $f_1(\text{Re})$ given by (8-209), the governing equation for ψ_1 can be obtained easily by substituting (8-158) into (8-140) and retaining all terms of $0(\text{Re})$. Not surprisingly, the result is Whitehead's equation (8-149), and the general solution of this equation is Whitehead's solution (8-151). Now, however, instead of trying to determine the coefficients in (8-151) by application of boundary conditions for $r \to \infty$ (which gave rise to Whitehead's paradox), we require matching between the first two terms in the inner and outer expansions, namely,

$$\boxed{\text{Re}^2 (\psi_0 + \text{Re}\psi_1) |_{r \gg 1} \iff (-\rho^2 Q_1 + \text{Re}\Psi_1) |_{\rho \ll 1} \text{ for } \text{Re} \to 0.} \tag{8-211}$$

The left-hand side is most conveniently expressed in outer variables as

$$\text{Re}^2 (\psi_0 + \text{Re}\psi_1) = -Q_1 \left(\rho^2 - \frac{3\text{Re}}{2} \rho + \frac{\text{Re}^3}{2\rho} \right)$$

$$+ \text{Re}^3 \left\{ \sum_{n=1}^{\infty} \left[A_n \frac{\rho^{n+3}}{\text{Re}^{n+3}} + B_n \frac{\rho^{n+1}}{\text{Re}^{n+1}} + C_n \frac{\rho^{2-n}}{\text{Re}^{2-n}} + D_n \frac{\text{Re}^n}{\rho^n} \right] Q_n(\eta) \right\}$$

$$+ \frac{3}{16} \left(2\text{Re}\rho^2 - 3\text{Re}^2 \rho - \frac{\text{Re}^4}{\rho} \right) Q_2(\eta). \tag{8-212}$$

The expansion of the right-hand side of (8-211) for small ρ is given by (8-210). If we recall the definitions

$$Q_1(\eta) \equiv \frac{1}{2}(\eta^2 - 1), \quad Q_2(\eta) \equiv \frac{1}{2}\eta(\eta^2 - 1), \tag{8-213}$$

the latter can be rewritten as

$$\Psi_{\rho \ll 1} = -\rho^2 Q_1 + \frac{3}{2} \text{Re}\rho Q_1 - \frac{3}{8} \text{Re}\rho^2 Q_1 + \frac{3}{8} \text{Re}\rho^2 Q_2 + 0(\text{Re}\rho^3) + o(\text{Re}). \tag{8-214}$$

Matching is now carried out by comparing (8-212) and (8-214) according to (8-211). We leave the details to the reader. The result is

$$A_n = 0, \quad n \geq 1,$$

$$B_n = 0, \quad n \geq 2,$$

$$B_1 = -\frac{3}{8}. \tag{8-215}$$

The coefficients C_n and D_n must be determined by application of the boundary conditions on ψ_1 at the sphere surface $r = 1$. Before giving the final result for ψ_1, it is perhaps useful to backtrack briefly to the matching condition (8–211) between (8–212) and (8–214). Specifically, we recall that it was the term $(3/8)\text{Re}\rho^2 Q_2(\eta)$ in (8–212) that led to Whitehead's paradox. Comparing (8–212) to (8–214), however, we see that this term now matches identically with one of the terms in the outer solution.

The result of applying the boundary conditions $\psi_1 = \partial\psi_1/\partial r = 0$ at $r = 1$ to Whitehead's solution (8–151), with A_n and B_n given by (8–215), is

$$\psi_1 = -\frac{3}{16}Q_1\left(2r^2 - 3r - \frac{1}{r}\right) + \frac{3}{16}Q_2\left(2r^2 - 3r + 1 - \frac{1}{r} + \frac{1}{r^2}\right). \qquad (8\text{–}216)$$

But this is just

$$\psi_1 = \frac{3}{8}\psi_0 + \frac{3}{16}Q_2\left(2r^2 - 3r + 1 - \frac{1}{r} + \frac{1}{r^2}\right). \qquad (8\text{–}217)$$

With the first two terms of the inner expansion now known, we can use (4–184) to calculate the drag to O(Re). The result is

$$\text{drag} = 6\pi a\mu U\left(1 + \frac{3}{8}\text{Re} + o(\text{Re})\right). \qquad (8\text{–}218)$$

The effect of inertia is to slightly increase the drag relative to Stokes's drag.

A large number of additional terms have actually been calculated in the present problem. The next term in the inner expansion and in the expression for the drag was found by Proudman and Pearson (1957) to be $0(\text{Re}^2\ln\text{Re})$,

$$\text{drag} = 6\pi a\mu U\left[1 + \frac{3}{8}\text{Re} + \frac{9}{40}\text{Re}^2\ln\text{Re} + 0(\text{Re}^2)\right]. \qquad (8\text{–}219)$$

Higher order terms to $0(\text{Re}^3)$ were obtained by Chester and Breach (1969).[13] It is disappointing, but fairly typical of asymptotic approximations, that the calculation of many terms achieves a relatively small increase in the range of Reynolds number where the drag can be evaluated accurately compared to (8–218). We may recall that asymptotic convergence is achieved by taking the limit Re $\to 0$ for a fixed number of terms in the expansion, rather than an increasing number of terms for some fixed value of Re.

Homework Problems

1. We consider heat transfer from a spherical body into a surrounding fluid that is completely motionless. If the temperature is denoted as T, then the equation governing the steady-state temperature distribution is

$$\nabla^2 T = 0.$$

a. Suppose the surface temperature of the sphere is constant with a value T_0 and the temperature of the fluid far from the sphere is T_∞ (also constant).

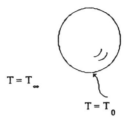

$T = T_\infty$

$T = T_0$

Show that $T = T_\infty + \dfrac{(T_0 - T_\infty)}{r} a$ where a is the sphere radius.

b. Now suppose that the surface of the sphere is insulating so that

$$\frac{\partial T}{\partial r} = 0 \quad \text{at} \quad r = a$$

and that there is a linear variation in the temperature of the fluid at infinity, that is,

$$T = \boldsymbol{\beta} \cdot \mathbf{x} \quad \text{as} \quad r \to \infty,$$

where $\mathbf{x}$ is a general position vector with origin at the sphere center. Determine the temperature distribution in the fluid for an arbitrary vector $\boldsymbol{\beta}$.

c. You should find that the disturbance produced by the sphere decreases as r^{-1} for case (a) and r^{-2} for case (b). Why does this happen?

2. Consider a heated cylinder whose centerline lies parallel to a cooled, infinite plane wall. The distance between the centerline and the wall is $a + h$, where a is the cylinder radius. We suppose that the surface of the cylinder is held at a constant prescribed temperature T_1, while the wall temperature is T_0, with $T_0 < T_1$. The fluid around the cylinder is assumed to remain motionless so that all heat transfer between the cylinder and wall occurs by conduction.

a. Assume that $h \ll a$, and obtain an approximation using lubrication ideas for the heat flux between the cylinder and wall.

b. Repeat the calculation for a heated sphere, instead of a heated cylinder, and comment (or explain) any difference between the results for a cylinder and for a sphere.

3. The following equation is a one-dimensional analogue of the Navier-Stokes equation:

$$\left(D^2 - \epsilon^2 \frac{1}{r} \frac{\partial \psi}{\partial r} \right) D^2 \psi = 0,$$

where

$$D^2 \equiv \frac{\partial^2}{\partial r^2} - \frac{2}{r^2},$$

$$1 \leq r < \infty.$$

Obtain an asymptotic solution of this equation for $\epsilon \ll 1$, including terms of $0(\epsilon)$, with the boundary conditions

$$\psi = \frac{\partial \psi}{\partial r} = 0 \quad \text{at} \quad r = 1,$$

$$\psi - \frac{1}{2} r^2 \to 0 \quad \text{as} \quad r \to \infty.$$

Note: The equation $(D^2 - \epsilon^2)\phi$ has solutions

$$e^{-\epsilon r}\left(1 + \frac{1}{\epsilon r}\right) \quad \text{and} \quad e^{\epsilon r}\left(1 - \frac{1}{\epsilon r}\right).$$

4. Let us consider the rate of heat transfer from a heated cylinder, whose central axis is perpendicular to the plane of a simple shear flow.

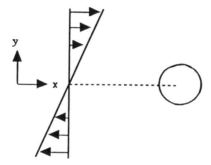

The ambient temperature of the fluid is denoted as T_∞, while the surface of the cylinder is T_0. You may also assume that the velocity field near the cylinder is given in terms of the streamfunction as

$$\psi = \frac{1}{4}(r^2 - 1) - \frac{1}{4}(r^2 - 2 + r^{-2})\cos 2\phi$$

when the cylinder is freely rotating. Here,

$$u_r \equiv \frac{1}{r}\frac{\partial \psi}{\partial \phi}, \quad u_\phi = -\frac{\partial \psi}{\partial r}$$

and

$$\mathbf{u}^\infty = y\mathbf{i}.$$

a. Solve the problem of heat transfer from the cylinder for small $\text{Pe} \ll 1$, and prove

$$\text{Nu} = 2\left(1.372 - \frac{1}{2}\log \text{Pe}\right)^{-1}.$$

What is the magnitude of the error in this expression for the Nusselt number?

Note: As part of the analysis for this problem, it is useful to have a solution in cylindrical coordinate of

$$\nabla^2 w - y \frac{\partial w}{\partial x} = 0$$

for a point source of w at the origin. This solution is

$$w = \int_0^\infty \frac{dS}{2S\left(1 + \frac{1}{12}S^2\right)^{1/2}} \exp\left[-\left\{\frac{\left(x - \frac{1}{2}yS\right)^2}{4S\left(1 + \frac{1}{12}S^2\right)} + \frac{y^2}{4S}\right\}\right].$$

You should derive this solution as part of the analysis of the overall problem.

b. Suppose there is a nonzero torque applied to the cylinder. How does this change the problem? Consider carefully, starting with a recalculation of the streamfunction field. Carry out calculations for enough cases to illustrate any qualitative trends in the heat transfer problem. (To be specific, you may wish to begin by considering the case in which the applied torque completely inhibits the rotation of the cylinder.)

5. A gas bubble rises through an infinite body of fluid under the action of buoyancy. The Reynolds number associated with this motion is very small but nonzero. Assume that the bubble remains *spherical*, and use the method of matched asymptotic expansions to calculate the drag on the bubble, including the first correction due to inertia at 0(Re). You may assume that the viscosity and density of the gas are negligible compared with that of the liquid so that you can apply the boundary conditions

$$u_r = 0 \quad \text{and} \quad \tau_{r\theta} = 0 \quad \text{at} \quad r = a,$$

where a is the bubble radius. The motion of the gas may thus be ignored in the calculation. Would the bubble actually remain spherical? If not, calculate the bubble shape through terms of 0(Re) by assuming that the shape is only slightly perturbed from spherical and using the method of domain perturbations. What condition (or conditions) must be satisfied to ensure that the near-sphere approximation is valid at 0(Re)?

6. A rigid sphere rotates with a constant angular velocity Ω in an infinite, quiescent fluid. It is desired to calculate the torque on the sphere to include any inertial correction at 0(Re), where

$$\text{Re} \equiv \frac{\Omega a^2 \rho}{\mu} \ll 1.$$

Can this be done by a regular perturbation solution, or is it necessary to use the method of matched asymptotic expansions? Calculate the torque to include the first inertial correction.

7. We consider uniaxial extensional flow exterior to a solid sphere whose center is located at the origin so that

$$\mathbf{u}^{\infty} = \begin{pmatrix} -E & 0 & 0 \\ 0 & -E & 0 \\ 0 & 0 & 2E \end{pmatrix} \cdot \mathbf{x}.$$

Here, $\mathbf{x}$ is zero at the sphere center. Let us suppose that the Reynolds number,

$$\text{Re} - \frac{\rho E a^2}{\mu} \ll 1,$$

is very small but nonzero. We desire to determine the first *inertial* contribution to the flow, beyond the classical creeping flow solution.

a. By considering the solution of the problem arising in a regular perturbation scheme at $0(\text{Re})$, show that a singular perturbation solution is necessary.

b. Develop appropriately scaled governing equations and boundary conditions for both the inner region nearest the sphere and the outer region in the far field. Using these equations and boundary conditions, determine the leading order approximation to the solution in the outer region, and use this solution, plus matching, to demonstrate that the creeping flow solution (obtained by assuming that the creeping flow approximation is uniformly valid) is the proper leading-order approximation in the inner region.

c. Demonstrate that the second term for the asymptotic approximation in the outer region is $0(\text{Re}^{3/2})$ in magnitude, and derive a governing equation for the streamfunction contribution at this order, namely, Ψ_1. This governing equation can be reduced to a simpler form by means of the substitution

$$E_\rho^2 \Psi_1 = \exp\left(\frac{1}{4}\rho^2 P_2(\eta)\right) \Phi(\rho, \eta),$$

but, so far as we know, this outer problem at $0(\text{Re}^{3/2})$ has not been solved analytically.

d. Derive governing equations and boundary conditions also for the second term in the expansion of ψ for the inner region by assuming that $f_1(\text{Re}) = \text{Re}$, that is

$$\psi = \psi_0 + f_1(\text{Re})\psi_1 + \cdots.$$

Obtain a solution for ψ_1, and demonstrate that the assumption $f_1(\text{Re}) = \text{Re}$ is required for matching. In fact, you should find that ψ_1 can be determined to within a single constant even though the lack of an analytic solution for Ψ_1 means that a detailed match of inner and outer solutions cannot be carried out.

8. Consider a *spherical* gas bubble and suppose that it contains a gaseous species A that is soluble in the liquid that surrounds the bubble and a second dominant species B that is insoluble in the liquid. We wish to calculate the rate of mass transfer in the creeping flow limit when the bubble is rising through an unbounded body of liquid due to buoyancy. The governing equation for

transport of a single solute A in the liquid is

$$\frac{\partial C}{\partial t} + \mathbf{u} \cdot \nabla C = D \nabla^2 C,$$

where D is the mutual diffusion coefficient for the species A in the liquid and C is the concentration of species A in the liquid. You may assume that $C \rightarrow 0$ far from the bubble. Obtain an expression for the rate of mass transfer in the limits $\mathrm{Re} \ll 1$ and $\mathrm{Pe} \ll 1$, where the Peclet number for mass transfer is defined as

$$\mathrm{Pe} \equiv \frac{Ua}{D},$$

with U being the rise velocity and a being the bubble radius. You may assume that any change in the volume of the bubble is negligible, and also neglect any change in the concentration of A inside the bubble. Hence, at equilibrium, the concentration of A on the liquid side of the gas-liquid interface is constant at a value C_0 that is given by Henry's law.

For this problem, you should assume that the mass transfer process occurs within the region surrounding the bubble where the Stokes approximation to the velocity field can be used. Thus, the solution will be valid provided $\mathrm{Re} \ll \mathrm{Pe}$. Explain the reason for this condition. Your calculation of mass transfer rate should be carried out to include the first correction due to convection.

References/Notes

1. Viscous dissipation effects can be important in a number of technological applications, either because velocity gradients are large, the fluid viscosity is large, or a combination of these. One class of problems of particular significance to chemical engineers occurs in many polymer processing systems. The interested reader may consult textbooks such as

a. Middleman, S. *Fundamentals of Polymer Processing.* McGraw-Hill: New York (1977).

b. Pearson, J.R.A., *Mechanics of Polymer Processing.* McGraw-Hill (Hemisphere): Washington (1983).

c. Pearson, J.R.A., and Richardson, S.M., *Computational Analysis of Polymer Processing.* Appl. Sci. Pub.: London and New York (1983).

2. The original analysis of this problem was published by

Acrivos, A., and Taylor, T.E., "Heat and Mass Transfer from Single Spheres in Stokes Flow," *Phys. Fluids* 5:387–394 (1962).

3. Goldstein, S., "The Steady Flow of Viscous Fluid Past a Fixed Obstacle at Small Reynolds Numbers," *Proc. Roy. Soc.* A123:225–235 (1929).

4. See Reference 2.

5. Brenner, H., "Forced Convection Heat and Mass Transfer at Small Peclet Numbers from a Particle of Arbitrary Shape," *Chem. Eng. Sci.* 18:109–122 (1963).

6. Elrick, D.E., "Source Functions for Diffusion in Uniform Shear Flow," *Australian J. Physics* 15:283–288 (1962).

7. Batchelor, G.K., "Mass Transfer from a Particle Suspended in Fluid with a Steady Linear Ambient Velocity Distribution," *J. Fluid Mech.* 95:369–400 (1979).

8. Acrivos, A., "A Note on the Rate of Heat or Mass Transfer from a Small Particle Freely Suspended in a Linear Shear Field," *J. Fluid Mech.* 98:299–304 (1980).

9. Whitehead, A.N., "Second Approximation to Viscous Fluid Motion," *Quart. J. Math.* 23:143–152 (1889).

10. a. Kaplan, S., "Low Reynolds Number Flow Past a Circular Cylinder," *J. Math. Mech.* 6:595–603 (1957).

   b. Kaplan, S., and Lagerstrom, P.A., "Asymptotic Expansions of Navier-Stokes Solutions for Small Reynolds Numbers," *J. Math. Mech.* 6:585–593 (1957).

   These and other published and unpublished works of Kaplan are reproduced in the book

   c. Kaplan, S., *Fluid Mechanics and Singular Perturbations*. P.A. Lagerstrom, L.N. Howard, and C.S. Lin (eds.). Academic Press: New York (1957).

11. Proudman, I., and Pearson, J.R.A., "Expansions at Small Reynolds Numbers for the Flow Past a Sphere and a Circular Cylinder," *J. Fluid Mech.* 2:237–262 (1957).

12. A reproduction of the work in this paper can be found in the textbook Oseen, C.W., *Hydrodynamik*. Akad. Verlagsgellschatt: Leipzig (1927).

13. Chester, W., and Breach, D.R., "On the Flow Past a Sphere at Low Reynolds Number," *J. Fluid Mech.* 37:751–760 (1969).

CHAPTER 9

Strong Convection Effects in Heat and Mass Transfer at Low Reynolds Number

In Chapter 8 we considered weak convection effects in the transfer of heat from a hot or cold body, and in the motion of a viscous fluid past a sphere. In the next several chapters we continue our study of convection effects in both heat (mass) transfer and fluid mechanics problems, but now we focus on the limits of $Pe \gg 1$ and/or $Re \gg 1$, where the convection of heat or vorticity (momentum) appear dominant in the governing equations (8–7) and (8–142) relative to conduction and diffusion, respectively.

There is no doubt of the practical need for solutions of fluid mechanics and transport problems outside the limiting cases of $Re \gg 1$ and $Pe \gg 1$. The majority of fluid motions, which do not involve very viscous fluids and/or very small particles, correspond to Reynolds numbers $Re \geq 0(1)$ and frequently $Re \gg 1$. For example, the Reynolds number for motion of a fluid characterized by a kinematic viscosity of 1 centistoke (for example, water), a length scale of $0(1 \text{ centimeter})$, and a velocity of only $0(1 \text{ centimeter/second})$ is 100. To achieve $Re \leq 0(10^{-1})$ where we might hope that creeping flow results would provide a reasonable approximation, we would need a body dimension $l \leq 10^{-3}$ centimeters (100 microns) even for the rather low velocity of 1 cm/sec.

Insofar as heat transfer is concerned, the analysis of Chapter 8 required both $Re \ll 1$ and $Pe \ll 1$. Since $Pe = RePr$, this condition is satisfied for gases, where $Pr \sim 0(1)$, and for relatively inviscid liquids such as water that have $Pr \sim 0(1)$ or slightly larger. On the other hand, viscous oils and greases can have $Pr \sim 10^3\text{--}10^6$, and for these fluids Pe may be large, even though Re is small. This provides one clear motivation for studying heat transfer for $Pe \geq 0(1)$. Another motivation is due to the fact that the same equation (8–7) also holds for transfer of a single solute by diffusion and convection in a two-component mass transfer system, provided only that Pr is replaced by the Schmidt number,

$$Sc \equiv \frac{\nu}{D},$$

where D is the diffusivity for the solute in the solvent. In this case, θ represents a dimensionless concentration of the solute, defined in such a way that the boundary conditions (8–8) still apply. A detailed discussion of the analogy between heat transfer by convection/conduction and mass transfer of a single solute by convection/diffusion is outside the scope of this book. The main point to note here is that the Schmidt number is $0(10^3)$ or larger for almost all solute/solvent systems in the liquid state and is $0(1)$ mainly for systems consisting of two gases (the one present at lowest concentration being designated as the solute). Thus, even for small Reynolds numbers, the Peclet number defined for mass transfer as

$$Pe = ReSc$$

is very often $0(1)$ or larger.

The preceding paragraphs demonstrate the importance of obtaining solutions for heat (or mass) transfer problems in the limit $Pe \gg 1$ and for fluid mechanics problems in the limit $Re \gg 1$. In general, however, exact analytical solutions are not possible in either of these limits. The Navier-Stokes equations are nonlinear. Aside from the special cases of unidirectional flow and of creeping flow in a bounded domain, this precludes exact solution for any $Re \neq 0$ except via numerical methods.* The forced convection heat transfer problem [equation (8–7) plus boundary conditions] is linear in θ, but it still cannot be solved exactly (except for special cases) for $Pe \geq 0(1)$ due to the complexity of the coefficient **u**. What may appear surprising at first is that simplifications arise in the limits $Re \gg 1$ or $Pe \gg 1$, which allow approximate solution even though no solutions (exact or approximate) are possible for intermediate values of Re or Pe. This is surprising because the importance of the troublesome convection term, which is the source of nonlinearity in the Navier-Stokes equations and of the complicated coefficient **u** in equation (8–7), would appear to increase as Re or Pe is increased.

To see how the problems simplify in the limit $Pe \gg 1$ or $Re \gg 1$, it is necessary to consider the solution of some specific prototype problems in detail. These solutions in the *strong-convection* limit are presented in the next several chapters. Specifically, in this chapter we begin with heat (or mass) transfer for $Pe \gg 1$; in Chapter 10 we discuss fluid mechanics problems for $Re \gg 1$; and in Chapter 11 we return to heat (or mass) transfer for $Pe \gg 1$, but in this case with $Re \gg 1$.

It will not surprise the reader who has come this far that we approach the two limits $Re \rightarrow \infty$ and $Pe \rightarrow \infty$ via the method of asymptotic approximation theory. In our view, however, the problems we pursue in Chapters 9, 10, and 11 provide the first real insight into the power and generality of the methods of asymptotic approximation and dimensional analysis. The asymptotic analyses of Chapter 8 were important in elucidating the role of dimensional analysis in the solution of nonlinear transport or fluid mechanics problems, especially in demonstrating how to deal with the fact that the temperature and/or velocity fields generally need not be characterized everywhere in the domain by the apparently obvious length scale imposed by the boundary geometry.

---

*Numerical solutions are themselves an approximate solution due to the discretization of the Navier-Stokes equations and boundary conditions to an equivalent set of algebraic equations. Current methods and computers allow accurate solution up to a maximum $Re \sim 0(10^2–10^3)$. The interested reader may wish to refer to references listed in Reference 1 of Chapter 4.

In practical terms, however, the results of Chapter 8 contribute only small corrections to the creeping flow and pure conduction correlations between the Nusselt number or drag coefficient and the independent dimensionless parameters Pe or Re. In contrast, the asymptotic analysis for strong convection plays a critical role in achieving the first approximation for correlations between independent and dependent dimensionless groups. Furthermore, we shall see that it is possible to deduce the complete functional form of this relationship, under extremely general conditions, by nothing more than a straightforward combination of nondimensionalization and an asymptotic formulation for the problem. For example, in Section A we consider the problem of heat transfer from a solid sphere in streaming flow at low Reynolds number but high Peclet number. In this case, we can show that

$$\text{Nu} = c\,\text{Pe}^{1/3} \qquad (9\text{--}1)$$

to leading order of approximation by simply formulating the asymptotics but without solving any differential equations! The constant c is independent of Pe and is guaranteed to have a numerical value of $0(1)$, which depends only on the geometry of the body. It is only if we wish to obtain a precise value for c that we must solve the leading-order set of differential equations. Although we would get different numerical values of c for different body shapes, we shall see that the general form of the correlation between Nu and Pe remains the same for bodies of arbitrary geometry at this leading order of approximation. The general form of a correlation like (9–1) is almost always more significant than the value of c. For example, with the form known, a rather small number of experiments could presumably be used to determine c for a quite arbitrarily shaped body. Without this knowledge, on the other hand, a much more extensive experimental program would be required to determine the functional dependence of Nu on Pe, and it would not presumably be evident without much more experimentation that the same form should hold for solid bodies of many different shapes.

But before extolling the virtues of the results and the approach, we must describe the method in detail. Let us therefore return to a systematic development of the ideas necessary to solve fluid mechanics and transport (heat or mass transfer) problems in the strong-convection limit. To do this, we begin with an already familiar problem from Chapter 8, namely, heat transfer from a solid sphere in a streaming flow at sufficiently low Reynolds number that the velocity field in the domain of interest can be approximated adequately by Stokes's solution. In the present case we consider the limit Pe $\gg$ 1. The resulting analysis will introduce us to the main ideas of boundary-layer theory.

A. Heat Transfer from a Solid Sphere in Streaming Flow for Small Reynolds Number and Large Peclet Number

We begin our analysis of strong convection effects by returning to the transfer of heat from a uniformly heated sphere in a uniform streaming flow at low Reynolds number, when the velocity field can be approximated by Stokes's solution. This problem was first considered by Acrivos and Goddard.[1] In dimensionless form, the problem we aim to solve is still given by equations (8–7) and (8–8). In these two equations, we have assumed that the characteristic length scale for variations in the temperature field is the sphere radius $l_c = a$. After substituting Stokes's solution for $\mathbf{u}$ and expressing (8–7) in spherical coordinates, we obtain (8–25), as before. In the present case, however, we consider the limit in which Pe $\gg 1$. To do this, we propose an asymptotic expansion for θ in the form

$$\theta = \theta_0 + \sum_{n=1}^{N} F_n(\text{Pe}^{-1})\theta_n. \qquad (9\text{–}2)$$

As usual, the governing equation for the first term in this expansion is simply generated by taking the limit Pe $\to \infty$ in the full equation, (8–7) or (8–25), namely,

$$\boxed{\mathbf{u} \cdot \nabla \theta_0 = 0.} \qquad (9\text{–}3)$$

We see that (9–3) involves a complete neglect of conduction relative to convection.

Equation (9–3) can be solved quite easily. The physical significance of (9–3) is that θ_0 must be constant along lines (or surfaces) parallel to $\mathbf{u}$. In other words, the projection of $\nabla \theta_0$ in the direction of $\mathbf{u}$ is zero. We may also note that the streamfunction ψ is, by *definition*, constant along lines parallel to $\mathbf{u}$, that is,

$$\mathbf{u} \cdot \nabla \psi = 0. \qquad (9\text{–}4)$$

Comparing (9–3) and (9–4), we conclude that

$$\boxed{\theta_0 = \theta_0(\psi),} \qquad (9\text{–}5)$$

that is, θ is constant on streamlines of the flow. The relationship (9–5) can be obtained in a more formal way by changing independent variables in (9–3) from (r, η) to (r, ψ), with the result

$$u_r \left(\frac{\partial \theta_0}{\partial r} \right)_\psi \equiv 0. \qquad (9\text{–}6)$$

Integrating (9–6), we again get the result (9–5). In Stokes's solution for streaming flow past a sphere, all streamlines begin "at infinity" in front of the sphere and terminate "at infinity" downstream, as we can see from the streamline sketch in Figure 4–13. When streamlines begin and end at infinity, we say that the streamlines are *open*. In contrast, if the fluid is recirculating in some part of the domain, the streamlines in that region are said to be *closed*.

Equation (9–5) is the general solution of (9–3), but in this case where all streamlines are open, we can go one step further since the boundary condition (8–8) for $r \to \infty$ implies that $\theta = 0$ sufficiently far upstream (or downstream) on every streamline. Hence, in combination with (9–5), we conclude that

$$\boxed{\theta_0 = 0} \qquad (9\text{–}7)$$

everywhere in the fluid domain. This "solution" satisfies (9–3) and the boundary condition for $r \to \infty$. However, it clearly cannot be uniformly valid throughout the domain because it does not satisfy the boundary condition $\theta = 1$ at the sphere surface.

The problem lies in the scaling inherent in (8–7) or (8–25); we assume that the sphere radius is an appropriate characteristic length scale for θ throughout the fluid domain. The resulting nondimensionalized form of the thermal energy balance leads to the conclusion that conduction terms can be neglected everywhere relative to convection for Pe $\to \infty$, and hence to the solution (9–7). However, it is clear that conduction terms *cannot* be negligible sufficiently near the surface of the sphere. In the first place, since $\mathbf{u} = 0$ at the sphere surface, the only mechanism for heat transfer from the sphere to the fluid is conduction. This is true *independent* of the magnitude of the Peclet number and is inherent in the formula (8–16) for the total heat flux. With conduction completely neglected, as in (9–3), it is not surprising that we end up with the solution (9–7). The dimensionless temperature upstream is $\theta = 0$, and we have thrown out the only possible mechanism for transfer of heat from the sphere (where $\theta = 1$) to the fluid. The only possible conclusion is that conduction *cannot* be negligible near the sphere surface, and thus that the sphere radius is *not* an appropriate characteristic length scale in this region. This may be contrasted with the Pe $\ll 1$ limit for the same problem, where we found that the sphere radius did not serve as an appropriate length scale at *large* distances from the sphere.

The fact that the temperature field apparently must be characterized by two different length scales (the sphere radius for the majority of the fluid domain and some other scale very near to the sphere surface) means that the asymptotic solution for Pe $\gg 1$ must be singular, as we have explained in Chapter 8. For this particular problem, the limiting process Pe $\to \infty$ applied to equation (8–8) or (8–25) provides a classic indicator of singular asymptotic behavior. Specifically, in the limit Pe $\to \infty$, the highest order derivatives are lost, and this means that the corresponding solution cannot satisfy all of the boundary conditions of the original problem. Thus, it will not be a valid first approximation in the whole domain. This is precisely the same type of behavior that we first observed in the high frequency limit of flow in a tube with a sinusoidal variation of the pressure gradient in time.

We have anticipated that the characteristic length scale $l_c = a$ breaks down in the neighborhood of the sphere surface. In the language of matched asymptotic expansions, then, the temperature field must be approximated by two asymptotic expansions for Pe $\gg 1$: one (termed *outer*) in the region within a distance of $0(a)$ from the sphere where $l_c = a$ and equations (9–2)–(9–7) are valid, and the other (termed *inner*) in some region much closer to the sphere surface where the details of the scaling and solution remain to be defined. Before attempting to determine an appropriate characteristic length scale and other features of the temperature field in the inner region by formal analysis of the governing equation and boundary condition, it is useful to discuss the expected qualitative behavior in this regime.

Beginning right at the sphere surface, heat is transferred radially outward by conduction. However, because Pe is large, very small convection velocities can overwhelm thermal conduction, and the heat transfer process becomes convection dominated at a very short distance from the sphere surface. Thus, before the heat released from the sphere can propagate very far outward in the radial direction, it is swept around the sphere and downstream into a wake. Very near the sphere surface where convection comes into play, the dominant velocity component is in the tangential direction. As a result, the region of heated fluid adjacent to the sphere is very thin. In fact, this region is known as the *thermal boundary layer*.

We can continue this discussion a little further to see how the dimension of this thin boundary layer should be expected to depend upon the dimensional parameters of the problem. Specifically, the time scale characteristic of the *conduction* process (conduction is simply the diffusion of heat) is

$$t^* = 0(L^2/\kappa), \tag{9-8}$$

where L is a length scale over which diffusion takes place and κ is the thermal diffusivity $k/\rho C_p$. On the other hand, the time scale characteristic of the motion of a fluid element completely around the sphere is

$$\hat{t} = 0(a/U_\infty). \tag{9-9}$$

Actually, $\hat{t}$ is a lower bound on the so-called convective time scale because the actual velocity of fluid elements very close to the sphere surface will be a small fraction of U_∞. Nevertheless, $\hat{t}$ provides an estimate of the time scale *available* for heating of a fluid element by conduction. Thus, substituting $\hat{t}$ for t^* in (9–8), we obtain an indication of the radial distance from the sphere over which we may expect the fluid to be heated by conduction, namely,

$$L^2 \sim \frac{a\kappa}{U_\infty}. \tag{9-10}$$

Thus, the distance nondimensionalized with respect to a is

$$l^2 \sim \left(\frac{\kappa}{aU_\infty}\right) = \mathrm{Pe}^{-1}. \tag{9-11}$$

We cannot take this estimate too seriously because of uncertainty about the magnitude of the velocity in the region where the heat transfer process occurs and the greatly simplified picture that is presumed to describe the heat transfer process (essentially a decoupled, sequential process of diffusion followed by convection). Nevertheless, (9–11) provides a strong indication that the thickness of the thermal boundary layer should decrease as Pe increases, i.e., as U_∞ increases or κ decreases. Furthermore, this estimate illustrates strongly the source of the difficulty in using the sphere radius everywhere in the domain as a characteristic length scale for nondimensionalization. In particular, as $\mathrm{Pe} \to \infty$, the preceding qualitative argument indicates that we should expect the temperature to decrease from $\theta = 1$ at the sphere surface (to $\theta = 0$) in a region near the sphere surface that gets thinner as the Peclet number increases. As a consequence,

$$\frac{\partial \theta}{\partial r} \to \infty \quad \text{as} \quad \mathrm{Pe} \to \infty, \tag{9-12}$$

and it is not at all evident that

$$\frac{1}{\mathrm{Pe}} \frac{\partial^2 \theta}{\partial r^2}$$

will be particularly small as Pe → ∞. On the other hand, in deriving (9–3) by letting Pe → ∞ in (8–7), it was intrinsically presumed that $(1/\text{Pe})\nabla^2\theta \to 0$, that is, that $\nabla^2\theta$ is $0(1)$ independent of Pe. The fact that $\partial\theta/\partial r$, with r nondimensionalized by a, increases in magnitude with Pe means that the radius is not an appropriate length scale for the thermal boundary-layer region. Thus, (8–7), and the corresponding expansion (9–2) with leading term $\theta_0 = 0$, is not applicable to this part of the domain, as we have stated already.

From a physical point of view, we can say that the physics introduces a new "intrinsic" length scale for Pe ≫ 1, which is precisely what is needed to maintain a balance between convection and conduction near the sphere surface *even as Pe → ∞*. As suggested by the sketch in Figure 9–1, this occurs by having the temperature drop occur in an ever-narrowing region as Pe increases; this region is what we have called the *thermal boundary layer.*

Governing Equations and Rescaling in the Thermal Boundary-Layer Region

We thus see from the preceding discussion that the limit Pe → ∞ must be analyzed as a singular perturbation problem with the length scale a and equations (8–8) and (9–2) holding in an outer region and the as yet to be determined thermal boundary-layer equations and scaling holding in an inner region immediately adjacent to the sphere surface.

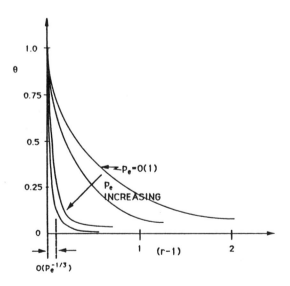

Figure 9–1 A schematic of the qualitative dependence of the temperature distribution on radial distance from the surface of a heated solid body ($\theta_s = 1.0$) for Re ≪ 1 and increasing Peclet number from $0(1)$ to asymptotically large values, Pe ≫ 1.

To obtain the nondimensionalized form of the thermal energy equations appropriate to the inner boundary-layer region, we must determine a length scale characteristic of the distance over which θ decreases by $0(1)$ from its surface value $\theta = 1$. As before, the process of determining the appropriate length scale might be accomplished by starting with the dimensional form of the thermal energy equation and simply looking for a combination of dimensional parameters with units of length that produces a dimensionless form of this equation in which at least some of the conduction terms are retained compared with convection terms in the limit as Pe $\rightarrow \infty$. In the weak convection problems considered in Chapter 8, this would, in fact, be a viable proposition because the new characteristic length scale turned out to equal the only combination of dimensional parameters, not involving a, that had units of length (namely, κ/U_∞ and ν/U_∞, respectively, for the thermal and fluid mechanics problems). However, we have pointed out that any combination of the form $(\kappa/U_\infty)^m a^{1-m}$ is an equally viable candidate as a characteristic length scale, and this suggests the need to use the formal *rescaling* procedure described in Chapter 8 to determine the length scale that is characteristic of the thermal boundary layer. This procedure begins with the previously nondimensionalized equation (8–7).

For the purposes of rescaling (8–7), it is convenient to redefine variables in the form

$$r - 1 = y \tag{9-13}$$

so that y is a radial variable that is zero at the sphere surface. Then, the rescaling to a new nondimensionalized radial variable appropriate to the boundary layer region can be rewritten as

$$\boxed{y = \text{Pe}^{-m}Y,} \tag{9-14}$$

where m remains to be determined. The rescaled variable, in this case Y, is simply the dimensional radial variable y' nondimensionalized with the length scale

$$l^* = a^{1-m}\left(\frac{\kappa}{U_\infty}\right)^m. \tag{9-15}$$

With the correct choice for m, this is the length scale characteristic of the inner (or boundary-layer) region. In the rescaled variables, the change in θ from $\theta = 1$ to approximately the freestream value $\theta = 0$ will occur over an increment $\Delta Y = 0(1)$ so that $\partial\theta/\partial Y = 0(1)$ independent of Pe. For $m > 0$, an increment $\Delta Y = 0(1)$ clearly corresponds to a very small increment in the radial distance Δy, scaled with respect to a. In particular, the thickness of the so-called thermal boundary layer is only $0(\text{Pe}^{-m})$ relative to the sphere radius a.

To determine m, we substitute the rescaling (9–14) into the governing equation (8–24) and choose m so that both convection and the largest conduction terms are retained in the inner region, even as Pe $\rightarrow \infty$. The only subtlety in carrying out this scheme is to remember that we are focusing on a very thin region $0(\text{Pe}^{-m})$, and in this region the magnitude of the velocity components will depend on Pe. To see that this must be true, and to facilitate the rescaling process, it is convenient to consider a

Taylor series approximation of the velocity components u_r and u_θ for $y \ll 1$. This is the region of interest for the boundary-layer analysis. Thus, starting with

$$u_r = \left(1 - \frac{3}{2r} + \frac{1}{2r^3}\right)\eta, \tag{9-16}$$

the Taylor series representation for $(r-1) = y \ll 1$ is

$$u_r = u_r|_{r=1} + \left(\frac{\partial u_r}{\partial r}\right)_{r=1}(r-1) + \left(\frac{\partial^2 u_r}{\partial r^2}\right)_{r=1}\frac{(r-1)^2}{2} + \cdots, \tag{9-17}$$

where, according to (9-16),

$$u_r|_{r=1} = 0,$$

$$\left(\frac{\partial u_r}{\partial r}\right)_{r=1} = 0,$$

and

$$\left(\frac{\partial^2 u_r}{\partial r^2}\right)_{r=1} = 3\eta.$$

Thus, for the region near the sphere's surface,

$$u_r \sim \frac{3}{2}(r-1)^2\eta + 0((r-1)^3). \tag{9-18}$$

Similarly, we can show that

$$u_\theta \sim \frac{-3}{2}(r-1)(1-\eta^2)^{1/2} + 0((r-1)^2). \tag{9-19}$$

Clearly, if we restrict our attention to a region $(r-1) = 0(\text{Pe}^{-m})$, the magnitudes of both u_r and u_θ in that region will depend on Pe. If we introduce the rescaling (9-14) into (9-18) and (9-19) and restrict our attention to $Y = 0(1)$, we confirm this fact,

$$u_r \sim \frac{3}{2}Y^2\eta\left(\frac{1}{\text{Pe}^{2m}}\right) + 0(\text{Pe}^{-3m}) \tag{9-20}$$

and

$$u_\theta \sim \frac{-3}{2}Y(1-\eta^2)^{1/2}\left(\frac{1}{\text{Pe}^m}\right) + 0(\text{Pe}^{-2m}). \tag{9-21}$$

Evidently, to leading order in Pe for Pe $\gg 1$, we can neglect all but the first term in the Taylor series approximation for u_r and u_θ.

Let us now consider rescaling the full thermal energy equation (8-25) using the results (9-20) and (9-21). When expressed in terms of the boundary-layer variable Y, we obtain

$$\frac{1}{Pe}\left\{\frac{\partial^2\theta}{\partial Y^2}Pe^{2m}+\frac{2Pe^m}{(1+Pe^{-m}Y)}\frac{\partial\theta}{\partial Y}+\frac{1}{(1+Pe^{-m}Y)^2}\frac{\partial}{\partial\eta}\left((1-\eta^2)\frac{\partial\theta}{\partial\eta}\right)\right\}=$$

$$\left\{\frac{3}{2}Y^2Pe^{-2m}+0(Pe^{-3m}Y^3)\right\}\eta Pe^m\frac{\partial\theta}{\partial Y}+\frac{1-\eta^2}{(1+Pe^{-m}Y)}\left\{\frac{3}{2}YPe^{-m}+0(Pe^{-2m}Y^2)\right\}\frac{\partial\theta}{\partial\eta}.$$

$$(9\text{--}22)$$

Clearly, for $Pe \gg 1$, the dominant conduction term is $\partial^2\theta/\partial Y^2$. Thus, retaining the largest terms on the two sides, we see that

$$Pe^{2m-1}\frac{\partial^2\theta}{\partial Y^2}+0(Pe^{m-1})=\left[\frac{3}{2}Y^2\eta\frac{\partial\theta}{\partial Y}+\frac{3}{2}(1-\eta^2)Y\frac{\partial\theta}{\partial\eta}\right]Pe^{-m}+0(Pe^{-2m}).$$

$$(9\text{--}23)$$

Hence, if conduction and convection are both to be retained in the boundary-layer region for $Pe \to \infty$, we see that

$$2m-1=-m$$

or

$$\boxed{m=\frac{1}{3}.}$$

$$(9\text{--}24)$$

Rewriting (9–23), we thus obtain

$$\boxed{\frac{\partial^2\theta}{\partial Y^2}=\frac{3}{2}Y^2\eta\frac{\partial\theta}{\partial Y}+\frac{3}{2}(1-\eta^2)Y\frac{\partial\theta}{\partial\eta}+0(Pe^{-1/3}).}$$

$$(9\text{--}25)$$

This is known as the *thermal boundary-layer equation* for this problem. Since we have obtained it by taking the limit $Pe \to \infty$ in the full thermal energy equation (9–22) with $m = 1/3$, we recognize that it governs only the first term in an asymptotic expansion similar to (9–2) for this inner region.

One interesting feature is that the operator $\nabla^2\theta$, which is expressed on the left-hand side of (9–22) in terms of rescaled spherical coordinate variables, takes a form in the limiting approximation (9–25) that appears to be just the normal derivative term in $\nabla^2\theta$ for a Cartesian coordinate system. In fact, we shall see that boundary-layer equations always can be expressed in terms of a local Cartesian coordinate system, with one variable normal to the body surface at each point (Y in this case) and the others tangent to it. This reduction of the equations in the boundary layer to a local, Cartesian form is due to the fact that the dimension of the boundary layer is so small relative to that of the body that surface curvature effects play no role.

To complete a solution for the leading-order approximation to the temperature field for $Pe \gg 1$, we need only solve (9–25) subject to appropriate boundary conditions. At the sphere surface, the condition

$$\boxed{\theta=1 \quad \text{at} \quad Y=0}$$

$$(9\text{--}26)$$

still holds. However, the second of the two conditions (8–8) cannot be applied directly since $r \to \infty$ is outside the domain of validity for this inner region. We should not

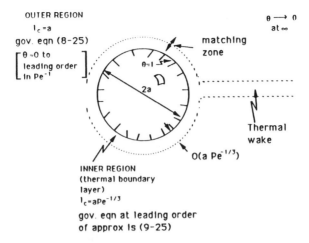

OUTER REGION
$l_c = a$
gov. eqn (8-25)
$\begin{bmatrix} \theta \sim 0 \text{ to} \\ \text{leading order} \\ \text{in Pe}^{-1} \end{bmatrix}$

$\theta = 1$

$2a$

$\theta \longrightarrow 0$
at ∞

matching zone

Thermal wake

$O(a \, Pe^{-1/3})$

INNER REGION
(thermal boundary layer)
$l_c = aPe^{-1/3}$

gov. eqn at leading order
of approx is (9-25)

Figure 9–2 A schematic representation of the solution domain for forced convection heat transfer due to uniform flow past a solid sphere with a uniform surface temperature for Re ≪ 1 but Pe ≫ 1.

forget, on the other hand, that the leading-order approximation in the outer region was already shown to be $\theta_0 = 0$. Thus, though we cannot directly apply the boundary condition $\theta \to 0$ as $r \to \infty$ to the solution (9–25), the matching condition with the first term in the outer region still produces the condition

$$\theta \to 0 \quad \text{for} \quad Y \gg 1 \quad \text{as} \quad Pe \to \infty, \tag{9–27}$$

which is equivalent. A summary sketch showing a qualitative representation of the solution domain for this problem in the limit of large Pe is shown in Figure 9–2.

We shall very shortly consider the solution of (9–25) subject to the conditions (9–26) and (9–27). First, however, there is a very important result that we can draw from the work we have done so far. This is the form of the relationship between the overall rate of heat transfer and the independent dimensional parameters of the system, which can now be determined in spite of the fact that we have not yet solved any differential equations, other than the trivial one (9–3). In the case of heat transfer for Re ≪ 1, the only independent dimensionless parameter is the Peclet number Pe, and in dimensionless terms the objective is thus the relationship

$$Nu = Nu(Pe),$$

where for the present spherical geometry

$$Nu \equiv \int_{-1}^{1} -\left(\frac{\partial \theta}{\partial r}\right)_{r=1} d\eta, \tag{9–28}$$

as shown in (8–20). But the formulation of the boundary-layer equations has shown that

$$\left(\frac{\partial\theta}{\partial r}\right)_{r=1} = \mathrm{Pe}^{1/3}\left(\frac{\partial\theta}{\partial Y}\right)_{Y=0}, \tag{9-29}$$

where $(\partial\theta/\partial Y)_{Y=0}$ is a function of η that is $0(1)$ in magnitude and completely independent of Pe. It follows from this and (9–28) that

$$\boxed{\mathrm{Nu} = c\mathrm{Pe}^{1/3},} \tag{9-30}$$

where

$$c \equiv \int_{-1}^{1} -\left(\frac{\partial\theta}{\partial Y}\right)_{Y=0} d\eta$$

is a numerical coefficient of magnitude $0(1)$.

   The correlation (9–30) is the most important result of the analysis of this section. To determine the coefficient c, we must solve (9–25)–(9–27). We will do this in the next subsection. However, we are guaranteed that c will be of $0(1)$, and we may not always wish to proceed further in an analysis of this kind. The fact that a result like (9–30) can be obtained without the necessity of solving any differential equations [apart from the trivial equation (9–3)] illustrates part of the power of the asymptotic method that we have been developing in the last several chapters.

Solution of the Thermal Boundary-Layer Equation

   The asymptotic formulation of the previous subsection has led to not only the important result given by (9–30) but also a very considerable simplification in the structure of the governing equation in the thermal boundary-layer region. As a consequence, it is now possible to obtain an analytic approximation for θ.

   To do this we introduce a similarity transformation of the form

$$\boxed{\theta = \theta(\zeta), \quad \text{where} \quad \zeta = \frac{Y}{g(\eta)}} \tag{9-31}$$

into equation (9–25). The function $g(\eta)$ determines the dependence of the boundary-layer thickness on η. In particular, assuming for the moment that a solution of the form (9–31) exists, we can associate the edge of the boundary-layer with the value of $\zeta = \zeta^*$, where θ has some arbitrary small value, say, 0,01. Then, in terms of Y, the edge will be located at $\zeta^* g(\eta)$. The derivatives of θ required in (9–25) are

$$\frac{\partial\theta}{\partial Y} = \frac{1}{g}\frac{\partial\theta}{\partial\zeta},$$

$$\frac{\partial^2\theta}{\partial Y^2} = \frac{1}{g^2}\frac{\partial^2\theta}{\partial\zeta^2},$$

and

$$\frac{\partial\theta}{\partial\eta} = -\frac{\zeta}{g}\frac{\partial g}{\partial\eta}\left(\frac{\partial\theta}{\partial\zeta}\right).$$

Substituting into (9–25), we thus obtain

$$\frac{\partial^2 \theta}{\partial \zeta^2} + \frac{3}{2}\zeta^2 \frac{\partial \theta}{\partial \zeta}\left\{\frac{(1-\eta^2)}{3}\frac{dg^3}{d\eta} - g^3\eta\right\} = 0 \qquad (9\text{–}32)$$

after multiplying through by g^2 to make the coefficient of the highest-order derivative equal to unity.

If a similarity solution exists, the coefficients of (9–32) must be either a constant or a function of ζ only. This means that

$$\frac{(1-\eta^2)}{3}\frac{dg^3}{d\eta} - g^3\eta = \text{const} = 2. \qquad (9\text{–}33)$$

The numerical value of the constant in (9–33) is arbitrary (provided it is nonzero) but is conveniently chosen as 2. However, the solution for θ would be unchanged if any other nonzero value were chosen. The constant cannot be zero because the resulting equation for θ does not have a solution that can satisfy both of the boundary conditions (9–26) and (9–27). Corresponding to (9–33), Equation (9–32) for θ now becomes

$$\frac{\partial^2 \theta}{\partial \zeta^2} + 3\zeta^2 \frac{\partial \theta}{\partial \zeta} = 0 \qquad (9\text{–}34)$$

and the problem has been reduced to the solution of (9–33) and (9–34) subject to appropriate boundary conditions.

Let us begin with (9–33). For this purpose, it is convenient to rewrite (9–33) in the form

$$\frac{dg^3}{d\eta} + \frac{3}{2}g^3 \frac{d\ln(1-\eta^2)}{d\eta} = \frac{6}{1-\eta^2}. \qquad (9\text{–}35)$$

Then, the left-hand side can be integrated directly to obtain the homogeneous solution $g^3 = c_1(1-\eta^2)^{-3/2}$. We leave it to the reader to verify that the general solution of (9–35) is

$$g^3 = \frac{c_1}{(1-\eta^2)^{3/2}} + \frac{6}{(1-\eta^2)^{3/2}}\int_{-1}^{\eta}(1-t^2)^{1/2}dt. \qquad (9\text{–}36)$$

The homogeneous solution is singular at both $\eta = 1$ and -1. The particular solution, on the other hand, blows up at $\eta = 1$ (that is, at the axis of symmetry downstream of the sphere) but is finite at $\eta = -1$. In view of the physical interpretation of $g(\eta)$ as representing the η dependence of the boundary-layer thickness, we expect that g must be finite at $\eta = -1$ (the upstream symmetry axis). Thus, we require that

$$c_1 = 0$$

and

$$g(\eta) = \frac{6^{1/3}}{(1-\eta^2)^{1/2}}\left[\int_{-1}^{\eta}(1-t^2)^{1/2}dt\right]^{1/3}. \qquad (9\text{–}37)$$

Even with this choice, $g(\eta) \to \infty$ as $\eta \to 1$. This means that the boundary-layer approximation breaks down in the limit as we approach the downstream axis of symmetry. We do not pursue this point here, except to comment that it is not very surprising that the boundary-layer solution should break down in this region. This is because the physical meaning of the rescaling used in deriving (9–25), as well as the qualitative discussion leading up to that rescaling, is that derivatives of θ in the radial direction are asymptotically large compared with derivatives along the sphere surface, that is,

$$\frac{\partial \theta}{\partial y} = Pe^{1/3} \frac{\partial \theta}{\partial Y} \gg \frac{\partial \theta}{\partial \eta}.$$

Near the rear stagnation point, however, this assumption cannot remain valid because the flow turns the corner and carries the thermal layer into a "wake" along the axis of symmetry. Mathematically, the fact that $g \to \infty$ means that the boundary-layer scaling is breaking down. Thus, if we were to completely analyze the temperature distribution in the whole fluid domain, it would be necessary to include at least two additional asymptotic regions: one along the downstream symmetry axis to analyze the structure of the thermal wake for $Pe \to \infty$ and at least one more within some small neighborhood of the rear-stagnation point to "connect" the boundary-layer on the sphere with the downstream wake.

We do not pursue solutions for these additional regions here. Careful research[2] has shown that the coefficient c in (9–30) can be evaluated with negligible error for $Pe \gg 1$ by evaluating the integral in (9–30) using the boundary-layer result for $(\partial \theta / \partial Y)_{Y=0}$ for all η in the range $-1 \leq \eta \leq 1$. In particular, the breakdown in the boundary-layer solution at $\eta = 1$ contributes an error in c_1 of only $O(Pe)$, which is asymptotically small for $Pe \to \infty$. Later, we will consider the form of the temperature distribution far downstream of a heated body in the thermal wake, but this has no direct bearing upon the present analysis.

To complete the present solution, we need only solve (9–34) subject to the conditions

$$\theta(0) = 1, \tag{9–38}$$

$$\theta(\infty) \to 0, \tag{9–39}$$

which are derived from (9–26) and (9–27). Integrating (9–34) once with respect to ζ, we obtain

$$\frac{\partial \theta}{\partial \zeta} = c_1 e^{-\zeta^3},$$

and after two integrations we get the general solution

$$\theta = c_2 + c_1 \int_0^\zeta e^{-t^3} dt.$$

Applying the boundary conditions (9–38) and (9–39) to determine c_1 and c_2, we find that

$$\boxed{\theta = 1 - \frac{\int_0^\zeta e^{-t^3} dt}{\int_0^\infty e^{-t^3} dt}.} \tag{9–40}$$

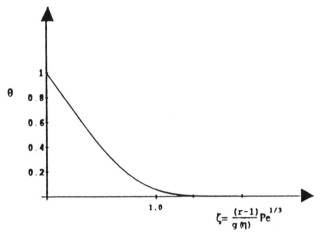

Figure 9–3 The self-similar temperature profile given by Equation (9–40) for forced convection heat transfer from a heated (or cooled) solid sphere in a uniform velocity field at small Re and large Pe. The function $g(\eta)$ represents the dependence of the thermal boundary-layer thickness on η and is given by (9–37).

This solution is plotted in Figure 9–3.

The denominator of (9–40) is the well-known gamma function $\Gamma(4/3)$,

$$\int_0^\infty e^{-t^3} dt = \int_0^\infty \frac{1}{3} e^{-u} u^{-2/3} du = \frac{1}{3}\Gamma\left(\frac{1}{3}\right) = \Gamma\left(\frac{4}{3}\right) = 0.8930.$$

The final step in the present analysis is to use the solutions (9–40) and (9–37) to evaluate c in (9–30). After some straightforward manipulation, we find that

$$c = \frac{3}{2(6)^{1/3}\Gamma(4/3)}\left(\frac{\pi}{2}\right)^{2/3} = 1.249. \tag{9–41}$$

Thus,

$$Nu = 1.249\,Pe^{1/3} + o(Pe^{1/3}). \tag{9–42}$$

It should be remembered that (9–25) and its solution (9–40) represent only the first term in an asymptotic expansion for Pe $\to \infty$ in the boundary-layer regime, while (9–3) and its solution (9–7) is the leading-order term in a corresponding expansion for the outer region. To obtain the next level of approximation, it is necessary to calculate an additional term in both of these expansions. We do not pursue this calculation here because of the complexity of the details. Nevertheless, it is worth recording the result obtained originally by Acrivos and Goddard:

$$\frac{Nu}{Pe^{1/3}} = 1.249 + .922\,Pe^{-1/3} + o(Pe^{-1/3}). \tag{9–43}$$

B. Generalization of the Correlation Between Nu and Pe to Solid Bodies of Nonspherical Shape in Uniform Streaming Flow

We have stated that the most important result of the previous section is the correlation, at the leading order of approximation, between the Nusselt and Peclet numbers, namely, equation (9–30):

$$\boxed{\text{Nu} = c\,\text{Pe}^{1/3}.}$$

However, all of the preceding analysis was for the special case of a solid, spherical body in uniform, steady streaming flow or, equivalently, for translation of a sphere, without rotation, through an otherwise quiescent fluid. What makes (9–30) especially significant is that it is valid for a much more general class of problems. In this section, we consider solid bodies of *arbitrary* shape for a uniform streaming flow in the creeping flow limit (or, equivalently, we consider translation of the body along a rectilinear path through an otherwise quiescent fluid). For this class of problems, the correlation (9–30) applies to all cases that involve smooth bodies without sharp corners or regions of extreme curvature that do not generate regions of closed streamlines* for Re ≪ 1. We shall shortly return to discuss the reasons for these limitations on allowable body shapes. It should be noted, however, that the restriction is rather mild. Almost all smooth particles will fall within the allowed class. In this case, only the coefficient c in (9–30) varies with the geometry of the body, and even then it is always guaranteed to be 0(1) in magnitude.

To understand the generality claimed for the correlation (9–30), it is well to remind ourselves that the general form (9–30) was deduced completely from the *asymptotic structure* of the heat transfer problem for the sphere at Pe ≫ 1. The fact that (9–30) can be applied for a much wider class of particle geometries is equivalent to stating that the asymptotic structure for Pe ≫ 1 is invariant to geometry within the limitations on shape that were stated above. The ease with which we can generalize the formulation to a body of arbitrary shape is a graphic demonstration of the power of the asymptotic method. We need only backtrack through the analysis for a sphere to see that the same asymptotic formulation, and thus the result (9–30), is still valid.

We begin by noting the obvious fact that the dimensionless form of the thermal energy equation (8–7) is preserved. Only the form of the velocity vector **u** will vary with the geometry of the body. Thus, the analysis for the outer region also will be preserved, leading to the solution (9–7), provided only that no regions of closed streamline or recirculating flow exist in the fluid domain. Since a fluid without inertia will tend to follow faithfully the contours of a body surface, one's intuition might suggest that recirculating flows would never occur for streaming motion at Re ≪ 1. However, it is necessary to be slightly cautious on this point, since recent research results have demonstrated the existence of counter examples.[3] For example, some body shapes have been found,[4] such as a hollow hemisphere moving along its symmetry

---

Streamline implies the existence of a streamfunction, which we have seen to be true only for axisymmetric and two-dimensional flows. In three dimensions, recirculating flows are associated with regions of closed *streamsurfaces*, or *pathlines*.

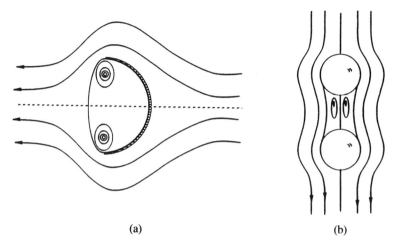

(a)   (b)

Figure 9–4 Creeping flows with regions of closed streamline motion: (a) streaming flow past a hollow hemispherical cap; (b) two spheres translating along their line of centers.

axis, as sketched in Figure 9–4a, that exhibit closed streamline "wakes" in zero Reynolds number flow. In addition, the region between two spheres that are translating with equal velocity along their line of centers has been found to contain closed streamlines if the distance between the spheres is smaller than one diameter, as sketched in Figure 9–4b. Nevertheless, for a single body in a zero Reynolds number streaming flow, the likelihood of a recirculating region is reasonably remote and does not occur for single bodies of "simple" shape, such as ellipsoids of arbitrary axis ratio. In all such cases, the high Peclet number heat transfer problem reduces to the analysis of a thermal boundary-layer region near the surface of the body.

Now, a key feature of the boundary-layer analysis in the case of a sphere was the fact that it was only the form of the velocity field very close to the sphere surface that determined the critical scaling parameter m, and thus the form of the correlation (9–30). At the level of approximation (9–18) and (9–19), we found that the velocity component tangent to the sphere surface varied linearly with distance from the surface y, while the velocity component normal to the surface varied quadratically as y^2. But it is a simple matter to show that this dependence of the tangential and normal velocity components on distance from the surface is actually typical of all bodies on which the no-slip condition is applied. To simplify the details, we restrict attention to axisymmetric or two-dimensional bodies (with the streaming motion parallel to the symmetry axis or perpendicular to the cylindrical axis, respectively) and adopt local Cartesian coordinates (x,y) parallel and perpendicular to the body surface, with corresponding velocity components (u,v). Then the no-slip condition requires that $u = 0$ at the body surface, which we denote as $y = 0$, and the first nonzero term in a Taylor series approximation for u must be linear in y, that is,

$$u \sim \left(\frac{\partial u}{\partial y}\right)_{y=0} y + 0(y^2), \qquad (9\text{-}44)$$

at all points x for which the shear stress is nonzero. The normal velocity component is also equal to zero on the body surface, $v = 0$ at $y = 0$, provided that there is no mass flux through the surface. But according to the continuity equation,

$$\frac{\partial u}{\partial x} + \frac{\partial v}{\partial y} = 0. \qquad (9\text{-}45)$$

Thus, if u depends linearly upon y, as in (9-44), it is evident from (9-45) that

$$v \sim -\frac{\partial}{\partial x}\left(\frac{\partial u}{\partial y}\right)_0 \frac{y^2}{2} + 0(y^3). \qquad (9\text{-}46)$$

Not surprisingly, the presence of a linear term in the tangential velocity component and a quadratic term in the normal velocity component is precisely what we found in the specific case of a solid sphere. However, we now see that the linear-quadratic combination is characteristic of any solid body on which a no-slip condition is applied. The only variation with changes in the geometry of the body will be in the dependence of the coefficients $(\partial u/\partial y)_0$ and $(\partial^2 u/\partial x\partial y)_0$ on x.

It follows from the invariance of the y dependence of the velocity components that the rescaled thermal energy equation for the inner boundary-layer region will have precisely the same form as (9-23) but with coefficients on the right-hand side that depend upon the geometry of the body, that is,

$$Pe^{2m-1}\frac{\partial^2 \theta}{\partial Y^2} = \left[\left(\frac{\partial u}{\partial y}\right)_{y=0} Y \frac{\partial \theta}{\partial x} - \frac{1}{2}\left(\frac{\partial^2 u}{\partial x\partial y}\right)_0 Y^2 \frac{\partial \theta}{\partial Y}\right]Pe^{-m} + 0(Pe^{-2m}).$$

Hence,

$$m = \frac{1}{3} \qquad (9\text{-}47)$$

as before, and it is clear that the general form of the correlation between Nussett number and Peclet number,

$$Nu = c\,Pe^{1/3},$$

is preserved for arbitrary solid, no-slip bodies of axisymmetric or two-dimensional shape, subject only to the assumption that the streamlines of the flow are open (that is, there exists no regions of recirculating motion). In fact, very similar arguments can be made without the restriction to axisymmetric or two-dimensional shapes, thus demonstrating that (9-30) is actually applicable to bodies of arbitrary shape (provided again that there is no region of recirculating flow and that the boundary is smooth, contains no sharp corners or regions of extreme curvature*). Of course, in this case,

---

*This latter restriction is necessary to preserve the ordering inherent in boundary-layer rescaling that requires $\partial/\partial y \gg \partial/\partial x$.

a more general form of the boundary-layer equation will be necessary. We leave it to the interested reader to fill in the details. For the two-dimensional or axisymmetric case, the coefficient c in equation (9–47) is given by

$$c \equiv \int_s - \left(\frac{\partial \theta}{\partial Y} \right)_0 \partial S, \tag{9–48}$$

where S denotes the body surface.

To determine c, we must solve the thermal boundary-layer equation, subject to the boundary conditions (9–26) and (9–27). Substituting $m = 1/3$ into (9–47), the latter takes the form

$$\frac{\partial^2 \theta}{\partial Y^2} = \left(\frac{\partial u}{\partial y} \right)_0 Y \frac{\partial \theta}{\partial x} - \frac{1}{2} \left(\frac{\partial^2 u}{\partial x \partial y} \right)_0 Y^2 \frac{\partial \theta}{\partial Y} + 0(\text{Pe}^{-1/3}). \tag{9–49}$$

It is thus evident that the numerical value of c will vary, depending on body geometry, since the coefficients in (9–49) depend upon the geometry.

For the particular case of a sphere, we found in Section **A** that the boundary-layer equation (9–25) could be solved by use of a similarity transformation. An interesting question is whether the general, boundary-layer equation (9–49) can be solved by a similar approach. For convenience in pursuing this question, we denote the surface velocity gradient as

$$\alpha(x) \equiv \left(\frac{\partial u}{\partial y} \right)_{y=0}$$

and its derivative as

$$\alpha'(x) = \frac{\partial \alpha}{\partial x} = \frac{\partial}{\partial x} \left(\frac{\partial u}{\partial y} \right)_0.$$

To see whether a similarity solution exists, we apply a similarity transformation of the form

$$\theta = \theta(\eta), \quad \eta \equiv \frac{Y}{g(x)}, \tag{9–50}$$

to equation (9–49). The result is

$$\theta'' + \left(\alpha g^2 g' + \frac{1}{2} \alpha' g^3 \right) \eta^2 \theta' = 0.$$

Thus, if a similarity solution exists,

$$\alpha g^2 g' + \frac{1}{2} \alpha' g^3 = \text{const} = 3 \tag{9–51}$$

and

$$\theta'' + 3\eta^2 \theta' = 0. \tag{9–52}$$

The constant coefficient that appears in (9–51) and (9–52) is arbitrary. The value 3 is chosen for convenience. In order to determine whether a similarity solution exists, we must determine whether solutions of (9–51) and (9–52) can be found that satisfy appropriate boundary conditions. Specifically, in order to satisfy boundary conditions on θ for $Y=0$ and $Y\to\infty$, we require that the similarity function satisfies

$$\boxed{\begin{aligned} \theta &= 1 \quad \text{at} \quad \eta = 0, \\ \theta &\to 0 \quad \text{as} \quad \eta \to \infty. \end{aligned}} \tag{9–53}$$

In addition, $g(x)$ is required to be *finite* except possibly at values of x corresponding to thermal wakes (for example, at the downstream stagnation point in the case of a sphere) where the assumption of a thin thermal layer is no longer valid.

Now, equation (9–52) is identical to equation (9–34), which was found earlier for the sphere, and we have already seen that it can be solved subject to the conditions (9–53). The solution for θ is given in (9–40). The existence of a similarity solution to (9–49) thus rests with equation (9–51). Specifically, for a similarity solution to exist, it must be possible to obtain a solution of (9–51) for $g(x)$, which remains finite for all x except possibly at a stagnation point where $\alpha = 0$, from which a thermal wake may emanate.

To seek a solution for (9–51), it is convenient to rewrite it in the form

$$\frac{\alpha}{3}\frac{dg^3}{dx} + \frac{\alpha'}{2}g^3 = 3, \tag{9–54}$$

of a first-order linear, ordinary differential equation for g^3. In this form, it is straightforward to write down a general solution

$$g^3 = \frac{c}{\alpha^{3/2}} + \frac{9}{\alpha^{3/2}}\int_{x_0}^{x}[\alpha(t)]^{1/2}dt,$$

where x_0 denotes the front stagnation point on the body surface (where $\alpha = 0$). Since x_0 is the upstream stagnation point, we must choose c such that g remains finite as $x\to x_0$. Thus, we require that $c = 0$ and

$$\boxed{g^3 = \frac{9}{\alpha^{3/2}}\int_{x_0}^{x}[\alpha(t)]^{1/2}dt.} \tag{9–55}$$

With this definition, $g(x)$ remains finite for all x (other than the rear stagnation point), and we claim to have constructed a similarity solution for the complete class of smooth axisymmetric or two-dimensional solid bodies (with no closed streamlines).

To verify this claim, we first note that the absence of a closed streamline region means that only two zeros are present in the function $\alpha(x)$ on the surface: one at the front stagnation point $x = x_0$ and the other at the rear stagnation point where the downstream symmetry axis of the flow and the body surface intersect. Thus, the solution (9–55), which was constructed to remain bounded at x_0, is now seen to remain bounded at all positions on the body surface (since $\alpha \neq 0$), except at the rear stagnation point where $\alpha \to 0$ and $g \to \infty$. But as we have already noted in the case of the sphere, the

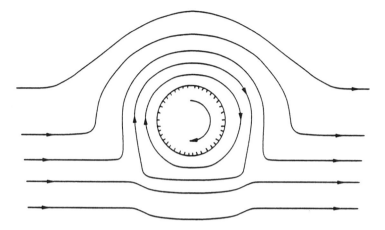

Figure 9–5 A qualitative sketch of the flow pattern for uniform, creeping flow past a rotating circular cylinder.

rear stagnation point is the point of departure for the thermal wake, in which heat is carried downstream by convection along the symmetry axis, and the boundary-layer assumptions clearly fail at this point because the thermal layer is not thin. Thus, with the exception of the rear stagnation point where the whole analysis breaks down, $g(x)$, as defined in (9–55), remains finite for all x and thus satisfies the critical requirement for existence of a similarity solution. Furthermore, this is true for axisymmetric and two-dimensional bodies of arbitrary shape so long as flow is characterized everywhere by open streamlines.

The reader may well ask whether anything can be said for bodies that generate regions of recirculating motion in uniform streaming flow. To answer this question, it is advantageous to consider two possible configurations of this general class. In the first configuration, an example of which is sketched in Figure 9–5, the body is completely surrounded by closed streamlines (or stream surfaces), as would happen if the body were rotating with an axis of rotation that is normal to the direction of the undisturbed flow; in the second configuration, a region of closed streamlines or recirculating flow is generated near the body (usually downstream) but it does not surround the body and is in contact with only a part of the body surface. Two examples of this case are sketched in Figure 9–4. In the first case, the boundary layer analysis of this section is not at all relevant, and we must discuss the problem on a completely different basis, which will be described in Section **E** of this chapter. For streaming flows (or equivalently, simple rectilinear translation through a quiescent fluid), the need to defer discussion of this case is not much of a limitation, because very few bodies rotate as a consequence of translation in the absence of an external torque. The second class of closed-streamline (or recirculating) flows has a higher probability of occurring, though a precise statement of particle geometries that lead to steady, closed-streamline (or stream surface) flows at low Reynolds number is not known at present. In any case, the boundary-layer results of this section still can be applied, at least qualitatively, to this second class of streaming flow problems even though they involve regions of recirculating motion adjacent to the body.

To do this, we must anticipate one result from the generel discussion of Section E on high Peclet number heat transfer in regions of closed streamlines: The dimensionless temperature gradient in such regions is determined primarily by its size. For example, in a closed streamline region of $0(1)$ in extent, dimensionless, steady-state temperature gradients will also be $0(1)$.

Suppose we consider a hypothetical situation where a region of closed streamline flow, with a linear dimension of $0(1)$ relative to the body, is generated downstream of some point A on the body surface. Upstream of A, the streamlines adjacent to the body surface are all open and the boundary-layer scaling is still relevant so that $\partial\theta/\partial y \sim 0(\text{Pe}^{1/3})$. Downstream of A, on the other hand, we have already noted that temperature gradients reflect the physical dimensions of the recirculating region. If we assume that these dimensions are $0(1)$, or, indeed, any size larger than $0(\text{Pe}^{-1/3})$, the temperature gradient at the body surface downstream of A will be asymptotically small compared with that upstream, and, as a consequence, almost all the heat transfer will occur in the upstream region. In this case, to leading order in Pe, the total heat flux will still be $0(\text{Pe}^{1/3})$, as it was in the absence of a closed streamline region, but the coefficient c in the correlation (9–30) will reflect the fact that the dominant heat transfer takes place on only the fraction of the body surface that is upstream of A. The total heat flux downstream of A, assuming the closed streamline region to be $0(1)$ in size, will be only $0(1)$—insignificant compared with that upstream of A for $\text{Pe} \gg 1$.

C. Boundary-Layer Analysis of Heat Transfer from a Solid Sphere in Generalized Shear Flows at Low Reynolds Number

It is perhaps timely to stop and reflect upon the nature of the thermal boundary-layer analysis to determine whether other generalizations of the basic result (9–30) may be possible. In particular, heat transfer from solid bodies occurs frequently when the fluid motion seen by the body cannot be approximated as a uniform streaming flow, and the reader may ask whether the correlation (9–30) can be applied in these cases with a proper choice for the characteristic velocity that appears in Pe. It is especially interesting, in this regard, to compare the present analysis to the corresponding low Peclet number problem of Chapter 8.

One immediately evident contrast between the limits $\text{Pe} \rightarrow 0$ and $\text{Pe} \rightarrow \infty$ is the nature of the dependence of the temperature field on the velocity field. In the low Peclet number limit, the temperature field near the body is dominated by conduction and is thus relatively insensitive to the details of the fluid motion. When convection effects do come into the low Peclet number problem, it is primarily the form of the velocity field at *large* distances from the body that determines the temperature field and the dependence of the Nusselt number on the Peclet number. On the other hand, at high Peclet number, the heat transfer process is confined by convective effects to a very thin region near the body surface, and it is only the local form of the velocity distribution in this region that matters. As a consequence, one might suppose that the qualitative nature of the high Peclet number process should be invariant to changes in the form of the flow at large distances from the body, in contrast with the low Peclet number case. Indeed, whatever the nature of the motion elsewhere in the fluid, the local forms (9–44) and (9–45) must be preserved in the immediate vicinity of the body surface, provided we describe the problem in a frame of reference that is fixed at the center of the body and rotates with it so that $u = v = 0$ on the body surface.

In view of these facts, it is tempting to suppose that the correlation

$$Nu \sim Pe^{1/3}$$

should hold for solid bodies, even when the fluid far from the body undergoes a more or less arbitrary motion. If true, this would be an extremely important result, because solid particles are very frequently subjected to motions that are much more complicated than a simple, uniform streaming flow.

Unfortunately, however, there are a large number of different types of flow conditions where the boundary-layer form of the heat transfer correlation (9–30) is *not applicable*. This applies, basically, to any flow configuration in which the body is completely surrounded by a region of closed streamlines (or pathlines, if the flow is not two-dimensional or axisymmetric). We will discuss high Peclet number heat transfer in such cases in Section **E**. Here, we consider only cases where the boundary-layer analysis can be used. In general, this requires that

1. the streamlines (or fluid pathlines) of the undisturbed flow at infinity must be open, as seen in a fixed, laboratory reference frame, and
2. the particle, when viewed in the same reference frame, cannot rotate.

A general discussion of flow types and particle shapes where these conditions are satisfied is beyond the scope of this book. As an alternative, which may impart at least a qualitative sense for the issues involved, we consider the special case of a *spherical* particle and the class of linear, two-dimensional flows described by

$$\mathbf{u} = \mathbf{\Gamma} \cdot \mathbf{x} = (\mathbf{E} + \mathbf{\Omega}) \cdot \mathbf{x}, \qquad (9\text{--}56)$$

where

$$\mathbf{\Gamma} = \frac{1}{2} \begin{bmatrix} 1+\lambda & 1+\lambda & 0 \\ -1+\lambda & -(1+\lambda) & 0 \\ 0 & 0 & 0 \end{bmatrix}$$

and $-1 \leq \lambda \leq 1$. A sketch of the streamlines for these flows is given in Figure 9–6. The limit $\lambda = 1$ is plane hyperbolic (extensional) flow, $\lambda = -1$ is a purely rotational flow with circular streamlines, and $\lambda = 0$ is simple shear flow. Viewed from another standpoint, flows with $\lambda > 0$ have a strain rate that exceeds vorticity—that is, $|E| > |\Omega|$—culminating at $\lambda = 1$ in a flow with no vorticity. On the other hand, flows with $\lambda < 0$ have a strain rate that is smaller than the vorticity, $|E| < |\Omega|$ culminating at $\lambda = -1$ in a flow that is purely rotational. For simple shear flow, $\lambda = 0$ and $|E| = |\Omega|$.

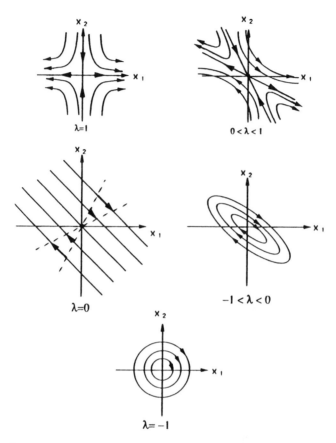

Figure 9–6 A sketch of streamlines for the class of linear, two-dimensional flows for $-1 \le \lambda \le 1$.

A special feature of simple shear flow is that it separates those flows with open stream-lines (hyperbola) from flows with closed streamlines (ellipses).

For the problem at hand, we assume that the sphere is neutrally buoyant and sub-ject to no body forces. Hence, because $\text{Re} \ll 1$, it will translate in a linear flow with its center moving as an element of the fluid, and there is no loss of generality in assum-ing that $x = 0$ corresponds to the center of the sphere.

The question at hand is whether circumstances exist, for this rather simple situa-tion in which the conditions (1) and (2) are satisfied so that boundary-layer analysis can be applied. So far as the first condition is concerned, the only flows of (9–56) that have open streamlines are those with $\lambda \ge 0$ (which includes simple shear flow). On the other hand, there is a nonzero hydrodynamic torque on the sphere that causes it to rotate for all flows in this subgroup except $\lambda = 1$. Thus, for a sphere in the general linear, two-dimensional flow, given by (9–56), there are only two cases that satisfy the conditions for applicability of boundary-layer theory:

1. *The sphere in pure extensional flow* ($\lambda = 1$). In this case, the streamlines are open (hyperbola at infinity), and the net hydrodynamic torque on the sphere is zero. Thus, if no external torque is applied to the sphere, it will not rotate.

2. *The sphere in one of the open streamline flows,* $0 \leq \lambda \leq 1$, but with an external torque applied so that the sphere cannot rotate. Note that the sphere would rotate, in the absence of an external torque, at the angular velocity that causes the hydrodynamic torque to vanish.

These are the only cases for a sphere in a linear velocity field where a boundary-layer analysis is expected to apply. In other cases, the sphere is surrounded by a zone of closed streamlines.

Similarly, if we consider the case of a circular cylinder with its axis at x = 0 and oriented normal to the plane of the two-dimensional flow, (9–56), the requirement of open streamline flow is satisfied only if $\lambda = 1$ or if the cylinder is constrained from rotation and $0 \leq \lambda < 1$.

Let us consider, then, the situation where the streamlines near a heated body are open and a boundary-layer structure is expected to hold for Pe $\gg 1$. How will the problem differ from that analyzed in Sections **A** and **B**? Since the leading-order approximation to θ is determined entirely by the form of the velocity field near the body surface, and this structure is invariant to changes both in body geometry and in the nature of the velocity field away from the surface, it is evident that there will be no qualitative change at all if the problem is one for which the boundary-layer structure can be expected. Indeed, if we examine the analysis in Section **B**, it should become apparent that it is only the functional form of the coefficients $\alpha(x) \equiv (\partial u / \partial y)_{y=0}$ and α' that should change from case to case and the general similarity solution should still apply.

The only changes required in these solutions are due to the fact that $\alpha(x)$ may be more complex than for uniform streaming flows. For example, a qualitative sketch of the flow structure for a nonrotating cylinder in simple shear flow at low Reynolds number is shown in Figure 9–7.[5] It is evident in this case that there are four stagnation points on the cylinder surface rather than two, as in the streaming flow problem. Two of the streamlines that lead to the stagnation points A and C are lines of inflow, and two from B and D are lines of outflow, where we should expect a thin thermal wake. In the limit as these outflow points are approached, we thus expect a breakdown

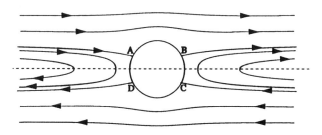

Figure 9–7 A qualitative sketch of the streamlines near a nonrotating circular cylinder in a simple, two-dimensional shear flow.

of the similarity solution with $g \to \infty$. At the inflow stagnation points, on the other hand, we require that g be finite. To accommodate these changes in the form of α, we can define $x = x_0 = 0$ in $g(x)$ as being coincident with point A in Figure 9–7. Then, g is defined by (9–55) for the interval between A and B with $x > 0$, and as

$$g^3 = -\frac{9}{(\alpha)^{3/2}} \int_0^x (\alpha)^{(1/2)} dt \quad \text{with} \quad \alpha \equiv -\frac{\partial u}{\partial y} \tag{9–57}$$

in the interval A to D with $x < 0$. At B and D, $g \to \infty$. The other regions of thermal boundary-layer structure between B and C and C and D do not need to be treated explicitly because of the symmetry of the problem. Among the problems at the end of this chapter, we pose additional examples involving nonuniform flows where the thermal boundary-layer analysis can be applied.

D. Mass Transfer from a Bubble or Drop that Translates Through a Quiescent Fluid at Low Re and Large Pe

The correlation (9–30) was shown to be valid for heat transfer from solid bodies of arbitrary shape in a variety of arbitrary undisturbed flows, subject only to the condition that the body not rotate nor be placed in an undisturbed flow that has closed streamlines at infinity.

Another question that the reader may ask is whether the restriction to solid bodies is really necessary for the validity of (9–30). In particular, an important problem closely related to that solved in Section A is the transport of a solute from a bubble or drop that translates through a quiescent fluid at low Re and Pe $\gg 1$. We have already noted that mass transfer in a liquid is almost always characterized by large values of the appropriate Peclet number (the Peclet number for mass transfer involves the product of the Schmidt number and Reynolds number instead of the Prandtl number and Reynolds number) and that the dimensionless form of the convection-diffusion equation governing transport of a single solute through a solvent is still (8–7) with θ now being a dimensionless solute concentration. For transfer of a solute from a bubble or drop into a liquid that previously contained no solute, the concentration θ at large distances from the bubble or drop will satisfy the condition

$$\theta \to 0 \quad \text{as} \quad |\mathbf{r}| \to \infty. \tag{9–58}$$

The condition at the bubble or drop surface (that is, in the liquid in the limit as we approach the surface) can be quite complicated. Here, for simplicity, we initially confine our attention to a bubble and suppose that the solute concentration inside the bubble remains constant (though this assumption will clearly break down eventually if we assume that there is no source for production of additional solute inside the bubble). We further assume that the bubble volume remains constant in the time interval of interest (again, the corresponding condition $\mathbf{u} \cdot \mathbf{n} = 0$ at the surface will never be precisely true). Finally, we assume that the solute concentration on the liquid side of the interface is a constant c_0 that can, in principle, be related to the constant concentration inside the bubble via the condition of thermodynamic equilibrium that is generally satisfied locally across the interface. Hence, in this case,

$$\theta \equiv \frac{c}{c_0}, \tag{9-59}$$

and the mass transfer problem has the same dimensionless form

$$\mathbf{u} \cdot \nabla \theta = \frac{1}{\text{Pe}} \nabla^2 \theta, \tag{9-60}$$

$$\theta = 1 \quad \text{at} \quad S,$$

and

$$\theta \to 0 \quad \text{at} \quad \infty \tag{9-61}$$

as solved in Section A for heat transfer from solid bodies.

Of course the feature that differs in this case is the form of the velocity field **u**. For simple translation through a quiescent fluid (that is, the uniform flow problem) at zero Reynolds number, this is solely a consequence of the change from no-slip conditions for a solid body to the condition of vanishing tangential stress at the surface of a clean bubble (recall that the shape remains spherical for Re ≪ 1; see Chapter 5).

We are concerned primarily with the form of the correlation between the dimensionless rate of mass transfer, which we again denote as Nu, and the Peclet number. To deduce this form, we again need to ascertain the asymptotic structure of the problem. In most respects, the analysis is very similar to that given in Section A for a solid sphere. Thus, as a first approximation, the analysis in the outer region [within ~$0(a)$ of the bubble] again leads to the estimate (9–7) for θ and the obvious need for a mass-transfer boundary-layer at the bubble surface. As before, the boundary-layer structure depends only on the form of the velocity field in the immediate vicinity of the bubble surface. But it is here that the present problem differs fundamentally from the case of a solid body. The change in boundary conditions to zero tangential stress now allows a nonzero tangential velocity at the surface, and the local approximations (9–44) and (9–46) are modified.

The consequences of this change can be explored first in rather general terms without the need for reference to specific details of the problem for a spherical bubble. After first examining the general consequences of ''slip'' at the boundary, we will then return to fill in the details for the bubble problem. To see the general situation, we return to the formulation of Section B in terms of a local, two-dimensional Cartesian coordinate system. In these terms, the only difference between the bubble and previous problems is in the Taylor series approximations for the velocity components (u, v). Here, the first nonzero term for the tangential component is the slip velocity, and the linear term is zero, that is,

$$u \sim u|_{y=0} + 0(y^2). \tag{9-62}$$

It follows from (9–62) and the continuity equation (9–45) that the normal velocity component is linear in y, that is,

$$v \sim -\frac{\partial}{\partial x}(u_{y=0})y + 0(y^3). \qquad (9\text{-}63)$$

These changes in the local form of the velocity components have a profound influence on the structure of the thermal boundary-layer.

To see this, we may first express (8–25) in terms of the local coordinates (x, y) and introduce the approximate forms, (9–62) and (9–63), for u and v at small values of y:

$$(u|_{y=0} + 0(y^2))\frac{\partial\theta}{\partial x} + \left[-\frac{\partial}{\partial x}(u_{y=0})y + 0(y^3)\right]\frac{\partial\theta}{\partial y} = \left(\frac{\partial^2\theta}{\partial x^2} + \frac{\partial^2\theta}{\partial y^2}\right)\frac{1}{Pe}.$$

Then, introducing a rescaling of the general form

$$Y = y Pe^m, \qquad (9\text{-}64)$$

we obtain

$$u_s(x)\frac{\partial\theta}{\partial x} - \frac{du_s}{dx}Y\frac{\partial\theta}{\partial Y} = Pe^{2m-1}\frac{\partial^2\theta}{\partial Y^2} + 0(Pe^{-2m}, Pe^{-1}). \qquad (9\text{-}65)$$

where we have denoted $u|_{y=0}$ as $u_s(x)$. Hence, in this case, the balance between conduction and convection in the limit $Pe \to \infty$ requires that

$$m = \frac{1}{2}. \qquad (9\text{-}66)$$

It follows that the thickness of the thermal boundary-layer is $0(Pe^{-1/2})$ rather than $0(Pe^{-1/3})$, and the correlation between Nu and Pe takes the general form

$$Nu \sim c Pe^{1/2}. \qquad (9\text{-}67)$$

We may note that the thermal boundary-layer in this case is asymptotically thin relative to the boundary-layer for a solid body. This is a consequence of the fact that the tangential velocity near the surface is larger, and hence convection is relatively more efficient.

Although we have approached (9–67) from the viewpoint of mass transfer from a gas bubble, it should be evident that a correlation of the same form should also be expected for liquid drops, or other systems in which a thermal boundary-layer forms on a "slip" surface. Of course, calculation of the coefficient c requires detailed solutions of the thermal boundary-layer equation(s), and this may be difficult for cases like a translating drop because of the necessity to simultaneously calculate concentration (or temperature) fields both inside and outside the drop. In spite of these difficulties of detail, however, it is clear that the general form (9–67) will be obtained in all such cases.

In the remainder of this section, we evaluate the coefficient c in (9–67) for the specific case of a translating gas bubble. In this case, the full creeping flow solution for the velocity field is

$$u_r = \left(1 - \frac{1}{r}\right)\cos\theta, \; u_\theta = -\left(1 - \frac{1}{2r}\right)\sin\theta, \; u_\phi = 0. \tag{9-68}$$

Thus, transforming from r to y according to (9–13) and expanding u_r and u_θ about $y = 0$, we find that

$$u_r \sim (\cos\theta)y + O(y^2) = (\cos x)y + O(y^2) \tag{9-69}$$

and

$$u_\theta \sim -\frac{1}{2}\sin\theta + O(y) = -\frac{1}{2}\sin x + O(y), \tag{9-70}$$

where $x = \theta$. Hence, the thermal boundary-layer equation (9–65) takes the specific form

$$(\cos x)\, Y\frac{\partial\theta}{\partial Y} - \frac{1}{2}\sin x\frac{\partial\theta}{\partial x} = \frac{\partial^2\theta}{\partial Y^2}. \tag{9-71}$$

In order to solve this equation, it is convenient to introduce the transformation $\eta = \cos x$ into (9–71):

$$\eta\, Y\frac{\partial\theta}{\partial Y} + \frac{1}{2}(1 - \eta^2)\frac{\partial\theta}{\partial\eta} = \frac{\partial^2\theta}{\partial Y^2}. \tag{9-72}$$

Following the example of previous problems in this chapter, we seek a similarity solution of the form

$$\theta = \theta(\xi), \quad \text{with} \quad \xi \equiv \frac{Y}{g(\eta)}. \tag{9-73}$$

Introducing (9–73) into (9–72), we find that

$$\left[\eta g^2 - \frac{1}{2}(1 - \eta^2)gg'\right]\xi\frac{\partial\theta}{\partial\xi} = \frac{\partial^2\theta}{\partial\xi^2}. \tag{9-74}$$

Thus, if a similarity solution exists, the coefficient in square brackets must be a non-zero constant that we choose, for convenience, as -2. It follows that

$$\frac{\partial^2\theta}{\partial\xi^2} + 2\xi\frac{\partial\theta}{\partial\xi} = 0, \tag{9-75}$$

with the boundary conditions

$$\theta = 1 \quad \text{at} \quad \xi = 0,$$
$$\theta \to 0 \quad \text{as} \quad \xi \to \infty. \tag{9-76}$$

A solution also must exist for $g(\eta)$ satisfying

$$\frac{1}{4}(1 - \eta^2)\frac{dg^2}{d\eta} - \eta g^2 = 2, \tag{9-77}$$

which is finite for all η except possibly $\eta = 1$, where a thermal wake may invalidate the boundary-layer assumptions.

It is a simple matter to solve (9–75), subject to (9–76). The result is

$$\theta = 1 - \frac{\int_0^\xi e^{-t^2} dt}{\int_0^\infty e^{-t^2} dt} = 1 - \frac{2}{\sqrt{\pi}} \int_0^\xi e^{-t^2} dt. \tag{9-78}$$

A general solution of (9–77) is also straightforward, namely,

$$g = \frac{c_1}{1 - \eta^2} + \frac{2\sqrt{2}}{(1 - \eta^2)} \left[\int_{-1}^\eta (1 - s^2) ds \right]^{1/2}. \tag{9-79}$$

This solution is finite for all η, except for the rear stagnation point at $\eta = 1$, provided $c_1 = 0$. Thus,

$$g(\eta) = \frac{2\sqrt{2}}{(1 - \eta^2)} \left[\int_{-1}^\eta (1 - s^2) ds \right]^{1/2}. \tag{9-80}$$

The constant c in (9–67) now can be calculated from the definition

$$c \equiv \int_{-1}^1 - \left(\frac{\partial \theta}{\partial Y} \right)_0 d\eta, \tag{9-81}$$

where

$$\left(\frac{\partial \theta}{\partial Y} \right)_0 = - \frac{2}{\sqrt{\pi}} \frac{1}{g(\eta)}. \tag{9-82}$$

Carrying out the indicated integration, the result is

$$c = \frac{4}{\sqrt{6\pi}} = 0.9213. \tag{9-83}$$

E.   Heat Transfer at High Peclet Number across Regions of Closed-Streamline Flow

In the first four sections of this chapter we considered heat (or mass) transfer at high Peclet number and low Reynolds number for a variety of circumstances in which the heated fluid region is confined to a thin thermal boundary-layer within $0(\text{Pe}^{-1/3})$ or $0(\text{Pe}^{-1/2})$ of the body surface. This limited radial extent of heated fluid is a consequence of the fact that convection is an intrinsically more efficient process than conduction for large Pe, in conjunction with the open streamline structure of the flow, which means that heat, initially transferred to the fluid at the body surface via conduction, can progress only a very short distance out from the surface before it is swept downstream around the body by convection and into the thermal wake. Because the thermal region is so thin in these cases, the temperature gradients at the body surface are large and the Nusselt number increases as $\text{Pe}^{1/3}$ or $\text{Pe}^{1/2}$.

We have already indicated in the preceding sections that this thermal boundary-layer structure does not occur when a particle (or body) is entirely surrounded by closed streamlines (or stream surfaces). In this case, the convection process near the body can no longer transfer heat directly to the streaming flow where it is carried into the wake but only circulates it in a closed path around the body. Thus, the heat transfer process is fundamentally altered, because heat can escape from the body only by *diffusing* slowly across the region of closed streamlines (or stream surfaces). Since the size of this region is independent of Pe, the steady-state temperature gradients will be 0(1), and we expect that

$$\boxed{\text{Nu} \sim 0(1) \quad \text{as} \quad \text{Pe} \to \infty.} \tag{9-84}$$

Assuming that this result is correct, it is very important because comparison with (9–30) and (9–67) shows that the rate of heat transfer from a particle (or heated body) to a surrounding fluid at high Peclet numbers depends critically on whether the streamlines (or stream surfaces) near the heated surface are open or closed.

In this section, we shall show how the temperature distribution can be calculated for heat transfer at high Pe across regions of closed streamline motion. However, the most important goal for the reader is to have a clear, qualitative understanding of the fundamental difference between open- and closed-streamline flows for convective heat transfer at high Peclet number. To this end, we begin by offering a somewhat more quantitative version of the time scale argument for the existence of thin boundary-layers that was originally introduced at the beginning of Section A. From this point of view, the thin thermal boundary-layer for *open*-streamline flows at high Pe is seen to exist because the time scale available for radial heat transfer by conduction is very short. In particular, if we use the free stream velocity as an estimate of the characteristic velocity of a fluid element near the body surface, then the time available for heating by conduction before the fluid passes around the body is

$$t^* = \frac{a}{U_\infty}. \tag{9-85}$$

On the other hand, conduction (or diffusion) occurs over a distance

$$l \sim \sqrt{\kappa t} \tag{9-86}$$

in a time increment t. Hence, combining (9–85) and (9–86), we obtain an estimate for the radial thickness of the heated, thermal layer,

$$l^* \sim \sqrt{\frac{\kappa a}{U_\infty}}. \tag{9-87}$$

In nondimensional terms, this is

$$\boxed{\bar{l}^* \sim \left(\frac{\kappa}{U_\infty a}\right)^{1/2} = \text{Pe}^{-1/2},} \tag{9-88}$$

and we see why the thermal layer is so thin for Pe $\gg$ 1. We may note that the estimate (9–88) is *precisely correct* for the boundary-layer on a gas bubble (or a drop) in spite of its simplistic view of the heat transfer process as decoupled conduction and convection. This is because the estimate U_∞ for the velocity of a fluid element near the body surface is reasonable in that case. For a solid body, on the other hand, the actual velocity in the region of interest is smaller,

$$0\left(U_\infty\left(\frac{\hat{l}}{a}\right)\right),$$

where $\hat{l}$ is the (dimensional) thermal boundary-layer thickness. As a consequence, the characteristic time available for conduction is longer,

$$\hat{t} = \frac{a^2}{U_\infty \hat{l}}, \tag{9-89}$$

and the predicted thickness of the thermal layer, nondimensionalized by the particle scale a, becomes

$$\frac{\hat{l}}{a} \sim (\text{Pe})^{-1/3}, \tag{9-90}$$

that is, the thermal boundary-layer thickness is increased relative to (9–88) because the available time for conduction of heat outward from the body surface is increased.

Now, we may consider how these time scale arguments change when heat is transferred across a region of *closed* streamlines or stream surfaces, still at large Peclet number. In this case, the convection process only circulates heat around a closed path and thus does not carry the heated fluid away from the body surface. As a consequence, the restriction on available time for conduction of heat outward from the body is no longer relevant, and the conduction process, though relatively inefficient for Pe $\gg$ 1, has an indefinite time to reach a steady-state configuration in which the whole of the closed-streamline (or stream surface) region is heated. Assuming that the size of this region is 0(1) relative to the heated body, it follows that the dimensionless temperature gradient at steady state must be 0(1), and the Nusselt number will approach an 0(1) value for Pe $\to \infty$, as indicated in equation (9–84). Indeed, at steady state, the *whole* of the temperature decrease from the surface value $\theta = 1$ to the free stream value $\theta = 0$ must occur across the recirculating flow region. If this were not true, the temperature on the outermost closed-streamline (or stream surface) would be 0(1), though obviously smaller than 1, and there would need to be a thermal boundary-layer on this surface of the type described in the preceding section. But then, the heat flux from the closed-streamline region to the bulk fluid stream would be 0(Pe$^{1/2}$), and this obviously could not be consistent with a temperature gradient and heat flux of 0(1) within the closed streamline region. Thus, we conclude that $\theta \to 0$ on the outermost closed streamline (or stream surface) of the recirculating zone. The qualitative arguments presented here clearly indicate that for Pe $\gg$ 1, the rate of heat transfer from a particle or heated body to a surrounding fluid depends critically on whether the streamlines near the heated surface are open or closed.

In the remainder of this section, we consider the details of solution for a particular problem of a heated body surrounded by a region of closed streamlines. The problem that is closest to that analyzed in previous sections of this chapter is a heated sphere in a flow where the sphere rotates [such as the general linear flows of equation (9–56) with $\lambda \neq 1$] and is thus surrounded by a closed region of recirculating fluid. In fact, however, the only problem of this general type that has been worked out in detail is the sphere in a simple shear flow ($\lambda = 0$) at low Reynolds number,[6] and even in that case, the flow is fully three-dimensional, the geometry of the recirculating region is complicated, and the analysis is more difficult in detail than we wish to present here. As an alternative, we consider a problem that retains the essential features of the heated sphere but is simpler to solve. This is the two-dimensional analogue of a freely rotating, heated circular cylinder in a simple shear flow.[7] Following the detailed analysis for the circular cylinder, we shall briefly return to the results for a sphere in shear flow.

To begin our analysis of the cylinder problem, it is necessary to consider briefly the velocity field. The fluid mechanics solution for a circular cylinder rotating with an imposed angular velocity Ω in a simple shear flow was given originally by Bretherton (1962)[8] for the creeping flow limit. In this case, the velocity field can be specified in terms of a streamfunction (see Chapter 4) such that

$$u_r \equiv \frac{1}{r}\frac{\partial\psi}{\partial\phi}, \quad u_\theta = -\frac{\partial\psi}{\partial r}. \tag{9-91}$$

A sketch of the physical problem and a definition of the cylindrical polar coordinates (r,ϕ) is given in Figure 9–8. For the creeping flow limit, the governing equations and boundary conditions are

$$\nabla^4\psi = 0 \tag{9-92}$$

with

$$\psi = 0, \quad \frac{\partial\psi}{\partial r} = \Omega \quad \text{at} \quad r = 1,$$

$$\psi \to \frac{1}{2}y^2 = \frac{1}{2}r^2\sin^2\phi \quad \text{as} \quad r \to \infty. \tag{9-93}$$

Here, the velocities are nondimensionalized with respect to Ga, where G is the shear rate of the undisturbed flow and a is the cylinder radius. The Bretherton solution of

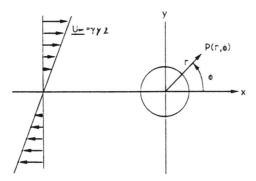

Figure 9–8 The coordinate axes and undisturbed flow for a cylinder immersed in an unbounded simple shear flow.

(9–92) and (9–93) is

$$\psi = \frac{1}{2}y^2 - \frac{1}{4}\left\{2(1-2\Omega)\ln r + 1 + \left[\frac{1}{r^2}-2\right]\cos 2\phi\right\}, \qquad (9\text{–}94)$$

and the corresponding torque on the cylinder is

$$T = -2\pi\left(\frac{v}{Ga^2}\right)(1-2\Omega). \qquad (9\text{–}95)$$

In the present case, there is no externally applied torque on the cylinder (i.e., it is free to rotate in the flow). Thus, the hydrodynamic torque must also vanish at steady state and

$$\Omega = \frac{1}{2}. \qquad (9\text{–}96)$$

It follows that

$$\psi = \frac{1}{2}y^2 - \frac{1}{4}\left[1 + \left(\frac{1}{r^2}-2\right)\cos 2\phi\right]. \qquad (9\text{–}97)$$

Alternatively,

$$\psi = \frac{1}{4}(r^2 - 1) - \frac{1}{4}(r^2 - 2 + r^{-2})\cos 2\phi. \qquad (9\text{–}98)$$

A full asymptotic analysis by Robertson and Acrivos (1970)[9] has shown that this solution of the creeping motion equations is a uniformly valid first approximation for Re $\ll$ 1.

A plot of the streamlines for (9–98) is shown in Figure 9–9. It is evident that all streamlines for $\psi < 1/4$ are closed, while those for $\psi > 1/4$ are open. The dividing streamline $\psi = 1/4$ can be thought of as "closing at infinity." The fact that $\psi = 1/4$ is the critical value can be seen by means of the following argument. First, if we evaluate ψ at $\phi = 0$ or π, we find that

$$\psi|_{\phi=0,\pi} = \frac{1}{4}\left(1 - \frac{1}{r^2}\right). \qquad (9\text{–}99)$$

Thus, $\psi \to 1/4$ as $r \to \infty$. On the other hand, for $\phi = \pi/2$,

$$\psi|_{\pi/2} = \frac{r^2}{2} - \frac{3}{4} + \frac{1}{4r^2}, \qquad (9\text{–}100)$$

and on this axis, $\psi = 1/4$ at $r \sim 1.32$. Thus, streamlines for $\psi > 1/4$ cross the y axis but do not cross the x axis, while those for $\psi < 1/4$ (including $\psi = 0$) cross both axes.

Now let us turn to the thermal energy equation

$$\mathbf{u}\cdot\nabla\theta = \frac{1}{Pe}\nabla^2\theta, \qquad (9\text{–}101)$$

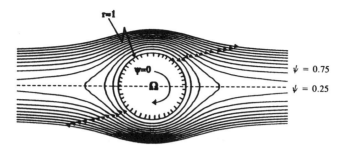

Figure 9–9 Streamlines for a freely rotating circular cylinder in simple, linear shear flow (9–98). Contours 0 to 0.75 in increments of 1/16.

which we must solve to determine the temperature distribution in the fluid. The boundary conditions are

$$\theta = 1 \quad \text{at} \quad r = 1$$

and

$$\theta \rightarrow 0 \quad \text{at} \quad r \rightarrow \infty. \tag{9-102}$$

In writing (9–101), we again have used Ga as the characteristic velocity, so that

$$\text{Pe} \equiv \frac{Ga^2}{\kappa}.$$

We seek an approximate solution in the form of an asymptotic expansion for the limit $\text{Pe} \rightarrow \infty$.

To obtain the governing equation for the first term in this expansion, we take the limit $\text{Pe} \rightarrow \infty$ in (9–101). The result is

$$\mathbf{u} \cdot \nabla \theta_0 = 0, \tag{9-103}$$

and we have already seen that a general solution of this equation is

$$\theta_0 = \theta_0(\psi). \tag{9-104}$$

In an open-streamline flow, the functional dependence of temperature on ψ can be obtained from the known dependence at ∞. This can be used, in the present case, to see that

$$\theta_0 = 0 \tag{9-105}$$

for $\psi \geq 1/4$. However, for the closed-streamline region, $\psi < 1/4$, (9–104) simply tells us that θ_0 is constant on any streamline but provides no basis to determine which value applies to which streamline. Generally, this apparent indeterminacy in the function $\theta_0(\psi)$ is characteristic of all problems involving closed streamlines or closed stream surfaces—we can say only that θ_0 is constant on streamlines for $\text{Pe} \gg 1$ but cannot determine the actual dependence of θ_0 on ψ without further analysis of some kind.

If we consider the qualitative description of heat transfer in closed-streamline regions presented at the beginning of this section, we may recognize that in order to obtain the temperature distribution, we need to somehow include conduction effects because this is the primary mode of heat transfer across streamlines for Pe $\gg$ 1, when the streamlines are closed. In the boundary-layer problems of the preceding sections of this chapter, the necessity for conduction signaled the need for rescaling to a different characteristic length scale and the development of a second distinct expansion with a new leading-order approximation. Here, on the other hand, the scaling and the general solution (9–104) are perfectly valid, and a different approach is necessary to see how the very slow and inefficient conduction process (acting over a very long period of time) can completely control the distribution of θ_0 over the closed-streamline region. For this purpose let us return to the full thermal energy equation (9–101) in the time dependent form that applies during the slow evolution toward a steady-state temperature distribution,

$$\frac{\partial \theta}{\partial t} + \mathbf{u} \cdot \nabla \theta = \frac{1}{\mathrm{Pe}} \nabla^2 \theta. \tag{9-106}$$

In writing (9–106), we have incorporated a characteristic time scale $t_c = a/U = G^{-1}$. This equation also can be written in the form

$$\boxed{\frac{\partial \theta}{\partial t} = \mathrm{div}\left[-\mathbf{u}\theta + \frac{1}{\mathrm{Pe}} \nabla\theta \right]} \tag{9-107}$$

since div $\mathbf{u} \equiv 0$. Now suppose that we integrate (9–107) over area A between any two arbitrarily chosen closed streamlines, say, ψ_1 and ψ_2. A sketch of a typical configuration is shown in Figure 9–10. Formally, this gives

$$\int_A \left(\frac{\partial \theta}{\partial t} \right) dA = \int_A \mathrm{div}\left[-\mathbf{u}\theta + \frac{1}{\mathrm{Pe}} \nabla\theta \right] dA, \tag{9-108}$$

which is nothing more than a macroscopic heat balance for the fluid region bounded by ψ_1 and ψ_2. The significance of the right-hand side of this balance is best exposed by applying the divergence theorem. This gives

$$\int_A \left(\frac{\partial \theta}{\partial t} \right) dA = \int_{\psi_1} \left[-\mathbf{u}\theta + \frac{1}{\mathrm{Pe}} \nabla\theta \right] \cdot \mathbf{n} dl - \int_{\psi_2} \left[-\mathbf{u}\theta + \frac{1}{\mathrm{Pe}} \nabla\theta \right] \cdot \mathbf{n} dl \tag{9-109}$$

where

$$\mathbf{n} = \mathbf{i}_\psi$$

is a unit vector that is orthogonal to the streamlines. But, by definition,

$$\mathbf{u} \cdot \mathbf{n} = 0 \tag{9-110}$$

on any streamline. Hence, (9–109) reduces to

$$\int_A \frac{\partial \theta}{\partial t} dA = \int_{\psi_1} \frac{1}{\mathrm{Pe}} \nabla\theta \cdot \mathbf{n} dl - \int_{\psi_2} \frac{1}{\mathrm{Pe}} \nabla\theta \cdot \mathbf{n} dl. \tag{9-111}$$

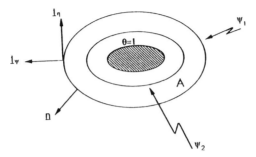

Figure 9–10 A sketch of a typical, two-dimensional closed-streamline configuration. Here, ψ_1 and ψ_2 are any two arbitrarily chosen closed streamlines. The area of the region between these streamlines is denoted as A. The unit normal is denoted as $\mathbf{n}$, and $\mathbf{i}_\psi$ and $\mathbf{i}_\eta$ are a pair of orthogonal unit vectors that are normal and tangential to the streamline (note that $\mathbf{i}_\psi \equiv \mathbf{n}$).

We see that in the transient period prior to establishing a steady state, heat accumulates in the region between any two arbitrarily chosen closed streamlines due solely to conduction of heat across the streamlines. In addition, in the limit $\text{Pe} \gg 1$, it is evident that the rate of change of the temperature distribution is very slow. Once $|\nabla\theta| \sim 0(1)$, the left-hand side of (9–111) changes on a dimensionless time scale $\Delta t \sim 0(\text{Pe})$ only. Eventually, however, steady state will be established, and it is this steady-state temperature distribution that we seek.

Clearly, whatever the final steady-state solution may be, it must satisfy the macroscopic condition (9–111), with the left-hand side set equal to zero. Since ψ_1 and ψ_2 are completely arbitrary, the resulting steady-state condition can be expressed in the form

$$\frac{1}{\text{Pe}} \int_\psi \nabla\theta \cdot \mathbf{n}\, dl = C, \qquad (9\text{–}112)$$

where C is a constant, independent of ψ. Physically, this condition states that the total heat flux across any streamline due to conduction must be independent of the particular streamline that we choose. Since one possible choice for a closed streamline in the present case is $\psi = 0$, coincident with the cylinder surface, we see that

$$C = \frac{1}{\text{Pe}} \int_{r=1} \frac{\partial\theta}{\partial r}\bigg|_{r=1} d\phi. \qquad (9\text{–}113)$$

But this integral is just the negative of the dimensionless heat flux per unit length of the cylinder, namely,

$$\text{Nu} = -\frac{1}{\pi} \int \frac{\partial\theta}{\partial r}\bigg|_{r=1} d\phi. \qquad (9\text{–}114)$$

Thus,

$$C = -\frac{\text{Nu}\,\pi}{\text{Pe}}, \tag{9-115}$$

and we can immediately determine the Nusselt number once C is known.

To see whether the condition (9–112) provides the added input that is required to determine $\theta_0(\psi)$ in the closed-streamline region, it is convenient to express it in terms of the indpendent variables ψ and η (which we choose as lines orthogonal to ψ) instead of cylindrical coordinates (r, ϕ). To do this, we need only introduce the appropriate scale or metric factors for the (ψ, η) coordinate system, h_ψ and h_η. These scale factors are defined such that the length of a differential line element, expressed in the (ξ, η) system, takes the form

$$(ds)^2 \equiv h_\psi^2(d\psi)^2 + h_\eta^2(d\eta)^2. \tag{9-116}$$

Equivalently,

$$h_\psi^2 \equiv \left(\frac{\partial x}{\partial \psi}\right)^2 + \left(\frac{\partial y}{\partial \psi}\right)^2, \quad h_\eta^2 \equiv \left(\frac{\partial x}{\partial \eta}\right)^2 + \left(\frac{\partial y}{\partial \eta}\right)^2. \tag{9-117}$$

Now,

$$\nabla\theta\cdot\mathbf{n} = \frac{1}{h_\psi}\frac{\partial\theta}{\partial\psi} \tag{9-118}$$

and

$$dl = h_\eta d\eta. \tag{9-119}$$

Thus, (9–112) can be written in the form

$$\frac{1}{\text{Pe}}\int_\psi \frac{h_\eta}{h_\psi}\frac{\partial\theta}{\partial\psi}\,d\eta = C. \tag{9-120}$$

But according to (9–104), $\theta \sim \theta_0(\psi)$ is a function of ψ only. Thus,

$$\frac{1}{\text{Pe}}\left[\int_\psi \frac{h_\eta}{h_\psi}\,d\eta\right]\frac{\partial\theta_0}{\partial\psi} = C,$$

where the coefficient in square brackets depends on ψ only. Let us denote this coefficient as $\Gamma(\psi)$. Then

$$\frac{1}{\text{Pe}}\Gamma(\psi)\frac{\partial\theta_0}{\partial\psi} = C. \tag{9-121}$$

Upon integration and applying the boundary condition $\theta_0 = 1$ at the cylinder surface, $\psi = 0$, we find that

$$\theta_0 = 1 + \text{Pe}\,C\int_0^\psi \frac{ds}{\Gamma(s)}. \tag{9-122}$$

The function $\Gamma(\psi)$ depends on the form of the streamfunction distribution and can be calculated once the flow is specified.

The solution (9–122) applies within the closed-streamline region. If we assume that it applies right up to the separating streamline $\psi = 1/4$ (as is true at the leading order of approximation), then the constant C can be determined from the second condition

$$\theta_0 = 0 \quad \text{at} \quad \psi = 1/4. \tag{9–123}$$

In this case,

$$C = -\frac{1}{\text{Pe} \int_0^{1/4} \frac{ds}{\Gamma(s)}} \tag{9–124}$$

and

$$\theta_0 = 1 - \frac{\int_0^\psi \frac{ds}{\Gamma(s)}}{\int_0^{1/4} \frac{ds}{\Gamma(s)}}. \tag{9–125}$$

In addition, we see from (9–115) that

$$\text{Nu}_0 = \frac{1}{\pi \int_0^{1/4} \frac{ds}{\Gamma(s)}}. \tag{9–126}$$

Numerical evaluation of the integral in (9–126), using the solution (9–98) for ψ and the definitions (9–117) for h_ψ and h_η to obtain $\Gamma(\psi)$, yields

$$\text{Nu}_0 = 5.73. \tag{9–127}$$

This is the primary result of the present calculation and confirms the general prediction that

$$\text{Nu} \to 0(1) \quad \text{as} \quad \text{Pe} \to \infty$$

for heat transfer from a body that is embedded within a region of closed-streamline flow.

The reader may note that the solution (9–125) for θ_0 approaches the same numerical value for $\psi \to 1/4$ as does the first-order approximation (9–105) for θ_0 in the region $\psi > 1/4$. It thus may seem that we have constructed a uniformly valid solution for the first term in a regular perturbation expansion for $\text{Pe} \gg 1$. Unfortunately, this is not the case. The governing equation for θ_0, (9–103), assumes that convection is dominant over conduction. But a careful analysis of the magnitude of conduction and convection terms, based on the leading-order solution (9–125) for θ_0, indicates that there is a region arbitrarily close to $\psi = 1/4$ where conduction is not small compared with convection as assumed. Thus, the limit $\text{Pe} \to \infty$ is singular, and a complete solution for $\text{Pe} \gg 1$ would require use of the method of matched asymptotic expansions. However, the "correction" necessary in the vicinity of $\psi = 1/4$ does not change (9–125) in the closed-streamline region nor the resulting estimate for the Nusselt number Nu_0. Indeed, the asymptotic solution in this case is analogous to Stokes's solution for flow past a sphere in the sense that the singular nature of the limit $Pe \gg 1$ does not appear explicitly in the leading-order approximation for θ but only when we attempt to generate higher-order approximations to the solution. We are content here with the leading-order estimates (9–105) and (9–125).

Before concluding the discussion of high Peclet number heat transfer in low Reynolds number flows across regions of closed streamlines (or stream surfaces), let us return briefly to the problem of heat transfer from a sphere in simple shear flow. This problem is qualitatively similar to the two-dimensional problem that we have just analyzed, and the physical phenomena are essentially identical. However, the details are much more complicated. The problem has been solved by Acrivos (1971),[10] and the interested reader may wish to refer to his paper for a complete description of the analysis. Here, we offer only the solution and a few comments. The primary difficulty is that an integral condition, similar to (9–112), which can be derived for the net heat transfer across an arbitrary isothermal stream surface, does not lead to any useful quantitative results for the temperature distribution because, in contrast with the two-dimensional case where the isotherms correspond to streamlines, the location of these stream surfaces is a priori unknown. To resolve this problem, Acrivos shows that the more general steady-state condition

$$\int_A \nabla^2 \theta \, dA = 0 \qquad (9\text{–}128)$$

must be used, which is obtained from (9–108) by applying the divergence theorem to only the first term on the right-hand side. After considerable effort, beginning with (9–104) and (9–128), Acrivos obtained the result

$$Nu = 8.9 \quad \text{for} \quad Pe \to \infty,$$

which corresponds qualitatively to the two-dimensional result (9–127).

It is important to recognize that the analysis presented in this section is generally applicable to any high Peclet number heat transfer process that takes place across a region of closed streamline flow. In particular, the limitation to small Reynolds number is not an intrinsic requirement for any of the development from equation (9–101) to equation (9–126). It is only in the specification of a particular form for the function $\Gamma(\psi)$ that we require an analytic solution for ψ and thus restrict our attention to the creeping flow limit. Indeed, all of (9–101)–(9–126) applies for any closed-streamline flow at any Reynolds number, provided only that $Pe \gg 1$. The same is true of the qualitative discussion of the heat transfer mechanism for closed-streamline (or stream surface) flows at the beginning of the section.

Homework Problems

1. We consider two unbounded flat surfaces that are parallel and separated by a distance d. Between the plates there is a flow component parallel to the boundaries that is driven by a pressure gradient

$$-\frac{\partial P}{\partial x} = G > 0.$$

The two plates are porous, and there is also a flow between the surfaces with a normally directed velocity component, V.

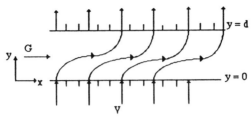

a. Show that the governing equation for the velocity component parallel to the walls is

$$\rho\left(V\frac{\partial u}{\partial y}\right) = G + \mu\left(\frac{\partial^2 u}{\partial y^2}\right),$$

with $u = 0$ at $y = 0$ and d. Nondimensionalize, and obtain an exact solution. (This solution is valid for arbitrary values of the Reynolds number $Re \equiv \rho V d/\mu$.)

b. Although an exact solution is available, it is instructive to consider an asymptotic solution of the above problem for $Re \gg 1$. Derive an approximate solution for this limiting case. Show that a boundary-layer exists at the upper wall, and determine the scaling of this region as $O(Re^{-\alpha})$. Show that this boundary-layer structure also can be determined by taking the limit $Re \to \infty$ in the exact solution.

2. For flow past a circular cylinder at low Reynolds number, $Re \ll 1$, a first (inner) approximation to the velocity field is

$$\psi = -\frac{1}{2\ln Re}\left(\frac{1}{r} - r + 2r\ln r\right)\sin\theta.$$

Suppose the cylinder is heated and $Pe \gg 1$. Find an appropriate thermal boundary-layer equation, and deduce the general form of the relationship between Nu and Pe.

3. In a classic paper, Payne and Pell (*J. Fluid Mech.* 7:529–549 [1960]) presented a general solution scheme for axisymmetric creeping flow problems. Among the specific examples that they considered was the uniform, axisymmetric flow past prolate and oblate ellipsoids of revolution (spheroids). This solution was obtained using prolate and oblate ellipsoidal coordinate systems, respectively. In the prolate case, the ellipsoidal coordinates (ξ, η) are related to circular cylindrical coordinates (r, z) via the transformation $z = c\cosh\xi\cos y$ and $r = c\sinh\xi\sin\eta$.

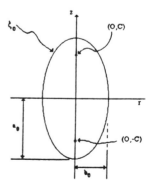

The coordinate surface $\xi = \xi_0$ (constant) defines the surface of a prolate spheroid, with major and minor semiaxes a_0 and b_0 given by

$$a_0 = c\cosh\xi_0 \quad , \quad b_0 = c\sinh\xi_0.$$

For convenience, we may note that

$$c^2 \equiv a_0^2 - b_0^2 \quad \text{and} \quad \xi_0 = \frac{1}{2}\ln\frac{a_0 + b_0}{a_0 - b_0}.$$

The line $\xi_0 = 0$ is a degenerate ellipse that reduces to a line segment $-c \leq z \leq c$ along the z axis. In these terms, Payne and Pell's solution for uniform streaming flow with undisturbed velocity U past a prolate spheroid with semimajor and semiminor axes a_0 and b_0 is

$$\psi = \frac{1}{2}Ur^2\left[1 - A\ln\frac{s+1}{s-1} + B\left(\frac{s}{s^2-1}\right)\right],$$

where

$$A = -\frac{\dfrac{1}{2}(s_0^2 + 1)}{s_0 - \dfrac{1}{2}(s_0^2 + 1)\ln\dfrac{s_0 + 1}{s_0 - 1}},$$

$$B = -\frac{s_0^2 - 1}{s_0 - \dfrac{1}{2}(s_0^2 + 1)\ln\dfrac{s_0 + 1}{s_0 - 1}},$$

and $s = \cosh\xi$, $s_0 = \cosh\xi_0$, with ξ_0 given in terms of a_0 and b_0 as indicated. In the oblate case, the coordinate transformation is

$$z = c\sinh\xi\cos\eta,$$

$$r = c\cosh\xi\sin\eta,$$

with $\xi = \xi_0$ corresponding to the surface of an oblate ellipsoid of revolution (spheroid) with major and minor axes

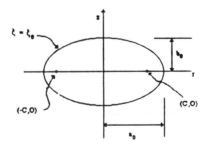

$$a_0 = c \cosh \xi_0,$$

$$b_0 = c \sinh \xi_0,$$

as before. In this case, Payne and Pell's solution for the uniform streaming flow problem is

$$\psi = \frac{1}{2} U r^2 \left\{ 1 - C \frac{\tau}{1 + \tau^2} - D \cot^{-1} \tau \right\},$$

with

$$A \equiv \frac{\left(1 + \tau_0^2\right)}{\tau_0 + \left(1 - \tau_0^2\right) \cot^{-1} \tau_0}$$

and

$$B \equiv -\frac{\left(1 - \tau_0^2\right)}{\tau_0 + \left(1 - \tau_0^2\right) \cot^{-1} \tau_0},$$

where $\tau = \sinh \xi$ and $\tau_0 = \sinh \xi_0$. One interesting special case is flow past a flat, circular disk, where $\tau_0 = 0$, and

$$\psi = \frac{1}{2} U r^2 \left\{ 1 - \frac{2}{\pi} \left(\cot^{-1} \tau + \frac{\tau}{1 + \tau^2} \right) \right\}.$$

a. With the preceding solutions given, develop thermal boundary-layer equations for a heated ellipsoid, and determine the relationship between Nu and Pe for each of the two cases of a prolate and oblate ellipsoid.

b. Discuss the influence of the ellipsoid geometry on the rate of heat transfer. In order to provide a fixed frame of reference, assume that a_0 and b_0 are varied in such a way that the aspect ratio of the ellipsoid changes but the *volume* is held fixed. Thus, as you vary the body shape, the surface area will change, as well as the heat flux distribution on the body surface. (You may find it convenient to define the Peclet number in terms of the length scale of a sphere of equal volume, so that its value remains fixed for fixed U/κ as the geometry is varied.) Include the limiting case of a circular disk in your discussion.

4. A heated circular cylinder is suspended in a fluid that is undergoing a simple shear flow. Assume that the cylinder does not rotate so that the streamfunction representing flow in the (inner) region near the cylinder is

$$\psi = \frac{1}{2}y^2 - \frac{1}{4}\left\{2\ln r + 1 + \left(\frac{1}{r^2} - 2\right)\cos 2\phi\right\}$$

in dimensionless terms [see equation (9–94)]. The temperature at the surface of the cylinder is held constant at a value T_0, while the ambient temperature of the fluid is T_1. Calculate the first approximation to the rate of heat transfer from the cylinder (per unit length) to the fluid, assuming that the Peclet number is large,

$$\text{Pe} = \frac{Ga^2}{\kappa} \gg 1.$$

5. We consider the same situation as in Problem 4, but the torque acting on the cylinder is now assumed to be too small to keep the cylinder from rotating. Hence, it rotates with an angular velocity Ω, which, in dimensionless terms, is in the range

$$0 < \Omega < \frac{1}{2}$$

between free rotation and no rotation. Thus, the velocity field is now given, in the region near the cylinder, by

$$\psi = \frac{1}{2}y^2 - \frac{1}{4}\left\{2(1 - 2\Omega)\ln r + 1 + \left(\frac{1}{r^2} - 2\right)\cos 2\phi\right\}$$

[that is, equation (9–94) with $\Omega \neq 0$ or $\Omega \neq 1/2$]. Obtain the relationship between Nu and Ω, assuming that Pe $\gg 1$.

6. Consider a *spherical* gas bubble that is suspended in a liquid that undergoes an axisymmetric extensional flow. The bubble contains a component A that is soluble in the exterior liquid and a major component B that is insoluble. Assume that the concentration of A is constant inside the bubble and that the bubble volume is fixed. Determine the concentration distribution in the liquid for species A and the mass transfer rate from the gas to the liquid (the Nusselt number), assuming Pe $\gg 1$.

  Note: In dimensional terms, the steady-state distribution of A in the liquid is governed by

$$\mathbf{u} \cdot \nabla C_A = D\nabla^2 C_A,$$

where D is the mutual diffusion coefficient for species A and C_A is the concentration of A in the liquid.

References/Notes

1. Acrivos, A., and Goddard, J.D., "Asymptotic Expansions for Laminar Convection Heat and Mass Transfer," *J. Fluid Mech.* 23:273–291 (1965).

2. See appendix 2 in Reference 1.

3. Among recent references that discuss closed-streamline patterns and eddies in low Reynolds number flows, the reader may wish to refer to Reference 11, Chapter 4, and

a. Jeffrey, D.J., and Sherwood, J.D., "Streamline Patterns and Eddies in Low-Reynolds-Number Flow," *J. Fluid Mech.* 96:315–334 (1980).

b. Davis, A.M.J., and O'Neill, M.E., "The Development of Viscous Wakes in a Stokes Flow When a Particle is Near a Large Obstacle," *Chem. Eng. Sci.* 32:899–906 (1977).

c. Davis, A.M.J., and O'Neill, M.E., "Separation in a Slow Linear Shear Flow Past a Cylinder and a Plane," *J. Fluid Mech.* 81:551–564 (1977).

4. See Reference 12, Chapter 4.

5. A general study of the streamlines for a circular cylinder in simple shear flow can be found in the following papers:

a. Robertson, C.R., and Acrivos, A., "Low Reynolds Number Shear Flow Past a Rotating Circular Cylinder, Part 1, Momentum Transfer, *J. Fluid Mech.* 40:685–704 (1970).

b. Cox, R.G., Zia, I.Y.Z., and Mason, S.G., "Particle Motions in Sheared Suspensions," 15, "Streamlines Around Cylinders and Spheres," *J. Coll. Interface Sci.* 27:7–18 (1968).

6. Acrivos, A., "Heat Transfer at High Peclet Number from a Small Sphere Freely Rotating in a Simple Shear Flow," *J. Fluid Mech.* 46:233–240 (1971).

7. Frankel, N.A., and Acrivos, A., "Heat and Mass Transfer from Small Spheres and Cylinders Freely Suspended in Shear Flow," *Phys. Fluids* 11:1913–1918 (1968).

8. Bretherton, F.P., "Slow Viscous Motion Around a Cylinder in Simple Shear," *J. Fluid Mech.* 12:591–613 (1962).

9. See Reference 5a.

10. See Reference 6.

CHAPTER 10

Laminar Boundary-Layer Theory

In Chapter 9 we considered strong convection effects in heat (or mass) transfer problems at low Reynolds numbers. The most important findings were the existence of a thermal boundary-layer for open-streamline flows at high Peclet numbers, and the fundamental distinction between open- and closed-streamline flows for heat or mass transfer processes at high Peclet numbers. An important conclusion in each of these cases is that conduction (or diffusion) plays a critical role in the transport process, even though $Pe \rightarrow \infty$. In open-streamline flows, this occurs because the temperature field develops increasingly large gradients near the body surface as $Pe \rightarrow \infty$. For closed streamline flows, on the other hand, the temperature gradients are $0(1)$—except possibly during some initial transient period—and conduction is important because it has an indefinite time to act.

In this chapter we continue the development of these ideas by considering their application to the approximate solution of fluid mechanics problems in the asymptotic limit $Re \rightarrow \infty$, with a particular emphasis on problems where boundary layers play a key role. Before embarking on this program, however, it is useful to highlight the expected goals and limitations of the analysis in which we formally require $Re \rightarrow \infty$ but still assume that the flow remains laminar. In practice, of course, most flows will become unstable at a large, but finite, value of Reynolds number and eventually undergo a transition to turbulence. However, the existence of an instability leading to a branch of unsteady and complicated solutions of the Navier-Stokes equations (e.g., a turbulent velocity field) does not preclude the existence of steady, laminar solutions the same range of Reynolds numbers. It is these latter solutions that we seek. In part, the practical significance of high Reynolds number laminar solutions is that they provide considerable insight and often even accurate results for large, but *finite*, Reynolds numbers below the instability limit. In addition, the high Reynolds number laminar flow solutions provide approximate base solutions whose stability can be studied to understand the mechanisms and critical conditions for instability.

As already indicated, our approach to the construction of approximate solutions for $Re \gg 1$ will be to use the method of matched asymptotic expansions. In view of this, it is worth noting that the earliest use of boundary-layer theory for fluid mechanics problems at high Reynolds number predated considerably the first *formal* use of asymptotic expansion procedures to solve partial differential equations. The initial applications of the method of matched asymptotic expansions appeared in the mid-1950s,[1] whereas the essentials of boundary-layer theory had already been presented in 1904 by Prandtl[2] in a paper that revolutionized fluid mechanics. Prandtl's approach, as well as that of many subsequent investigators in the intervening thirty-five to forty years, was an ad hoc, but physically motivated, simplification of the Navier-Stokes equations for large Reynolds number. The interested reader may wish to refer to the classic textbook by Schlichting[3] for a derivation of boundary-layer theory based upon this physical approach.

Prandtl's theory was revolutionary because it provided for the first time a theoretical understanding of the critical role played by viscous effects in determining fluid motions in the limit Re → ∞. Prior to Prandtl, inviscid flow theory had dominated attempts to describe fluid motions for Re ≫ 1 but was in serious disagreement with experimental observation[4] since it could not deal in any way with *separation phenomena*—namely, the existence of recirculating wakes and large *form* drag in flow past nonstreamlined bodies. It was only with the advent of Prandtl's boundary-layer theory that a theoretical basis existed to predict whether fluid motion past a body of given shape would remain attached (with correspondingly "small" drag) or separate (with a large form drag). The importance of this advance in the development of rational aerodynamic design of airfoil shapes with large lift/drag ratios cannot be overemphasized.[5] We shall return to the problem of predicting separation phenomena later in this chapter.

A potential advantage of the "physical" approach to boundary-layer theory is that it forces an emphasis on the underlying *physical* description of the flow. However, unlike the asymptotic approach presented here, the physically derived theory provides no obvious means to improve the solution beyond the first level of approximation. Provided that the physical picture underlying the analysis is properly emphasized, the asymptotic approach can incorporate the principal positive aspect of the earlier theories within a rational framework for systematic improvement of the approximation scheme.

A. Potential Flow Theory

The first part of this chapter will be concerned largely with a specific prototype problem in which a stationary solid body is immersed in an unbounded, incompressible fluid that is undergoing a steady, uniform translational motion at large distances from the body ("at infinity"). For simplicity, we shall assume in most instances that the body is two-dimensional; namely, that it extends indefinitely in the third direction, z, without change of shape so that its geometry can be specified completely by its cross section in the x, y plane. The streaming motion at infinity is then assumed to be parallel to the xy plane.

To analyze streaming flow at high Reynolds number past a two-dimensional body, the starting point is the full, steady-state Navier-Stokes and continuity equations, nondimensionalized using the streaming velocity U_∞ as a characteristic velocity scale and a scalar length of the body in the xy plane, say, a, as the characteristic length scale, namely,

$$\mathbf{u} \cdot \nabla \mathbf{u} = -\nabla p + \frac{1}{Re} \nabla^2 \mathbf{u} \qquad (10\text{-}1)$$

and

$$\nabla \cdot \mathbf{u} = 0. \qquad (10\text{-}2)$$

Pressure has been nondimensionalized using ρU_∞^2, as is appropriate for flow at large Reynolds number.

A convenient way to discuss some aspects of flow at high Reynolds number is in terms of the *transport of vorticity* rather than directly in terms of velocity and pressure. We recall that the vorticity is defined as the curl of the velocity,

$$\boldsymbol{\omega} \equiv \nabla \wedge \mathbf{u}. \qquad (10\text{-}3)$$

Hence, physically, it represents the local rotational motion of the fluid. An equation for transport of vorticity is obtained directly by taking the curl of all terms in equation (10–1). Since

$$\nabla \wedge (\nabla \phi) \equiv 0 \qquad (10\text{-}4)$$

for any scalar ϕ, we see that the pressure is eliminated, and we have remaining

$$\mathbf{u} \cdot \nabla \boldsymbol{\omega} = \boldsymbol{\omega} \cdot \nabla \mathbf{u} + \frac{1}{Re} \nabla^2 \boldsymbol{\omega}. \qquad (10\text{-}5)$$

The left-hand side represents the advection (or convection) of vorticity by the velocity $\mathbf{u}$, while the second term on the right represents the transport of vorticity by diffusion (with diffusivity = the kinematic viscosity ν). These two terms are "familiar" in the sense that they resemble the convection and diffusion terms appearing in the transport equation for any passive scalar. A counterpart to the third term does not appear in these transport equations, however. Known as the *production* term, it is associated with the intensification of vorticity due to stretching of vortex lines. It is not a true production term, however, since it cannot produce vorticity where none exists. Indeed, since (10–5) contains $\boldsymbol{\omega}$ linearly in every term, it is clear that vorticity can be neither created nor destroyed in the interior of an isothermal, incompressible fluid: It can only be convected, diffused, or changed in magnitude once it is already present.*

The main value of the vorticity transport equation, in the present context, is that a direct analogy exists for two-dimensional motions between this equation and the thermal energy equation of Chapter 9. Specifically, for a two-dimensional flow,

$$\boldsymbol{\omega} = \omega \mathbf{k},$$

where $\mathbf{k}$ is a unit vector normal to the plane of flow, so that

$$\boldsymbol{\omega} \cdot \nabla u \equiv 0$$

---

*An important assumption inherent in the derivation of (10–5) is that $\rho = $ constant. If $\rho \neq$ constant, vorticity can be generated internally when $\mathbf{g} \wedge \nabla \rho \neq 0$. This is one reason why motions of density stratified fluids are often qualitatively different from the motion of a constant density fluid in the same domain.[6]

and the general equation (10–5) reduces to the form

$$\boxed{\mathbf{u}\cdot\nabla\omega = \frac{1}{\mathrm{Re}}\,\nabla^2\omega,} \tag{10–6}$$

where ω is the scalar magnitude of $\boldsymbol{\omega}$. If we compare (10–6) with the thermal energy equation (8–7), we see that the forms are identical, and Re plays the same role as Pe. In particular, in a two-dimensional motion, vorticity, like heat, is convected and diffused and it is produced only at the boundaries of the flow. The main differences between the thermal and vorticity transport problems are that θ and $\mathbf{u}$ are truly independent variables in the forced convection approximation, whereas ω and $\mathbf{u}$ are intimately connected via (10–3). In addition, the magnitude of the vorticity cannot be specified on the boundaries since $\mathbf{u}$ is already given, and it is almost never independent of position on the surface as we assumed for θ. Nevertheless, there is considerable similarity between heat transfer at high Pe and vorticity transport at high Re, which can be used to advantage now that we have solved the thermal problem for Pe $\gg 1$.

In particular, let us start with the nondimensionalized vorticity transport equation (10–6) and attempt to obtain an approximate solution for Re $\gg 1$. We expect an asymptotic expansion with Re^{-1} as the small parameter, but we restrict our attention here to the leading-order term in this expansion, which can be obtained by solving the limiting form of (10–6) for Re $\to \infty$, namely,

$$\mathbf{u}\cdot\nabla\omega = 0. \tag{10–7}$$

But this is precisely the same form as equation (9–3), which we have already shown to have the solution

$$\omega = \omega(\psi), \tag{10–8}$$

that is, the vorticity is constant along a streamline. Hence, if the flow is characterized by open streamlines, it follows that

$$\omega \equiv 0 \tag{10–9}$$

on any streamline that begins upstream in the uniform flow region (where $\omega \equiv 0$). Flows in which $\omega = 0$ are known as *irrotational*.

Although $\omega = 0$ is a valid, leading-order approximation for the solution of (10–6) as Re $\to \infty$, we cannot tell without further analysis if it corresponds to a uniformly valid solution of the original problem (10–1) and (10–2) for $\mathbf{u}$. For this, we need to determine the corresponding velocity fields. This is most easily done by expressing (10–9) in terms of $\mathbf{u}$,

$$\nabla \wedge \mathbf{u} = 0. \tag{10–10}$$

Then, introducing the streamfunction, via the usual definition (see Chapter 4, Section C),

$$\mathbf{u} \equiv \nabla \wedge (\psi\mathbf{k}), \tag{10–11}$$

and combining (10–10) and (10–11), we find that

$$\nabla^2\psi = 0. \tag{10–12}$$

Thus, we see that ψ in a two-dimensional, irrotational flow satisfies Laplace's equation. Such motions are therefore also known as *potential flows*.

If we compare (10–12) with the full two-dimensional Navier-Stokes equation, expressed in terms of the streamfunction, we note that the latter is fourth order (the viscous terms generate $\nabla^4\psi$), whereas (10–12) is only second order. As a result, it is clear that the velocity field obtained from (10–12) will, at most, be able to satisfy only one of the boundary conditions of the original problem at the body surface. Intuitively, we may anticipate that the kinematic condition on the normal component of velocity,

$$\mathbf{u}\cdot\mathbf{n} = 0 \quad \text{on} \quad S, \tag{10–13}$$

is the more crucial of the two original boundary conditions (the second being the no-slip condition) because it insures that the predicted flow field moves *around* the body and not through it. From a physical viewpoint, no-slip at the wall can be established only when the fluid viscosity is nonzero. Hence, for an inviscid fluid, we also should expect that the no-slip condition must be abandoned. These intuitive arguments are, in fact, correct. It is generally possible to solve (10–12) subject to the kinematic condition (10–13), but it is *not* possible to simultaneously satisfy the no-slip condition. Thus, the approximate solution (10–9) is clearly not valid in the immediate vicinity of the body surface.

In spite of this, we shall see that *potential flow theory* plays an important part in the development of asymptotic solutions for Re $\gg$ 1. Indeed, if we compare the assumptions and analysis leading to (10–9) and then to (10–12), with the early steps in analysis of heat transfer at high Peclet number, it is clear that the solution $\omega = 0$ is a valid first approximation for Re $\gg$ 1 everywhere except in the immediate vicinity of the body surface. There, the body dimension, a, that was used to nondimensionalize (10–1) is not a relevant characteristic length scale. In this region, we shall see that the flow develops a boundary-layer in which viscous forces remain important even as Re $\rightarrow \infty$, and this allows the no-slip condition to be satisfied.

Before discussing the flow structure in the boundary-layer region, we digress briefly to say a few words about the historical development of boundary-layer theory. The early years of theoretical studies of high Reynolds number flows were, in fact, dominated by attempts to obtain solutions for the so-called potential-flow problems satisfying (10–12) and (10–13).[7] The reader may well wonder how early researchers could overlook the obvious flaw of not satisfying the no-slip condition. However, before being too critical, it should be remembered that the no-slip condition is not a consequence of a general conservation principle such as Cauchy's equations of motion but is similar to constitutive equations in representing a mathematical hypothesis consistent with observed behavior.[8] To early workers in fluid mechanics, the necessity of satisfying the no-slip condition was not so obvious, especially for very large values of the Reynolds number (or, equivalently, small viscosities). Indications that the potential-flow theory was incomplete, even as an approximation for Re $\rightarrow \infty$, were mainly macroscopic and based upon comparisons between experimental observations and predictions of the potential-flow theory. The most dramatic evidence of a fundamental flaw in the theory was known as D'Alembert's Paradox. This paradox was simply that the drag on a two-dimensional body of arbitrary shape was predicted to be zero by the potential-flow theory but was found to increase monotonically with Re in experimental observations.

A second, equally dramatic difference between the potential-flow theory and experimental observation was that the flow patterns were often completely different. In the case of streaming flow past a circular cylinder, for example, the potential-flow solution is fore-aft symmetric with no indication of a wake downstream of the body. To show this we simply solve the potential-flow equation in cylindrical coordinates,

$$\frac{1}{r}\frac{\partial}{\partial r}\left(r\frac{\partial\psi}{\partial r}\right) + \frac{1}{r^2}\frac{\partial^2\psi}{\partial\theta^2} = 0, \tag{10-14}$$

subject to the boundary conditions

$$\psi \rightarrow r\sin\theta \quad \text{as} \quad r \rightarrow \infty \tag{10-15}$$

and

$$\psi = 0 \quad \text{at} \quad r = 1. \tag{10-16}$$

The condition (10-16) is just the kinematic condition (10-13) expressed in terms of the streamfunction, while (10-15) requires that the velocity field approach a uniform streaming motion at large distances from the cylinder. Equation (10-14), subject to (10-15) and (10-16), is solved easily via separation of variables, or other standard transform methods, with the resulting solution

$$\boxed{\psi = r\sin\theta\left(1 - \frac{1}{r^2}\right).} \tag{10-17}$$

This solution is clearly symmetric with respect to the fore-aft (upstream and downstream) directions. A sketch of streamlines (10-17) is shown in Figure 10-1. For comparison we show streakline photographs of flow past a cylinder at two values of Reynolds number, Re = 13.1 and 26, in Figure 10-2. The most obvious difference

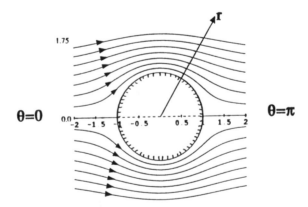

$\theta=0$ $\theta=\pi$

Figure 10-1 Streamlines for uniform streaming flow past a circular cylinder in the potential flow limit, according to (10-17). Contours shown range from $\psi = 0$ to $\psi = 1.75$ in increments of 0.25.

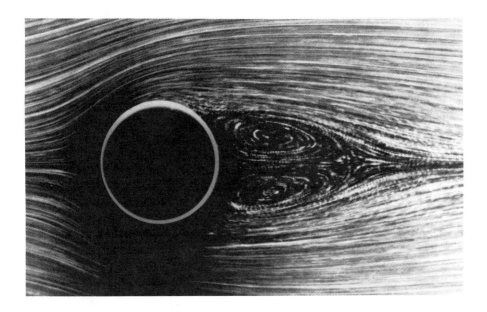

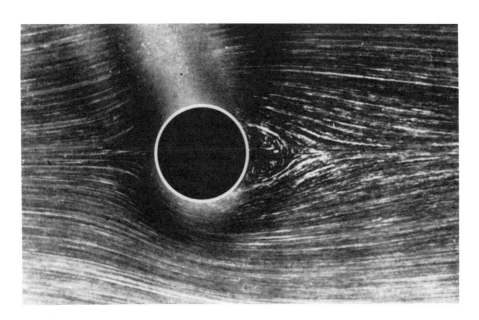

Figure 10–2 Streakline visualization of uniform streaming flow of a Newtonian fluid past a circular cylinder at Reynolds numbers R = 13.1 and R = 26. These photos were taken using a long exposure time to record the pathways of very small tracer particles in the flow. (Source: Van Dyke, M.D., *An Album of Fluid Motion*. Parabolic Press: Stanford, 1982; original photos were taken by S. Taneda. The photo depicting R = 26 was originally published by Taneda in 1956 in the *Journal of the Physical Society of Japan*, 11, 302–307.)

between the photographs and the predicted streamlines is the asymmetric flow pattern, with the large region of recirculation immediately to the rear of the cylinder. Of course, the theory is a limiting result for Re $= \infty$, whereas the photographs are for *finite* Reynolds number. The main reason for the moderate experimental Reynolds number is, in fact, that the flow pictured in Figure 10–2 becomes unsteady at somewhat larger Reynolds numbers, and there is a complex series of transitions until eventually the motion becomes highly turbulent. Thus, it is not necessarily true that the photographs are indicative of what the laminar flow would look like if we could take the limit Re $\rightarrow \infty$ without encountering instabilities. Nevertheless, there is nothing to suggest that the flow pattern would revert to one resembling (10–17). In fact, the unsteady flows encountered experimentally at large Reynolds numbers continue to show a strong degree of asymmetry and a clear remnant of the closed-streamline pattern in the form of a downstream pattern of vortices that is known as the Karman vortex street. All this input suggests strongly that the potential-flow pattern, (10–17) and Figure 10–1, is highly unrealistic for Re $\gg 1$.

One way to think of the difference between (10–17) and the actual velocity field near a cylinder for Re $\gg 1$, which allows us to take advantage of the close analogy to the heat transfer problem of Section **A** in Chapter 9, is that the potential-flow solution lacks the vorticity characteristic of the motions of real fluids. It is particularly instructive to think of the transient physical process of establishing a steady-state velocity (or vorticity) field, beginning with the fluid initially at rest relative to the body and then suddenly imposing a uniform motion on the fluid at $t = 0$. Initially, $\omega \equiv 0$ everywhere, and we have already seen that vorticity is not created within the fluid, so that the motion must remain irrotational [and precisely of the form (10–17) if the body is a circular cylinder] unless vorticity is produced at the bounding surfaces of the fluid and transported into the fluid via diffusion or convection according to (10–6). The mechanism for vorticity production at a *solid* bounding surface is the rotational motion imparted by the no-slip condition. Initially, in a real, viscous fluid, the motion will be very close to the irrotational potential flow except in the immediate vicinity of the body surface where the no-slip condition produces a large gradient in the tangential velocity component.[9] If we think in terms of vorticity, this corresponds to $\omega = 0$ almost everywhere, but with a very strong source at the body surface. In the real fluid, this vorticity is transported radially outward into the fluid by diffusion, until eventually a steady-state configuration is achieved when production is balanced by convection and diffusion. From this viewpoint, it is obvious that the deficiency in the approximation (10–7)—or, equivalently, the potential-flow solution—is that it neglects vorticity diffusion completely and thus allows no mechanism for vorticity to ''escape'' from the body surface. As a consequence, $\omega \equiv 0$ everywhere in the fluid, according to (10–9).

The conclusion to be drawn from the preceding discussion is that the potential-flow theory (10–9) [or, equivalently, (10–12) and (10–13)] does *not* provide a uniformly valid first approximation to the solution of the Navier-Stokes and continuity equations (10–1) and (10–2) for Re $\gg 1$. Furthermore, our experience in Chapter 9 with the thermal boundary-layer structure for large Peclet number would lead us to believe that this is because the velocity field near the body surface is characterized by a length scale $O(a \mathrm{Re}^{-n})$, instead of the body dimension a that was used to nondimensionalize (10–1). As a consequence, the terms $\nabla^2\omega$ and $\mathbf{u}\cdot\nabla\omega$ in (10–6), which are nondimensionalized using a, are not $O(1)$ and independent of Re everywhere in the domain, as was assumed in deriving (10–7), but instead are increasing functions of Re in the region very close to the body surface. Thus, in this region it is *not* true that $(1/\text{Re})\nabla^2\omega$ becomes negligible compared with $\mathbf{u}\cdot\nabla\omega$ in the limit Re $\rightarrow \infty$, as equation (10–6) suggests.

We see then that (10–7)—and potential-flow theory—provides a leading-order approximation for Re $\gg 1$ in an outer region away from the body but is *not* valid very near the body surface. *Physically, in order to retain the essential effects of viscosity in the vicinity of the body, there is an internal readjustment in the flow that produces very large gradients in ω (or $\mathbf{u}$) near the surface.* This is not reflected in equations (10–1), (10–2), or (10–6), which are scaled with respect to the length scale of the body. In order to determine the form of the Navier-Stokes equation that holds near the body, we must rescale (10–1) and (10–2), or (10–6), to introduce a characteristic length scale that is appropriate to this inner boundary-layer region, and then employ the method of matched asymptotic expansions to obtain a uniformly valid approximation to the solution for Re $\gg 1$.

B. The Boundary-Layer Equations

The discussion in the previous section leads to the following qualitative description of the flow domain for Re $\gg 1$:

1. *Outer region*, where variations of velocity are characterized by the length scale a of the body and *potential-flow theory* provides a valid first approximation in an asymptotic expansion of the solution for Re $\rightarrow \infty$.
2. *Inner region*, a boundary layer, of thickness $O(a\text{Re}^{-\alpha})$ near the body surface (with α yet to be determined), where *viscous effects* must be included, even in the limit Re $\rightarrow \infty$.

Since this inner boundary-layer region is infinitesimal in thickness relative to a, all curvature terms that appear when the equations of motion are expressed in curvilinear coordinates will drop out to first order in Re^{-1}, thus leaving boundary-layer equations that are effectively expressed in terms of a local, Cartesian coordinate system.

To illustrate this latter point, we first derive equations for the inner boundary-layer region for the specific problem of streaming flow past a circular cylinder, starting from the equations of motion expressed in a cylindrical coordinate system. These are

$$\left(u_r \frac{\partial u_\theta}{\partial r} + \frac{u_\theta}{r}\frac{\partial u_\theta}{\partial \theta} + \frac{u_r u_\theta}{r} \right) = -\frac{1}{r}\frac{\partial p}{\partial \theta} + \frac{1}{\text{Re}}\left[\frac{\partial^2 u_\theta}{\partial r^2} + \frac{\partial}{\partial r}\left(\frac{u_\theta}{r}\right) + \frac{1}{r^2}\frac{\partial^2 u_\theta}{\partial \theta^2} + \frac{2}{r^2}\frac{\partial u_r}{\partial \theta} \right],$$

$$(10\text{–}18)$$

$$\left(u_r \frac{\partial u_r}{\partial r} + \frac{u_\theta}{r}\frac{\partial u_r}{\partial \theta} - \frac{u_\theta^2}{r} \right) = -\frac{\partial p}{\partial r} + \frac{1}{\text{Re}}\left[\frac{\partial^2 u_r}{\partial r^2} + \frac{\partial}{\partial r}\left(\frac{u_r}{r}\right) + \frac{1}{r^2}\frac{\partial^2 u_r}{\partial \theta^2} - \frac{2}{r^2}\frac{\partial u_\theta}{\partial \theta} \right],$$

$$(10\text{–}19)$$

and

$$\frac{1}{r}\frac{\partial}{\partial r}(ru_r) + \frac{1}{r}\frac{\partial u_\theta}{\partial \theta} = 0. \tag{10-20}$$

For convenience, we change from r to y,

$$r - 1 \equiv y,$$

so that y is a radial variable (normal to the body surface) that is *zero* at the surface, and we also change from θ to x,

$$\theta \equiv x.$$

Then rewriting (10–18)–(10–20), we have

$$\left(v\frac{\partial u}{\partial y} + \frac{u}{1+y}\frac{\partial u}{\partial x} + \frac{uv}{1+y}\right) =$$

$$-\frac{1}{1+y}\frac{\partial p}{\partial x} + \frac{1}{\mathrm{Re}}\left[\frac{\partial^2 u}{\partial y^2} + \frac{\partial}{\partial y}\left(\frac{u}{1+y}\right) + \frac{1}{(1+y)^2}\frac{\partial^2 u}{\partial x^2} + \frac{2}{(1+y)^2}\frac{\partial v}{\partial x}\right], \tag{10-21}$$

$$\left(v\frac{\partial v}{\partial y} + \frac{u}{1+y}\frac{\partial v}{\partial x} - \frac{u^2}{1+y}\right) =$$

$$-\frac{\partial p}{\partial y} + \frac{1}{\mathrm{Re}}\left[\frac{\partial^2 v}{\partial y^2} + \frac{\partial}{\partial y}\left(\frac{v}{1+y}\right) + \frac{1}{(1+y)^2}\frac{\partial^2 v}{\partial x^2} - \frac{2}{(1+y)^2}\frac{\partial u}{\partial x}\right], \tag{10-22}$$

and

$$\frac{1}{1+y}\frac{\partial}{\partial y}\left[(1+y)v\right] + \frac{1}{1+y}\frac{\partial u}{\partial x} = 0, \tag{10-23}$$

where we have denoted $u_r \equiv v$ and $u_\theta \equiv u$ in anticipation of a reduction, locally, to a Cartesian form for these equations in the boundary-layer.

To obtain the appropriate form of the equations of motion for the inner, boundary-layer region, we must rescale with a new characteristic length scale. Following the precedent of previous analyses, we do this by rescaling the radial variable y in the equations of motion and continuity, (10–21)–(10–23), according to

$$Y = \mathrm{Re}^\alpha y, \tag{10-24}$$

since it is the direction normal to the surface that is expected to exhibit a length scale much smaller than a and dependent on Re. If we introduce (10–24) into the continuity equation (10–23), we find that

$$\frac{1}{1+\dfrac{Y}{\mathrm{Re}^\alpha}}\mathrm{Re}^\alpha\frac{\partial}{\partial Y}\left[\left(1+\frac{Y}{\mathrm{Re}^\alpha}\right)v\right] + \frac{1}{1+\dfrac{Y}{\mathrm{Re}^\alpha}}\frac{\partial u}{\partial x} = 0. \tag{10-25}$$

Thus, approximating $1 + Y/\mathrm{Re}^\alpha$ as 1, as is appropriate at the leading order of approximation, we have

$$\mathrm{Re}^\alpha \frac{\partial v}{\partial Y} + \frac{\partial u}{\partial x} = 0. \tag{10-26}$$

Clearly, the magnitude of either u or v must depend on Re in this region if the form of the continuity equation is to be preserved in the limit $\mathrm{Re} \to \infty$, and this condition is necessary to insure that the boundary-layer solution is consistent with conservation of mass. The tangential velocity component u will be required to *match* the tangential velocity component from the potential-flow solution in the region of overlapping validity between the inner (boundary-layer) and outer (potential-flow) regions. But the potential-flow solution, say, (10–17) for the circular cylinder, clearly yields a tangential velocity in the vicinity of the body that is $0(1)$, and this implies that $u = 0(1)$ in the inner boundary-layer domain if the two solutions are to match. The normal velocity component, on the other hand, approaches zero in the potential-flow approximation as the body surface is approached. Thus, a normal velocity component of $0(\mathrm{Re}^{-\beta})$ in the boundary-layer region can be matched to within an arbitrarily small error for $\mathrm{Re} \to \infty$ with the potential-flow solution. This suggests that the rescaling

$$\boxed{V = \mathrm{Re}^\beta v} \tag{10-27}$$

is allowable, and we see immediately from (10–26) that $\alpha = \beta$ so that

$$\boxed{\frac{\partial u}{\partial x} + \frac{\partial V}{\partial Y} = 0} \tag{10-28}$$

in terms of the rescaled inner variables Y and V. The reader should recall that α is to be chosen so that derivatives with respect to Y are $0(1)$—independent of Re—in this inner region. The dimension of the inner region is $0(a\mathrm{Re}^{-\alpha})$ according to (10–24), and provided that $\alpha > 0$, as we have assumed, this is extremely small for $\mathrm{Re} \gg 1$. Accordingly, it is not surprising that the continuity equation reduces to a local Cartesian form (10–28). If we go back to equation (10–25), we see that the "curvature" terms are $0(\mathrm{Re}^{-\alpha})$ compared with the terms that we have retained. It must be remembered, however, that equation (10–28) governs only the first term in an asymptotic expansion of the full solution for the inner region.* If we wished to calculate higher order contributions to this expansion, we would have to retain the curvature terms in equation (10–25) as they would appear in the governing equations at $0(\mathrm{Re}^{-\alpha})$. In the present analysis, we will be content with obtaining only the first term in both the inner and outer regions (the latter being the potential-flow solution).

To determine α and the appropriate form of the equations of motion, (10–21) and (10–22), in the inner region, we substitute the rescaled variables (10–24) and (10–27) into these equations. The result for the tangential component (10–21) is

---

*It was obtained from equation (10–25) by taking the limit $\mathrm{Re} \to \infty$.

$$\left(V\frac{\partial u}{\partial Y} + \frac{u}{\left(1+\dfrac{Y}{\mathrm{Re}^\alpha}\right)}\frac{\partial u}{\partial x} + \frac{1}{\mathrm{Re}^\alpha}\frac{uV}{\left(1+\dfrac{Y}{\mathrm{Re}^\alpha}\right)} \right) =$$

$$= -\frac{1}{1+\dfrac{Y}{\mathrm{Re}^\alpha}}\frac{\partial p}{\partial x} + \mathrm{Re}^{2\alpha-1}\frac{\partial^2 u}{\partial Y^2} + \frac{\mathrm{Re}^\alpha}{\mathrm{Re}}\frac{\partial}{\partial Y}\left(\frac{u}{1+\dfrac{Y}{\mathrm{Re}^\alpha}}\right) +$$

$$\frac{1}{\mathrm{Re}\left(1+\dfrac{Y}{\mathrm{Re}^\alpha}\right)^2}\left\{\frac{\partial^2 u}{\partial x^2} + \frac{2}{\mathrm{Re}^\alpha}\frac{\partial V}{\partial x}\right\}. \qquad (10\text{--}29)$$

The governing equation for the first approximation to u and V can be obtained by taking the limit $\mathrm{Re} \to \infty$,

$$u\frac{\partial u}{\partial x} + V\frac{\partial u}{\partial Y} + 0(\mathrm{Re}^{-\alpha}) = -\frac{\partial p}{\partial x} + \mathrm{Re}^{2\alpha-1}\left[\frac{\partial^2 u}{\partial Y^2} + 0(\mathrm{Re}^{-\alpha})\right] + 0(\mathrm{Re}^{-1}). \quad (10\text{--}30)$$

Hence, the largest viscous term is $0(\mathrm{Re}^{2\alpha-1})$, and it is clear that

$$\alpha = \frac{1}{2} \qquad (10\text{--}31)$$

if viscous effects are to be retained in the limit $\mathrm{Re} \to \infty$. It follows from (10–31) that the dimensionless boundary-layer thickness is $0(\mathrm{Re}^{-1/2})$.

The reader may note that we have thus far considered only the tangential component of the equations of motion. To see whether any additional information can be obtained from the normal component equation (10–22), we substitute (10–24) and (10–27) with $\beta = \alpha = 1/2$. The result is

$$\frac{\partial p}{\partial Y} = 0(\mathrm{Re}^{-1/2}), \qquad (10\text{--}32)$$

from which we conclude that the pressure is constant across the boundary-layer to a first approximation; that is, p is a function of x only. Equations (10–28), (10–30), and (10–32) are sufficient to determine the three unknowns in the boundary-layer region, u, V, and p. We shall return shortly to the method of solution.

First, however, it is important to recognize that the *form* of equations (10–28), (10–30), and (10–32) is *independent* of the geometry of the body. Although we started our analysis with the specific problem of flow past a circular cylinder, and thus with the equations of motion in cylindrical coordinates, the equations for the leading-order approximation in the inner (boundary-layer) region reduce to a local, Cartesian form with Y being normal to the body surface and x tangential. The geometry of the body does not directly enter these so-called boundary-layer equations at all. Indeed, we shall see shortly that it is only through the requirement of matching with the outer potential-flow solution that the boundary-layer solution depends on the body geometry.

Let us then consider an *arbitrary* two-dimensional body. In all cases, the governing equations at leading order in the boundary-layer are

$$\frac{\partial u}{\partial x} + \frac{\partial V}{\partial Y} = 0, \qquad (10\text{--}33)$$

and

$$u\frac{\partial u}{\partial x} + V\frac{\partial u}{\partial Y} = -\frac{\partial p}{\partial x} + \frac{\partial^2 u}{\partial Y^2}, \qquad (10\text{--}34)$$

$$\frac{\partial p}{\partial Y} = 0. \qquad (10\text{--}35)$$

These are known as the *boundary-layer equations*. There is one further simplification that we can always introduce. In particular, if we integrate (10–35), we see that the pressure distribution in the boundary-layer is a function of x only. Thus, the pressure gradient in the boundary-layer ($\partial p/\partial x$) is also independent of Y and must have the same form as the pressure gradient in the outer potential flow, evaluated in the limit as we approach the body surface, namely,

$$\left.\frac{\partial p}{\partial x}\right|_{\text{boundary layer}} = \lim_{y \to 0}\left(\frac{\partial p}{\partial x}\right)_{\text{potential flow}}. \qquad (10\text{--}36)$$

In other words, the pressure distribution in the boundary-layer is completely determined at this level of approximation by the limiting form of the pressure distribution impressed at its outer edge by the potential flow. It is convenient to express this distribution in terms of the potential-flow velocity distribution. In particular, let us define the tangential velocity function $u_e(x)$ as

$$u_e(x) \equiv (\mathbf{u} \cdot \mathbf{t})|_{y \to 0}, \qquad (10\text{--}37)$$

where $y = 0$ denotes the body surface and $\mathbf{t}$ is a unit tangent vector (that is, $\mathbf{n} \cdot \mathbf{t} \equiv 0$). In terms of the velocity and pressure, the leading-order equation of motion in the potential-flow region is

$$\mathbf{u} \cdot \nabla \mathbf{u} = -\nabla p. \qquad (10\text{--}38)$$

But the normal velocity component is zero at the surface of the body, and (10–38) is reduced in the limit $y \to 0$ to an equation for the pressure gradient that is required in (10–36),

$$\left.\frac{\partial p}{\partial x}\right|_{y \to 0} = -u_e\frac{du_e}{dx}. \qquad (10\text{--}39)$$

This is often referred to as *Bernoulli's equation*. Once the potential-flow velocity field is known, it provides a means to determine $\partial p/\partial x$ at the outer edge of the boundary-layer region and hence, via (10–36), the pressure gradient $\partial p/\partial x$ in the boundary-layer itself.

Thus, insofar as the boundary-layer is concerned, the pressure gradient may be regarded as "known"—it is actually obtained by solving the outer potential-flow problem—and the boundary-layer equations can be written as

$$u \frac{\partial u}{\partial x} + V \frac{\partial u}{\partial Y} = u_e(x) \frac{du_e}{dx} + \frac{\partial^2 u}{\partial Y^2}$$ (10–40)

and

$$\frac{\partial u}{\partial x} + \frac{\partial V}{\partial Y} = 0,$$ (10–41)

with u and V being unknown. In order to demonstrate that these equations and the potential-flow equation (10–12) actually provide a basis to determine a uniformly valid approximation to the solution of the equations of motion for $\text{Re} \to \infty$, we must consider boundary conditions and matching conditions for the solutions in the two regions.

The boundary conditions are straightforward. The outer, potential-flow solution is required to take the form of the undisturbed velocity field at infinity; that is, in the case of uniform streaming motion,

$$\mathbf{u} \to \mathbf{e} \quad \text{as} \quad |\mathbf{r}| \to \infty,$$ (10–42)

where $\mathbf{e}$ is a unit vector in the direction of the free-stream motion. At the body surface, the no-slip and kinematic conditions give boundary conditions for the inner fields:

$$u = V = 0 \quad \text{at} \quad Y = 0.$$ (10–43)

However, the conditions (10–42) and (10–43) are *not* sufficient to obtain unique solutions in either the inner or outer domains. The boundary-layer equations are third order with respect to Y and first order with respect to x [this can be seen clearly if (10–40) and (10–41) are expressed in terms of the streamfunction]. Hence, to specify completely the velocity profiles in the boundary-layer, we require one additional boundary condition in Y and an initial profile at the leading edge of the boundary-layer (x is usually defined so that this point corresponds to $x = 0$). In addition, the potential-flow equations are second order and thus require at least one boundary condition in addition to (10–42). In Section **A**, it was suggested that an appropriate condition was

$$\mathbf{u} \cdot \mathbf{n} = 0 \quad \text{on the body surface, } S_B.$$ (10–44)

However, this was predicated on the assumption that the potential-flow solution was to be applied in the whole of the domain, whereas we now see that it should apply only up to a distance of $0(\text{Re}^{-1/2})$ from S_B, but not right at the surface itself. In order to determine the additional conditions that must be imposed to obtain a well-posed problem for the velocity fields in the inner and outer domains, we must examine the *matching* conditions at the region of overlap between them. These matching conditions can be written in general form as

$$\lim_{y \ll 1} u(x, y) \Longleftrightarrow \lim_{Y \gg 1} u(x, Y) \quad \text{as} \quad \text{Re} \to \infty$$ (10–45)

and

$$\lim_{y \ll 1} v(x, y) \Longleftrightarrow \lim_{Y \gg 1} \frac{1}{\sqrt{\text{Re}}} V(x, Y) \quad \text{as} \quad \text{Re} \to \infty,$$ (10–46)

where $y = 0$ is still used to denote the body surface in outer variables. The condition (10–46) includes $\sqrt{Re}$ so that both sides can be expressed in equivalent terms according to (10–27) with $\beta = 1/2$.

Let us now examine the matching conditions in more detail. Turning first to (10–46), we see that the right-hand side is arbitrarily small, $0(Re^{-1/2})$ for $Re \to \infty$, relative to the left-hand side, provided only that $V(x, \infty)$ is finite. Thus, subject to a posteriori verification of this latter fact, we see that the matching condition (10–46) is precisely equivalent to the boundary condition (10–44), which can thus be used for the normal velocity component in the outer solution. But this condition, together with (10–42), is sufficient to completely determine the outer-flow solution, which we see is just the potential-flow solution of the previous section. The only difference is that (10–44) is now seen as a *matching* condition, valid only as a leading-order approximation that is applied at the outer edge of the boundary-layer, rather than an exact condition that is applied right at the body surface, as before.

Let us suppose that we have solved the potential flow problem. Then the left-hand side of (10–45) becomes a *known* function of x, which we have denoted previously as $u_e(x)$, and this matching condition becomes a boundary condition for the boundary-layer solution

$$\lim_{Y \gg 1} u(x, Y) = u_e(x). \tag{10–47}$$

Together with (10–43) and an initial profile at $x = 0$, this condition is sufficient to completely specify the solution of (10–40) and (10–41).

It would thus appear that the potential-flow and boundary-layer equations (10–12), (10–40), and (10–41), plus the boundary conditions (10–42) and (10–43) and the matching conditions (10–44) and (10–47), provide a self-contained basis to determine the leading-order approximation to the velocity and pressure fields for steady, laminar streaming flows past two-dimensional bodies of arbitrary shape. First, we solve the potential-flow problem using (10–42) and (10–44). Then, with $u_e(x)$ known, we calculate the pressure gradient in the boundary-layer using (10–39) and solve the boundary-layer equations (10–40) and (10–41) with the boundary conditions (10–43) and (10–47), plus an initial condition at $x = 0$. This prescription works for a wide variety of *streamlined* body shapes. However, there is one intrinsic approximation in the formulation that can invalidate the procedure (or at least decrease its domain of validity) for bodies of *nonstreamlined* shape (such as the circular cylinder that we originally began to analyze). This is the assumption, inherent in (10–44) and the corresponding matching condition (10–46), that all regions of nonzero vorticity remain confined to a very thin layer, with a thickness of $0(Re^{-1/2})$, immediately adjacent to the body surface. For the horizontal flat plate and streamlined airfoil shapes, which we shall consider in the next section, this assumption is actually satisfied. However, assuming that the photographs in Figure 10–2 for large, but finite, Reynolds number are even qualitatively indicative of the asymptotic flow structure for $Re \to \infty$, it is patently obvious that the assumption is drastically violated for the circular cylinder (and all nonstreamlined bodies) due to the large, recirculating region of flow immediately downstream. In such cases, a boundary layer still exists over the front part of the body, but at some point (called the *separation point*), this boundary-layer departs radically from the surface, leading to the formation of the recirculating region downstream.

The reader may wonder why this change in flow structure invalidates the boundary-layer theory. Upstream of the separation point, the flow still divides into a boundary-layer of the usual kind on the surface and an outer, potential flow. However, the potential-flow approximation applies only outside any region of nonzero vorticity. When these regions all remain attached to the body and are of $0(Re^{-1/2})$ in thickness, it is adequate for at least the first term in an asymptotic solution to approximate this region as encompassing the whole of the domain external to the body itself. When the boundary-layer separates, however, it almost always leads to the formation of a large region of recirculating flow downstream in which the vorticity is nonzero. Thus, in these cases, the potential-flow solution should be applied to the region outside both the body and this attached recirculating wake, with (10–44) applied to this "composite" body rather than the solid body alone. Furthermore, since the potential-flow equations are elliptic in type, we should expect the solution everywhere in the domain to be influenced by the modification in boundary shape to include the recirculating wake. Thus, in order to obtain the pressure distribution in the boundary-layer prior to the separation point, it is necessary, in principle, to solve the potential-flow problem including any recirculating flow regions with dimensions of $0(1)$ as part of the "effective" body. But now we encounter a severe difficulty. The shape of the recirculating wake region is a priori unknown, and it can be calculated only in the context of a complete asymptotic theory in the whole domain—appropriate scaling and governing equations in the recirculating flow region and any boundary-layers adjacent to this region, plus, in all likelihood, a wake region downstream of the recirculating zone. At present, and in spite of intensive effort for many years, such a theory does not exist.[10] One problem is that the wake region becomes unstable at quite modest Reynolds numbers [$\sim 0(40)$ for the circular cylinder], and, thus, experimental studies do not provide a clear indication of the scale of the recirculating region or the magnitude of velocities within it in a range of Reynolds number that can be extrapolated to $Re \sim \infty$. In the absence of this kind of input, it has been extremely difficult to develop an asymptotic theory for the recirculating wake. Recently, numerical solutions have been obtained for steady flow past a circular cylinder up to $Re \sim 300$.[11] These appear to be suggestive of the asymptotic behavior for $Re \to \infty$ and may eventually lead to a satisfactory asymptotic solution for laminar, separated flows at $Re \gg 1$. For the moment, however, the available proposals for the flow structure in the limit $Re \to \infty$ remain controversial, and the steady, separated-flow problem is still regarded as one of the two most important unresolved problems in fluid mechanics (the other being a satisfactory theoretical description of turbulence).

What is amazing about boundary-layer theory is that it still has been incredibly useful for precisely the class of problems where separation occurs. While it does *not* provide a uniformly valid description of velocity and pressure fields in such cases, it can be used to determine when (that is, for what body geometries) and where the boundary-layer will separate. Beyond the separation point, the boundary-layer assumptions breakdown. For example, the dimension of the region of nonzero vorticity is no longer $0(Re^{-1/2})$. Hence, a prediction of separation is tantamount to the statement that the boundary-layer theory is capable of providing an accurate prediction of the point on the body surface where it becomes invalid! This prediction is obtained, as we shall see, by calculating the pressure distribution for potential flow around the body *without* taking into account the possibility of a closed streamline wake and using this pressure distribution in the boundary-layer equations. The existence of a separation point is signaled if $\partial u/\partial Y|_{Y=0} \to 0$ at some point on the body surface, and when this occurs, the boundary-layer solution develops a singularity.[12] This scheme, though clearly an ad hoc application of the asymptotic, boundary-layer approach, works because the *actual* pressure distribution in the presence of the recirculating wake and the predicted distribution from potential flow theory in absence of a wake are reasonably similar almost up to the point of actual separation. This is illustrated in Figure 10–3, where we show a comparison between an experimental pressure distribution on a circular cylinder for relatively small Reynolds numbers, $30 \le \text{Re} \le 40$, and the predicted pressure distribution from potential-flow theory. The ability of the boundary-layer theory to predict separation is probably its most important characteristic.

We have stated repeatedly that the boundary-layer and potential-flow equations apply to only the leading term in an asymptotic expansion of the solution for $\text{Re} \gg 1$.

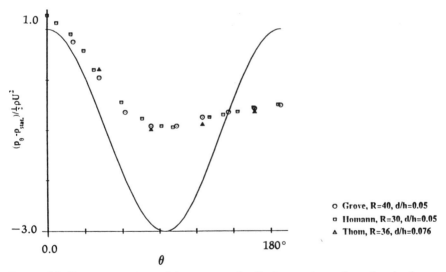

o Grove, R=40, d/h=0.05
□ Homann, R=30, d/h=0.05
▲ Thom, R=36, d/h=0.076

Figure 10–3 A comparison of the pressure distribution on the surface of a circular cylinder according to potential-flow theory (---------) and as measured experimentally for $30 \le \text{Re} \le 40$. The experimental data is taken from Grove, A.S., et al., "Steady Separated Flow Past a Circular Cylinder," *J. Fluid Mech.* 19:60–80 (1964), where original references are given to Homann and Thom's papers. The offset in stagnation point pressure at $\theta - 0$ is a result of the finite Reynolds number, as shown in Grove et al. (1964). The difference in the pressure between $\theta = 0$ and the point of separation (that occurs at $\theta \sim 135°$ at this relatively low Reynolds number) is partly due to finite Reynolds number effects within the boundary-layer, and partly due to the influence of the recirculating wake on the external flow, which is felt significantly upstream of the separation point.

This is clear from the fact that both were derived in their respective domains of validity by simply taking the limit Re $\rightarrow \infty$ in the appropriately nondimensionalized Navier-Stokes equations. Frequently, in the analysis of laminar flow at high Reynolds number, we do not proceed beyond these leading-order approximations because they already contain the most important information: a prediction of whether or not the flow will separate and, if not, an analytic approximation for the drag. Nevertheless, the reader may be interested in how we would proceed to the next level of approximation, and this is described briefly in the remainder of this section.[13]

At the next level, we first calculate the second term in the outer solution. If we examine the equation of motion for the outer region,

$$\mathbf{u} \cdot \nabla \mathbf{u} + \nabla p = \frac{1}{\mathrm{Re}} \nabla^2 \mathbf{u}, \quad \nabla \cdot \mathbf{u} = 0, \tag{10-48}$$

we see that the first neglected term in the preceding analysis was $0(\mathrm{Re}^{-1})$. On the other hand, examination of the matching conditions (10–45) and (10–46) shows that there is a mismatch between the first term in the outer region that satisfies the boundary condition (10–44) and the boundary-layer solution that takes a value

$$\lim_{Y \gg 1} \frac{1}{\sqrt{\mathrm{Re}}} V(x, Y) = 0(\mathrm{Re}^{-1/2}) \tag{10-49}$$

at the outer edge of the boundary-layer. Taken together, (10–48) and (10–49) imply that the next term in the asymptotic expansion for the outer region must be $0(\mathrm{Re}^{-1/2})$, that is,

$$\mathbf{u} = \mathbf{u}_0 + \frac{1}{\sqrt{\mathrm{Re}}} \mathbf{u}_1 + o(\mathrm{Re}^{-1/2}) \tag{10-50}$$

and

$$p = p_0 + \frac{1}{\sqrt{\mathrm{Re}}} p_1 + o(\mathrm{Re}^{-1/2}). \tag{10-51}$$

To obtain the governing equations for the first and second terms in this expansion, we substitute (10–50) and (10–51) into (10–48):

$$\mathbf{u}_0 \cdot \nabla \mathbf{u}_0 + \nabla p_0 + \frac{1}{\sqrt{\mathrm{Re}}} (\mathbf{u}_0 \cdot \nabla \mathbf{u}_1 + \mathbf{u}_1 \cdot \nabla \mathbf{u}_0 + \nabla p_1) = 0(\mathrm{Re}^{-1}), \tag{10-52}$$

$$\nabla \cdot \mathbf{u}_0 + \frac{1}{\sqrt{\mathrm{Re}}} (\nabla \cdot \mathbf{u}_1) + o(\mathrm{Re}^{-1/2}) = 0. \tag{10-53}$$

Hence, at $0(1)$,

$$\mathbf{u}_0 \cdot \nabla \mathbf{u}_0 + \nabla p_0 = 0, \quad \nabla \cdot \mathbf{u}_0 = 0, \tag{10-54}$$

which, together with the boundary conditions (10–42) and (10–44), is just the potential-flow problem we discussed in Section **A** of this chapter. At $0(\mathrm{Re}^{-1/2})$,

$$\mathbf{u}_0 \cdot \nabla \mathbf{u}_1 + \mathbf{u}_1 \cdot \nabla \mathbf{u}_0 + \nabla p_1 = 0, \quad \nabla \cdot \mathbf{u}_1 = 0, \tag{10-55}$$

and, at this level of approximation, the boundary condition at infinity is

$$\mathbf{u}_1 \to 0 \quad \text{as} \quad |\mathbf{x}| \to \infty, \tag{10-56}$$

while the matching condition (10–46), together with (10–49), yields

$$\mathbf{u}_1 \cdot \mathbf{n} = V(x, \infty) \quad \text{at} \quad y = 0. \tag{10-57}$$

At this second level, the outer solution is modified to account for the weak outward displacement of the potential flow that is caused by the presence of the boundary-layer at the body surface. It can be seen in (10–57) that this displacement manifests itself in the outer flow as a weak "blowing" velocity normal to the body surface.

In the inner boundary-layer region, there are neglected terms of $0(\mathrm{Re}^{-1/2})$ in the first-order equation obtained from (10–29) by taking the limit $\mathrm{Re} \to \infty$. In addition, the $0(\mathrm{Re}^{-1/2})$ velocity field $\mathbf{u}_1$ in the outer region requires an $0(\mathrm{Re}^{-1/2})$ correction in the inner region due both to the matching condition (10–45) and to the modification of the pressure gradient imposed on the boundary-layer from the outer flow. To obtain governing equations for the higher order terms in the boundary-layer regime, it follows that we would again use an expansion of the form

$$u = u_0(x, Y) + \mathrm{Re}^{-1/2} u_1(x, Y) + o(\mathrm{Re}^{-1/2}),$$

$$V = V_0(x, Y) + \mathrm{Re}^{-1/2} V_1(x, Y) + o(\mathrm{Re}^{-1/2}) \tag{10-58}$$

in (10–29), with $\alpha = 1/2$.

C. Streaming Flow Past a Horizontal Flat Plate—The Blasius Solution

In this section we apply the general boundary-layer analysis of the previous sections to the problem of laminar streaming flow past a horizontal flat plate at high Reynolds number. We assume, at the outset, that the plate length is finite (denoted as L) and that the flow at infinity is parallel to the plate surface. A schematic representation of the flow configuration is given in Figure 10–4. In this case, we define the Reynolds number as $\mathrm{Re} \equiv U_\infty L/\nu$. Later, we shall see that the plate can be either finite or semi-infinite in length with no change in the solution structure.

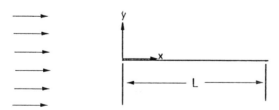

Figure 10–4 A schematic of the geometric configuration for streaming flow past a horizontal flat plate of finite length L.

Following the prescription for solving boundary-layer problems, outlined in the previous section, we begin with the outer potential flow problem. For this purpose, it is convenient to adopt a fixed system of Cartesian coordinates (x, y), with y normal to the body surface and x tangent to the surface with $x = 0$ denoting the upstream (or *leading*) edge of the plate. Then, the potential flow equation is just

$$\frac{\partial^2 \psi}{\partial x^2} + \frac{\partial^2 \psi}{\partial y^2} = 0, \tag{10-59}$$

which is to be solved subject to the free-stream boundary condition

$$\psi \to y \quad \text{as} \quad y \to \infty \tag{10-60}$$

and the matching condition (10–44) at $y = 0$, which becomes just

$$\psi = 0 \quad \text{at} \quad y = 0 \tag{10-61}$$

when expressed in terms of the streamfunction ψ. The solution of (10–59)–(10–61) is trivial, namely,

$$\boxed{\psi = y, \quad \text{for all } x \text{ and } y.} \tag{10-62}$$

This solution simply implies that the velocity everywhere in the outer region is just the undisturbed free-stream velocity. The flat plate is idealized mathematically as having zero thickness, and since the no-slip condition does not apply to the potential flow regime, the fluid moves past the plate with no disturbance whatsoever.

The next step is to solve the boundary-layer equations (10–40) and (10–41) with $\partial p / \partial x$ calculated from (10–39). But in this case,

$$u_e = 1,$$

and thus

$$\partial p / \partial x = 0 \tag{10-63}$$

to leading order in the boundary-layer. It follows that the boundary-layer problem reduces to

$$\boxed{u \frac{\partial u}{\partial x} + V \frac{\partial u}{\partial Y} = \frac{\partial^2 u}{\partial Y^2}, \quad \frac{\partial u}{\partial x} + \frac{\partial V}{\partial Y} = 0,} \tag{10-64}$$

with boundary conditions, from (10–43) and (10–47),

$$u = V = 0 \quad \text{at} \quad Y = 0 \quad x > 0, \tag{10-65}$$

and an initial profile

$$u \to 1 \quad \text{as} \quad Y \to \infty \quad x > 0, \tag{10-66}$$
$$u = 1 \quad \text{at} \quad x = 0. \tag{10-67}$$

The latter condition corresponds to the physical fact that the fluid is not disturbed at all by the plate until it actually encounters the no-slip condition for $x \geq 0$.

The problem (10–64)–(10–67) was first solved by Blasius in his doctoral thesis (1908) using the similarity transformation[14]

$$\boxed{u \equiv f'(\eta) \quad \text{with} \quad \eta \equiv \frac{Y}{g(x)}.} \tag{10-68}$$

It is convenient to define f so that $u = \partial f/\partial\eta$, since otherwise V must involve an integral of f. As it is, we see from the continuity equation and the boundary condition on V at $Y=0$ that

$$V = -\int_0^Y \frac{\partial u}{\partial x}\, dY = g'(x)\int_0^\eta t f''(t)\, dt.$$

Integrating by parts,

$$\boxed{V \equiv g'(x)\left[\eta f' - f\right],} \tag{10-69}$$

where $f(0) = 0$ so that $V=0$ at $Y=\eta=0$ and $f'(0)=0$ so that $u=0$ at $\eta=0$. Substituting (10–68) and (10–69) into (10–64) yields

$$\boxed{f''' + gg'ff'' = 0.} \tag{10-70}$$

Boundary conditions on f can be derived from (10–65) and (10–66):

$$\boxed{f = f' = 0 \quad \text{at} \quad \eta = 0,} \tag{10-71}$$

$$\boxed{f' = 1 \quad \text{as} \quad \eta \to \infty.} \tag{10-72}$$

Since the differential equation for f is third order, this is all the boundary conditions that we can impose. Thus, if a similarity solution of the form (10–68) is to be possible, the condition (10–72) must represent both the boundary condition for $Y \to \infty$ and the initial condition at $x=0$. This is possible, in the present problem, provided that

$$g(0) = 0, \tag{10-73}$$

and this is a necessary condition for existence of a similarity solution. A second necessary condition is that the coefficient gg' must be either zero or a nonzero constant. However, it cannot be zero because no solution of $f''' = 0$ exists to satisfy (10–71) and (10–72).

Since $gg' \neq 0$, we choose

$$\boxed{gg' = 1} \tag{10-74}$$

for convenience, and then

$$\boxed{f''' + ff'' = 0.} \tag{10-75}$$

The latter is known as the *Blasius equation*. In order to complete the similarity solution, we must solve (10–74) subject to (10–73), and (10–75) subject to (10–71) and (10–72).

The solution for $g(x)$ is trivial. Equation (10–74) can be written in the alternative form

$$\frac{dg^2}{dx} = 2. \tag{10-76}$$

Hence, integrating once and applying the boundary condition $g(0) = 0$, we obtain

$$\boxed{g = \sqrt{2x}.}$$ (10–77)

The solution for $f(\eta)$ cannot be obtained analytically. Though the similarity transformation has reduced the set of partial differential equations, (10–64), to a single ordinary differential equation, (10–75), the latter is still *nonlinear.* In fact, Blasius originally solved (10–75) using a numerical method, but with the algebra carried out by hand! Fortunately, today accurate numerical solutions can be obtained using a computer. The main difficulty in solving (10–75) numerically is that most methods for solving ordinary differential equations are set up for initial value problems. To convert a nonlinear boundary-value problem like (10–71), (10–72), and (10–75) to an initial value form suitable for solution via standard numerical methods, the most common approach is to use a "shooting" method. In this method, we first guess a value of f'' at $\eta = 0$, say,

$$f''(0) = A$$

and solve (10–74) as an initial value problem, with $f(0) = f'(0) = 0$ and $f''(0) = A$. Of course, it is highly unlikely that $f'(\infty)$ will be equal to 1 for an arbitrary choice of A. However, if A is not too far from the correct value, we may find that

$$f'(\infty) \to B \quad \text{as} \quad \eta \to \infty.$$

Then, if the problem is solved again for $f''(0) = A + \epsilon$, we should find $f'(\eta) \to B + \delta$ as $\eta \to \infty$. Provided $\delta \ll 1$ when $\epsilon \ll 1$, a new guess for $f''(0)$ can be obtained since we have effectively calculated

$$\Delta = \frac{\Delta f'(\infty)}{\Delta f''(0)}.$$

For example, a new guess for $f''(0)$ could be

$$f''(0) = A + \left(\frac{\Delta f''(0)}{\Delta f'(\infty)}\right)(1 - B) = A + \frac{1 - B}{\Delta}.$$

Although "shooting" works well for the Blasius equation, an unusual feature of this particular equation is that we can transform directly from the first solution for $f''(0) = A$ to the exact solution once we have calculated $f'(\infty) = B$. To see how this works, let us define an auxiliary function $F(z)$ that satisfies the following equations and boundary conditions:

$$F''' + FF'' = 0,$$

$$F(0) = 0,$$

$$F'(0) = 0,$$

$$F''(0) = A,$$ (10–78)

where the independent variable in this case has been denoted as z. But this is just the initial value problem that was solved as first step in the "shooting" method. Thus, the solution is known to yield

$$F'(z) = B \quad \text{as} \quad z \to \infty. \tag{10-79}$$

But now let us introduce a new function $f(\eta)$, where

$$f = \frac{1}{\sqrt{B}} F \quad \text{and} \quad \eta = \sqrt{B} z. \tag{10-80}$$

Then, transforming from $F(z)$ to $f(\eta)$ in the problem (10–78), using (10–80), we find that

$$f''' + ff'' = 0,$$
$$f(0) = 0,$$
$$f'(0) = 0,$$
$$f''(0) = A/B^{3/2}.$$

Furthermore, since the function $F(z)$ satisfies (10–79), the new function f must satisfy

$$f'(\infty) = 1.$$

But this is just the *Blasius* problem, and $f(\eta)$ is therefore the Blasius function. Clearly, we need solve only a *single* trial problem with initial value $F''(0) = A$ to obtain the asymptotic value $F'(\infty) = B$, and we can transform directly to the Blasius solution which is obtained as an initial value problem with $f''(0) = A/B^{3/2}$.

A plot for $f'(\eta)$ versus η was obtained by numerical solution of the Blasius equation and is shown in Figure 10–5. We find that

$$\boxed{f''(0) = 0.469} \tag{10-81}$$

and that $f'(\eta) \sim 0.99$ for $\eta \sim 3.6$. The latter result provides a qualitative estimate of the width of the boundary-layer. In terms of the original variables, nondimensionalized

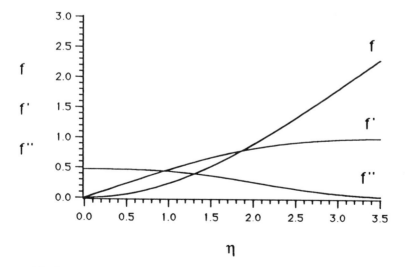

Figure 10–5 The Blasius function and its derivatives as a function of the similarity variable η.

by the plate length, $\eta = 3.6$ corresponds to

$$y \sim 3.6\sqrt{\frac{2x}{\text{Re}}}. \tag{10-82}$$

We recognize that the numerical coefficient 3.6 is arbitrary and dependent on the value of $f'(\eta) = u$, which we choose to represent an effective "edge" of the boundary-layer. However, the dependence on x and Re are independent of this choice and are thus of considerable physical interest. In particular, the dependence of the boundary-layer thickness on $\sqrt{x}$ indicates a singularity in the solution in the limit as $x \to 0$. This shows up most directly in the shear stress at the plate surface,

$$\boxed{\tau_{xy} = \mu\left(\frac{\partial u}{\partial y}\right)_{y=0} = \frac{\mu U_\infty}{L}\sqrt{\frac{\text{Re}}{2x}}f''(0).} \tag{10-83}$$

Obviously, $\tau_{xy} \to \infty$ as $x^{-1/2}$ for $x \to 0$. We may note also that the normal velocity component diverges as $x^{-1/2}$ for all η other than $\eta = 0$, where $V \equiv 0$ for all x.

This *singularity* in the solution as $x \to 0$ indicates that the boundary-layer approximation breaks down as we approach the leading edge of the plate. This is not surprising. The boundary-layer approximation is based on the assumption that derivatives with respect to y exceed those with respect to x by a large amount, proportional to $\sqrt{\text{Re}}$. However, this assumption clearly breaks down near $x = 0$, where there is a discontinuity in boundary conditions on the axis $y = 0$. Fortunately, the error inherent in the boundary-layer solution for $x \to 0$ does not have a serious effect on the solution for other values of x, nor does it introduce an appreciable error in the predicted total force on the plate since τ_{xy} is still integrable for $x \to 0$. The reader interested in the correct flow structure for $x \ll 1$ may wish to refer to a paper by Carrier and Lin (1948).[15] Here, we shall simply ignore the singularity in the boundary-layer solution at $x = 0$ since it does not affect the calculated force on the plate. The latter is obtained by integrating the local shear stress (10–83) over the top and bottom surfaces of the plate. The result, per unit length of the plate in the third direction z, is

$$D = 2\mu U_\infty\sqrt{\frac{\text{Re}}{2}}f''(0)\int_0^1 \frac{dx}{\sqrt{x}},$$

$$D = 4\mu U_\infty\sqrt{\frac{\text{Re}}{2}}f''(0). \tag{10-84}$$

It is common practice to report the dimensionless drag coefficient (in analogy to Nu) rather than D, that is,

$$\boxed{C_D \equiv \frac{D}{\rho U_\infty^2 L} = \frac{4}{\sqrt{2}}\cdot\frac{f''(0)}{\sqrt{\text{Re}}}.} \tag{10-85}$$

The functional dependence of C_D on $\text{Re}^{-1/2}$ is a feature common to all streamlined bodies where the boundary-layer analysis is valid over the complete body surface (i.e., the flow does not separate). The correlation

$$C_D = \frac{\text{const}}{\sqrt{\text{Re}}} \qquad (10\text{-}86)$$

is, in this sense, analogous to the correlations between Nu and Pe that were derived in Chapter 9.

Two features of the Blasius solution are worth mentioning here. The first is the asymptotic behavior of the normal velocity component for large Y. It was noted earlier that the matching condition between the inner and outer regions was satisfied to $0(\text{Re}^{-1/2})$ by the boundary-layer and potential-flow solutions, provided conditions (10–44) and (10–47) were satisfied and the normal velocity component $V(x, Y)$ remained finite for $Y \to \infty$. The latter condition is necessary because the boundary-layer equations are only third order in Y, and thus the behavior of V at large Y cannot be specified in advance, while retaining the boundary conditions (10–43) and the matching condition (10–47). By examining the numerical solution of the Blasius equation and/or developing a series expansion of the solution for large η, (see Schlichting, 1968), it can be shown that

$$f(\eta) \to \eta - 1.217 \quad \text{for} \quad \eta \to \infty.$$

Hence,

$$V = \frac{1}{\sqrt{2x}} \left[\eta f' - f \right] \to \frac{1}{\sqrt{2x}} (1.217) \quad \text{for} \quad \eta \to \infty.$$

Except in the limit $x \to 0$, which we have already discussed, we see that V is finite as required. If we transform back to outer variables, we find that

$$\frac{V}{\sqrt{\text{Re}}} = \frac{1}{\sqrt{2}} \left(\frac{1.217}{\sqrt{x}\sqrt{\text{Re}}} \right). \qquad (10\text{-}87)$$

Thus, so far as the flow in the outer region is concerned, the presence of the boundary layer acts like a weak vertical flow at the plate surface which tends to displace streamlines outward. This feature would appear in the solution for the outer region at $0(\text{Re}^{-1/2})$, as a correction to the $0(1)$ solution given by (10–62). The physical reason for the outward displacement of streamlines in the outer flow is the deceleration of fluid that occurs in the boundary-layer due to the no-slip condition at the plate surface.

A second property of the Blasius solution is worth mentioning because it reflects a general property of the boundary-layer equations. The fact is that the Blasius solution is actually independent of the length of the plate. This may seem strange or even incorrect at first because we have obviously used the plate length L to nondimensionalize the equations leading to (10–59), and the plate length L also appears in the Reynolds number that was used to rescale the boundary layer equations (10–64). However, we have shown that

$$u = f'(\eta),$$

where

$$\eta = \frac{Y}{\sqrt{2x}},$$

and it can be shown that η is completely independent of L. To see this, we can rewrite η in terms of the original dimensional variables (x', y'):

$$\eta = \frac{1}{\sqrt{2}}\frac{Y}{\sqrt{x}} = \frac{1}{\sqrt{2}}\frac{y\sqrt{Re}}{\sqrt{x}} = \frac{1}{\sqrt{2}}\frac{(y'/L)}{(x'/L)^{1/2}}\left(\frac{U_\infty L}{\nu}\right)^{1/2} = \frac{y'}{\sqrt{2x'}}\sqrt{U_\infty/\nu}.$$

It is in fact the distance from the leading edge of the plate that provides the relevant measure of the relative importance of viscous and inertia terms in the equations,

$$\mathrm{Re}_x = \frac{U_\infty x'}{\nu}. \tag{10--88}$$

Indeed, we have already seen that the solution breaks down as $x' \to 0$.

The fact that the solution is independent of L is a reflection of the fact that the boundary-layer equations are *parabolic*, with characteristics that proceed in the direction of increasing x. This means that the solution of the boundary-layer equations at a given position x^* depends only on conditions within the boundary-layer for $x < x^*$ (upstream) but is not at all influenced directly by conditions downstream. This is a general characteristic of the boundary-layer equations (10–40) and (10–41). In the problem of flow past a flat plate, it is reflected by the fact that the solution at any point $x' < L$ is completely unaltered by the proximity of x' to the end of the plate. Indeed, the solution for $x' < L$ is precisely the same as if the plate were *semi-infinite* in extent. We already have seen that the boundary-layer equations are local in form, with no direct dependence on the geometry of the body except through the form of the pressure gradient. In other words, the solution in the boundary-layer depends upon the geometry of the body only because the solution in the outer region depends on the geometry. In the case of a flat plate, however, the first approximation in the outer region is completely independent of the length of the plate. In fact, the free-stream motion is not changed by the plate at all. Thus, the boundary-layer solution is also independent of L.

D. Streaming Flow Past a Semi-Infinite Wedge— The Falkner-Skan Solution

We saw in the previous section and in Chapter 9 that the boundary-layer equations very frequently have similarity solutions. In fact, in every case that we have studied, we have been able to find a self-similar solution. In Chapter 3, where we first introduced the concept of self-similar solutions, it was suggested that they should exist only for problems that are not characterized by a physical length scale. The local nature of the boundary-layer equations might lead to the assumption that the physical length scale of the problem, namely, the length scale of the body, would never be relevant and thus that similarity solutions should always be possible for these equations. However, this is not true. The form of the *momentum* boundary-layer equations depends on the form of the pressure gradient $-\partial p/\partial x$, which is imposed from the outer, potential flow around the body, and the potential flow solution will generally introduce a length scale into the problem. Thus, in general, we should not expect to be able to obtain similarity solutions of the momentum boundary-layer equations. The flat plate problem of the preceding section is a special case in that the potential flow solution introduces no geometric parameters. In heat transfer problems, the geometry of the body enters through the velocity field, but the thermal boundary-layer equations depend on only the local form of the velocity components at the body surface. These local approximations are independent of the length scale of the body, and thus it is not surprising that similarity solutions can be obtained for *arbitrarily* shaped bodies, as shown in Chapter 9.

We have indicated above that the similarity solutions of the momentum boundary-layer equations should not generally be expected. An interesting question is whether similarity solutions can be obtained in any case other than the flat plate problem in the previous section. To answer this question, we start with the boundary-layer equations in their most general form:

$$u \frac{\partial u}{\partial x} + V \frac{\partial u}{\partial Y} = u_e \frac{du_e}{dx} + \frac{\partial^2 u}{\partial Y^2} \tag{10-89}$$

and

$$\frac{\partial u}{\partial x} + \frac{\partial V}{\partial Y} = 0, \tag{10-90}$$

with

$$u = V = 0 \quad \text{at} \quad Y = 0,$$
$$u \to u_e(x) \quad \text{as} \quad Y \to \infty. \tag{10-91}$$

The only distinction between bodies of different geometry is the function $u_e(x)$. Thus, the question we wish to answer is whether similarity solutions exist for functions $u_e(x)$ other than $u_e(x) = \text{const}$ (the flat plate problem).

To answer this question, let us attempt a similarity solution of the general form

$$u = u_e(x) f'(\eta) \quad \text{with} \quad \eta = \frac{Y}{g(x)}. \tag{10-92}$$

The function $u_e(x)$ must be included in (10–92), otherwise $f'(\eta)$ would have to be a function of x as $\eta \to \infty$ [see (10–91)]. The form for V corresponding to (10–92) can be obtained using (10–90) and the boundary condition $V = 0$ at $Y = 0$, that is,

$$V = - \int_0^Y \left(\frac{\partial u}{\partial x} \right) dY = - \int_0^\eta g \left(\frac{\partial u}{\partial x} \right) d\eta. \tag{10-93}$$

The result of substituting for $(\partial u / \partial x)$ from (10–92) and carrying out the integration is

$$V = - \frac{d}{dx} (g u_e) f + g' u_e \eta f', \tag{10-94}$$

where $f(0) = 0$. If we now substitute for u and V in the boundary-layer equation (10–89), we obtain

$$f''' + \left[g \frac{d}{dx} (g u_e) \right] f f'' + g^2 u_e' \left[1 - (f')^2 \right] = 0. \tag{10-95}$$

The boundary conditions (10–91) become

$$\boxed{f = f' = 0 \quad \text{at} \quad \eta = 0}$$

and

(10–96)

$$\boxed{f' \to 1 \quad \text{as} \quad \eta \to \infty.}$$

(10–97)

In general, (10–89) requires an initial profile at $x = 0$ (corresponding to the most upstream point in the boundary-layer). Evidently, if a similarity solution does exist, the boundary condition (10–97) must represent this initial condition, as well as the boundary condition for $Y \to \infty$. This implies that $g(0) = 0$; that is, the boundary-layer thickness must go to zero as $x \to 0$. This can occur only for a body that has a *pointed* leading edge (such as the flat plate).

If we examine (10–95), the existence of a similarity solution requires that the coefficients involving $g(x)$ and $u_e(x)$ must be either zero or a nonzero constant. Let us denote these constants as α and β, that is,

$$\boxed{g \frac{d}{dx}(u_e g) = \alpha}$$

(10–98)

and

$$\boxed{g^2 u_e' = \beta.}$$

(10–99)

We suppose that both of these constants are nonzero. Equations (10–98) and (10–99) should then lead to a unique combination of functions $u_e(x)$ and $g(x)$ for which similarity solutions exist. Of course, in the approach that we have adopted, the problem of determining a function $u_e(x)$ that allows similarity solutions is a purely mathematical question. The related physical problem is to determine whether the resulting forms for $u_e(x)$ correspond to any realizable body shape.

Let us first resolve the mathematical question by solving (10–98) and (10–99). To do this, substitute (10–99) into (10–98). This gives

$$2\beta + u_e \frac{dg^2}{dx} = 2\alpha.$$

(10–100)

Now, adding (10–99) and (10–100), we obtain

$$\frac{d}{dx}(g^2 u_e) = 2\alpha - \beta$$

(10–101)

and, integrating this equation,

$$g^2 u_e = (2\alpha - \beta)x + c,$$

(10–102)

where c is an arbitrary constant of integration. Now, substituting for g^2 from (10–99), we find that

$$\frac{1}{u_e}\frac{du_e}{dx} = \frac{\beta}{[(2\alpha - \beta)x + c]},$$

(10–103)

and integration of this equation gives

$$u_e = c_1 \left[(2\alpha - \beta)x + c \right]^{\beta/(2\alpha - \beta)}, \tag{10-104}$$

where c_1 is another arbitrary constant. Finally, with u_e known, we obtain g^2 from (10–102). The result is

$$g^2 = \frac{1}{c_1} \left[(2\alpha - \beta)x + c \right]^{(2\alpha - 2\beta)/(2\alpha - \beta)}. \tag{10-105}$$

Evidently, $(2\alpha - \beta)$ cannot be zero, but we shall return to this point momentarily.

Now, the remaining questions are whether the requirements for similarity impose additional constraints on the constants α, β, c, and c_1, and then whether the resulting form for u_e corresponds to any physically realizable geometry for the body. Let us first consider the case $\alpha, \beta \neq 0$, which turns out to be the most interesting. In this case, either α or β can be set equal to unity with no loss of generality. We choose $\alpha = 1$. Thus,

$$u_e = c_1 \left[(2 - \beta)x + c \right]^{\beta/(2 - \beta)} \tag{10-106}$$

and

$$g^2 = \frac{1}{c_1} \left[(2 - \beta)x + c \right]^{(2 - 2\beta)/(2 - \beta)}. \tag{10-107}$$

One of the remaining conditions for existence of a similarity solution is that $g(0) = 0$. Thus, we can see from (10–107) that

$$c = 0.$$

What remains for u_e is

$$u_e = c_1 \left[(2 - \beta)x \right]^{\beta/(2 - \beta)}.$$

However, the constant $c_1(2 - \beta)^{\beta/(2 - \beta)}$ is superfluous since it always can be made equal to unity by a proper choice of characteristic velocity U_c. Thus, finally, we can write

$$\boxed{u_e = x^{\beta/(2 - \beta)} = x^m,} \tag{10-108}$$

where

$$m = \frac{\beta}{2 - \beta} \quad \text{or} \quad \beta = \frac{2m}{m + 1}, \tag{10-109}$$

and

$$c_1 = (2 - \beta)^{-\beta/(2 - \beta)}. \tag{10-110}$$

Substituting for β and c_1 in (10–107), we see that the corresponding form for g is

$$\boxed{g = \left[\frac{2}{m + 1} x^{1 - m} \right]^{1/2}.} \tag{10-111}$$

Thus, a necessary condition for the existence of similarity solutions for $\alpha, \beta \neq 0$ is that $u_e = x^m$. The sufficient condition is that $u_e = x^m$ and a solution $f(\eta)$ exists that satisfies (10–95) in the form

$$\boxed{f''' + ff'' + \beta\{1 - (f')^2\} = 0} \tag{10-112}$$

and the boundary conditions (10–96) and (10–97). Equation (10–112) is known as the *Falkner-Skan* equation.

Assuming that (10–112) can be solved subject to (10–96) and (10–97), the question is whether $u_e = x^m$ corresponds to any physically realizable body shapes. To answer from first principles, we would have to solve an *inverse problem* in potential flow theory. We do not propose to do that here. Rather, we simply state the result, which is that

$$u_e = x^m$$

only for streaming flow past a semi-infinite wedge of included angle $\pi\beta$, with the motion at infinity parallel to the bisector of the wedge.[16] The flow configuration is sketched in Figure 10–6. A special case is $\beta = 0$, when the definition of g and the governing Falkner-Skan equation reduce to the Blasius equations for flow past a parallel flat plate.

When $0 < \beta < 1$, we have the wedge geometry. A plot of the solutions for $f'(\eta)\left[\equiv u/u_e(x)\right]$ is given for various β in Figure 10–7. The constant m increases monotonically in the range $0 < m < 1$ as β increases. Further, since $u_e = x^m$ and $dp/dx = -u_e(du_e/\partial x)$, it follows that the magnitude of the pressure gradient in the streamwise direction

$$\frac{\partial p}{\partial x} = -mx^{2m-1} \tag{10-113}$$

increases as the wedge angle increases. We will refer to $\partial p/\partial x < 0$ as a *favorable* pressure gradient since the pressure decreases in the direction of flow, and this tends to accelerate the fluid in the boundary-layer relative to the velocity it would have in the absence of the pressure gradient (i.e., for the flat plate $m = 0$). This is reflected in the velocity profiles in Figure 10–7.

The case of $\beta = 1$ ($m = 1$) deserves special mention. In this case, the geometry reduces to flow directly toward a perpendicular flat plate. The position $x = 0$ is chosen in this instance to be the midpoint, so that flow near the surface is diverted either left or right as it approaches the plate from $x < 0$ or $x > 0$. The resulting motion is known as the *two-dimensional stagnation flow* and corresponds to the potential flow $u = x$, $v = -y$. However, if we recall that the leading edge of the body should be sharp in order for the condition $g(0) = 0$ to make sense, we may be skeptical that a solution of similarity form should exist for this problem. Indeed, if we examine the function $g(x)$ for $\beta = m = 1$, we find that

$$g = 1, \tag{10-114}$$

and thus the general similarity form (10–92) reduces to

$$\boxed{u = xf'(Y),} \tag{10-115}$$

which is not a similarity solution at all. The transformation (10–115) reduces the laminar boundary-layer equation to the same ordinary differential equation that is obtained

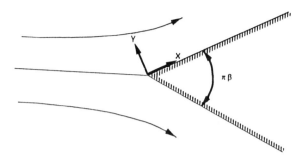

Figure 10–6 A sketch of the flow configuration that corresponds to the Falkner-Skan equation. The body is a semi-infinite wedge with included angle $\pi\beta$. The boundary-layer coordinates (x, Y) are shown for the upper surface, with $x = 0$ corresponding to the leading edge of the wedge.

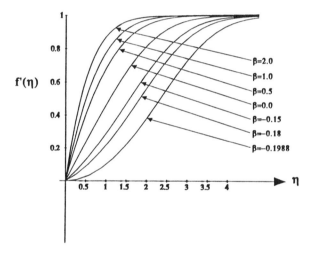

Figure 10–7 Solutions of the Falkner-Skan equation, plotted as f' versus η for several different values of β. The physically significant values of β that correspond to the semi-infinite wedge configuration of Figure 10–6 are $0 \leq \beta \leq 1$. The solutions for $\beta > 1$ and $\beta < 0$ are discussed on page 588.

from the Falkner-Skan equation for $\beta = 1$, namely,

$$f''' + ff'' + (1 - f'^2) = 0, \qquad (10\text{--}116)$$

but in this case f is a function of Y alone. The reader may recall that the function g provides a measure of the boundary-layer thickness as a function of x. In this limiting case, it reverts to a constant. Physically, the tendency of vorticity to diffuse out from the body surface (and thus to increase the thickness of the boundary layer as we move in the downstream direction) is precisely compensated in this instance by the normal motion of fluid toward the boundary, which convects vorticity *toward* the boundary at a rate that just balances diffusion in the opposite direction.

We have shown from a mathematical viewpoint that similarity solutions of the boundary-layer equations exist whenever

$$u_e = x^m$$

for *arbitrary m*. From this infinity of possible cases, we have identified a class of problems for $0 \leq m \leq 1$ where the functional dependence of u_e on x^m corresponds to a realizable, physical problem. However, this limitation on m does not mean that solutions of the Falkner-Skan equations may not exist for $m < 0$ or $m > 1$. Indeed, such solutions have been found numerically by Hartree (1937) and Stewartson (1954).[17] As m increases above 1.0, the solution of the Falkner-Skan equation smoothly continues the trend indicated in Figure 10–7 for increasing values of m. However, for $m > 1.0$, it has been pointed out by Hartree and Stewartson that these solutions do not lead to real values of u and V and thus are not of any physical significance or interest. Hartree also obtained a family of solutions for β between 0 and -0.1988 that were physically acceptable in the sense that $f' \to 1$ *from below* as $\eta \to \infty$. At $\beta = -0.1988$, $f''(0) = 0$. For $\beta < -0.1988$, all solutions of the Falkner-Skan equation are unacceptable on physical grounds because $f' \to 1$ *from above* as $\eta \to \infty$, and this would correspond to velocities within the boundary-layer that *exceed* the outer potential-flow value at the same streamwise position, x. The reader may well ask why we bother to even mention these solutions of the Falkner-Skan equation for m outside the range $0 \leq m \leq 1$, since there is evidently no corresponding physical problem. The main justification is the *qualitative* connection between solutions for $m < 0$ and boundary-layer *separation*. In particular, it will be noted from (10–113) that

$$\frac{\partial p}{\partial x} > 0$$

for $m < 0$. This means that the pressure increases in the direction of motion, and the resulting pressure gradient tends to decelerate the fluid in the boundary-layer relative to its velocity in the absence of a pressure gradient. A pressure gradient acting in this sense is known as an *adverse* pressure gradient. This deceleration is reflected by the fact that $f'(\eta)$ is smaller for any $\eta \neq 0$ for $m < 0$ than it is for $m = 0$. Furthermore, when the pressure gradient increases only a little above zero, corresponding to $\beta = -0.1988$, the solution exhibits zero shear stress, $f''(0) = 0$. Stewartson[18] has suggested as a qualitative model that the Falkner-Skan equation might be viewed as describing the velocity profile in a boundary-layer where the constant β is treated as a parameter that determines the local pressure gradient. Thus, for example, if we consider a circular cylinder, β starts out as equal to unity at the front stagnation point, decreases monotonically as we pass around the body, and finally passes through zero at the equator. In this framework, the point of separation would occur at the point where the local pressure gradient gives $\beta = -0.1988$. Of course, this model is a gross oversimplification, and we should not expect more than the qualitative indication that separation may be expected when a boundary-layer encounters a region with only a very weakly increasing pressure (i.e.

a small adverse pressure gradient).[18] We shall see in the next section that this is, in fact, the case and we will discuss the "mechanism" of separation in detail.

E. Streaming Flow Past Cylindrical Bodies— Boundary-Layer Separation

In the previous two sections we studied boundary-layer problems that yield similarity solutions. Here, we turn to a general class of problems that cannot be solved via similarity transform, namely, flow past cylindrical bodies, with the circular cylinder serving as a convenient prototype. This class of problems is interesting not only because it requires a new approach to obtain solutions but also because many cylindrically shaped bodies exhibit boundary-layer separation. Experimentally, steady separated flows are generally identified by the presence of a pair of large recirculating eddies downstream of the body. Another indication of separation that is more important from a practical point of view is that the drag on the body is much larger than it is for the flat plate or other streamlined bodies that do not lead to a separated flow.

The cause of large drag in the case of a body like a circular cylinder is the asymmetry in the velocity and pressure distributions at the cylinder surface that results from separation. All bodies in laminar streaming flow at large Reynolds number are subjected to viscous stresses that boundary-layer analysis shows must be

$$\tau \sim \frac{\partial u}{\partial y} \sim \sqrt{\text{Re}} \left(\frac{\partial u}{\partial Y} \right) \sim 0(\sqrt{\text{Re}}). \tag{10-117}$$

These viscous stresses lead to a contribution to the drag coefficient (that is, the dimensionless drag per unit length for a two-dimensional body):

$$\boxed{C_D \equiv \frac{\text{drag}}{\rho U_\infty^2 l_c} \sim 0(\text{Re}^{-1/2}) \quad \text{for} \quad \text{Re} \to \infty.} \tag{10-118}$$

In addition, there are pressure forces at the body surface. We have seen from boundary-layer theory that the pressure does not vary to leading order across the boundary-layer. Hence, the pressure on the body surface can be approximated by the pressure distribution imposed upon the boundary-layer by the outer potential flow. The pressure variations in this outer region according to (10–1) are $0(\rho U_\infty^2 / l_c)$. However, so long as the boundary-layer does not separate, the *net* contribution to the drag due to pressure variations over the body surface will be zero. This is, in fact, nothing more than D'Alembert's paradox revisited. Even though there are large pressure variations over the body surface, the downstream force due to pressure forces acting over the front portion of the body is precisely balanced by an equal and opposite upstream force due to pressure forces on the rear of the body. On the other hand, when the boundary-layer separates, the pressure at the rear of the body is sharply reduced and the pressure distribution becomes very asymmetric.* Thus, there is a net force on the body due to

---

*An example of the pressure distribution predicted by potential flow theory and the experimentally measured distribution for a case where the boundary-layer separates was shown in Figure 10–3.

the imbalance between the pressure forces on the front and rear of the body. Since the pressure variations are $0(\rho U^2/l_c)$, the net force on the body will be $0(\rho U^2 l_c)$, and the corresponding contribution to the drag coefficient will be

$$\boxed{C_D \sim 0(1) \quad \text{as} \quad \text{Re} \to \infty.}$$

(10–119)

The drag contribution associated with an asymmetric pressure distribution is sometimes known as *form* drag, while that due to the viscous stresses is called *friction* drag. Comparing (10–118) and (10–119), we see that the form drag on a body in separated flow is asymptotically large compared with the drag on a streamlined body without separation. Indeed, the very large drag penalty associated with separation explains the emphasis in aerodynamics on the design of streamlined cross-sectional shapes for aircraft wings that achieve maximum lift without inducing separation. Regardless of the specific application, the relative magnitudes of the form and friction contributions to the drag show why it is critically important to have an *a priori* method to predict when separation can be expected.

Let us now turn to the application of boundary-layer theory to streaming flow past cylindrical bodies using the circular cylinder to illustrate the details of the calculation. We have already noted that experimental studies show the existence of separated flow in the case of a circular cylinder. Thus, this should provide an interesting test case for prediction of separation via boundary-layer theory. As we noted earlier, the analysis is carried out initially ignoring the possibility that the flow may separate. In particular, we calculate the pressure gradient in the boundary-layer using potential flow theory for streaming flow around the cylinder alone (that is, without any attempt to incorporate the recirculating wake as part of a composite body). The presence of a separation point is then signaled in the boundary-layer solution by the existence of a point on the body surface where the velocity gradient $\partial u/\partial Y$ vanishes, namely,

$$\boxed{\left.\frac{\partial u}{\partial Y}\right|_{Y=0} \Rightarrow 0 \quad \text{as} \quad x \to x_s,}$$

(10–120)

where x_s is used to denote the position of the separation point.

The potential flow solution for streaming motion past a circular cylinder was obtained earlier and given in terms of the streamfunction in (10–17). To calculate the pressure gradient in the boundary-layer, we first determine the tangential velocity function, u_e, as defined in (10–37):

$$u_e = \lim_{r \to 1} \left(\frac{\partial \psi}{\partial r}\right).$$

(10–121)

Adopting the local Cartesian coordinates appropriate to the boundary-layer equations, the result can be expressed as

$$\boxed{u_e = 2\sin x.}$$

(10–122)

The corresponding pressure gradient according to (10–40) is

$$\boxed{\frac{\partial p}{\partial x} = -2\sin 2x.}$$ (10-123)

The velocity function $u_e(x)$ and the pressure gradient dp/dx are sufficient to completely specify the boundary-layer problem, but it is of interest to compare the predicted pressure distribution on the cylinder surface with experimentally measured results, and for this purpose it is convenient to proceed one step beyond (10–123) to calculate p. To do this, we integrate with respect to x,

$$p = c + \cos 2x,$$ (10-124)

where c is a constant of integration that is determined by the choice of a reference pressure. For experimental work it is convenient to use the pressure in the free stream as the reference value. Let us denote this reference pressure as p_∞. To relate p_∞ to p on the body surface, we can integrate (10–19) with u_r evaluated from the potential-flow solution:

$$u_r = -\left(1 - \frac{1}{r^2}\right)\cos\theta$$ (10-125)

along the front symmetry axis $\theta = 0$ from $r = 1$ to $r \to \infty$. The result is

$$p\big|_{r=1,\theta=0} = p_\infty + \frac{1}{2}.$$ (10-126)

Hence, comparing (10–126) and (10–124) at $x = \pi$, we see that

$$c = p_\infty - \frac{1}{2},$$ (10-127)

and (10–124) can be rewritten as

$$\boxed{p - p_\infty = -\frac{1}{2} + \cos 2x.}$$ (10-128)

This quantity was plotted in Figure 10–3.* Also shown is the measured pressure distribution for Re in the range 30 to 40. It is evident that the measured and predicted distributions are in close agreement over the front portion of the cylinder. Thus, it is not surprising that boundary-layer theory, based on the potential-flow pressure distribution, should be quite accurate up to the vicinity of the separation point. It is this fact that explains the ability of boundary-layer theory to provide a reasonable estimate of the onset point for separation, as we shall demonstrate shortly.

It will be noted that the pressure gradient over the front half of the cylinder is *favorable*—that is, the pressure decreases in the direction of motion. Beyond the half-way point ($\theta = x = \pi/2$), on the other hand, the pressure begins to increase; in this

---

*Note that the scaling factor used in Figure 10–3 is that usually adopted by experimentalists, $(1/2)\rho u_\infty^2$, and thus differs by a factor of 2 from (10–128).

region the pressure gradient is *adverse*. Our experience with the solutions of the Falkner-Skan equation suggests that this latter region is a candidate for boundary-layer separation. Indeed, experimental observation corroborates this fact. If we stop for a moment and think qualitatively about cylindrical bodies of other cross-sectional shapes, it should be clear that the potential flow pressure distribution will *always* have the same *qualitative* appearance as shown in Figure 10–3. At the front stagnation point, there is always a pressure maximum. Furthermore, as the fluid moves around the cylinder, its velocity tangent to the surface will increase, and thus the pressure will decrease until the point of maximum body width is reached. Beyond this point, the tangential velocity in the potential-flow region will decrease, and the pressure will then increase. Thus, for every cylindrical body, the boundary-layer will experience a region of increasing pressure (an adverse pressure gradient) and thus have the potential to separate. Experience has shown that separation can be avoided only by reducing the magnitude of this adverse pressure gradient by streamlining the body shape beyond the point of maximum width. Not surprisingly, the resulting cross-sectional profile resembles an airfoil. Obviously, to distinguish between which cases will and which will not separate, it is extremely important to have a method to solve the boundary-layer equations for cylindrical bodies of quite general shapes. One such method is based upon the so-called *Blasius series* and is illustrated below.

The boundary-layer problem for the specific case of a circular cylinder is (10–40), (10–41), (10–43), and (10–47), with u_e and $\partial p/\partial x$ given by (10–122) and (10–123). The first point to note is that a similarity solution does not exist for this problem. Furthermore, in view of the qualitative similarity of the pressure distributions for cylinders of arbitrary shape, it is obvious that similarity solutions do not exist for any problems of this general class. The Blasius series solution developed here is nothing more than a power-series approximation of the boundary-layer solution about $x = 0$.

We begin by noticing that the velocity function $u_e(x)$ can be expressed, for any cylindrical body that has a stagnation point at $x = 0$ and is symmetrical about an axis parallel to the free stream, in the general form

$$u_e(x) = u_1 x + u_3 x^3 + u_5 x^5 + \cdots . \tag{10–129}$$

Here, u_1, u_3, u_5, and so forth are the numerical coeffients of a power-series expansion of $u_e(x)$ and are assumed to be known. In the case of a circular cylinder, the first few are

$$u_1 = 2, \tag{10–130}$$

$$u_3 = -\frac{1}{3}, \tag{10–131}$$

$$u_5 = +\frac{1}{60}, \tag{10–132}$$

and

$$u_n = -(-1)^{(n+1)/2}\left(\frac{2}{n!}\right). \tag{10–133}$$

It may be noted that the radius of convergence for this power-series representation of $\sin x$ is $|x| < \infty$. In general, the number of terms required in the series (10–129) to give a good approximation of $u_e(x)$ will vary depending on the form of $u_e(x)$ and thus the geometry of the body. Asymmetric shapes resembling airfoils with long tapering rear sections will generally require many more terms than a symmetric shape such as a circular or elliptic cylinder. Since the boundary-layer velocity u must match $u_e(x)$ for large Y, it seems reasonable to assume that u can be represented throughout the boundary-layer by a similar power series in x but with coefficients that depend on Y since $u = 0$ at $Y = 0$. Thus, we propose an approximation:

$$u = F_1'(Y)x + F_3'(Y)x^3 + F_5'(Y)x^5 + \cdots,$$

(10–134)

with

$$F_1'(0) = F_3'(0) = F_5'(0) = F_n'(0) = 0$$

(10–135)

and

$$F_n'(Y) = u_n \quad \text{as} \quad Y \to \infty \quad \text{for all } n.$$

(10–136)

But with this form for u, it is evident from the continuity equation that a power-series representation also must exist for V,

$$V = -\int_0^Y \frac{\partial u}{\partial x}\, dY = -\left(F_1 + 3F_3 x^2 + 5F_5 x^4 + \cdots\right),$$

(10–137)

where

$$F_1(0) = F_3(0) = F_5(0) = F_n(0) = 0.$$

(10–138)

If we substitute for u, V, and u_e in the boundary-layer equation (10–40) using the power-series approximations in (10–129), (10–134), and (10–137), and collect all terms of like power in x, we obtain a set of ordinary differential equations for the coefficients $F_1(Y), F_3(Y)$, and so forth:

$$F_1''' + F_1 F_1'' + \left(u_1^2 - F_1'^2\right) = 0,$$

(10–139)

$$F_3''' + F_1 F_3'' + 4\left(u_1 u_3 - F_1' F_3'\right) + 3F_1'' F_3 = 0,$$

(10–140)

and

$$F_5''' + F_1 F_5'' + 6\left(u_1 u_5 + \frac{1}{2} u_3^2 - F_1' F_5'\right) + 5F_1'' F_5 =$$

$$3\left((F_3')^2 - F_3 F_3''\right).$$

(10–141)

These ordinary differential equations are to be solved subject to (10–135), (10–136), and (10–138). The first is just the Falkner-Skan equation for a two-dimensional stagnation flow. This is not at all surprising since the geometry of a cylinder near $x = 0$ is locally just the stagnation point geometry as sketched in Figure 10–8. Subsequent equations are linear in F_3, F_5, and so forth. There is, therefore, no doubt that solutions exist for the functions $F_n(Y)$, though the equations will have to be solved numerically in view of the fact that (10–139) is nonlinear.

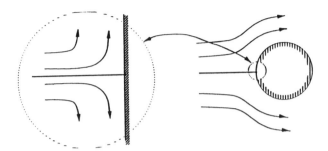

Figure 10–8 A sketch showing streaming flow past a circular cylinder. On the right is the flow as seen with a course level of resolution. On the left is the flow in the immediate vicinity of the front stagnation point, as seen from a much finer resolution. It is evident that the *local* flow examined in a region close enough to the stagnation point reduces to a classic stagnation point flow as described by the Falkner-Skan equation for $\beta = 1$.

The only problem with the approached outlined here is that the ordinary differential equations (10–139)–(10–141), as well as the equations for the higher order functions $F_n(Y)$, all depend explicitly on the numerical coefficients u_1, u_3, and u_5 from the expansion (10–129). This is not surprising. These coefficients are the only source of dependence of the boundary-layer equations on the geometry of the cylinder. Thus, it would seem inevitable that they should appear explicitly in the analysis. However, this means that the solution for every new problem—that is, every new cylinder geometry—will require a new numerical integration of the set of ordinary differential equations for $F_n(Y)$. The astonishing fact, however, is that a slight reformulation of the power-series approximation yields a set of ordinary differential equations and boundary conditions that do not contain the coefficients u_i at all. This transformation was originally discovered by the English mathematician Leslie Howarth (1935).[19] Howarth proposed to replace the power series (10–134) by a power series of the form

$$u = u_1 x f_1'\left(Y\sqrt{u_1}\right) + 4u_3 x^3 f_3'\left(Y\sqrt{u_1}\right) + 6u_5 x^5 f_5'\left(Y\sqrt{u_1}\right) + \cdots, \quad (10\text{–}142)$$

where

$$u_5 f_5 = u_5 g_5 + \frac{u_3^2}{u_1} h_5,$$

$$u_7 f_7 = u_7 g_7 + \frac{u_3 u_5}{u_1} h_7 + \frac{u_3^3}{u_1^2} k_7,$$

and so on. This is known as the *Blasius series*. With this form for u, we obtain in place of (10–139)–(10–141) the alternative set

$$f_1'^2 - f_1 f_1'' = 1 + f_1''' , \qquad (10\text{–}143)$$

$$4f_1' f_3' - 3f_1'' f_3 - f_1 f_3'' = 1 + f_3''' , \qquad (10\text{–}144)$$

$$6f_1'g_5' - 5f_1''g_5 - f_1g_5'' = 1 + g_5''', \tag{10-145}$$

$$6f_1'h_5' - 5f_1''h_5 - f_1h_5'' = \frac{1}{2} + h_5''' - 8\left(f_3'^2 - f_3f_3''\right), \tag{10-146}$$

and so on. The boundary conditions for these equations are

$$f_1(0) = f_1'(0) = 0, \quad f_1'(\infty) = 1,$$

$$f_3(0) = f_3'(0) = 0, \quad f_3'(\infty) = \frac{1}{4}$$

$$g_5(0) = g_5'(0) = 0, \quad g_5'(\infty) = \frac{1}{6},$$

$$h_5(0) = h_5'(0) = 0, \quad h_5'(\infty) = 0. \tag{10-147}$$

The key point to note is that (10–143)–(10–147) are completely independent of the geometric coefficients u_1, u_3, u_5, and so on. Similarly, if we continued the series to higher order terms, the corresponding equations and boundary conditions would all be independent of the u_i. This means that the functions $f_i(Y\sqrt{u_1})$ are *universal functions* that apply to arbitrarily shaped cylindrical bodies or to any boundary-layer problem for which the function $u_e(x)$ can be expressed as either a power series or polynomial in odd powers of x. Thus, we can integrate (10–143)–(10–146) once and for all and tabulate the results. Application to a specific problem then involves only the algebra of inserting the coefficients u_n into the series for u, V, or any other quantity that we may wish to calculate. Tables of values for the functions $f_n'(z)$ for $n \leq 11$ are given in books on boundary-layer theory, such as Schlichting's, as well as in the paper by Howarth (1935).

It is of primary interest in many cases to calculate $\partial u/\partial Y$ at $Y = 0$. This quantity is proportional to the shear stress and can be used in the absence of separation to calculate the frictional force on a body. More importantly, however, condition (10–120) indicates that separation will occur if $\partial u/\partial Y \to 0$ at any point x on the body surface (other than the stagnation point $x = 0$). Furthermore, the point x where this occurs should provide an estimate of the position of the separation point. To calculate $(\partial u/\partial Y)_{Y=0}$ from the Blasius series solution, we require numerical values for the second derivative of $f_n(Y)$ at $Y = 0$, namely,

$$\frac{1}{\sqrt{u_1}}\left(\frac{\partial u}{\partial Y}\right)_{Y=0} = u_1 x f_1''(0) + 4u_3 x^3 f_3''(0) + 6x^5\left[u_5 g_5''(0) + \frac{u_3^2}{u_1} h_5''(0)\right]$$

$$+ 8x^7\left[u_7 g_7''(0) + \frac{u_3 u_5}{u_1} h_7''(0) + \frac{u_3^3}{u_1^2} k_7''(0)\right] + 0(x^9). \tag{10-148}$$

The numerical values of the coefficients required to evaluate (10–148) are

$$\begin{array}{|l|}
\hline
f_1''(0) = 1.2326, \\
f_3''(0) = 0.7244, \\
g_5''(0) = 0.6347, \\
h_5''(0) = 0.1192, \\
g_7''(0) = 0.5792, \\
h_7''(0) = 0.1829, \\
k_7''(0) = 0.0076. \\
\hline
\end{array} \qquad (10\text{-}149)$$

At the beginning of this section, we started out to analyze the boundary-layer solution for the specific case of a circular cylinder. Thus, it is of interest to apply (10–148) and (10–149) to that problem using the u_n coefficients given by (10–133). The result is

$$\frac{1}{\sqrt{u_1}}\left(\frac{\partial u}{\partial Y}\right)_{Y=0} = 2x(1.2326) - \frac{4}{3}x^3(0.7244) + 6x^5\left(\frac{0.6347}{60} + \frac{0.1192}{18}\right) + 0(x^7).$$
$$(10\text{-}150)$$

A plot of the right-hand side of (10–150) is shown in Figure 10–9 for the two-term approximation in which we neglect the contribution of the x^5 term and for the four-term approximation truncated after x^7, as indicated in (10–150). At both levels of approximation, $\partial u/\partial Y \to 0$ for $x > \pi/2$. With the first two terms only, (10–150) yields

$$x_s \sim 1.597,$$

which is barely beyond the point $x = \pi/2 = 1.57$, where the pressure begins to increase. If we were to retain all terms in (10–148) up to order x^{11}, we would find a predicted separation point at

$$\theta_s = 109.6°.$$

Thus, the boundary-layer solution, obtained via the Blasius series, indicates that separation should be expected at about $\theta_s \sim 110°$. Careful numerical solution of the boundary-layer equations in the form (10–41) with u_e given by (10–122) yields a predicted separation point at

$$\theta_s = 104.5°.$$

Experimental observations of the flow past a circular cylinder show that separation does indeed occur, with a separation point at $\theta_s \sim 110°$. It should be noted, however, that steady recirculating wakes can be achieved, even with artificial stabilization,[20] only up to Re ~ 200, and it is not clear that the separation angle has yet achieved an asymptotic (Re $\to \infty$) value at this large, but finite, Reynolds number. In any case, we should not expect the separation point to be predicted too accurately because it is based on the pressure distribution for an unseparated potential flow, and this becomes increasingly inaccurate as the separation point is approached. The important fact is that the boundary-layer analysis does provide a method to predict whether separation

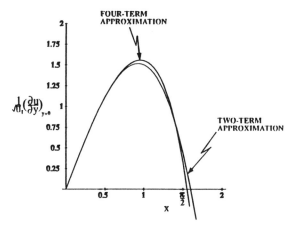

Figure 10-9 The dimensionless shear stress as a function of position on the surface of a circular cylinder as calculated using the approximate Blasius series solution. Note that x is measured in radians from the front stagnation point. The predicted point of boundary-layer separation corresponds to the second zero of $\partial u / \partial Y|_0$ and is predicted to occur just beyond the minimum pressure point at $x = \pi/2$.

should be expected for a body of specified shape. This is a major accomplishment, as we have already pointed out.

Before turning to other topics, it is worth considering the physical events that lead to separation. There are two plausible ways to explain the phenomenon. A common feature of these two mechanisms is that viscous effects play a critical role. Experimentally, we find that separation (or at least the downstream recirculating wake that we associate with separation) usually occurs for Reynolds numbers larger than some critical value, and we infer from this that separation is basically a large Reynolds number phenomenon. Indeed, this point of view is consistent with the fact that separation can be predicted by boundary-layer theory, which is an asymptotic theory for Re → ∞. In spite of this, separation is a phenomenon that can be explained only by considering the consequences of *viscous* contributions to the motion of the fluid.

The most commonly accepted explanation for separation can be paraphrased in the following way. In a potential flow with no viscosity, there is an exact interchange of kinetic energy and work due to the action of the pressure gradient for a fluid element that is near the body surface. On the front portion of the body, the pressure gradient is favorable and the fluid element gains kinetic energy. On the backside, it encounters an adverse pressure gradient causing it to slow down, but the acceleration process on the front side has provided just enough kinetic energy for the fluid element to reach the rear stagnation point before it is brought to rest by the pressure forces at the back. The presence of viscous effects within the boundary-layer alters this picture in a profound way. Some of the work done by the favorable pressure gradient at the front is lost to viscous dissipation, and the kinetic energy of the fluid element is decreased. Thus, when it encounters the adverse pressure gradient at the rear of the body, it no longer has enough kinetic energy to make it to the rear stagnation point, and its tangential motion is completely arrested at some intermediate point. Since fluid mass cannot simply accumulate at this point, the only possibility is for the fluid element to depart from the surface; that is, the boundary-layer separates from the body. In this view of separation, it is the local dynamics within the boundary-layer that trigger the separation process, and the large recirculating region of high vorticity downstream of the body is seen to exist as a consequence of the separation of the boundary-layer.

An alternative point of view is that vorticity accumulates at the rear of the body, which leads to a large recirculating eddy structure, and as a consequence, the flow in the vicinity of the body surface is forced to detach from the surface. This is quite a different mechanism from the first one since it assumes that the primary process leading to separation is the production and accumulation of vorticity rather than the local dynamics within the boundary-layer.[21] However, viscosity still plays a critical role for a solid body in the *production* of vorticity. In fact, for any finite Reynolds number, there is probably some element of truth in both explanations. Furthermore, it is unlikely that experimental evidence (or evidence based on numerical solutions of the complete Navier-Stokes equations) will be able to distinguish between these ideas, because such evidence for steady flows will inevitably be limited to moderate Reynolds numbers.

F. An Approximate Method to Estimate Shear Stress in Boundary-Layer Flows

We have seen in the preceding sections that a primary objective of solving the boundary-layer equations is the shear stress distribution $\partial u/\partial Y|_{Y=0}$. One reason, in the absence of separation, is to predict the force acting on a body. A second is that the primary criteria for determining whether separation will occur is the condition $\partial u/\partial Y|_{Y=0} = 0$. Similarly, in the analogous convective heat-transfer problems, one is usually interested primarily in the local dimensionless heat flux at the surface of a body, namely, $\partial \theta/\partial Y|_{Y=0}$. As a consequence, methods have been developed that can provide a rapid and relatively accurate estimate of $\partial u/\partial Y|_{Y=0}$ or $\partial \theta/\partial Y|_{Y=0}$, without requiring a complete solution of the boundary-layer equations. In this section we illustrate one of the best-known methods for estimating $\partial u/\partial Y|_{Y=0}$ from the momentum boundary-layer equations. This is the so-called *Karman-Pohlhausen method*. It will become evident that the method is easily generalized to estimate $\partial \theta/\partial Y|_{Y=0}$ from the thermal boundary-layer equations, though we do not pursue that generalization here.

The basic idea underlying the Karman-Pohlhausen method is to find an approximation to the velocity profile that satisfies the momentum boundary-layer equations in an overall, integral sense and that is designed to provide a direct estimate of $\partial u/\partial Y|_{Y=0}$. To do this, we first derive an integral condition from the boundary-layer equation in the general form

$$u\frac{\partial u}{\partial x} + V\frac{\partial u}{\partial Y} = u_e\frac{du_e}{dx} + \frac{\partial^2 u}{\partial Y^2} \qquad (10\text{--}151)$$

by integrating across the boundary-layer from $Y=0$ to $Y=\infty$. This yields

$$\left(\frac{\partial u}{\partial Y}\right)_{Y=\infty} - \left(\frac{\partial u}{\partial Y}\right)_{Y=0} = \int_0^\infty \left[u\frac{\partial u}{\partial x} + V\frac{\partial u}{\partial Y} - u_e\frac{du_e}{dx}\right]dY. \tag{10-152}$$

The second term on the right-hand side can be expressed in the alternate form

$$\int_0^\infty V\frac{\partial u}{\partial Y}dY = -u|_{Y=\infty}\int_0^\infty \left(\frac{\partial u}{\partial x}\right)dY + \int_0^\infty u\frac{\partial u}{\partial x}dY \tag{10-153}$$

by applying the continuity equation for V and integrating by parts. But $u_{Y=0} = u_e(x)$. Thus, combining (10-152) and (10-153),

$$\frac{\partial u}{\partial Y}\bigg|_{Y=0} = \left(\frac{\partial u}{\partial Y}\right)_{Y=\infty} + \int_0^\infty \left[u_e\frac{du_e}{dx} - 2u\frac{\partial u}{\partial x} + u_e\frac{\partial u}{\partial x}\right]dY. \tag{10-154}$$

Finally, noting that

$$2u\frac{\partial u}{\partial x} = \frac{\partial u^2}{\partial x} \quad \text{and} \quad u_e\frac{\partial u}{\partial x} = \frac{\partial(uu_e)}{\partial x} - u\frac{du_e}{dx},$$

(10-154) can be rewritten in the final form

$$\boxed{\left(\frac{\partial u}{\partial Y}\right)_{Y=0} = \left(\frac{\partial u}{\partial Y}\right)_{Y=\infty} + \frac{du_e}{dx}\int_0^\infty (u_e - u)dY + \int_0^\infty \frac{\partial}{\partial x}\left[u(u_e - u)\right]dY.} \tag{10-155}$$

This equation, known as the *Karman-Pohlhausen equation*, is an exact deduction from the boundary-layer equation (10-151). In particular, if we put the exact profile for $u(x, Y)$ into the right-hand side, we get the exact form for $(\partial u/\partial Y)_{Y=0}$. Of course, if we knew the exact solution for the boundary-layer velocity profile, we could calculate $(\partial u/\partial Y)_{Y=0}$ by simply differentiating. Here, we propose using simple guesses for $u(x, Y)$—sometimes called *trial functions*—with the hope that (10-155) will then yield a reasonable estimate for $(\partial u/\partial Y)_{Y=0}$.

The obvious question is whether these trial velocity profiles need to satisfy any constraints. It seems clear that, at minimum, they should satisfy the boundary conditions at $Y=0$ and $Y=\infty$, namely,

$$\boxed{u = 0 \quad \text{at} \quad Y = 0} \tag{10-156}$$

and

$$\boxed{u \to u_e(x) \quad \text{for} \quad Y \to \infty.} \tag{10-157}$$

Clearly, if the second of these conditions is to be satisfied, it is also necessary that

$$\left(\frac{\partial u}{\partial Y}\right) \to 0 \quad \text{as} \quad Y \to \infty. \tag{10-158}$$

Furthermore, the differential equation (10-151) can be used to generate additional conditions at $Y=0$. For example, if we require $V=0$, then

$$\frac{\partial^2 u}{\partial Y^2} = -u_e\frac{du_e}{dx} \quad \text{at} \quad Y = 0. \tag{10-159}$$

Generally speaking, we shall see that the estimate of $(\partial u/\partial Y)_0$ is improved as the trial function is forced to satisfy additional conditions at $Y=0$ and $Y=\infty$. We have also stated that the Karman-Pohlhausen method is based on the use of a velocity profile that satisfies the momentum balance in an overall, integral sense [represented by (10–155)]. To see how this additional constraint is imposed upon the trial functions, it is simplest to demonstrate the method by applying it to a specific problem.

We choose for illustration purposes the simplest of all boundary-layer problems, namely, streaming flow past a horizontal flat plate. This problem is advantageous for two reasons. First, we have already obtained an exact solution, and hence we know that

$$\left.\frac{\partial u}{\partial Y}\right|_{Y=0} = \frac{0.332}{\sqrt{x}}. \tag{10–160}$$

Second, since $u_e = 1$, the Karman-Pohlhausen equation is somewhat simplified,

$$\boxed{\left(\frac{\partial u}{\partial Y}\right)_{Y=0} = \left(\frac{\partial u}{\partial Y}\right)_{Y=\infty} + \int_0^\infty \frac{\partial}{\partial x}\left[u(1-u)\right]dY.} \tag{10–161}$$

Now, the simplest possible trial velocity profile is one that increases linearly from $Y=0$ to its "asymptotic" value, $u=1$ at some $Y=\delta(x)$, and is then constant for all $y \geq \delta(x)$. The introduction of $\delta(x)$ is necessary if this simple trial function is to satisfy the integral constraint (10–161), as we shall see shortly. In functional form, the proposed trial function for u is

$$\boxed{u = \begin{cases} Y/\delta(x) & Y \leq \delta(x) \\ 1 & Y > \delta(x) \end{cases}.} \tag{10–162}$$

This is the simplest "guess" that satisfies the three boundary constraints (10–156)–(10–158). But does it satisfy the integral condition (10–161)? To see, let us substitute (10–162) into (10–161). The result is

$$\frac{1}{\delta(x)} = \int_0^{\delta(x)} \frac{\partial}{\partial x}\left[\frac{Y}{\delta(x)}\left(1 - \frac{Y}{\delta(x)}\right)\right]dY. \tag{10–163}$$

For convenience, we denote $Y/\delta(x)$ as ξ. In terms of ξ, (10–163) becomes

$$\frac{1}{\delta(x)} = \int_0^1 \frac{\partial\xi}{\partial x} \frac{\partial}{\partial\xi}\left[\xi(1-\xi)\right]\delta(x)d\xi. \tag{10–164}$$

Since

$$\frac{\partial\xi}{\partial x} = -\frac{\xi}{\delta(x)}\frac{d\delta}{dx},$$

this becomes

$$\boxed{\frac{1}{\delta(x)} = -\frac{d\delta}{dx}\int_0^1 \xi(1-2\xi)d\xi.} \tag{10–165}$$

Clearly, then, (10–161) is only satisfied by the trial function (10–162) if the function $\delta(x)$ satisfies (10–165).

This is a first-order ordinary differential equation for $\delta(x)$. Carrying out the integration with respect to ξ,

$$\frac{1}{\delta(x)} = \frac{1}{6} \frac{d\delta}{dx}, \tag{10-166}$$

and thus

$$\delta(x) = \sqrt{12x + c}.$$

We choose the integration constant $c = 0$ so that the boundary-layer "thickness" goes to zero at the leading edge of the plate. Hence,

$$\delta(x) = \sqrt{12x}. \tag{10-167}$$

The trial velocity profile (10–162), with $\delta(x)$ given by (10–167), now satisfies both the boundary conditions (10–156)–(10–157) and the integral momentum constraint (10–161). Of course it is *not* a solution in any sense of the differential boundary-layer equations (10–151). The critical question is whether it gives a reasonable estimate for $(\partial u / \partial Y)_0$. To see, we need simply note that

$$\left(\frac{\partial u}{\partial Y}\right)_{Y=0} = \frac{1}{\delta(x)} = \frac{1}{\sqrt{12x}} = \frac{0.289}{\sqrt{x}}. \tag{10-168}$$

Comparing (10–168) with the exact answer (10–160), we see that the approximate Karman-Pohlhausen result is remarkably accurate, especially given the crude straight-line guess for the velocity profile.

The next question we may ask is whether we can achieve a better result with a more sophisticated trial function for u. Now that we have gone through the analysis for one particular trial function, it is a simple matter to investigate other choices. The only question that arises is how to choose a better trial function? Here, we cannot give hard-and-fast rules. The general idea is to try to make the proposed profile satisfy as many conditions at the boundaries $Y = 0$ and $Y = \infty$ as possible while still satisfying the integral momentum constraint. Intuition, based on the objective of more accurate results for $\partial u / \partial Y$ at $Y = 0$, would suggest that the most effective additional conditions would be those derived from the boundary-layer equation at $Y = 0$. For $u_e = 1$, (10–159) becomes

$$\frac{\partial^2 u}{\partial Y^2} = 0 \quad \text{at} \quad Y = 0, \tag{10-169}$$

and if we differentiate (10–151) with respect to Y, we find additionally that

$$\frac{\partial^3 u}{\partial Y^3} = 0 \quad \text{at} \quad Y = 0. \tag{10-170}$$

However, the most general polynomial form for u that satisfies $u = \partial^2 u / \partial Y^2 = \partial^3 u / \partial Y^3 = 0$ is just (10–162).

Suppose that instead of trying to use both (10–169) and (10–170), we drop (10–170) for the moment but then require that

$$\frac{\partial u}{\partial Y} = 0 \quad \text{at} \quad Y = \delta(x). \tag{10-171}$$

Physically, this requires that the trial function u and its first derivative be continuous across the boundary-layer. We thus propose a trial function of the form

$$u = a + b\eta + c\eta^2 + d\eta^3, \tag{10-172}$$

with $\eta = Y/\delta(x)$ and $\delta(x)$ a function to be determined such that the integral constraint (10–161) is satisfied. The boundary conditions that we would like our trial function to satisfy are (10–156), (10–157), (10–169), and (10–171). Applying these conditions to (10–172) yields algebraic equations for the coefficients, $a, b, c,$ and d, namely,

$$0 = a,$$

$$1 = a + b + c + d,$$

$$0 = 2c,$$

and

$$0 = b + 2c + 3d.$$

Hence,

$$a = 0, \ c = 0,$$

and

$$b = \frac{3}{2}, \ d = -\frac{1}{2}.$$

To apply the integral constraint (10–161), we substitute for u using (10–172). The result is again a differential condition for $\delta(x)$,

$$\frac{3}{2}\frac{1}{\delta(x)} = \frac{39}{280}\frac{d\delta}{dx}. \tag{10-173}$$

Hence,

$$\delta(x) = \sqrt{\frac{280}{13}} x \tag{10-174}$$

and

$$\boxed{\frac{\partial u}{\partial Y}\bigg|_0 = \frac{3}{2\delta(x)} = \frac{0.323}{\sqrt{x}}.} \tag{10-175}$$

Clearly, the result (10–175) is better than (10–168).

It must be emphasized, however, that the Karman-Pohlhausen method is an ad hoc application of (10–161). As such, we should not expect too much in the way of accuracy, nor should we be lulled by the apparent success of the preceding calculation into a belief that the incorporation of additional conditions and/or complexity into the trial function will always give an improved result. Indeed, if we start with a fourth-order polynomial in the preceding calculation instead of (10–172) and add the additional "boundary condition" (10–170), we find that

$$\left(\frac{\partial u}{\partial Y}\right)_0 = \frac{0.343}{\sqrt{x}} . \tag{10-176}$$

Clearly, (10–176) is no better than (10–175). In fact, the procedure outlined above is truly ad hoc, and there is no sense in which it should be expected to converge to the exact answer by the addition of more conditions on the trial functions. Further, beyond the essential conditions (10–156) and (10–157), there is no guarantee that the conditions we have used are necessarily the best choice at any given level of approximation.

In spite of the limitations of the method, however, the Karman-Pohlhausen technique does provide a simple and powerful method to obtain an estimate of $(\partial u/\partial Y)_{Y=0}$ for any given function $u_e(x)$. The method also can be generalized quite easily to obtain estimates of $(\partial \theta/\partial Y)_{Y=0}$, based on an integrated form of the thermal boundary-layer equation. We leave this latter generalization to the reader.

G. Streaming Flow Past Axisymmetric Bodies—A Generalization of the Blasius Series

Essentially all of the preceding developments of this chapter have assumed that the geometry of the flow can be approximated as two-dimensional. Although we have suggested that the same general principles will apply for bodies of arbitrary, but smooth, shape, it is worth considering at least the generalization to axisymmetric flows, since this includes the important practical problem of streaming flow past a sphere. We do not need to reiterate every detail of the analysis, since some parts are virtually identical to the two-dimensional problems discussed earlier. Instead, we concentrate on those features where it is not clear that the same general analysis should apply.

For convenience, we focus our discussion initially on streaming flow past a sphere. The configuration is sketched in Figure 10–10, where we also indicate the

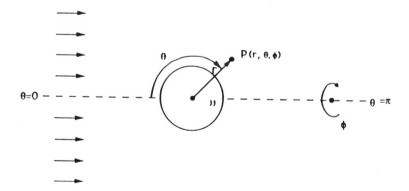

Figure 10–10 A schematic representation for streaming flow past a sphere. The flow is axisymmetric, and thus is independent of the azimuthal angle ϕ. The polar angle θ is measured from the symmetry axis and is equal to zero on the upstream axis.

direction of motion relative to the spherical coordinates (r, θ). The dimensionless equations of motion for the outer region, in which $l_c = a$ (the sphere radius), are identical to (4–97), and it is clear that the leading-order approximation for $\text{Re} \gg 1$ is thus

$$E^2\psi = 0,$$ (10–177)

with

$$\psi = 0 \quad \text{at} \quad r = 1$$ (10–178)

and

$$\psi \to -\frac{1}{2}r^2\sin^2\theta \quad \text{for} \quad r \to \infty.$$ (10–179)

This is equivalent to the potential-flow problem in two dimensions. The solution is

$$\psi = -\frac{1}{2}\sin^2\theta\left(r^2 - \frac{1}{r}\right),$$ (10–180)

with corresponding velocity components

$$u_\theta = \sin\theta\left(1 + \frac{1}{2r^3}\right)$$ (10–181)

and

$$u_r = -\cos\theta\left(1 - \frac{1}{r^3}\right).$$ (10–182)

It will be noted that this solution again exhibits fore-aft symmetry and satisfies the condition $u_r = 0$ at the sphere surface $r = 1$. However, like the potential-flow solutions in two dimensions, this first outer solution does not satisfy the no-slip condition at $r = 1$, and this points clearly to the necessity for a thin boundary-layer near the body surface where viscous effects are important.

Thus, in a manner entirely equivalent to the two-dimensional analysis, we seek rescaled equations in the inner region very near to the sphere surface where the tangential velocity goes from the potential-flow value $(3/2)\sin\theta$ to 0 at the body surface. The only difference from the previous analysis is in the detailed form of the Navier-Stokes and continuity equations for axisymmetric geometries. When expressed in terms of spherical coordinates, these equations are

$$\frac{1}{r^2}\frac{\partial}{\partial r}(r^2 u_r) + \frac{1}{r\sin\theta}\frac{\partial}{\partial\theta}(u_\theta\sin\theta) = 0,$$ (10–183)

$$\frac{1}{\text{Re}}\left[\frac{1}{r^2}\frac{\partial}{\partial r}\left(r^2\frac{\partial u_r}{\partial r}\right) + \frac{1}{r^2\sin\theta}\frac{\partial}{\partial\theta}\left(\sin\theta\frac{\partial u_r}{\partial\theta}\right) - \frac{2u_r}{r^2} - \frac{2}{r^2}\frac{\partial u_\theta}{\partial\theta} - \frac{2u_\theta\cot\theta}{r^2}\right] =$$

$$\frac{\partial p}{\partial r} + u_r\frac{\partial u_r}{\partial r} + \frac{u_\theta}{r}\frac{\partial u_r}{\partial\theta} - \frac{u_\theta^2}{r},$$ (10–184)

and

$$\frac{1}{Re}\left[\frac{1}{r^2}\frac{\partial}{\partial r}\left(r^2\frac{\partial u_\theta}{\partial r}\right) + \frac{1}{r^2\sin\theta}\frac{\partial}{\partial\theta}\left(\sin\theta\frac{\partial u_\theta}{\partial\theta}\right) + \frac{2}{r^2}\frac{\partial u_r}{\partial\theta} - \frac{u_\theta}{r^2\sin^2\theta}\right] =$$

$$\frac{1}{r}\frac{\partial p}{\partial\theta} + u_r\frac{\partial u_\theta}{\partial r} + \frac{u_\theta}{r}\frac{\partial u_\theta}{\partial\theta} + \frac{u_\theta u_r}{r}. \quad (10\text{--}185)$$

Thus, when we apply the usual boundary-layer rescaling to the continuity equation,

$$\boxed{u_r = \frac{1}{\sqrt{Re}}V, \quad r-1 = \frac{Y}{\sqrt{Re}},} \quad (10\text{--}186)$$

we obtain

$$\frac{1}{\left(\frac{Y}{\sqrt{Re}}+1\right)^2}\frac{\partial}{\partial Y}\left[\left(\frac{Y}{\sqrt{Re}}+1\right)^2 V\right] + \frac{1}{\left(1+\frac{Y}{\sqrt{Re}}\right)\sin\theta}\frac{\partial}{\partial\theta}(u\cdot\sin\theta) = 0, \quad (10\text{--}187)$$

where the tangential component u_θ has been denoted as u. Hence, in the limit $Re\to\infty$,

$$\boxed{\frac{\partial V}{\partial Y} + \frac{1}{\sin\theta}\frac{\partial}{\partial\theta}(u\cdot\sin\theta) = 0(Re^{-1/2})\to 0.} \quad (10\text{--}188)$$

Similarly, the tangential component of the momentum equations becomes

$$\boxed{\begin{array}{l}\dfrac{\partial^2 u}{\partial Y^2} + \dfrac{2}{\sqrt{Re}}\dfrac{\partial u}{\partial Y} + 0(Re^{-1}) = \\[2ex] \dfrac{\partial p}{\partial\theta} + V\dfrac{\partial u}{\partial Y} + u\dfrac{\partial u}{\partial\theta} + \dfrac{1}{\sqrt{Re}}uV - \dfrac{Y}{\sqrt{Re}}\dfrac{\partial p}{\partial\theta} - \dfrac{Y}{\sqrt{Re}}u\dfrac{\partial u}{\partial\theta} + 0(Re^{-1}),\end{array}} \quad (10\text{--}189)$$

so that, in the limit $Re\to\infty$,

$$\boxed{\frac{d^2 u}{\partial Y^2} = \frac{\partial p}{\partial\theta} + V\frac{\partial u}{\partial Y} + u\frac{\partial u}{\partial\theta}.} \quad (10\text{--}190)$$

The normal component gives

$$\boxed{\frac{\partial p}{\partial Y} = +\frac{1}{\sqrt{Re}}u^2 + 0(Re^{-1}),} \quad (10\text{--}191)$$

so that

$$\boxed{\frac{\partial p}{\partial Y} = 0} \quad (10\text{--}192)$$

for $Re\to\infty$. As usual, the governing equations for the leading-order approximation in the inner boundary-layer region are the limit of the full rescaled equations for $Re\to\infty$, namely, (10–188), (10–190), and (10–192). It is important to note that the same rescaling in (10–186), as deduced originally for two-dimensional flows, still applies for this axisymmetric problem (and, indeed, would apply to any axisymmetric geometry). Furthermore, with the exception of the continuity equation, the form of the boundary-layer equations at leading order is identical to the two-dimensional case. To obtain a solution for the leading approximation in the inner region, we need to solve (10–188), (10–190), and (10–192) subject to the boundary conditions

$$u = Y = 0 \quad \text{at} \quad Y = 0 \tag{10-193}$$

and the matching condition

$$u \to \frac{3}{2} \sin\theta \quad \text{as} \quad Y \to \infty. \tag{10-194}$$

The latter, taken with (10–192) and the Bernoulli equation in the outer region, implies that

$$\frac{\partial p}{\partial \theta} = \frac{9}{8} \sin 2\theta \tag{10-195}$$

in the boundary layer.

Although the preceding governing equations have been derived for the specific case of a sphere, Schlichting has shown that the same general form would apply to any axisymmetric body. In particular, if we define the geometry in terms of a function $r(x)$, as shown in Figure 10–11, the governing boundary-layer equations for a general axisymmetric body can be shown to take the form

$$\frac{\partial(ru)}{\partial x} + \frac{\partial(rV)}{\partial Y} = 0, \tag{10-196}$$

$$u\frac{\partial u}{\partial x} + V\frac{\partial u}{\partial Y} = -\frac{\partial p}{\partial x} + \frac{\partial^2 u}{\partial Y^2}, \tag{10-197}$$

$$\frac{\partial p}{\partial Y} = 0, \tag{10-198}$$

$$u = V = 0 \quad \text{at} \quad Y = 0, \tag{10-199}$$

and

$$u \to u_e(x) \quad \text{as} \quad Y \to \infty. \tag{10-200}$$

Clearly, the governing equations for the sphere are a special case with

$$r(x) = \sin\theta = \sin x \tag{10-201}$$

and

$$u_e = \frac{3}{2} \sin\theta = \frac{3}{2} \sin x. \tag{10-202}$$

Now, the general problem (10–196)–(10–200) can be solved by a generalization of the Blasius series approximation from Section **E**.

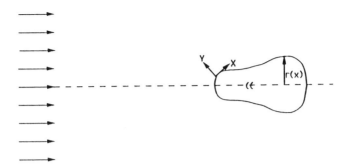

Figure 10–11 A schematic representation for streaming flow past an arbitrary axisymmetric body. The body geometry is specified by the function $r(x)$, which measures the distance from the symmetry axis to the body surface as a function of the position x.

Let us suppose that the two functions $r(x)$ and $u_e(x)$, which differentiate between bodies with different geometries, can be approximated in the form

$$r = r_1 x + r_3 x^3 + r_5 x^5 + \cdots \tag{10-203}$$

and

$$u_e = u_1 x + u_3 x^3 + u_5 x^5 + \cdots . \tag{10-204}$$

Then, a power-series approximation also can be obtained for u within the boundary-layer of the form

$$u = u_1 x f_1'(\eta) + 2u_3 x^3 f_3'(\eta) + 3u_5 x^5 f_5'(\eta) + O(x^7). \tag{10-205}$$

Furthermore, Froessling (1940)[22] has followed the same general approach pioneered by Howarth (1935) to show that the governing equations and boundary conditions for the functional coefficients in (10–205) can be made independent of the coefficients u_i and r_i from the expansions (10–203) and (10–204). This requires the following substitutions:

$$f_3 = g_3 + \frac{r_3 u_1}{r_1 u_3} h_3, \tag{10-206}$$

$$f_5 = g_5 + \frac{r_5 u_1}{r_1 u_5} h_5 + \frac{u_3^2}{u_1 u_5} k_5 + \frac{r_3 u_3}{r_1 u_5} j_5 + \frac{r_3^2 u_1}{r_1^2 u_5} m_5, \tag{10-207}$$

and so forth. The resulting equations for the coefficients at $O(x)$ and $O(x^3)$ are

$$f_1''' + f_1 f_1'' - \frac{1}{2}(f'^2 - 1) = 0, \tag{10-208}$$

$$g_3''' + f_1 g_3'' = 2f_1' g_3' - 2f_1'' g_3 - 1, \tag{10-209}$$

and

$$h_3''' + f_1 h_3'' = 2f_1' h_3' - 2f_1'' h_3 - \frac{1}{2} f_1 f_1'', \tag{10-210}$$

with boundary conditions

$$f_1(0) = g_3(0) = h_3(0) = 0,$$

$$f_1'(\infty) \to 1, \; g_3'(\infty) \to \frac{1}{2}, \; h_3'(\infty) \to 0,$$

$$f_1'(0) = g_3'(0) = h_3'(0) = 0. \tag{10-211}$$

Numerical solutions for all of the coefficients through $0(x^5)$ in (10–205) were actually obtained originally by Froessling, and later Scholkemeier (1949)[23] solved for the ten additional functions required to completely specify (10–205) to $0(x^7)$. The results from Froessling have been reproduced in the book by Schlichting (1968).

Most important of these are the numerical values

$$f_1''(0) = 0.9277,$$

$$g_3''(0) = 1.0475,$$

$$h_3''(0) = 0.0448,$$

$$g_5''(0) = 0.9054,$$

$$h_5''(0) = 0.0506,$$

$$k_5''(0) = 0.1768,$$

$$j_5''(0) = 0.0291,$$

$$m_5''(0) = -0.0244, \tag{10-212}$$

since these can be used to determine the shear stress at the body surface, and thus test for the existence of a separation point. In particular,

$$\left.\frac{\partial u}{\partial Y}\right|_0 = u_1 x f_1''(0)\sqrt{2u_1} + 2u_3 x^3 \sqrt{2u_1} f_3''(0) \cdots . \tag{10-213}$$

Now, for the case of a sphere,

$$r_1 = 1, \quad r_3 = -\frac{1}{6}, \cdots$$

and

$$u_1 = \frac{3}{2}, \; u_3 = -\frac{1}{4}, \cdots . \tag{10-214}$$

If we truncate after only two terms in (10–213), we therefore find that

$$\left.\frac{\partial u}{\partial Y}\right|_0 = \frac{3\sqrt{3}}{2}(0.9277)[1 - 0.3925x^2 + \cdots]. \tag{10-215}$$

Thus, a separation point exists, according to this approximation, at

$$x_s = 1.596 \quad (\theta_s \sim 91.5°). \tag{10-216}$$

This is very close to the result for the similar two-term approximation for a circular cylinder. According to Schlichting, if we retain all terms through $0(x^7)$, we find that

$$x_s = 1.913 \quad (\theta_s \sim 109.6°). \tag{10-217}$$

Again, this is almost identical to the result for the circular cylinder.

The most important point of the preceding analysis is that the same scaling laws and the same asymptotic solution structure applies for both axisymmetric and two-dimensional geometries. Indeed, though we do not pursue the subject here, the same general solution scheme applies for arbitrary three-dimensional geometries. The only complication in solving the general equations for axisymmetric bodies is the shape function $r(x)$ that appears in the continuity equation, and we have seen that a slight generalization of the Blasius series yields a universal power-series solution for arbitrary $r(x)$.

H. The Boundary-Layer on a Spherical Bubble

By now the reader may have come to associate the momentum boundary-layer as a thin region of $0(Re^{-1/2})$ adjacent to a body surface across which the dimensionless tangential velocity changes by $0(1)$ in order to satisfy the no-slip condition at the body surface. To be sure, many boundary-layers do exhibit this structure, but it is overly restrictive as a general description. A more accurate and generally applicable description is that the boundary-layer is a thin region of $0(Re^{-1/2})$ adjacent to a body surface inside of which the vorticity generated at the body surface is confined.

In this regard, it is of interest to contrast the two problems of the streaming motion of a fluid at large Reynolds number past a solid sphere and a spherical bubble. In the case of a solid sphere, the potential flow solution (10–181)–(10–182) does not satisfy the no-slip condition at the sphere surface, and the necessity for a boundary-layer in which viscous forces are important is transparent. For the spherical bubble, on the other hand, the no-slip condition is replaced by the condition of zero tangential stress, $\tau_{r\theta} = 0$, and it may not be immediately obvious that a boundary-layer is needed. However, in this case, the potential flow solution does not satisfy the zero tangential stress condition (as we shall see shortly), and a boundary-layer in which viscous forces are important still must exist. We shall see that the detailed features of the boundary-layer are different from those of a no-slip, solid body. However, in both cases, the surface of the body acts as a source of vorticity, and this vorticity is confined at high Reynolds number to a thin $0(Re^{-1/2})$ region near the surface.

The source of vorticity at a solid, no-slip surface is the velocity gradient that is generated in satisfying the no-slip condition. This mechanism yields vorticity of $0(Re^{1/2})$ at the body surface. At an interface where the tangential velocity is *not* zero, on the other hand, vorticity is produced by rotation of fluid elements caused by the surface curvature. This latter mechanism generates vorticity of magnitude proportional to the local curvature of the surface in the direction of the motion of the fluid. As an example of vorticity production in the latter case, we may consider the condition of zero tangential stress at the surface of a bubble whose shape we assume, for simplicity, to be spherical. In this case, for an axisymmetric motion,

$$\tau_{r\theta} = \frac{\partial u_\theta}{\partial r} - \frac{u_\theta}{r} = 0, \qquad (10\text{–}218)$$

and if this condition is expressed in terms of vorticity,

$$\omega \equiv \frac{1}{r} \frac{\partial}{\partial r}(ru_\theta) - \frac{\partial u_r}{\partial \theta}, \qquad (10\text{–}219)$$

we find that

$$\omega|_{r=1} = \frac{2u_\theta}{r}\bigg|_{r=1} = 2u_\theta. \qquad (10\text{–}220)$$

The corresponding condition for motion past an axisymmetric bubble of arbitrary shape is

$$\omega|_{r=1} = 2\kappa u_s, \qquad (10\text{–}221)$$

that is, the vorticity is equal to twice the product of the tangential velocity u_s and the local curvature in the direction of motion, κ.

Regardless of the source of the vorticity, it remains confined near the body surface for $\mathrm{Re} \gg 1$ because the time scale for radial diffusion is limited by convection around the body to $0(L/U_\infty)$, where L is a characteristic linear dimension of the body. This restricted time scale for radial diffusion then leads to the same estimate of the thickness of the boundary-layer, $0(L\mathrm{Re}^{-1/2})$, in all cases. However, the *strength* of the boundary-layer, as indicated either by the magnitude of the vorticity or the magnitude of velocity gradients, differs dramatically between the cases of no-slip and slip boundaries. When the body surface is solid, the velocity must change by $0(U_\infty)$ across this layer, and this requires that the velocity gradients be $0[(U_\infty/L)\mathrm{Re}^{1/2}]$, as we have already seen in the preceding sections of this chapter. For the bubble, on the other hand, it is the gradient of tangential velocity normal to the surface that must change by $0(U_\infty/L)$; that is, the dimensionless velocity gradient must change by $0(1)$ in order for the shear stress to go to zero at the bubble surface. Thus, the tangential velocity must change by $0[(U_\infty/L)\cdot L(\mathrm{Re})^{-1/2}] = 0(U_\infty\mathrm{Re}^{-1/2})$. In other words, the change in the velocity across the boundary-layer is *very small* compared with the characteristic velocity in the potential-flow region. This suggests that the velocity distribution in the boundary-layer will be the potential flow profile plus a $0(\mathrm{Re}^{-1/2})$ correction to satisfy the zero shear stress condition at the bubble surface.

In order to verify these predictions of dimensional reasoning, we now turn to a detailed analysis for a bubble moving through a quiescent fluid with velocity U_∞. An alternative description of the problem for the point of view of a reference system fixed at the center of the bubble is that the fluid undergoes a uniform motion with velocity U_∞ past a stationary bubble. We will adopt this latter viewpoint for our analysis because it is the same as that used in the earlier sections of this chapter. At the beginning, we will assume that the bubble is *spherical*, though we may anticipate, by direct observation or by examination of the pressure distribution imposed on a bubble in a potential flow, that a real bubble will tend to deform to a flattened, axisymmetric shape.[24] We shall return at the end of this section to consider the necessary conditions for a bubble to remain approximately spherical when the Reynolds number is not small.

In addition, we will actually calculate the bubble shape when the deformation is small. However, the primary goal here is to examine the change in boundary-layer structure for a curved surface when the surface boundary condition is changed from no-slip to zero shear stress. For this purpose it is adequate to concentrate on the limiting case of a perfectly spherical bubble.

As usual, we begin by nondimensionalizing the equations of motion using the bubble radius as a characteristic length scale. Thus, we obtain (10–183)–(10–185), which are to be solved subject to the boundary conditions

$$u_r = 0 \quad \text{at} \quad r = 1, \tag{10-222}$$

$$\tau_{r\theta} = r \frac{\partial}{\partial r}\left(\frac{u_\theta}{r}\right) = 0 \quad \text{at} \quad r = 1, \tag{10-223}$$

and

$$u_r \to -\cos\theta, \quad u_\theta \to \sin\theta \quad \text{as} \quad r \to \infty. \tag{10-224}$$

Clearly, in the limit Re $\gg$ 1, the leading-order approximation for the solution to this problem is identical to the inviscid flow problem for a solid sphere. Although the no-slip boundary-layer has been replaced in the present problem by the zero shear stress condition (10–223), this has no influence on the leading-order inviscid flow approximation since the potential fluid solution can, in any case, only satisfy the kinematic condition $\mathbf{u} \cdot \mathbf{n} = 0$ at $r = 1$. Hence, the first approximation in the outer part of the domain where the bubble radius is an appropriate characteristic length scale is precisely the same as for the no-slip sphere, namely, (10–181) and (10–182). However, this solution does *not* satisfy the zero shear stress condition (10–223) at the bubble surface, and thus it is clear that the inviscid flow equations do not provide a uniformly valid approximation to the Navier-Stokes equations even for Re $\gg$ 1. The problem is the implicit assumption, inherent in the nondimensionalized equation (10–182)–(10–185), that viscous terms can be neglected everywhere, including the region near the bubble surface. Both the source of this incorrect assumption (that is, the use of the body radius as a characteristic length scale everywhere in the domain) and its remedy are the same as in the preceding sections of this chapter, though the details are different, as we shall see.

Thus, to obtain governing equations appropriate to the inner region near the surface, we introduce the usual boundary-layer rescaling,

$$u_r = \frac{1}{\sqrt{\text{Re}}} V, \quad r - 1 = \frac{Y}{\sqrt{\text{Re}}}, \tag{10-225}$$

into (10–183)–(10–185). The resulting equations are (10–187), (10–189), and (10–191), in which we retain terms of both $0(1)$ and $0(\text{Re}^{-1/2})$. The boundary conditions (10–222) and (10–223) in terms of inner variables are

$$V = 0 \quad \text{at} \quad Y = 0 \tag{10-226}$$

and

$$\frac{\partial u}{\partial Y} - \frac{u}{\sqrt{\text{Re}}} + 0(\text{Re}^{-1}) = 0 \quad \text{at} \quad Y = 0. \tag{10-227}$$

As in the previous section we denote the tangential velocity component within the boundary-layer as u to distinguish it from the tangential component u_θ in the outer, inviscid flow region. The boundary-layer solution is also required to match the outer inviscid solution, namely,

$$\lim_{Y \gg 1} u \iff \frac{3}{2}\sin\theta - \frac{3}{2}Y\sin\theta\left(\frac{1}{\sqrt{Re}}\right) + 0(Re^{-1}) \tag{10-228}$$

and

$$\lim_{Y \gg 1}\frac{1}{\sqrt{Re}}V \iff \lim_{(r-1)\ll 1}(u_r) \tag{10-229}$$

(both as $Re \to \infty$). The second of these produces the boundary condition $u_r = 0$ at $r = 1$, which was already used for the leading-order approximation in the inviscid flow region (10–181) and (10–182). The matching condition (10–228) incorporates the limiting form of (10–181) for $y \equiv (r-1) \ll 1$ on the right-hand side and provides a boundary condition for the leading approximation (at least) to u in the boundary-layer,

$$u \to \frac{3}{2}\sin\theta \quad \text{for} \quad Y \to \infty. \tag{10-230}$$

Now, let us suppose, following the scaling arguments at the beginning of this section, that the first two terms in the solution for the inner boundary-layer region can be expressed in the form

$$u = u_0 + \frac{1}{\sqrt{Re}}u_1 + \cdots,$$

$$V = V_0 + \frac{1}{\sqrt{Re}}V_1 + \cdots,$$

$$p = p_0 + \frac{1}{\sqrt{Re}}p_1 + \cdots. \tag{10-231}$$

Thus, substituting into (10–187), (10–189), and (10–191), we find the governing equations at $0(1)$ for the first term in (10–231), namely,

$$\frac{\partial V_0}{\partial Y} + \frac{1}{\sin\theta}\frac{\partial}{\partial\theta}(\sin\theta\, u_0) = 0, \tag{10-232}$$

$$\frac{\partial^2 u_0}{\partial Y^2} = \frac{\partial p_0}{\partial\theta} + V_0\frac{\partial u_0}{\partial Y} + u_0\frac{\partial u_0}{\partial\theta}, \tag{10-233}$$

$$\frac{\partial p_0}{\partial Y} = 0. \tag{10-234}$$

The corresponding boundary conditions from (10–226)–(10–228) are

$$V_0 = 0$$

$$\left.\frac{\partial u_0}{\partial Y} = 0\right\} \quad \text{at} \quad Y = 0 \qquad (10\text{–}235)$$

and

$$u_0 \to \frac{3}{2}\sin\theta \quad \text{as} \quad Y \to \infty. \qquad (10\text{–}236)$$

Not surprisingly, these equations and boundary conditions are just the limiting forms of (10–187), (10–189), (10–191), and (10–226)–(10–228) for Re → ∞.

The solution for (10–232)–(10–236) is just

$$u_0 = \frac{3}{2}\sin\theta \quad \text{and} \quad V_0 = -3Y\cos\theta, \qquad (10\text{–}237)$$

that is, the boundary (matching) condition for $Y \gg 1$ is a uniformly valid solution. This is *not* a surprising result. The scaling arguments at the beginning of this section suggest that the change in the velocity profile in the boundary-layer that is necessary to satisfy the zero shear stress boundary condition is only an $0(\text{Re}^{-1/2})$ correction to the potential-flow solution. The $0(1)$ solution in the boundary-layer (10–237) is nothing but the largest term in an expansion of the potential flow solution for $y \equiv (r - 1) \ll 1$. Indeed, if we simply express the potential-flow solution in terms of inner variables, we find that

$$u_\theta \to \frac{3}{2}\sin\theta - \frac{3}{2}Y\sin\theta\,\frac{1}{\sqrt{\text{Re}}} + 0(\text{Re}^{-1}), \qquad (10\text{–}238)$$

$$u_r\sqrt{\text{Re}} = V \to -3Y\cos\theta + 6Y^2\cos\theta\,\frac{1}{\sqrt{\text{Re}}} + 0(\text{Re}^{-1}), \qquad (10\text{–}239)$$

and the first terms are identical to (10–237).

Normally, the next step in the method of matched asymptotic expansions would be to seek a second approximation in the outer region, followed by a second approximation in the inner region and so on. However, if we examine the governing equations (10–183)–(10–185) for the outer region, we see that the largest neglected term is $0(\text{Re}^{-1})$. Furthermore, since the first approximation to the velocity distribution in the inner region turned out to be just the first term of the outer solution for $(r - 1) \ll 1$ and Re → ∞, there is no mismatch at this level to generate a nonzero second approximation in the outer region. Instead, the next term is the $0(\text{Re}^{-1/2})$ correction in the expansion (10–231) for the inner boundary-layer region.

To determine the form of this term, we must solve the governing equations and boundary conditions that derive from (10–187), (10–189), (10–191), (10–226), and (10–227) at $0(\text{Re}^{-1/2})$. These are

$$\frac{\partial V_1}{\partial Y} + \frac{1}{\sin\theta}\frac{\partial}{\partial\theta}(u_1\sin\theta) = -2V_0 + \frac{Y}{\sin\theta}\frac{\partial}{\partial\theta}(u_0\sin\theta), \qquad (10\text{–}240)$$

$$\frac{\partial^2 u_1}{\partial Y^2} = \frac{\partial p_1}{\partial \theta} + V_0 \frac{\partial u_1}{\partial Y} + V_1 \frac{\partial u_0}{\partial Y} +$$

$$u_1 \frac{\partial u_0}{\partial \theta} + u_0 \frac{\partial u_1}{\partial \theta} + u_0 V_0 - Y \frac{\partial p_0}{\partial \theta} - Y u_0 \frac{\partial u_0}{\partial \theta} - 2 \frac{\partial u_0}{\partial Y}, \quad (10\text{-}241)$$

and

$$\frac{\partial p_1}{\partial Y} = u_0^2, \quad (10\text{-}242)$$

with boundary conditions

$$V_1 = 0, \quad \text{at} \quad Y = 0, \quad (10\text{-}243)$$

$$\frac{\partial u_1}{\partial Y} - u_0 = 0 \quad \text{at} \quad Y = 0, \quad (10\text{-}244)$$

and

$$u_1 \to -\frac{3}{2} Y \sin \theta \quad \text{as} \quad Y \to \infty \, (\text{match}). \quad (10\text{-}245)$$

The latter condition requires that the boundary-layer solution match to $0(\text{Re}^{-1/2})$ with the inner limit of the outer solution (10-238).

Considerable simplification occurs when we substitute for u_0 and V_0 from (10-237). First, the right-hand side of (10-240) is identically zero, and we recover the usual form of the continuity equation for (u_1, V_1), namely,

$$\frac{\partial V_1}{\partial Y} + \frac{1}{\sin \theta} \frac{\partial}{\partial \theta} (u_1 \sin \theta) = 9 Y \cos \theta. \quad (10\text{-}246)$$

Similarly, the right-hand side of (10-241) also simplifies. First, note that

$$\frac{\partial p_0}{\partial \theta} = -u_0 \frac{\partial u_0}{\partial \theta}$$

since u_0 is independent of Y. Thus,

$$\frac{\partial^2 u_1}{\partial Y^2} = \frac{\partial p_1}{\partial \theta} + V_0 \frac{\partial u_1}{\partial Y} + V_1 \frac{\partial u_0}{\partial Y} + u_1 \frac{\partial u_0}{\partial \theta} + u_0 \frac{\partial u_1}{\partial \theta} + u_0 V_0.$$

But according to (10-242) and the form of p_0 in the outer, inviscid flow,

$$\frac{\partial p_1}{\partial \theta} = 2 u_0 \frac{\partial u_0}{\partial \theta} Y.$$

Hence, substituting for u_0 from (10-237),

$$\frac{\partial p_1}{\partial \theta} = \frac{9}{2} Y \sin \theta \cos \theta. \quad (10\text{-}247)$$

But this is just $-u_0 V_0$, and so finally we obtain

$$\boxed{\frac{\partial^2 u_1}{\partial Y^2} = V_0 \frac{\partial u_1}{\partial Y} + V_1 \frac{\partial u_0}{\partial Y} + u_1 \frac{\partial u_0}{\partial \theta} + u_0 \frac{\partial u_1}{\partial \theta}.} \qquad (10\text{-}248)$$

Thus, the problem at $0(\mathrm{Re}^{-1/2})$ is to solve the *linear* equations (10–246) and (10–248), subject to the boundary and matching conditions (10–243)–(10–245). We may note that (10–248) is actually independent of V_1 because $(\partial u_0 / \partial Y) \equiv 0$. Hence, (10–248) can be solved first using the boundary conditions (10–244) and (10–245), and then V_1 can be obtained from the continuity equation (10–246) and the boundary condition (10–243).

A solution of (10–248), (10–243), and (10–244) can be obtained by similarity transformation. For this purpose, however, it is convenient to formulate the problem in terms of the *difference* velocity

$$\hat{u}_1 \equiv u_1 - \left(-\frac{3}{2} Y \sin\theta \right). \qquad (10\text{-}249)$$

Substituting (10–249) into (10–244), (10–245), and (10–248), and using the $0(1)$ solution (10–237) to evaluate the coefficients, we obtain

$$\boxed{\frac{\partial^2 \hat{u}_1}{\partial Y^2} = -3Y\cos\theta \frac{\partial \hat{u}_1}{\partial Y} + \frac{3}{2}\cos\theta\,\hat{u}_1 + \frac{3}{2}\sin\theta \frac{\partial \hat{u}_1}{\partial \theta},} \qquad (10\text{-}250)$$

$$\boxed{\frac{\partial \hat{u}_1}{\partial Y} = 3\sin\theta \quad \text{at} \quad Y=0,} \qquad (10\text{-}251\text{a})$$

and

$$\boxed{\hat{u}_1 \to 0 \quad \text{as} \quad Y \to \infty.} \qquad (10\text{-}251\text{b})$$

To solve (10–250) and (10–251), we propose a similarity transformation of the general form

$$\boxed{\hat{u}_1 = h(\theta) f(\eta), \quad \text{where} \quad \eta = \frac{Y}{g(\theta)}.} \qquad (10\text{-}252)$$

It is necessary to include the multiplying coefficient $h(\theta)$ because of the form of the boundary condition (10–251a). In particular, if we substitute (10–252) into (10–251a), we find that

$$h(\theta) \frac{1}{g(\theta)} f'(0) = 3\sin\theta, \qquad (10\text{-}253)$$

so that

$$\boxed{\frac{h(\theta)}{g(\theta)} = 3\sin\theta.} \qquad (10\text{-}254)$$

Although it might at first seem that (10–253) could be satisfied with $h = $ const by proper choice of $g(\theta)$, the resulting similarity transform would not reduce (10–250) to an ordinary differential equation. With the present formulation, $g(\theta)$ still can be chosen to reduce (10–250) to a similarity form. To see that this is possible, we substitute for $\hat{u}_1$ in the form

$$\hat{u}_1 = 3\sin\theta\, g(\theta) f(\eta), \quad \text{where} \quad \eta = \frac{Y}{g(\theta)}, \tag{10–255}$$

into the governing equation (10–250) and boundary conditions (10–251). The result is

$$\frac{d^2 f}{d\eta^2} = \left[-3g^2 \cos\theta - \frac{3}{2} gg' \sin\theta \right] \eta \frac{df}{d\eta} + \left[3g^2 \cos\theta + \frac{3}{2} \sin\theta gg' \right] f, \tag{10–256}$$

with

$$f'(0) = 1,$$

$$f(\eta) \to 0 \quad \text{as} \quad \eta \to \infty. \tag{10–257}$$

But the two coefficients involving g in (10–256) are identical, and the condition for existence of a similarity solution thus reduces to a single, first-order ordinary differential equation for $g(\theta)$,

$$\frac{3}{2} gg' \sin\theta + 3g^2 \cos\theta = 2, \tag{10–258}$$

where the constant 2 is arbitrary (provided only that the constant is nonzero). Hence, the problem reduces to solving (10–258) for g, with the requirement that g remains bounded for all θ (except possibly $\theta = \pi$, where a wake may be expected), plus the ordinary differential equation

$$\frac{d^2 f}{d\eta^2} + 2\eta \frac{df}{d\eta} - 2f = 0, \tag{10–259}$$

with boundary conditions (10–257) for f.

Solving (10–259) is straightforward. We let

$$s = \frac{f}{\eta}, \tag{10–260}$$

and (10–259) is transformed to

$$\frac{d^2 s}{d\eta^2} + 2\left(\frac{1+\eta^2}{\eta}\right)\frac{ds}{d\eta} = 0. \tag{10–261}$$

Integrating twice with respect to η,

$$s(\eta) = k_1 \int_\eta^\infty \frac{1}{t^2} e^{-t^2} dt + k_2 \tag{10–262}$$

or

$$\boxed{f(\eta) = k_1 \eta \int_\eta^\infty \frac{1}{t^2} e^{-t^2} dt + k_2 \eta.} \tag{10–263}$$

The two integration constants are evaluated by application of the boundary conditions (10–257). The condition

$$f(\eta) \to 0 \quad \text{as} \quad \eta \to \infty$$

requires $k_2 = 0$. (The reader may wish to verify that the first term goes to zero as $\eta \to \infty$ for arbitrary k_1.) Application of the second condition is facilitated by expressing $f(\eta)$ in the alternate form

$$f(\eta) = k_1 \left[e^{-\eta^2} - 2\eta \int_\eta^\infty e^{-t^2} dt \right], \tag{10-264}$$

which is obtained from (10–263) by integrating by parts. It follows from (10–264) that

$$\frac{df}{d\eta} = -2k_1 \int_\eta^\infty e^{-t^2} dt. \tag{10-265}$$

Therefore, applying (10–257),

$$1 = -2k_1 \int_0^\infty e^{-t^2} dt = -2k_1 \sqrt{\pi/2}$$

or

$$k_1 = -\frac{1}{\sqrt{\pi}}. \tag{10-266}$$

Thus, finally,

$$\boxed{f(\eta) = -\frac{1}{\sqrt{\pi}} e^{-\eta^2} + \eta \operatorname{erfc}(\eta),} \tag{10-267}$$

where

$$\operatorname{erfc}(\eta) \equiv \frac{2}{\sqrt{\pi}} \int_\eta^\infty e^{-t^2} dt \tag{10-268}$$

is the complementary error function, whose values are tabulated in most mathematics handbooks (see Abramowitz and Stegun, 1965).[25] To solve for $g(\theta)$, we can multiply (10–258) by $\sin^3\theta$ and then rewrite the resulting equation in the form

$$\frac{\partial}{\partial\theta} \left(\frac{3}{4} \sin^4\theta g^2 \right) = 2\sin^3\theta. \tag{10-269}$$

Integrating and applying the condition that $g(\theta)$ is finite at $\theta = 0$, it can be shown easily that

$$\boxed{g(\theta) = \frac{\sqrt{8/3}}{\sin^2\theta} \left[\frac{2}{3} - \cos\theta + \frac{\cos^3\theta}{3} \right]^{1/2}.} \tag{10-270}$$

Hence, combining the results of our analysis, we find that

$$\boxed{u_1 = -\frac{3}{2} Y\sin\theta + 3g(\theta)\sin\theta f(\eta),} \tag{10-271}$$

where $f(\eta)$ and $g(\theta)$ are given in (10–267) and (10–270), respectively. Thus, the general solution for the inner boundary-layer region through terms of $0(\mathrm{Re}^{-1/2})$ is

$$u = \frac{3}{2}\sin\theta + \frac{1}{\sqrt{Re}}\left[-\frac{3}{2}Y\sin\theta + 3g(\theta)\sin\theta f(\eta)\right] + 0(Re^{-1}). \quad (10\text{–}272)$$

From (10–271) and the continuity equation (10–246), the normal velocity component V_1 can be obtained, if so desired.

This completes the solution to $0(Re^{-1/2})$. It should be noted that the first two terms in (10–272) are, in fact, nothing but the first two terms in the inviscid solution, evaluated in the inner region, namely, (10–238). Thus, to $0(Re^{-1/2})$, we see that the solution in the complete domain consists of the inviscid solution (10–181) and (10–182), with an $0(Re^{-1/2})$ viscous correction in the inner boundary-layer region to satisfy the zero shear stress boundary condition at the bubble surface. Because the viscous correction in the inner region is only $0(Re^{-1/2})$, the governing equation for it is *linear*. Hence, unlike the no-slip boundary-layers considered earlier in this chapter, it is possible to obtain an analytic solution for the leading-order departure from the inviscid flow solution.

It is of particular interest to calculate the force acting on the bubble as a complement to the drag law obtained earlier for the force on a spherical bubble in *creeping* flow. One way to determine the force is to calculate the normal stress and pressure forces on the bubble surface and integrate in the usual manner. However, this approach turns out to be somewhat complicated because it is necessary to include the leading-order departures in u, V, and p from the inviscid flow solution to obtain the first non-zero contribution to the drag (note that the inviscid flow solution alone gives zero drag when used in this manner). Here, we follow an alternative approach pioneered by Levich (1949),[26] based on an overall mechanical energy balance, in the general form

$$\mathbf{U}\cdot\mathbf{F}_{drag} = \int_V \phi \, dV, \quad (10\text{–}273)$$

where $\Phi \equiv \nabla\mathbf{u}:\mathbf{T}$ is the rate of viscous dissipation in the fluid. Physically, the condition (10–273) states that there is a precise balance at steady state between the rate of working on the fluid due to the bubble motion and the rate of dissipation of mechanical energy in the fluid to heat. This macroscopic balance applies to an arbitrary body undergoing an arbitrary (though steady) motion, but it is particularly useful, as we shall see, for the special case of bubble motion at high Re.

Before considering the application of (10–273), it is instructive to outline its derivation. To do this, we consider a so-called "control volume" of fluid lying within an arbitrarily large spherical surface in the fluid that circumscribes the particle or body of interest and is fixed with respect to the center of the particle, so that $\mathbf{u} \to \mathbf{U}_\infty$ at its surface as the radius is increased to very large values. We begin with the equation of motion in the form

$$\rho\frac{D\mathbf{u}}{Dt} = \nabla\cdot\mathbf{T}, \quad (10\text{–}274)$$

which applies at all points within the fluid. A differential mechanical energy balance can be obtained directly from (10–274) by taking its inner (scalar) product with **u** (see Chapter 2 Section **D**), and can be written in the form

$$\frac{\rho}{2}\frac{D}{Dt}(\mathbf{u}\cdot\mathbf{u}) = \nabla\cdot(\mathbf{u}\cdot\mathbf{T}) - \nabla\mathbf{u}:\mathbf{T}. \tag{10–275}$$

Hence, at steady state,

$$\frac{\rho}{2}\nabla\cdot\left[(\mathbf{u}\cdot\mathbf{u})\mathbf{u}\right] = \nabla\cdot(\mathbf{u}\cdot\mathbf{T}) - \nabla\mathbf{u}:\mathbf{T} \tag{10–276}$$

where we have utilized the continuity condition $\nabla\cdot\mathbf{u}=0$ to rewrite the left-hand side. This mechanical energy equation is satisfied at every point in the fluid. Hence, it also applies in the integral sense,

$$\int_V\nabla\cdot\left[\frac{\rho}{2}(\mathbf{u}\cdot\mathbf{u})\,\mathbf{u}\right]dV = \int_V\nabla\cdot(\mathbf{u}\cdot\mathbf{T})\,dV - \int_V(\nabla\mathbf{u}:\mathbf{T})\,dV, \tag{10–277}$$

where V represents the whole of the volumetric region within the control volume. But now the divergence theorem can be applied to both the left-hand side of (10–277) and to the first term on the right. The integral on the left-hand side is equal to zero, that is,

$$\int_A\frac{\rho}{2}(\mathbf{u}\cdot\mathbf{n})(\mathbf{u}\cdot\mathbf{u})\,dA \equiv -\int_{A_{particle}}\frac{\rho}{2}(\mathbf{u}\cdot\mathbf{n})(\mathbf{u}\cdot\mathbf{u})\,dA + \int_{A_\infty}\frac{\rho}{2}U_\infty^3\cos\theta\sin\theta\,d\theta\,d\phi \equiv 0, \tag{10–278}$$

where the symbol A_∞ is used to denote the surface of the large circumscribed sphere in the fluid. The integral at the particle surface is zero because $\mathbf{u}\cdot\mathbf{n}=0$ at $A_{particle}$. The first integral on the right-hand side of (10–277) becomes

$$\int_A\mathbf{n}\cdot(\mathbf{u}\cdot\mathbf{T})\,dA = \int_A\mathbf{u}\cdot(\mathbf{n}\cdot\mathbf{T})\,dA = U_\infty\cdot\int_{A_\infty}\mathbf{T}\cdot\mathbf{n}\,dA - \int_{A_{particle}}\mathbf{u}\cdot(\mathbf{n}\cdot\mathbf{T})\,dA. \tag{10–279}$$

For the spherical bubble, the second term here is zero because the integrand is proportional to the shear stress $\tau_{r\theta}$, which is zero at the bubble surface. The first term on the right is just

$$U_\infty\cdot\int_{A_\infty}\mathbf{T}\cdot\mathbf{n}\,dA = U_\infty\cdot\int_{A_{particle}}\mathbf{T}\cdot\mathbf{n}\,dA = U_\infty\cdot\mathbf{F}_{drag}. \tag{10–280}$$

The equality of the integral of $\mathbf{T}\cdot\mathbf{n}$ over A_∞ and $A_{particle}$ is true only to $0(Re^{-1/2})$ because it does not account for the deficit in the momentum flux in the wake region behind the bubble, but this approximation is adequate for present purposes. Combining (10–277)–(10–280), we obtain (10–273).

Let us now apply (10–273) to calculate the drag on the spherical bubble, following the analysis of Levich (1949). First, it is convenient to nondimensionalize:

$$U_\infty F_{drag} = \mu U_\infty^2 a\int_V[\nabla\mathbf{u}:\mathbf{T}]\,dV. \tag{10–281}$$

Thus, expressing the drag in terms of the dimensionless drag coefficient,

$$C_D \equiv \frac{F_{\text{drag}}}{\pi a^2 \left(\frac{1}{2}\rho U_\infty^2\right)},$$

we find that

$$C_D = \frac{4}{\text{Re}}\left[\int_0^\pi \int_1^\infty r^2 [\nabla \mathbf{u}:\mathbf{T}] \sin\theta d\theta dr\right], \tag{10-282}$$

where

$$\Phi \equiv \nabla \mathbf{u}:\mathbf{T} = 2\left(\frac{\partial u_r}{\partial r}\right)^2 + 2\left(\frac{1}{r}\frac{\partial u_\theta}{\partial \theta} + \frac{u_r}{r}\right)^2 + 2\left(\frac{u_r}{r} + \frac{u_\theta \cot\theta}{r}\right)^2 +$$

$$\left[r\frac{\partial}{\partial r}\left(\frac{u_\theta}{r}\right) + \frac{1}{r}\frac{\partial u_r}{\partial \theta}\right]^2. \tag{10-283}$$

But now, a first approximation to the integral term in (10–282) can be obtained using the *inviscid* flow solution. In particular, we have shown that

$$u_r = u_r^p + 0(\text{Re}^{-1/2}),$$

$$u_\theta = u_\theta^p + 0(\text{Re}^{-1/2}), \tag{10-284}$$

throughout the domain $1 \le r\infty$, where u_r^p and u_θ^p are the inviscid velocity components given in (10–161) and (10–162), and the $0(\text{Re}^{-1/2})$ terms correspond to the viscous correction within the boundary-layer. We have not, of course, proven that the boundary-layer is the only region where $0(\text{Re}^{-1/2})$ corrections exist to the inviscid flow solution. In fact, Moore (1963)[27] has demonstrated that velocity corrections of this order also must exist in a thin wake of width $0(\text{Re}^{-1/4})$ around the axis of symmetry.

Now, substituting (10–284) into (10–283) and integrating according to (10–282), we find that

$$\int_V \Phi dV = 6 + 0(\text{Re}^{-1/2}). \tag{10-285}$$

Thus,

$$C_D = \frac{24}{\text{Re}}\left[1 + 0(\text{Re}^{-1/2})\right]. \tag{10-286}$$

Although we have determined the $0(\text{Re}^{-1/2})$ velocity corrections within the boundary layer, we cannot determine the numerical coefficient of the $0(\text{Re}^{-1/2})$ correction for C_D without a detailed analysis of the downstream wake. This is beyond the scope of the present work, but Moore (1963) carried out the necessary analysis to show that

$$C_D = \frac{24}{\text{Re}}\left[1 - \frac{2.2}{\text{Re}^{1/2}} + 0(\text{Re}^{-1})\right]. \tag{10-287}$$

The amazing feature of (10–286) is that it is obtained entirely from the inviscid flow solution—the boundary-layer analysis does play an important role in demonstrating that the volume integral of Φ, based on the inviscid velocity field, will provide a valid first approximation to the total viscous dissipation but does not enter directly.

An obvious question that may occur to the reader is why the very simple method of integrating the viscous dissipation function has not been used earlier for calculation of the force on a solid body. The answer is that the method provides no real advantage except for the motion of a shear stress-free bubble because the easily attained inviscid or potential-flow solution does not generally yield a correct first approximation to the dissipation. For the bubble, $\nabla\mathbf{u}:\mathbf{T} = 0(1)$ everywhere to leading order, including the viscous boundary-layer where the deviation from the inviscid solution yields only a correction of $0(\text{Re}^{-1/2})$. For bodies with no-slip boundaries, on the other hand, $\nabla\mathbf{u}:\mathbf{T}$ is still $0(1)$ outside the boundary-layer, but inside the boundary-layer $\nabla\mathbf{u}:\mathbf{T} = 0(\text{Re})$. When integrated over the boundary-layer, which is $0(\text{Re}^{-1/2})$ in radial thickness, this produces a $0(\sqrt{\text{Re}})$ contribution to the total dissipation,

$$\text{B.L. contribution viscous dissipation} \sim \sqrt{\text{Re}} \cdot \text{const},$$

and the corresponding estimate for C_D is

$$\boxed{C_D \sim \frac{\text{const}}{\sqrt{\text{Re}}}}$$

(assuming that the flow does not separate). Thus, in the more common no-slip problem, the dominant contribution to the dissipation comes from the boundary-layer. Although the estimate of dissipation due to the inviscid or potential flow does not change, it is of little use, and an estimation of drag via the rate of viscous dissipation still requires that we determine the complete boundary-layer solution.

One other aspect of the present problem that we promised to reconsider is the bubble shape. So far in this section we have assumed that the bubble is exactly spherical. However, in general we would expect a bubble to deform in a flow, and we have shown earlier that the shape can be determined from the normal stress condition

$$\mathbf{n} \cdot \mathbf{n} \cdot \mathbf{T} - \mathbf{n} \cdot \mathbf{n} \cdot \hat{\mathbf{T}} = \sigma(\nabla \cdot \mathbf{n}), \tag{10–288}$$

which represents a balance between the normal stress components due to the fluid motion and capillary forces due to interfacial or surface tension. For a bubble with negligible density and viscosity relative to the surrounding fluid, the normal stress inside is just the pressure, that is,

$$\mathbf{n} \cdot \mathbf{n} \cdot \hat{\mathbf{T}} = -p_{\text{bubble}}. \tag{10–289}$$

In *general,* if we regard the bubble shape as unknown, we must solve for the velocity, pressure, and bubble shape *simultaneously,* using the equations of motion, the usual boundary conditions

$$\left. \begin{array}{c} \mathbf{u} \cdot \mathbf{n} = 0 \\ \mathbf{n} \cdot \mathbf{t} \cdot \mathbf{T} = 0 \end{array} \right\} \quad \text{at} \quad r = r_s(\theta), \tag{10–290}$$

and

$$u_r \to -\cos\theta, \ u_\theta \to \sin\theta \quad \text{as} \quad r \to \infty, \tag{10–291}$$

plus the normal stress condition (10–288). Unless the bubble is only weakly deformed from spherical, this requires a numerical approach because the problem is highly nonlinear, not only in the equations of motion for Re $\neq 0$ but also in all of the boundary conditions, since these involve the unit normal and unit tangent vectors at an interface of unknown shape. If we assume that the shape is approximately spherical, however, we can analytically calculate the first departure from sphericity using the solution for the velocity and pressure fields for a spherical bubble. At the same time, we can obtain a criteria in terms of dimensionless parameters for the bubble to remain nearly spherical.

To do this, let us suppose that the bubble is nearly spherical, with a shape described in dimensionless terms by

$$\boxed{r_s = 1 + \epsilon f(\theta) + \cdots,} \qquad (10\text{–}292)$$

where ϵ is a small parameter, to be determined, and $f(\theta)$ is an unknown function that describes the departure from sphericity at $0(\epsilon)$, assuming that the shape will be axisymmetric. We showed in Chapter 5 that the unit normal and tangent vectors, calculated in terms of $f(\theta)$, are

$$\mathbf{n} = (1 + 0(\epsilon^2))\mathbf{i}_r - \left(\epsilon \frac{\partial f}{\partial \theta} + 0(\epsilon^3)\right)\mathbf{i}_\theta \qquad (10\text{–}293)$$

and

$$\mathbf{t} = \left(\epsilon \frac{\partial f}{\partial \theta} + 0(\epsilon^3)\right)\mathbf{i}_r + (1 + 0(\epsilon^2))\mathbf{i}_\theta, \qquad (10\text{–}294)$$

while the curvature function $\nabla \cdot \mathbf{n}$ appearing in the normal stress balance is

$$\nabla \cdot \mathbf{n} = 2 - \epsilon \left[2f + \frac{1}{\sin\theta} \frac{\partial}{\partial \theta}\left(\sin\theta \frac{\partial f}{\partial \theta}\right)\right] + 0(\epsilon^2). \qquad (10\text{–}295)$$

Thus, the boundary conditions (10–288)–(10–290) become

$$\boxed{T_{rr} + p_{\text{bubble}} + 0(\epsilon) = \frac{\sigma}{\rho U_\infty^2 a}\left[2 - \epsilon\left(2f + \frac{1}{\sin\theta} \frac{\partial}{\partial \theta}\left(\sin\theta \frac{\partial f}{\partial \theta}\right)\right) + 0(\epsilon^2)\right]} \qquad (10\text{–}296)$$

and

$$\left. \begin{aligned} u_r + 0(\epsilon) = 0 \\ \tau_{r\theta} + 0(\epsilon) = 0 \end{aligned} \right\} \quad \text{at} \quad r = 1 + 0(\epsilon). \qquad (10\text{–}297)$$

It can be seen from (10–297) that the boundary conditions (10–290) revert, at the leading order of approximation, to the boundary conditions (10–222) and (10–223). It follows that the velocity and pressure fields at this same level of approximation are simply the solutions for a spherical bubble that were obtained earlier. The normal-stress condition has not been required at all to obtain these results, but now, with the shape assumed to be unknown, the normal-stress condition provides a means to determine the $0(\epsilon)$ departure from sphericity. The dimensionless parameter in (10–296) is known as the Weber number,

$$\text{We} \equiv \frac{\rho U_\infty^2 a}{\sigma}. \qquad (10\text{–}298)$$

With the initial estimates of the velocity and pressure fields for a spherical bubble, we can calculate the left-hand side of (10–296). The normal-stress component is

$$T_{rr} = -p + \frac{1}{Re}\left(2\frac{\partial u_r}{\partial r}\right). \tag{10–299}$$

To leading order for $Re \gg 1$, we have seen already that the velocity field is given by the potential flow solution [see (10–284)], and this is also true of the velocity gradients, except in the boundary-layer region where the $0(Re^{-1/2})$ correction to u_r and u_θ produces a $0(1)$ modification in the gradient of the tangential velocity normal to the wall. In this region, however,

$$\left(\frac{\partial u_r}{\partial r}\right)_{r=1} = \frac{\partial V}{\partial Y}\bigg|_0 = -3\cos\theta + 0(Re^{-1/2}). \tag{10–300}$$

The pressure in the boundary-layer is

$$p = p_0(\theta) + 0(Re^{-1/2}) \tag{10–301}$$

since $\partial p_0/\partial Y = 0$ [see (10–234)], and thus $p_0(\theta)$ is determined from the pressure distribution in the outer, inviscid flow in the limit $(r-1) \ll 1$. In general, this pressure distribution consists of both a dynamic and a static part, as was also true in the corresponding problem at low Reynolds number (see Chapter 5, Section **A**). To calculate the dynamic part, we utilize the leading-order, inviscid flow equations for the outer region:

$$\frac{\partial p}{\partial r} = -u_r\frac{\partial u_r}{\partial r} - \frac{u_\theta}{r}\frac{\partial u_r}{\partial \theta} + \frac{u_\theta^2}{r} \tag{10–302}$$

and

$$\frac{1}{r}\frac{\partial p}{\partial \theta} = -u_r\frac{\partial u_\theta}{\partial r} - \frac{u_\theta}{r}\frac{\partial u_\theta}{\partial \theta} - \frac{u_r u_\theta}{r}, \tag{10–303}$$

with u_r and u_θ given by the inviscid flow solution (10–181) and (10–182). If we denote the far-field pressure on the horizontal plane through the center of the bubble as p_∞, we obtain upon integration

$$p_{\text{fluid}} = p_\infty + \frac{1}{2} - \frac{9}{8}\sin^2\theta + p_{\text{static}}. \tag{10–304}$$

The hydrostatic pressure variation at the bubble surface, in dimensional terms, is

$$p'_{\text{static}} = -(\rho g a)\cos\theta. \tag{10–305}$$

Hence,

$$p_{\text{fluid}} = p_\infty + \frac{1}{2} - \frac{9}{8}\sin^2\theta - \frac{\rho g a}{\rho U_\infty^2}\cos\theta. \tag{10–306}$$

For the spherical bubble, however, we have calculated the hydrodynamic drag (10–286). Hence, at steady state, the balance between buoyancy and drag is

$$\frac{4}{3}\pi a^3 \rho g = 12\pi\mu a U_\infty,$$

(10–307)

and it follows that the coefficient of the hydrostatic term in (10–306) is

$$\frac{a\rho g}{\rho U_\infty^2} = \frac{9}{\mathrm{Re}}.$$

(10–308)

The implication of (10–308) is that the hydrostatic variation in pressure at the bubble surface is negligible for $\mathrm{Re} \gg 1$ relative to the dynamic pressure variation in (10–304). Finally, collecting the results (10–299)–(10–308), we can rewrite the normal-stress condition (10–296) in the form

$$\left[\left[p_B - p_\infty - \frac{1}{2} + \frac{9}{8}\sin^2\theta + 0(\mathrm{Re}^{-1})\right] = \frac{1}{\mathrm{We}}\left[2 - \epsilon\left[2f + \frac{1}{\sin\theta}\frac{\partial}{\partial\theta}\left(\frac{\partial f}{\partial\theta}\sin\theta\right)\right]\right]\right].$$

(10–309)

It is now evident that for f to be $0(1)$ as assumed in writing (10–292), the small parameter must be

$$\boxed{\epsilon \equiv \mathrm{We}.}$$

(10–310)

In some ways, this is the most important result of the analysis. To ensure that the shape is near spherical, we require $\mathrm{We} \ll 1$. To determine the bubble shape at $0(\mathrm{We})$, we rewrite (10–309) to include (10–310) and change independent variables from θ to $\mu = \cos\theta$ to obtain

$$\frac{d}{d\mu}\left[(1-\mu^2)\frac{df}{d\mu}\right] + 2f = \left(-p_B + p_\infty + \frac{1}{2} + \frac{2}{\mathrm{We}} - \frac{9}{8}\sin^2\theta\right).$$

(10–311)

Rewriting the right-hand side in terms of Legendre polynomials,

$$\frac{d}{d\mu}\left[(1-\mu^2)\frac{df}{d\mu}\right] + 2f = \frac{3}{4}P_2(\mu) + \left[\frac{2}{\mathrm{We}} - p_B + p_\infty - \frac{1}{4}\right]P_0(\mu).$$

(10–312)

We require a solution of (10–312), which is bounded at $\mu = \pm 1$ and satisfies the two constraints

$$\int_{-1}^{1} f(\mu)d\mu = 0$$

(10–313)

and

$$\int_{-1}^{1} f(\mu)\mu d\mu = 0.$$

(10–314)

The first of these derives from the condition of constant bubble volume, while the second requires that the center of volume of the deformed bubble remains fixed at the origin of the coordinate system. Solving (10–312), subject to (10–313) and (10–314), we find that

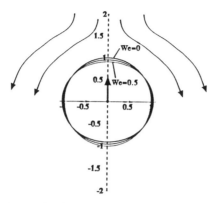

Figure 10–12 The shape of a rising gas bubble as a function of the Weber number, as predicted by small deformation, boundary-layer theory. We = 0, 0.25, 0.5.

$$f = -\frac{3}{16} P_2(\mu) \qquad (10\text{--}315)$$

and

$$p_B - p_\infty + \frac{1}{4} = \frac{2}{\text{We}}. \qquad (10\text{--}316)$$

The latter determines the pressure rise across the bubble surface expressed in terms of p_∞. Combining (10–315) with (10–292), we see that the bubble shape is

$$r_s = 1 + \text{We}\left(-\frac{3}{16} P_2(\mu)\right) + 0(\text{We}^2). \qquad (10\text{--}317)$$

Closer examination of (10–317) shows that the bubble is deformed into an oblate ellipsoid of revolution. A sketch of the bubble shape for several small values of We is given in Figure 10–12.

Moore (1965)[28] generalized the boundary-layer analysis for a spherical bubble to include the deformation to an oblate ellipsoidal shape. We do not reproduce his analysis, but it is worth recording the final result for the drag coefficient, which takes the form

$$C_D = \frac{24}{\text{Re}} G(\chi)\left[1 + \frac{H(\chi)}{\text{Re}^{-1/2}}\right],$$

where χ is the ratio of major to minor axes of the ellipsoid and $G(\chi)$ and $H(\chi)$ are functions of χ. Numerical values for G and H were tabulated by Harper (1972)[29] for $1.0 \le \chi \le 4.0$ and are reproduced here in Table 10–1.

Table 10.1 The Functions $G(\chi)$ and $H(\chi)$*

χ	$G(\chi)$	$H(\chi)$	χ	$G(\chi)$	$H(\chi)$
1.0	1.000	−2.211	2.6	4.278	1.499
1.1	1.137	−2.129	2.7	4.565	1.884
1.2	1.283	−2.025	2.8	4.862	2.286
1.3	1.437	−1.899	2.9	5.169	2.684
1.4	1.600	−1.751	3.0	5.487	3.112
1.5	1.772	−1.583	3.1	5.816	3.555
1.6	1.952	−1.394	3.2	6.155	4.013
1.7	2.142	−1.186	3.3	6.505	4.484
1.8	2.341	−0.959	3.4	6.866	4.971
1.9	2.549	−0.714	3.5	7.237	5.472
2.0	2.767	−0.450	3.6	7.620	5.987
2.1	2.994	−0.168	3.7	8.013	6.517
2.2	3.231	+0.131	3.8	8.418	7.061
2.3	3.478	+0.448	3.9	8.834	7.618
2.4	3.735	+0.781	4.0	9.261	8.189
2.5	4.001	+1.131			

*Reproduced from reference 29.

Homework Problems

1. Consider inviscid, potential flow past the half cylinder depicted below. Calculate the force (lift and drag) on the object assuming that the flow does not separate. If the flow does separate at 90°, what happens to the lift and drag?

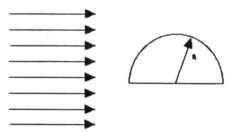

2. The shape of a gas bubble rising through a liquid at large Reynolds and Weber numbers (We $\equiv \rho U^2 a / \sigma \gg 1$) is very nearly a perfect spherical cap, with an upper surface that is spherical and a lower surface that is flat.

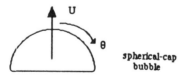

spherical-cap
bubble

Suppose that σ is sufficiently small; i.e., We is sufficiently large, that surface tension plays no role in determining the bubble shape, except possibly locally in the vicinity of the rim where the spherical upper surface and the flat lower surface meet. Further, suppose that the Reynolds number is sufficiently large that the motion of the liquid can be approximated to a first approximation, via the potential-flow theory. Denote the radius of curvature at the nose of the bubble as R (at $\theta = 0$). Show that a self-consistency condition for existence of a spherical shape with radius R in the vicinity of the stagnation point, $\theta = 0$, is that the velocity of rise of the bubble is

$$U = \frac{2}{3} \sqrt{gR}.$$

(Hint: Do not forget that p at the bubble surface is *total* pressure.)

3. Consider flow past a flat plate in the throat of a two-dimensional channel as depicted below. If the free-stream velocity is given by $u_\infty = \lambda x$, where x is the distance from the leading edge, show that the flow in the boundary-layer on the plate is governed by an ordinary differential equation. How does the boundary-layer thickness grow with x? How does the shear stress vary with x?

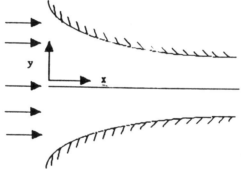

4. A flat plate is pulled through a wall with a constant velocity, U_0, into a Newtonian fluid of kinematic viscosity ν. Assume that $U_0/\nu \gg 1$.

   a. Derive the governing differential equations and boundary or matching conditions for the boundary-layer motion that occurs in the immediate vicinity of the plate surface. Show that this problem can be reduced to solving an ordinary differential equation by similarity transformation.

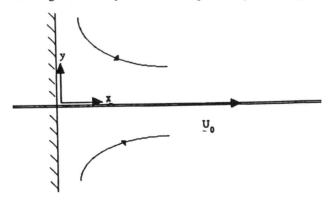

b. Use the Karman-Pohlhausen method and the velocity profile

$$u = \begin{cases} 1 - \sin\left(\dfrac{\pi}{2}\dfrac{Y}{\delta}\right) & \text{for } Y \le \delta(x) \\ \\ 0 & \text{for } Y > \delta(x) \end{cases}$$

to evaluate the boundary-layer thickness, the surface shear stress on the plate, and the drag coefficient.

5.* Consider converging flow (that is, flow inward toward the vertex) between two infinite plane walls with an included angle 2α as shown below.

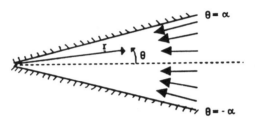

The flow is best described in cylindrical coordinates (r,θ,z) with z being normal to the plane of flow. We assume $u_z = 0$ and there is no z dependence, $\partial/\partial z(\ ) = 0$. Also, assume that the flow is entirely radial in the sense that fluid moves along rays $\theta = $ constant toward the vertex, that is, $u_\theta = 0$. (Obviously a solution of this form will only apply for *large* r, away from the opening at the vertex.)

a. Show that $u_r = f(\theta)/r$.

b. Obtain the governing differential equation for $f(\theta)$:

$$\frac{2}{\nu}f\frac{df}{d\theta} + \frac{d^3f}{d\theta^3} + 4\frac{df}{d\theta} = 0.$$

c. Assume no-slip conditions apply at the walls, and obtain a sufficient number of boundary conditions or other constraints on $f(\theta)$ to have a well-posed problem for the determination of $f(\theta)$.

d. Introduce a normalized angle ϕ and a normalized flow variable $F(\phi)$, such that

$$\phi = \theta/\alpha \quad \text{and} \quad F(\phi) = \alpha f/q,$$

where $q \equiv$ volume flow rate per unit width in the z direction, to obtain the dimensionless equation

$$\text{Re}\,F\frac{dF}{d\phi} + \frac{d^3F}{d\phi^3} + 4\alpha^2\frac{dF}{d\phi} = 0$$

plus boundary conditions and auxiliary conditions on $F(\phi)$. (Re $\equiv$ $2q\alpha/\nu$.)

*This problem is attributed to Denn, M.M., *Process Fluid Mechanics*. Prentice-Hall: Englewood Cliffs, NJ (1980).

e. Obtain an approximate solution for the creeping flow limit $Re \rightarrow 0$.
f. Obtain an approximate solution for $\alpha \rightarrow 0$ (the lubrication limit).
g. Consider the solution for $Re \rightarrow \infty$. Show that the solution reduces in this case to

$$F(\phi) = -\frac{1}{2}.$$

The solution obviously does not satisfy the no-slip condition on the solid boundaries, $\phi = \pm 1$. Explain why. Outline how a solution could be obtained for $Re \rightarrow \infty$, using the method of matched asymptotic expansions, which satisfies the boundary conditions on the walls. Be as detailed and explicit as possible, including actually setting up the equations and boundary conditions for the region near the wall. An exact solution is possible. The following integral may be useful:

$$\int \frac{d\xi}{\left(\xi - \frac{1}{2}\right)\sqrt{\xi + 1}} = \sqrt{\frac{2}{3}} \log \frac{\sqrt{\xi + 1} - \sqrt{3/2}}{\sqrt{\xi + 1} + \sqrt{3/2}}.$$

Calculate the drag on the plates.

One may also obtain an approximate solution via similarity transform using local Cartesian variables x and y, parallel and perpendicular to either of the walls. For this purpose, the solution $F(\phi) = -1/2$ provides an expression for $u_e(x)$. Use the following formulas if necessary:

$$\int \frac{d\xi}{\sqrt{(\xi - 1)^2 (\xi + 2)}} = \frac{2}{\sqrt{3}} \tanh^{-1} \left[\frac{\sqrt{2 + \xi}}{\sqrt{3}} \right] + C,$$

$$\tanh^{-1} \left(\sqrt{\frac{2}{3}} \right) = 1.146.$$

6. Consider the boundary-layer problem for a flat plate when the potential flow yields a velocity

$$u_e(x) = U_0 (1 - \alpha x^n)$$

A self-similar solution is not possible. Why? However, a solution can be generated as a power series in x with coefficients that are functions that depend on a similiarity variable.
a. Obtain the governing equations and boundary conditions for these coefficient functions. (The first should be the Blasius equation.)
b. Use integral methods to estimate the point of boundary-layer separation for $x > 0$.
7. The Karman-Pohlhausen method can be applied to estimate the shear stress distribution for any general potential velocity function $u_e(x)$. Determine $\delta(x)$ for a general function $u_e(x)$ using both the linear and cubic velocity profiles. Assume that the no-slip condition applies at $Y = 0$. In the case of the linear profile, calculate $\delta(x)$ for the special cases of
a. a circular cylinder ($u_e = 2 \sin x$),

b. a semi-infinite wedge $(u_e = x^m)$,

c. linearly decelerated flow $(u_e = 1 - x)$.

Compare with exact answers in each case. In the case of the cubic profile, stop when you have obtained an ordinary differential equation for $\delta^2(x)$ in terms of $u_e(x)$.

8. The classical boundary-layer theory represents only the first term in an asymptotic approximation for Re $\gg 1$. However, in cases involving separation, we do not seek additional corrections because the existence of a separation point signals the breakdown of the whole theory. When the flow does not separate, we can calculate higher order corrections, and these provide useful insight and results. In this problem, we reconsider the familiar Blasius problem of streaming flow past a semi-infinite flat plate that is oriented parallel to a uniform flow.

a. The Blasius function $f(\eta)$ for large η has the asymptotic form

$$f(\eta) \sim \eta - \beta_1 + \text{terms that are exponentially small.}$$

Show that the second term in the outer region is $0(\text{Re}^{-1/2})$ in magnitude and that the second-order outer problem is equivalent to potential flow past a semi-infinite, parabolically shaped body.

b. In general, the second-order *inner* solution satisfies a boundary-layer–like equation in which account is taken of the fact that the pressure distribution in the boundary-layer is modified by the second-order correction to the *outer* flow. Derive governing equations and boundary-conditions for the second-order *inner* solution for the Blasius problem. Show that the solution for the second-order term in the outer region, corresponding to a semi-infinite flat plate, yields the matching condition

$$\frac{\partial \Psi_2}{\partial Y}(x, \infty) = 0,$$

where Ψ_2 is the streamfunction for the second-order inner problem, and thus demonstrate that the second-order *inner* solution is

$$\Psi_2(x, Y) \equiv 0.$$

(Note that $u \equiv \partial \Psi/\partial Y$ and $V \equiv -\partial \Psi/\partial x$, where Y and V are the usual stretched boundary-layer variables. The above solution for Ψ_2 implies that the $0(\text{Re}^{-1/2})$ change in the pressure distribution for a semi-infinite flat plate produces *no* change in the boundary-layer region.)

c. For a *finite* flat plate, the second-order problem is modified because the condition

$$\psi_2(x, 0) = -\beta_1 \sqrt{2x}$$

(here ψ_2 is the streamfunction for the second-order *outer* problem) applies only for $0 < x \le 1$ ($x = 1$ is the end of the plate if the plate length is used as a characteristic length scale). For $x > 1$, an approximation is that the *displacement thickness* associated with the wake is a *constant* rather than a parabolic function of x as in the region $0 < x \le 1$. Thus, a solution can be obtained for ψ_2 in this case, which leads to the matching condition

$$\lim_{Y \gg 1} \Psi_2(x, Y) = \frac{\beta_1}{\sqrt{2\pi}} \frac{1}{\sqrt{x}} \log \frac{1 + \sqrt{x}}{1 + \sqrt{x}}.$$

Show that an approximate solution can be obtained for $\Psi_2(x, Y)$ in the form of a series expansion in x with coefficients that are functions of the Blasius similarity variable η. Determine the governing differential equations and boundary conditions for the first two terms in this expansion. It can be shown that

$$C_D \equiv \frac{\text{drag}}{\frac{1}{2}\rho U_\infty^2 L} \sim \frac{1.328}{\sqrt{\text{Re}}} + \frac{C_1}{\text{Re}} + \cdots,$$

and if enough terms are retained in the approximate solution for Ψ_2, the constant C_1 can be approximated as $C_1 = 4.12$. You need not calculate C_1 in detail, but you should explain the result above for C_D and explain in principle how to calculate C_1. What value does C_1 have for a semi-infinite flat plate? Does this surprise you? Explain.

9. The boundary-layer approximation, in which derivatives in the primary flow direction are assumed to be small compared with derivatives across the flow, finds application in a number of problems far from a source of mass or momentum in an unbounded fluid. One example is in the analysis of the wake far downstream of a body in a uniform streaming flow. As an example, let us consider a two-dimensional cylindrical body that has a cross section that is symmetric about the $y = 0$ axis, but is otherwise arbitrary, placed in a Newtonian fluid that is moving with a constant velocity $\mathbf{U} = U\mathbf{i}$, as depicted below.

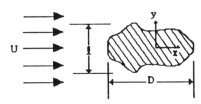

Very near to the body, the velocity field is very complicated due, in part, to the complex geometry of the body and differs considerably from the free-stream form $\mathbf{U}$, which would occur everywhere if the body were not present. However, far downstream ($x \gg 1$) the velocity field gradually returns to its undisturbed state; that is, the deviations from the undisturbed velocity field $\mathbf{U}$ become small. Under these circumstances, the actual velocity field, which we denote as $\mathbf{u}$, can be expressed as

$$\mathbf{u} = \mathbf{U} - \mathbf{u'} \text{ with } |\mathbf{u'}| \ll |\mathbf{U}| \qquad (x \text{ large}). \tag{1}$$

The field $\mathbf{u'}$ is called the disturbance velocity field since it represents the difference between the undisturbed field $\mathbf{U}$, which would exist if the body were not present, and the actual velocity in the presence of the body. It is observed that the velocity field $\mathbf{u}$ is completely symmetric about $y = 0$ (all x). Furthermore, when x is large, the major deviations from the free-stream velocity $\mathbf{U}$ occur in the vicinity of the x axis ($y = 0$); that is, $\mathbf{u'}$ is nonzero only near $y = 0$ for large x, and this region is called the *wake flow region*.

a. One interesting feature of the *wake flow region* is that one can directly determine the force acting on the body by knowing (measuring) the disturbance velocity component $u'_x \equiv \mathbf{i} \cdot \mathbf{u'}$ as a function of y. Employ the appropriate macroscopic balance equation(s) to derive an expression for the streamwise (x) component of the force (F_x) acting on the body in terms of $U = \mathbf{i} \cdot \mathbf{U}$ and u'_x at a distant downstream station. Assume that $\mathbf{j} \cdot \mathbf{u} \equiv 0$ for large values of x. Show all of your work, including your choice of a control volume surface, and carefully state all assumptions that you may have had to employ. You should find that

$$F_x = c_1 \int_{y=0}^{\infty} u'_x (2U - u'_x) dy, \tag{2}$$

in which c_1 is a constant. Note that the y component of the net force acting on the body is zero because of the symmetry imposed about the x axis ($y = 0$). Show that $F_y \equiv 0$, using the macrobalance approach. Since the drag (F_x) is a fixed number for a given body and a given value of U, it follows from (2) that

$$\int_{y=0}^{\infty} u'_x (2U - u'_x) dy = \text{const, independent of } x \text{ for large values of } x.$$

Hence, neglecting the small term that is quadratic in $|u'_x|$,

$$\int_{y=0}^{\infty} 2u'_x U dy = \text{const (independent of } x); \quad x \gg 1,$$

or

$$\int_{y=0}^{\infty} 2u'_x dy = \text{const} = \frac{F_x}{c_1 U}. \tag{3}$$

b. Observations lead us to believe, in addition to the other features of the flow described above, that

$$\left| \frac{\partial^2 u'_x}{\partial y^2} \right| \gg \left| \frac{\partial^2 u'_x}{\partial x^2} \right| \tag{4}$$

in the wake flow region. Use this fact, as well as (1), to show that an approximate solution for u'_x can be obtained from

$$U \frac{\partial u'_x}{\partial x} = \nu \frac{\partial^2 u'_x}{\partial y^2} \qquad (x \gg 1), \tag{5}$$

subject to boundary conditions

$$u'_x \to 0, \qquad |y| \to \infty \qquad (x \gg 1), \tag{6}$$

$$\frac{\partial u_x'}{\partial y} = 0, \qquad y = 0 \qquad (x \gg 1),$$

and condition (3).

c. Suppose that you decide to nondimensionalize equations (5) and (6). What length and velocity scales would you use? Do both of these choices continue to make sense as you move far away from the body? (Note that the body would appear as a thin line as we move to very large values of x.) Assume that

$$u_x' = c_2 x^m f(\eta),$$

where

$$\eta = \frac{y}{x^n} c_3$$

and c_2 and c_3 are constants. Determine the coefficients m and n, and derive the governing equation for f. With appropriate choices for the constants c_2 and c_3, you should find that

$$f'' + \frac{1}{2}\eta f' + \frac{1}{2}f = 0. \tag{7}$$

(Hint: The coefficient m cannot be zero. Why not?) What boundary conditions must f satisfy? Why should you have expected a similarity solution in this problem?

d. Solve for the similarity function f.

e. Show that the constant that appears in the similarity function $f(\eta)$ can be related simply to the force magnitude F_x.

10. Consider the two-dimensional flow created when fluid is forced with high velocity through a very narrow slit in a wall as pictured below (this flow is referred to as a two-dimensional symmetric jet).

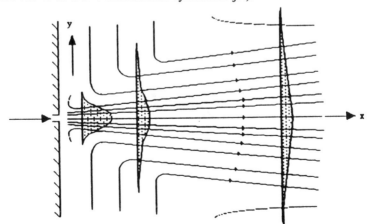

As the fluid leaving the slit contacts the surrounding fluid, the action of viscosity slows it down while causing some of the surrounding fluid to be accelerated. Hence, with increasing x, the maximum velocity decreases while the width of the jet increases.

a. Show that this process must occur so that the total momentum in the x direction is conserved; that is, show that

$$J = \rho \int_{-\infty}^{\infty} [u(x,y)]^2 dy = \text{constant (independent of } x).$$

b. When the average velocity of the fluid leaving the slit divided by the kinematic viscosity is large, the jet is observed to be long and narrow. Under these conditions the equations governing the flow in the jet can be reduced to the boundary-layer–like equations. What is the pressure gradient in the jet? What are the appropriate boundary conditions? (Note that it is more convenient to use conditions at the center of the jet and at one edge rather than conditions at both edges of the jet.)

c. Rewrite your equation in terms of the nondimensionalized streamfunction Ψ defined as

$$\frac{\partial \Psi}{\partial Y} = u \quad , \quad \frac{\partial \Psi}{\partial x} = -V,$$

where variables are considered to be nondimensionalized as in the boundary-layer theory in the limit of $\text{Re} \to \infty$.

d. Show that by introducing the similarity transformation

$$\Psi = x^m f(\eta) \quad , \quad \eta = \frac{Y}{x^n},$$

the governing equation can be reduced to

$$f''' + 2ff'' + 2f'^2 = 0.$$

Find m and n as part of your solution. What are the boundary conditions and momentum constraint in terms of $f(\eta)$? What is the rate of decrease of the centerline velocity with x, and how fast does the width of the jet increase?

e. This equation can be integrated twice to find $f(\eta)$; after applying the boundary conditions we obtain

$$f(\eta) = \alpha \tanh(\alpha\eta).$$

What are the corresponding velocity components u and V? Relate α to J, the total momentum of the fluid leaving the slit.

f. Find an expression for the half width of the jet defined as the distance from the centerline to the point where u equals 1 percent of its centerline value. Find an expression for the local Reynolds number as a function of x.

References/Notes

1. See reference 10 of Chapter 8.

2. Prandtl, L., "Über Flüssigkeitsbewegung bei sehr kleiner Reibung," *Proc. Third Intern. Math. Congr.*, Heidelberg (1904). Reprinted in *Collected Works*, vol. II, 575–584 (1961).

3. Schlichting, H., *Boundary-Layer Theory*, 6th ed. McGraw-Hill: New York (1968). See chapter 7, pp. 118–121.

4. See Reference 3, Chapter 1, section f.

5. Goldstein, S., "Fluid Mechanics in the First Half of This Century," *Ann. Rev. of Fluid Mech.* 1:1–28 (1969).

6. The subject of geophysical fluid dynamics encompasses two effects not usually considered in other areas of fluid mechanics: rotation of the fluid and the existence of density stratifications in the oceans and atmosphere. Together, these lead to very important and fascinating qualitative changes in the fluid motions. A useful account of some of the phenomena encountered in rotating and stratified fluids can be found in chapters 15 and 16 of Tritton, D.J., *Physical Fluid Dynamics.* Van Nostrand Reinhold: London (1977).

7. A complete compilation of results from the subject of theoretical hydrodynamics—the motion of an inviscid fluid satisfying Euler's equations [namely, Equations (10–1) and (10–2), but with the viscous term deleted]—can be found in the classic book

Milne-Thompson, L.M., *Theoretical Hydrodynamics*, 5th ed. MacMillan: London (1967).

8. See Reference 2 of Chapter 2, for a discussion of the historical evolution of the no-slip boundary condition.

9. The problem of start-up flow for a circular cylinder has received a great deal of attention over the years because of its role in understanding the inception and development of boundary-layer separation. A relatively recent publication with a comprehensive reference list of both analytical and numerical studies is Cowley, S.J., "Computer Extension and Analytic Continuation of Blasius' Expansion for Impulsive Flow Past a Circular Cylinder," *J. Fluid Mech.* 135:389–405 (1983).

10. One recent proposal, which also contains references to earlier work on this problem is Smith, F.T., "A Structure for Laminar Flow Past a Bluff Body at High Reynolds Number," *J. Fluid Mech.* 155:175–191 (1985).

11. Fornberg, B., "Steady Viscous Flow Past a Circular Cylinder up to Reynolds Number 600," *J. Comp. Phys.* 61:297–320 (1985).

12. The original publications describing the behavior of solutions of the boundary-layer equations at a separation point are:

a. Goldstein, S., "On Laminar Boundary-Layer Flow Near a Position of Separation," *Quart. J. Mech. Appl. Math.* 1:43–69 (1948).

b. Stewartson, K., "On Goldstein's Theory of Laminar Separation," *Quart. J. Mech. Appl. Math.* 11:399–410 (1958).

A more modern view of boundary-layer separation including the structure of boundary-layer solutions near a separation point can be found in

c. Williams, J.C., III, "Incompressible Boundary-Layer Separation," *An. Rev. of Fluid Mech.* 9:113–144 (1977).

d. Smith, F.T., "Steady and Unsteady Boundary-Layer Separation," *An. Rev. of Fluid Mech.* 18:197–220 (1986).

13. A very readable account of higher order approximations in boundary-layer theory may be found in the textbook by Van Dyke, cited earlier as Reference 1c in Chapter 6, or in a review paper by the same author,

Van Dyke, M., "Higher-Order Boundary-Layer Theory," *An. Rev. Fluid. Mech.* 1:265–292 (1969).

14. Blasius, H., "Grenzschichten in Flüssigkeiten mit Kleiner Reibung," *Zeib. Math. u. Phys.* 56:1–37 (1908). English translation in *NACA TM* 1256.

15. Carrier, G.F., and Lin, C.C., "On the Nature of the Boundary-Layer Near the Leading Edge of a Flat Plate," *Quart. Appl. Math.* 6:63–68 (1948).

16. See Reference 7.

17. a. Hartree, D.R., "On an Equation Occurring in Falkner and Skan's Approximate Treatment of the Equations of the Boundary-Layer," *Proc. Cambr. Phil. Soc.* 33:223–239 (1937).

b. Stewartson, K., "Further Solutions of the Falkner-Skan Equation," *Proc. Camb. Phil. Soc.* 50:454–465 (1954).

18. An interesting extension of these ideas was discovered by Stewartson (Reference 17), who showed that there is another branch of solutions for $0 > \beta > -0.1998$ that is physically acceptable in the sense that $f' \to 1$ from below but which contain a region for η near zero where $f'(\eta) < 0$. It was originally suggested by Stewartson that these solutions may be relevant to the postseparation region of the flow, assuming that the separation process (which occurs at $\beta = -0.1998$) produces a decrease in the adverse pressure gradient so that $\beta > -0.1998$ is applicable in this region. Later analysis, by Stewartson and others, has analyzed the details of interaction between the boundary layer and the exterior flow in the vicinity of a separation point. Much of this work is summarized in a fascinating review paper by Stewartson, "D'Alembert's Paradox," *SIAM Rev.* 23:308–343 (1981).

19. Howarth, L., "On the Calculation of Steady Flow in the Boundary-Layer Near the Surface of a Cylinder in a Stream," *ARC RM* 1632 (1935).

20. A series of experiments was carried out by Acrivos and co-workers that showed that the transition from a steady symmetric recirculating wake flow downstream of a cylindrical body to an unsteady wake with periodic vortex shedding could be delayed from an expected Reynolds number of about 40 to a value of $R \approx 200$, by insertion of a thin plate (known as a splitter plate) on the downstream axis of symmetry at the end of the circulating region. See Acrivos, A., Leal, L.G., Snowden, D.D., and Pan, F., "Further Experiments on Steady Separated Flows Past Bluff Objects," *J. Fluid Mech.* 34:25–48 (1968).

21. These ideas are developed in more detail in a recent paper

a. Leal, L.G., "Vorticity Transport and Wake Structure for Bluff Bodies at Finite Reynolds Number," *Phys. Fluids* A1:124–131 (1989).

Additional insight into the evolution of a separated wake for a solid, circular cylinder may be obtained at large, but finite, Reynolds number from numerical solutions such as

b. Collins, W.M., and Dennis, S.C.R., "Flow Past an Impulsively Started Circular Cylinder," *J. Fluid Mech.* 60:105–127 (1973).

22. The original reference to this work (in German) is given by Schlichting (Reference 3), p. 148. The presentation here is similar to that given by Schlichting.

23. Scholkemeier, F.W., "Die Laminare Reibungsschicht an Rotationssymmetrischen Körpen," *Arch. d. Math.* 1:270–277 (1949).

24. Experimental observations of bubble shape and of other aspects of bubble motions is given in

a. Bhaga, D., and Weber, M.E., "Bubbles in Viscous Liquids: Shapes, Wakes and Velocities," *J. Fluid Mech.* 105:61–85 (1981).

This and other work on this problem is summarized in the book

b. Clift, R., Grace, J.R., and Weber, M.E., *Bubbles, Drops and Particles.* Academic Press: New York (1978).

25. See Reference 6 of Chapter 3.

26. Levich, V.G., "Motion of Gas Bubbles with High Reynolds Numbers" (in Russian), *Zhn. Eksp. Teor. Fiz.* 19:18–24 (1949).

An excellent summary of theoretical and experimental work on the motion of bubbles and drops, including the preceding reference, is found in Reference 1 of Chapter 5.

27. Moore, D.W., "The Boundary Layer on a Spherical Gas Bubble," *J. Fluid Mech.* 16:161–176 (1963).

28. Moore, D.W., "The Velocity of Rise of Distorted Gas Bubbles in a Liquid of Small Viscosity," *J. Fluid Mech.* 23:749–766 (1965).

29. Harper, J.F., "The Motion of Bubbles and Drops Through Liquids," *Adv. in Applied Mech.* 12:59–129 (1972).

CHAPTER 11

Thermal Boundary-Layer Theory at Large Reynolds Number

We saw in Chapter 10 that the boundary-layer structure, which arises naturally in flows past bodies at large Reynolds numbers, provides a basis for approximate analysis of the flow. In this chapter, we consider heat transfer (or mass transfer for a single solute in a solvent) in the same high Reynolds number limit for problems in which the velocity field takes the boundary-layer form. We saw previously that the thermal energy equation in the absence of significant dissipation, and at steady state, takes the dimensionless form

$$\mathbf{u} \cdot \nabla \theta = \frac{1}{\text{Pe}} \nabla^2 \theta, \tag{11–1}$$

where

$$\theta \equiv \frac{T - T_\infty}{T_0 - T_\infty}$$

and

$$\text{Pe} \equiv \text{Re} \cdot \text{Pr}.$$

In Chapters 8 and 9, we considered the solution of this equation in the limit $\text{Re} \ll 1$, where the velocity distribution could be approximated via solutions of the creeping flow equations. When $\text{Pe} \gg 1$, we found that the fluid was heated (or cooled) significantly only in a very *thin thermal boundary-layer* of $0(\text{Pe}^{-1/3})$ in thickness, immediately adjacent to the body surface.

Here, we consider heat transfer from heated (or cooled) bodies in the dual limit $\text{Re} \gg 1$ and $\text{Pe} \gg 1$.[1] Because $\text{Re} \gg 1$, we assume that the velocity field in the vicinity of the body is characterized by a momentum boundary-layer. In addition, since $\text{Pe} \gg 1$, we also expect the temperature distribution to be characterized by a thin thermal boundary-layer. Although both boundary-layers will be thin, we should not necessarily expect the scaling of Chapters 9 and 10 to be preserved. Note that the condition $\text{Pe} \gg 1$ does not limit the range of possible values for the Prandtl number, Pr. Indeed, if Pr is small, the Peclet number can still be large provided that the Reynolds number is

637

large enough. However, the magnitude of Pr does determine the relative size of Re and Pe. When $Pr \gg 1$, then $Pe \gg Re \gg 1$. On the other hand, when $Pr \ll 1$, then $Re \gg Pe \gg 1$. Although the dependence of the thermal boundary-layer thickness on the independent parameters Re and Pr (or Pe) remains to be determined, we may anticipate that the magnitudes of Re and Pe will determine the *relative* dimensions of the two boundary-layers. If $Pe \gg Re \gg 1$, both the momentum and thermal layers will be thin, but it seems likely that the thermal layer will be much the thinner of the two. Likewise, if $Re \gg Pe \gg 1$, we can guess that the momentum boundary-layer will be thinner than the thermal layer. In the analysis that follows in later sections of this chapter, we consider both of the asymptotic limits $Pr \to \infty (Pe \gg Re \gg 1)$ and $Pr \to 0(Re \gg Pe \gg 1)$. We shall see that the relative dimensions of the thermal and momentum layers, anticipated above on purely heuristic grounds, will play an important and natural role in the theory.

The primary focus of our analysis, as in Chapters 8 and 9, is a prediction of the correlation between the Nusselt number, Nu, representing the dimensionless total heat flux, and the independent dimensionless parameters Re and Pr (or Pe). The engineering literature abounds with such correlations, determined empirically from experimental data.[2] The general form of these correlations is

$$Nu = c Re^a Pr^b. \tag{11–2}$$

For laminar flow conditions and $Re \gg 1$, the coefficient a is found to be

$$a = 0.5,$$

while the coefficient b varies from approximately 0.3 to 0.5, depending upon the range of values considered for Pr. The coefficient c is generally $0(1)$ and *dependent* only on the geometry of the body. We shall prove, from the asymptotic *formulation* of the problem (that is, without solving any detailed equations) that

$$b = \frac{1}{2} \text{ in the limit } Pr \to 0$$

and

$$b = \frac{1}{3} \text{ in the limit } Pr \to \infty.$$

Let us begin by considering the governing equation and boundary conditions for $Re \gg 1$ and $Pe \gg 1$.

A. Governing Equations ($Re \gg 1$, $Pe \gg 1$, with Pr Arbitrary)

We start with the nondimensionalized thermal energy equation, in the form (11–1), where spatial variables are scaled using the dimension of the body as a characteristic length scale. The geometry of the body is assumed to be arbitrary, but two-dimensional for simplicity. However, if the flow separates, the boundary-layer analysis that we eventually obtain applies only to the portion of the body surface where the momentum boundary-layer remains attached. It is simplest to restrict attention to cases where the boundary-layer does not separate at all. The Prandtl number may be either large or small, but we assume Pe $\gg$ 1 as stated above. In this limit, (11-1) reduces to the familiar form

$$\boxed{\mathbf{u}\cdot\nabla\theta = 0,}$$

(11-3)

in which conduction terms are neglected altogether, and the solution in regions where the streamlines are open is

$$\boxed{\theta = \theta(\psi) = 0.}$$

(11-4)

It is evident from our previous analyses of heat transfer at large Pe (Chapter 9) that this solution and the governing equation (11-3) is valid as a first approximation in some *outer* region away from the body surface.

This outer solution $\theta \equiv 0$ must be supplemented near the body surface by a solution that includes the effects of conduction, and thus satisfies the condition $\theta = 1$ at the boundary. For large Peclet numbers, Pe $\gg$ 1, such a solution will be valid in a thin thermal boundary-layer. Following the procedures of earlier chapters, the characteristic length scale for this region is determined by *rescaling*. The primary change from the preceding analyses is that the thermal boundary-layer thickness depends *independently* upon the two parameters Re and Pr rather than depending on the combined parameter Pe = RePr only. This occurs because the velocity distribution in the region near the body surface is characterized by a length scale, $l_c = a\mathrm{Re}^{-1/2}$, that depends explicitly on the Reynolds number. Thus, the Reynolds number and the Prandtl number "appear" *independently* in the thermal energy equation rather than only in the form of the combined parameter Pe. In contrast, for the low Reynolds number limit considered earlier, the velocity field in the region relevant to the thermal energy equation is characterized by the length scale of the body, and it is only Pe that appears in the thermal problem.

We begin by considering the inner thermal boundary-layer limit for arbitrary Pr $\sim$ 0(1), but with Pe $\gg$ 1. The velocity distribution in the vicinity of the body surface is characterized by the length scale

$$l \sim a(\mathrm{Re})^{-1/2}$$

independent of the form of the temperature distribution. Thus, as a first step, we rescale the thermal energy equation in a manner consistent with the momentum boundary-layer scaling; that is, we introduce

$$\boxed{Y = y\sqrt{\mathrm{Re}}, \ V = v\sqrt{\mathrm{Re}}}$$

(11-5)

into the thermal energy equation (11-1), now expressed in terms of local Cartesian, boundary-layer coordinates. The result is

$$\boxed{u(x,Y)\frac{\partial\theta}{\partial x} + V(x,Y)\frac{\partial\theta}{\partial Y} = \frac{1}{\mathrm{Pr}}\frac{\partial^2\theta}{\partial Y^2} + 0(\mathrm{Re}^{-1}\mathrm{Pr}^{-1}).}$$

(11-6)

We wish to solve this equation subject to the boundary conditions

$$\theta = 1 \text{ at } Y = 0, \ x > x^*, \tag{11-7a}$$

$$\theta \to 0 \text{ for } Y \gg 1, \tag{11-7b}$$

and

$$\theta = 0 \text{ at } x = x^*, \ Y > 0. \tag{11-7c}$$

The conditions (11-7a) and (11-7c) correspond to the assumption that the surface of the body is heated (or cooled) to a constant temperature T_o beyond a certain position denoted as x^*. Upstream of x^*, the body is not heated. Hence, at steady state in the present boundary-layer limit, both the body surface and the fluid remain at the ambient temperature T_∞ for $x < x^*$. If the body is heated over its whole surface, then $x^* = 0$. In this case, the "leading edge" of the thermal and momentum boundary-layers are coincident.

Before going any further, there is one aspect of the rescaling (11-5) that requires discussion. We have stated that this rescaling is relevant to $\mathrm{Pr} \sim 0(1)$, and this is clearly consistent with our earlier discussion of the role of the Prandtl number in determining the relative dimensions of the thermal and momentum boundary-layers. In particular, for $\mathrm{Pr} \sim 0(1)$, the two parameters Re and Pe are of comparable magnitude, and we should expect the width of the momentum and thermal layers to be similar. It is not surprising in this case that the momentum boundary-layer scaling should apply to the thermal energy equation. However, our goal is to solve the thermal problem for a wide range of Prandtl numbers, and it is not so obvious that the rescaling (11-5) and governing equation (11-6) should apply to the limiting cases $\mathrm{Pr} \to 0$ or $\mathrm{Pr} \to \infty$, where our previous arguments suggested that the scales of the two boundary-layers might be quite different.

The fact is that the thermal boundary-layer equation (11-6) is relevant for the whole range of possible values for Pr, provided only that Re, Pe $\gg 1$, as stated earlier. In the case of $\mathrm{Pr} \to \infty$, our earlier discussion suggested that the thermal layer (the region in which θ decreases from its surface value, $\theta = 1$, to its free-stream value, $\theta = 0$) would be thin compared with the dimensions of the momentum layer [the region in which u increases from its surface value, $u = 0$, for a no-slip body to its potential flow limit, $u_e(x)$]. But if this is true, the whole of the thermal layer will lie "inside" the momentum layer, and it should not be surprising if the scaling (11-5) is relevant. The less obvious case is the limit $\mathrm{Pr} \ll 1$. In this case, we suggested earlier that the extent of the thermal layer would be large relative to the momentum layer. Nevertheless, the boundary-layer velocity distributions are still relevant in the following sense. As Pr decreases in magnitude, the thermal layer lies increasingly in the part of the momentum boundary-layer that corresponds to large values of Y. To put it another way, the thermal layer is increasingly controlled by the potential-flow velocity distribution. However, regardless of the magnitude of Pr, we assume Pe $\gg 1$! Hence, no matter how wide the thermal region is compared with the momentum boundary-layer, it always remains asymptotically thin compared with the dimension of the body. Thus, no matter how small the Prandtl number becomes, it is only *the limit of the potential-flow velocity profile as we approach the body surface* that is relevant to the thermal energy equation. However, at leading order in Re^{-1}, the matching condition for the momentum boundary-layer solution,

$$\lim_{Y \gg 1} u(x, Y) \Longleftrightarrow \lim_{y \to 0} u(x, y) = u_e(x) \text{ for } Re \to \infty, \tag{11-8}$$

ensures that this limit of the potential flow solution is *identical* to the outer limit $(Y \gg 1)$ of the boundary-layer solution. Thus, for all Pr, no matter how small, the use of the momentum boundary-layer scaling remains valid in (11–6), and there is a smooth transition to the limiting case $\text{Pr} \to 0$, provided only that $\text{Pe} \gg 1$.

We deduce from these arguments that the Reynolds number dependence of the Nusselt number can be determined directly from (11–5). We recall that the definition of the Nusselt number, in terms of the dimensionless temperature gradient, is

$$\boxed{\text{Nu} \equiv \int_{x=x^*}^{x=x_L} -\left(\frac{\partial \theta}{\partial y}\right)_{y=0} dx,} \tag{11-9}$$

where x_L is the end of the heated surface (usually the "trailing edge" of the body) and $y = 0$ corresponds to the body surface. Hence, as shown earlier, it follows from dimensional analysis of the thermal energy equation (11–1) that

$$\text{Nu} = \text{Nu}(x^*, x_L; \text{Re}, \text{Pr}). \tag{11-10}$$

For all Re, Pr $\sim 0(1)$, the value of the number on the right-hand side of (11–10) is also $0(1)$. Here, however, we consider the limit Re $\gg 1$. Hence, applying the momentum boundary-layer scaling (11–5) to the definition (11–9), we see that

$$\boxed{\text{Nu} = \sqrt{\text{Re}} \int_{x^*}^{x_L} -\left(\frac{\partial \theta}{\partial Y}\right)_{Y=0} dx.} \tag{11-11}$$

The temperature gradient $(\partial \theta / \partial Y)_{Y=0}$ is determined by the governing equation and boundary conditions (11–6) and (11–7). Thus,

$$\frac{\partial \theta}{\partial Y} = \frac{\partial \theta}{\partial Y}(x, Y; \text{Pr}),$$

and it follows that

$$\boxed{\text{Nu} = \sqrt{\text{Re}} F(x^*, x_L; \text{Pr}),} \tag{11-12}$$

where

$$\boxed{F(x^*, x_L; \text{Pr}) \equiv -\int_{x^*}^{x_L} \left(\frac{\partial \theta}{\partial Y}\right)_{Y=0} dx.} \tag{11-13}$$

B. Exact Solutions for Pr ~ 0(1)

To proceed further, it is necessary to solve the thermal boundary-layer problem defined by (11–6) and (11–7). However, in spite of the fact that we have a *linear* differential equation for θ, it is generally difficult to solve due to the complex and variable coefficients $u(x, Y)$ and $V(x, Y)$. We shall see shortly that we can obtain approximate closed-form solutions irrespective of the geometry of the body (provided, of course, that the basic assumptions of an "attached" boundary-layer remain valid) for the two asymptotic limits $\text{Pr} \to \infty$ and $\text{Pr} \to 0$. For arbitrary $\text{Pr} \sim 0(1)$, it is possible to solve (11–6) in analytic form only for a few very specific bodies.

In particular, let us consider the class of body geometries for which the momentum boundary-layer equations allow similarity solutions to be obtained. Thus, we consider streaming flow past a semi-infinite wedge with included angle $\pi\beta$, as sketched in Figure 11–1. Then, provided that $x^* = 0$, so that the leading edge of the thermal layer is coincident with the leading edge of the momentum layer, it can be proven that similarity solutions also can be obtained for θ in the form

$$\boxed{\theta = \theta_s(x) H(\eta),} \tag{11–14}$$

where η is the similarity variable from the fluid mechanics problem ($\eta \equiv Y/g(x)$), and the surface temperature distribution is of the form

$$\theta_s(x) \approx x^m. \tag{11–15}$$

Here we consider, in detail, only the simplest case with $m = 0$ and $\beta = 0$, namely, a semi-infinite flat plate with $\theta_s = 1$ for all x. In this case, the velocity field is given by the Blasius solution

$$u = f'(\eta), \quad V = g'(\eta f' - f), \tag{11–16}$$

with

$$\eta \equiv \frac{Y}{g} \quad \text{and} \quad g(x) = \sqrt{2x}. \tag{11–17}$$

Now, let us assume that a similarity solution also exists for θ in the form

$$\boxed{\theta = H(\eta).} \tag{11–18}$$

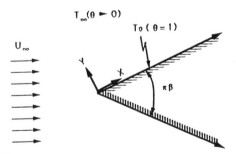

Figure 11–1 A schematic representation for heat transfer from a semi-infinite wedge with surface temperature T_0 (dimensionless $\theta \equiv 1$) and ambient temperature $T_\infty (\theta = 0)$.

The similarity variable η is the same, (11–17), as for the Blasius solution. Substituting (11–18) into (11–6) leads to the similarity equation

$$\frac{d^2H}{d\eta^2} + \Pr f(\eta)\frac{dH}{d\eta} = 0,$$ (11–19)

where we have used the fact that $gg' = 1$. As expected, the similarity transformation (11–18) leads to a second-order ordinary differential equation for H. Hence, in order for the similarity solution to be workable, it is evident that the boundary condition (11–7b) and the initial condition (11–7c) must collapse into a single condition for H. It is for this reason that we require the leading edge of the thermal and momentum boundary layers to coincide (that is, $x^* = 0$). In this case, the boundary conditions on $H(\eta)$ become

$$H(0) = 1,$$
$$H(\eta) = 0 \quad \text{as} \quad \eta \to \infty.$$ (11–20)

A formal solution satisfying (11–19) and (11–20) is obtained easily, namely,

$$H(\eta) = 1 - \frac{\int_0^\eta e^{-\Pr\int_0^t f(s)\,ds}\,dt}{\int_0^\infty e^{-\Pr\int_0^t f(s)\,ds}\,dt}.$$ (11–21)

However, the Blasius function $f(\eta)$ is available only as the numerical solution of the Blasius equation, and it is thus inconvenient to evaluate this formula for $H(\eta)$. A simple alternative is to numerically integrate the Blasius equation and the thermal energy equation (11–19) simultaneously. The function $H(\eta)$, obtained in this manner, is plotted in Figure 11–2 for several different values of the Prandtl number, $0.01 \le \Pr \le 100$. As suggested earlier, it can be seen that the thermal boundary-layer thickness depends strongly on Pr. For $\Pr \gg 1$, the thermal layer is increasingly thin relative to the Blasius layer (recall that $f \to .99$ for $\eta \sim 4$). The opposite is true for $\Pr \ll 1$.

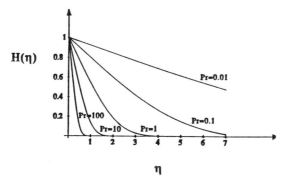

Figure 11–2 The temperature profile for heat transfer from a horizontal flat plate at large Reynolds number (large Peclet number) for several values of the Prandtl number, $0.01 \le \Pr \le 100$.

Apart from the special cases discussed in this section, the thermal boundary-layer equation (11–6) can be solved analytically only for the two asymptotic limits $\Pr \ll 1$ and $\Pr \gg 1$. This is the subject of the next two sections.

C. The Asymptotic Limit, $\Pr \gg 1$

Let us begin with the limit $\Pr \gg 1$. The problem is now to obtain an approximate asymptotic solution of (11–6) and (11–7). Although the problem defined by (11–6) and (11–7) is derived in Section A as the *inner* boundary-layer approximation of the overall problem for $\mathrm{Pe} \gg 1$, $\mathrm{Re} \gg 1$, its solution for arbitrary geometry and Pr either large or small is itself a singular perturbation problem. In other words, for these limiting cases, the domain consists of the *global* outer region where the governing equation and solution at the leading order of approximation are (11–3) and (11–4), plus the boundary-layer region governed by (11–6) and (11–7), which itself splits into two parts.

In the high Pr case, the leading-order term in an asymptotic expansion, for the part of the domain where the momentum boundary-layer scaling is applicable, is obtained by taking the limit $\Pr \to \infty$ in (11–6). The result is

$$\mathbf{u} \cdot \nabla \theta = 0, \tag{11–22}$$

now expressed in terms of momentum boundary-layer variables, but with the same solution as before,

$$\theta = 0 \tag{11–23}$$

for all parts of the momentum boundary-layer domain that contain open streamlines. Clearly, this approximate solution still does not satisfy the boundary condition $\theta = 1$ at the body surface. In the limit $\Pr \to \infty$, the conduction term in (11–6) is neglected, and the resulting solution pertains to only the *outer portion* of the momentum boundary-layer, away from the body surface. Thus, as suggested earlier, we see that the solution of the thermal boundary-layer equation (11–6) for large Pr is itself a singular perturbation (matched asymptotic expansion) problem.

From a physical point of view, the fact that conduction is neglected in (11–6) for $\Pr \to \infty$ is indicative of the fact that the momentum boundary-layer length scale is *not* relevant in the vicinity of the body surface for this limit. This is, in fact, demonstrated nicely by the exact similarity solution of the preceding section. Turning again to Figure 11–2, we see that the thermal region of heated fluid becomes thinner and thinner in the momentum boundary-layer variables as Pr is increased. Clearly, the momentum boundary-layer dimension is not an appropriate measure of the distance over which the temperature changes from its surface value to the free-stream value $\theta = 0$. Viewed in another way, it is evident from Figure 11–2 that the temperature gradient for large Pr is not independent of Pr when calculated using momentum boundary-layer variables. In order to obtain the form of (11–6) that applies to the region in the immediate vicinity of the body surface, it is necessary to rescale so that the dimension of the thermal region in terms of the new independent variables is $0(1)$.

We proceed formally. Thus, we introduce the rescaled variable

$$\boxed{\tilde{Y} = Y \mathrm{Pr}^{\alpha}} \qquad (11\text{--}24)$$

(where $\alpha > 0$) into (11–6). As before, the change of variables from Y to $\tilde{Y}$ corresponds to the introduction of a new characteristic length scale. Equation (11–6) becomes

$$\mathrm{Pr}^{2\alpha-1} \frac{\partial^2 \theta}{\partial \tilde{Y}^2} = u\left(x, \frac{\tilde{Y}}{\mathrm{Pr}^{\alpha}}\right) \frac{\partial \theta}{\partial x} + \mathrm{Pr}^{\alpha} V\left(x, \frac{\tilde{Y}}{\mathrm{Pr}^{\alpha}}\right) \frac{\partial \theta}{\partial \tilde{Y}}. \qquad (11\text{--}25)$$

The objective is to choose α such that the conduction term is retained in this rescaled equation, even in the limit $\mathrm{Pr} \to \infty$ ($\tilde{Y} \sim 0(1)$). Examining (11–25), we see formally that it is the limit of u and V for small values of the second argument that is relevant to the limit $\mathrm{Pr} \gg 1$. Thus, to determine the appropriate forms for u and V, we consider a Taylor series expansion for small values of the momentum boundary-layer variable Y, namely,

$$u\left(x, \frac{\tilde{Y}}{\mathrm{Pr}^{\alpha}}\right) \sim \left(\frac{\partial u}{\partial Y}\right)_{Y=0} \frac{\tilde{Y}}{\mathrm{Pr}^{\alpha}} + \frac{1}{2}\left(\frac{\partial^2 u}{\partial Y^2}\right)_{Y=0} \frac{\tilde{Y}^2}{\mathrm{Pr}^{2\alpha}} + \cdots. \qquad (11\text{--}26)$$

Here, we have set $u|_{Y=0} = 0$ because of the no-slip condition at the body surface. The coefficient of the second term, $(\partial u/\partial Y)_{Y=0}$, is the dimensionless shear stress, which is a function of position on the body surface. We denote this function as $\lambda(x)$, that is,

$$\left.\frac{\partial u}{\partial Y}\right|_{Y=0} = \lambda(x). \qquad (11\text{--}27)$$

The coefficient of the second term is

$$\left.\frac{\partial^2 u}{\partial Y^2}\right|_{Y=0} = -u_e(x) \frac{du_e}{dx}. \qquad (11\text{--}28)$$

Thus, for $\mathrm{Pr} \gg 1$, we can approximate u in the form

$$u\left(x, \frac{\tilde{Y}}{\mathrm{Pr}^{\alpha}}\right) \sim \lambda(x) \frac{\tilde{Y}}{\mathrm{Pr}^{\alpha}} - \frac{1}{2} u_e(x) \frac{du_e}{dx} \frac{\tilde{Y}^2}{\mathrm{Pr}^{2\alpha}} + \cdots. \qquad (11\text{--}29)$$

The corresponding form for V can then be determined from (11–29) using the continuity equation

$$V\left(x, \frac{\tilde{Y}}{\mathrm{Pr}^{\alpha}}\right) \sim -\frac{1}{2} \lambda'(x) \frac{\tilde{Y}^2}{\mathrm{Pr}^{2\alpha}} + \frac{1}{6} \frac{d}{dx}\left(u_e \frac{du_e}{dx}\right) \frac{\tilde{Y}^3}{\mathrm{Pr}^{3\alpha}} + \cdots. \qquad (11\text{--}30)$$

The geometry of the body has not been specified, except for the restriction to two dimensions and the assumption that the velocity field in the vicinity of the body is given in terms of a boundary-layer solution (that is, no separation). Indeed, the only way that the geometry is reflected in (11–25), (11–29), or (11–30) is via the functions $\lambda(x)$ and $u_e(x)$. We shall see that the leading-order approximation to the solution of (11–25) can be obtained for *arbitrary* $\lambda(x)$—that is, for *arbitrary* geometry.

To obtain α, we substitute (11–29) and (11–30) into (11–25). The result is

$$\Pr^{3\alpha-1}\frac{\partial^2\theta}{\partial\tilde{Y}^2} = \left\{\lambda\tilde{Y} - \frac{1}{2}u_e\frac{du_e}{dx}\frac{\tilde{Y}^2}{\Pr^\alpha} + \cdots\right\}\frac{\partial\theta}{\partial x} -$$

$$\left\{\frac{1}{2}\lambda'\tilde{Y}^2 - \frac{1}{6}\frac{d}{dx}\left(u_e\frac{du_e}{dx}\right)\frac{\tilde{Y}^3}{\Pr^\alpha} + \cdots\right\}\frac{\partial\theta}{\partial\tilde{Y}}. \qquad (11\text{–}31)$$

We pick α so that the conduction term on the left balances the largest convection terms on the right, namely,

$$\alpha = \frac{1}{3}.$$

Thus,

$$\boxed{\tilde{Y} = Y\Pr^{1/3},} \qquad (11\text{–}32)$$

and the resulting form for the leading-order approximation to (11–31) is

$$\boxed{\frac{\partial^2\theta}{\partial\tilde{Y}^2} = \lambda(x)\tilde{Y}\frac{\partial\theta}{\partial x} - \frac{\lambda'(x)}{2}\tilde{Y}^2\frac{\partial\theta}{\partial\tilde{Y}} + 0(\Pr^{-1/3}).} \qquad (11\text{–}33)$$

This is the governing equation for the leading-order approximation in an asymptotic expansion for θ in the innermost region near the body surface. The boundary condition (11–7a) applies to this leading approximation for θ,

$$\boxed{\theta = 1 \quad \text{at} \quad \tilde{Y} = 0.} \qquad (11\text{–}34a)$$

The solution is also required to match (11–23), which is the leading-order approximation in the outer part of the momentum boundary-layer.

$$\boxed{\theta = 0 \quad \text{for} \quad \tilde{Y} \gg 1 \quad (\Pr \to \infty).} \qquad (11\text{–}34b)$$

A schematic showing the relative scales of the thermal and momentum layers in this limit of $\Pr \gg 1$ is shown in Figure 11–3.

The form of the correlation between Nu, Re, and Pr can now be obtained for the present case $\text{Re} \gg 1$ and $\Pr \gg 1$. The starting point is (11–11). The rescaling (11–32) now provides an asymptotic estimate for $\partial\theta/\partial Y|_{Y=0}$, namely,

$$\frac{\text{Nu}}{\text{Re}^{1/2}} = -\Pr^{1/3}\int_{x^*}^{x_L}\left(\frac{\partial\theta}{\partial\tilde{Y}}\right)_{\tilde{Y}=0}dx. \qquad (11\text{–}35)$$

The dimensionless gradient $(\partial\theta/\partial\tilde{Y})_{\tilde{Y}=0}$ is a function of x that is independent of Re or Pr. Hence,

$$\boxed{\frac{\text{Nu}}{\text{Re}^{1/2}\Pr^{1/3}} = F(x^*, x_L), \quad \text{Re} \gg 1, \ \Pr \gg 1.} \qquad (11\text{–}36)$$

This is the primary result of the present section. The constant on the right-hand side depends on only the geometry of the body, that is, $\lambda(x)$ at the leading order of approximation. We shall see shortly that the equation and boundary conditions (11–33) and (11–34) can be solved analytically for arbitrary $\lambda(x)$.

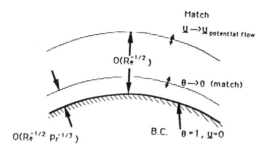

Figure 11–3 A schematic showing the two-layer structure of the thermal-momentum boundary-layer for large Pe, with Re ≫ 1 and Pr ≫ 1. In this case, the distance over which θ changes to its free-stream value is vanishingly small compared with the dimension of the momentum boundary-layer.

First, however, it is important to note that the form of the correlation (11–36), though *independent* of the geometry of the body, is *dependent* on the boundary conditions at its surface. In particular, let us suppose that the no-slip boundary condition is replaced by a condition that allows $u(x,0) \neq 0$ at the body surface. In this case, instead of (11–29), the form for u is

$$u\left(x, \frac{\tilde{Y}}{\mathrm{Pr}^\alpha}\right) = u(x,0) + \lambda(x)\frac{\tilde{Y}}{\mathrm{Pr}^\alpha} + \cdots. \tag{11–37}$$

Hence, when this form (and the corresponding expression for V) is substituted into (11–25), the value of α required to balance conduction with the largest convection term now becomes

$$\alpha = \frac{1}{2}.$$

Thus, for any problem with $u(x,0) \neq 0$, the form of the correlation for Nu changes to

$$\boxed{\frac{\mathrm{Nu}}{\mathrm{Re}^{1/2}\mathrm{Pr}^{1/2}} = G(x^*, x_L)} \tag{11–38}$$

for Re ≫ 1, Pr ≫ 1. We consider an example of this type in the exercises at the end of this chapter.

Let us now return to (11–33), which is valid for any arbitrary (no-slip) two-dimensional body in the limit Re ≫ 1, Pr ≫ 1. As noted earlier, the geometry of the body enters at this leading order of approximation only through the function $\lambda(x)$. The somewhat surprising fact is that we can solve (11–33) for *arbitrary* $\lambda(x)$. Hence, we can obtain a *universal* first approximation for θ and for the Nusselt number correlation (11–36).

In order to solve (11–33) for arbitrary $\lambda(x)$, we introduce a similarity transformation of the form

$$\theta = \theta(\eta), \quad \text{where} \quad \eta = \frac{\tilde{Y}}{g(x)}. \tag{11–39}$$

The function $g(x)$ is to be determined as part of the solution and must depend explicitly on $\lambda(x)$. In order to satisfy the boundary conditions (11–34), as well as the initial condition (11–7c), it is necessary that

$$g(x^*) = 0. \tag{11–40}$$

When the similarity transformation (11–39) is introduced into (11–33), the result is

$$\frac{d^2\theta}{d\eta^2} + \eta^2 \frac{d\theta}{d\eta} \left\{ \lambda g^2 g' + g^3 \frac{\lambda'}{2} \right\} = 0. \tag{11–41}$$

Hence, if a similarity solution of the form (11–39) exists, it must be possible to determine $g(x)$ such that the coefficient of the second term in (11–41) is a constant. As usual, the value of this constant can be chosen for convenience (without any effect on the solution).

Let us suppose that the constant is equal to 3. Then, the constraint on g for existence of a similarity solution can be written in the form of a first-order ordinary differential equation for g^3, namely,

$$\frac{\lambda}{3} \frac{dg^3}{dx} + \frac{1}{2} \frac{d\lambda}{dx} g^3 = 3. \tag{11–42}$$

To see whether this equation has a solution satisfying the condition (11–40), we can rewrite it in the form

$$\frac{dg^3}{dx} + g^3 \left(\frac{d \ln \lambda^{3/2}}{dx} \right) = \frac{9}{\lambda}. \tag{11–43}$$

The homogeneous equation has a solution

$$\frac{c}{\lambda^{3/2}}.$$

Hence, the general solution of (11–42) is

$$g^3 = \frac{c}{\lambda^{3/2}} + \frac{9}{\lambda^{3/2}} \int_{x^*}^{x} \sqrt{\lambda(t)} \, dt. \tag{11–44}$$

To satisfy the condition (11–40),

$$c = 0$$

so that

$$g(x) = \frac{9^{1/3}}{\sqrt{\lambda(x)}} \left[\int_{x^*}^{x} \sqrt{\lambda(t)} \, dt \right]^{1/3}. \tag{11–45}$$

The governing equation for the similarity function $\theta(\eta)$ now takes the form

$$\frac{d^2\theta}{d\eta^2} + 3\eta^2 \frac{d\theta}{d\eta} = 0,$$

(11–46)

with boundary (matching) conditions

$$\theta(0) = 1,$$

(11–47a)

$$\theta(\infty) = 0.$$

(11–47b)

Equation (11–46) can be integrated twice, with respect to η, to obtain

$$\theta = c_1 \int_\eta^\infty e^{-t^3} dt + c_2.$$

Hence, applying the boundary conditions (11–47), we find that

$$\theta = \frac{\int_\eta^\infty e^{-t^3} dt}{\int_o^\infty e^{-t^3} dt} = \frac{1}{\Gamma(4/3)} \int_\eta^\infty e^{-t^3} dt.$$

(11–48)

This completes the similarity solution for arbitrary $\lambda(x)$. The temperature profile given by (11–48) is plotted in Figure 11–4.

The only remaining task is to use the solution (11–48) with $g(x)$ given by (11–45) to evaluate the geometry-dependent coefficient on the right-hand side of (11–36). For this purpose, we first evaluate $(\partial\theta/\partial\tilde{Y})_{\tilde{Y}=0}$:

$$-\left(\frac{\partial\theta}{\partial\tilde{Y}}\right)\bigg|_{\tilde{Y}=0} = -\frac{d\theta}{d\eta}\bigg|_{\eta=0} \left(\frac{\partial\eta}{\partial\tilde{Y}}\right) = -\frac{1}{g(x)} \frac{d\theta}{d\eta}\bigg|_{\eta=0}.$$

(11–49)

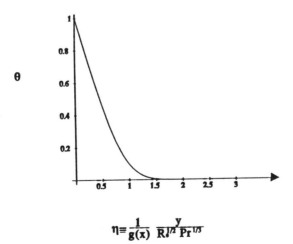

$$\eta \equiv \frac{1}{g(x)} \frac{y}{Re^{1/2} Pr^{1/3}}$$

Figure 11–4 The self-similar temperature profile for the thermal boundary-layer in the limit $Re \gg 1$, $Pr \gg 1$.

Now,

$$\left(\frac{d\theta}{d\eta}\right)_{\eta=0} = -\frac{1}{\Gamma(4/3)}. \tag{11-50}$$

Hence,

$$-\left(\frac{\partial\theta}{\partial\tilde{Y}}\right)_{\tilde{Y}=0} = \frac{\sqrt{\lambda(x)}}{9^{1/3}\Gamma(4/3)}\left[\int_{x*}^{x}\sqrt{\lambda(t)}dt\right]^{-1/3}. \tag{11-51}$$

To obtain $F(x^*, x_L)$, this must be integrated over the body surface. Thus

$$F(x^*, x_L) \equiv -\int_{x*}^{x_L}\left(\frac{\partial\theta}{\partial\tilde{Y}}\right)_{\tilde{Y}=0}dx = \frac{3^{1/3}}{2\Gamma(4/3)}\left[\int_{x*}^{x_L}\sqrt{\lambda(t)}dt\right]^{2/3} \tag{11-52}$$

The results in (11–51) and (11–52) are, of course, completely general for arbitrary $\lambda(x)$. Here, for illustration purposes, we evaluate them only for the special case of a flat plate (parallel to the free stream) where, according to the Blasius solution from Chapter 10,

$$\lambda(x) = \frac{0.332}{\sqrt{x}}.$$

In this case,

$$-\left(\frac{\partial\theta}{\partial\tilde{Y}}\right)_{\tilde{Y}=0} = \frac{(0.332)^{1/3}}{9^{1/3}\Gamma(4/3)}\frac{1}{x^{1/4}}\frac{1}{(4/3)^{1/3}(x^{3/4}-x^{*3/4})^{1/3}} = \frac{0.339}{x^{1/2}}\left(\frac{1}{1-\left(\frac{x^*}{x}\right)^{3/4}}\right)^{1/3}. \tag{11-53}$$

Hence,

$$\frac{\text{Nu}}{\text{Re}^{1/2}\text{Pr}^{1/3}} = F(x^*, x_L) = 0.677(x_L^{3/4} - x^{*3/4})^{2/3}. \tag{11-54}$$

Here, x_L is the dimensionless position corresponding to the end of the heated portion of the plate surface, while x^* is the position of the beginning of the heated zone. If the plate length is finite, so that the plate length is used as a characteristic length scale, and the whole of the plate surface is heated, then

$$x^* = 0 \quad \text{and} \quad x_L = 1$$

so that

$$\text{Nu} = 0.677\,\text{Re}^{1/2}(\text{Pr}^{1/3} + 0(1)). \tag{11-55}$$

The geometry-dependent coefficient in this case is 0.677. The reader will probably no longer be surprised to find that this coefficient is 0(1) in magnitude.

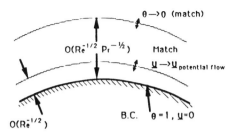

Figure 11–5 A schematic representation showing the two-layer structure of the thermal-momentum boundary-layer for large Pe, with Re $\gg$ 1 and Pr $\ll$ 1. In this limit, the distance required for θ to approach its free-stream value is much larger than the distance required for the velocity to approach the potential-flow form. Thus, convection within the thermal boundary-layer is dominated by the outer limit of the momentum boundary-layer velocity distribution (or equivalently, the inner limit as we approach the body surface of the potential-flow distribution).

D. The Asymptotic Limit, Pr $\ll$ 1

Now let us turn to the solution of the thermal boundary-layer equation (11–6) for the limit Pr $\rightarrow$ 0. Although the thermal and momentum boundary-layers both remain very thin with the dimension of the body, this is the limit, as seen in Figure 11–5, where the thermal layer (the region where θ decreases from the surface value $\theta = 1$ to the free-stream value $\theta = 0$) is of much greater extent than the momentum layer (the region where u depends explicitly on Y). From a physical point of view, the limit Pr $\ll$ 1 can be thought of as corresponding to increasing thermal conductivity, for fixed values of the other parameters. Hence, as Pr $\rightarrow$ 0, the relative efficiency of heat transfer by conduction goes up, and we require larger velocities before the transport by convection becomes as important as conduction. Hence, the thermal region increases in width relative to the momentum boundary-layer scaling as the Prandtl number is decreased.

If we proceed formally, then the governing equation for the first term in an asymptotic expansion for small Prandtl number is obtained for the region in which the momentum boundary-layer scaling is appropriate by taking the limit Pr $\rightarrow$ 0 in (11–6). The result is

$$\boxed{\frac{\partial^2 \oplus}{\partial Y^2} = 0.}$$

(11–56)

For convenience, we denote the dimensionless temperature as $\oplus$ in this part of the domain to differentiate it from the temperature distribution elsewhere. It can be seen from (11–56), that the heat transfer process in this region is dominated at leading order by pure conduction. If we integrate (11–56), the temperature profile is thus predicted to take the linear form

$$\oplus = a(x) Y + b(x).$$

Hence, applying the boundary condition (11-7a),

$$b \equiv 1$$

and

$$\boxed{\oplus = 1 + a(x) Y.} \tag{11-57}$$

It is evident, upon further examination, that the solution (11-57) *cannot* be uniformly valid within the thermal boundary-layer because it cannot be made to satisfy the matching condition (11-7b) for any choice of the coefficient $a(x)$. Hence, (11-56) and the solution (11-57) are only valid in the part of the thermal layer that lies closest to the body surface, and in order to obtain $a(x)$, we must seek a second solution, valid in the outer part of the thermal boundary layer, with $a(x)$ determined from matching.

To obtain an asymptotic approximation in the outer part of the thermal region, we seek a rescaling of (11-6)—that is, a new nondimensionalization—in which both conduction and convection terms are retained in the governing equation for the leading-order term in the asymptotic expansion for Pr $\ll 1$. We have seen, by examination of Figure 11-2, that the extent of the thermal layer for decreasing values of Pr depends strongly on Pr. Thus, so too does $\nabla \theta$, $\nabla^2 \theta$, and so on. From this perspective, the objective of rescaling is to find a new characteristic length scale such that the dimensionless distance over which θ changes, in terms of the newly scaled variables, is $O(1)$, that is, independent of Pr. To rescale, we introduce the formal change of variables

$$\boxed{\bar{Y} = \text{Pr}^\alpha Y \quad (\alpha > 0).} \tag{11-58}$$

Hence, for Pr $\ll 1$, changes in $\bar{Y}$ of $O(1)$ can be seen to correspond to very large changes in Y of $O(\text{Pr}^{-\alpha})$!

In the preceding section for Pr $\gg 1$, we rescaled both u and V also in recognition of the fact that these quantities have a characteristic magnitude proportional to $\text{Pr}^{-\beta}$ ($\beta > 0$) when we restrict ourselves to a region within $O(\text{Pr}^{-1/3})$ of the body surface. This rescaling, however, was not applied directly to u and V. Instead, a Taylor series approximation was introduced for u and V at small values of Y, and the rescaling of u and V was accomplished implicitly by rescaling Y. It should be noted, however, that the result was equivalent to a *direct* rescaling of u and V of the form

$$u = \tilde{u} \, \text{Pr}^{-1/3},$$

$$V = \tilde{V} \, \text{Pr}^{-2/3}.$$

Here, for illustrative purposes, we proceed in the other way, by directly rescaling u and V in accord with (11-58), that is,

$$\bar{u} = \text{Pr}^\beta u \left(x, \frac{\bar{Y}}{\text{Pr}^\alpha} \right) \tag{11-59a}$$

and, hence, by continuity,

$$\bar{V} = \mathrm{Pr}^{\alpha+\beta} V\left(x, \frac{\bar{Y}}{\mathrm{Pr}^{\alpha}}\right). \tag{11-59b}$$

Now, in the limit $\mathrm{Pr} \to 0$,

$$\bar{u} = \mathrm{Pr}^{\beta} u(x, \infty) = \mathrm{Pr}^{\beta} u_e(x). \tag{11-60}$$

But since $u_e(x)$ is already a function of $0(1)$, it appears that the appropriate scale for u must be

$$\beta \equiv 0. \tag{11-61}$$

Let us suppose that this result is correct. In this case, the rescaling of (11–6) requires only the transformations (11–58) and (11–59b), with $\alpha \neq 0$. Expressing (11–6) in terms of $\bar{Y}$ and $\bar{V}$, we find that

$$u\left(x, \frac{\bar{Y}}{\mathrm{Pr}^{\alpha}}\right)\frac{\partial \theta}{\partial x} + \bar{V}\left(x, \frac{\bar{Y}}{\mathrm{Pr}^{\alpha}}\right)\frac{\partial \theta}{\partial \bar{Y}} = \mathrm{Pr}^{2\alpha-1}\frac{\partial^2 \theta}{\partial \bar{Y}^2}. \tag{11-62}$$

Now, in the limit $\mathrm{Pr} \to 0$,

$$u\left(x, \frac{\bar{Y}}{\mathrm{Pr}^{\alpha}}\right) \to u_e(x) \tag{11-63}$$

and, thus, via continuity,

$$\bar{V}\left(x, \frac{\bar{Y}}{\mathrm{Pr}^{\alpha}}\right) \to -\frac{du_e}{dx}\, Y. \tag{11-64}$$

The value of the rescaling parameter α is determined by the condition that the limiting form of (11–62) should retain both conduction and convection in the limit $\mathrm{Pr} \to 0$. Hence, in view of (11–63) and (11–64), it can be seen from (11–62) that

$$\boxed{\alpha = \frac{1}{2}.} \tag{11-65}$$

Thus, the governing equation for the leading-order term in an asymptotic expansion for this outer part of the thermal boundary-layer is

$$\boxed{u_e(x)\frac{d\theta}{dx} - u_e' \bar{Y}\frac{\partial \theta}{\partial \bar{Y}} = \frac{\partial^2 \theta}{\partial \bar{Y}^2}.} \tag{11-66}$$

It may be noted that the scaling parameter α and the form of the governing equation (11–66) in this case are independent of whether the boundary condition at the body surface involves slip or no slip.

The boundary conditions for (11–66) are (11–7c), plus the condition

$$\boxed{\theta \to 0 \quad \text{for} \quad \bar{Y} \gg 1} \tag{11-67a}$$

as a matching condition with the global outer solution (11–4), and the matching condition

$$\lim_{Y \gg 1} \Theta(x, Y) \iff \lim_{\bar{Y} \ll 1} \theta(x, \bar{Y}) \quad \text{for} \quad \mathrm{Pr} \to 0. \tag{11-67b}$$

The left-hand side of this condition contains, at leading order, just the solution (11–57). It is convenient to express this solution in terms of the rescaled variables for the *outer* part of the thermal boundary-layer, namely,

$$\oplus(x, Y) = a(x)\frac{\bar{Y}}{Pr^{1/2}} + 1. \tag{11-68}$$

But in this form, it is clear that $a(x) \equiv 0$. If $a(x) \neq 0$, then it would be necessary that the solution for $\theta(x, \bar{Y})$ contain a term of $0(Pr^{-1/2})$. But $\theta(x, \bar{Y}) = 0(1)$ at the leading order of approximation. Hence, the first approximation in the asymptotic expansion for $\oplus$ in the innermost region of the thermal boundary-layer reduces to just

$$\oplus = 1 + \text{higher order terms in Pr}, \tag{11-69}$$

and the matching condition (11–67b) can be treated as a boundary condition for the leading-order approximation to θ, namely,

$$\boxed{\theta(x, 0) = 1.} \tag{11-67c}$$

Physically, in the limit $Pr \rightarrow 0$, the width of the thermal layer is so much greater than the momentum layer that θ does not vary at all across the momentum layer at the leading order of approximation. A sketch showing the multiscale structure of the thermal boundary region in this limit of $Pr \ll 1$ is shown in Figure 11–5.

The problem for the first term in an asymptotic solution for the temperature distribution θ in the outer part of the thermal layer is thus to solve (11–66) subject to the conditions (11–67a), (11–67c), and (11–7c). Again, we see that the geometry of the body enters implicitly through the function $u_e(x)$ only. As in the high Pr limit, a general solution of (11–66) is possible even for an arbitrary functional form for $u_e(x)$. Before moving forward to obtain this solution, however, a few comments are probably useful about the solution (11–69) for the innermost part of the boundary-layer immediately adjacent to the body surface.

The most important point to notice is that the leading-order solution $\oplus = 1$ does not imply that the heat flux at the body surface is zero, even though this might at first appear to be the case. In general, the condition of matching between the inner and outer approximations within the thermal boundary-layer requires not only (11–67c) but also that the spatial derivatives of θ and $\oplus$ should match at each level of approximation for $Pr \rightarrow 0$. Thus, in particular,

$$\lim_{\bar{Y} \ll 1} Pr^{1/2} \frac{\partial\theta}{\partial\bar{Y}} \Longleftrightarrow \lim_{Y \gg 1} \left(\frac{\partial\oplus}{\partial Y}\right) \quad \text{for} \quad Pr \rightarrow 0. \tag{11-70}$$

Now, $\partial\theta/\partial\bar{Y} = 0(1)$ at the leading order of approximation. Hence, it follows that the dimensionless temperature gradient within the innermost region should be $0(Pr^{1/2})$. But the solution (11–69) represents only the first term in an asymptotic expansion. The condition (11–70) suggests strongly that

$$\oplus_{\text{inner}} = 1 + Pr^{1/2}\oplus_1(x, Y) + \cdots. \tag{11-71}$$

But if this is true, then it can be seen from (11–6) that the governing equation for $\oplus_1$

is still

$$\frac{d^2 \oplus_1}{\partial Y^2} = 0,$$ (11-72)

with the solution

$$\oplus_1 = c(x)Y + d(x).$$ (11-73)

The boundary condition (11-7a) at $Y=0$ now becomes

$$\oplus_1 = 0 \quad \text{at} \quad Y = 0$$ (11-74)

so that $d(x) \equiv 0$. The matching condition (11-70) then requires that

$$\frac{\partial \theta}{\partial \bar{Y}}\bigg|_{\bar{Y} \ll 1} \Longleftrightarrow \left(\frac{\partial \oplus_1}{\partial Y}\right)_{Y \gg 1} \quad \text{for} \quad \text{Pr} \to 0.$$ (11-75)

Thus, $c(x) \equiv (\partial\theta/\partial\bar{Y})_{\bar{Y} \ll 1}$, and it follows from (11-71) that

$$\boxed{\oplus_{\text{inner}} = 1 + \text{Pr}^{1/2} \left(\frac{\partial\theta}{\partial Y}\right)_{\bar{Y} \ll 1} Y + \cdots.}$$ (11-76)

Hence, to evaluate the heat flux at the body surface, it is sufficient to determine the solution $(\partial\theta/\partial\bar{Y})|_{\bar{Y} \ll 1}$ in the *outer* part of the thermal layer. It also follows from this discussion that

$$\frac{\text{Nu}}{\text{Re}^{1/2}} \equiv -\int_{x^*}^{x_L} \left(\frac{\partial\oplus}{\partial Y}\right)_{Y=0} dx = -\text{Pr}^{1/2}\int_{x^*}^{x_L} \left(\frac{\partial\theta}{\partial\bar{Y}}\right)_{\bar{Y} \ll 1} dx.$$

Hence,

$$\boxed{\frac{\text{Nu}}{\text{Re}^{1/2}\text{Pr}^{1/2}} = -\int_{x^*}^{x_L} \left(\frac{\partial\theta}{\partial\bar{Y}}\right)_{\bar{Y} \ll 1} dx = K(x^*, x_L)}$$ (11-77)

for $\text{Re} \gg 1$, $\text{Pr} \ll 1$, but with $\text{Pe} = \text{Re}\,\text{Pr} \gg 1$. Thus, to obtain a leading-order approximation for the Nusselt number, we must solve the differential equation (11-66) subject to the boundary conditions (11-67).

The coefficient $K(x^*, x_L)$ depends upon the geometry of the body, at least as specified via the function $u_e(x)$. However, as in the high Pr limit, it is possible to obtain a general solution of (11-66) and (11-67) for arbitrary geometry [that is, arbitrary $u_e(x)$]. The method of analysis is once again the similarity transformation

$$\theta = \theta(\eta) \quad \text{with} \quad \eta = \frac{\bar{Y}}{g(x)}.$$ (11-78)

Again, the function $g(x)$ must be determined subject to the condition $g(x^*) = 0$.

If the similarity transformation (11-78) is introduced into (11-66), the result is

$$\frac{d^2\theta}{d\eta^2} + \eta\frac{d\theta}{d\eta}\{u_e g g' + u_e' g^2\} = 0,$$ (11-79)

and a necessary condition for existence of a similarity solution is that it be possible to determine $g(x)$ in terms of $u_e(x)$ such that the coefficient in brackets is constant. For convenience, we choose this constant equal to 2. Examination of (11–79) indicates that the condition on g can be written in the form of a first-order ordinary differential equation for g^2,

$$\frac{dg^2}{dx} + 2g^2 \frac{d \ln u_e}{dx} = \frac{4}{u_e}. \tag{11–80}$$

Hence, a necessary condition for existence of a similarity solution is that g^2 satisfy (11–80), i.e.,

$$g^2 = \frac{c}{u_e^2} + \frac{4}{u_e^2} \int_{x*}^{x} u_e(t) dt. \tag{11–81}$$

The constant c is determined from the boundary condition that $g^2(x*) = 0$. Thus,

$$c = 0$$

and

$$g(x) = \frac{2}{u_e} \left[\int_{x*}^{x} u_e(t) dt \right]^{1/2}. \tag{11–82}$$

With this form for $g(x)$, the boundary condition (11–67a) and the initial condition (11–7c) are both expressed in terms of the single condition

$$\theta(\infty) = 0. \tag{11–83}$$

The governing, similarity form of the equation for $\theta(\eta)$ is now

$$\frac{d^2\theta}{d\eta^2} + 2\eta \frac{d\theta}{d\eta} = 0. \tag{11–84}$$

Hence, integrating twice with respect to η and applying the condition (11–83), plus the condition (11–67b) in the form

$$\theta(0) = 1, \tag{11–85}$$

we obtain the general solution

$$\theta = \frac{\int_{\eta}^{\infty} e^{-t^2} dt}{\int_{0}^{\infty} e^{-t^2} dt}. \tag{11–86}$$

The integral coefficient in the denominator is just $\Gamma(3/2)$, which has a numerical value $\sqrt{\pi}/2$. Hence,

$$\theta = \frac{2}{\sqrt{\pi}} \int_{\eta}^{\infty} e^{-t^2} dt \equiv \operatorname{erfc}(\eta), \tag{11–87}$$

where $\operatorname{erfc}(\eta)$ is the *complimentary error function*,

$$\operatorname{erfc}(\eta) = 1 - \operatorname{erf}(\eta),$$

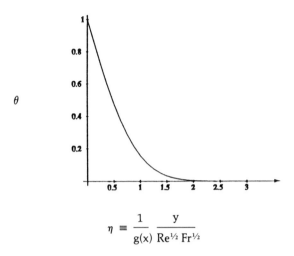

$$\eta \equiv \frac{1}{g(x)} \frac{y}{\mathrm{Re}^{\frac{1}{2}} \, \mathrm{Fr}^{\frac{1}{2}}}$$

Figure 11-6 The self-similar temperature profile for the thermal boundary-layer with $\mathrm{Re} \gg 1$, $\mathrm{Pr} \ll 1$.

where

$$\mathrm{erf}\,(\eta) \equiv \frac{2}{\sqrt{\pi}} \int_0^\eta e^{-t^2} dt.$$

This temperature profile is plotted in Figure 11-6.

The only remaining task is to use the solution (11-87) and (11-82) to evaluate the coefficient $K(x^*, x_L)$ in the correlation (11-77) for the Nusselt number as a function of Re and Pr. The integrand is

$$\left(\frac{\partial \theta}{\partial \bar{Y}}\right)_{\bar{Y} \ll 1} = \frac{d\theta}{d\eta}\bigg|_{\eta=0} \frac{\partial \eta}{\partial \bar{Y}} = \frac{1}{g(x)} \frac{d\theta}{d\eta}\bigg|_{\eta=0}.$$

But

$$\frac{d\theta}{d\eta}\bigg|_{\eta=0} = -\frac{2}{\sqrt{\pi}},$$

and thus

$$\left|\frac{\partial \theta}{\partial \bar{Y}}\right|_{\bar{Y}=0} = -\frac{u_e}{2}\left[\int_{x^*}^x u_e(t)\,dt\right]^{-1/2} \cdot \frac{2}{\sqrt{\pi}}. \tag{11-88}$$

Hence,

$$K(x^*, x_L) \equiv \int_{x^*}^{x_L} -\left(\frac{\partial \theta}{\partial \bar{Y}}\right)_{\bar{Y}=0} dx = \frac{2}{\sqrt{\pi}}\left[\int_{x^*}^{x_L} u_e(t)\,dt\right]^{1/2} \tag{11-89}$$

The results in (11–88) and (11–89) is for general geometries. Here, we again consider explicitly only the specific case of a horizontal flat plate for which

$$u_e(x) \equiv 1. \tag{11-90}$$

In this case, the local heat flux is

$$\left| \frac{\partial \theta}{\partial \bar{Y}} \right|_{\bar{Y}=0} = -\frac{0.564}{\sqrt{x - x^*}} = -\frac{0.564}{\sqrt{x}} \quad \text{if } x^* \equiv 0. \tag{11-91}$$

Then

$$K(x^*, x_L) = \frac{2}{\sqrt{\pi}} (x_L - x^*)^{1/2}, \tag{11-92}$$

or in case

$$x^* \equiv 0 \quad \text{and} \quad x_L \equiv 1,$$

we find that

$$K(x^*, x_L) \equiv \frac{2}{\sqrt{\pi}} = 1.128. \tag{11-93}$$

E. Discussion of the Asymptotic Results for Pr ≪ 1 and Pr ≫ 1

Before proceeding to other topics, it is perhaps worthwhile to reflect briefly on the results of Sections 11.C and 11.D. In particular, let us focus our attention on the two results in (11–53) and (11–91) for the local temperature gradients in the case of a horizontal flat plate, with $x^* \equiv 0$. These correspond to correlations for the *local* dimensionless heat flux of the forms

$$\frac{\mathrm{Nu}_x}{\mathrm{Re}^{1/2} \mathrm{Pr}^{1/2}} = \frac{0.564}{\sqrt{x}} \quad \text{for } \mathrm{Pr} \ll 1 \tag{11-94}$$

and

$$\frac{\mathrm{Nu}_x}{\mathrm{Re}^{1/2} \mathrm{Pr}^{1/3}} = \frac{0.339}{\sqrt{x}} \quad \text{for } \mathrm{Pr} \gg 1. \tag{11-95}$$

It may be noted that the functional dependence of the local heat flux on x is the same as the shear stress. This is a consequence of the fact that the thickness of the thermal boundary-layer varies with x in the same way as the momentum boundary-layer thickness. Furthermore, the form of the correlations for large and small Prandtl numbers are also quite similar. However, this latter observation may be somewhat misleading. In the case $\mathrm{Pr} \gg 1$, the heat flux increases to very large values, proportional to $\mathrm{Pr}^{1/3}$ as Pr increases. On the other hand, the heat flux for $\mathrm{Pr} \ll 1$ decreases as $\mathrm{Pr}^{1/2}$ for $\mathrm{Pr} \to 0$. Clearly, the heat transfer process is much less efficient for small Pr than for large Pr.

From a practical point of view, it frequently will be the case that the Prandtl number is neither very small nor very large. For example, for most gases, Pr ~ 0(1), while the Prandtl number for relatively simple, low molecular weight liquids like water is Pr ~ 0(5). Thus, an obvious question is whether the asymptotic results for Pr → 0 and Pr → ∞ can provide any useful guidance for such cases. The answer is that a useful estimate of the rate of heat transfer can be obtained for the whole range of values of Pr by simple interpolation between the two asymptotic results. For example, let us again consider the case of a horizontal flat plate, for which the local Nusselt number is given by (11–94) and (11–95) for the two limiting cases Pr → 0 and Pr → ∞, respectively. In this case, we plot $\mathrm{Nu}_x \sqrt{x}/\mathrm{Re}^{1/2}$, corresponding to (11–94) and (11–95), versus Pr on a log-log graph. The result shown in Figure 11-7 is two straight lines, one with slope 1/2 and the other with slope 1/3, that intersect at Pr ~ 0(1). Also shown on the same plot is the *exact* result for this problem, obtained from the similarity solution in Section **B**,

$$-\frac{\partial \theta}{\partial Y}\sqrt{x} = -\left.\frac{dH}{d\eta}\right|_{\eta=0} = \frac{1}{\sqrt{2}\int_0^\infty e^{-\mathrm{Pr}\int_0^t f(s)\,ds}\,dt}. \qquad (11\text{–}96)$$

We see that the maximum difference between this exact result and the nearest of the two asymptotes is about 20 percent at the point where the two asymptotes intersect. We derive two basic conclusions from the comparison between the exact and asymptotic results in Figure 11-7. First, the transition region between the low to the high Pr number limits is fairly narrow, and the asymptotes thus provides a good approximation of the exact results for a surprisingly wide range of values for Pr. Second, the comparison suggests strongly the possibility of obtaining a reasonable estimate for Nu at intermediate values of Pr for more complex problems by a smooth interpolation between the two asymptotes.

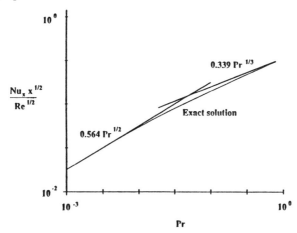

Figure 11–7 A comparison of the asymptotic forms for the local Nusselt number correlation for Pr ≪ 1 and Pr ≫ 1, with the result from the exact similarity solution for arbitrary Prandtl number. (Notice: All results are for the special case of a horizontal flat plate.)

F. Approximate Results for Surface Temperature with Specified Heat Flux or Mixed Boundary Conditions

All of the results obtained in the preceding sections of this chapter have been based on the assumption that the temperature at the body surface is uniform. However, there are many important cases where this condition does not hold. For example, in many practical applications it is the heat flux at the surface that is held constant, rather than the surface temperature. One example of such a problem is a wire (resistance) heater. For this type of problem, the objective of theory is no longer the heat flux (or total rate of heat transfer), since this is already known, but the temperature distribution in the fluid and especially at the surface of the body (where one might wish to determine the maximum temperature, for example). Other problems exist where neither the temperature nor the heat flux is known, but some combination of these. One example is the flow of a gas containing a condensable component past a cooled solid surface, when a liquid film is deposited on the body. In this case, an exact solution for the temperature distribution in the gas (and thus the rate of heat transfer between the body and the gas), would require solving a coupled problem in which the velocity and temperature fields are determined in both the liquid film and the gas. However, an approximate solution is possible, provided the liquid film is thin enough so that the transfer of heat across the film is dominated by conduction. In this case, at steady state, the rate of heat transfer across the film is balanced by the transport of heat from the outer surface of the film to the gas, and this leads to a *mixed-type boundary condition* for the gas-phase heat transfer process. To see this, let us suppose that the temperature at the surface of the cooled body is fixed at a value T_o and that the width of the liquid film is l. Then, the steady-state heat flux balance is

$$k_L \frac{T_o - T_s}{l} = -k_G \left. \frac{\partial T}{\partial \eta} \right|_s ,\qquad (11\text{--}97)$$

where k_L and k_G are the liquid and gas-phase thermal conductivities, T_s denotes the surface temperature (unknown) at the outer surface of the liquid film, and $(\partial T/\partial \eta)_s$ is the temperature gradient normal to this surface in the gas. Rearranging,

$$\boxed{\left. -\frac{\partial T}{\partial \eta} \right|_s + \frac{k_L}{k_G l} T_s = \frac{k_L}{k_G l} T_o.}\qquad (11\text{--}98)$$

The right-hand side is known from the specified temperature at the solid. Hence, (11–98) is a boundary condition of the so-called *mixed-type* for the temperature distribution to the gas. A schematic representation of the situation is shown in Figure 11–8. Both T_s and $(\partial T/\partial \eta)_s$ are unknown, and we thus seek both the surface temperature and the local heat flux from the solution. Note, however, that it is enough to determine one of these two quantities directly in our solution. The other can then be obtained from (11–98).

There are at least two approaches that we can take to solve problems in which either the heat flux or the mixed-type condition is specified as a boundary condition.

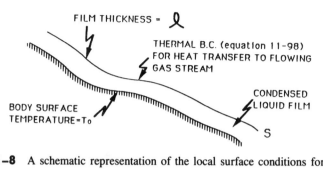

Figure 11-8 A schematic representation of the local surface conditions for heat transfer from a solid body with surface temperature T_0 to a gas stream when there is a condensed liquid film on the body surface. As explained in the text, this leads to an approximate boundary condition on the surface S of mixed type (11-98).

If it is desired to determine the temperature distribution throughout the fluid, then we must return to the governing thermal boundary-layer equation (11-6)—assuming Re, Pe $\gg$ 1—and develop new asymptotic solutions for large and small Pr, with either $\partial T/\partial \eta|_{y=0}$ or a condition of the mixed type specified at the body surface. The problem for a constant, specified heat flux is relatively straightforward, and we pose such a case as one of the exercises at the end of this chapter. On the other hand, in many circumstances, we might be concerned only with determining the temperature distribution on the body surface [and thus $\partial T/\partial \eta|_s$ from (11-98) for the mixed-type problem], and for this, there is an even simpler approach that can be derived directly from the results of Sections **C** and **D**. This involves the derivation of an integral equation for the surface temperature.

In order to see how this can be done, let us pose a generalization of (11-6) and (11-7) as follows:

$$u\frac{\partial \theta}{\partial x} + V\frac{\partial \theta}{\partial Y} = \frac{1}{\text{Pr}}\frac{\partial^2 \theta}{\partial Y^2},\tag{11-99}$$

with

$$\theta = 0 \quad \text{at} \quad Y = \infty,$$

$$\theta = 0 \quad \text{for} \quad Y > 0, \ x = x^*,\tag{11-100}$$

and

$$\theta = \theta_s(x) \quad \text{for} \quad Y = 0, \ x > x^*.$$

Of course, if our objective were to calculate the complete temperature distribution in the fluid, we would never pose the problem in this form since $\theta_s(x)$ is unknown and thus not satisfactory as a boundary condition. However, our goal here is to determine $\theta_s(x)$ for cases in which $\partial \theta/\partial Y|_{Y=0}$ or $\left[(\partial \theta/\partial Y) + (k_L/k_G l)\theta\right]_{Y=0}$ is specified, and the objective in writing the problem in the form (11-99)–(11-100) is to show that we can evaluate $\theta_s(x)$ directly from the asymptotic solutions obtained earlier for a constant surface temperature, $\theta_s = 1$. The key to transforming from our original solutions to a formula for $\theta_s(x)$ satisfying (11–99)–(11–100) is that the governing equations and boundary conditions for θ are all linear.

It is convenient to proceed formally in a series of steps.

1. First, we consider the problem in which $\theta_s(x) = 0$ for $0 \le x \le x^*$, and $\theta_s(x) = 1$ for $x > x^*$. For arbitrary two-dimensional bodies, we have already solved this problem for the two limits $\Pr \to 0$ and $\Pr \to \infty$. In particular, let us define $K(x; x^*, \Pr)$ as

$$\boxed{-\left.\frac{\partial \theta}{\partial Y}\right|_{Y=0} \equiv K(x; x^*, \Pr).} \qquad (11\text{–}101)$$

Then, it can be seen from (11–51) and (11–88), that

$$\boxed{K(x, x^*, \Pr) = \begin{cases} \dfrac{\Pr^{1/3}\sqrt{\lambda(x)}}{9^{1/3}\Gamma(4/3)}\left[\displaystyle\int_{x^*}^{x}\sqrt{\lambda(t)}\,dt\right]^{-1/3} & \text{for } \Pr \to \infty \\[4mm] \dfrac{\Pr^{1/2}u_e(x)}{\sqrt{\pi}}\left[\displaystyle\int_{x^*}^{x}u_e(t)\,dt\right]^{-1/2} & \text{for } \Pr \to 0 \end{cases}} \qquad (11\text{–}102)$$

2. Second, if we consider the trivial generalization in which the surface temperature distribution is

$$\theta_s(x) = 0, \; 0 \le x < x^*,$$

$$\theta_s(x) = \theta_s(x^*) \quad \text{for } x \ge x^*,$$

where $\theta_s(x^*)$ is a constant value not equal to unity, the solution for $(\partial\theta/\partial Y)_{Y=0}$ is now

$$\boxed{-\left(\frac{\partial\theta}{\partial Y}\right)_{Y=0} = K(x, x^*, \Pr)\theta_s(x^*).} \qquad (11\text{–}103)$$

3. Third, we consider the case of a surface temperature distribution of the form

$$\theta_s(x) = 0, \quad 0 \le x < x^*,$$

$$\theta_s(x) = \theta_s(x^*), \quad x^* \le x \le x^* + \Delta x^*,$$

$$\theta_s(x) = 0, \quad x > x^* + \Delta x^*.$$

The solution, in terms of $(\partial\theta/\partial Y)_{Y=0}$ for this problem is just a linear combination of the solutions to two problems of the type considered in the second step, namely,

Problem 1: $\theta_s(x) = 0, \; 0 \le x < x^*,$

$$\theta_s(x) = \theta_s(x^*), \; x \ge x^*;$$

Problem 2: $\theta_s(x) = 0$, $0 \le x < x^* + \Delta x^*$,

$$\theta_s(x) = -\theta_s(x^*), \quad x \ge x^* + \Delta x^*.$$

Hence,

$$-\left(\frac{\partial \theta}{\partial Y}\right)_{Y=0} = K(x, x^*, \mathrm{Pr})\theta_s(x^*) - K(x, x^* + \Delta x^*, \mathrm{Pr})\theta_s(x^*)$$

and

$$-\left(\frac{\partial \theta}{\partial Y}\right)_{Y=0} = \{K(x, x^*, \mathrm{Pr}) - K(x, x^* + \Delta x^*, \mathrm{Pr})\}\,\theta_s(x^*). \qquad (11\text{–}104)$$

In the limit as the increment $\Delta x^* \to 0$, this can be written in the equivalent form

$$-\left(\frac{\partial \theta}{\partial Y}\right)_{Y=0} = -\frac{\partial K}{\partial x^*}\theta_s(x^*)dx^* \qquad (\Delta x^* \to 0). \qquad (11\text{–}105)$$

4. Finally, let us suppose that $\theta_s(x)$ is an arbitrary but smooth function of x, and that this function can be approximated by breaking the surface region into a large number of discrete intervals in which $\theta_s(x)$ can be approximated as a constant, corresponding, for example, to the average value over the interval. Then the solution, in terms of $-(\partial\theta/\partial Y)_{Y=0}$, for each individual interval is of the type shown in Step 3. Thus, the *discrete* approximation to the whole solution is a sum of solutions of the type from Step 3 for each subinterval, namely,

$$-\frac{\partial \theta}{\partial Y}\bigg|_{Y=0} = \sum_k \{K(x, k\Delta x^*, \mathrm{Pr}) - K(x, (k+1)\Delta x^*, \mathrm{Pr})\}\,\overline{\theta}_s(k\Delta x^*),$$

$$(11\text{–}106)$$

where the sum on k is such that $(k+1)\Delta x^* = x_L$ for the largest value of k.

Now, in the limit $\Delta x^* \to 0$, the discrete approximation to the surface temperature distribution passes smoothly to the continuous function $\theta_s(x)$, and the summation in (11–106) becomes an integral over x. Thus,

$$-\left(\frac{\partial \theta}{\partial Y}\right)_{Y=0} = -\int_0^x \left(\frac{\partial K}{\partial t^*}\right)\theta_s(t^*)dt^*. \qquad (11\text{–}107)$$

Now let us suppose that we are given the heat flux on the body surface as a function of x,

$$\left(\frac{\partial \theta}{\partial Y}\right)_{Y=0} = g(x), \qquad (11\text{–}108)$$

Substituting this into (11–107), we obtain

$$\int_0^x \frac{\partial K}{\partial t^*}(x,t^*,\mathrm{Pr})\theta_s(t^*)dt^* = g(x).$$
(11–109)

This is a *linear integral equation* (of the first kind) that now can be solved directly, in principle, to determine the unknown surface temperature distribution $\theta_s(x)$. The main advantage of this formulation, relative to solving the whole problem (11–6) with (11–108) as a boundary condition, is that (11–109) can be solved to determine $\theta_s(x)$ directly, without any need to determine the temperature distribution elsewhere in the domain. This latter problem is only one-dimensional, in spite of the fact that the original problem was fully two-dimensional.

If instead of (11–108) we have boundary conditions of the mixed type, a similar result is obtained. In particular, let us suppose that

$$\frac{\partial \theta}{\partial Y}\bigg|_{Y=0} + c_1\theta_s = c_2 f(x),$$
(11–110)

where $f(x)$ is known. Then substituting into (11–107), we obtain

$$\int_0^x \frac{\partial K}{\partial t^*}(x,t^*,\mathrm{Pr})\theta_s(t^*)dt^* + c_1\theta_s(x) = c_2 f(x).$$
(11–111)

In this case, we again have a linear integral equation (of the second kind) for the unknown surface temperature distribution.

The solutions to (11–109) or (11–111) can be carried out either numerically or, since the equations are linear, by an assortment of approximate analytical techniques. We do not attempt to present these solution methods here. Instead, to preserve space, we refer the interested reader to any of the standard textbooks on linear integral equations. A classic is the book *Linear Integral Equations* by W.V. Lovitt, which is available from Dover in paperback.[3]

Let us conclude by briefly summarizing the material of this section in the following manner. If we wish to determine θ everywhere in the domain, we would have to solve the differential equation (11–6)—either exactly or via an asymptotic approximation—subject to either of the boundary conditions (11–108) or (11–110) at the body surface. If, on the other hand, we wish only to determine $\theta_s(x)$, then it is advantageous to solve the *boundary integral* (11–109) or (11–111). This converts the original two-dimensional problem into a one-dimensional problem and allows $\theta_s(x)$ to be calculated directly without the necessity of determining θ everywhere in the domain.

Homework Problems

1. If $u_e = x^m$, find the most general functional form for the surface temperature, $\theta_s = \theta_s(x)$, that allows a similarity solution of the thermal boundary-layer equation in two dimensions for large $\mathrm{Re} \gg 1$ and $\mathrm{Pr} = 0(1)$.

2. We have considered the development of thermal boundary-layer theory for a

two-dimensional body with a constant surface temperature θ_0. We have also discussed a method to determine the *surface* temperature when the heat flux is specified. In this problem, we wish to solve for the leading-order approximation to the temperature distribution in the fluid for an arbitrary two-dimensional body when the heat flux is specified as a *constant*.

    a. Solve this problem *directly* using matched asymptotic expansions for $\text{Pr} \gg 1$ and $\text{Pr} \ll 1$.

    b. Suppose the heat flux is $Q = Q(x)$. Is there any combination of $u_e(x)$ and $Q(x)$ for which similarity solutions can be obtained? Prove your answer.

3. Consider Problem 4 from Chapter 10. If you have not solved part (a) of this problem, you will need to do so prior to tackling the following additional questions.

    a. Assume that the plate surface temperature is T_0, which is larger than the ambient temperature of the fluid, T_∞. Assume that $U/(k/\rho C_p) \gg 1$. Use the method of matched asymptotic expansions to obtain the governing differential equations and boundary conditions in the limit, $\text{Pr} \gg 1$. Solve the resulting equation for the temperature distribution and determine the relationship between Nu and Pr.

    b. Consider the same problem, but for $\text{Pr} \ll 1$. If necessary, you may assume that the *tangential* velocity u in the outer potential-flow region is $0(\text{Re}^{-1/2})$ for $\text{Re} \gg 1$. The normal velocity at the outer edge of the boundary-layer can be obtained from the solution of Problem 4 from Chapter 10.

4. We have stated in the text that there is an *analogy* between heat transfer and mass transfer of a single component, say A, in a fluid (gas or liquid) B. In particular, if we denote the concentration of A in weight percent as ω_A, then transport of A by convective diffusion is governed by

$$\mathbf{u} \cdot \nabla \omega_A = D \nabla^2 \omega_A \qquad (1)$$

at steady state, where D is the binary diffusion coefficient. However, a complete analogy between heat and mass transfer is possible only in certain special situations. In particular, the rates of heat and mass transfer between a stationary surface and a moving fluid may be appreciably different even for systems with identical external flow configurations and transport parameters (that is, $D \equiv k/\rho C_p$). One reason for this is that in mass transfer the surface can act either as a source or sink of material (the solute) with the result that a net hydrodynamic velocity is established normal to the solid-fluid interface. If this velocity, v_0, is appreciable in magnitude, it can cause a distortion in the velocity profile that would normally exist in the absence of mass exchange and thus substantially affect the mass transfer rate. Indeed, even if the velocity profile is not altered substantially, the rate of mass transfer still can be altered by the presence of a nonzero normal velocity at the boundary since this contributes a convection mode normal to the boundary in addition to the mass flux by diffusion that exists when $v_0 = 0$.

To be specific, consider a system that is composed of a gas consisting of one inert component and one diffusing component. The body is assumed to be a two-dimensional cylinder of arbitrary cross section, and the flow far from the body is a uniform stream, with characteristic velocity U_∞. The mass flux of the inert component at the body surface is zero; the diffusing species can be produced either at the solid (consider, for example, a porous body impregnated with fuel, which releases a gaseous product as the fuel burns), or it can be used up at the surface, which then acts as a sink for the diffusing species.

a. Derive an expression for the normal velocity at the body surface in terms of the binary *diffusivity* for A in B and the concentration distribution ω_A in the vicinity of the body surface.

b. Derive a *dimensionless* boundary-layer form of the governing differential equations and boundary conditions, assuming $\mathrm{Re} \gg 1$, so that boundary-layer equations hold for both the momentum equation and the species mass-balance equation (1). You may take the fluid properties ρ, μ, and D to be independent of composition. Further, you may assume that the velocity v_0 is large enough to affect the velocity profile inside the momentum boundary-layer but is not so large as to cause a change in the pressure distribution around the surface; that is, u_e is still given by potential flow theory with v approximated as zero at the body surface. What dimensionless parameter (or parameters) must be small if this condition is true? What parameters would have to be small if the mass flux at the body surface is to have negligible influence on the velocity distribution even in the boundary-layer?

c. Consider the case of $\mathrm{Sc} \gg 1$, where Sc is the Schmidt number ν/D. Derive the limiting form of the mass transfer boundary-layer equation for this case. You should be able to show that

$$\frac{\mathrm{Nu}_x}{\mathrm{Re}^{1/2}\,\mathrm{Sc}^{1/3}} = F(x;\lambda(x),B),$$

where $\lambda(x)$ is the dimensionless shear stress at the body surface and

$$B \equiv \frac{\omega_{A_S} - \omega_{A_\infty}}{1 - \omega_{A_S}},$$

with ω_{A_S} being the weight fraction of the diffusing species at the surface, $y = 0$, and x is the local cartesian coordinate parallel to the body surface.

References/Notes

1. The analysis in this chapter largely follows the original developments of Goddard, J.D., and Acrivos, A., "Asymptotic Expansions for Laminar Forced-Convection Heat and Mass Transfer, Part 2, Boundary-Layer Flows," *J. Fluid Mech.* 24:339–366 (1966).

2. Perry, R.H., and Chilton, C.H., *Chemical Engineers Handbook,* 5th Ed. McGraw-Hill: New York (1973).

3. Lovitt, W.V., *Linear Integral Equations.* Dover: New York (1950); originally published in 1924 by McGraw-Hill.

CHAPTER 12

Natural and Mixed Convection Flows

All of the analysis of the preceding chapters has either assumed that the fluids are isothermal or that the fluid motion is completely independent of the temperature distribution (the *forced convection approximation*). In reality, however, when the temperature is nonuniform, there will be gradients in all of the material properties including the density, and it is impossible (with the exception of a zero gravity environment) to have heat transfer without some contribution of buoyancy forces to the fluid motion. A critical question that we address in this chapter is when this buoyancy-driven contribution to the fluid's motion can actually be neglected as we have assumed all along.

In addition, there are many important problems where the motion of the fluid is entirely due to density gradients. These motions are commonly known as *natural convection*.[1] Although the velocities associated with natural convection are sometimes quite small, the mechanism of heat transfer by conduction alone (due to random thermal motion at the molecular level) is quite inefficient, and thus even *weak* convection motions can dominate conduction in the heat transfer process. Hence, for heat transfer processes that involve temperature gradients in a nominally quiescent gas or liquid, it can be extremely important to include natural convection contributions. We consider the analysis of such problems in this chapter.

The first step for all problems involving nonuniform density distributions is to reconsider the form of the equations of motion. Clearly, any contribution to fluid motion that is due to density gradients must be a consequence of the body-force term $\rho \mathbf{g}$ in these equations, but in all of the preceding chapters we have assumed that ρ is constant so that the body force contribution consists wholly of a hydrostatic pressure gradient that can be suppressed in the governing equations by introduction of the so-called dynamic pressure [see (2–114)]. On the other hand, for motions that are driven by density variations in the fluid, it seems logical that we should consider the equations of motion in a form that explicitly displays the body force contribution. Thus, in the first section, we reconsider the effect of the body force term in the equations of motion and discuss the so-called Boussinesq approximation that is introduced to make those equations tractable. Following this, we first consider problems involving

669

natural convection alone and then the so-called mixed problems where natural and forced convection contributions are both present.

A. The Boussinesq Equations

We begin with the complete dimensional form of the continuity, momentum, and thermal energy equations for an incompressible Newtonian fluid, simplified only to the extent that we limit our attention to *steady-state* problems and neglect viscous dissipation:

$$\rho(\mathbf{u}'\cdot\nabla'\mathbf{u}') = -\nabla'p' + \nabla'\cdot(\mu\nabla'\mathbf{u}') + \rho\mathbf{g}, \tag{12-1}$$

$$\nabla'\cdot(\rho\mathbf{u}') = 0, \tag{12-2}$$

and

$$\rho C_p(\mathbf{u}'\cdot\nabla'T) = \nabla'\cdot(k\nabla'T). \tag{12-3}$$

The primes on $\mathbf{u}'$, r', and ∇' simply indicate that the variables are dimensional. Since the temperature, T, is assumed to be nonuniform, all of the material properties, ρ, μ, k, and C_p also depend upon spatial position and thus $\nabla'\mu$, $\nabla'\rho$, $\nabla'k \neq 0$.

It is convenient at the outset to consider a specific problem. Thus, we imagine a heated (or cooled) body that has a uniform surface temperature T_o immersed in a fluid that has a uniform ambient temperature T_∞. Apart from any motion that is generated by density gradients, we suppose that the fluid is motionless.

The first step, following the examples of previous chapters, should be to nondimensionalize the equations (12-1)–(12-3). However, this immediately presents some difficulties. In all previous problems it has been assumed that the fluid is undergoing an undisturbed motion that defines an obvious characteristic velocity, u_c. Here, however, the motion is due to density variations in the fluid, and it is not so obvious what choice to make for u_c. In addition, since the fluid motion is completely dependent on the density (that is, temperature) distribution in the fluid, which in turn depends on the velocity field, all three equations are completely coupled. Furthermore, the equations are very strongly nonlinear; even the viscous and conduction terms in (12-1) and (12-3) are now nonlinear in view of the dependence of μ and k on T and the coupling between $\mathbf{u}'$ and T.

In order to make progress, and coincidentally point the way toward meaningful nondimensionalization, we need to consider carefully the role of the body force term in (12-1). It is clear from the preceding discussion that natural convection is due to the body force term $\rho\mathbf{g}$. However, we have already seen that nonzero $\rho\mathbf{g}$ does not necessarily produce motion. In particular, when ρ is constant, the constant body force can be balanced completely by a hydrostatic variation of pressure with no motion. It is only the *variation* in ρ *with spatial position* that leads to fluid motion, and in the problem considered here this variation is due to the local heating of the fluid by the body. In order to account for these physical facts, it is convenient to express (12-1) in an alternative form by subtracting the hydrostatic terms, namely,

$$0 = -\nabla'p_h' + \rho_\infty\mathbf{g}. \tag{12-4}$$

Then, expressing p' in (12–1) as the sum of a hydrostatic and dynamic contribution,

$$p' = p'_h + p'_d, \tag{12-5}$$

and subtracting (12–4) from (12–1), we obtain

$$\boxed{\rho(\mathbf{u}' \cdot \nabla' \mathbf{u}') = -\nabla' p'_d + \nabla' \cdot (\mu \nabla' \mathbf{u}') + (\rho - \rho_\infty)\mathbf{g}.} \tag{12-6}$$

Now, if $\rho - \rho_\infty \neq 0$ (or a constant), then $\mathbf{u}'$ must be *nonzero* to satisfy (12–6). It is evident in this form that the magnitude of the driving force for fluid motion is proportional to the magnitude of gradients of the density difference $\rho - \rho_\infty$.

Of course, the introduction of (12–4) and (12–5) into (12–1) to obtain (12–6) is analogous to the introduction of dynamic pressure to derive equation (2–115) for systems of *constant* density. Indeed, the commonly used description of p'_d as the dynamic pressure contribution derives from this previous analysis. In the present case, however, this designation is somewhat misleading. The contribution p'_d is actually the difference between the total pressure at a point and the hydrostatic pressure that would occur at the same point in the fluid if the density were ρ_∞ everywhere. Thus, it includes both a *dynamic* contribution and a *static* contribution due to the fact that the density actually differs from ρ_∞.

Of course, (12–6) is still exact, and the system of equations (12–6), (12–2), and (12–3) is no easier to solve than the original system of equations. To produce a tractable problem for analytic solution, it is necessary to introduce the so-called *Boussinesq approximation*, which has been used for virtually all existing analyses of natural and mixed convection problems. The essence of this approximation is the assumption that the temperature variations in the fluid are small enough that the material properties ρ, μ, k, and C_p can be approximated by their values at the ambient temperature T_∞, except in the body force term in (12–6), where the approximation $\rho = \rho_\infty$ would mean that the fluid remains motionless.

The Boussinesq approximation can be formalized in the following way.[2] Let us assume that the maximum value of $T - T_\infty$ is small in the sense that the various material properties can be approximated in the linear forms

$$\frac{\rho_\infty}{\rho} = 1 + \beta(T - T_\infty) + \cdots,$$

$$\frac{\mu_\infty}{\mu} = 1 + \alpha(T - T_\infty) + \cdots,$$

$$\frac{k_\infty}{k} = 1 + \gamma(T - T_\infty) + \cdots,$$

and

$$\frac{C_{p\infty}}{C_p} = 1 + \delta(T - T_\infty) + \cdots. \tag{12-7}$$

If we substitute these expressions into (12–6), (12–2), and (12–3), we obtain

$$\rho_\infty\left[1-\beta\Delta T+0(\Delta T)^2\right]\mathbf{u}'\cdot\nabla'\mathbf{u}'=$$

$$-\nabla'p'_d+\mu_\infty\nabla'\cdot\left[(1-\alpha\Delta T+0(\Delta T)^2)\nabla'\mathbf{u}'\right],-\rho_\infty(\beta\Delta T+0(\Delta T)^2)\mathbf{g}, \quad (12\text{-}8)$$

$$\nabla'\cdot\left[(1-\beta\Delta T+0(\Delta T)^2)\mathbf{u}'\right]=0, \quad (12\text{-}9)$$

and

$$\rho_\infty C_{p\infty}\left[(1-\beta\Delta T+0(\Delta T)^2)(1-\delta\Delta T+0(\Delta T)^2)\right]\mathbf{u}'\cdot\nabla'T=$$

$$k_\infty\nabla'\cdot\left[(1-\gamma\Delta T+0(\Delta T)^2)\nabla'T\right], \quad (12\text{-}10)$$

where $\Delta T=T-T_\infty$. Now, let us suppose that

$$\beta\Delta T,\ \gamma\Delta T,\ \alpha\Delta T,\ \delta\Delta T\ll 1. \quad (12\text{-}11)$$

The coefficients β, γ, α, and δ are generally quite small ($\beta\sim 10^{-3}$ per degree centigrade for liquids), and thus the conditions (12-11) do not necessarily imply very small values for ΔT. In any case, if (12-11) is satisfied, it is clear that a first approximation to (12-9) and (12-10) is

$$\boxed{\nabla'\cdot\mathbf{u}'=0} \quad (12\text{-}12)$$

and

$$\boxed{\rho_\infty C_{p\infty}(\mathbf{u}'\cdot\nabla'T)=k_\infty\nabla'^2T.} \quad (12\text{-}13)$$

In each case, the error in the terms neglected is $0(\Delta T)$ times the appropriate material coefficient α, β, γ, or δ, relative to the terms retained. The corresponding approximation to (12-8) requires a little more care. Superficially, the prescription leading to (12-12) and (12-13) would seem to suggest *neglecting all* terms of $0(\Delta T)$ including the buoyancy term, thus leaving only the inertia, viscous, and dynamic pressure terms evaluated using the material properties at the ambient temperature.

However, a little thought will indicate that the inertia, viscous, and dynamic pressure terms in (12-8) cannot be *larger* than the buoyancy term because the flows that are responsible for the *existence* of these terms are *driven* by the buoyancy forces. On the other hand, they cannot all be smaller either and still satisfy (12-8). Thus, to approximate (12-8) at the same level that is inherent in (12-12) and (12-13), we keep the first approximation to the inertia and viscous terms (that is, the terms independent of ΔT), plus the buoyancy term at $0(\Delta T)$, which *must* be the same magnitude as the *largest* of the inertia and/or viscous terms. The result is

$$\boxed{\rho_\infty(\mathbf{u}'\cdot\nabla'\mathbf{u}')=-\nabla p'_d+\mu_\infty\nabla'^2\mathbf{u}'-\rho_\infty\beta\Delta T\mathbf{g}.} \quad (12\text{-}14)$$

Although the approximation leading to (12-14) may at first seem quite arbitrary, this is only a consequence of trying to obtain approximations to the equations in *dimensional* form rather than first nondimensionalizing as we have systematically done in previous chapters. Thus, it appears that we have kept $0(1)$ approximations to the viscous, inertia, and pressure gradient terms but an $0(\Delta T)$ approximation to the buoyancy term. However, this is misleading. Because the motion in natural convection is driven by the buoyancy term, the characteristic magnitude of the velocities is actually proportional to $\beta g\Delta T$ raised to some power. Hence, it will turn out that the terms retained in (12-14) are all of the same order of magnitude in general, and the terms neglected are all $0(\Delta T)^m$ smaller, where m is some positive exponent.

The equations (12-12)-(12-14) are known as the *Boussinesq equations of motion* and will form the basis for analysis of natural convection effects in this chapter. In fact, the Boussinesq approximation has been used for almost all published theoretical work on natural convection flows. Although one should expect quantitative deviations from the Boussinesq predictions for systems in which the temperature differences are large (greater than $\sim 10\text{-}20\,^{\circ}\text{C}$), it is likely that the Boussinesq equations remain qualitatively useful over a considerably larger range of temperature differences. In any case, although the Boussinesq equations represent a very substantial simplification of the exact equations, the essential property of *coupling* between the thermal and velocity fields is preserved, and even in the Boussinesq approximation, the solution of natural convection problems is more complicated than the forced convection heat transfer problems that we encountered earlier.

B. Natural Convection Boundary-Layers at High Grasshof Numbers

In order to illustrate the preceding point and to introduce an important prototype problem, let us begin by considering the motion induced in the immediate vicinity of a body that is either heated or cooled with respect to the ambient temperature of a surrounding fluid. For simplicity, we assume that the fluid would be motionless in the absence of buoyancy-driven natural convection and that the temperature of the body surface is a constant that we denote as T_o. The general problem is thus to solve (12-12)-(12-14) subject to the conditions of uniform ambient temperature T_∞ and no motion at large distances from the body, plus constant temperature T_o and the kinematic and no-slip boundary conditions at the body surface. Not surprisingly, this general problem is too difficult for analytic solution—the complete system of Navier-Stokes and thermal energy equations is nonlinear and highly coupled [the nonlinearity of (12-13) is a consequence of the fact that **u** is a function of T through (12-14), and vice versa]. To make any progress, it is necessary to seek further simplification of the governing equations by limiting our considerations to special ranges of parameter space and then applying the asymptotic techniques that have been developed in the preceding chapters.

A first step toward this goal is to nondimensionalize the governing equations. The main uncertainly or difficulty in doing this is that it is not immediately obvious what choice to make for a characteristic velocity. Unlike all of the problems that we have seen in previous chapters, the magnitude of the velocity is *not* imposed by setting the magnitude of **u** at a boundary (or at infinity) but is determined by other independent parameters including, presumably, the magnitude of the temperature differential $(T_o - T_\infty)$. In order to proceed, let us formally designate the characteristic velocity as u_c and see whether we can determine the dependence of u_c on the independent parameters of the problem during the course of analysis. Thus, we define dimensionless variables as

$$\Theta \equiv \frac{T - T_\infty}{T_o - T_\infty}, \quad \mathbf{u} = \frac{\mathbf{u}'}{u_c}, \quad p = \frac{p'_d}{\rho_\infty u_c^2}, \text{ and } \mathbf{x} = \frac{\mathbf{x}'}{l} \tag{12–15}$$

and substitute into the Boussinesq equations (12–12)–(12–14). Note that l denotes a characteristic linear dimension of the heated (or cooled) body. Furthermore, the use of $\rho_\infty u_c^2$ as a characteristic pressure suggests that we will ultimately be concerned with a parameter range that is equivalent, in some sense, to a finite Reynolds number regime.

Beginning with the Navier-Stokes equation, (12–14), the resulting nondimensionalized equation can be expressed in the form

$$\mathbf{u} \cdot \nabla \mathbf{u} = -\nabla p + \frac{\nu_\infty}{u_c l} \nabla^2 \mathbf{u} - \frac{l \beta g (T_o - T_\infty)}{u_c^2} \Theta \mathbf{i}_g, \tag{12–16}$$

where we have expressed the gravitational acceleration vector $\mathbf{g}$ in the form $\mathbf{g} \equiv g \mathbf{i}_g$. The two groups of dimensional parameters that appear in (12–16) are, of course, dimensionless. However, because u_c remains to be specified and must depend in some way on the remaining dimensional parameters in (12–16), only one of these dimensionless groups is actually independent.

The form of (12–16) suggests one of two possible choices for u_c, depending on the dominant physical balance. We have noted earlier that the inertia, pressure, and viscous terms in (12–16) exist only because of the motion produced by the buoyancy term. Hence, there must be a balance between one or more of these terms and buoyancy.

If the dominant balance is between *inertia* and buoyancy, then it is clear from (12–16) that the appropriate choice for u_c is

$$\boxed{u_c = \sqrt{l \beta g (T_o - T_\infty)}} \tag{12–17}$$

(assuming $T_o > T_\infty$) so that (12–16) becomes

$$\boxed{\mathbf{u} \cdot \nabla \mathbf{u} = -\nabla p + \frac{1}{\mathrm{Gr}^{1/2}} \nabla^2 \mathbf{u} - \Theta \mathbf{i}_g.} \tag{12–18}$$

The dimensionless parameter that remains in (12–18) is known as the *Grasshof number* and is defined as

$$\boxed{\mathrm{Gr} \equiv \frac{\beta (T_o - T_\infty) g l^3}{\nu_\infty^2}.} \tag{12–19}$$

Clearly, for natural convection flows, $\mathrm{Gr}^{1/2}$ plays the role of a Reynolds number, and the scaling in (12–17) is relevant for $\mathrm{Gr} \gg 1$, where inertia dominates viscous effects.

On the other hand, it might occur that the dominant balance in (12–16) actually should be between *viscous* forces and the buoyancy term. In this case, an appropriate choice for u_c is the one corresponding to

$$\frac{\nu_\infty}{u_c l} = \frac{l \beta g (T_o - T_\infty)}{u_c^2}$$

or

$$u_c = \frac{l^2 \beta g (T_o - T_\infty)}{\nu_\infty}. \tag{12-20}$$

Then, (12–16) takes the form

$$\text{Gr} \left[\mathbf{u} \cdot \nabla \mathbf{u} \right] = -\nabla \bar{p} + \nabla^2 \mathbf{u} - \theta \mathbf{i}_g, \tag{12-21}$$

where

$$\bar{p} = \text{Gr} p. \tag{12-22}$$

The scaling in (12–20) is appropriate for *low* Grasshof numbers, and the rescaling in (12–22) is just equivalent to the change from a characteristic pressure ρu_c^2, as assumed originally, to $\mu_\infty u_c / l$, as is appropriate for the viscous-dominated limit for natural convection.

In practice, the Grasshof number is usually rather large, and the nondimensionalization in (12–17)–(12–18) is more commonly used for theoretical analysis. The reason is that the combination of parameters gl^3 / ν_∞^2 tends to be large. For example, let us suppose that we have a rather small heated body with a linear dimension of order 1 centimeter and that the fluid of interest is water at an ambient temperature of 20 °C. Then,

$$\text{Gr} \equiv \frac{\beta g l^3}{\nu_\infty^2} (T_o - T_\infty) = \left(\frac{2 \times 10^3}{°C} \right) (T_o - T_\infty).$$

Clearly, even for rather small temperature differences, the Gr number is quite large.

Thus, in the analysis that follows, we consider only the limiting case of large Gr.[3] The dimensionless form of the equation of motion is thus (12–18), and when the scaling in (12–15) and (12–17) is introduced into the continuity and thermal energy equations (12–12) and (12–13), these become

$$\nabla \cdot \mathbf{u} = 0 \tag{12-23}$$

and

$$\mathbf{u} \cdot \nabla \theta = \frac{1}{\text{Pr} \text{Gr}^{1/2}} \nabla^2 \theta. \tag{12-24}$$

Recall that the problem we have set out to solve is that of a heated (or cooled) body in an unbounded ambient fluid that is motionless at large distances from the body surface. Thus, we seek to solve (12–18), (12–23), and (12–24), subject to the boundary conditions

$$\mathbf{u} \to 0, \ \theta \to 0 \quad \text{far from the body}, \tag{12-25a}$$
$$\mathbf{u} = 0, \ \theta = 1 \quad \text{on the body surface}, \tag{12-25b}$$

in the limit Gr $\gg 1$.

The Outer Region

We have already seen that $\text{Gr}^{1/2}$ plays a role in the nondimensionalized Boussinesq equations that is similar to Reynolds number in the forced convection problems studied earlier. Thus, we should not be surprised if the natural convection problem for $Gr \gg 1$ exhibits some type of boundary-layer structure. If this is true, however, the nondimensionalized form of (12–18), (12–23), and (12–24) can be relevant only to the portion of the domain that is not too close to the body surface, since these equations were nondimensionalized assuming that the length scale characteristic of changes in $\mathbf{u}$ or θ was the linear dimension of the body, l. In particular, if we take the limit $Gr \rightarrow \infty$ in (12–18) and (12–24), we obtain the inviscid and convection dominated limits

and

$$\mathbf{u} \cdot \nabla \mathbf{u} = -\nabla p - \theta \mathbf{i}_g \qquad (12–26)$$

$$\mathbf{u} \cdot \nabla \theta = 0. \qquad (12–27)$$

Assuming that the streamlines of the natural convection flow are open (or at least only close very far from the body surface), the solution of (12–27) is

$$\theta = \theta(\psi),$$

which implies that $\theta = 0$ everywhere in the domain where (12–27) is valid. Hence, there is no body force in (12–26), and we thus obtain the homogeneous equations

$$\mathbf{u} \cdot \nabla \mathbf{u} = -\nabla p \quad \text{and} \quad \nabla \cdot \mathbf{u} = 0.$$

Since the boundary conditions (12–25) in $\mathbf{u}$ are homogeneous, it would appear that the leading-order solution for $\mathbf{u}$ in the limit $Gr \rightarrow \infty$ might be $\mathbf{u} \equiv 0$, $p = \text{const}$. Of course, this solution,

$$\theta \equiv 0, \quad \mathbf{u} \equiv 0, \quad p = \text{const},$$

cannot be valid in the immediate vicinity of the body surface because it does not satisfy the boundary condition $\theta \equiv 1$.

The question is whether it is a plausible result for the leading-order approximation in the rest of the domain. On the one hand, if $\theta = 0$, there is no body force to induce motion. On the other hand, in the *inner* region immediately adjacent to the body surface, it is obvious from the boundary conditions (12–25) that $\theta \neq 0$, and thus there is a buoyancy force that *will* induce motion. Furthermore, it will be shown by matching between the two regions that $\mathbf{u} \neq 0$ in the outer part of the domain in spite of the fact that $\theta \equiv 0$. The important question is whether $\mathbf{u}$ is nonzero in the outer region at $O(1)$ or only at higher orders of approximation $O(Gr^{-m})$. A definitive answer to this question can only be obtained by examination of the solution in the inner region, and thereby of the matching conditions between the two parts of the domain.

The Inner Region—
The Natural Convection Boundary-Layer

To proceed further, we must consider the solution in the inner part of the domain, immediately adjacent to the body surface, and for this we must specify the geometry of the heated body. For convenience, we consider a horizontal, two-dimensional, cylindrical body. The cylindrical cross section is assumed to contain a central plane of symmetry that is parallel to $\mathbf{g}$, but the shape is otherwise arbitrary but smooth (no sharp

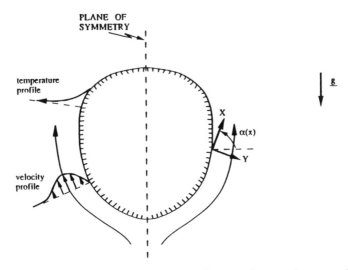

Figure 12-1 A sketch of the geometric configuration for natural convection near a heated, two-dimensional body.

corners, cusps, or so on). In view of the fact that the cylinder is horizontal (while **g** is vertical), there will be no flow induced in the axial direction, and the problem is reduced to a *two-dimensional* flow in the plane at any arbitrary axial position. For definition purposes, we sketch a portion of the arbitrary cross section of our body in Figure 12-1. The central symmetry plane ensures only that the motion induced on one side of the body is the same as that on the other side and that the division of fluid going to the left or right occurs at the lowest point on the body surface.

For the present analysis, we are considering the *limiting* case in which $Gr \gg 1$. We have already seen that $Gr^{1/2}$ plays a role in the governing equations (12–18) and (12–24) that is equivalent to the Reynolds number in the forced convection problems studied earlier. Hence, we may anticipate, by analogy, that the structure of the solutions for **u** and θ in this inner region will exhibit a boundary-layer form. The necessity for a boundary-layer is, in fact, corroborated by the leading-order approximation to the solution of (12–18) and (12–23)–(12–24). Following the precedent of previous examples, it should be possible to develop this solution using local pseudo-Cartesian coordinates x and y, as shown in Figure 12-1. In this case, however, it is necessary to keep track of the orientation of the x and y axes relative to the vertical since this determines the relative importance of the buoyancy term in the two directions. For this purpose, we introduce the function $\alpha(x)$, which is defined as the angle between the x axis (that is locally parallel to the body surface) and the horizontal plane as shown in Figure 12-1.

The governing equations (12–18), (12–23), and (12–24), expressed in terms of the local x and y coordinates but neglecting all curvature terms in anticipation of the boundary-layer-like analysis are then

$$u\frac{\partial u}{\partial x} + v\frac{\partial u}{\partial y} = -\frac{\partial p}{\partial x} + \frac{1}{Gr^{1/2}}\left(\frac{\partial^2 u}{\partial x^2} + \frac{\partial^2 u}{\partial y^2}\right) + \theta\sin\alpha, \tag{12–28}$$

$$u\frac{\partial v}{\partial x} + v\frac{\partial v}{\partial y} = -\frac{\partial p}{\partial y} + \frac{1}{Gr^{1/2}}\left(\frac{\partial^2 v}{\partial x^2} + \frac{\partial^2 v}{\partial y^2}\right) - \theta\cos\alpha, \tag{12–29}$$

$$\frac{\partial u}{\partial x} + \frac{\partial v}{\partial y} = 0, \tag{12–30}$$

and

$$u\frac{\partial \theta}{\partial x} + v\frac{\partial \theta}{\partial y} = \frac{1}{Pr\,Gr^{1/2}}\left(\frac{\partial^2 \theta}{\partial x^2} + \frac{\partial^2 \theta}{\partial y^2}\right). \tag{12–31}$$

To obtain a *rescaled* version of these equations, suitable to the boundary-layer region, we proceed at first by *analogy* with the forced-convection boundary-layer analysis of Chapter 11. Thus, since $Gr^{1/2}$ plays the role of a Reynolds number, we introduce a rescaling of variables in the form

$$Y = y\,Gr^{1/4}, \quad V = v\,Gr^{1/4}. \tag{12–32}$$

The objective of this rescaling is to introduce a length scale appropriate to the boundary-layer regime so that the leading-order approximation to (12–28)–(12–31) retains both conduction and viscous terms in the limit $Gr \to \infty$. The assumption inherent in (12–32) is that the region in which $\theta \neq 0$ is limited to a thin boundary-layer immediately adjacent to the body surface. Introducing (12–32) into (12–28), we obtain

$$u\frac{\partial u}{\partial x} + V\frac{\partial u}{\partial Y} = -\frac{\partial p}{\partial x} + \frac{\partial^2 u}{\partial Y^2} + \theta\sin\alpha + 0(Gr^{-1/2}). \tag{12–33}$$

Clearly, in the limit $Gr \to \infty$, the viscous term is retained. The component of the equation of motion normal to the body surface takes the form

$$\frac{1}{Gr^{1/4}}\left[u\frac{\partial V}{\partial x} + V\frac{\partial V}{\partial Y}\right] = -Gr^{1/4}\frac{\partial p}{\partial Y} + \frac{1}{Gr^{1/4}}\frac{\partial^2 V}{\partial Y^2} - \theta\sin\alpha + 0(Gr^{-1/2}). \tag{12–34}$$

Hence,

$$\frac{\partial p}{\partial Y} = -\frac{1}{Gr^{1/4}}\theta\cos\alpha + 0(Gr^{-1/2}). \tag{12–35}$$

Now, provided $\alpha \neq 0$, except possibly at discrete points along the body surface, the *dominant* contribution of buoyancy is clearly the term $\theta\sin\alpha$ in (12–33), and we can approximate $\partial p/\partial Y$ as

$$\frac{\partial p}{\partial Y} = 0(Gr^{-1/4}) \to 0 \quad \text{as} \quad Gr \to \infty. \tag{12–36}$$

The situation where $\alpha = 0$ is a special case since the direct body force contribution to (12–33) is zero and the influence of buoyancy comes from the *streamwise* pressure gradient, $\partial p / \partial x$, that arises from integrating (12–35) with respect to Y and then differentiating with respect to x. In this special case, if $\partial \theta / \partial x \neq 0$, the effect of buoyancy is weaker by $0(\mathrm{Gr}^{-1/4})$ but is still the dominant influence on the dynamics of the boundary-layer for $\alpha = 0$. We limit ourselves here to situations in which $\alpha \neq 0$ so that the approximation (12–36) can be used. This implies, as in the forced convection case, that the pressure is constant across the boundary-layer so that

$$\frac{\partial p}{\partial x} = \frac{\partial p_\infty}{\partial x} + 0(\mathrm{Gr}^{-1/4}), \tag{12–37}$$

where p_∞ is the pressure imposed on the boundary-layer from the outer inviscid flow region. In the forced convection problems that we have studied previously, the nature of the boundary-layer flow is actually determined by the pressure gradient imposed from the outer region. However, in the present case, the arguments of the preceding section suggest that $u = v = 0$ to leading order in the outer regime as $\mathrm{Gr} \to \infty$, and $\partial p_\infty / \partial x = 0$. Hence,

$$\frac{\partial p}{\partial x} = 0(\mathrm{Gr}^{-1/4}) \sim 0, \tag{12–38}$$

and (12–33) can be simplified to the form

$$u \frac{\partial u}{\partial x} + V \frac{\partial u}{\partial Y} = \frac{\partial^2 u}{\partial Y^2} + \theta \sin \alpha + 0(\mathrm{Gr}^{-1/4}). \tag{12–39}$$

Finally, introducing the rescaled variables (12–32) into the continuity and thermal energy equations (12–30) and (12–31), we obtain

$$\frac{\partial u}{\partial x} + \frac{\partial V}{\partial Y} = 0 \tag{12–40}$$

and

$$u \frac{\partial \theta}{\partial x} + V \frac{\partial \theta}{\partial Y} = \frac{1}{\mathrm{Pr}} \frac{\partial^2 \theta}{\partial Y^2} + 0(\mathrm{Gr}^{-1/2}\mathrm{Pr}^{-1}). \tag{12–41}$$

The equations (12–39)–(12–41) are known as the *natural convection boundary-layer equations*. For arbitrary Prandtl number, Pr, the leading-order asymptotic approximation to the solution in the natural convection boundary-layer is obtained by solving (12–39)–(12–41) subject to the conditions

$$\theta = 1, \ u = V = 0 \quad \text{at} \quad Y = 0 \tag{12–42}$$

and

$$\theta \to 0, \ u \to 0 \quad \text{as} \quad Y \to \infty. \tag{12–43}$$

The latter conditions represent a matching condition with the leading-order approximation to the solution in the outer part of the domain based on the assumption that $u = v = 0$ in this region.

Clearly, as $Gr \to \infty$, we retain both conduction and viscous terms in the natural convection boundary-layer region. Furthermore, based upon the scaling in (12–32), we see that

$$\boxed{Nu = Gr^{1/4} f(Pr) \quad \text{for} \quad Gr \gg 1,}$$ (12–44)

where Nu is the Nusselt number, defined previously as

$$Nu \equiv \int_S -\left(\frac{\partial \theta}{\partial y}\right) \partial S,$$

where S represents the body surface.

In the remainder of this section we consider the solution of (12–39)–(12–43). Although the problem has been considerably simplified, it is still not possible to obtain closed-form analytic solutions except in special circumstances. These follow exactly the pattern from the forced convection problem in Chapter 11; for arbitrary Pr we can in principle obtain *similarity solutions,* but only for very special geometries. On the other hand, for $Pr \gg 1$ or $Pr \ll 1$ we can obtain asymptotic approximations of the solution for bodies of arbitrary shape. The scaling involved in these limiting cases will be shown to lead to the additional results

$$\boxed{f(Pr) = \begin{cases} C_1 Pr^{1/4} & \text{as} \quad Pr \to \infty \\ C_2 Pr^{1/2} & \text{as} \quad Pr \to 0. \end{cases}}$$ (12–45)

Нερε, C_1 and C_2 are two $0(1)$ constants that depend on the shape of the body.

It should be evident from the arguments of Chapter 11 that exact analytic solutions of (12–39)–(12–43) are possible only via similarity transformation and, further, that a similarity transformation only exists for $\sin \alpha(x) \cong x^m$. As far as we can see, the only problems satisfying this condition correspond to a flat plate at some fixed orientation so that $\sin \alpha$ is constant, independent of x, and $m = 0$. An example, which is considered as a homework problem at the end of this chapter, is the vertical flat plate in which $\sin \alpha = 1$. We focus our attention here on the two limiting cases $Pr \to \infty$ and $Pr \to 0$.

The Limiting Case $Pr \to \infty$.

We begin with the limiting case $Pr \to \infty$. In this limit, the natural convection boundary-layer splits again into two parts because the solution of (12–39)–(12–43) for large Pr is, itself, a singular perturbation problem. To see this, we attempt to solve (12–39)–(12–43) directly via an asymptotic expansion for large Pr, with the governing equations for the first terms in this expansion derived from (12–39)–(12–43) by applying the limit $Pr \to \infty$.

We need consider only (12–41) to demonstrate our point. Clearly, in the limit $Pr \to \infty$, this equation reduces to the form

$$\mathbf{u} \cdot \nabla \theta = 0,$$

and the solution is

$$\theta \equiv 0. \tag{12-46}$$

Since this solution does *not* satisfy the boundary condition on θ at the body surface and completely throws out the body-force term in (12–39), it is evident that (12–39)–(12–41) apply to only the outer part of the natural convection boundary-layer, away from the body surface. In view of the absence of a body force term in this region, it is tempting to guess that the solution for u and V will again be zero at the leading order. However, we shall see shortly that this is not true. Instead, the motion induced in the *inner* part of the boundary-layer, where the buoyancy force is *nonzero*, causes a motion of *equal magnitude* in the outer part of the boundary-layer where the buoyancy force is *zero*.

To obtain a rescaled form of (12–39)–(12–41) appropriate to the *inner* part of the boundary-layer for $Pr \gg 1$, we introduce the following change of variables:

$$\boxed{Y_1 = Y Pr^a, \quad u_1 = u Pr^b, \quad \text{and} \quad V_1 = V Pr^{a+b}.} \tag{12-47}$$

The objective is to determine a and b such that the conduction term is retained in the thermal energy equation as well as the viscous and buoyancy terms (at least) in (12–39). The motivation for rescaling Y in the form shown is the same as in previous examples; for $Pr \gg 1$, we "stretch" the normal variable Y so that changes that occur over a very short distance in terms of Y now occur over a distance ΔY_1 of $0(1)$. The rescaling of u suggested in (12–47) is not so obvious. We note simply that u is driven by θ through the buoyancy term, and θ depends on Pr so that it is quite plausible to assume that the magnitude of u also depends on Pr. Of course, it could happen that $b = 0$. The rescaling of V follows directly from this scaling for u and Y, plus the requirement of preserving the form of the continuity equation in terms of the rescaled variables, namely,

$$\boxed{\frac{\partial u_1}{\partial x} + \frac{\partial V_1}{\partial Y_1} = 0.} \tag{12-48}$$

To determine a and b, we rescale (12–39) and (12–41) according to (12–47). The result for the thermal energy equation is

$$u_1 \frac{\partial \theta}{\partial x} + V_1 \frac{\partial \theta}{\partial Y_1} = Pr^{2a+b-1} \frac{\partial^2 \theta}{\partial Y_1^2}. \tag{12-49}$$

Since the objective of rescaling is to express the equations in a form that retains the balance between convection and conduction in the limit $Pr \to \infty$, it follows that

$$\boxed{2a + b - 1 = 0.} \tag{12-50}$$

The equation of motion, on the other hand, takes the rescaled form

$$\frac{1}{Pr^{2b}} \left\{ u_1 \frac{\partial u_1}{\partial x} + V_1 \frac{\partial u_1}{\partial Y_1} \right\} = \theta \sin \alpha + Pr^{2a-b} \left(\frac{\partial^2 u_1}{\partial Y_1^2} \right). \tag{12-51}$$

Clearly, we must retain the body force in the limit $Pr \to \infty$ and at least one of the other two terms in (12–51). Since these rescaled equations are intended to be valid in the innermost region very close to the body surface, it would be surprising if the dominant balance in (12–51) did not include the viscous rather than inertia term. However, let us suppose that we did not possess this insight and decided that it was the inertia terms that should balance the body force term in this innermost region. Then we would require that

$$2b = 0,$$

and thus, from (12–50), it would follow that $a = 1/2$. But with $b = 0$ and $a = 1/2$, the rescaled form of (12–51) is

$$u_1 \frac{\partial u_1}{\partial x} + V_1 \frac{\partial u_1}{\partial Y_1} = \theta \sin \alpha + \mathrm{Pr} \left(\frac{\partial^2 u_1}{\partial Y_1^2} \right).$$

However, this is *inconsistent* with the assumption that led to the condition $2b = 0$ in the first place, because it implies that both the inertia and buoyancy terms are smaller than the viscous term for $\mathrm{Pr} \gg 1$.

The correct assumption is that the dominant balance in equation (12–51) is between the buoyancy and viscous terms. This requires that

$$\boxed{2a - b = 0} \tag{12–52}$$

and, in conjunction with (12–50), leads to the values

$$\boxed{a = 1/4, \quad b = 1/2.} \tag{12–53}$$

The governing equations (12–49) and (12–51) then take the forms

$$\boxed{u_1 \frac{\partial \theta}{\partial x} + V_1 \frac{\partial \Theta}{\partial Y_1} = \frac{\partial^2 \theta}{\partial Y_1^2}} \tag{12–54}$$

and

$$\boxed{\frac{\partial^2 u_1}{\partial Y_1^2} + \theta \sin \alpha = 0(\mathrm{Pr}^{-1}),} \tag{12–55}$$

and this is consistent with the assumption leading to (12–52). Furthermore, the asymptotic result for the Nusselt number correlation for $\mathrm{Gr} \gg 1$ and $\mathrm{Pr} \gg 1$ is now seen to be

$$\boxed{\mathrm{Nu} = C_1 \mathrm{Pr}^{1/4} \mathrm{Gr}^{1/4},} \tag{12–56}$$

as suggested earlier.

Although the rescaling in (12–47) with $a = 1/4$ and $b = 1/2$ is clearly self-consistent and leads to a form of the governing equations in which both conduction and viscous terms are retained in the limit $\mathrm{Pr} \to \infty$, it is worthwhile to pause for a moment to explore the physical significance of the rescaling. In terms of the original natural convection boundary-layer variables, the rescaling is

$$u = \frac{u_1}{\mathrm{Pr}^{1/2}}, \quad Y = \frac{Y_1}{\mathrm{Pr}^{1/4}}, \quad \text{and} \quad V = \frac{V_1}{\mathrm{Pr}^{3/4}}.$$

One obvious question, since $\theta = O(1)$ and the body force term in (12–39) is $O(1)$, is why the tangential velocity component turns out to be only $O(\text{Pr}^{-1/2})$ instead of $O(1)$ for $\text{Pr} \gg 1$. The answer is twofold: First, the magnitude of the velocity component is restricted by the fact that we are considering only a very thin layer of $O(\text{Pr}^{-1/4})$ in the region immediately adjacent to a no-slip boundary; second, although the magnitude of the body force is $O(1)$, it is confined to an increasingly thinner region with increase of Pr.

It is also worth noting the critical role played by Pr in determining the size of the regions where the various physical effects are important within the natural convection boundary-layer. In this regard, the Prandtl number plays a similar role to that found earlier in conjunction with the analysis of the *forced* convection thermal boundary-layer problem. In particular, for $\text{Pr} \gg 1$, the whole of the so-called thermal layer in which $\theta \neq 0$ lies sufficiently close to the body surface so that inertia plays no role. In this sense, the thermal boundary-layer lies *inside* the momentum layer just as it did in the forced convection case for $\text{Pr} \gg 1$.

It is, perhaps, worthwhile to summarize the asymptotic structure of the natural convection boundary-layer, as determined by the preceding analysis. This is summarized by the schematic representation in Figure 12–2. In the outer part of the *boundary-layer*, the characteristic scale $l\text{Gr}^{-1/4}$ is applicable, and we have seen that the governing equations are

$$
\begin{aligned}
u\frac{\partial u}{\partial x} + V\frac{\partial u}{\partial Y} &= \frac{\partial^2 u}{\partial Y^2}, \\
\frac{\partial u}{\partial x} + \frac{\partial V}{\partial Y} &= 0, \\
\theta &= 0,
\end{aligned}
\tag{12-57}
$$

with the single boundary condition

$$
u \to 0 \quad \text{as} \quad Y \to \infty
\tag{12-58}
$$

from matching with the leading-order approximation in the region outside the boundary-layer. In the inner (thermal) region, characterized by the length scale $l\text{Gr}^{-1/4}\text{Pr}^{-1/4}$, the governing equations are

$$
\begin{aligned}
0 &= \frac{\partial^2 u_1}{\partial Y_1^2} + \theta \sin\alpha, \\
\frac{\partial u_1}{\partial x} + \frac{\partial V_1}{\partial Y_1} &= 0, \\
u_1\frac{\partial \theta}{\partial x} + V_1\frac{\partial \theta}{\partial Y_1} &= \frac{\partial^2 \theta}{\partial Y_1^2},
\end{aligned}
\tag{12-59}
$$

and the boundary conditions at the body surface are applicable, namely,

$$
u_1 = V_1 = 0, \ \theta_1 = 1 \quad \text{at} \quad Y_1 = 0.
\tag{12-60}
$$

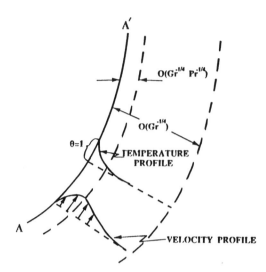

Figure 12–2 A pictorial representation of the asymptotic boundary-layer structure for natural convection flow in the limit $Gr \gg 1$, $Pr \gg 1$.

Clearly, to solve (12–57) and (12–59), additional boundary conditions must be specified, and these come from matching between the asymptotic approximations in the two regions (namely, $Y_1 \rightarrow \infty$ and $Y \rightarrow 0$).

One of the matching conditions is obvious,

$$\boxed{\theta \rightarrow 0 \quad \text{as} \quad Y_1 \rightarrow \infty,} \tag{12–61}$$

since the solution for θ in the outer part of the boundary-layer is $\theta = 0$. Additional conditions must come from matching between the velocity components and (if necessary) their derivatives with respect to Y and Y_1. Examining (12–57)–(12–60), it is evident that we need three such conditions from matching—two for the solution from (12–57) and one for the solution from (12–59).

In order to formulate these conditions, it is useful to examine the qualitative nature of the solution for u_1 in the inner region. In particular, we see that

$$\frac{\partial^2 u_1}{\partial Y_1^2} = -\theta \sin\alpha < 0 \quad \text{for all } Y_1,$$

since θ varies from $\theta = 1$ at $Y_1 = 0$, to $\theta = 0$ in the asymptotic limit $Y_1 \rightarrow \infty$. Thus, $(\partial u_1 / \partial Y_1)$, which is positive at $Y_1 = 0$, decreases monotonically to zero as $Y_1 \rightarrow \infty$, and u_1, which is zero at $Y_1 = 0$, asymptotes to a constant, nonzero value as $Y_1 \rightarrow \infty$, that is,

$$\boxed{u_1 \rightarrow \text{const} \sim 0(1) \quad \text{as} \quad Y_1 \rightarrow \infty.} \tag{12–62}$$

Now, the formal matching condition between u and u_1 is

$$\boxed{\lim_{Y_1 \gg 1} u_1 \Longleftrightarrow \lim_{Y \ll 1} \mathrm{Pr}^{1/2} u \quad \text{as} \quad \mathrm{Pr} \to \infty.} \qquad (12\text{-}63)$$

But we have just shown that the left-hand side has a value of $O(1)$ according to (12–62). Thus, $\mathrm{Pr}^{1/2} u$ must be asymptotic to the same $O(1)$ values as $Y \to 0$, and it thus follows that u must be $O(\mathrm{Pr}^{-1/2})$ for this large Pr limit, rather than $O(1)$ as assumed in (12–57) and (12–58). The source of the original estimate $u = O(1)$ was the nondimensionalization (12–17) which was based on the assumption that the buoyancy force acted in a region of $O(l)$—that is, comparable in size to the linear dimension of the body. However, we now see that the buoyancy force for $\mathrm{Pr} \gg 1$ is restricted to a very much smaller region (indeed, asymptotically small for $\mathrm{Pr} \gg \infty$), and the velocity induced by buoyancy is smaller by $O(\mathrm{Pr}^{-1/2})$ than the original estimate (12–17).

The preceding argument suggests that to complete the formulation of the problem, we need to go back and rescale u in the governing equations (12–57) for the outer part of the boundary-layer. However, in rescaling u, we need to be careful not to change the basic form of (12–57). This requires rescaling Y and V in addition to u. In particular, let us suppose that

$$\tilde{u} = u \mathrm{Pr}^{1/2},$$

$$\tilde{V} = V \mathrm{Pr}^{\alpha},$$

and

$$\tilde{Y} = Y \mathrm{Pr}^{\alpha - 1/2}.$$

This preserves the form of the continuity equation,

$$\frac{\partial \tilde{u}}{\partial x} + \frac{\partial \tilde{V}}{\partial \tilde{Y}} = 0.$$

The equation of motion (12–57) then becomes

$$\frac{1}{\mathrm{Pr}} \left[\tilde{u} \frac{\partial \tilde{u}}{\partial x} + \tilde{V} \frac{\partial \tilde{u}}{\partial \tilde{Y}} \right] = \frac{1}{\mathrm{Pr}^{3/2 - 2\alpha}} \frac{\partial^2 \tilde{u}}{\partial \tilde{Y}^2}.$$

In order to preserve the form of this equation, we require that $\alpha = 1/4$. Hence,

$$\tilde{u} = u \mathrm{Pr}^{1/2},$$

$$\tilde{V} = V \mathrm{Pr}^{1/4},$$

$$\tilde{Y} = Y \mathrm{Pr}^{-1/4},$$

and the governing equations in the outer part of the natural convection layer (often called the *inertial layer*) remain in the form

$$\frac{\partial \tilde{u}}{\partial x} + \frac{\partial \tilde{V}}{\partial \tilde{Y}} = 0,$$

$$\tilde{u}\frac{\partial \tilde{u}}{\partial x} + \tilde{V}\frac{\partial \tilde{u}}{\partial \tilde{Y}} = \frac{\partial^2 \tilde{u}}{\partial \tilde{Y}^2},$$

$$\theta = 0. \tag{12-64}$$

But now the velocity component $\tilde{u}$ is rescaled to be consistent with the magnitude of the velocity induced by buoyancy in the inner region.

To complete the problem formulation, we must now use the matching conditions between the solutions in the two parts of the natural convection boundary-layer to obtain additional boundary conditions for (12–59) and (12–64). As noted earlier, we require three such conditions. Two come from matching of the velocity components

$$\lim_{Y_1 \gg 1}(u_1) \Longleftrightarrow \lim_{\tilde{Y} \ll 1}(\tilde{u}) \quad \text{as} \quad \text{Pr} \to \infty \tag{12-65}$$

and

$$\lim_{Y_1 \gg 1}\left(\frac{V_1}{\text{Pr}^{3/4}}\right) \Longleftrightarrow \lim_{\tilde{Y} \ll 1}\left(\frac{\tilde{V}}{\text{Pr}^{1/4}}\right) \quad \text{as} \quad \text{Pr} \to \infty. \tag{12-66}$$

A third condition can be obtained by requiring that the first derivatives of u also match

$$\lim_{Y_1 \gg 1}\left(\frac{\partial u_1}{\partial Y_1}\right) \Longleftrightarrow \lim_{\tilde{Y} \ll 1}\left(\frac{1}{\text{Pr}^{1/2}}\frac{\partial \tilde{u}}{\partial \tilde{Y}}\right) \quad \text{as} \quad \text{Pr} \to \infty. \tag{12-67}$$

In the rescaled forms shown, all of the nondimensionalized variables are $0(1)$. Hence, in the limit $\text{Pr} \to \infty$, the last two conditions reduce to equivalent boundary conditions,

$$\tilde{V}(o,x) = 0 \tag{12-68}$$

and

$$\frac{\partial u_1}{\partial Y_1} \to 0 \quad \text{as} \quad Y_1 \to \infty. \tag{12-69}$$

The second of these conditions is sufficient to completely specify the solution of (12–59) and (12–60) for the inner (*thermal-viscous*) region. Thus, in principle, we can obtain the *limiting* form of u_1 for $Y_1 \gg 1$ by solving this inner problem, and the matching condition (12–62) then provides an additional boundary condition for the solution of (12–64), namely,

$$\tilde{u}(0,x) = u_1(x,\infty). \tag{12-70}$$

This condition, together with (12–68) and (12–58), expressed in the form

$$\tilde{u} \to 0 \quad \text{as} \quad \tilde{Y} \to \infty, \tag{12-71}$$

is then sufficient to completely specify the solution of (12–64) for the outer (inertial) layer.

In fact, although the leading-order asymptotic solution in the thermal boundary-layer is completely specified by the formulation above, the details of actually solving for the velocity and temperature fields are difficult, except for a flat plate. The problem is to obtain closed-form, analytic solutions for the thermal energy equation in the inner region and for the nonlinear boundary-layer equation (12–64) in the outer region. In any case, provided that solutions *exist,* it should be clear from previous examples that the essential features of the problem are inherent in the asymptotic *formulation.* In particular, we can see from the formulation alone that

$$\mathrm{Nu} = C_1 \mathrm{Gr}^{1/4} \mathrm{Pr}^{1/4} \quad (\mathrm{Gr} \gg 1,\ \mathrm{Pr} \gg 1),$$

where C_1 is a geometry-dependent constant of $0(1)$. To obtain C_1 it is necessary to determine the temperature distribution in the boundary-layer, but for this we need only solve the (inner) thermal-layer problem. The special case of a vertical flat plate, which can be solved via similarity transformation is posed at the end of this chapter as a homework problem. The coefficient C_1 for this case is

$$\frac{\mathrm{Nu}}{\mathrm{Gr}^{1/4}\mathrm{Pr}^{1/4}} = 0.65. \tag{12–72}$$

In other cases, the only change is the numerical value of C_1, and the problem is reduced to obtaining this constant either by solving the thermal-layer problem numerically or by carrying out a small number of experiments.

The Limiting Case Pr → 0

Let us now turn to the other limiting case, $\mathrm{Pr} \to 0$. The starting point is again the natural convection boundary-layer equations (12–39)–(12–41) plus boundary conditions (12–42) and matching conditions (12–43). With the detailed discussion of the limit $\mathrm{Pr} \gg 1$ now behind us, along with the evident analogies between forced convection at high Re and natural convection at large Gr, we can limit the level of detail in the present discussion.

It is perhaps a useful starting point to remind ourselves of the expected role of Pr in determining the relative scales of the regions in which conduction and viscous diffusion play a dominant role in the transport of heat and momentum, respectively. In particular, for $\mathrm{Pr} \gg 1$, we have seen in both the forced and natural convection problems that the region in which conduction effects are important is very much thinner than the region in which viscous effects are important, and in this sense the thermal layer is seen to be very much thinner than the momentum layer. For the limit $\mathrm{Pr} \ll 1$, the behavior in the forced convection limit suggests that the opposite will be true—namely, that the region in which viscous effects are important will now be much thinner than the region in which conduction is important. To see whether this expectation is borne out, we turn to (an outline of) the asymptotic analysis for $\mathrm{Pr} \ll 1$.

We again start with the governing equations (12–39)–(12–41), expressed in terms of the natural convection boundary-layer coordinates. The leading-order approximation for $\mathrm{Pr} \ll 1$ is obtained by simply taking the limit $\mathrm{Pr} \to 0$ in these equations. The continuity and momentum equations (12–39) and (12–40) are unchanged. However, the result from (12–41) is

$$\boxed{\frac{\partial^2 \theta}{\partial Y^2} = 0,}$$

(12–73)

and the solution of this equation satisfying the boundary condition (12–42) is

$$\boxed{\theta = 1 + a(x)Y.}$$

(12–74)

We shall see shortly that $a(x) \equiv 0$ in order that this solution matches the temperature distribution in the outer part of the boundary-layer.

The natural convection boundary-layer equations (12–39)–(12–41) are, in effect, expressed in terms of the cross-stream variable Y scaled using a length scale that characterizes the width of the momentum boundary-layer (that is, the region where viscous effects are in balance with inertia). Thus, the significance of the reduction from (12–41) to (12–73) in the limit $\text{Pr} \to 0$ is that the thermal layer where *convection* of heat balances conduction becomes very thick compared with this momentum-layer length scale. However, since θ is expected to decrease from its surface value, $\theta = 1$, to its ambient value, $\theta = 0$, over the whole of this thermal-layer thickness, it is evident that θ will hardly vary at all over increments in the momentum-layer variable Y of $0(1)$ for $\text{Pr} \to 0$. Clearly, the solution (12–74) cannot satisfy the matching condition (12–43), $\theta \to 1$ as $Y \to \infty$, and still satisfy the boundary condition (12–42), $\theta = 1$ at $Y = 0$. It is thus evident that (12–73), plus (12–39) and (12–40), can only apply to an *inner* region in the vicinity of the body surface. In the *outer* region (still in the natural convection boundary-layer), both conduction and convection terms must be important, even in the limit $\text{Pr} \to 0$.

To obtain the governing equations for this *outer* region, we must rescale (12–39)–(12–41) by introducing

$$\boxed{\bar{Y} = \text{Pr}^\alpha Y.}$$

(12–75)

The physical significance of this change of variables is that we now nondimensionalize in such a way that changes that occur over very large distances in Y occur over distances of $0(1)$ in $\bar{Y}$. Alternatively, we may say that we nondimensionalize in such a way that the change of θ from 1 to 0 occurs over a distance in $\bar{Y}$ of $0(1)$ as $\text{Pr} \to 0$. To preserve the form of the continuity equation, we must simultaneously rescale V, and for completeness, we rescale both V and u according to

$$\boxed{\begin{aligned} \bar{u} &= \text{Pr}^\beta u, \\ \bar{V} &= \text{Pr}^{\alpha+\beta} V. \end{aligned}}$$

(12–76)

As usual, it may happen that $\beta = 0$.

To determine α and β, we introduce (12–75) and (12–76) into the thermal energy and momentum equations. The result for the thermal energy equation is formally identical to (12–49), that is,

$$\bar{u}\frac{\partial\theta}{\partial x} + \bar{V}\frac{\partial\theta}{\partial\bar{Y}} = \Pr^{2\alpha+\beta-1}\frac{\partial^2\Theta}{\partial\bar{Y}^2}, \tag{12-77}$$

except now $\Pr \ll 1$. Clearly, in view of (12–73) and (12–74), the limiting form of this equation must retain both convection and conduction terms. This requires that

$$\boxed{2\alpha + \beta - 1 = 0.} \tag{12-78}$$

Similarly, the rescaled form of the momentum equation (12–39) is identical to (12–51), that is,

$$\frac{1}{\Pr^{2\beta}}\left\{\bar{u}\frac{\partial\bar{u}}{\partial x} + \bar{V}\frac{\partial\bar{u}}{\partial\bar{Y}}\right\} = \theta\sin\alpha + \Pr^{2\alpha-\beta}\frac{\partial^2\bar{u}}{\partial\bar{Y}}. \tag{12-79}$$

Now, the choice of α and β depends on whether the dominant balance in this equation for the *outer* region in the natural convection boundary-layer is between buoyancy and inertia or buoyancy and the viscous term.

Since we are considering the portion of the boundary-layer that is farthest from the body, it should be intuitively evident that the correct choice is a balance between the buoyancy and inertia terms. However, as before, if this choice were not apparent and we were to guess that the balance should be between viscous and buoyancy terms, an obvious inconsistency would arise in the analysis. To see this, let us suppose that

$$2\alpha - \beta = 0, \tag{12-80}$$

which implies that the dominant balance is between the viscous and buoyancy terms. Then, combining (12–79) and (12–80), we would find that

$$\alpha = \frac{1}{4}, \quad \beta = \frac{1}{2},$$

which is the same result that we obtained earlier for the large Pr case. In this case, however, (12–79) becomes

$$\frac{1}{\Pr}\left\{\bar{u}\frac{\partial\bar{u}}{\partial x} + \bar{V}\frac{\partial\bar{u}}{\partial\bar{Y}}\right\} = \theta\sin\alpha + \frac{\partial^2\bar{u}}{\partial\bar{Y}^2}.$$

But, in the limit $\Pr \to 0$, this implies that the *inertia* term dominates *both* buoyancy and inertia terms. This is clearly inconsistent with the assumption leading to (12–80). This fundamental inconsistency implies that the correct choice must be the balance between inertia and buoyancy, as suggested originally. From (12–79), we see that this choice is equivalent to

$$\boxed{2\beta = 0.}$$

Thus, from (12–78),

$$\boxed{\alpha = \frac{1}{2}.} \tag{12-81}$$

In this case, (12–79) becomes

$$\bar{u}\frac{\partial \bar{u}}{\partial x} + \bar{V}\frac{\partial \bar{u}}{\partial \bar{Y}} = \theta \sin\alpha + 0(\text{Pr}),\qquad (12\text{–}82)$$

and we see that this is consistent with the assumption of a dominant inertia-buoyancy balance. The thermal energy and continuity equations in this *outer* region then take the forms

$$\bar{u}\frac{\partial \theta}{\partial x} + \bar{V}\frac{\partial \theta}{\partial \bar{Y}} = \frac{\partial^2 \theta}{\partial \bar{Y}^2}.\qquad (12\text{–}83)$$

and

$$\frac{\partial \bar{u}}{\partial x} + \frac{\partial \bar{V}}{\partial \bar{Y}} = 0.\qquad (12\text{–}84)$$

To complete the formulation for this case, $\text{Pr} \ll 1$, we require boundary and matching conditions. The boundary conditions are (12–42) for the *inner* region and (12–43) for the *outer* region. To complete the specification of the velocity and thermal fields in the inner and outer regions, we must use matching conditions in the overlap region of common validity. For the inner solution, we require one additional condition on u, plus a condition on θ to determine $a(x)$ in (12–74). For the outer region, we require one additional condition for the inviscid flow equation (12–82), plus a condition on θ for $\bar{Y}\to 0$.

Let us first consider the matching condition on θ, namely,

$$\lim_{\bar{Y} \ll 1} \theta(x, \bar{Y}) \Leftrightarrow \lim_{Y \gg 1}\left(1 + a(x)\frac{\bar{Y}}{\text{Pr}^{1/2}}\right)\text{as Pr}\to 0.\qquad (12\text{–}85)$$

The right-hand side is the inner solution, (12–74), expressed in terms of outer variables. Clearly, since the left-hand side is $0(1)$, the matching condition (12–85) requires $a(x) \equiv 0$, as suggested earlier. The matching condition (12–85) then becomes a boundary condition for the leading-order approximation to θ in the outer region, namely,

$$\theta(x,\bar{Y}) = 1 \quad\text{at}\quad \bar{Y} = 0.\qquad (12\text{–}86)$$

The matching conditions on u and V are

$$\lim_{\bar{Y} \ll 1}\left(\bar{u}(x,\bar{Y})\right) \Leftrightarrow \lim_{Y \gg 1}\left(u(x,Y)\right)\text{as Pr}\to 0\qquad (12\text{–}87)$$

and

$$\lim_{\bar{Y} \ll 1}\left(\bar{V}(x,\bar{Y})\right) \Leftrightarrow \lim_{Y \gg 1}\left(\text{Pr}^{1/2}V(x,Y)\right)\text{as Pr}\to 0.\qquad (12\text{–}88)$$

Since we are concerned here with only the leading-order approximation to the solution for $\text{Pr}\to 0$, the last of these conditions is equivalent to a boundary condition for $\bar{V}$ in the outer region, namely,

$$\bar{V}(x, \bar{Y}) = 0 \quad \text{at} \quad \bar{Y} = 0. \tag{12-89}$$

This condition, together with (12-86), is sufficient to completely specify the solution of (12-82)-(12-84) for the velocity and temperature field in the outer region. Thus, in principle at least, $\bar{u}(x, \bar{Y})$ at $\bar{Y} = 0$ can be calculated, and the matching condition (12-87) then becomes a boundary condition for the solution in the inner region. The full inner problem is then

$$u \frac{\partial u}{\partial x} + V \frac{\partial u}{\partial Y} = \sin\alpha + \frac{\partial^2 u}{\partial Y^2}, \tag{12-90}$$

$$\frac{\partial u}{\partial x} + \frac{\partial V}{\partial Y} = 0, \tag{12-91}$$

and

$$\theta \equiv 1$$

with

$$u = V = 0 \quad \text{at} \quad Y = 0$$

and

$$u = \bar{u}(x, 0) \quad \text{as} \quad Y \to \infty. \tag{12-92}$$

To determine the complete leading-order solution for the temperature and velocity fields, we first solve the *outer* problem to obtain $\bar{u}$, $\bar{V}$, and $\theta(x, \bar{Y})$ and then solve the *inner* problem (12-90)-(12-92) using $\bar{u}$ at $\bar{Y} = 0$ to specify the form of u for $Y \to \infty$. If, on the other hand, the prime objective is either the local or overall rate of heat transfer, we require only the temperature distribution, and for this we need only solve the *outer* problem. The reason is simply that the limiting form for $\partial\theta/\partial\bar{Y}|_{\bar{Y}\to 0}$ from the outer problem also yields a correct first approximation to the temperature gradient at the body surface. The proof of this fact is virtually identical to the argument given earlier for the *forced* convection analysis at high Reynolds number and low Prandtl number, and we leave it to the reader to fill in the details. The net result of these arguments is that the rescaling of the *outer* problem directly yields an asymptotic estimate for the overall Nusselt number,

$$\text{Nu} = C_2 \text{Gr}^{1/4} \text{Pr}^{1/2}. \tag{12-93}$$

As usual, the details of the outer problem must be solved if we wish to evaluate the constant C_2, but in any case, we know that this constant will be $0(1)$ and dependent only on the geometry of the body.

C. The Laminar Plume Above a Line or Point Source of Heat

In the previous section, we analyzed the structure of natural convection flows in the immediate vicinity of a heated (or cooled) cylindrical body in the limit $\text{Gr} \gg 1$. Here, we turn to the structure of the temperature and velocity fields far above (or below) a heated (or cooled) body in a fluid that is otherwise stationary. This is known as the *thermal plume problem*. We consider only the special case in which the motion remains laminar.

Of course, one would generally like to determine the velocity and temperature fields everywhere in the fluid rather than just near the body and far from the body in the thermal plume. Unfortunately, however, the structure of the intermediate transition region is quite complex, and there are no obvious approximations to allow asymptotic solutions in this zone. On the other hand, if we move sufficiently far from the body, the structure of the velocity and temperature fields becomes increasingly simplified. In particular, the structure of these fields becomes less sensitive to the details of the body geometry, which appears increasingly as a line (or point) source of heat and as a line (or point) force on the fluid. In this case, analytic solutions again become possible, and this is the subject of the present section.

It is worth noting, before proceeding to the details, that the thermal plume problem is at least qualitatively representative of a much wider class of problems that are related to earlier parts of this book. In particular, let us reconsider briefly the *forced* convection flow past a body that is heated relative to the ambient fluid. In this case, the presence of the body is manifested downstream in a so-called *wake* in which the velocity and temperature fields differ from ambient conditions. Again, in the region sufficiently far from the body, the structure of these fields is simplified as the body appears either as a line (or point) source of heat and as a line (or point) force on the fluid. There is a substantial analogy between the methods of analysis introduced in this section for the thermal plume and the usual approach for forced convection wakes. Indeed, as illustrated by some of the exercises at the end of this chapter, the natural convection problem provides a general framework that can be adapted to the whole class of wake, jet and plume-type problems.

To begin, we assume that the heated body is two-dimensional (that is, a horizontal, cylindrical geometry as discussed in the preceding section) and that we are sufficiently far away that it can be approximated as a line source of heat in a fluid that is isothermal and motionless in its ambient state. The starting point for analysis is again the Boussinesq equations written in dimensional form:

$$\rho_\infty (\mathbf{u} \cdot \nabla \mathbf{u}) = -\nabla p_d + \mu_\infty \nabla^2 \mathbf{u} + \rho_\infty \beta (T - T_\infty) g \mathbf{i}_g, \qquad (12\text{--}94)$$

$$\nabla \cdot \mathbf{u} = 0, \qquad (12\text{--}95)$$

and

$$\rho_\infty C_{p\infty} (\mathbf{u} \cdot \nabla T) = k_\infty \nabla^2 T. \qquad (12\text{--}96)$$

We have seen that the natural first step in the analysis of any physical problem is to nondimensionalize the governing equations. However, the present problem differs from most of those analyzed previously in the sense that the approximation of the body as a line (or point) *disturbance* of heat means that there is no characteristic length scale inherent in the geometry. Furthermore, it is not easy to guess at a characteristic temperature difference, since it is the net heat flux from the body to the fluid that is specified rather than the temperature of the body surface.

As far as the lack of a characteristic length scale is concerned, the implication is that the solution will be self-similar (see Chapter 3, Section **D**). Nevertheless it is convenient initially to formulate the problem as though there were a length scale, l_c, with the expectation that l_c will cancel in the solution as it did in previous self-similar examples that we have studied (refer to the Blasius problem for flow past a semi-infinite flat plate).

The appropriate choice for definition of a nondimensional temperature also is not obvious for the reasons just cited. Thus, for the moment, we simply adopt the formal definition

$$\oplus = \frac{T - T_\infty}{\Delta T^*}, \qquad (12\text{-}97)$$

where ΔT^* remains to be specified in terms of the total heat flux Q and other dimensional parameters. Finally, the characteristic velocity is conveniently defined in the same manner as in (12-17),

$$u_c = \sqrt{\beta l_c \Delta T^* g}. \qquad (12\text{-}98)$$

Then, the dimensionless form of the governing equations is formally identical to (12-18), (12-23), and (12-24) but with

$$\mathrm{Gr} \equiv \frac{\beta g l_c^3 \Delta T^*}{\nu_\infty^2}. \qquad (12\text{-}99)$$

The boundary conditions, however, are no longer the same because we consider the body only as a line source of heat with a specified heat flux Q per unit length. Thus, though the solution must still satisfy the far-field boundary conditions (12-25a), the boundary conditions at the body surface are not directly relevant. Instead, we require only that the solution be consistent with the specified rate of heat release from the body. We shall see that this constraint is sufficient to obtain a unique solution.

We formally seek a solution for large Grashof number, $\mathrm{Gr} \gg 1$. The structure of the plume, in this limit, is very reminiscent of the boundary-layer solutions obtained earlier. In particular, derivatives *across* the plume become asymptotically large *relative* to derivatives *along* the plume axis. However, the dimensions across the plume are *not* small compared to the scale of the heated body—rather, the distances over which u and θ change *along* the plume become extremely large compared with the body scale.

The leading-order approximation for the temperature and velocity fields in most of the fluid domain is obtained by taking the limit $\mathrm{Gr} \to \infty$ in (12-18) and (12-23)-(12-24). Clearly, the viscous and thermal conduction terms are lost, and the result is

$$\boxed{\oplus = \mathbf{u} = p = 0.} \qquad (12\text{-}100)$$

This "solution" is consistent with the far-field boundary conditions and at first may seem generally satisfactory in the sense that it violates no obvious constraints (recall that this solution is intended to apply far from the heated body so that the *nonhomogeneous* boundary conditions at the body surface are not directly applicable). However, one physical condition *is* violated in that a certain heat flux is released from the body, which should be conserved as we move outward away from the body. However, the net heat flux associated with (12–100) is clearly *zero* across any surface surrounding the line source.

From a physical point of view, if the body is heated, it is evident that the lighter, heated fluid will rise upward in the vicinity of the vertical plane that passes through the center of the body. Indeed, when we are sufficiently far away so that the body appears only as a line source of heat, we should expect the temperature and velocity fields to be symmetric on either side of the vertical plane that contains the line source, with maximum values of $\oplus$ and u occurring at the plane. Evidently, in view of (12–100), we cannot capture the buoyant plume for Gr $\gg 1$ unless we retain conduction and/or viscous terms in the equations. This can only mean that a rescaling of the relative magnitudes of the length scales is necessary in the inner region on either side of the symmetry plane to suit the equations to the importance of viscous and conduction effects for large Gr. Intuitively, it is clear that derivatives of **u** or $\oplus$ across the plume will tend to be larger than derivatives along the plume, and this must be reflected in the nondimensionalization.

In view of the arguments of the preceding paragraph and the fact that Gr is large, we try the usual boundary-layer rescaling,

$$Y = y\text{Gr}^{1/4}, \quad V = v\text{Gr}^{1/4}, \tag{12-101}$$

in which Y and V are the coordinate and velocity components normal to the symmetry plane. (The velocity and coordinate components parallel to this plane are denoted as u and x.) Substituting (12–101) into (12–18) and (12–23)–(12–24), we obtain

$$u\frac{\partial u}{\partial x} + V\frac{\partial u}{\partial Y} = -\frac{\partial p}{\partial x} + \frac{\partial^2 u}{\partial Y^2} + \oplus + 0(\text{Gr}^{-1/2}), \tag{12-102}$$

$$\frac{\partial p}{\partial Y} = 0(\text{Gr}^{-1/4}) \to 0 \quad \text{as} \quad \text{Gr} \to \infty, \tag{12-103}$$

$$u\frac{\partial\oplus}{\partial x} + V\frac{\partial\oplus}{\partial Y} = \frac{1}{\text{Pr}}\frac{\partial^2\oplus}{\partial Y^2} + 0(\text{Pr}^{-1}\text{Gr}^{-1/2}), \tag{12-104}$$

and

$$\frac{\partial u}{\partial x} + \frac{\partial V}{\partial Y} = 0. \tag{12-105}$$

Since the dynamic pressure in the outer part of the domain is constant (zero), it follows from (12–103) that $\partial p/\partial x$ in (12–102) is *zero*.

The thermal equation is second order in Y and first order in x, while the coupled pair (12–102) and (12–105) is third order in Y and first order in x. The boundary conditions required in Y come from matching with the trivial solution (12–100),

$$u \to 0, \ \oplus \to 0 \quad \text{as} \quad Y \to \infty, \tag{12-106}$$

and symmetry conditions at the central plane,

$$V = \frac{\partial u}{\partial Y} = \frac{\partial\oplus}{\partial Y} = 0 \quad \text{at} \quad Y = 0. \tag{12-107}$$

Ordinarily, one would also expect to apply initial conditions at some small x to complete the specification of the problem, but of course no such condition is available in the present case. Instead, for problems of this general type (plumes and wakes), the x dependence of the solution is specified in terms of an *integral constraint* that derives from an overall (macroscopic) heat or momentum balance.

In the present case, the constraint derives from an overall heat balance. Let us suppose that the heat flux per unit length at the line source is Q. According to the present boundary-layer approximation, heat can only be advected with the buoyancy-driven motion of the fluid, or conducted outward in the Y direction. Thus, the total heat flux across any plane surface above the line source must be Q at steady state. In conjunction with (12–106) and (12–107), we shall see that this constraint is sufficient to completely specify the solution of (12–102), (12–104), and (12–105).

Of course, any macroscopic constraint based on thermal energy conservation must be derivable from the differential thermal energy equation since this is simply the pointwise version of the same principle. In particular, the constraint on total heat flux across a horizontal plane surface can be derived directly by integrating (12–104) over the surface. We consider the heat flux per unit length, and thus it is only necessary to integrate (12–104) over Y from $-\infty$ to $+\infty$:

$$\int_{-\infty}^{\infty} u \frac{\partial \oplus}{\partial x} \, dY + \int_{-\infty}^{\infty} V \frac{\partial \oplus}{\partial Y} \, dY = \frac{1}{\mathrm{Pr}} \int_{-\infty}^{\infty} \frac{\partial^2 \oplus}{\partial Y^2} \, dY. \tag{12–108}$$

The right-hand side is

$$\int_{-\infty}^{\infty} \frac{\partial^2 \oplus}{\partial Y^2} \, dY = 2 \int_0^{\infty} \frac{\partial^2 \oplus}{\partial Y^2} \, dY = 2 \left[\left(\frac{\partial \oplus}{\partial Y} \right)_{Y=\infty} - \left(\frac{\partial \oplus}{\partial Y} \right)_{Y=0} \right] \equiv 0 \tag{12–109}$$

because of the boundary conditions at $Y = 0$ and $Y = \infty$. The second term on the left side of (12–108) is evaluated by integrating by parts:

$$2 \int_0^{\infty} V \frac{\partial \oplus}{\partial Y} \, dY = 2 V \oplus \Big|_0^{\infty} - 2 \int_0^{\infty} \oplus \left(\frac{\partial V}{\partial Y} \right) dY. \tag{12–110}$$

The first term, $V \oplus |_0^{\infty}$, is zero because of the boundary conditions at $Y = 0$ and $Y = \infty$. The second term in (12–110) can be combined with the first term on the left side of (12–108) rewritten in the form

$$2 \int_0^{\infty} u \frac{\partial \oplus}{\partial x} \, dY = 2 \int_0^{\infty} \left[\frac{\partial}{\partial x} (u \oplus) - \oplus \frac{\partial u}{\partial x} \right] dV. \tag{12–111}$$

Thus, combining (12–108)–(12–111), together with the continuity equation (12–105), we find that

$$\boxed{\frac{d}{dx} \left[2 \int_0^{\infty} u \oplus dY \right] \equiv 0.} \tag{12–112}$$

This condition is equivalent to the statement

$$2\int_0^\infty u \oplus dY = \bar{Q}$$

(12–113)

for any arbitrary plane $x > 0$, where $\bar{Q}$ is the dimensionless heat flux per unit length at the line source,

$$\bar{Q} \equiv \frac{Q}{\mathrm{Gr}^{1/4}(\Delta T^* \nu_\infty \rho_\infty C_{p\infty})}.$$

(12–114)

By choosing a convenient numerical value for $\bar{Q}$ (for example, $\bar{Q} = 1$), the definition (12–114) provides a basis for defining ΔT^* in terms of Q. However, this is not essential, and we retain the form (12–113) for the following analysis.

Thus, finally, we are in a position to solve for the velocity and temperature fields in a buoyant plume above a line source of heat. The governing equations at the leading order of approximation are (12–102), (12–104), and (12–105), with the integral constraint (12–113) and the boundary conditions (12–106) and (12–107). To proceed in the most convenient fashion, we eliminate (12–105) by introducing a streamfunction,

$$u \equiv \frac{\partial \psi}{\partial Y}, \quad V \equiv -\frac{\partial \psi}{\partial x}.$$

Then, since the physical statement of the problem does not contain an obvious length scale, our prior experience suggests seeking a similarity solution. We try the very general forms

$$\oplus = \frac{A}{x^n} \theta(\eta), \quad \psi = Bx^m f(\eta); \quad \eta = \frac{CY}{x^p}.$$

(12–115)

This choice is motivated by the expectation that the magnitude of θ and u at the central symmetry plane, $Y = 0$ (that is, $\eta = 0$), will decrease with increasing x. Although the form (12–115) may turn out to be more complex than is actually necessary, the only harm in beginning with a too general form is the possibility of some unnecessary algebra.

To determine the unspecified constants in (12–115), we substitute into the governing equations and boundary conditions. The reader may recall from earlier chapters that the best first step in any similarity solution is to see whether the proposed form is general enough to satisfy all of the boundary conditions. Here, however, the problem is not the boundary conditions but the integral constraint (12–113). To see whether the proposed forms (12–115) can satisfy this constraint for some choice of the constants, we substitute into (12–113). The result is

$$2ABx^{m-n}\int_0^\infty \theta \frac{df}{d\eta} d\eta = \bar{Q}.$$

Thus, this constraint can be satisfied by the proposed similarity form provided that

$$m = n$$

(12–116)

and the coefficient AB is chosen such that

$$2AB \int_0^\infty \theta \frac{df}{d\eta} \, d\eta = \bar{Q}. \qquad (12\text{–}117)$$

To determine n and p, we next substitute the forms (12–115) into the thermal energy equation (12–104) and the equation of motion (12–102) expressed in terms of ψ (recall that $\partial p / \partial Y = 0$). The result from the thermal energy equation is

$$\frac{B}{x^{1+p}} \left[-n\theta \frac{df}{d\eta} - nf \frac{d\theta}{d\eta} \right] = \frac{1}{\text{Pr}} \frac{C}{x^{n+2p}} \frac{d^2\theta}{d\eta^2}.$$

If a similarity solution exists, the powers on x must be equal on the two sides. Hence,

$$1 + p = n + 2p, \qquad (12\text{–}118)$$

and then

$$-n \frac{B}{C} \text{Pr} \left[\theta \frac{df}{d\eta} + f \frac{d\theta}{d\eta} \right] = \frac{d^2\theta}{d\eta^2}. \qquad (12\text{–}119)$$

The result from (12–102) is

$$\frac{B^2 C^2}{x^{2p+1-2n}} \left[(n-p) \left(\frac{df}{d\eta} \right)^2 - nf \frac{d^2 f}{d\eta^2} \right] = \frac{BC^3}{x^{3p-n}} \frac{d^3 f}{d\eta^3} + \frac{A}{x^n} \theta.$$

Thus, again, the powers on x must be the same in the various terms so that

$$2p + 1 - 2n = 3p - n = n. \qquad (12\text{–}120)$$

Combining this with (12–118) we find that

$$m = n = \frac{3}{5} \quad \text{and} \quad p = \frac{2}{5}. \qquad (12\text{–}121)$$

With these values for m, n, and p, the governing equations for θ and f now appear as

$$-\frac{3}{5} \frac{B \text{Pr}}{C} \left[\theta \frac{df}{d\eta} + f \frac{d\theta}{d\eta} \right] = \frac{d^2\theta}{d\eta^2} \qquad (12\text{–}122)$$

and

$$B^2 C^2 \left[\frac{1}{5} \left(\frac{df}{d\eta} \right)^2 - \frac{3}{5} f \frac{d^2 f}{d\eta^2} \right] = BC^3 \frac{d^3 f}{d\eta^3} + A\theta. \qquad (12\text{–}123)$$

The choice for A, B, and C is somewhat arbitrary. However, we follow the lead of Yih (1952, 1956)[4] in choosing these constants to be

$$A = 1, \ B = 5^{3/4}, \ \text{and} \ C = \left(\frac{1}{5} \right)^{1/4},$$

so that

$$\theta f' + f\theta' = -\frac{\theta''}{3\,\text{Pr}}$$

(12–124)

and

$$(f')^2 - 3ff'' = f''' + \theta.$$

With ΔT^* chosen so that $\bar{Q} = 5^{3/4}$, the integral constraint becomes

$$2\int_0^\infty \theta \frac{df}{d\eta}\, d\eta = 1.$$

(12–126)

The boundary conditions, written in terms of θ and f, are

$$f(0) = f''(0) = \theta'(0) = 0,$$

$$f'(\eta) \to 0, \quad \theta(\eta) \to 0 \quad \text{as} \quad \eta \to \infty.$$

(12–127)

Yih showed many years ago that the nonlinear problem defined by (12–124)–(12–127) has the curious property that a closed form analytic solution exists, but only for two specific values of $\text{Pr} = 5/9$ and $\text{Pr} = 2$. For the specific case of $\text{Pr} = 2$, the solutions for f and θ take the form

$$f = \frac{2\beta}{3}\tanh\beta\eta, \quad \theta = \frac{16}{9}\beta^4\text{sech}^4\beta\eta, \quad \text{with } \beta \equiv \frac{3}{2}\left(\frac{5}{48}\right)^{1/5} = 0.9542,$$

(12–128)

and for $\text{Pr} = 5/9$,

$$f = \frac{6\beta}{5}\tanh\beta\eta, \quad \theta = \frac{96}{25}\beta^4\text{sech}^2\beta\eta, \quad \text{with } \beta \equiv \frac{1}{2}\left(\frac{375}{72}\right)^{1/5} = 0.6955.$$

(12–129)

Numerical results for θ and u are plotted in Figure 12–3 for these two cases, $\text{Pr} = 5/9$ and $\text{Pr} = 2$. Clearly, as Pr is increased, there is a sharper focusing of the temperature distribution than of the velocity distribution. Lines of constant θ are also plotted in Figure 12–4. The familiar plumelike image is unmistakable.

All of the preceding analysis of this section pertains to the plume created by a two-dimensional, line source of heat. This is the obvious far-field counterpart of the boundary-layer analysis for cylindrical bodies from the previous section. However, it is also possible to obtain a corresponding solution for velocity and temperature fields above a *point* source of heat. This is the far-field problem, corresponding to a heated three-dimensional body at the origin. We pose this problem as one of the questions at the end of this chapter, and thus leave the details to the reader. In cylindrical coordinates, with z along the axis of the plume and r perpendicular to it, the leading order approximation within the plume again reduces to a self-similar form

$$\oplus = \frac{1}{z}\theta(\eta), \quad \psi = 4zf(\eta); \quad \eta = \frac{R}{\sqrt{2}z},$$

(12–130)

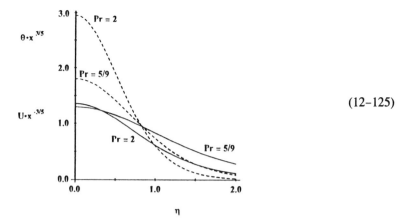

$$(12\text{–}125)$$

Figure 12–3 The vertical velocity distribution (solid line) and the temperature profile (dashed line) for natural convection flow above a two-dimensional line source of heat at Pr = 5/9 and Pr = 2.

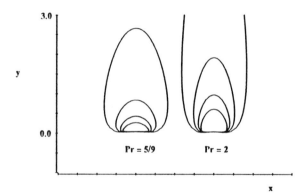

Figure 12–4 Isotherms in natural convection flow above a two-dimensional, line source of heat at Pr = 5/9 and Pr = 2. Contour levels are 1 (outermost) to 4 (innermost).

where

$$W = \frac{1}{R}\frac{\partial \psi}{\partial R}, \quad V = -\frac{1}{R}\frac{\partial \psi}{\partial z},$$

and

$$R = r\,\mathrm{Gr}^{1/4}, \quad V = v\,\mathrm{Gr}^{1/4}.$$

The governing equations and boundary conditions, with

$$\Delta T^* \equiv \frac{Q}{2\pi\rho C_p^\infty l_c v_\infty} \tag{12–131}$$

are

$$-\theta \frac{df}{d\eta} - f\frac{d\theta}{d\eta} = \frac{1}{4\,\mathrm{Pr}}\left(\eta\frac{d^2\theta}{d\eta^2} + \frac{d\theta}{d\eta}\right) \tag{12-132}$$

and

$$4\left[-f\frac{d^2f}{d\eta^2} + \frac{1}{\eta}f\frac{df}{d\eta}\right] = \eta\left[\frac{d^3f}{d\eta^3} - \frac{1}{\eta}\frac{d^2f}{d\eta^2} + \frac{1}{\eta^2}\frac{df}{d\eta}\right] + \theta\eta^2, \tag{12-133}$$

with

$$\int_0^\infty \theta\frac{df}{d\eta}\,d\eta = \frac{1}{4} \tag{12-134}$$

and

$$f = \frac{d^2f}{d\eta^2} = \frac{d\theta}{d\eta} = 0 \quad \text{at} \quad \eta = 0, \tag{12-135}$$

$$\theta \to 0, \quad \frac{df}{d\eta} \to 0 \quad \text{as} \quad \eta \to \infty. \tag{12-136}$$

Curiously, this problem also has a closed-form analytic solution, discovered by Yih (1950), for two specific values of Pr, namely, $\mathrm{Pr} = 1$ and $\mathrm{Pr} = 2$.[5]

The form of the solution for $\mathrm{Pr} = 2$ is

$$f = \frac{A_1\eta^2}{A_2 + \eta^2}, \quad \theta = \frac{B_1}{(A_2 + \eta^2)^4}, \tag{12-137}$$

where

$$A_1 = 1, \quad A_2 = \frac{8}{\sqrt{5}}, \quad \text{and} \quad B_1 = \frac{2(8)^3}{5},$$

and for $\mathrm{Pr} = 1$ is

$$f = \frac{A_1\eta^2}{A_2 + \eta^2}, \quad \theta = \frac{B_1}{(A_2 + \eta^2)^3}, \tag{12-138}$$

with

$$A_1 = \frac{3}{2}, \quad A_2 = 6, \quad \text{and} \quad B_1 = 4(6)^2.$$

The axial velocity component, u, and the temperature profile are both plotted as a function of η in Figure 12–5 for $\mathrm{Pr} = 1$ and $\mathrm{Pr} = 2$. The effect of increased Prandtl number is again to produce narrower and more sharply peaked velocity and temperature distributions, but the effect is stronger on θ than on u.

D. Combined Forced and Free Convection

An important question that has not yet been addressed in this book is the accuracy of the forced convection approximation. Until now, we have assumed that natural

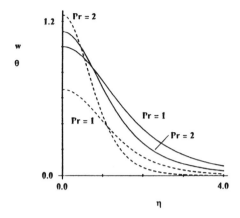

Figure 12–5 The vertical velocity distribution (solid line) and the temperature profile (dashed line) for natural convection flow above a point source of heat at $Pr = 1$ and $Pr = 2$.

convection effects will be entirely negligible whenever there is motion of the ambient fluid that is not driven by density gradients. But, of course, this is an approximation. Whenever the temperature of the fluid is nonuniform, there will be density gradients, and these will alter the motion of the fluid. The question is when the *changes* in motion from this mechanism can be neglected compared to the motions that result from the imposed ambient velocity field. Alternatively, if the ambient motions are very weak, we might ask whether these motions can be neglected so that the problem is approximated as pure natural convection. All of these questions are addressed in this section.

To begin, let us adopt a viewpoint in which the problem under consideration is one in which we would previously have adopted the forced convection approximation. Specifically, we consider a heated body with surface temperature T_o, immersed in an ambient fluid that is undergoing a uniform translational motion far from the body that is characterized by a velocity u_∞ and an ambient temperature T_∞. For simplicity, we assume that the body is two-dimensional, with a straight "cylindrical" axis that is oriented normal to **g**. Further, we suppose that the undisturbed fluid motion is perpendicular to this axis but that the direction relative to **g** is arbitrary.

Following the methodology from previous chapters, we begin by nondimensionalizing the governing equations and boundary conditions using u_∞ as a characteristic velocity, a linear dimension of the cross-sectional profile of the body l as a characteristic length scale, and a dimensionless temperature $\theta \equiv (T - T_\infty)/(T_o - T_\infty)$. In other words, we proceed as though the problem were to be analyzed via the forced convection approximation except that the governing equations are the Boussinesq equations, including the body force term that was neglected in the earlier analyses for this class of problems. The result for the steady-state problem is

$$\mathbf{u} \cdot \nabla \mathbf{u} = -\nabla p + \frac{1}{\mathrm{Re}} \nabla^2 \mathbf{u} - \frac{\mathrm{Gr}}{\mathrm{Re}^2} \theta \mathbf{i}_g, \tag{12-139}$$

$$\mathbf{u} \cdot \nabla \theta = \frac{1}{\mathrm{Pr}\,\mathrm{Re}} \nabla^2 \theta, \tag{12-140}$$

and

$$\nabla \cdot \mathbf{u} = 0, \tag{12-141}$$

where

$$\mathrm{Re} \equiv \frac{u_\infty l}{\nu} \quad \text{and} \quad \mathrm{Gr} = \frac{l^3 \beta g (T_o - T_\infty)}{\nu^2}.$$

The combination $\mathrm{Gr}/\mathrm{Re}^2$ is often called the Rayleigh number,

$$\mathrm{Ra} \equiv \frac{l \beta g (T_o - T_\infty)}{u_\infty^2}.$$

The boundary conditions corresponding to (12–139)–(12–141) are

$$\theta \to 0, \quad \mathbf{u} \to \mathbf{e} \quad \text{as} \quad |x| \to \infty \tag{12-142}$$

and

$$\theta = 1, \quad \mathbf{u} = 0 \quad \text{for} \quad |x| \, \epsilon \, s. \tag{12-143}$$

Here $\mathbf{e}$ is a unit vector in the direction of the undisturbed fluid motion and s signifies the body surface.

By examining (12–139), we see from dimensional analysis alone that a general criteria for neglect of natural convection effects for Re and Pr $\sim O(1)$, but arbitrary, is

$$\frac{\mathrm{Gr}}{\mathrm{Re}^2} \ll 1.$$

This result, in itself, is very useful. However, we can get even more insight for the limiting cases involving Re $\gg 1$ and Pr either very large or very small by continuing further with the analysis.

In particular, for Re $\gg 1$, we proceed with a conventional forced convection boundary-layer analysis. The leading-order equations where the scaling $l_c = l$ is relevant are obtained by letting Re $\to \infty$ in (12–139) and (12–140). The resulting solution, following the earlier analysis, is clearly

$$\theta \equiv 0, \tag{12-144}$$

and from this we see that the fluid mechanics problem reduces exactly to the potential-flow limit with no body force. As before, we denote the limit of the velocity component parallel to the body surface as $u_e(x)$ for $\mathbf{x} \to s$. Neither the solution (12–144) nor the potential-flow solution satisfies the boundary conditions (12–143). For this, we must analyze the boundary-layer region.

In the following analyis, we use the same notation as in previous chapters. Thus, we introduce local Cartesian coordinates (x, y) that are parallel and perpendicular, respectively, to the body surface. As stated earlier, the body and the fluid motion are both two-dimensional. The boundary-layer equations are obtained by rescaling, using the standard boundary-layer change of variables,

$$Y = y\mathrm{Re}^{1/2}, \quad V = v\mathrm{Re}^{1/2}. \tag{12-145}$$

The resulting equations for the boundary-layer (inner) region are

$$\frac{\partial u}{\partial x} + \frac{\partial V}{\partial Y} = 0, \tag{12-146}$$

$$u\frac{\partial u}{\partial x} + V\frac{\partial u}{\partial Y} = -\frac{\partial p}{\partial x} + \frac{\partial^2 u}{\partial Y^2} + \frac{\mathrm{Gr}}{\mathrm{Re}^2}\theta\sin\alpha + 0(\mathrm{Gr}^{-1}), \tag{12-147}$$

$$\frac{\partial p}{\partial Y} = -\frac{\mathrm{Gr}}{\mathrm{Re}^{5/2}}\theta\cos\alpha + 0(\mathrm{Re}^{-1}), \tag{12-148}$$

and

$$u\frac{\partial\theta}{\partial x} + V\frac{\partial\theta}{\partial Y} = \frac{1}{\mathrm{Pr}}\frac{\partial^2\theta}{\partial Y^2} + 0(\mathrm{Pr}^{-1}\mathrm{Re}^{-1}). \tag{12-149}$$

Here α is again the angle between the x axis and the horizontal. The boundary and matching conditions for the solution in this *inner* region are

$$u = V = 0, \quad \theta = 1 \quad \text{at} \quad Y = 0,$$
$$u \rightarrow u_e(x), \quad \theta \rightarrow 0 \quad \text{for} \quad Y \gg 1. \tag{12-150}$$

Of course, the solution of these equations only represents the first *inner* approximation in a singular perturbation expansion. Since we are primarily interested in the effects of *natural convection* on the forced convection problem, we have retained the terms of $0(\mathrm{Gr}/\mathrm{Re}^2)$ and $0(\mathrm{Gr}/\mathrm{Re}^{5/2})$, respectively, in (12–147) and (12–148).

As before, there are two types of problems, depending on whether $\alpha \neq 0$ (except possibly at isolated points) so that the dominant buoyancy effect is the direct buoyancy force in (12–147),[6] or $\alpha = 0$ (a horizontal flat surface) so that the dominant effect of buoyancy is a consequence of the pressure gradient in (12–148).[7] In the latter case, we retain the term of $0(\mathrm{Gr}/\mathrm{Re}^{5/2})$ and integrate (12–148) to obtain

$$\frac{\partial p}{\partial x} = \frac{\partial p_\infty}{\partial x} + \frac{\mathrm{Gr}}{\mathrm{Re}^{5/2}}\int_Y^\infty\left(\frac{\partial\theta}{\partial x}\right)\partial Y. \tag{12-151}$$

Now, $p_\infty(x)$ is the imposed pressure at the outer edge of the momentum boundary-layer, and in this case we can see from the outer potential-flow equations (or Bernoulli's equation) that

$$\frac{\partial p_\infty}{\partial x} = -u_e\frac{du_e}{dx}. \tag{12-152}$$

The second term in (12–151) is the contribution to the streamwise pressure gradient due, essentially, to hydrostatic pressure variations along the body surface. When the surface is heated, Gr > 0 as defined, and since $\partial\theta/\partial x < 0$ (note x increases in the direction of positive u_e), it follows that the natural convection effect on an upper facing flat surface (for example, a horizontal flat plate) is equivalent to a *favorable* streamwise pressure gradient. On the other hand, if the surface were cooled, Gr as defined would be negative while $\partial\theta/\partial x$ is still negative, and the buoyancy effect on the same surface would be equivalent to an *adverse* streamwise pressure gradient. Indeed, Robertson and co-workers[8] used numerical solutions of the full Boussinesq equation some time ago to show that the boundary-layer on the upper surface of a cooled, horizontal flat plate could actually *separate* due to this effective adverse gradient. In the remaining discussion, we limit our considerations to the case in which $\alpha \neq 0$.

In this case, we can combine (12–147), (12–148), (12–151), and (12–152) to obtain

$$u\frac{\partial u}{\partial x} + V\frac{\partial u}{\partial Y} = u_e\frac{du_e}{dx} + \frac{\partial^2 u}{\partial Y^2} + \frac{Gr}{Re^2}\theta\sin\alpha + 0(Re^{-1}) + 0\left(\frac{Gr}{Re^{5/2}}\right). \qquad (12\text{–}153)$$

As before, Gr > 0 for a heated surface, and Gr < 0 if the surface is cooled relative to the ambient fluid. It can be seen that the buoyancy term in (12–153) plays a role that is analogous to an extra pressure gradient, which is favorable for Gr > 0 and adverse for Gr < 0 (note $\sin\alpha \geq 0$).

The problem now is to solve the boundary-layer equations (12–146), (12–149), and (12–153) for the case in which $\alpha \neq 0$, subject to the conditions (12–150), plus whatever initial conditions are appropriate to the particular problem. There are a number of interesting cases where this is possible. First, for arbitrary Pr and Gr/Re^2, we can obtain a similarity solution if the surface temperature is constant over the whole body surface, as already assumed, and if

$$u_e = x^m, \quad \sin\alpha = Bx^{2m-1}. \qquad (12\text{–}154)$$

We have already seen that the first of these conditions means that the geometry of the body must be a semi-infinite wedge with included angle $\pi\beta$, where $\beta = 2m/(m+1)$. But in this case, $\sin\alpha$ must be a constant independent of x. Hence, the *only* possibility is $m = 1/2$, which corresponds to a wedge with included angle $2\pi/3$, so that $B = \sqrt{2}/3$. The governing equations for this somewhat peculiar case are

$$f''' + ff'' + \frac{2}{3}(1 - (f')^2) = -\frac{\sqrt{2}}{4}\frac{Gr}{Re^2}\theta(\eta) \qquad (12\text{–}155)$$

and

$$\frac{1}{Pr}\theta'' + \frac{1}{3}f\theta' = 0. \qquad (12\text{–}156)$$

A second case of the general problem that we can consider is the asymptotic limit $Gr/Re^2 \ll 1$. Although this limit of the boundary-layer problem is regular, the problem is rather uninteresting because it automatically requires natural convection to be weak compared with forced convection. Finally, a third class of problems that is amenable to analytic solution is the limit Pe $\gg 1$ or Pr $\ll 1$ with arbitrary values of

Gr/Re^2. We consider these latter cases in the remainder of this section, starting with the limit $Pr \gg 1$.

The Limit $Pr \gg 1$

We begin with the thermal boundary-layer equations in the dimensionless form (12–146), (12–149), and (12–153). The leading-order term in an asymptotic solution for $Pr \gg 1$ satisfies the limiting form of these equations for $Pr \to \infty$. It is a trivial matter to see that the conduction term is neglected in the thermal energy equation, which thus has the trivial solution

$$\boxed{\theta \equiv 0,} \tag{12–157}$$

and in this case, the buoyancy term is zero in (12–153), so that the fluid mechanics problem becomes

$$\frac{\partial u}{\partial x} + \frac{\partial V}{\partial Y} = 0, \tag{12–158}$$

with

$$u\frac{\partial u}{\partial x} + V\frac{\partial u}{\partial Y} = u_e\frac{du_e}{dx} + \frac{\partial^2 u}{\partial Y^2}, \tag{12–159}$$

$$u \to u_e(x) \quad \text{for} \quad Y \gg 1. \tag{12–160}$$

The boundary condition for $Y \ll 1$ will be obtained from matching with the solution in the *inner* region, where conduction and buoyancy terms are both important. Clearly, the equations in this *outer* part of the boundary-layer are unchanged from the pure forced convection case.

The equations for the *inner* part of the thermal boundary-layer are obtained by simply following the prescription of Chapter 11 for *forced* convection heat transfer in the limit $Re \gg 1$ and $Pr \gg 1$. Hence we rescale variables according to (11–32) so that the inner variables are

$$\hat{Y} = Y Pr^{1/3}, \quad \hat{u} = u Pr^{1/3}, \quad \text{and} \quad \hat{V} = V Pr^{2/3}. \tag{12–161}$$

The governing equations are again identical to the pure forced convection case, except now the buoyancy force is retained in the equation of motion:

$$\frac{\partial^2 \hat{u}}{\partial \hat{Y}^2} + \frac{Gr}{Re^2 Pr^{1/3}} \sin\alpha(x)\theta = 0(Pr^{-1/3}) \to 0, \tag{12–162}$$

$$\frac{\partial \hat{u}}{\partial x} + \frac{\partial \hat{V}}{\partial \hat{Y}} = 0, \tag{12–163}$$

$$\hat{u}\frac{\partial \theta}{\partial x} + \hat{V}\frac{\partial \theta}{\partial \hat{Y}} = \frac{\partial^2 \theta}{\partial \hat{Y}^2}. \tag{12–164}$$

The boundary conditions are

$$\boxed{u = V = 0, \quad \theta = 1 \quad \text{at} \quad \hat{Y} = 0.}$$

(12–165)

In order to solve either the *inner* or *outer* problems, we require matching conditions. These again follow the prescription from Chapter 11. In particular,

$$\boxed{\begin{aligned} \lim_{Y \ll 1} u &\Longleftrightarrow \lim_{\hat{Y} \gg 1} \frac{1}{\mathrm{Pr}^{1/3}} \hat{u} \to 0 \quad \text{as} \quad \mathrm{Pr} \to \infty, \\[2mm] \lim_{Y \ll 1} V &\Longleftrightarrow \lim_{\hat{Y} \gg 1} \frac{1}{\mathrm{Pr}^{2/3}} \hat{V} \to 0 \quad \text{as} \quad \mathrm{Pr} \to \infty, \\[2mm] \lim_{Y \ll 1} \frac{\partial u}{\partial Y} &\Longleftrightarrow \lim_{\hat{Y} \gg 1} \frac{\partial \hat{u}}{\partial \hat{Y}} \quad \text{as} \quad \mathrm{Pr} \to \infty. \end{aligned}}$$

(12–166)

(12–167)

and

(12–168)

The first two conditions are sufficient to solve the outer problem, (12–158)–(12–160). Once the outer velocity field is known, the last condition can then be used as a boundary condition for the inner problem, (12–162)–(12–165) with $\partial u / \partial Y = \lambda(x)$ now known from the outer solution. The physical explanation of (12–166) and (12–167) is simply that the thermal (*inner*) part of the boundary-layer is so thin for $\mathrm{Pr} \to \infty$ relative to the overall dimension of the momentum boundary-layer that the leading-order approximations to u and V remain zero at the inner edge of the outer region. This is, in fact, the same as in the forced convection limit.

One important result that has already been obtained without actually solving all of these equations is the relative magnitude of the buoyancy term for $\mathrm{Pr} \gg 1$. In particular, comparison of the two remaining terms in (12–162) shows clearly that natural convection effects will play a negligible role whenever

$$\frac{\mathrm{Gr}}{\mathrm{Re}^2 \mathrm{Pr}^{1/3}} \ll 1 \qquad (\mathrm{Pr} \gg 1).$$

(12–169)

For $\mathrm{Pr} \gg 1$, this condition on $\mathrm{Gr}/\mathrm{Re}^2$ is much less severe than the condition $\mathrm{Gr}/\mathrm{Re}^2 \ll 1$ that was obtained earlier for $\mathrm{Pr} \sim 0(1)$. From a physical point of view, this is not because the buoyancy force is any weaker for $\hat{Y} \approx 0$, but because it is only active in a very thin region immediately adjacent to the body surface for $\mathrm{Pr} \gg 1$.

It is important to recognize that (12–162)–(12–164) make no assumption about the magnitude of the buoyancy effect. Hence, for *arbitrary* Gr, it follows from the rescaling leading to these equations and their form that the local Nusselt number,

$$\mathrm{Nu}_x \equiv -\left.\frac{\partial \theta}{\partial y}\right|_{y=0},$$

(12–170)

will take the general form

$$\boxed{\frac{\mathrm{Nu}_x}{\mathrm{Re}^{1/2} \mathrm{Pr}^{1/3}} = F\left(x; \frac{\mathrm{Gr}}{\mathrm{Re}^2 \mathrm{Pr}^{1/3}}\right).}$$

(12–171)

Of course, the details of the function $F(x; Gr/Re^2 Pr^{1/3})$ can be obtained only by solving the full thermal boundary-layer problem. However, it is significant to note that the correlation in (12-171) is sufficiently general to reproduce the full spectrum of results, ranging from the forced convection limit $(Gr \ll 1)$, where

$$Nu_x = a_1(x) Re^{1/2} Pr^{1/3}, \qquad (12\text{-}172)$$

to the natural convection limit $(Gr \gg 1)$, where

$$Nu_x = a_2(x) Gr^{1/4} Pr^{1/4}. \qquad (12\text{-}173)$$

Obviously, by comparison of (12-171), (12-172), and (12-173), it must follow that

$$F\left(x; \frac{Gr}{Re^2 Pr^{1/3}}\right) = \begin{cases} a_1(x) & \text{for } \dfrac{Gr}{Re^2 Pr^{1/3}} \to 0 & (12\text{-}174a) \\[4mm] a_2(x)\left(\dfrac{Gr}{Re^2 Pr^{1/3}}\right)^{1/4} & \text{for } \dfrac{Gr}{Re^2 Pr^{1/3}} \to \infty. & (12\text{-}174b) \end{cases}$$

The key question is whether these limiting forms could have been deduced from (12-171) and the governing equations (12-162)–(12-164) if we had not already known the answers (12-172) and (12-173). In fact, the first of the two limiting forms is straightforward. In particular, with $\partial u / \partial Y \equiv \lambda(x)$ known from the momentum boundary-layer solution in the *outer* region, the *(inner)* thermal-layer problem can be solved as a regular perturbation expansion for $Gr/Re^2 Pr^{1/3} \ll 1$. As usual, the governing equations for the leading terms in this expansion can be obtained by taking the limit $Gr/Re^2 Pr^{1/3} \to 0$ in (12-162)–(12-164), and it is evident from this that the leading order solution will be *independent* of $Gr/Re^2 Pr^{1/3}$. It follows that the function F (that is, $-\partial\theta/\partial\hat{Y}|_{\hat{Y}=0}$) will be a function only of x, as indicated in (12-174).

A formal proof that F will adopt the form indicated in (12-174b) for $Gr/Re^2 Pr^{1/3} \to \infty$ can be constructed as follows. Note that the form for F suggests that $\hat{Y}$ must be rescaled for $(Gr/Re^2 Pr^{1/3}) \gg 1$ as

$$Y_1 = \hat{Y}\left(\frac{Gr}{Re^2 Pr^{1/3}}\right)^{1/4} \qquad (12\text{-}175)$$

since this rescaling gives

$$Nu_x = -\left.\frac{\partial\theta}{\partial y}\right|_{y=0} = Re^{1/2} Pr^{1/3}\left(\frac{Gr}{Re^2 Pr^{1/3}}\right)^{1/4}\left(-\frac{\partial\theta}{\partial Y_1}\right)_{Y_1=0}, \qquad (12\text{-}176)$$

and if $(\partial\theta/\partial Y_1)$ is independent of $(Gr/Re^2 Pr^{1/3})$, this is identical to (12-147)b. Not surprisingly, if we combine (12-175) with (12-161) and (12-145), we see that the result is

$$Y_1 = \left(\frac{Gr}{Re^2 Pr^{1/3}}\right)^{1/4} Re^{1/2} Pr^{1/3} y = Gr^{1/4} Pr^{1/4} y, \qquad (12\text{-}177)$$

and this is precisely the rescaling found earlier for the natural convection boundary-layer. To complete our proof of (12–174b), we introduce the additional changes of variables

$$u_1 = \hat{u} \left(\frac{\mathrm{Gr}}{\mathrm{Re}^2 \mathrm{Pr}^{1/3}} \right)^{-1/2}, \quad V_1 = \hat{V} \left(\frac{\mathrm{Gr}}{\mathrm{Re}^2 \mathrm{Pr}^{1/3}} \right)^{-1/4}, \qquad (12\text{–}178)$$

and the *inner* problem (12–162)–(12–164) is then transformed to the form

$$\frac{\partial^2 u_1}{\partial Y_1^2} + \sin \alpha(x)\theta = 0, \qquad (12\text{–}179)$$

$$\frac{\partial u_1}{\partial x} + \frac{\partial V_1}{\partial Y_1} = 0, \qquad (12\text{–}180)$$

$$u_1 \frac{\partial \theta}{\partial x} + V_1 \frac{\partial \theta}{\partial Y_1} = \frac{\partial^2 \theta}{\partial Y_1^2}, \qquad (12\text{–}181)$$

with boundary and matching conditions

$$u_1 = V_1 = 0, \quad \theta = 1 \quad \text{at} \quad Y_1 = 0 \qquad (12\text{–}182)$$

and

$$\theta \to 0, \quad \frac{\partial u_1}{\partial Y_1} \sim \lambda(x) \left(\frac{\mathrm{Gr}}{\mathrm{Re}^2 \mathrm{Pr}^{1/3}} \right)^{-3/4} \quad \text{for} \quad Y_1 \gg 1. \qquad (12\text{–}183)$$

For large $(\mathrm{Gr}/\mathrm{Re}^2 \mathrm{Pr}^{1/3})$, this problem can be solved as a regular perturbation expansion. The first term in this solution satisfies the limits of (12–179)–(12–183) for $(\mathrm{Gr}/\mathrm{Re}^2 \mathrm{Pr}^{1/3}) \to \infty$. The only change that occurs in the limiting process is that the matching condition on $\partial u_1 / \partial Y_1$ becomes

$$\frac{\partial u_1}{\partial Y_1} \sim 0 \quad \text{for} \quad Y_1 \gg 1. \qquad (12\text{–}184)$$

This leading-order problem for large Gr is identical to the inner (thermal) part of the natural convection boundary-layer that was analyzed in Section **B**. Clearly, since $(\mathrm{Gr}/\mathrm{Re}^2 \mathrm{Pr}^{1/3})$ does not appear in the limiting problem, the leading-order approximation to the solution is independent of $\mathrm{Gr}/\mathrm{Re}^2 \mathrm{Pr}^{1/3}$, and thus

$$-\left(\frac{\partial \theta}{\partial Y_1} \right)_{Y_1 = 0} = a_2(x). \qquad (12\text{–}185)$$

This completes our proof. The critical conclusion from this analysis is that the problem (12–158)–(12–168) encompasses the complete spectrum of asymptotic results from $\mathrm{Gr} \to 0$ to $\mathrm{Gr} \to \infty$.

Although the preceding demonstration is completely rigorous, it may well strike the reader as somewhat artificial in the sense that we start knowing that the result for large Gr should take the form (12–173), and this "guides" our choice of rescaling in (12–175). However, it is not necessary to approach the problem in this ad hoc manner. Instead, we can start with the thermal energy equation for arbitrary Pr,

$$u\frac{\partial\theta}{\partial x} + V\frac{\partial\theta}{\partial Y} = \frac{1}{\text{Pr}}\frac{\partial^2\theta}{\partial Y^2}, \tag{12–186}$$

and proceed in a completely formal manner to the same result based on the observation that the dimension of the thermal layer, where both conduction and convection are important, is much thinner for large Pr than the dimension of the momentum layer in which u changes from its surface value to the free-stream condition (12–142). Thus, the thermal energy equation depends only on the limiting forms of u and V for small values of the momentum boundary-layer variable Y, namely,

$$u \sim \left(\frac{\partial u}{\partial Y}\right)_{Y=0} Y + \frac{1}{2}\left(\frac{d^2 u}{\partial Y^2}\right)_{Y=0} Y^2 + \cdots \tag{12–187}$$

and

$$V \sim -\frac{\partial}{\partial x}\left(\frac{\partial u}{\partial Y}\bigg|_{Y=0}\right)\frac{Y^2}{2} - \frac{1}{6}\frac{\partial}{\partial x}\left(\frac{\partial^2 u}{\partial Y^2}\bigg|_{Y=0}\right)Y^3 + \cdots. \tag{12–188}$$

But the momentum boundary-layer equation (12–168) takes the limiting form for $Y \to 0$,

$$0 = u_e\frac{du_e}{dx} + \frac{\partial^2 u}{\partial Y^2}\bigg|_{Y=0} + \frac{\text{Gr}}{\text{Re}^2}\theta\sin\alpha.$$

Thus, denoting $(\partial u/\partial Y)_{Y=0}$ as $\lambda(x)$,

$$\boxed{u \sim \lambda(x)Y + \left(-\frac{1}{2}u_e\frac{du_e}{dx} - \frac{1}{2}\frac{\text{Gr}}{\text{Re}^2}\theta\sin\alpha\right)Y^2 + \cdots \tag{12–189}}$$

and

$$\boxed{V \sim -\frac{d\lambda}{dx}\frac{Y^2}{2} + \frac{1}{6}\left[\frac{d}{dx}\left(u_e\frac{du_e}{dx}\right) + \frac{\text{Gr}}{\text{Re}^2}\frac{d}{dx}(\theta\sin\alpha)\right]Y^3 + \cdots. \tag{12–190}}$$

Now let us reexamine the rescaling of (12–186) for Pr $\gg 1$, keeping in mind that the governing equation in the inner region should retain both conduction and convection terms in the limit Pr $\to \infty$. Thus we introduce the change of variables

$$\hat{Y} = Y\text{Pr}^\alpha. \tag{12–191}$$

The result for (12–189) and (12–190) is

$$u \sim \lambda(x)\frac{\hat{Y}}{\text{Pr}^\alpha} + \left(-\frac{1}{2}u_e\frac{du_e}{dx} - \frac{1}{2}\frac{\text{Gr}}{\text{Re}^2}\theta\sin\alpha\right)\frac{\hat{Y}^2}{\text{Pr}^{2\alpha}} + \cdots. \tag{12–192}$$

and

$$V \sim -\frac{1}{2}\frac{d\lambda}{dx}\frac{\hat{Y}^2}{\text{Pr}^{2\alpha}} + \frac{1}{6}\left[\frac{d}{dx}\left(u_e\frac{du_e}{dx}\right) + \frac{\text{Gr}}{\text{Re}^2}\frac{d}{dx}(\theta\sin\alpha)\right]\frac{\hat{Y}^3}{\text{Pr}^{3\alpha}} + \cdots. \tag{12–193}$$

Now, if Gr $\ll 1$, then (12–186) becomes

$$\left(\lambda(x)\frac{\hat{Y}}{\Pr^\alpha} + \cdots\right)\frac{\partial\theta}{\partial x} - \left(\frac{d\lambda}{dx}\frac{\hat{Y}^2}{2\Pr^{2\alpha}} + \cdots\right)\Pr^\alpha\frac{\partial\theta}{\partial\hat{Y}} = \Pr^{2\alpha-1}\frac{\partial^2\theta}{\partial\hat{Y}^2}. \quad (12\text{-}194)$$

Hence,

$$\boxed{\alpha = \frac{1}{3},} \quad (12\text{-}195)$$

and this is the same result already obtained for the forced convection limit. On the other hand, if $\mathrm{Gr}/\mathrm{Re}^2\Pr^{2\alpha} \gg 1/\Pr^\alpha$—that is, if $\mathrm{Gr} \gg 1$—then (12–186) can be expressed in the form

$$-\frac{1}{2}\frac{\mathrm{Gr}}{\mathrm{Re}^2\Pr^{2\alpha}}\theta\sin\alpha\,\hat{Y}^2\frac{\partial\theta}{\partial x} + \frac{1}{6}\frac{\mathrm{Gr}}{\mathrm{Re}^2}\frac{\partial}{\partial x}(\theta\sin\alpha)\frac{\hat{Y}^3}{\Pr^{3\alpha}}\Pr^\alpha\frac{\partial\theta}{\partial\hat{Y}} =$$

$$\Pr^{2\alpha-1}\frac{\partial^2\theta}{\partial\hat{Y}^2}. \quad (12\text{-}196)$$

It follows from this that

$$\boxed{\alpha = \frac{1}{4}} \quad (12\text{-}197)$$

and

$$\boxed{\frac{\mathrm{Gr}}{\mathrm{Re}^2}\left[-\frac{1}{2}\theta\sin\alpha\,\hat{Y}^2\frac{\partial\theta}{\partial x} + \frac{1}{6}\frac{\partial}{\partial x}(\theta\sin\alpha)\,\hat{Y}^3\frac{\partial\theta}{\partial\hat{Y}}\right] = \frac{\partial^2\theta}{\partial\hat{Y}^2}.} \quad (12\text{-}198)$$

To remove $\mathrm{Gr}/\mathrm{Re}^2$, we can use

$$\bar{Y} = \hat{Y}\left(\frac{\mathrm{Gr}}{\mathrm{Re}^2}\right)^{-1/4}, \quad (12\text{-}199)$$

and this leads again to the rescaling in (12–177).

As usual, the main information from the preceding analysis is the scaling inherent in the formulation leading to the outer problem (12–158)–(12–160) and the two limiting versions of the inner problem: (12–162)–(12–164), which turns out to be particularly relevant for $\mathrm{Gr} \ll 1$, and (12–179)–(12–181) for $\mathrm{Gr} \gg 1$. Specifically, this leads to the general form (12–171) for the local heat flux and the two limiting forms for F given by (12–174) for $\mathrm{Gr}/\mathrm{Re}^2\Pr^{1/3} \ll 1$ and $(\mathrm{Gr}/\mathrm{Re}^2\Pr^{1/3}) \gg 1$, respectively. Nevertheless, it is of some interest to demonstrate the derivation of *detailed solutions* for the case of a *vertical* flat plate, and we do this in the next section.

The Vertical Flat Plate for $\Pr \gg 1$

Although the most important insights about the mixed convection problem for large $\Pr$ are contained in the preceding section, it is worthwhile to consider the detailed solutions for at least one case. We select the vertical flat plate (that is, its surface is parallel to **g**), with the external flow parallel to the plate. For this problem, an approximate analytic solution is possible for $\Pr \gg 1$.[9]

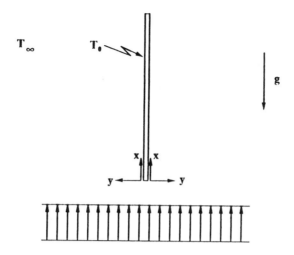

Figure 12–6 A schematic representation of the mixed convection problem for a heated (or cooled) vertical flat plate in a uniform flow that is parallel to gravity.

The starting point in this case is the *outer* problem (12–158)–(12–160), plus matching conditions (12–166) and (12–167). The imposed streamwise pressure gradient is *zero* because $u_e = 1$, and thus the problem is, in fact, just the classical Blasius problem of streaming flow past a semi-infinite flat plate. The solution of this problem is well known to be

$$u = f'(\eta) \quad \text{with} \quad \eta = \frac{Y}{\sqrt{x}},$$

with f being the Blasius similarity function satisfying

$$2f''' + ff'' = 0 \quad \text{with} \quad f(0) = f'(0) = 0, \; f'(\infty) = 1.$$

A schematic representation of the problem is shown in Figure 12-6. As shown, we assume that the external flow is vertically upward with $x = 0$ at the leading edge of the plate.

Now the inner problem is equations (12–162)–(12–164) with boundary conditions (12–165) and the matching condition (12–168), which can now be written in the form

$$\frac{\partial \hat{u}}{\partial \hat{Y}} \to \frac{0.332}{\sqrt{x}} \quad \text{as} \quad \hat{Y} \to \infty, \tag{12-200}$$

where $f''(0) = 0.332$. The nature of the solution to the problem depends upon whether the buoyancy force is in the same direction as the external, ambient flow ('aiding'), or in the opposite direction ('opposing'). Some insight about the nature of the solution can be obtained by examining the shear stress in the inner region. The govening equation for $\hat{u}$ is (12–162),

$$\frac{\partial^2 \hat{u}}{\partial \hat{Y}^2} = -\frac{\text{Gr}}{\text{Re}^2 \text{Pr}^{1/3}} \sin\alpha(x)\theta$$

but with $\sin\alpha(x) \equiv 1$. Thus, integrating,

$$\frac{\partial \hat{u}}{\partial \hat{Y}} = \left(\frac{\partial \hat{u}}{\partial \hat{Y}}\right)_\infty + \left(\frac{\text{Gr}}{\text{Re}^2 \text{Pr}^{1/3}}\right)\int_{\hat{Y}}^{\infty} \theta \, dY. \tag{12–201}$$

At the surface of the plate,

$$\boxed{\frac{\partial \hat{u}}{\partial \hat{Y}}\bigg|_0 = \lambda(x) + \frac{\text{Gr}}{\text{Re}^2 \text{Pr}^{1/3}}\int_0^{\infty} \theta \, dY.} \tag{12–202}$$

Now, for the case of a heated plate ('aiding'), $(T_o - T_\infty) > 0$ and $\text{Gr} > 0$. Thus, the shear stress at the plate surface is increased relative to the forced convection case. The effect of buoyancy is to accelerate the fluid in the same direction as the external flow. For a cooled plate ('opposing'), $\text{Gr} < 0$. In this case, the shear stress contributions of the external flow and buoyancy have opposite signs; that is, the effect of buoyancy is to decelerate the flow. Clearly, at some point, it is possible for $\partial \hat{u}/\partial \hat{Y}|_0$ to equal zero. Let us suppose that the natural convection contribution is weak enough that $\partial \hat{u}/\partial \hat{Y} > 0$ for small x. Then, since $\lambda(x)$ is a decreasing function of x, it is possible for $\partial \hat{u}/\partial \hat{Y}|_0$ to change signs from positive to negative at some point x_s on the plate surface. In this case, there is a buoyancy-induced separation as sketched schematically in Figure 12–7. Clearly, if the external flow were in the opposite direction, the role of buoyancy would be reversed; that is, the cooled plate would be aiding the external flow and the heated plate would be opposing, possibly leading to separation or reversed flow, as explained above.

Now, if we examine the inner problem, (12–162)–(12–164), (12–165), and (12–200), it is clear that a similarity solution does *not* exist. In particular, if we seek a solution in the form

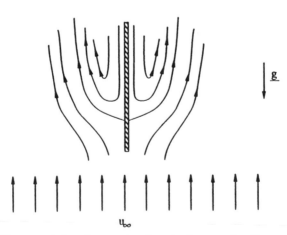

Figure 12–7 Buoyancy-induced separation for mixed convection flow near a cooled vertical plate in a uniform flow.

$$\hat{u} = f'(\eta), \quad \theta = \theta(\eta), \quad \eta = \frac{Y}{g(x)}, \tag{12-203}$$

then it can be seen from the matching condition (12-200) that $g(x) = \sqrt{x}$. But then (12-162) becomes

$$\frac{1}{x}\frac{d^3 f}{d\eta^3} + \frac{Gr}{Re^2 Pr^{1/3}} \theta(\eta) = 0, \tag{12-204}$$

and it is obvious that (12-203) cannot be correct. However, the simple form of (12-204) is suggestive that the solution could be expressed approximately as a series in x for small x, but with the coefficient of each term being a function of η.

Thus, we seek an approximate solution of the form

$$\boxed{\theta = \theta_0(\eta) + x\theta_1(\eta) + \cdots} \tag{12-205}$$

and

$$\boxed{\hat{u} = f_0'(\eta) + x f_1(\eta) + \cdots .} \tag{12-206}$$

It follows from the continuity equation (12-163) that

$$\hat{V} = \frac{1}{2\sqrt{x}}(\eta f_0' - f_0) + \sqrt{x}\left(\eta f_1' - \frac{3}{2}f_1\right) + \cdots . \tag{12-207}$$

Then, the equation of motion (12-162) becomes

$$\frac{1}{x}\frac{d^3 f_0}{d\eta^3} + \frac{d^3 f_1}{d\eta^3} + \cdots = -\frac{Gr}{Re^2 Pr^{1/3}}(\theta_0 + x\theta_1 + \cdots), \tag{12-208}$$

and the thermal energy equation is

$$-\frac{1}{2x}f_0\frac{d\theta_0}{d\eta} + \left(f_0'\theta_1 - \frac{1}{2}f_0\frac{d\theta_1}{d\eta} - \frac{3}{2}f_1\frac{d\theta_0}{d\eta}\right) + 0(x) = \frac{1}{x}\frac{d^2\theta_0}{d\eta^2} + \frac{d^2\theta_1}{d\eta^2} + 0(x). \tag{12-209}$$

Hence, collecting terms of like powers in x, it follows that

$$\boxed{\begin{aligned} \frac{d^3 f_0}{d\eta^3} &= 0, & (12\text{-}210) \\[2mm] \frac{d^2\theta_0}{d\eta^2} + \frac{1}{2}f_0\frac{d\theta_0}{d\eta} &= 0, & (12\text{-}211) \\[2mm] \frac{d^3 f_1}{d\eta^3} &= -\frac{Gr}{Re^2 Pr^{1/3}}\theta_0, & (12\text{-}212) \end{aligned}}$$

and

$$\boxed{\frac{d^2\theta_1}{d\eta^2} + \frac{1}{2}f_0\frac{d\theta_1}{d\eta} - \frac{df_0}{d\eta}\theta_1 = -\frac{3}{2}f_1\frac{d\theta_0}{d\eta}.} \tag{12-213}$$

Equations for the higher-order terms in (12–205) and (12–206) can be obtained by extending (12–208) and (12–209) to higher orders in x. The boundary conditions derived from (12–165) and the matching condition (12–200), plus the condition $\theta \to 0$ as $\hat{Y} \to \infty$, are

$$f_0 = f_0' = 0 \quad \text{at} \quad \eta = 0, \quad f_0'' \to 0.332 \quad \text{as} \quad \eta \to \infty, \qquad (12\text{–}214)$$

$$\theta_0 = 1 \quad \text{at} \quad \eta = 0, \quad \theta_0 \to 0 \quad \text{as} \quad \eta \to \infty, \qquad (12\text{–}215)$$

$$f_1 = f_1' = 0 \quad \text{at} \quad \eta = 0, \quad f_0'' \to 0 \quad \text{as} \quad \eta \to \infty, \qquad (12\text{–}216)$$

and

$$\theta_1 = 0 \quad \text{at} \quad \eta = 0, \quad \theta_1 = 0 \quad \text{as} \quad \eta \to \infty. \qquad (12\text{–}217)$$

Analytic solutions can be obtained for the functions f_0, θ_0, f_1, θ_1, and so on, though these become increasingly complex as we proceed to higher order terms. It should be noted that a power-series solution in x does not exist in terms of the original independent variables x and Y. To see this, we need only note that the first term in a series for θ would have to be independent of x in order to satisfy the boundary condition (12–165), but the first term in $\hat{u}$ must be proportional to $1/\sqrt{x}$ in view of equation (12–200). These two choices are incompatible with the thermal equation, as can be seen by seeking a solution in the form

$$u = \frac{1}{\sqrt{x}} (f_0'(Y) + x f_1'(Y) + \cdots),$$

$$\theta = \theta_0(Y) + x \theta_1(Y) + \cdots.$$

The idea of obtaining approximate solutions by expanding as a power series in the streamwise spatial variable with *coefficients* that are functions of a similarity variable has been used to advantage in a number of fluid mechanics problems.[10] Examples include the boundary-layer on a flat plate with an accelerating or decelerating potential flow given by $u_e(x) = U_0 - ax^n$, and the related problem of the boundary-layer development at the inlet of a straight wall channel.

The Limit Pr ≪ 1

The second asymptotic limit of interest in the mixed convection problem is $Pr \to 0$. Again, the task is to solve the *thermal* boundary-layer equations (12–161), (12–164), and (12–168). We may recall from our earlier analysis of the forced and natural convection problems for large Re or Gr, respectively, that the relative scales of the thermal and momentum layers is reversed for $Pr \ll 1$ compared to the case of $Pr \gg 1$. Thus, in the region where viscous and inertia terms are both important, the thermal energy equation reduces to pure conduction, and the leading-order approximation to the temperature is $\theta \approx 1$. On the other hand, where convection and conduction are both important in the thermal energy balance, the equation of motion is reduced to an inviscid form. The thermal layer is very much thicker than the momentum layer in the limit $Pr \to 0$.

To carry out the analysis, we begin by following the development outlined in Chapter 11 for forced convection heat transfer at low Pr but now retain the buoyancy term in equation (12–168). Hence, a leading-order approximation to the thermal energy equation, scaled using momentum boundary-layer variables, is obtained by letting $Pr \to 0$ in (12–168), and the solution of this equation satisfying the boundary condition $\theta = 1$ at $Y = 0$ is simply

$$\theta \equiv 1. \tag{12-218}$$

Clearly, this is the leading-order approximation in some inner region, nearest the body surface, and it is in this region that the momentum boundary-layer scaling is apparently relevant. Before considering the remainder of the equations and boundary-conditions in this region, let us turn to the equations and boundary-conditions for the outer region, away from the body surface.

Again, we follow the analysis for the forced convection approximation and rescale according to (11–58)–(11–60), with $\beta = 0$ and $\alpha = 1/2$,

$$Y^* = Y \mathrm{Pr}^{1/2}, \quad V^* = V \mathrm{Pr}^{1/2}. \tag{12-219}$$

This transforms (12–161), (12–164), and (12–168) to the forms

$$\frac{\partial u}{\partial x} + \frac{\partial V^*}{\partial Y^*} = 0, \tag{12-220}$$

$$u\frac{\partial u}{\partial x} + V^*\frac{\partial u}{\partial Y^*} - u_e\frac{du_e}{dx} - \frac{\mathrm{Gr}}{\mathrm{Re}^2}\sin\alpha(x)\theta = 0(\mathrm{Pr}) = 0 \quad \text{as} \quad \mathrm{Pr} \to 0, \tag{12-221}$$

and

$$u\frac{\partial\theta}{\partial x} + V^*\frac{\partial\theta}{\partial Y^*} = \frac{\partial^2\theta}{\partial Y^{*2}}. \tag{12-222}$$

The solution in this region should match the potential-flow, isothermal solution in the region outside the boundary-layer, as indicated by (12–165). Thus, as $Y^* \to \infty$,

$$u \to u_e(x), \quad \theta \to 0 \quad \text{as} \quad Y^* \to \infty. \tag{12-223}$$

The remaining conditions come from matching with the solution in the innermost part of the boundary-layer where (12–218) is the first approximation to θ. Thus,

$$\lim_{Y^* \ll 1} \theta = 1 \quad \text{for} \quad \mathrm{Pr} \to 0, \tag{12-224}$$

$$\lim_{Y^* \ll 1} u^* \Longleftrightarrow \lim_{Y \gg 1} u \quad \text{for} \quad \mathrm{Pr} \to 0, \tag{12-225}$$

and

$$\lim_{Y^* \ll 1} V^* \Longleftrightarrow \lim_{Y \gg 1} \sqrt{\mathrm{Pr}}\,V \to 0 \quad \text{for} \quad \mathrm{Pr} \to 0. \tag{12-226}$$

Note that we have denoted the velocity component u in the outermost region as u^* to distinguish it from u in the inner region, though no rescaling is involved in going from

u to $u*$. Clearly, the condition (12–226) is sufficient, together with (12–223) and (12–224), to determine the solution in the outer (thermal layer) part of the boundary-layer.

To determine the velocity field closer to the body surface, we must solve the remainder of the problem in the innermost region, with θ approximated according to (12–218). This is a standard fluid mechanics boundary-layer problem, namely,

$$u\frac{\partial u}{\partial x} + V\frac{\partial u}{\partial Y} = u_e^*\frac{du_e^*}{dx} + \frac{Gr}{Re^2}\sin\alpha(x) + \frac{\partial^2 u}{\partial Y^2} \tag{12–227}$$

and

$$\frac{\partial u}{\partial x} + \frac{\partial V}{\partial Y} = 0, \tag{12–228}$$

with

$$u = V = 0 \quad \text{at} \quad Y = 0 \tag{12–229}$$

and

$$u \rightarrow u_e^*(x) \quad \text{for} \quad Y \gg 1. \tag{12–230}$$

The latter condition comes from the matching condition (12–225) with $u_e^*(x) \equiv \lim_{Y^* \rightarrow 0} u^*(x, Y^*)$.

The rescaling inherent in the thermal-layer equations and the form of these equations lead directly to the general form of the heat transfer correlation for mixed convection problems in the low Pr limit. In particular, we see that

$$\frac{Nu_x}{Re^{1/2}Pr^{1/2}} = G\left(x; \frac{Gr}{Re^2}\right). \tag{12–231}$$

Furthermore, since the limits of the correlation for pure forced and natural convection are known to be

$$Nu_x = a_1(x)Re^{1/2}Pr^{1/2} \quad \left(\begin{array}{c}\text{forced}\\\text{convection}\end{array}\right) \tag{12–232}$$

and

$$Nu_x = a_2(x)Gr^{1/4}Pr^{1/4} \quad \left(\begin{array}{c}\text{natural}\\\text{convection}\end{array}\right), \tag{12–233}$$

it is evident that the function G must adopt the limiting forms

$$G\left(x, \frac{Gr}{Re^2}\right) = \begin{cases} a_1(x) & \text{for } Gr/Re^2 \rightarrow 0 \tag{12–234a}\\ \\ a_2(x)\left(\dfrac{Gr}{Re^2}\right)^{1/2} & \text{for } Gr/Re^2 \gg 1. \tag{12–234b} \end{cases}$$

Clearly, in the $\Pr \ll 1$ limit, the strength of the natural convection contribution is measured by $\mathrm{Gr}/\mathrm{Re}^2$ alone (that is, it is independent of $\Pr$).

The proposed form (12–234a) for G in the limit $\mathrm{Gr}/\mathrm{Re}^2 \to 0$ is evident directly from (12–220)–(12–222) plus (12–223), (12–224), and (12–226) since it is clear that a solution can be found for this problem in the form of a regular expansion with $\mathrm{Gr}/\mathrm{Re}^2$ as the small parameter, in which the first term is completely independent of $\mathrm{Gr}/\mathrm{Re}^2$. To show that the form (12–234b) appears for $\mathrm{Gr}/\mathrm{Re}^2 \gg 1$, we can introduce the change of variables

$$Y_1 = \left(\frac{\mathrm{Gr}}{\mathrm{Re}^2}\right)^{1/4} Y^*, \quad u_1 = \left(\frac{\mathrm{Gr}}{\mathrm{Re}^2}\right)^{-1/2} u^*, \quad V_1 = \left(\frac{\mathrm{Gr}}{\mathrm{Re}^2}\right)^{-1/4} V^* \quad (12\text{–}235)$$

into the outer problem (12–220)–(12–222), plus boundary conditions, to obtain

$$\frac{\partial u_1}{\partial x} + \frac{\partial V_1}{\partial Y_1} = 0, \quad (12\text{–}236)$$

$$u_1 \frac{\partial u_1}{\partial x} + V_1 \frac{\partial u_1}{\partial Y_1} - \theta \sin\alpha(x) - \left(\frac{\mathrm{Gr}}{\mathrm{Re}^2}\right)^{-1} u_e \frac{du_e^*}{dx} = 0, \quad (12\text{–}237)$$

and

$$u_1 \frac{\partial \theta}{\partial x} + V_1 \frac{\partial \theta}{\partial Y_1} = \frac{\partial^2 \theta}{\partial Y^2}, \quad (12\text{–}238)$$

with boundary conditions

$$u_1 \to \left(\frac{\mathrm{Gr}}{\mathrm{Re}^2}\right)^{-1/2} u_e(x), \quad \theta \to 0 \quad \text{as} \quad Y_1 \to \infty \quad (12\text{–}239)$$

and

$$\theta = 1, \quad V_1 = 0 \quad \text{at} \quad Y_1 = 0. \quad (12\text{–}240)$$

This problem can be solved as a perturbation expansion for $\epsilon \equiv (\mathrm{Gr}/\mathrm{Re}^2)^{-1/2} \to 0$, and the leading-order term in this expansion is clearly independent of $\mathrm{Gr}/\mathrm{Re}^2$ (as can be seen by taking the limit $\mathrm{Gr}/\mathrm{Re}^2 \to \infty$ in the governing equations plus boundary conditions). Thus,

$$G\left(x; \frac{\mathrm{Gr}}{\mathrm{Re}^2}\right) \equiv -\left.\frac{\partial \theta}{\partial Y^*}\right|_{Y^*=0} = a_2(x)\left(\frac{\mathrm{Gr}}{\mathrm{Re}^2}\right)^{-1/4} \quad \text{for} \quad \frac{\mathrm{Gr}}{\mathrm{Re}^2} \gg 1,$$

where $a_2(x) \equiv -\partial\theta/\partial Y_1|_{Y_1=0}$.

It can be seen in both the limits $\Pr \to \infty$ and $\Pr \to 0$ that the thermal boundary-layer equations provide a description of mixed-convection problems that is uniformly valid over the whole range of $\mathrm{Gr}/\mathrm{Re}^2$, from the forced convection limit $\mathrm{Gr} \to 0$ to natural convection-dominated flow $\mathrm{Gr} \to \infty$. In particular, the form of the heat transfer correlation has been shown to pass smoothly between these two limits to reproduce the correlations found earlier for the pure forced or free convection problems.

E. Natural Convection in a Horizontal Fluid Layer Heated from Below— The Rayleigh-Benard Problem

In all of the problems considered until now in this chapter, we have assumed that the heated (or cooled) boundaries are not horizontal, except possibly in a local region. Thus, there has always been a nonzero body force component parallel to the boundary, and the fluid undergoes natural convection motion for any *arbitrarily* small temperature nonuniformity. In practice, this is the most common situation, but it is by no means unique. In particular, if the temperature gradient at every point in the fluid is in the same direction as gravity, it is possible to have a pure conduction mode of heat transfer with *no* induced motion in the fluid. The best-known prototype occurs in an "unbounded" layer of fluid between two horizontal plane surfaces with heating from below. This is the famous Rayleigh-Benard problem.[11] Convection occurs in this case only because the basic pure conduction state is unstable above some critical value of the temperature gradient. For small temperature differences between the two horizontal boundaries, the fluid does not move, and heat transfer is by conduction. However, as the temperature difference is increased, a critical condition is eventually reached where the motionless base state becomes unstable, and there is an abrupt and spontaneous transition to a convective mode of heat transfer.

The Rayleigh-Benard configuration is sketched in Figure 12-8. We assume that the fluid is between two infinite plane surfaces that are separated by a distance d. The lower surface is at a constant temperature T_0, and the upper one is at a lower temperature T_1. Our starting point is the time-dependent Boussinesq equations, in dimensionless form,

$$\nabla \cdot \mathbf{u} = 0, \tag{12-241}$$

$$\frac{\partial \theta}{\partial t} + \mathbf{u} \cdot \nabla \theta = \frac{1}{\text{Pr}} \left(\frac{v_0}{u_c d} \right) \nabla^2 \theta, \tag{12-242}$$

and

$$\frac{\partial \mathbf{u}}{\partial t} + \mathbf{u} \cdot \nabla \mathbf{u} = -\nabla p + \left(\frac{v_0}{u_c d} \right) \nabla^2 \mathbf{u} - \frac{\beta T_c g d}{u_c^2} \theta \mathbf{i}_z, \tag{12-243}$$

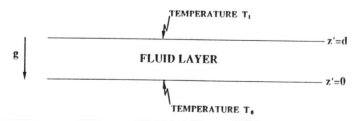

Figure 12-8 The Rayleigh-Benard configuration for buoyancy-driven convection in a horizontal fluid layer that is heated from below.

where the characteristic lengthscale is d, the characteristic velocity and temperature scales are denoted as u_c and T_c but are left unspecified for the moment, and

$$\theta \equiv \frac{T_0 - T}{T_c}. \tag{12-244}$$

The first thing to notice is that there is a steady-state solution for *arbitrary* values of the dimensionless parameters, namely,

$$\mathbf{u} \equiv 0, \tag{12-245}$$

$$\theta = \bar{\theta} \equiv \frac{T_0 - T_1}{T_c} z, \tag{12-246}$$

and

$$\nabla p = \nabla \bar{p} = -\frac{T_0 - T_1}{T_c} \left(\frac{\beta T_c g d}{u_c^2} \right) z \mathbf{i}_z. \tag{12-247}$$

In this mode, heat transfer is strictly by *conduction*. Because the temperature gradient is vertical, the body force can be completely balanced by a hydrostatic pressure gradient, and the solution (12-245)–(12-247) is valid over the whole range of possible values for $T_0 - T_1$. This is in sharp contrast with all of the natural convection systems studied earlier, where arbitrarily small temperature gradients led to motion.

In fact, *experimental* studies show that the static layer is actually maintained up to some finite, critical value of $T_0 - T_1$, but beyond this point there is an *abrupt* and *spontaneous* transition to convective motion. The most clear-cut evidence of this transition, in addition to flow visualization, is an abrupt increase in the slope of the heat transfer rate versus the temperature difference $T_0 - T_1$. It is notable that the abrupt transition from conduction to convection occurs, for a given fluid and given fluid layer depth, at a value of $T_0 - T_1$ that is independent of any precautions that may be taken to isolate the system from external disturbances. This suggests that the static fluid layer is *unstable* to disturbances of arbitrarily small magnitude.

The most important feature to try and predict theoretically is the critical condition for instability, since this marks the boundary between relatively inefficient heat transfer by conduction and the onset of convection. To do this, we consider the fate of an arbitrary, fully three-dimensional initial perturbation of the base state (12-245)–(12-247), namely,

$$\mathbf{u} = (u', v', w'),$$

$$\theta = \bar{\theta} + \theta',$$

and

$$p = \bar{p} + p', \tag{12-248}$$

where the primed variables are assumed to be arbitrarily small. If the system is unstable to infinitesimal disturbances, we should find that these primed quantities increase in magnitude with time. If it is stable, on the other hand, they should decrease in time.

The governing equations for the disturbance variables are just the Boussinesq equations, (12-241)–(12-243). However, in view of the fact that $|\mathbf{u}'|, \theta', p' \ll 1$, these equations can be *linearized* by neglecting all quantities that are quadratic in the primed variables. Thus, substituting (12–248) into the Boussinesq equations and linearizing, we obtain (in Cartesian component form)

$$\frac{\partial u'}{\partial x} + \frac{\partial v'}{\partial y} + \frac{\partial w'}{\partial z} = 0, \tag{12–249}$$

$$\frac{\partial \theta'}{\partial t} + w' \frac{\partial \bar{\theta}}{\partial z} = \frac{1}{Pr\sqrt{Gr}} \nabla^2 \theta', \tag{12–250}$$

$$\frac{\partial u'}{\partial t} = -\frac{\partial p'}{\partial x} + \frac{1}{\sqrt{Gr}} \nabla^2 u', \tag{12–251}$$

$$\frac{\partial v'}{\partial t} = -\frac{\partial p'}{\partial y} + \frac{1}{\sqrt{Gr}} \nabla^2 v', \tag{12–252}$$

and

$$\frac{\partial w'}{\partial t} = -\frac{\partial p'}{\partial z} + \frac{1}{\sqrt{Gr}} \nabla^2 w' - \theta', \tag{12–253}$$

where we have adopted the usual form for u_c,

$$u_c = \sqrt{\beta g d T_c}.$$

The most convenient choice for T_c, which has heretofore been left unspecified, is

$$T_c = T_0 - T_1,$$

since then

$$\frac{\partial \bar{\theta}}{\partial z} = 1 \quad \text{and} \quad Gr = \frac{\beta g d^3 (T_0 - T_1)}{v_\infty^2}.$$

The boundary conditions depend, of course, on the nature of the flat surfaces at $z = 0$ and 1. We shall return to conditions for the velocity components shortly.

In order to analyze (12–249)–(12–253), it is convenient to combine them in order to reduce the number of dependent variables. This can be done in a straightforward way. First, (12–251) and (12–252) can be combined by differentiating with respect to x and y, respectively, and adding. The result is

$$\frac{\partial}{\partial t} \left(\frac{\partial u'}{\partial x} + \frac{\partial v'}{\partial y} \right) = -\nabla_1^2 p' + \frac{1}{\sqrt{Gr}} \nabla^2 \left(\frac{\partial u'}{\partial x} + \frac{\partial v'}{\partial y} \right), \tag{12–254}$$

where

$$\nabla_1^2 \equiv \frac{\partial^2}{\partial x^2} + \frac{\partial^2}{\partial y^2}.$$

However, using continuity, this can be written as

$$\left(\frac{\partial}{\partial t} - \frac{1}{\sqrt{Gr}} \nabla^2 \right) \frac{\partial w'}{\partial z} = \nabla_1^2 p'. \tag{12–255}$$

We have thus reduced the original set of five equations and five unknowns to three equations, (12–250), (12–253), and (12–255), for three variables, w', p', and θ'. To eliminate p', we can combine (12–253) and (12–255). First, differentiate (12–255) with respect to z,

$$\left(\frac{\partial}{\partial t} - \frac{1}{\sqrt{Gr}}\nabla^2\right)\frac{\partial^2 w'}{\partial z^2} = \nabla_1^2\left(\frac{\partial p'}{\partial z}\right), \tag{12–256}$$

and then operate on (12–253) with ∇_1^2,

$$\left(\frac{\partial}{\partial t} - \frac{1}{\sqrt{Gr}}\nabla^2\right)\nabla_1^2 w' = -\nabla_1^2\left(\frac{\partial p'}{\partial z}\right) - \nabla_1^2\theta'. \tag{12–257}$$

Then, adding (12–156) and (12–257), we obtain

$$\left(\frac{\partial}{\partial t} - \frac{1}{\sqrt{Gr}}\nabla^2\right)\nabla^2 w' = -\nabla_1^2\theta'. \tag{12–258}$$

We now have two equations, (12–250) and (12–258), for θ' and w'. Finally, these two equations can be combined to eliminate θ'. To do this, we operate on (12–258) with $[\partial/\partial t - (1/Pr\sqrt{Gr})\nabla^2]$,

$$\left(\frac{\partial}{\partial t} - \frac{1}{Pr\sqrt{Gr}}\nabla^2\right)\left(\frac{\partial}{\partial t} - \frac{1}{\sqrt{Gr}}\nabla^2\right)\nabla^2 w' = -\nabla_1^2\left(\frac{\partial}{\partial t} - \frac{1}{Pr\sqrt{Gr}}\nabla^2\right)\theta'.$$

Hence, substituting on the right-hand side using (12–250), we obtain

$$\boxed{\left(\frac{\partial}{\partial t} - \frac{1}{Pr\sqrt{Gr}}\nabla^2\right)\left(\frac{\partial}{\partial t} - \frac{1}{\sqrt{Gr}}\nabla^2\right)\nabla^2 w' = +\nabla_1^2 w'.} \tag{12–259}$$

Thus, the problem of describing the time-dependent evolution of an arbitrary initial disturbance reduces to the solution of a single sixth-order partial differential equation for the disturbance velocity component w'. Once w' is known, the other disturbance quantities can be calculated. For example, θ' can be obtained from (12–250), and p', u', and v' can then be determined via (12–249), (12–251), and (12–252). The objective here is to use (12–259) to examine the conditions for growth or decay of an arbitrary initial disturbance in w'. Since the governing equation is linear and separable, it will clearly have a solution of the form

$$\boxed{w' = W(x,y,z)e^{\sigma t},} \tag{12–260}$$

with

$$\boxed{\left(\sigma - \frac{1}{Pr\sqrt{Gr}}\nabla^2\right)\left(\sigma - \frac{1}{\sqrt{Gr}}\nabla^2\right)\nabla^2 W = +\nabla_1^2 W.} \tag{12–261}$$

The analysis of stability is then equivalent to determining the *sign* of the real part of the growth rate σ. Positive values of $Re(\sigma)$ indicate instability via exponential growth of an initial disturbance, whereas $\text{Re}(\sigma) < 0$ corresponds to a stable system. The so-called neutral stability condition is $\text{Re}(\sigma) = 0$. It is clear, upon examination of (12-261), that σ generally depends on Gr, Pr, and the spatial form of the disturbance function $W(x, y, z)$.

There are two questions that need to be addressed in order to analyze (12-261) to determine the stability of the system. First, we need to consider the representation of an *arbitrary* initial disturbance. Second, we require boundary conditions for the disturbance function W. The question of representing an arbitrary initial disturbance is best considered in the context of an overall solution scheme for (12-261). Since this equation is linear and separable, the solution W can be expressed in the general form

$$\boxed{W(x, y, z) = f(z) F(x, y).} \tag{12-262}$$

Further, it is clear upon examination of (12-161) that any solution for F must satisfy the constraint

$$\nabla_1^2 F = \text{const } F.$$

In fact, such a solution is just

$$\boxed{F = e^{i(\alpha_x x + \alpha_y y)},} \tag{12-263}$$

in which case

$$\nabla_1^2 F = -\alpha^2 F \tag{12-264}$$

where

$$\alpha^2 = \alpha_x^2 + \alpha_y^2. \tag{12-265}$$

But (12-163) is nothing more than a normal Fourier component with wave numbers α_x and α_y.

Now, as long as the initial disturbance is continuous, then any arbitrary variation with respect to x and y can be expressed as a Fourier series (or integral), and because the governing equation is linear, the dynamic behavior of each Fourier mode is *independent* of the others. Thus, it is sufficient simply to consider the dynamics of a single Fourier mode with the wave number α as a parameter. Although the growth rate σ will clearly depend on α, a sufficient condition for instability can be obtained by finding the maximum value of the Grashof number for which $\text{Re}(\sigma) \equiv 0$ over all possible values of α, $0 \leq \alpha \leq \infty$.

The functional form for $f(z)$ must satisfy the ordinary differential equation obtained by substituting (12-262) and (12-264) into (12-261), that is,

$$\left[\left(\sigma - \frac{1}{\text{Pr}\sqrt{\text{Gr}}} (D^2 - \alpha^2) \right) \left(\sigma - \frac{1}{\sqrt{\text{Gr}}} (D^2 - \alpha^2) \right) (D^2 - \alpha^2) - \alpha^2 \right] f = 0 \tag{12-266}$$

as well as boundary conditions for f that are to be specified at $z = 0.1$. Note that D^2 in (12-266) is shorthand for the second derivative, $\partial^2/\partial z^2$. The boundary conditions on f depend on the physical assumptions that are made about the bounding surfaces at $z = 0, 1$. Since these surfaces are assumed to be nondeforming, the kinematic condition is simply

$$\boxed{f=0 \quad \text{at} \quad z=0, 1.}$$ (12–267)

However, the differential equation for f is sixth order. Thus, we require at least four additional boundary conditions. One physically plausible assumption is that the boundaries at $z=0,1$ are isothermal. In this case,

$$\theta' = 0 \quad \text{at} \quad z = 0, 1,$$

and we can see from (12–258) that this implies

$$\boxed{\left(\sigma - \frac{1}{\sqrt{Gr}}(D^2 - \alpha^2)\right)(D^2 - \alpha^2)f = 0 \quad \text{at} \quad z = 0, 1.}$$ (12–268)

The third pair of boundary conditions on f depends on whether the boundaries at $z=0$ and 1 are solid or free surfaces. At a solid surface, $u' = v' = 0$, and it therefore follows from continuity that

$$\frac{\partial w'}{\partial z} = 0, \quad z = 0, 1.$$

From (12–260) and (12–262), we see that this implies

$$\boxed{Df = 0 \text{ at a solid boundary.}}$$ (12–269)

On the other hand, at a free surface, where the fluid layer is bounded above (or below) by an inviscid fluid, the boundary condition of shear stress continuity is approximated as

$$\tau'_{xz} = \tau'_{yz} = 0.$$

Using the definition for τ for a Newtonian fluid, these conditions become

$$\frac{\partial u'}{\partial z} = \frac{\partial v'}{\partial z} = 0$$

or

$$\frac{\partial}{\partial z}\left(\frac{\partial u'}{\partial x} + \frac{\partial v'}{\partial y}\right) = 0.$$

Hence, by continuity,

$$\frac{\partial^2 w'}{\partial z^2} = 0,$$

and this implies that

$$\boxed{D^2 f = 0 \text{ at a free surface.}}$$ (12–270)

The most common configuration for a fluid layer that is heated from below is probably to have a solid boundary at $z = 0$ and a free surface (gas-liquid interface) at $z = 1$. However, there is a major simplification of the analysis if both surfaces are free, and the results are qualitatively similar regardless of which of the boundary conditions, (12–269) or (12–270), is applied at which boundary.

The simplification for two free surfaces is that f and *all* even derivations of f are zero at $z = 0, 1$. To see this, we note that the conditions (12–267) and (12–270), when substituted into the condition (12–268), require that

$$\boxed{D^4f = 0 \quad \text{at} \quad z = 0, 1.} \tag{12–271}$$

Hence, from the governing equation (12–266), it can be seen that $D^6f = 0$ at $z = 0, 1$, and so on, by differentiating (12–266) to higher order derivatives of f with respect to z.

Thus, for two free surfaces, the problem is reduced to the governing equation (12–266) and the six boundary conditions (12–267), (12–270), and (12–271), This is a classic *eigenvalue problem*. In particular, let us suppose that we specify Pr, Gr and α^2 (the wave number of the normal model of perturbation). Then, if σ is left arbitrary, the only solution of (12–266) satisfying the boundary conditions is $f \equiv 0$. In general, there is a single value of σ such that $f \neq 0$, and this is called the *eigenvalue*. If Real $(\sigma) < 0$ for all α, the system is stable to infinitesimal disturbances. On the other hand, if Real $(\sigma) > 0$ for *any* α, it is unconditionally unstable. In reality, an infinitesimal disturbance of arbitrary wave number α is impossible to avoid, and the system will exhibit a spontaneous instability if Real $(\sigma) > 0$ for *any* α. Stated in another way, the above statements imply that for any Pr there will be a certain value of Gr such that all disturbances of any α decay. The largest such value of Gr is called the *critical* value for linear stability.

The stability problem with two free surfaces was first solved in 1916 by Lord Rayleigh. Since f and all even derivatives of f are zero at $z = 0$ and $z = 1$, a fundamental solution of the problem is

$$\boxed{f_n = \sin (n\pi z),} \tag{12–272}$$

where n is any integer, $n \geqslant 1$. This satisfies all of the boundary conditions (12–267), (12–269), and (12–270). The governing equation (12–266) becomes

$$\left(\sigma + \frac{1}{\text{Pr}\sqrt{\text{Gr}}} (n^2\pi^2 + \alpha^2) \right) \left(\sigma + \frac{1}{\sqrt{\text{Gr}}} (n^2\pi^2 + \alpha^2) \right) (n^2\pi^2 + \alpha^2) - \alpha^2 = 0.$$

Multiplying by PrGr, this equation can be rewritten in the form first derived by Rayleigh:

$$\boxed{(\hat{\sigma}\text{Pr} + \gamma)(\hat{\sigma} + \gamma)\gamma - \alpha^2\text{Ra} = 0,} \tag{12–273}$$

where the product Pr·Gr is known as the Rayleigh number, that is,

$$\text{Ra} \equiv \text{Pr}\cdot\text{Gr}$$

$$\hat{\sigma} = \sigma\sqrt{\text{Gr}}$$

and

$$\gamma \equiv (n^2\pi^2 + \alpha^2) \tag{12–274}$$

The change from σ to $\hat{\sigma}$ is precisely equivalent to using d^2/v_0 as the characteristic time-scale, instead of d/u_c. The equation (12–273) shows that a natural parameter to use in discussing the stability of a fluid layer for fixed Pr is Ra rather than $\sqrt{\text{Gr}}$.

Clearly, for Pr, γ, α, and Ra fixed, (12–273) is a *characteristic* equation for the eigenvalues $\hat{\sigma}$. For Ra $\neq 0$, the solution is

$$\hat{\sigma} = \frac{1}{2\mathrm{Pr}} \left\{ -\gamma(\mathrm{Pr}+1) \pm \sqrt{\gamma^2(\mathrm{Pr}-1)^2 + \frac{4\alpha^2\mathrm{Ra}\,\mathrm{Pr}}{\gamma}} \right\}. \qquad (12\text{--}275)$$

In the limit $\mathrm{Ra} \equiv 0$, there is no buoyancy to sustain an initial disturbance, and we see that there are two real, negative values for $\hat{\sigma}$:

$$\hat{\sigma}_1 = -\frac{\gamma}{\mathrm{Pr}} \quad \text{and} \quad \hat{\sigma}_2 = -\gamma \quad \text{for} \quad \mathrm{Ra} \equiv 0. \qquad (12\text{--}276)$$

As expected, in this case the system is stable.

   More generally, since $\mathrm{Pr} > 0$ and $\gamma > 0$, it can be seen from (12–275) that the two eigenvalues are always *real*, and this means that an initial disturbance will either grow or decay monotonically with time. Thus, the neutral state, separating stable and unstable conditions, corresponds to

$$\boxed{\hat{\sigma} = 0.} \qquad (12\text{--}277)$$

It is interesting to note that the growth rates predicted from (12–275) generally depend on Pr, but the condition of neutral stability is seen from (12–273) to be independent of Pr in the present case. Going back to (12–273), it follows from (12–277) that the value of Rayleigh number that gives neutral stability is

$$\boxed{\mathrm{Ra}^* = \frac{(n^2\pi^2 + \alpha^2)^3}{\alpha^2} \quad \text{for} \quad n = 1, 2, 3, \ldots .} \qquad (12\text{--}278)$$

For a given wave number, the *least* stable mode corresponds to $n=1$ (this yields the smallest value of Ra^*). If we plot Ra^* versus α for $n=1$, as in Figure 12–9, we obtain

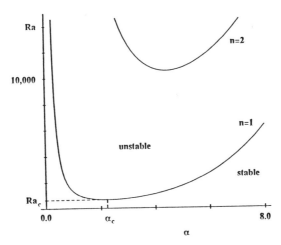

Figure 12–9 Stability criteria for the Rayleigh-Benard problem. The two curves shown are the neutral stability curves for the modes $n=1$ and $n=2$. The region above the curve for $n=1$ is unstable, while that below is stable. The critical Rayleigh number is 657.511 at a critical wave number of 2.2214.

the so-called neutral stability curve. For a given α, any value of Ra that exceeds Ra*(α) corresponds to an *unstable* system, while any smaller value is stable. The *critical* Rayleigh number, Ra_{crit}, for linear instability is the minimum value of Ra* for all possible values of α, and the corresponding value of $\alpha = \alpha_{min}$ is known as the critical wave number.

An analytic expression for Ra_{crit} can be obtained from (12–278) by finding the value of α such that $d Ra*/d\alpha^2 = 0$. Upon differentiating (12–278), we find that

and

$$\alpha_c = \sqrt{\frac{\pi^2}{2}} = 2.2214$$

$$Ra_c = \frac{27}{4}\pi^4 = 657.511. \qquad (12\text{–}279)$$

Thus, for Ra < 657.511, the system is stable to arbitrary small disturbances. However, any real system that satisfies the assumptions of the analysis will be unstable for Ra > 657.511, since it is impossible to avoid an infinitesimal disturbance that contains the α_c component.

It should be remarked that we have considered the stability of the base state, (12–245)–(12–247), only to an infinitesimal disturbance. Any system that is unstable to an infinitesimal disturbance is unconditionally unstable. On the other hand, a system that is stable to infinitesimal disturbances could turn out to be unstable to disturbances of *finite amplitude* at lower values of the critical parameter (i.e., Ra < 657.511). In the present problem, however, the infinitesimal disturbance is the most unstable, and the linear stability criteria thus constitutes an absolute stability bound on Ra.

The reader may well ask whether the linear stability analysis can tell us anything about the state to which the system evolves in time for Ra $> Ra_c$. In particular, we have seen that the most unstable infinitesimal disturbance is the one with wave number α_c. Further, it can be shown from (12–275) that this is also the fastest growing disturbance for Ra $\sim Ra_c$, and one might therefore expect to see a convection pattern evolve with this wave number in the xy plane. It should be noted however, that the exponential growth rate predicted soon yields disturbances of finite amplitude, and the linear analysis quickly breaks down. Any prediction of the ultimate fate of the system must involve more complex analysis of finite amplitude effects. This is beyond the scope of the present work.

We have noted earlier that the case of a fluid layer with two free surfaces is qualitatively representative of the general nature of the solution for more realistic boundary conditions. Of course, the change from a slip to no-slip boundary condition does tend to stabilize the flow somewhat so that the critical Rayleigh number increases in the order

$$Ra_c \quad < \quad Ra_c \quad < \quad Ra_c \ .$$

| Two Free | One Free | Two no-slip |
| Boundaries | Boundary | Boundaries |

A detailed analysis in the latter case shows, in fact, that

$$\text{Ra}_c = 1707.762 \quad \text{with } \alpha_c = 3.117.$$

Two no-slip
Boundaries

One may also question the assumption of isothermal boundaries. In reality, rather than strictly isothermal conditions, we might expect to have a finite heat flux to the surroundings and a mixed boundary condition of the type

$$\frac{\partial \theta}{\partial z} + \text{Nu}\theta = 0.$$

In this case, the isothermal limit corresponds to $\text{Nu} \to \infty$, and it turns out that the system is less stable (lower Ra_c) for finite Nu. Another effect for free surfaces is that nonisothermal boundaries are subject to tensile stress due to surface-tension gradients, and these can also drive a convective motion, known as Marangoni instability.

There has been a great deal of work on both the buoyancy-driven convection problem discussed here and on other related phenomena such as Marangoni instability and double-diffusive convection since the early analysis of Rayleigh almost seventy-five years ago. Many useful texts exist[12] that describe this work, and the present discussion is just a brief introduction. The main conclusion from our analysis is that a horizontal fluid layer can actually sustain a *finite* (subcritical) temperature gradient without any buoyancy-driven convection. This, then, identifies a second basic class of natural convection phenomena that occur only as a consequence of *instability* at finite Gr or Ra numbers instead of occurring for arbitrarily small values of these parameters, as was the case in earlier examples.

Homework Problems

1. Consider the limit of small Grashof number, $\text{Gr} \ll 1$, in the natural convection equations. Use asymptotic methods to formulate governing equations and boundary conditions for a heated two-dimensional body. "Solve" for the leading-order approximations for a semi-infinite, vertical flat plate in an unbounded fluid.

2. a. We consider a point source of heat of strength Q located in an unbounded, quiescent fluid. Derive the governing equations (12–132)–(12–136) for this problem, taking care to state all assumptions. Verify the solutions (12–137) and (12–138).

   b. Consider a heated jet, oriented vertically upward (with **g** acting vertically downward), that is produced when a heated fluid is forced through an orifice into a quiescent body of the same fluid that is at a lower temperature. Derive governing equations and solutions (if possible) for large distances above the orifice using a boundary-layer–like analysis similar to part (a), but including the momentum flux out of the orifice. When can natural convection effects be ignored? When can forced convection be ignored so that the problem reverts to part (a)?

3. Start with the natural convection boundary-layer equations (12–39)–(12–43). Prove that similarity solutions are possible only if $\sin\alpha(x) = x^m$. Apply this analysis to a vertical flat plate that is heated compared to an adjacent, ambient fluid. Solve the resulting equations if possible. What happens if the plate is tilted to some finite angle γ from the vertical?

4. Verify that similarity solutions of the mixed convection problem with a constant surface temperature are possible only if the conditions (12–154) are satisfied, and derive the governing equations (12–155) and (12–156) for the special case outlined in the text where these conditions are satisfied.

5. It is observed experimentally that a fluid layer with an upper free surface may sometimes exhibit cellular convection patterns when heated from below, even though the Rayleigh number is quite small in magnitude, that is, Ra ~ 0(1). Further, paint films often display steady cellular circulatory flow of the same general type as that observed in the cases of fluid layers heated from below, and this may occur whether the paint layer is on the top or on the underside of the solid, heated boundary. Both of these observations suggest that there must be an additional mechanism for instability, in addition to buoyancy-driven motion. This was shown by Sterling and Scriven [*AIChE J.* 5:514 (1959)] and Pearson (*J. Fluid Mech.* 4:489 (1958)] to be flow driven by surface tension gradients—called Marangoni instabilities. In this problem, we work out the linear stability analysis for a fluid layer heated from below, taking account of surface tension variations on the upper surface, which is assumed to be a gas-liquid interface, due to temperature variations along the surface.

a. Let us first consider the problem from a qualitative point of view. Suppose we have a layer of fluid that is resting without motion on a solid heated surface of temperature T_0 that exceeds the ambient temperature T_1 of the overlaying gas. Gravity is assumed to be zero (we may imagine that this

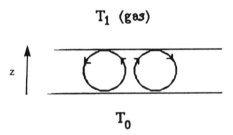

system is being studied in space). Suppose at $t = 0$, a disturbance is introduced into the fluid that takes the form of a pair of two-dimensional roll cells as pictured. Assume that the surface tension increases as the liquid temperature at the surface decreases. What, qualitatively, do you expect the temperature distribution on the interface to look like, assuming there is heat transfer between the liquid and the gas. Does this produce a surface tension distribution that could sustain the cellular motion? What mechanisms or effects would you expect to be present that would cause the motion to decay with time? Presumably, spontaneous motion would occur if the driving force were large enough to overcome the dissipative mechanism. What factors would you expect to control the magnitude of the two competing effects?

b. Let us now consider the governing equations and boundary conditions that govern the dynamics of the disturbance in the system described above. We assume that the Rayleigh number is identically zero ($Ra \equiv 0$). Further, let us suppose that the neutral stability state corresponds to the growth parameter $\sigma \equiv 0$. Based on these assumptions and the analysis of the Rayleigh-Benard problem in this chapter, what are the governing equations in the neutral state (note that the thermal equation is uncoupled from the dynamical equation since we have neglected buoyancy). Assume that the lower surface is a no-slip, isothermal flat wall. The upper surface is assumed to remain flat. The heat exchange between the liquid and gas is assumed to be at local equilibrium. Hence,

$$-k \frac{\partial T}{\partial z} = h(T - T_1)$$

at the upper surface $z = d$, where h is a heat transfer coefficient and k is the the thermal conductivity of the liquid. The shear stress components at this interface are linearly related to the temperature gradients,

$$\tau_{xz} = -\beta \frac{\partial T}{\partial x}, \quad \tau_{yz} = -\beta \frac{\partial T}{\partial y},$$

where γ is the temperature coefficient of interfacial tension, $\beta \equiv -(\partial \gamma / \partial T)$. From this, you should be able to derive a condition relating the normal velocity component W to T. State the boundary conditions in a dimensionless form that is consistent with that used to nondimensionalize the governing equations. You should have six conditions altogether for f and θ, three at each boundary. Two new dimensionless parameters should appear in the boundary conditions

$$Nu_g \equiv \frac{hd}{k} \quad \text{and} \quad M \equiv \frac{d\beta(T_0 - T_1)}{\rho \nu \kappa},$$

where $\kappa = k/\rho C_p$ and $\nu = \mu/\rho$. The latter is known as the Marangoni number.

c. We have already put $\sigma = 0$. Hence the problem reduces to a so-called eigenvalue problem for determining the magnitude of M that corresponds to this neutral state. The governing equations are linear, and you should be able to solve them easily. After applying all of the boundary conditions except the one involving the parameter M, you should have

$$f = C \left(\sinh \alpha z + \frac{\alpha \cosh \alpha - \sinh \alpha}{\sinh \alpha} z \sinh \alpha z - \alpha z \cosh \alpha z \right)$$

and a corresponding expression for $\theta(z)$,

$$\theta = C \left[\frac{3}{4\alpha} z \cosh \alpha z + \left(\frac{\alpha \cosh \alpha - \sinh \alpha}{4\alpha \sinh \alpha} \right) z^2 \cosh \alpha z \right.$$

$$-\frac{1}{4} z^2 \sinh \alpha z - \frac{\alpha \cosh \alpha - \sinh \alpha}{4\alpha^2 \sinh \alpha} z \sinh \alpha z$$

$$\left. - \sinh \alpha z \left[\frac{\alpha^2 \cosh^2 \alpha + \alpha \sinh \alpha \cosh \alpha + \sinh^2 \alpha + Nu_g(\alpha^2 + \alpha \sinh \alpha \cosh \alpha + \sinh^2 \alpha)}{4\alpha^2 \sinh \alpha (\alpha \cosh \alpha + Nu_g \sinh \alpha)} \right] \right],$$

where C is an arbitrary constant.

d. Use the last boundary condition to derive an equation for $M(\sigma = 0)$ in terms of the parameters α and Nu_g. Sketch the relationship α versus M for $Nu_g = 0$. The critical value of M is the minimum value for all possible α. Determine M_{crit} for $Nu_g = 0$, 2, and 4. What are the corresponding values of α? Does the critical M increase or decrease as Nu_g increases? Explain qualitatively (it might help to think about the two limiting cases $Nu_g = 0$ and $Nu_g \to \infty$).

6. The analysis of Rayleigh-Benard instability was based on the assumption that the density of the liquid depends on temperature only. In that case, we obtain the apparently obvious result that a fluid layer can only undergo a convective instability if it consists of heavy (cold) fluid on top of lighter (hot) fluid. In many circumstances, however, the fluid system will be more complicated in the sense that it may contain a solute (or several solutes) that also influence the fluid density. One common example is seawater, where the density depends on both temperature and salinity. In this case, it is possible to obtain the apparently paradoxical result of an instability due to buoyancy even when the overall density distribution is apparently stable, with the dense fluid on the bottom and the less dense fluid on the top.

Convection arising spontaneously via instability in a fluid system involving two "components"—salt and heat, or two solutes—is known as *double-diffusive convection*. Although a number of regimes of instability are possible, we shall concentrate on the so-called "finger regime" of warm, salty water overlaying cold, fresh water. The density distribution associated with the temperature in this case is stable, while that associated with salt is unstable. Let us suppose that the total density profile actually has lighter water over heavier water. This system may still be unstable. The explanation lies in the difference in the thermal and mass diffusivities. A fluid element displaced upward as a consequence of some disturbance will quickly equilibrate with the local temperature but will still have less salt. Hence, it will be lighter than its surroundings and continue to move up. In this instability, it is the rapid diffusion of heat relative to salt that is the essential destabilizing process. In this problem, we generalize the Rayleigh-Benard analysis to consider the case of double diffusive convection.

a. We assume that we have a fluid layer of depth d that is cooler at the bottom but saltier at the top. Thus, we suppose that the basic steady state is

$$u = v = w = 0,$$

$$\bar{T} = T_0 + \left(\frac{T_1 - T_0}{d}\right)z \quad \text{with } T_o < T_1,$$

$$\bar{C} = C_0 + \left(\frac{C_1 - C_0}{d}\right)z \quad \text{with } C_o < C_1,$$

and

$$\frac{\partial \bar{p}}{\partial z} = -\bar{\rho}g,$$

with C the concentration of salt.

$$C_1, \ T_1$$

———————————————————————————— $z = d$

FLUID LAYER

———————————————————————————— $z = d$

$$C_1, \ T_1$$

Assume that the equation of state for the fluid is

$$\frac{1}{\rho} = \frac{1}{\rho_0}\{1 + \beta(T - T_0) - \beta'(C - C_0)\},$$

and derive linearized disturbance equations. The equations should be non-dimensionalized as in the Benard problem, but with a characteristic temperature $T_c = \Pr(T_0 - T_1)$ and concentration $C_c = \mathrm{Sc}(C_0 - C_1)$, where $\Pr$ and Sc are the Prandtl and Schmidt numbers, respectively. These equations should contain four dimensionless parameters, $\Pr$, Sc, and the two Rayleigh numbers

$$\mathrm{Ra} \equiv \frac{g\beta(T_1 - T_0)d^3}{\nu^2}\Pr \ ; \quad \hat{\mathrm{Ra}} \equiv \frac{g\beta'(C_1 - C_0)d^3}{\nu^2}\mathrm{Sc}.$$

In the present configuration $\mathrm{Ra}, \ \hat{\mathrm{Ra}} > 0$.

b. Reduce the governing equations to a coupled triad for $f(z)$, $\theta(z)$, and $\hat{\theta}(z)$—the latter being the z-dependent parts of the dimensionless vertical velocity, temperature, and concentration, namely,

$$\left.\begin{array}{l} \hat{\theta} = e^{\sigma t}F(x,y) \ \hat{\theta}(z) \\[4pt] \theta = e^{\sigma t}F(x,y) \ \theta(z) \\[4pt] W = e^{\sigma t}F(x,y) \ f(z) \end{array}\right\} \quad \text{with } \nabla_1^2 F = -\alpha^2 F.$$

    c. Consider the case of two free surfaces that are isothermal. State the boundary conditions, and derive a characteristic equation for σ in a manner analogous to the analysis for the Rayleigh-Benard case. You should obtain a cubic, algebraic equation for σ with coefficients that depend on Sc, Pr, Ra, $\hat{Ra}$, α, and $k^2 = (n^2\pi^2 + \alpha^2)$.

    d. Use the fact that the neutral state for this configuration of hot, salty fluid over fresh, cold fluid has $\sigma \equiv 0$ to obtain an analytic expression for the critical condition for instability.

References/Notes

1. A thorough summary of the broad spectrum of buoyancy effects in fluid mechanics, including the problems that we identify here as *natural convection* can be found in
    Turner, J.S., *Buoyancy Effects in Fluids.* Cambridge Univ. Press: Cambridge (1973).
2. The reader who is interested in a more detailed account of the development of the Boussinesq approximation may refer to
    a. Spiegel, E.A., and Veronis, G., "On the Boussinesq Approximation for a Compressible Fluid," *Astrophys. J.* 131:442–447 (1960).
    b. Mihaljan, J.M., "A Rigorous Exposition of the Boussinesq Approximations Applicable to a Thin Layer of Fluid," *Astrophys. J.* 136:1126–1133 (1962).
3. A useful summary of work on large Grasshof number boundary-layer theory is Eshghy, S., "Free Convection Layers at Large Prandtl Number," *J. of Appl. Math. and Physics* (*ZAMP*) 22:275–292 (1971).
4. Yih, C.S., "Laminar Free Convection Due to a Line Source of Heat," *Trans. Am. Geophys. Union* 33:699–672 (1952).
    Also (with corrections) in Yih, C.S., "Free Convection Due to Boundary Sources," presented at the Symposium on the Use of Models in Geophysical Fluid Mechanics, Johns Hopkins University, Baltimore, 1953. Proceedings published by Government Printing Office: Washington, DC, 117–133 (1956).
5. Yih, C.S., "Free Convection Due to a Point Source of Heat," Proceedings of the First U.S. National Congress of Applied Mechanics, Chicago, pp. 941–947 (1950).
    Reproduced in the proceedings cited in Reference 4.
6. An analysis of this case was originally presented in Acrivos, A., "On the Combined Effect of Forced and Free Convection in Laminar Boundary-Layer Flows," *Chem. Eng. Sci.* 21:343–352 (1966).
7. The problem of combined forced and free convection for a horizontal flat plate was analyzed via a boundary-layer, asymptotic solution for $Pr \gg 1$ and $Pr \ll 1$, in Leal, L.G., "Combined Forced and Free Convection Heat Transfer from a Horizontal Flat Plate," *J. Appl. Math. and Physics* (*ZAMP*) 24:20–42 (1973).
8. Robertson, G.E., Seinfeld, J.H., and Leal, L.G., "Combined Forced and Free Convection Flow Past a Horizontal Flat Plate," *AIChE J.* 19:998–1008 (1973).
9. A similar analysis for $Pr = 1$ was given by Merkin, J.H., "The Effect of Buoyancy Forces on the Boundary-Layer Flow Over a Semi-Infinite Vertical Flat Plate in a Uniform Free Stream," *J. Fluid Mech.* 35:439–450 (1969).
10. A number of examples from boundary-layer theory can be found in the classic textbook by Schlichting, cited as Reference 3, Chapter 10.
11. Chandrasekhar, S., *Hydrodynamic and Hydromagnetic Stability.* Oxford Univ. Press: London (1961). Also reprinted by Dover Publications: New York (1981).
12. An excellent, brief summary of many linear analyses of hydrodynamic stability problems is given in chapter 9 of Yih, C.S., *Fluid Mechanics.* McGraw-Hill: New York (1969).
    A discussion of double-diffusive convection is given in Chapter 8 of Reference 1. Additional references to hydrodynamic stability theory were listed as Reference 4 in Chapter 3.

Index